Oncolytic Virus Immunotherapy

Oncolytic Virus Immunotherapy

Editors

Antonio Marchini
Carolina S. Ilkow
Alan Melcher

MDPI • Basel • Beijing • Wuhan • Barcelona • Belgrade • Manchester • Tokyo • Cluj • Tianjin

Editors

Antonio Marchini
Department of Oncology
Luxembourg Institute of Health
L-1526 Luxembourg
Luxembourg, and
Laboratory of Oncolytic Virus
Immuno-Therapeutics (LOVIT)
German Cancer Research Center
69120 Heidelberg
Germany

Carolina S. Ilkow
Cancer Therapeutics Program
Ottawa Hospital
Research Institute
Ottawa, ON K1H 8L6, and
Department of Biochemistry,
Microbiology and Immunology
University of Ottawa
Ottawa, ON K1H 8M5
Canada

Alan Melcher
Translational Immunotherapy
Team
Division of Radiotherapy
and Imaging
Institute of Cancer Research
London SW3 6JB
United Kingdom

Editorial Office
MDPI
St. Alban-Anlage 66
4052 Basel, Switzerland

This is a reprint of articles from the Special Issue published online in the open access journal *Cancers* (ISSN 2072-6694) (available at: www.mdpi.com/journal/cancers/special_issues/OVI).

For citation purposes, cite each article independently as indicated on the article page online and as indicated below:

LastName, A.A.; LastName, B.B.; LastName, C.C. Article Title. *Journal Name* **Year**, *Volume Number*, Page Range.

ISBN 978-3-0365-2549-5 (Hbk)
ISBN 978-3-0365-2548-8 (PDF)

Contents

About the Editors

Antonio Marchini

Dr. Antonio Marchini is Head of the Laboratory of Oncolytic Virus Immuno Therapeutics (LOVIT), a binational research unit at the Luxembourg Institute of Health (LIH, Luxembourg) and German Cancer Research Center (DKFZ, Heidelberg, Germany). His Team focuses on the development of new anticancer strategies based on oncolytic autonomous parvoviruses (PVs). Major areas of research are: i) generation of innovative oncolytic vectors: retargeted PVs, chimeric vectors of adenovirus-PV genomes, PVs expressing sh/miRNAs; ii) design and assessment of novel combination therapies using PVs together with other anticancer agents such as apoptosis inducers, immune checkpoint blockade and/or epigenetic modulators of gene expression, iii) identification of H-1PV cellular modulators and predictive biomarkers of viral oncolysis. He earned his Ph.D degree at the University of Heidelberg in 2001, working on the identification of new cellular proteins interacting with human papillomavirus 16 E7 oncoprotein.

Carolina S. Ilkow

Carolina S. Ilkow is originally from Buenos Aires, Argentina, where she obtained her bachelor's degree in Science. She then decided to move to Edmonton, Canada, to continue her graduate studies at the University of Alberta, where she obtained her PhD in cell biology, after which she joined Dr. John Bell's lab as a post-doctoral fellow. Carolina's work in the Bell lab aimed at developing novel and tailored virotherapies to fight Pancreatic cancer. Her discoveries in this field led Carolina to win a prestigious Researcher in Training Award and to publish impactful papers.

In 2016, Carolina was recruited as a scientist at the Ottawa Hospital Research Institute, and an Assistant Professor in the department of Biochemistry, Microbiology and Immunology at the University of Ottawa. Her research is focused on the development of innovative biotherapeutics for cancer treatment.

Alan Melcher

Professor Melcher is a clinician scientist, and currently Professor of Translational Immunotherapy at The Institute of Cancer Research, London, and Honorary Consultant Oncologist at The Royal Marsden NHS Foundation Trust. He graduated in medicine from the University of Oxford and trained in Clinical Oncology (Radiotherapy and Chemotherapy) in Cardiff, London and Leeds. Following completion of his PhD at the Imperial Cancer Research Fund (now Cancer Research UK) in London, he was a post-doctoral research fellow at the Mayo Clinic, Minnesota, before returning to the UK, where in 2007, he became Professor of Clinical Oncology and Biotherapy at Leeds, before moving to London in 2016. Professor Melcher's laboratory and translational research is focused on oncolytic viruses, radiotherapy and immunotherapy for the treatment of cancer. He and his team are investigating how to improve the ability of oncolytic viruses to trigger immunogenic cell death in tumours.

Preface to "Oncolytic Virus Immunotherapy"

Dear Readers,

Oncolytic Viruses (OV) are self-propagating agents that can selectively induce the lysis of cancer cells while sparing normal tissues. OV-mediated cancer cell death is often immunogenic and triggers robust anticancer immune responses and immunoconversion of tumor microenvironments. This makes oncolytic virotherapy a promising new form of immunotherapy and OVs ideal candidates for combination therapy with other anticancer agents, including other immunotherapeutics. There are more than 40 OVs from nine different families in clinical development, and many more at the preclinical stage. Each OV has its own unique characteristics, its pros and cons. Although herpes simplex virus is currently the lead clinical agent, a real champion among the OVs has not yet emerged, justifying the continuous development and optimization of these agents. In this book, our goal was to compile reviews that summarize the state-of-the-art and give a comprehensive overview of the OV arena with a particular focus on new trends, directions, challenges, and opportunities.

Antonio Marchini, Carolina S. Ilkow, Alan Melcher
Editors

Editorial

Oncolytic Virus Immunotherapy

Antonio Marchini [1,2,*], **Carolina S. Ilkow** [3,4] **and Alan Melcher** [5]

1 Laboratory of Oncolytic Virus Immuno-Therapeutics, Luxembourg Institute of Health, 84 Val Fleuri, L-1526 Luxembourg, Luxembourg

2 German Cancer Research Centre, Laboratory of Oncolytic Virus Immuno-Therapeutics, Im Neuenheimer Feld 242, 69120 Heidelberg, Germany

3 Cancer Therapeutics Program, Ottawa Hospital Research Institute, Ottawa, ON K1H 8L6, Canada; cilkow@ohri.ca

4 Department of Biochemistry, Microbiology and Immunology, University of Ottawa, Ottawa, ON K1H 8M5, Canada

5 Institute of Cancer Research, London SW3 6JB, UK; alan.melcher@icr.ac.uk

* Correspondence: antonio.marchini@lih.lu or a.marchini@dkfz.de; Tel.: +352-26-970-856 or +49-6221-424969

Citation: Marchini, A.; Ilkow, C.S.; Melcher, A. Oncolytic Virus Immunotherapy. *Cancers* **2021**, *13*, 3672. https://doi.org/10.3390/cancers13153672

Received: 12 May 2021
Accepted: 7 July 2021
Published: 22 July 2021

Publisher's Note: MDPI stays neutral with regard to jurisdictional claims in published maps and institutional affiliations.

Oncolytic viruses (OVs) were originally developed as direct cytotoxic agents but have been increasingly recognised as a form of immunotherapy. Oncolytic viruses have now reached the stage of significant widespread clinical testing, with more than 40 OVs currently being evaluated for the treatment of various tumour entities [1]. The majority of past and ongoing clinical trials have been phase I studies evaluating the safety of the treatment as the primary (and most of the time only) end-point. Very few OVs entered more advance stages in the clinical development pipeline, and only one agent has been FDA- and EMA-approved, talimogene laherperepvec (T-VEC, an HSV encoding GMCSF), for intratumoural administration in advanced melanoma [2]. Whilst there is a wealth of encouraging early trial data confirming the safety of OVs across a number of viruses, tumour types and administration routes [1], more recent data from emerging larger, randomised studies have not been so encouraging. The last (and only) positive randomised phase 3 trial of an OV, testing T-VEC against subcutaneous granulocyte–macrophage colony-stimulating factor in melanoma, was published back in 2015 [2], and that study predated immunotherapy, which is now standard of clinical care in this disease. The next logical steps with T-VEC, combining the virus with checkpoint blockade, were initially encouraging with ipilimumab (an anti-CTLA4 antibody) [3], but the recent discontinuation of the randomised phase 3 of pembrolizumab (an anti-PD1) +/− T-Vec due to futility (Thousand Oaks, Calif., accessed on 2 February 2021 https://investors.amgen.com/news-releases/news-release-details/amgen-reports-fourth-quarter-and-full-year-2020-financial) has raised significant concerns about the long-term potential for the OV field in the clinic.

There have been other disappointing large, randomised trials. Vocimagene amiretrorepvec (TOCA 511) is a replicating retrovirus encoding a transgene for cytosine deaminase, which converts the prodrug 5-fluorocytosine into 5-fluorouracil. This failed in a study of over 400 patients, where viral injection into the resection cavity on first or second resection for high-grade glioma was randomised against standard of care treatment [4]. Then, pexastimogene devacirepvec (Pexa-Vec), a vaccinia virus again encoding GM-CSF, was also unsuccessful when tested after [5] or first line in combination with sorafenib in hepatocellular carcinoma.

Whilst there is no hiding from the disappointments of these studies, rather than abandoning the field, now is the time to reconsider and regroup. There are many drugs that fail on progression from early to randomised studies, but OVs represent an immune strategy rather than a single therapeutic, and so should not be abandoned en masse. Their greatest promise lies in 'heating up' an immunologically 'cold' tumour to prime for checkpoint blockade, and there are good translational clinical data that suggest that this can happen in patients [6,7]. There are a number of reasons why the large studies to date have been unsuccessful, including wrong choice of tumour type (for Pexa-Vec, advanced liver cancer patients' often poor performance status makes altering the course of the disease notoriously difficult), clinical stage targeting (single agent pembrolizumab is too effective a single agent in limited metastatic melanoma for the addition of T-VEC to make a significant difference), and common problems seen on transitioning from early to later phase testing (in the TOCA511 study, patients had fewer cycles of treatment than in earlier trials). The key to further progress now is to better understand the immunobiology of OVs in patients via in-depth translational studies and to use this knowledge to inform careful development and progression of clinical trials in the most appropriate patient context.

We are proud and pleased to present this Special Issue of Cancers, "Oncolytic Virus immunotherapy" which includes 16 reviews written by many of the top oncolytic virus experts. There is a common message that comes across reading this issue, and it is one of justified optimism. The efficacy of OVs can be further improved, and the "second-generation" OV-based therapies once in the clinic may become a game changer in the history of cancer therapies for, e.g., pancreatic carcinoma, glioblastoma or lung cancer that still await effective treatment options. As our knowledge of the tumour cell, its microenvironment and components improves, the ideal of a "one size fits all" OV-based therapy becomes less real or attainable. Different individuals or cancers will need different approaches. Genetic engineering and arming of OVs and development of optimal combinatorial treatments must be carefully evaluated, taking into consideration the intra/inter patient heterogeneity of cancer and the complex interactions that cancer cells have with other components of the tumour microenvironment (resident and infiltrating non-transformed cells, secreted factors and extracellular matrix proteins). This complexity needs to be understood for every tumour entity and the (epi)genetic characteristics that distinguish it.

Oncolytic viruses are a very diverse group of "living drugs", comprising viruses with very different biology and unique features. Every OV platform has strengths and weaknesses. Developing them further will exploit their positive aspects and mitigate the negative ones, considering the type and stage of cancer patients, route, schedule of administration, and the insurgence of neutralising antiviral immune responses that can reduce efficacy. To this end, it is also crucial to identify predictive biomarkers of response that suggest the most opportune OV treatment for each patient.

This Special Issue shows that while a real champion among OVs has not yet emerged, there are many great advances in the field that could lead to an improvement in therapeutic outcomes in the near future. A new wave of OV platforms are being developed thanks to the advances in genetic engineering and our improved understanding of the tumour ecosystem, which is allowing for the rational combination of OVs with other anti-cancer therapies. The importance of developing combination strategies that synergise against the tumour without leading to unwanted off-tumour effects is a common theme across the reviews. Müller et al. describe the community efforts for reovirus [8]; Burman et al. for Newcastle disease virus [9]; Engeland and Ungerechts for measles virus [10]; Angelova et al. for parvovirus [11]; and Malin and Kühnel [12] and Cunliffe et al. [13] for the adenovirus platforms. In addition to being "lysing machines", OVs are "vehicles" that can deliver and express transgenes in the tumour ecosystem. Examples are given for the HSV platform by Vannini et al. [14]; for the adenovirus platform by Cunliffe et al. [13]; and, more generally, for the treatment of solid tumours by Jin et al. [15]. It is clear now that one avenue for improving the success of virotherapy resides in maximising the ability of OVs to harness the immune system to act against cancer, for example, through combination with other immunotherapies—especially immunocheckpoint blockers and adoptive cell therapy—or through the insertion of immunomodulatory transgenes into the virus genome. Combinatorial therapies of OVs and other treatment modalities are an active area of development, and, herein, Evgin and Vile [16], Kuryk et al. [17], Holbrook et al. [18] and Spiesschaert et al. [19] review the recent advances in this exciting field of research. Recent advances in genetic profiling of tumours are changing the way that we treat cancer patients. The latter is also impacting the way that we foresee the use of OVs in the near future. Both

Fisher et al. [20] and Enrilich and Bacharach [21] describe the importance of understanding the tumour and its microenvironment for selecting the right OV platform for each cancer patient. Similarly, Stavrakaki et al. [22] discuss the importance of finding biomarkers to "personalise OVs" based on the tumour-specific characteristics. Finally, Kock et al. provide us with a comprehensive summary of oncolytic HSV-1 in its journey through the clinical arena [23].

We hope that the readers enjoy this Special Issue and that the OV scientific community continues working together towards the development of virotherapeutics that could positively impact the life of people living with cancer.

Author Contributions: Conceptualisation: A.M. (Antonio Marchini), C.S.I. and A.M. (Alan Melcher); writing, review and editing, A.M. (Antonio Marchini), C.S.I. and A.M. (Alan Melcher). All authors have read and agreed to the published version of the manuscript.

Funding: A.M. (Antonio Marchini) is a recipient of research grants from the Luxembourg Cancer Foundation, Télévie and the Cooperational Research Program of the German Cancer Research Center (DKFZ), Heidelberg with the Ministry of Science and Technology, Israel. C.S.I (Carolina S. Ilkow) is a recipient of grants from the Canadian Institutes of Health Research (#377104), BioCanRx, and the Canadian Cancer Society Innovation (#705973) and Impact (#IMP-14 and # 706162) as well as the generous support from the Ontario Institute for Cancer Research, the Ottawa Regional Cancer Foundation and the Ottawa hospital foundation.

Conflicts of Interest: A.M. (Antonio Marchini) is the inventor on several H-1PV-related patents/patent applications. No other conflict of interest are declared by the authors.

References

1. Macedo, N.; Miller, D.M.; Haq, R.; Kaufman, H.L. Clinical landscape of oncolytic virus research in 2020. *J. Immunother. Cancer* **2020**, *8*. [CrossRef] [PubMed]
2. Andtbacka, R.H.; Kaufman, H.L.; Collichio, F.; Amatruda, T.; Senzer, N.; Chesney, J.; Delman, K.A.; Spitler, L.E.; Puzanov, I.; Agarwala, S.S.; et al. Talimogene Laherparepvec Improves Durable Response Rate in Patients with Advanced Melanoma. *J. Clin. Oncol.* **2015**, *33*, 2780–2788. [CrossRef] [PubMed]
3. Chesney, J.; Puzanov, I.; Collichio, F.; Singh, P.; Milhem, M.M.; Glaspy, J.; Hamid, O.; Ross, M.; Friedlander, P.; Garbe, C.; et al. Randomized, Open-Label Phase II Study Evaluating the Efficacy and Safety of Talimogene Laherparepvec in Combination with Ipilimumab Versus Ipilimumab Alone in Patients with Advanced, Unresectable Melanoma. *J. Clin. Oncol.* **2018**, *36*, 1658–1667. [CrossRef]
4. Cloughesy, T.F.; Petrecca, K.; Walbert, T.; Butowski, N.; Salacz, M.; Perry, J.; Damek, D.; Bota, D.; Bettegowda, C.; Zhu, J.J.; et al. Effect of Vocimagene Amiretrorepvec in Combination with Flucytosine vs Standard of Care on Survival Following Tumor Resection in Patients with Recurrent High-Grade Glioma: A Randomized Clinical Trial. *JAMA Oncol.* **2020**, *6*, 1939–1946. [CrossRef] [PubMed]
5. Moehler, M.; Heo, J.; Lee, H.C.; Tak, W.Y.; Chao, Y.; Paik, S.W.; Yim, H.J.; Byun, K.S.; Baron, A.; Ungerechts, G.; et al. Vaccinia-based oncolytic immunotherapy Pexastimogene Devacirepvec in patients with advanced hepatocellular carcinoma after sorafenib failure: A randomized multicenter Phase IIb trial (TRAVERSE). *Oncoimmunology* **2019**, *8*, 1615817. [CrossRef]
6. Ribas, A.; Dummer, R.; Puzanov, I.; VanderWalde, A.; Andtbacka, R.H.I.; Michielin, O.; Olszanski, A.J.; Malvehy, J.; Cebon, J.; Fernandez, E.; et al. Oncolytic Virotherapy Promotes Intratumoral T Cell Infiltration and Improves Anti-PD-1 Immunotherapy. *Cell* **2017**, *170*, 1109–1119.e1110. [CrossRef]
7. Samson, A.; Scott, K.J.; Taggart, D.; West, E.J.; Wilson, E.; Nuovo, G.J.; Thomson, S.; Corns, R.; Mathew, R.K.; Fuller, M.J.; et al. Intravenous delivery of oncolytic reovirus to brain tumor patients immunologically primes for subsequent checkpoint blockade. *Sci. Transl. Med.* **2018**, *10*. [CrossRef]
8. Muller, L.; Berkeley, R.; Barr, T.; Ilett, E.; Errington-Mais, F. Past, Present and Future of Oncolytic Reovirus. *Cancers* **2020**, *12*, 3219. [CrossRef] [PubMed]
9. Burman, B.; Pesci, G.; Zamarin, D. Newcastle Disease Virus at the Forefront of Cancer Immunotherapy. *Cancers* **2020**, *12*, 3552. [CrossRef]
10. Engeland, C.E.; Ungerechts, G. Measles Virus as an Oncolytic Immunotherapy. *Cancers* **2021**, *13*, 544. [CrossRef]
11. Angelova, A.; Ferreira, T.; Bretscher, C.; Rommelaere, J.; Marchini, A. Parvovirus-Based Combinatorial Immunotherapy: A Reinforced Therapeutic Strategy against Poor-Prognosis Solid Cancers. *Cancers* **2021**, *13*, 342. [CrossRef]
12. Peter, M.; Kuhnel, F. Oncolytic Adenovirus in Cancer Immunotherapy. *Cancers* **2020**, *12*, 3354. [CrossRef] [PubMed]
13. Cunliffe, T.G.; Bates, E.A.; Parker, A.L. Hitting the Target but Missing the Point: Recent Progress towards Adenovirus-Based Precision Virotherapies. *Cancers* **2020**, *12*, 3327. [CrossRef] [PubMed]

14. Vannini, A.; Leoni, V.; Sanapo, M.; Gianni, T.; Giordani, G.; Gatta, V.; Barboni, C.; Zaghini, A.; Campadelli-Fiume, G. Immunotherapeutic Efficacy of Retargeted oHSVs Designed for Propagation in an Ad Hoc Cell Line. *Cancers* **2021**, *13*, 266. [CrossRef] [PubMed]
15. Jin, K.T.; Du, W.L.; Liu, Y.Y.; Lan, H.R.; Si, J.X.; Mou, X.Z. Oncolytic Virotherapy in Solid Tumors: The Challenges and Achievements. *Cancers* **2021**, *13*, 588. [CrossRef]
16. Evgin, L.; Vile, R.G. Parking CAR T Cells in Tumours: Oncolytic Viruses as Valets or Vandals? *Cancers* **2021**, *13*, 1106. [CrossRef]
17. Kuryk, L.; Bertinato, L.; Staniszewska, M.; Pancer, K.; Wieczorek, M.; Salmaso, S.; Caliceti, P.; Garofalo, M. From Conventional Therapies to Immunotherapy: Melanoma Treatment in Review. *Cancers* **2020**, *12*, 3057. [CrossRef]
18. Holbrook, M.C.; Goad, D.W.; Grdzelishvili, V.Z. Expanding the Spectrum of Pancreatic Cancers Responsive to Vesicular Stomatitis Virus-Based Oncolytic Virotherapy: Challenges and Solutions. *Cancers* **2021**, *13*, 1171. [CrossRef]
19. Spiesschaert, B.; Angerer, K.; Park, J.; Wollmann, G. Combining Oncolytic Viruses and Small Molecule Therapeutics: Mutual Benefits. *Cancers* **2021**, *13*, 3386. [CrossRef]
20. Fisher, K.; Hazini, A.; Seymour, L.W. Tackling HLA Deficiencies Head on with Oncolytic Viruses. *Cancers* **2021**, *13*, 719. [CrossRef]
21. Ehrlich, M.; Bacharach, E. Oncolytic Virotherapy: The Cancer Cell Side. *Cancers* **2021**, *13*, 939. [CrossRef] [PubMed]
22. Stavrakaki, E.; Dirven, C.M.F.; Lamfers, M.L.M. Personalizing Oncolytic Virotherapy for Glioblastoma: In Search of Biomarkers for Response. *Cancers* **2021**, *13*, 614. [CrossRef] [PubMed]
23. Koch, M.S.; Lawler, S.E.; Chiocca, E.A. HSV-1 Oncolytic Viruses from Bench to Bedside: An Overview of Current Clinical Trials. *Cancers* **2020**, *12*, 3514. [CrossRef] [PubMed]

Review

Past, Present and Future of Oncolytic Reovirus

Louise Müller [†] [iD], Robert Berkeley [†], Tyler Barr, Elizabeth Ilett [‡] and Fiona Errington-Mais [*,‡]

Leeds Institute of Medical Research (LIMR), University of Leeds, Leeds LS9 7TF, UK;
lme.muller@outlook.com (L.M.); raberkeley@outlook.com (R.B.); um17tkb@leeds.ac.uk (T.B.);
e.ilett@leeds.ac.uk (E.I.)
* Correspondence: f.errington@leeds.ac.uk; Tel.: +44-113-3438410
† Authors contributed equally to this review article.
‡ Joint Senior Author.

Received: 18 September 2020; Accepted: 30 October 2020; Published: 31 October 2020

Simple Summary: Within this review article the authors provide an unbiased review of the oncolytic virus, reovirus, clinically formulated as pelareorep. In particular, the authors summarise what is known about the molecular and cellular requirements for reovirus oncolysis and provide a comprehensive summary of reovirus-induced anti-tumour immune responses. Importantly, the review also outlines the progress made towards more efficacious combination therapies and their evaluation in clinical trials. The limitations and challenges that remain to harness the full potential of reovirus are also discussed.

Abstract: Oncolytic virotherapy (OVT) has received significant attention in recent years, especially since the approval of talimogene Laherparepvec (T-VEC) in 2015 by the Food and Drug administration (FDA). Mechanistic studies of oncolytic viruses (OVs) have revealed that most, if not all, OVs induce direct oncolysis and stimulate innate and adaptive anti-tumour immunity. With the advancement of tumour modelling, allowing characterisation of the effects of tumour microenvironment (TME) components and identification of the cellular mechanisms required for cell death (both direct oncolysis and anti-tumour immune responses), it is clear that a "one size fits all" approach is not applicable to all OVs, or indeed the same OV across different tumour types and disease locations. This article will provide an unbiased review of oncolytic reovirus (clinically formulated as pelareorep), including the molecular and cellular requirements for reovirus oncolysis and anti-tumour immunity, reports of pre-clinical efficacy and its overall clinical trajectory. Moreover, as it is now abundantly clear that the true potential of all OVs, including reovirus, will only be reached upon the development of synergistic combination strategies, reovirus combination therapeutics will be discussed, including the limitations and challenges that remain to harness the full potential of this promising therapeutic agent.

Keywords: reovirus; oncolytic virus; immunotherapy

---

1. Oncolytic Virotherapy (OVT)

Advancements in virology and molecular biology techniques over recent decades have allowed us to exploit the anti-tumour potential of oncolytic viruses (OVs) [1]. The unique ability of OVs to exploit oncogenic signalling pathways provides a significant advantage over traditional treatment modalities. OVs are specifically defined as viruses which: (i) preferentially infect and kill malignant cells through viral replication and oncolysis, and (ii) engage the immune system to promote anti-tumour immunity. Additional mechanisms of action have also been reported, including disruption of tumour-associated vasculature or stroma and modulation of the tumour microenvironment (TME) [2–4].

An array of OVs—naturally occurring, attenuated, and genetically modified—have been investigated in pre-clinical models and clinical trials but only two have received approval for clinical use: (i) a genetically engineered adenovirus H101, approved in China in 2005 [5], and (ii) the Food and Drug Administration (FDA)-approved talimogene laherparepvec (T-VEC)—a herpes simplex virus type 1 (HSV-1) genetically engineered to limit neurovirulence and promote an immunostimulatory environment [6,7]. This review will provide an overview of what we have learnt about oncolytic mammalian orthoreovirus since its rise as a clinically applicable agent, we will discuss areas of active pre-clinical and clinical research and consider the challenges that exist to harness its full therapeutic potential.

2. The Emergence of Reovirus as a Therapeutic Agent

The Reoviridae family of viruses has found hosts in mammals, fish, birds and plants [8,9]. Three serotypes of mammalian orthoreovirus have been identified: type one Lang, type two Jones, and type three Abney and Dearing [10]. Each differs in its in vivo tropism, despite a high degree of genetic similarity [11]. Type-specific diversity occurs in the S1 gene, encoding the outer capsid σ1 attachment protein, which has undergone significant evolutionary divergence [12]. Orthoreovirus type two *Jones* was the first serotype observed to replicate specifically in malignant cell lines [13]; however, it is the mammalian orthoreovirus type three Dearing strain (T3D)—now manufactured as pelareorep but previously known as Reolysin®—that has made progress as a therapeutic agent. Mammalian orthoreovirus T3D (hereafter referred to as reovirus) is typically isolated from human gastrointestinal and upper respiratory tracts [14,15]. In most individuals, infection proceeds asymptomatically causing mild enteric or respiratory illness in young children and being relatively non-pathogenic in adults, in line with its designation as a respiratory enteric orphan virus (reovirus) [10]. There have been sporadic reports of severe pathology associated with reovirus infection in infants and immunocompromised individuals [9,16–21] and more recently, reovirus has been implicated in coeliac disease by promoting a T_H1 immune response, a response that bodes well for its use as an immunotherapeutic tool although oral delivery should be avoided to limit these potential unwanted side effects [22].

Reovirus is a non-enveloped, double-stranded (ds) RNA virus approximately 85 nm in diameter, with two concentric icosahedral protein capsids [23]. The outer and inner capsids protect the dsRNA genome which comprises 23.5 kbp in ten segments termed large (L1-3), medium (M1-3), or small (S1-4) according to size [23–25]. The gene segments encode eight structural proteins (λ1-3, μ1-2, and σ1-3) and the non-structural proteins, μNS and σNS [26]. μ1 and σ3 form part of the outer capsid, λ3 forms a subunit of the RNA polymerase and σ1 and λ2 are important for viral attachment, although σ1 initiates target cell entry [23]. The proteins also protect the virus from immune-surveillance by preventing a host anti-viral interferon (IFN) response; σ3 binds to dsRNA and prevents its binding to dsRNA-dependent protein kinase R (PKR; a dsRNA sensor) [27] and μNS sequesters the IFN transcription factor (interferon regulatory factor 3; IRF3) and inhibits its translocation to the nucleus [28].

3. Tumour Specificity and Replication

The reovirus life-cycle is shown in Figure 1. Viral entry occurs over multiple steps, the first being a low-affinity "tethering" of the reovirus σ1 protein to cell surface sialic acid [29,30]. Subsequently, σ1 engages junctional adhesion molecule A (JAM-A), the canonical reovirus receptor [31–33], which is ubiquitously expressed throughout the body and has several important roles in normal cellular processes including tight junction formation, leukocyte migration, and angiogenesis [34]. Fortuitously, JAM-A is also overexpressed in several cancers, including both haematological and solid malignancies [35–41]. Following reovirus engagement with JAM-A and receptor-mediated endocytosis, the viral particle undergoes acid-dependent cathepsin-mediated proteolysis within the endosome [42,43] to form an intermediate subviral particle (ISVP) characterised by the loss of σ3 and cleavage of μ1 [44]. The proteolytic uncoating, principally by cathepsins L and B, is critical for penetration of the endosome membrane by μ1; ISVPs undergo a conformational change causing autocleavage of μ1 into μ1N which triggers pore formation in the endocytic membrane [45] and delivers transcriptionally active reovirus into the cytosol [46,47] for replication. Capped, positive-sense single stranded (ss) RNA serves as mRNA for protein translation and provides a template for replication of nascent dsRNA genomes [48]. Transcription and translation occur in cytoplasmic "viral factories" [49,50], with packaging of the segmented genome into virions occurring concomitantly with RNA synthesis [51,52]. Viral egress can be non-cytolytic in the absence of transformation; however, the release of progeny virus is typically lytic in permissive, transformed cells [53,54].

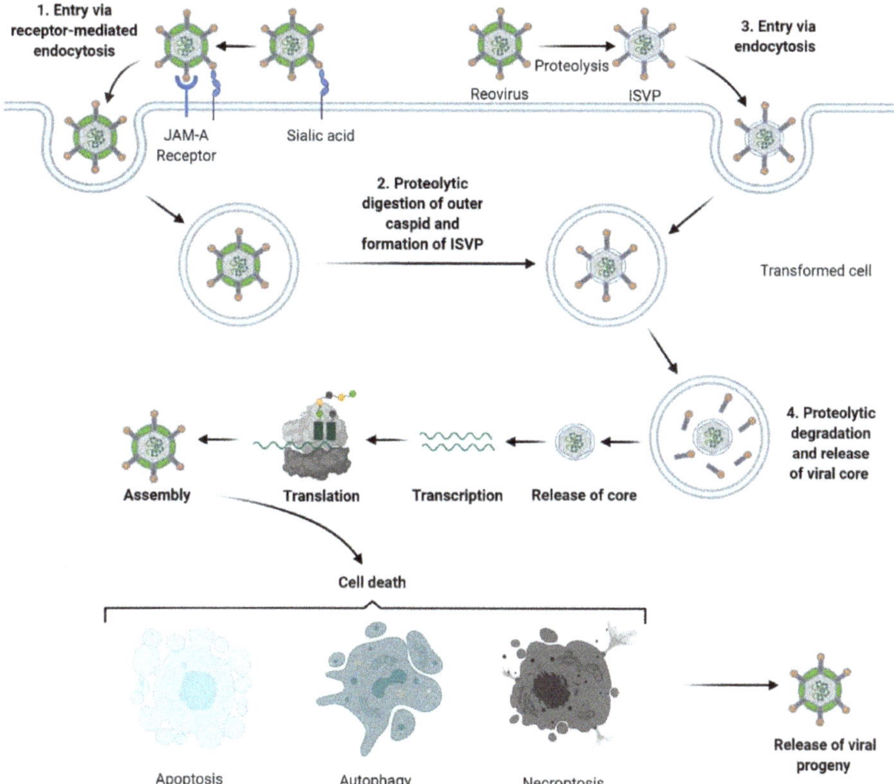

Figure 1. Reovirus replication: **1.** Reovirus is first tethered via a weak interaction between σ1 and cell surface sialic acid; σ1 then binds with high affinity to junctional adhesion molecule A (JAM-A) resulting in internalization of the virus via receptor-mediated endocytosis. **2.** Once internalized, the virus is transported to early and late endosomes where it undergoes proteolytic digestion to remove the outer capsid protein σ3 resulting in the formation of infectious subvirion particles (ISVPs). **3.** Alternatively, ISVPs may be formed by extracellular proteases within the tumour environment allowing direct entry into cells via membrane penetration. **4.** After further proteolytic degradation a transcriptionally active viral core is released into the cytoplasm. Transcription and translation occur ultimately leading to the assembly of new viral progeny, host cell death and progeny release. Figure created using Biorender (https://biorender.com/).

The molecular features associated with the oncolytic capacity of reovirus have been the subject of decades of research. Initially, an association between reovirus permissiveness and epidermal growth factor receptor (EGFR) status was revealed [55,56], along with evidence that activation of downstream signalling pathways, induced after transfection with the oncogene *v-erb*, are important [57]. Subsequent transfection of cells with constitutively active elements of the RAS pathway, a group of small GTP-binding proteins that regulate cell fate and growth, identified a role for RAS in reovirus permissiveness [58]. Therefore, although JAM-A is important for host cell entry, gain-of-function mutations activating RAS signalling [59] could promote reovirus replication and the release of virus progeny [60]. *RAS* mutations are prevalent in cancer [61], supporting the use of reovirus as a potential therapeutic agent [58,62]. The link between reovirus and cellular RAS status was further strengthened by observations that tumour cell susceptibility could be influenced by modulating RAS and/or its downstream effectors using short-hairpin RNA or small-molecule inhibitors [63,64]. Mechanistically, modulation of RAS signalling may promote susceptibility via inhibition of PKR [58]. In healthy cells, binding of dsRNA by PKR results in its dimerization, autophosphorylation and activation. Activated PKR subsequently phosphorylates the translation initiation factor, eIF2, rendering it inactive, which prevents the translation of viral transcripts [65]; however, in *RAS*-transformed cells PKR remains inactive and viral replication can occur [58,66,67]. Currently, the mechanism that coordinates *RAS*-transformation and PKR inactivation remains unclear [68].

Although the RAS–PKR axis provides a plausible explanation for the susceptibility of cancer cells to reovirus, the true molecular mediator has been the subject of debate, with doubt being cast by the survival of some infected *RAS*-transformed cells [69,70]. Moreover, the absence of a correlation between total or phospho-PKR with RAS expression or cell death contradicts previous studies [71], as does the lack of association between oncolysis and EGFR signalling [72]. It has become increasingly apparent that viral replication and cell death are not inextricably linked. Indeed, it is possible that RAS activation does not underlie viral replication but rather sensitivity to apoptosis which can occur independently of replication [53,64]. Sensitivity to reovirus oncolysis is likely to be dependent on multiple cellular and molecular determinants, many of which may yet be undiscovered.

4. Mechanisms of Oncolysis

Reovirus was originally considered to operate predominantly by apoptosis (reviewed in [73]). The apoptotic signalling often displayed by infected cells includes the generation of IFN and activation of NF-κB, either through detection of cytoplasmic dsRNA via PKR, retinoic acid-inducible gene I (RIG-I) or melanoma differentiation-associated protein 5 (MDA5), or following σ1 and μ1 receptor engagement or membrane penetration [53,74–77]. In response to NF-κB and/or IRF3 signalling, inflammatory cytokines such as TNF-related apoptosis-inducing ligand (TRAIL) are secreted, which bind to surface death receptors and trigger activation of caspase-3 and -7 [78–80]. While IFN is a potent promoter of cell death, it can be dispensable for reovirus-induced apoptosis, which explains the ability of infected, IFN-deficient tumour cells to undergo apoptosis [79,81]. Blockade of apoptotic caspases does not always abrogate reovirus-induced cell death, indicating that other modes of cell death can also occur [82]. Necroptosis, contingent on recognition of viral dsRNA and induction of a type I IFN response [83], and autophagy following acute endoplasmic reticulum (ER) stress [84] have both been identified as alternative modes of reovirus-induced cell death. Thus, reovirus-induced death is exquisitely linked to the phenotype of the target cell and the surrounding TME; indeed, our recent unpublished data suggest that modulation of pro- vs. anti-apoptotic proteins upon co-culture with stromal cell support can abrogate reovirus-induced apoptosis in malignant B cells. Therefore, examination of viral replication and/or oncolysis in multiple cancer models, and in the context of TME support, will be essential to identify mechanisms of cancer-selective activity and cell death.

5. Reovirus Modulation of the Immune System

5.1. Reovirus-Induced Innate Anti-Tumour Immunity

Immune cells and infected tumour cells secrete pro-inflammatory cytokines and chemokines in response to reovirus treatment [85–88]. This occurs via engagement of pathogen-associated molecular patterns (PAMPs; e.g., viral RNA, DNA or proteins) or damage-associated molecular patterns (DAMPs; e.g., heat-shock proteins, calreticulin, uric acid and ATP released from infected cells) with pattern recognition receptors (PRRs) [89]. As with most viral infections, the secretion of type I IFN is a key component of the innate response to reovirus [90]. Viral dsRNA in the cytoplasm of infected cells is detected by PRRs such as RIG-I, MDA5, PKR or Toll-like receptor-3 [91,92] and triggers the transcription of type I IFNs from both infected tumour cells and immune cells; dendritic cells (DCs) and monocytes are important in the detection of reovirus and secretion of IFN-α [35,85,93]. Indeed, specific roles for RIG-I and mitochondrial antiviral-signaling protein (MAVS) but not MDA5 have been reported for reovirus activation of IRF3/IRF7, whilst reovirus activation of an NF-κB was dependent on MDA5 [76]. Moreover, it has been suggested that long reovirus dsRNA gene segments activate MDA5 while short dsRNA segments activate RIG-I [76]. Importantly, a role for RIG-I signaling has also been implicated in reovirus permissiveness of *RAS*-transformed cells; the MEK/ERK pathway—downstream of RAS—blocks signaling from RIG-I and inhibits IFN production, thus enabling reovirus replication [94]. In addition, a role for TLR-3 has been described for reovirus detection within the TME. Here, reovirus inhibited the immunosuppressive activity of myeloid-derived suppressor cells (MDSCs) in a TLR-3-dependent manner [95].

The generation of a pro-inflammatory environment reverses the immunosuppressive state of the TME, induces cytotoxic bystander cytokine killing of tumour cells, activates and recruits innate immune effector cells to kill neoplastic cells, and facilitates the generation of an adaptive anti-tumour immune response [96–99]. Reciprocal cell-to-cell interactions between DCs and natural killer (NK) cells within the TME or tumour-draining lymph nodes, can stimulate both NK cell activation and DC maturation [85,100]; NK cell anti-tumour immunity within peripheral blood mononuclear cells (PBMCs) is mediated by type I IFN secretion from monocytes [35]. In addition to the recruitment and activation of NK cells, reovirus also activates innate T cells which are capable of eliminating tumour cells via the release of cytolytic granules [85,101]; this remains a poorly understood mechanism of action.

5.2. Adaptive Anti-Tumour Immunity

In addition to PAMPs and DAMPs, tumour-associated antigens (TAAs) are also released into the TME during oncolysis. TAAs are phagocytosed by antigen presenting cells (APCs), such as DCs, and the cytokine-rich milieu stimulates DC maturation [102]. Reovirus-activated DCs cross-present TAAs via major histocompatibility complex (MHC) class I to naive CD8^{+ve} T cells [102,103]. These processes facilitate the priming of tumour-specific cytotoxic T lymphocytes (CTLs) [89,102,103]. Interestingly, direct reovirus oncolysis is not essential to generate adaptive anti-tumour immunity, as tumour-specific CTLs have been successfully generated against reovirus-resistant melanoma cells in vivo [104]. Thus, even if a particular cancer is not killed directly by the lytic effects of reovirus, reovirus treatment may offer immunotherapeutic value for patients. By contrast, a recent study by Martin et al., [105] suggested that reovirus was ineffective at priming a systemic immune response compared to alternative OVs, despite effective eradication of the primary tumour. These conflicting data are difficult to interpret; however, the discrepancies observed could be due to the different mouse strains; previous studies [103,104] have utilized T_H1-dominant C57BL/6 mouse models, whilst this later study used T_H2 dominant Balb/c mice. Of note, Martin et al., did not examine the induction of tumour specific CTLs but eradication of a secondary tumour. Therefore, it is possible that reovirus did prime effector CTLs which were inhibited due to the upregulation of immune checkpoint molecules, such as programmed death-ligand 1 (PD-L1), or the induction of regulatory T cells (T_{regs}) within the TME. Indeed, it is important to note that reovirus can promote the accumulation of T_{regs} and MDSCs [106–108] and also upregulates immune checkpoint molecules [108–110], which could impede both effector NK cell and CTL responses. Figure 2 (the inner circle) provides an overview of known reovirus mechanisms of action, including oncolysis and the induction of innate and adaptive anti-tumour immunity.

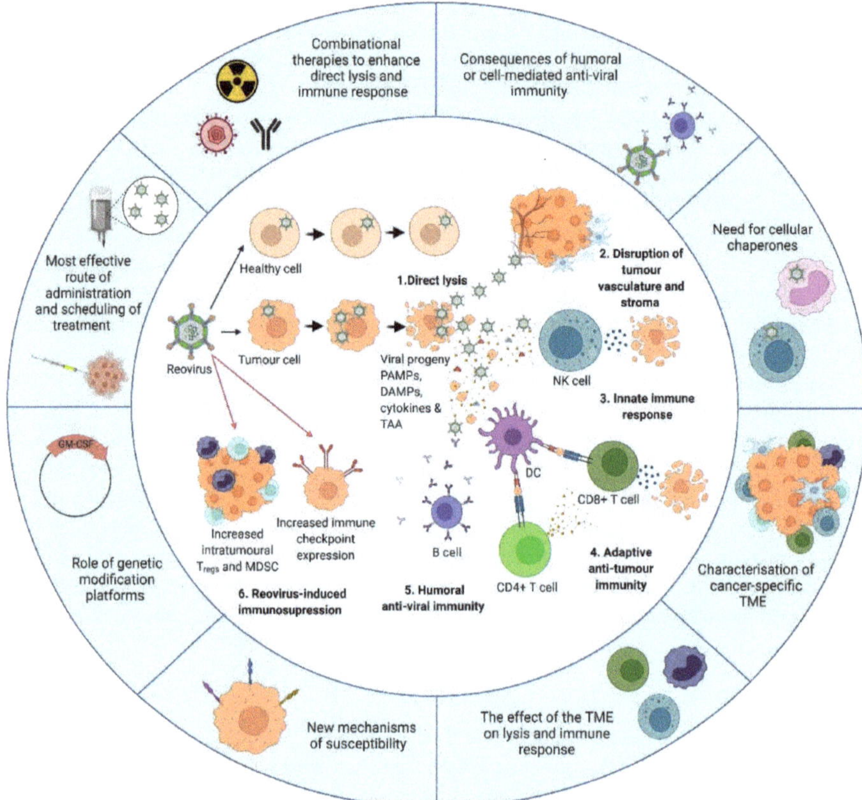

Figure 2. Overview of reovirus mechanisms of action and the developments required. The **inner circle** illustrates what is currently known about reovirus. **1.** In healthy cells, anti-viral immune responses limit reovirus replication and prevent lytic killing. By contrast, oncogenic signalling pathways render tumour cells susceptible to reovirus replication and direct oncolysis. **2.** Reovirus replicates in the tumour vasculature and stroma due to reciprocal cell:cell interactions which alter anti-viral signalling. **3.** Infection of tumour cells leads to the release of viral progeny, cytokines and tumour-associated antigens (TAAs), which initiates innate anti-tumour immunity including cytokine-mediated bystander killing and natural killer (NK) cell-mediated cytotoxicity. **4.** Adaptive anti-tumour immunity is generated following the phagocytosis of TAAs by dendritic cells (DCs) and presentation of TAAs to CD4^{+ve} and CD8^{+ve} T cells, which facilitates priming of tumour-specific cytotoxic T lymphocytes (CTLs). **5.** In addition to innate and adaptive anti-tumour immune responses, humoral anti-viral immunity is induced, leading to the production of reovirus-specific neutralising antibodies (NAbs). **6.** Following induction of anti-tumour/anti-viral immune responses, regulatory immune mechanisms are "switched-on" to control ongoing immune responses, including upregulation of immune checkpoints and increased levels of regulatory T cells (T$_{regs}$) and/or myeloid-derived suppressor cells (MDSCs). The **outer circle** highlights priority research areas to improve reovirus efficacy. These include gaining a greater understanding of: (i) the consequence of humoral and/or cell-mediated anti-viral immunity on reovirus efficacy which would inform the development of, or the requirement for, cellular chaperones; (ii) the tumour microenvironment (TME) and how it influences reovirus oncolysis and anti-tumour immunity; (iii) the cellular determinants utilized by reovirus for direct oncolysis, including mechanisms of reovirus resistance; (iv) the potential benefits of genetically-modified reovirus platforms; (v) reovirus scheduling to maximize virus delivery and efficacy including the best route of virus administration; and (vi) combinatorial approaches that are designed to boost both direct oncolysis and anti-tumour immune responses. PAMPs: pathogen-associated molecular patterns; DAMPs: damage-associated molecular patterns; GM-CSF: granulocyte-macrophage colony stimulating factor. Figure created using Biorender.com.

In a recent phase I study of intravenous (i.v.) reovirus there was an increase in transcripts of the pro-recruitment chemokines macrophage inflammatory protein (MIP)-1α and MIP-1β in tumour RNA and in the expression of the intracellular adhesion molecule 1 (ICAM-1) by T cells 48–72 h after infusion [109]. Along with CD68[+ve] myeloid cells, tumours of reovirus-treated vs. control patients appeared to contain a higher number of CD8[+ve] T cells [109], whose presence is strongly associated with superior outcomes [111]. Moreover, pro-inflammatory cytokines and IFN were upregulated in the serum of reovirus-treated patients [109,112], which can promote APC maturation and activate NK and T cells, as evidenced by the increased expression of CD69 [113]. Collectively, the evidence suggests that, as an immune adjuvant, reovirus can promote leukocyte infiltration into tumours and support tumour immune surveillance. However, to promote and sustain reovirus-induced anti-tumour immunity it is essential that long-term characterisation of the TME after reovirus treatment is carried out and that combination strategies are developed to counteract any inhibitory/regulatory mechanisms that develop.

5.3. The Antiviral Immune Response

The "antiviral" immune response is designed to combat the invading pathogen; however, it could also be fundamental to OV efficacy because of the overlap with "anti-tumour" processes. The humoral arm of adaptive immunity plays an important role in preventing reovirus infection through the generation of neutralising antibodies (NAbs) and there is evidence that circulating reovirus-specific antibodies can impair viral persistence and access to tumours [114]. As reovirus is ubiquitous in the environment [115], the global seroprevalence among adults is commonly above 50% and typically closer to 100% [116–120]. While NAbs may have a positive effect in protecting against reovirus infection, their effect on reovirus therapeutic activity remains controversial. Interesting, but generally less considered in relation to OV therapy, is the fact that viral antigens also prime virus-specific T cells [98,121,122]. These could either potentiate anti-cancer activity through eradication of virally infected tumour cells or abrogate anti-cancer activity by abrogating viral replication and direct oncolysis.

6. Reovirus Delivery—Systemic vs. Intra-Tumoural

Although the mechanisms by which reovirus exerts its cytotoxic effects have been the subject of some debate, the fact that it can reliably do so against malignant targets remains unquestioned. Reovirus has oncolytic activity against the vast majority of solid tumour types in vitro (lung, breast, ovarian, prostate, colorectal, pancreatic, glioma, melanoma, and head and neck squamous cell carcinoma (HNSCC)) [72,87,93,123–126] and has shown promise in haematological models, such as multiple myeloma and both lymphoid and myeloid leukaemias [35,37,127].

When first used as a cancer therapeutic in pre-clinical in vivo models, reovirus was delivered by the intra-tumoural (i.t.) route [128] and induced regression of established subcutaneous B16 melanomas [129], colorectal liver metastases [70] and subcutaneous and orthotopic gliomas [130]. Interestingly, i.t.-administered UV-inactivated reovirus also controlled tumour growth via immune-mediated mechanisms in a liver cancer model [131]. However, the systemic administration of virus into the bloodstream would appear to have the greatest potential to access disseminated tumour cells within the vasculature or distant organs. This is of clinical importance given that metastasis causes ~90% of all cancer-related deaths [132]. Oral intake, by far the most convenient route of systemic drug administration, is not suited to OV therapy as the virus is a gastrointestinal pathogen and is contained within the gastrointestinal system. Vascular injection is therefore the preferred systemic delivery route, being less invasive than locoregional administration. Unfortunately, the impact of i.v. reovirus upon tumour growth is often limited in comparison to i.t. injection; this could be due to: (i) limited delivery to the tumour; (ii) the generation of NAbs resulting in virus neutralisation prior to tumour access; and/or iii) reduced recruitment of immune effector cells to the tumour site.

Because of the size of the typical therapeutic OV infusion (10^9–10^{10} pfu), B cell mobilisation and antibody production occurs rapidly. From a not-insubstantial baseline, anti-reovirus antibody titres commonly increase ~1000-fold [133] and is greater in response to i.v. than i.t. injection [112]. Strategies to reduce and/or counteract reovirus NAbs have involved the use of immunosuppressive chemotherapy, particularly cyclophosphamide (CPA). CPA can deplete T_{regs} and boost T cell anti-tumour immunity [134]; however, at higher doses, it can suppress the effector functions of all lymphocytes, including B cell antibody production [135,136]. In preclinical models, CPA successfully curtailed B cell responses and enhanced the persistence of reovirus and delivery to tumours [114,137,138]. CPA and other chemotherapy agents have been used successfully alongside i.v. reovirus in clinical trials to reduce NAbs [139,140], with the exception of one phase I trial where CPA did not attenuate anti-viral responses [141].

In patients, reovirus persists in the bloodstream of seropositive individuals in association with immune cells after i.v. infusion and can gain access to the tumour tissue [133,141]. In a reovirus brain trial (EudraCT) 2011-005635-10), reovirus was found in six of nine brain tumours by immunohistochemistry (IHC) and nine of nine tumours by electron microscopy [109] after a single viral infusion. In its predecessor REO-013, reovirus protein was also found in nine of 10 colorectal cancer liver metastases by IHC [133]. Remarkably, in REO-020, it was in patients exhibiting some of the highest NAb titres that reovirus was successfully detected in the tumour [142]. Therefore, it appears that elimination of circulating NAbs is not essential for effective viral delivery. In fact, NAbs may play an important role in controlling toxicity, a phenomenon highlighted in mice with reduced NAbs (due to CPA treatment), and mirrored in B cell-deficient mice, where reovirus replication occurring in the heart and other organs proved lethal [114]. Although not severe, the identification of occasional hepatic and cardiac toxicities in some trials combining reovirus with chemotherapy emphasises the importance of NAbs in systemic virotherapy [140]. Perhaps a more important consideration in this matter is that immunosuppressive agents such as CPA could also dampen cell-mediated immunity [136] and compromise the development of long-term anti-tumour immune responses. Thus, identifying appropriate dosing schedules is essential. For example, low-dose CPA effectively enhances reovirus delivery to tumours while maintaining protective NAb levels [114] and, crucially, has the potential to promote the development of anti-tumour immunity [143,144], although in the context of reovirus this remains unknown

Given the initial belief that NAbs would be detrimental to efficacy, the concept of using cellular chaperones to deliver reovirus to tumours was explored. Immune cells have excellent tumour trafficking potential, and also have the potential to enhance anti-tumour immune effects. When administered i.v., reovirus naturally associates with a number of immune cells in the blood and can be detected on monocytes, NK cells, B cells and granulocytes [109]. Moreover, replication-competent reovirus associates with PBMC in seropositive patients [133,141] and strategies using human PBMC as reovirus carriers have demonstrated that DCs, T cells, and monocytes can act as protective cell carriers with efficient "hand-off" to tumour cells, despite pre-existing antiviral immunity [145–148]. Similarly, a heterogeneous population of lymphokine-activated killer cells and DCs can deliver reovirus to ovarian cancer cells in the presence of NAbs [149]. Of particular significance is the fact that mice co-treated with reovirus and granulocyte-macrophage colony stimulating factor (GM-CSF) were dependent on NAbs to achieve effective therapy, indicating that NAbs may in fact promote reovirus efficacy [147].

7. Unlocking the Potential of Reovirus with Combination Therapeutics

No matter which route of delivery is chosen, it remains clear that combination therapies will be necessary to optimise reovirus efficacy. Combination with radiotherapy has been investigated on the basis that activating mutations in *RAS* are associated with resistance to radiotherapy but confer sensitivity to reovirus. Twigger et al. reported that this treatment combination increased cell death in a number of cancer cell lines in vitro and in vivo, particularly in cell lines that showed only moderate reovirus sensitivity [150]. Similarly, the combination of reovirus with radiotherapy enhanced therapeutic outcomes in two models of paediatric sarcoma [151]. In both studies, the enhanced therapeutic outcome appeared to be due to increased direct cytotoxicity.

Multiple studies have investigated the combination of reovirus with chemotherapeutic agents, with synergy being frequently observed. As with radiotherapy, the enhanced treatment effect appeared to be due to increased oncolysis. For example, treatment of a range of prostate cancer cell lines with reovirus plus docetaxel, paclitaxel, vincristine, cisplatin or doxorubicin led to increased apoptosis/necrosis in vitro and reovirus improved docetaxel therapy in a xenograft prostate cancer model [152]. Increased apoptosis and/or necrosis has also been demonstrated by the combination of reovirus with: cisplatin in a melanoma model [153]; cisplatin, gemcitabine or vinblastine in non-small cell lung cancer cell lines [125]; and cisplatin plus paclitaxel in both in vitro and in vivo models of head and neck cancer [154]. Collectively, this evidence suggests that the beneficial outcomes resulting from combining reovirus with chemotherapy agents are generally mediated through oncolysis rather than immune-mediated mechanisms. However, Gujar et al. suggested that improved survival following reovirus plus gemcitabine treatment in an ovarian cancer model was at least partly immune-mediated, with reduced numbers of MDSC in tumours and improved anti-tumour CTL responses [155].

The majority of chemotherapeutic agents induce apoptosis, though the mechanisms by which they do this differ: Paclitaxel utilizes different apoptotic pathways depending on its concentration [156]; tamoxifen and gemcitabine activate mitogen-activated protein kinase (MAPK) and p53-dependent pathways and upregulate pro-apoptotic factors [157,158]; while docetaxel induces a non-apoptotic mode of death [159]. Reovirus itself induces apoptosis but can also induce necroptosis, which requires later stages of infection [83]. The reported synergy between reovirus and chemotherapy agents may be due to the induction of this additional form of cell death; however, it could also be due to the ability of reovirus to increase the expression of pro-apoptotic Bcl-2 family proteins [160]. Of particular significance is the dependence of reovirus on apoptosis, which may make it sensitive to resistance mechanisms utilized by cancer cells to escape chemotherapy cytotoxicity. Indeed, our studies have shown that stromal cell support of malignant B cells and multiple myeloma cells can inhibit reovirus sensitivity, in line with that observed for standard of care (SOC) chemotherapy agents (data not shown).

More recently, reovirus has been combined successfully with more targeted cancer therapies. The majority of malignant melanomas carry activating mutations in the RAS-RAF-MEK-ERK signalling pathway, with *NRAS* and *BRAF* mutations being most common. Although inhibition of this pathway might be expected to antagonize reovirus-induced cytotoxicity, the combination of reovirus with small molecule inhibitors of BRAF or MEK actually enhanced ER stress-induced apoptosis [161]. Similarly, the combination of reovirus with bortezomib, a proteasome inhibitor that increases ER stress, increased apoptosis in multiple myeloma cell lines in vitro and improved outcomes in vivo [160]. Energy metabolism within cancer cells is now emerging as an important element for OV susceptibility. OVs, such as reovirus, utilise host metabolic pathways to provide essential nucleotides, lipids, and amino acids for virus propagation and as such, metabolic reprogramming has been considered as a strategy to potentiate OV efficacy [162]. In the context of reovirus, susceptibility has been reported to correlate with pyruvate metabolism and oxidative stress, with a central role for pyruvate dehydrogenase (PDH). Specifically, the early oxidative stress response following reovirus treatment inhibits pyruvate dehydrogenase (PDH), via PDH kinase (PDK) phosphorylation, and induces a metabolic state that does not support reovirus replication. However, reactivation of PDH, using the PDK inhibitors dichloroacetate and AZD7545 enhanced reovirus efficacy in vitro and in vivo.

Therefore, metabolic reprogramming is a promising approach to increase the therapeutic potential of reovirus in cancer patients [163]. Another interesting study found that pre-conditioning tumours with bevacizumab—a vascular endothelial growth factor (VEGF) inhibitor—and then withdrawing treatment, rendered endothelial cells susceptible to reovirus infection, induced vascular collapse and promoted immune-mediated tumour clearance [164]. Similar effects were also observed following withdrawal of paclitaxel-mediated inhibition of VEGF signalling [165].

Other combination strategies have focused on boosting immune-mediated anti-tumour effects. For instance, combining reovirus with oncolytic vesicular stomatitis virus (VSV) in a dual-OV "prime-boost" regimen led to improved melanoma therapy via induction of different arms of the immune response; VSV induced a melanoma-specific T_H17 response which augmented the T_H1 response induced by reovirus [166]. As discussed above, cell carriage of reovirus by circulating myeloid cells has been potentiated by pre-conditioning the host with GM-CSF to expand immune effector populations [147]. Another strategy that has demonstrated successful results in several cancer models is the combination of reovirus with immune checkpoint inhibitors. Rajani et al. showed that the combination of i.t. reovirus with systemic anti-programmed cell death protein 1 (PD-1) enhanced survival in melanoma-bearing mice compared to either therapy alone [106]. Addition of checkpoint blockade to the dual OV "prime-boost" approach described above also enhanced survival [166]. Three studies have also demonstrated that reovirus can "sensitize" tumours to subsequent checkpoint blockade: (i) reovirus treatment of multiple myeloma cells in vitro increased PD-L1 expression, with systemic reovirus treatment followed by anti-programmed death-ligand 1 (PD-L1) increasing survival in a syngeneic model of multiple myeloma [110]; (ii) increased PD-L1 expression was observed in high grade glioma patients following reovirus treatment and systemic reovirus/anti-PD-1 therapy improved survival in a syngeneic, orthotopic murine glioma model [109]; and (iii) i.t. reovirus increased both PD-L1 expression on tumour cells and the number of intra-tumoral T_{regs} in a murine breast cancer model, while combination reovirus/anti-PD-1 treatment enhanced survival by reducing T_{reg} numbers and improving tumour-specific CTL responses [167]. More recently, reovirus has also been used in combination with CD3-bispecific antibodies. Reovirus-induced IFN stimulated the recruitment of NK cells and reovirus-specific $CD8^+$ T cells to the tumour site. Non-exhausted reovirus-specific effector T cells acted in synergy with CD3- bispecific antibodies to reduce the in vivo growth of multiple tumour types including pancreas, melanoma and breast; moreover, reovirus preconditioning was required for maximal efficacy. Importantly, combination treatment was also effective at distant lesions, not injected with reovirus, demonstrating the potential of this strategy for the treatment of metastatic disease.

8. Reovirus Clinical Trials

Reovirus T3D is the subject of one of the largest clinical trial programmes in oncolytic virotherapy (OVT). The clinical grade formulation of reovirus is now marketed as pelareorep (formerly Reolysin®) by Oncolytics Biotech Inc. (Calgary, AB, Canada). The virus is listed in 26 trials identified on www.clinicaltrials.gov. As of 2018, reovirus holds orphan drug status from the FDA for glioma, ovarian, pancreatic, peritoneal and gastric cancers, and from the European Medicines Agency (EMA, Amsterdam, The Netherlands) for ovarian and pancreatic cancer.

The first-in-man phase I study of reovirus, REO-001, enrolled 19 patients with accessible, advanced malignancies, who were treated intra-tumourally with ascending doses of the virus. No dose-limiting toxicities were observed, all being grade two or below, with nausea, headache or vomiting being the most common [168]. Tumour responses were apparent in 37% of patients. Based on this and its promising safety profile in animal models, reovirus progressed quickly into trials of systemic treatment. Intravenous delivery was first tested in REO-004. Eighteen patients with advanced solid tumours received virus doses of up to 3×10^{10} TCID$_{50}$ without identifying dose-limiting toxicity. In fact, only two patients experienced grade two events, even when multiple doses were given on successive days [169]. When corroborated by other phase I trials [112,170], these results demonstrated that when delivered by infusion as a very large, non-physiological bolus, reovirus is remarkably well tolerated. Interestingly, i.v. administration of reovirus in a phase I trial of heavily pre-treated patients with advanced cancers increased the number of CD4^{+ve} T cells, CD8^{+ve} T cells and NK cells, as well as cytokine levels, in the blood, suggesting the onset of an immune response. Significantly, i.v. administration of reovirus in brain tumours also led to a local IFN response with recruitment of CTLs [109].

Reovirus has now undergone further evaluation in phase I and II clinical trials across a range of indications; summarised in Table 1. Historically, the tumours most heavily targeted within the reovirus programme have been melanoma, myeloma and glioma [142,171,172], although trials have also included pancreatic, lung, breast, colorectal, prostate, and head and neck cancers [108,109,173–176]. Initial trials deployed reovirus as a monotherapy, the majority utilising i.v. administration; safety was established in the almost total absence of serious adverse events [177], with equivocal outcomes reported in phase II trials [142,178]. The mixed outcomes of patient response in clinical trials have made the therapeutic potential of reovirus a topic of debate. It is accurate to state that i.v. reovirus has often shown very modest activity, particularly as a monotherapy [142]. However, it reliably gains access to tumour lesions when administered systemically [109,133]. Currently, the virus is no longer under active investigation as a monotherapy and Oncolytics Biotech Inc. is instead developing combination programmes (www.oncolyticsbiotech.com).

Table 1. Summary of Reovirus Clinical Trials.

Disease	Combinations	Phase	Trial ID	Route	Dose(s) TCID$_{50}$	Results
Gliomas	N/A	I	NCT00528684	I.T	1×10^7, 1×10^8, 1×10^9	No DLT, 10/12 patients had PD, 1/12 SD and 1/12 patients unevaluable for response, but alive >4.5 years post treatment [171].
	N/A	I	EudraCT 2011-005635-10	I.V	1×10^{10}	Reovirus detected in within tumours and increased CTL infiltration [109].
Brain cancer	Sargramostim (GM-CSF)	I	NCT02444546	I.V	MTD	Ongoing
Pancreatic cancer	Carboplatin and Paclitaxel	II	NCT01280058	I.V	3×10^{10}	No significant enhancement of PFS with reovirus combination therapy ($n = 36$) vs. Carboplatin/Paclitaxel alone ($n = 37$) (4.9 vs. 5.2 months) [108].
	Pembrolizumab and 5 Fluorouracil or gemcitabine or irinotecan	I	NCT02620423	I.V	4.5×10^{10}	Well tolerated. 3/10 evaluable patients had SD, 1 of which had PR for 17.4 months. Biopsies show reovirus infection in tumour cells and immune infiltrates [179].
	Pembrolizumab	II	NCT03723915	I.V	Not reported	Ongoing
	Gemcitabine	II	NCT00998322	I.V	1×10^{10}	Well tolerated. 1/29 patients had PR, 23/29 SD, 5/29 PD. Single patient with SD had upregulated expression of PD-L1 following treatment [180].
Colorectal cancer	Irinotecan and Leucovorin and 5-Fluorouracil	I	NCT01274624	I.V	1×10^{10}– 3×10^{10}	2/21 patients had DLT. 18/21 evaluable for response. 1/18 PR, 9/18 SD, 8/18 PD [174].
	Leucovorin and 5-Fluorouracil and Oxaliplatin and Bevacizumab	II	NCT01622543	I.V	3×10^{10}	Poorer PFS with reovirus combination therapy (7 months vs. 9 months). No significant difference in OS [181].
Head and Neck Cancers	Carboplatin and Paclitaxel	II	NCT00753038	I.V	3×10^{10}	Well tolerated. 4/13 evaluable patients had PR, 2/13 had SD for >12 weeks [173].
	Carboplatin and Paclitaxel	III	NCT01166542	I.V	3×10^{10}	Interim results reported (www.oncolyticsbiotech.com). 118 evaluable patients, reovirus increased PFS from 48 to 95 days. Significantly increased OS. Curtailed to larger phase II trial.
Melanoma	N/A	II	NCT00651157	I.V	3×10^{10}	Well tolerated, viral replication was detected in 2/15 patients despite NAb, average PFS 45 days [142].
	Carboplatin and Paclitaxel	II	NCT00984464	I.V	3×10^{10}	Well tolerated. 3/14 patients had PR, 9/14 SD, 2/14 PD. ORR of 21%, no complete responses [182].
Multiple Myeloma	N/A	I	NCT01533194	I.V	3×10^9, 3×10^{10}	No DLT reported, reovirus localization to BM, SD for up to 8 months [172].
	Lenalidomide or Pomalidomide	I	NCT03015922	I.V	3×10^{10}	Ongoing
	Dexamethasone and Carfilzomib	I	NCT02101944	I.V	MTD	Recruiting
	Dexamethasone and Bortezomib	I	NCT02514382	I.V	MTD up to 4.5×10^{10}	Ongoing
	Dexamethasone and Carfilzomib and Nivolumab	I	NCT03605719	I.V	MTD	Recruiting

Table 1. *Cont.*

Disease	Combinations	Phase	Trial ID	Route	Dose(s) TCID$_{50}$	Results
Lung Cancer	Carboplatin or Paclitaxel	II	NCT00861627	I.V	3×10^{10}	11/37 of patients PR, 20/37 SD, PFS 4 months [175].
	Carboplatin or Paclitaxel	II	NCT00998192	I.V	3×10^{10}	Treatment well tolerated, 12/25 patients had PR, 10/25 SD, 3/25 PD [183].
	Pemetrexed or Docetaxel	II	NCT01708993	I.V	4.5×10^{10}	Virus was well tolerated, no enhancement of PFS with reovirus vs. drugs alone (2.96 vs. 2.83 months) [184].
Prostate cancer	Docetaxel and Prednisone	II	NCT01619813	I.V	3×10^{10}	Poorer OS in virus and drug combination arm, vs. drug alone [185].
Breast cancer	Paclitaxel	II	NCT01656538	I.V	3×10^{10}	Combination arm showed improved OS vs. drug alone arm (17.4 vs. 10.4 months) [176].
	Avelumab and Paclitaxel	II	NCT04215146	I.V	4.5×10^{10}	Recruiting
	Retifanlimab	II	NCT04445844	I.V	MTD	Recruiting
Ovarian cancer	Paclitaxel	II	NCT01166542	I.V	3×10^{10}	Median PFS 4.3 months and ORR 20% for patients receiving Pacitaxel alone vs. 4.4 months and 17.4%, for combination treatment. Addition of reovirus to treatment does not reduce the hazard of progression or death [186].
Bone and soft tissue sarcoma	N/A	II	NCT00503295	I.V	3×10^{10}	Well tolerated. 14/33 patients had SD for >2 months, including 5 patients which had SD for >6 months [178].
Advanced cancer	Radiotherapy	I		I.T	$1 \times 10^{8} - 1 \times 10^{10}$	No DLT. Low dose radiation arm 2/7 PR and 5/7 SD. High dose radiation arm 5/7 PR and 2/7 SD [187].
	Carboplatin and Paclitaxel	I		I.V	3×10^{9}, 1×10^{10}, 3×10^{10}	No DLT. 1/26 patients had CR, 6/26 PR, 9/26 SD, 2/25 major clinical response, and 9/25 PD [139].
	Docetaxel	I		I.V	3×10^{9}, 1×10^{10}, 3×10^{10}	MTD not reached. 1/16 patients had CR, 3/16 PR, 3/16 minor response, 7/16 SD, 2/16 PD [188].
	Gemcitabine	I		I.V	1×10^{9} 3×10^{9}, 1×10^{10}, 3×10^{10}	3/16 patients had DLT. 10/16 patients evaluable for response, 1/10 PR, 6/10 SD, 3/10 PD [140].

DLT: dose-limiting toxicity, PFS: progression-free survival, PR: partial response, ORR: overall response rate, SD: stable disease, PD: progressive disease, IT: intra-tumoural, I.V: intravenous, MTD: maximum tolerated dose, PD-L1: anti-programmed death-ligand 1, BM: bone marrow, OS: overall survival.

9. The Future for Reovirus—Pre-Clinical Requirements and Clinical Considerations

In spite of its efficacy in pre-clinical models, reovirus treatment (as with other OVs) has benefited only a minority of patients. Figure 2 highlights some possible reasons for this and summarizes what we currently know about reovirus (the inner circle) along with some priority areas of research which should aid the development of more effective reovirus therapies (the outer circle). Currently it remains unclear how best to administer reovirus in order to obtain optimal therapeutic responses while maintaining safety. The route designed to maximize efficacy via oncolysis may differ from that designed to facilitate immune-mediated tumour clearance. Although translational studies reliably demonstrate that reovirus can access tumours after i.v. administration [109,133], a greater understanding of the effect of anti-reovirus immunity, both humoral and cell-mediated, is pivotal to maximize its clinical efficacy.

Born of the desire to accelerate clinical application, reovirus has generally been combined with SOC therapies. This has generally led to improved efficacy due to increased cytotoxicity but a more strategic approach, based on a complete understanding of the mechanisms of death induced by each therapy and the challenges faced within defined TMEs, would generate further improvements.

An important aspect of combination therapies is the dosing regimen employed. How many reovirus administrations are required? How frequent should they be? Should they be administered before, after or simultaneously with other agents? Currently, the treatment regimens employed in clinical trials reveal no consensus on what the optimum dosing schedule might be. The planned regimen for the most recent trial is 4.5×10^{10} TCID$_{50}$ reovirus i.v. on days 1/2/8/9/15/16 of a 28-day cycle, but other regimens have been used including delivery on days 1/2/3/4/5 of a 28-day cycle or days 1/2/3/8 for the first 21-day cycle and days 1/8 thereafter. These regimens may be pragmatic to facilitate combination with SOC therapies but they may not be the most efficacious. Going forward, it will be important to optimise chemotherapy-induced cytotoxicity while maintaining reovirus-mediated anti-tumour immunity. For example, chemotherapy agents that induce lymphopenia might abrogate immune responses, therefore careful selection of complementary chemotherapies is essential. Indeed, combination of reovirus with gemcitabine can improve anti-tumour immune responses [155] indicating that the two mechanisms can be compatible. Consideration of treatment regimens will be particularly important for combination with immune checkpoint inhibitors because anti-cytotoxic T lymphocyte-associated 4 (CTLA-4) antibodies are likely to potentiate early stages of T cell priming, whilst anti-PD1/anti-PD-L1 antibodies would act to reverse T cell exhaustion within the TME.

Whilst murine pre-clinical models will be essential to identify and validate novel reovirus combinations with improved efficacy, it is important to recognise, and reflect on, the limitations of many commonly used in vivo models. In particular, xenograft models utilizing immunocompromised mice do not consider OV-induced anti-tumour immune responses; moreover, syngeneic tumour models, in immunocompetent mice, do not always model tumour progression at the correct anatomical site. Although more advanced in vivo modes are available (e.g., spontaneous cancer models), which more accurately reflect disease progression, these are expensive and time consuming, restricting their use for many cancer researchers. Importantly, these models do not represent the heterogenous nature of patient tumours. Therefore, it is it imperative that clinical trials are designed to gain as much information as possible. Specifically, clinical trials should allow downstream interrogation of the tumour and the TME, including cancer-associated fibroblasts, immune cell components and soluble factors/extracellular vesicles. Ideally, multiple patient samples (e.g., blood and primary/secondary tumour tissue) should be obtained pre- and post-treatment to gain insight into why some patients may respond, whilst others do not. Detailed characterization of these samples will facilitate the development of more complex combination regimes to counteract resistance mechanisms and allow predictive biomarkers of response to be identified.

While genetic modification of other OV has improved efficacy in pre-clinical models, this approach has not been widely used with reovirus because the segmented RNA backbone makes it difficult to modify. Nevertheless, recent identification and characterisation of reovirus mutants isolated from human U118MG glioblastoma cells has revealed the capacity of JAM-A-independent (jin) mutants to infect JAM-A^{-ve} cells, which are usually resistant to wild-type virus [189]. Following this, a reverse genetics approach was developed to allow genetic modification of expanded-tropism jin mutants [190] and small transgenes including reporter constructs have been inserted [191,192]. This yields tremendous scope to develop novel, genetically engineered reovirus platforms, with enhanced tropism, increased infectivity and replication, and improved immune stimulation. Indeed, reovirus has recently been armed with functional GM-CSF to boost anti-tumour immunity [193]. In addition to the reovirus jin mutants, reassorted reovirus platforms are also undergoing pre-clinical development. Co-infection and serial passage of MDA-MB-231 cells with the prototype laboratory strains for reovirus (type one *Lang*, type two *Jones*, and type three *Dearing*) generated a reassorted virus with a predominant type one genetic composition and some type three gene segments which displayed enhanced infectivity and cytotoxicity in triple-negative breast cancer cells [194]. Moreover, the advancement of reovirus engineering has enabled mutations to be made that can counteract inhibitory mechanisms within the TME. In particular, mutations within σ1 have been incorporated to prevent proteolytic cleavage of σ1 by breast cancer-associated proteases, which abrogated binding to sialic acid; infectivity was restored in the σ1 mutants [195]. These innovations suggest a new and exciting era of reovirus research is emerging.

No single OV has emerged as the undisputed leader in terms of efficacy and it is unlikely that a "one size fits all" OV exists. Having demonstrated some clinical activity, reovirus remains a promising weapon in the cancer therapy arsenal where viral modifications, allied with informed scheduling and strategic combination with other treatments, should pay dividends for cancer patients.

Author Contributions: L.M., R.B., E.I., and F.E.-M., wrote and reviewed the manuscript. L.M. designed the graphical abstract and T.B., prepared the figures, the clinical trial table and reviewed the manuscript. E.I. and F.E.-M. were responsible for the conceptual design of the article. All authors have read and agreed to the published version of the manuscript.

Funding: Studentships for L.M. and R.B. were provided by the University of Leeds and T.B. is currently funded by the Bone Cancer Research Trust, grant number 113852. No addition external funding was received for this research.

Conflicts of Interest: The authors declare no conflict of interest.

References

1. Kelly, E.; Russell, S.J. History of oncolytic viruses: Genesis to genetic engineering. *Mol. Ther. J. Am. Soc. Gene Ther.* **2007**, *15*, 651–659. [CrossRef] [PubMed]
2. Breitbach, C.J.; Arulanandam, R.; De Silva, N.; Thorne, S.H.; Patt, R.; Daneshmand, M.; Moon, A.; Ilkow, C.; Burke, J.; Hwang, T.H.; et al. Oncolytic vaccinia virus disrupts tumor-associated vasculature in humans. *Cancer Res.* **2013**, *73*, 1265–1275. [CrossRef] [PubMed]
3. Ilkow, C.S.; Marguerie, M.; Batenchuk, C.; Mayer, J.; Ben Neriah, D.; Cousineau, S.; Falls, T.; Jennings, V.A.; Boileau, M.; Bellamy, D.; et al. Reciprocal cellular cross-talk within the tumor microenvironment promotes oncolytic virus activity. *Nat. Med.* **2015**, *21*, 530–536. [CrossRef] [PubMed]
4. Bauzon, M.; Hermiston, T. Armed therapeutic viruses—A disruptive therapy on the horizon of cancer immunotherapy. *Front. Immunol.* **2014**, *5*, 74. [CrossRef] [PubMed]
5. Garber, K. China approves world's first oncolytic virus therapy for cancer treatment. *J. Natl. Cancer Inst.* **2006**, *98*, 298–300. [CrossRef] [PubMed]
6. Randazzo, B.P.; Tal-Singer, R.; Zabolotny, J.M.; Kesari, S.; Fraser, N.W. Herpes simplex virus 1716, an ICP 34.5 null mutant, is unable to replicate in CV-1 cells due to a translational block that can be overcome by coinfection with SV40. *J. Gen. Virol* **1997**, *78 Pt 12*, 3333–3339. [CrossRef]
7. Liu, B.L.; Robinson, M.; Han, Z.Q.; Branston, R.H.; English, C.; Reay, P.; McGrath, Y.; Thomas, S.K.; Thornton, M.; Bullock, P.; et al. ICP34.5 deleted herpes simplex virus with enhanced oncolytic, immune stimulating, and anti-tumour properties. *Gene Ther.* **2003**, *10*, 292–303. [CrossRef] [PubMed]
8. Lelli, D.; Beato, M.S.; Cavicchio, L.; Lavazza, A.; Chiapponi, C.; Leopardi, S.; Baioni, L.; De Benedictis, P.; Moreno, A. First identification of mammalian orthoreovirus type 3 in diarrheic pigs in Europe. *Virol. J.* **2016**, *13*, 139. [CrossRef]

9.   Steyer, A.; Gutierrez-Aguire, I.; Kolenc, M.; Koren, S.; Kutnjak, D.; Pokorn, M.; Poljsak-Prijatelj, M.; Racki, N.; Ravnikar, M.; Sagadin, M.; et al. High similarity of novel orthoreovirus detected in a child hospitalized with acute gastroenteritis to mammalian orthoreoviruses found in bats in Europe. *J. Clin. Microbiol.* **2013**, *51*, 3818–3825. [CrossRef]

10.   Rosen, L.; Hovis, J.F.; Mastrota, F.M.; Bell, J.A.; Huebner, R.J. Observations on a newly recognized virus (Abney) of the reovirus family. *Am. J. Hyg.* **1960**, *71*, 258–265. [CrossRef]

11.   Gaillard, R.K., Jr.; Joklik, W.K. Quantitation of the relatedness of reovirus serotypes 1, 2, and 3 at the gene level. *Virology* **1982**, *123*, 152–164. [CrossRef]

12.   Cashdollar, L.W.; Chmelo, R.A.; Wiener, J.R.; Joklik, W.K. Sequences of the S1 genes of the three serotypes of reovirus. *Proc. Natl. Acad. Sci. USA* **1985**, *82*, 24–28. [CrossRef] [PubMed]

13.   Hashiro, G.; Loh, P.C.; Yau, J.T. The preferential cytotoxicity of reovirus for certain transformed cell lines. *Arch. Virol.* **1977**, *54*, 307–315. [CrossRef] [PubMed]

14.   Ramos-Alvarez, M.; Sabin, A.B. Characteristics of poliomyelitis and other enteric viruses recovered in tissue culture from healthy American children. *Proc. Soc. Exp. Biol. Med.* **1954**, *87*, 655–661. [CrossRef]

15.   Ramos-Alvarez, M.; Sabin, A.B. Enteropathogenic viruses and bacteria; role in summer diarrheal diseases of infancy and early childhood. *J. Am. Med. Assoc.* **1958**, *167*, 147–156. [CrossRef]

16.   Chua, K.B.; Crameri, G.; Hyatt, A.; Yu, M.; Tompang, M.R.; Rosli, J.; McEachern, J.; Crameri, S.; Kumarasamy, V.; Eaton, B.T.; et al. A previously unknown reovirus of bat origin is associated with an acute respiratory disease in humans. *Proc. Natl. Acad. Sci. USA* **2007**, *104*, 11424–11429. [CrossRef]

17.   Tillotson, J.R.; Lerner, A.M. Reovirus type 3 associated with fatal pneumonia. *N. Engl. J. Med.* **1967**, *276*, 1060–1063. [CrossRef]

18.   Ouattara, L.A.; Barin, F.; Barthez, M.A.; Bonnaud, B.; Roingeard, P.; Goudeau, A.; Castelnau, P.; Vernet, G.; Paranhos-Baccala, G.; Komurian-Pradel, F. Novel human reovirus isolated from children with acute necrotizing encephalopathy. *Emerg. Infect. Dis.* **2011**, *17*, 1436–1444. [CrossRef]

19.   Tyler, K.L.; Barton, E.S.; Ibach, M.L.; Robinson, C.; Campbell, J.A.; O'Donnell, S.M.; Valyi-Nagy, T.; Clarke, P.; Wetzel, J.D.; Dermody, T.S. Isolation and molecular characterization of a novel type 3 reovirus from a child with meningitis. *J. Infect. Dis.* **2004**, *189*, 1664–1675. [CrossRef]

20.   Morecki, R.; Glaser, J.H.; Cho, S.; Balistreri, W.F.; Horwitz, M.S. Biliary atresia and reovirus type 3 infection. *N. Engl. J. Med.* **1982**, *307*, 481–484. [CrossRef]

21.   Richardson, S.C.; Bishop, R.F.; Smith, A.L. Reovirus serotype 3 infection in infants with extrahepatic biliary atresia or neonatal hepatitis. *J. Gastroenterol. Hepatol.* **1994**, *9*, 264–268. [CrossRef] [PubMed]

22.   Bouziat, R.; Hinterleitner, R.; Brown, J.J.; Stencel-Baerenwald, J.E.; Ikizler, M.; Mayassi, T.; Meisel, M.; Kim, S.M.; Discepolo, V.; Pruijssers, A.J.; et al. Reovirus infection triggers inflammatory responses to dietary antigens and development of celiac disease. *Science* **2017**, *356*, 44–50. [CrossRef]

23.   Dermody, T.; Parker, J.; Sherry, B. *Fields Virology*, 6th ed.; Knipe, D.M., Howley, P.M., Eds.; Lippincott, Williams & Wilkins: Philadelphia, PA, USA, 2013; Volume 2, pp. 1304–1346.

24.   Lemay, G. Transcriptional and translational events during reovirus infection. *Biochem. Cell Biol.* **1988**, *66*, 803–812. [CrossRef] [PubMed]

25.   Shatkin, A.J.; Sipe, J.D.; Loh, P. Separation of ten reovirus genome segments by polyacrylamide gel electrophoresis. *J. Virol.* **1968**, *2*, 986–991. [CrossRef] [PubMed]

26.   Chandran, K.; Zhang, X.; Olson, N.H.; Walker, S.B.; Chappell, J.D.; Dermody, T.S.; Baker, T.S.; Nibert, M.L. Complete in vitro assembly of the reovirus outer capsid produces highly infectious particles suitable for genetic studies of the receptor-binding protein. *J. Virol.* **2001**, *75*, 5335–5342. [CrossRef]

27.   Yue, Z.; Shatkin, A.J. Double-stranded RNA-dependent protein kinase (PKR) is regulated by reovirus structural proteins. *Virology* **1997**, *234*, 364–371. [CrossRef] [PubMed]

28.   Stanifer, M.L.; Kischnick, C.; Rippert, A.; Albrecht, D.; Boulant, S. Reovirus inhibits interferon production by sequestering IRF3 into viral factories. *Sci. Rep.* **2017**, *7*, 10873. [CrossRef]

29.   Barton, E.S.; Connolly, J.L.; Forrest, J.C.; Chappell, J.D.; Dermody, T.S. Utilization of sialic acid as a coreceptor enhances reovirus attachment by multistep adhesion strengthening. *J. Biol. Chem.* **2001**, *276*, 2200–2211. [CrossRef]

30.   Chappell, J.D.; Duong, J.L.; Wright, B.W.; Dermody, T.S. Identification of carbohydrate-binding domains in the attachment proteins of type 1 and type 3 reoviruses. *J. Virol.* **2000**, *74*, 8472–8479. [CrossRef]

31. Antar, A.A.; Konopka, J.L.; Campbell, J.A.; Henry, R.A.; Perdigoto, A.L.; Carter, B.D.; Pozzi, A.; Abel, T.W.; Dermody, T.S. Junctional adhesion molecule-A is required for hematogenous dissemination of reovirus. *Cell Host Microbe* **2009**, *5*, 59–71. [CrossRef]

32. Barton, E.S.; Forrest, J.C.; Connolly, J.L.; Chappell, J.D.; Liu, Y.; Schnell, F.J.; Nusrat, A.; Parkos, C.A.; Dermody, T.S. Junction adhesion molecule is a receptor for reovirus. *Cell* **2001**, *104*, 441–451. [CrossRef]

33. Shmulevitz, M.; Gujar, S.A.; Ahn, D.G.; Mohamed, A.; Lee, P.W. Reovirus variants with mutations in genome segments S1 and L2 exhibit enhanced virion infectivity and superior oncolysis. *J. Virol.* **2012**, *86*, 7403–7413. [CrossRef] [PubMed]

34. Sugano, Y.; Takeuchi, M.; Hirata, A.; Matsushita, H.; Kitamura, T.; Tanaka, M.; Miyajima, A. Junctional adhesion molecule-A, JAM-A, is a novel cell-surface marker for long-term repopulating hematopoietic stem cells. *Blood* **2008**, *111*, 1167–1172. [CrossRef]

35. Parrish, C.; Scott, G.B.; Migneco, G.; Scott, K.J.; Steele, L.P.; Ilett, E.; West, E.J.; Hall, K.; Selby, P.J.; Buchanan, D.; et al. Oncolytic reovirus enhances rituximab-mediated antibody-dependent cellular cytotoxicity against chronic lymphocytic leukaemia. *Leukemia* **2015**, *29*, 1799–1810. [CrossRef]

36. Kelly, K.R.; Espitia, C.M.; Zhao, W.; Wendlandt, E.; Tricot, G.; Zhan, F.; Carew, J.S.; Nawrocki, S.T. Junctional adhesion molecule-A is overexpressed in advanced multiple myeloma and determines response to oncolytic reovirus. *Oncotarget* **2015**, *6*, 41275–41289. [CrossRef] [PubMed]

37. Hall, K.; Scott, K.J.; Rose, A.; Desborough, M.; Harrington, K.; Pandha, H.; Parrish, C.; Vile, R.; Coffey, M.; Bowen, D.; et al. Reovirus-mediated cytotoxicity and enhancement of innate immune responses against acute myeloid leukemia. *BioRes. Open Access* **2012**, *1*, 3–15. [CrossRef]

38. Xu, P.P.; Sun, Y.F.; Fang, Y.; Song, Q.; Yan, Z.X.; Chen, Y.; Jiang, X.F.; Fei, X.C.; Zhao, Y.; Leboeuf, C.; et al. JAM-A overexpression is related to disease progression in diffuse large B-cell lymphoma and downregulated by lenalidomide. *Sci. Rep.* **2017**, *7*, 7433. [CrossRef] [PubMed]

39. Zhao, C.; Lu, F.; Chen, H.; Zhao, X.; Sun, J.; Chen, H. Dysregulation of JAM-A plays an important role in human tumor progression. *Int. J. Clin. Exp. Pathol.* **2014**, *7*, 7242–7248.

40. Zhang, M.; Luo, W.; Huang, B.; Liu, Z.; Sun, L.; Zhang, Q.; Qiu, X.; Xu, K.; Wang, E. Overexpression of JAM-A in non-small cell lung cancer correlates with tumor progression. *PLoS ONE* **2013**, *8*, e79173. [CrossRef]

41. McSherry, E.A.; McGee, S.F.; Jirstrom, K.; Doyle, E.M.; Brennan, D.J.; Landberg, G.; Dervan, P.A.; Hopkins, A.M.; Gallagher, W.M. JAM-A expression positively correlates with poor prognosis in breast cancer patients. *Int. J. Cancer* **2009**, *125*, 1343–1351. [CrossRef]

42. Alain, T.; Kim, T.S.; Lun, X.; Liacini, A.; Schiff, L.A.; Senger, D.L.; Forsyth, P.A. Proteolytic disassembly is a critical determinant for reovirus oncolysis. *Mol. Ther. J. Am. Soc. Gene Ther.* **2007**, *15*, 1512–1521. [CrossRef]

43. Golden, J.W.; Linke, J.; Schmechel, S.; Thoemke, K.; Schiff, L.A. Addition of exogenous protease facilitates reovirus infection in many restrictive cells. *J. Virol.* **2002**, *76*, 7430–7443. [CrossRef] [PubMed]

44. Ebert, D.H.; Deussing, J.; Peters, C.; Dermody, T.S. Cathepsin L and cathepsin B mediate reovirus disassembly in murine fibroblast cells. *J. Biol. Chem.* **2002**, *277*, 24609–24617. [CrossRef] [PubMed]

45. Chandran, K.; Farsetta, D.L.; Nibert, M.L. Strategy for nonenveloped virus entry: A hydrophobic conformer of the reovirus membrane penetration protein micro 1 mediates membrane disruption. *J. Virol.* **2002**, *76*, 9920–9933. [CrossRef] [PubMed]

46. Nibert, M.L.; Odegard, A.L.; Agosto, M.A.; Chandran, K.; Schiff, L.A. Putative autocleavage of reovirus mu1 protein in concert with outer-capsid disassembly and activation for membrane permeabilization. *J. Mol. Biol.* **2005**, *345*, 461–474. [CrossRef]

47. Odegard, A.L.; Chandran, K.; Zhang, X.; Parker, J.S.; Baker, T.S.; Nibert, M.L. Putative autocleavage of outer capsid protein micro1, allowing release of myristoylated peptide micro1N during particle uncoating, is critical for cell entry by reovirus. *J. Virol.* **2004**, *78*, 8732–8745. [CrossRef] [PubMed]

48. Li, J.K.; Scheible, P.P.; Keene, J.D.; Joklik, W.K. The plus strand of reovirus gene S2 is identical with its in vitro transcript. *Virology* **1980**, *105*, 282–286. [CrossRef]

49. Fields, B.N.; Raine, C.S.; Baum, S.G. Temperature-sensitive mutants of reovirus type 3: Defects in viral maturation as studied by immunofluorescence and electron microscopy. *Virology* **1971**, *43*, 569–578. [CrossRef]

50. Kobayashi, T.; Chappell, J.D.; Danthi, P.; Dermody, T.S. Gene-specific inhibition of reovirus replication by RNA interference. *J. Virol.* **2006**, *80*, 9053–9063. [CrossRef]

51. Antczak, J.B.; Joklik, W.K. Reovirus genome segment assortment into progeny genomes studied by the use of monoclonal antibodies directed against reovirus proteins. *Virology* **1992**, *187*, 760–776. [CrossRef]

52. McDonald, S.M.; Patton, J.T. Assortment and packaging of the segmented rotavirus genome. *Trends Microbiol.* **2011**, *19*, 136–144. [CrossRef] [PubMed]

53. Connolly, J.L.; Barton, E.S.; Dermody, T.S. Reovirus binding to cell surface sialic acid potentiates virus-induced apoptosis. *J. Virol.* **2001**, *75*, 4029–4039. [CrossRef] [PubMed]

54. Lai, C.M.; Mainou, B.A.; Kim, K.S.; Dermody, T.S. Directional release of reovirus from the apical surface of polarized endothelial cells. *mBio* **2013**, *4*, e00049-13. [CrossRef] [PubMed]

55. Strong, J.E.; Tang, D.; Lee, P.W. Evidence that the epidermal growth factor receptor on host cells confers reovirus infection efficiency. *Virology* **1993**, *197*, 405–411. [CrossRef]

56. Tang, D.; Strong, J.E.; Lee, P.W. Recognition of the epidermal growth factor receptor by reovirus. *Virology* **1993**, *197*, 412–414. [CrossRef]

57. Strong, J.E.; Lee, P.W. The v-erbB oncogene confers enhanced cellular susceptibility to reovirus infection. *J. Virol.* **1996**, *70*, 612–616. [CrossRef]

58. Strong, J.E.; Coffey, M.C.; Tang, D.; Sabinin, P.; Lee, P.W. The molecular basis of viral oncolysis: Usurpation of the Ras signaling pathway by reovirus. *EMBO J.* **1998**, *17*, 3351–3362. [CrossRef]

59. Downward, J. Targeting RAS signalling pathways in cancer therapy. *Nat. Rev. Cancer* **2003**, *3*, 11–22. [CrossRef]

60. Marcato, P.; Shmulevitz, M.; Pan, D.; Stoltz, D.; Lee, P.W. Ras transformation mediates reovirus oncolysis by enhancing virus uncoating, particle infectivity, and apoptosis-dependent release. *Mol. Ther. J. Am. Soc. Gene Ther.* **2007**, *15*, 1522–1530. [CrossRef]

61. Prior, I.A.; Lewis, P.D.; Mattos, C. A comprehensive survey of Ras mutations in cancer. *Cancer Res.* **2012**, *72*, 2457–2467. [CrossRef]

62. Shmulevitz, M.; Marcato, P.; Lee, P.W. Unshackling the links between reovirus oncolysis, Ras signaling, translational control and cancer. *Oncogene* **2005**, *24*, 7720–7728. [CrossRef] [PubMed]

63. Norman, K.L.; Hirasawa, K.; Yang, A.D.; Shields, M.A.; Lee, P.W. Reovirus oncolysis: The Ras/RalGEF/p38 pathway dictates host cell permissiveness to reovirus infection. *Proc. Natl. Acad. Sci. USA* **2004**, *101*, 11099–11104. [CrossRef]

64. Smakman, N.; van den Wollenberg, D.J.; Borel Rinkes, I.H.; Hoeben, R.C.; Kranenburg, O. Sensitization to apoptosis underlies KrasD12-dependent oncolysis of murine C26 colorectal carcinoma cells by reovirus T3D. *J. Virol.* **2005**, *79*, 14981–14985. [CrossRef]

65. Sadler, A.J.; Williams, B.R. Structure and function of the protein kinase R. *Curr. Top. Microbiol. Immunol.* **2007**, *316*, 253–292. [CrossRef]

66. Christian, S.L.; Zu, D.; Licursi, M.; Komatsu, Y.; Pongnopparat, T.; Codner, D.A.; Hirasawa, K. Suppression of IFN-induced transcription underlies IFN defects generated by activated Ras/MEK in human cancer cells. *PLoS ONE* **2012**, *7*, e44267. [CrossRef] [PubMed]

67. de Haro, C.; Mendez, R.; Santoyo, J. The eIF-2alpha kinases and the control of protein synthesis. *FASEB J.* **1996**, *10*, 1378–1387. [CrossRef] [PubMed]

68. Gong, J.; Mita, M.M. Activated ras signaling pathways and reovirus oncolysis: An update on the mechanism of preferential reovirus replication in cancer cells. *Front. Oncol.* **2014**, *4*, 167. [CrossRef]

69. Kim, M.; Egan, C.; Alain, T.; Urbanski, S.J.; Lee, P.W.; Forsyth, P.A.; Johnston, R.N. Acquired resistance to reoviral oncolysis in Ras-transformed fibrosarcoma cells. *Oncogene* **2007**, *26*, 4124–4134. [CrossRef]

70. Smakman, N.; van der Bilt, J.D.; van den Wollenberg, D.J.; Hoeben, R.C.; Borel Rinkes, I.H.; Kranenburg, O. Immunosuppression promotes reovirus therapy of colorectal liver metastases. *Cancer Gene Ther.* **2006**, *13*, 815–818. [CrossRef]

71. Song, L.; Ohnuma, T.; Gelman, I.H.; Holland, J.F. Reovirus infection of cancer cells is not due to activated Ras pathway. *Cancer Gene Ther.* **2009**, *16*, 382. [CrossRef]

72. Twigger, K.; Roulstone, V.; Kyula, J.; Karapanagiotou, E.M.; Syrigos, K.N.; Morgan, R.; White, C.; Bhide, S.; Nuovo, G.; Coffey, M.; et al. Reovirus exerts potent oncolytic effects in head and neck cancer cell lines that are independent of signalling in the EGFR pathway. *BMC Cancer* **2012**, *12*, 368. [CrossRef]

73. Clarke, P.; Richardson-Burns, S.M.; DeBiasi, R.L.; Tyler, K.L. Mechanisms of apoptosis during reovirus infection. *Curr. Top. Microbiol. Immunol.* **2005**, *289*, 1–24. [CrossRef] [PubMed]

74. Deb, A.; Haque, S.J.; Mogensen, T.; Silverman, R.H.; Williams, B.R. RNA-dependent protein kinase PKR is required for activation of NF-kappa B by IFN-gamma in a STAT1-independent pathway. *J. Immunol.* **2001**, *166*, 6170–6180. [CrossRef] [PubMed]

75. Goubau, D.; Schlee, M.; Deddouche, S.; Pruijssers, A.J.; Zillinger, T.; Goldeck, M.; Schuberth, C.; Van der Veen, A.G.; Fujimura, T.; Rehwinkel, J.; et al. Antiviral immunity via RIG-I-mediated recognition of RNA bearing 5′-diphosphates. *Nature* **2014**, *514*, 372–375. [CrossRef]

76. Sherry, B. Rotavirus and reovirus modulation of the interferon response. *J. Interferon Cytokine Res.* **2009**, *29*, 559–567. [CrossRef] [PubMed]

77. Smith, J.A.; Schmechel, S.C.; Williams, B.R.; Silverman, R.H.; Schiff, L.A. Involvement of the interferon-regulated antiviral proteins PKR and RNase L in reovirus-induced shutoff of cellular translation. *J. Virol.* **2005**, *79*, 2240–2250. [CrossRef] [PubMed]

78. Clarke, P.; Meintzer, S.M.; Gibson, S.; Widmann, C.; Garrington, T.P.; Johnson, G.L.; Tyler, K.L. Reovirus-induced apoptosis is mediated by TRAIL. *J. Virol.* **2000**, *74*, 8135–8139. [CrossRef]

79. Kominsky, D.J.; Bickel, R.J.; Tyler, K.L. Reovirus-induced apoptosis requires both death receptor- and mitochondrial-mediated caspase-dependent pathways of cell death. *Cell Death Differ.* **2002**, *9*, 926–933. [CrossRef]

80. Kominsky, D.J.; Bickel, R.J.; Tyler, K.L. Reovirus-induced apoptosis requires mitochondrial release of Smac/DIABLO and involves reduction of cellular inhibitor of apoptosis protein levels. *J. Virol.* **2002**, *76*, 11414–11424. [CrossRef]

81. Knowlton, J.J.; Dermody, T.S.; Holm, G.H. Apoptosis induced by mammalian reovirus is beta interferon (IFN) independent and enhanced by IFN regulatory factor 3- and NF-kappaB-dependent expression of Noxa. *J. Virol.* **2012**, *86*, 1650–1660. [CrossRef]

82. Berger, A.K.; Danthi, P. Reovirus activates a caspase-independent cell death pathway. *mBio* **2013**, *4*, e00178-13. [CrossRef] [PubMed]

83. Berger, A.K.; Hiller, B.E.; Thete, D.; Snyder, A.J.; Perez, E., Jr.; Upton, J.W.; Danthi, P. Viral RNA at Two Stages of Reovirus Infection Is Required for the Induction of Necroptosis. *J. Virol.* **2017**, *91*. [CrossRef]

84. Thirukkumaran, C.M.; Shi, Z.Q.; Luider, J.; Kopciuk, K.; Gao, H.; Bahlis, N.; Neri, P.; Pho, M.; Stewart, D.; Mansoor, A.; et al. Reovirus modulates autophagy during oncolysis of multiple myeloma. *Autophagy* **2013**, *9*, 413–414. [CrossRef]

85. Errington, F.; Steele, L.; Prestwich, R.; Harrington, K.J.; Pandha, H.S.; Vidal, L.; de Bono, J.; Selby, P.; Coffey, M.; Vile, R.; et al. Reovirus activates human dendritic cells to promote innate antitumor immunity. *J. Immunol.* **2008**, *180*, 6018–6026. [CrossRef] [PubMed]

86. Douville, R.N.; Su, R.C.; Coombs, K.M.; Simons, F.E.; Hayglass, K.T. Reovirus serotypes elicit distinctive patterns of recall immunity in humans. *J. Virol.* **2008**, *82*, 7515–7523. [CrossRef]

87. Errington, F.; White, C.L.; Twigger, K.R.; Rose, A.; Scott, K.; Steele, L.; Ilett, L.J.; Prestwich, R.; Pandha, H.S.; Coffey, M.; et al. Inflammatory tumour cell killing by oncolytic reovirus for the treatment of melanoma. *Gene Ther.* **2008**, *15*, 1257–1270. [CrossRef]

88. Steele, L.; Errington, F.; Prestwich, R.; Ilett, E.; Harrington, K.; Pandha, H.; Coffey, M.; Selby, P.; Vile, R.; Melcher, A. Pro-inflammatory cytokine/chemokine production by reovirus treated melanoma cells is PKR/NF-kappaB mediated and supports innate and adaptive anti-tumour immune priming. *Mol. Cancer* **2011**, *10*, 20. [CrossRef]

89. Kaufman, H.L.; Kohlhapp, F.J.; Zloza, A. Oncolytic viruses: A new class of immunotherapy drugs. *Nat. Rev. Drug Discov.* **2015**, *14*, 642–662. [CrossRef] [PubMed]

90. Johansson, C.; Wetzel, J.D.; He, J.; Mikacenic, C.; Dermody, T.S.; Kelsall, B.L. Type I interferons produced by hematopoietic cells protect mice against lethal infection by mammalian reovirus. *J. Exp. Med.* **2007**, *204*, 1349–1358. [CrossRef]

91. Levy, D.E.; Marie, I.J.; Durbin, J.E. Induction and function of type I and III interferon in response to viral infection. *Curr. Opin. Virol.* **2011**, *1*, 476–486. [CrossRef] [PubMed]

92. Rojas, M.; Arias, C.F.; Lopez, S. Protein kinase R is responsible for the phosphorylation of eIF2alpha in rotavirus infection. *J. Virol.* **2010**, *84*, 10457–10466. [CrossRef]

93. Adair, R.A.; Scott, K.J.; Fraser, S.; Errington-Mais, F.; Pandha, H.; Coffey, M.; Selby, P.; Cook, G.P.; Vile, R.; Harrington, K.J.; et al. Cytotoxic and immune-mediated killing of human colorectal cancer by reovirus-loaded blood and liver mononuclear cells. *Int. J. Cancer* **2013**, *132*, 2327–2338. [CrossRef] [PubMed]

94. Shmulevitz, M.; Pan, L.Z.; Garant, K.; Pan, D.; Lee, P.W. Oncogenic Ras promotes reovirus spread by suppressing IFN-beta production through negative regulation of RIG-I signaling. *Cancer Res.* **2010**, *70*, 4912–4921. [CrossRef] [PubMed]

95. Katayama, Y.; Tachibana, M.; Kurisu, N.; Oya, Y.; Terasawa, Y.; Goda, H.; Kobiyama, K.; Ishii, K.J.; Akira, S.; Mizuguchi, H.; et al. Oncolytic Reovirus Inhibits Immunosuppressive Activity of Myeloid-Derived Suppressor Cells in a TLR3-Dependent Manner. *J. Immunol* **2018**, *200*, 2987–2999. [CrossRef]

96. Cai, J.; Lin, Y.; Zhang, H.; Liang, J.; Tan, Y.; Cavenee, W.K.; Yan, G. Selective replication of oncolytic virus M1 results in a bystander killing effect that is potentiated by Smac mimetics. *Proc. Natl. Acad. Sci. USA* **2017**, *114*, 6812–6817. [CrossRef]

97. Arulanandam, R.; Batenchuk, C.; Varette, O.; Zakaria, C.; Garcia, V.; Forbes, N.E.; Davis, C.; Krishnan, R.; Karmacharya, R.; Cox, J.; et al. Microtubule disruption synergizes with oncolytic virotherapy by inhibiting interferon translation and potentiating bystander killing. *Nat. Commun.* **2015**, *6*, 6410. [CrossRef]

98. Diaz, R.M.; Galivo, F.; Kottke, T.; Wongthida, P.; Qiao, J.; Thompson, J.; Valdes, M.; Barber, G.; Vile, R.G. Oncolytic immunovirotherapy for melanoma using vesicular stomatitis virus. *Cancer Res.* **2007**, *67*, 2840–2848. [CrossRef]

99. Benencia, F.; Courreges, M.C.; Conejo-Garcia, J.R.; Mohamed-Hadley, A.; Zhang, L.; Buckanovich, R.J.; Carroll, R.; Fraser, N.; Coukos, G. HSV oncolytic therapy upregulates interferon-inducible chemokines and recruits immune effector cells in ovarian cancer. *Mol. Ther. J. Am. Soc. Gene Ther.* **2005**, *12*, 789–802. [CrossRef]

100. Prestwich, R.J.; Errington, F.; Steele, L.P.; Ilett, E.J.; Morgan, R.S.; Harrington, K.J.; Pandha, H.S.; Selby, P.J.; Vile, R.G.; Melcher, A.A. Reciprocal human dendritic cell-natural killer cell interactions induce antitumor activity following tumor cell infection by oncolytic reovirus. *J. Immunol* **2009**, *183*, 4312–4321. [CrossRef]

101. Spiotto, M.T.; Rowley, D.A.; Schreiber, H. Bystander elimination of antigen loss variants in established tumors. *Nat. Med.* **2004**, *10*, 294–298. [CrossRef] [PubMed]

102. Prestwich, R.J.; Errington, F.; Ilett, E.J.; Morgan, R.S.; Scott, K.J.; Kottke, T.; Thompson, J.; Morrison, E.E.; Harrington, K.J.; Pandha, H.S.; et al. Tumor infection by oncolytic reovirus primes adaptive antitumor immunity. *Clin. Cancer Res.* **2008**, *14*, 7358–7366. [CrossRef] [PubMed]

103. Gujar, S.A.; Pan, D.A.; Marcato, P.; Garant, K.A.; Lee, P.W. Oncolytic virus-initiated protective immunity against prostate cancer. *Mol. Ther. J. Am. Soc. Gene Ther.* **2011**, *19*, 797–804. [CrossRef] [PubMed]

104. Prestwich, R.J.; Ilett, E.J.; Errington, F.; Diaz, R.M.; Steele, L.P.; Kottke, T.; Thompson, J.; Galivo, F.; Harrington, K.J.; Pandha, H.S.; et al. Immune-mediated antitumor activity of reovirus is required for therapy and is independent of direct viral oncolysis and replication. *Clin. Cancer Res.* **2009**, *15*, 4374–4381. [CrossRef] [PubMed]

105. Martin, N.T.; Roy, D.G.; Workenhe, S.T.; van den Wollenberg, D.J.M.; Hoeben, R.C.; Mossman, K.L.; Bell, J.C.; Bourgeois-Daigneault, M.C. Pre-surgical neoadjuvant oncolytic virotherapy confers protection against rechallenge in a murine model of breast cancer. *Sci. Rep.* **2019**, *9*, 1865. [CrossRef]

106. Rajani, K.; Parrish, C.; Kottke, T.; Thompson, J.; Zaidi, S.; Ilett, L.; Shim, K.G.; Diaz, R.M.; Pandha, H.; Harrington, K.; et al. Combination Therapy With Reovirus and Anti-PD-1 Blockade Controls Tumor Growth Through Innate and Adaptive Immune Responses. *Mol. Ther. J. Am. Soc. Gene Ther.* **2016**, *24*, 166–174. [CrossRef]

107. Clements, D.R.; Sterea, A.M.; Kim, Y.; Helson, E.; Dean, C.A.; Nunokawa, A.; Coyle, K.M.; Sharif, T.; Marcato, P.; Gujar, S.A.; et al. Newly recruited CD11b+, GR-1+, Ly6C(high) myeloid cells augment tumor-associated immunosuppression immediately following the therapeutic administration of oncolytic reovirus. *J. Immunol.* **2015**, *194*, 4397–4412. [CrossRef]

108. Noonan, A.M.; Farren, M.R.; Geyer, S.M.; Huang, Y.; Tahiri, S.; Ahn, D.; Mikhail, S.; Ciombor, K.K.; Pant, S.; Aparo, S.; et al. Randomized Phase 2 Trial of the Oncolytic Virus Pelareorep (Reolysin) in Upfront Treatment of Metastatic Pancreatic Adenocarcinoma. *Mol. Ther. J. Am. Soc. Gene Ther.* **2016**, *24*, 1150–1158. [CrossRef]

109. Samson, A.; Scott, K.J.; Taggart, D.; West, E.J.; Wilson, E.; Nuovo, G.J.; Thomson, S.; Corns, R.; Mathew, R.K.; Fuller, M.J.; et al. Intravenous delivery of oncolytic reovirus to brain tumor patients immunologically primes for subsequent checkpoint blockade. *Sci. Transl. Med.* **2018**, *10*. [CrossRef]

110. Kelly, K.R.; Espitia, C.M.; Zhao, W.; Wu, K.; Visconte, V.; Anwer, F.; Calton, C.M.; Carew, J.S.; Nawrocki, S.T. Oncolytic reovirus sensitizes multiple myeloma cells to anti-PD-L1 therapy. *Leukemia* **2018**, *32*, 230–233. [CrossRef]

111. Gooden, M.J.; de Bock, G.H.; Leffers, N.; Daemen, T.; Nijman, H.W. The prognostic influence of tumour-infiltrating lymphocytes in cancer: A systematic review with meta-analysis. *Br. J. Cancer* **2011**, *105*, 93–103. [CrossRef]

112. White, C.L.; Twigger, K.R.; Vidal, L.; De Bono, J.S.; Coffey, M.; Heinemann, L.; Morgan, R.; Merrick, A.; Errington, F.; Vile, R.G.; et al. Characterization of the adaptive and innate immune response to intravenous oncolytic reovirus (Dearing type 3) during a phase I clinical trial. *Gene Ther.* **2008**, *15*, 911–920. [CrossRef] [PubMed]

113. El-Sherbiny, Y.M.; Holmes, T.D.; Wetherill, L.F.; Black, E.V.; Wilson, E.B.; Phillips, S.L.; Scott, G.B.; Adair, R.A.; Dave, R.; Scott, K.J.; et al. Controlled infection with a therapeutic virus defines the activation kinetics of human natural killer cells in vivo. *Clin. Exp. Immunol.* **2015**, *180*, 98–107. [CrossRef] [PubMed]

114. Qiao, J.; Wang, H.; Kottke, T.; White, C.; Twigger, K.; Diaz, R.M.; Thompson, J.; Selby, P.; de Bono, J.; Melcher, A.; et al. Cyclophosphamide facilitates antitumor efficacy against subcutaneous tumors following intravenous delivery of reovirus. *Clin. Cancer Res.* **2008**, *14*, 259–269. [CrossRef] [PubMed]

115. Matsuura, K.; Ishikura, M.; Nakayama, T.; Hasegawa, S.; Morita, O.; Uetake, H. Ecological studies on reovirus pollution of rivers in Toyama Prefecture. *Microbiol. Immunol.* **1988**, *32*, 1221–1234. [CrossRef] [PubMed]

116. Minuk, G.Y.; Paul, R.W.; Lee, P.W. The prevalence of antibodies to reovirus type 3 in adults with idiopathic cholestatic liver disease. *J. Med. Virol.* **1985**, *16*, 55–60. [CrossRef]

117. Lerner, A.M.; Cherry, J.D.; Klein, J.O.; Finland, M. Infections with reoviruses. *N. Engl. J. Med.* **1962**, *267*, 947–952. [CrossRef]

118. Selb, B.; Weber, B. A study of human reovirus IgG and IgA antibodies by ELISA and western blot. *J. Virol. Methods* **1994**, *47*, 15–25. [CrossRef]

119. Pal, S.R.; Agarwal, S.C. Sero-epidemiological study of reovirus infection amongst the normal population of the Chandigarh area—Northern India. *J. Hyg. (Lond.)* **1968**, *66*, 519–529. [CrossRef]

120. Tai, J.H.; Williams, J.V.; Edwards, K.M.; Wright, P.F.; Crowe, J.E., Jr.; Dermody, T.S. Prevalence of reovirus-specific antibodies in young children in Nashville, Tennessee. *J. Infect. Dis.* **2005**, *191*, 1221–1224. [CrossRef]

121. Li, X.; Wang, P.; Li, H.; Du, X.; Liu, M.; Huang, Q.; Wang, Y.; Wang, S. The Efficacy of Oncolytic Adenovirus Is Mediated by T-cell Responses against Virus and Tumor in Syrian Hamster Model. *Clin. Cancer Res.* **2017**, *23*, 239–249. [CrossRef]

122. Zamarin, D.; Holmgaard, R.B.; Subudhi, S.K.; Park, J.S.; Mansour, M.; Palese, P.; Merghoub, T.; Wolchok, J.D.; Allison, J.P. Localized oncolytic virotherapy overcomes systemic tumor resistance to immune checkpoint blockade immunotherapy. *Sci. Transl. Med.* **2014**, *6*, 226ra32. [CrossRef] [PubMed]

123. Hirasawa, K.; Nishikawa, S.G.; Norman, K.L.; Alain, T.; Kossakowska, A.; Lee, P.W. Oncolytic reovirus against ovarian and colon cancer. *Cancer Res.* **2002**, *62*, 1696–1701.

124. Norman, K.L.; Coffey, M.C.; Hirasawa, K.; Demetrick, D.J.; Nishikawa, S.G.; DiFrancesco, L.M.; Strong, J.E.; Lee, P.W. Reovirus oncolysis of human breast cancer. *Hum. Gene Ther.* **2002**, *13*, 641–652. [CrossRef] [PubMed]

125. Sei, S.; Mussio, J.K.; Yang, Q.E.; Nagashima, K.; Parchment, R.E.; Coffey, M.C.; Shoemaker, R.H.; Tomaszewski, J.E. Synergistic antitumor activity of oncolytic reovirus and chemotherapeutic agents in non-small cell lung cancer cells. *Mol. Cancer* **2009**, *8*, 47. [CrossRef]

126. Thirukkumaran, C.M.; Nodwell, M.J.; Hirasawa, K.; Shi, Z.Q.; Diaz, R.; Luider, J.; Johnston, R.N.; Forsyth, P.A.; Magliocco, A.M.; Lee, P.; et al. Oncolytic viral therapy for prostate cancer: Efficacy of reovirus as a biological therapeutic. *Cancer Res.* **2010**, *70*, 2435–2444. [CrossRef] [PubMed]

127. Thirukkumaran, C.M.; Shi, Z.Q.; Luider, J.; Kopciuk, K.; Gao, H.; Bahlis, N.; Neri, P.; Pho, M.; Stewart, D.; Mansoor, A.; et al. Reovirus as a viable therapeutic option for the treatment of multiple myeloma. *Clin. Cancer Res.* **2012**, *18*, 4962–4972. [CrossRef]

128. Coffey, M.C.; Strong, J.E.; Forsyth, P.A.; Lee, P.W. Reovirus therapy of tumors with activated Ras pathway. *Science* **1998**, *282*, 1332–1334. [CrossRef]

129. Qiao, J.; Kottke, T.; Willmon, C.; Galivo, F.; Wongthida, P.; Diaz, R.M.; Thompson, J.; Ryno, P.; Barber, G.N.; Chester, J.; et al. Purging metastases in lymphoid organs using a combination of antigen-nonspecific adoptive T cell therapy, oncolytic virotherapy and immunotherapy. *Nat. Med.* **2008**, *14*, 37–44. [CrossRef] [PubMed]

130. Wilcox, M.E.; Yang, W.; Senger, D.; Rewcastle, N.B.; Morris, D.G.; Brasher, P.M.; Shi, Z.Q.; Johnston, R.N.; Nishikawa, S.; Lee, P.W.; et al. Reovirus as an oncolytic agent against experimental human malignant gliomas. *J. Natl. Cancer Inst.* **2001**, *93*, 903–912. [CrossRef]

131. Samson, A.; Bentham, M.J.; Scott, K.; Nuovo, G.; Bloy, A.; Appleton, E.; Adair, R.A.; Dave, R.; Peckham-Cooper, A.; Toogood, G.; et al. Oncolytic reovirus as a combined antiviral and anti-tumour agent for the treatment of liver cancer. *Gut* **2018**, *67*, 562–573. [CrossRef]

132. Mehlen, P.; Puisieux, A. Metastasis: A question of life or death. *Nat. Rev. Cancer* **2006**, *6*, 449–458. [CrossRef]

133. Adair, R.A.; Roulstone, V.; Scott, K.J.; Morgan, R.; Nuovo, G.J.; Fuller, M.; Beirne, D.; West, E.J.; Jennings, V.A.; Rose, A.; et al. Cell carriage, delivery, and selective replication of an oncolytic virus in tumor in patients. *Sci. Transl. Med.* **2012**, *4*, 138ra77. [CrossRef] [PubMed]

134. Scurr, M.; Pembroke, T.; Bloom, A.; Roberts, D.; Thomson, A.; Smart, K.; Bridgeman, H.; Adams, R.; Brewster, A.; Jones, R.; et al. Low-Dose Cyclophosphamide Induces Antitumor T-Cell Responses, which Associate with Survival in Metastatic Colorectal Cancer. *Clin. Cancer Res.* **2017**, *23*, 6771–6780. [CrossRef] [PubMed]

135. Hurd, E.R.; Giuliano, V.J. The effect of cyclophosphamide on B and T lymphocytes in patients with connective tissue diseases. *Arthritis Rheum* **1975**, *18*, 67–75. [CrossRef]

136. Varkila, K.; Hurme, M. The effect of cyclophosphamide on cytotoxic T-lymphocyte responses: Inhibition of helper T-cell induction in vitro. *Immunology* **1983**, *48*, 433–438.

137. Ikeda, K.; Ichikawa, T.; Wakimoto, H.; Silver, J.S.; Deisboeck, T.S.; Finkelstein, D.; Harsh, G.R.t.; Louis, D.N.; Bartus, R.T.; Hochberg, F.H.; et al. Oncolytic virus therapy of multiple tumors in the brain requires suppression of innate and elicited antiviral responses. *Nat. Med.* **1999**, *5*, 881–887. [CrossRef]

138. Peng, K.W.; Myers, R.; Greenslade, A.; Mader, E.; Greiner, S.; Federspiel, M.J.; Dispenzieri, A.; Russell, S.J. Using clinically approved cyclophosphamide regimens to control the humoral immune response to oncolytic viruses. *Gene Ther.* **2013**, *20*, 255–261. [CrossRef] [PubMed]

139. Karapanagiotou, E.M.; Roulstone, V.; Twigger, K.; Ball, M.; Tanay, M.; Nutting, C.; Newbold, K.; Gore, M.E.; Larkin, J.; Syrigos, K.N.; et al. Phase I/II trial of carboplatin and paclitaxel chemotherapy in combination with intravenous oncolytic reovirus in patients with advanced malignancies. *Clin. Cancer Res.* **2012**, *18*, 2080–2089. [CrossRef]

140. Lolkema, M.P.; Arkenau, H.T.; Harrington, K.; Roxburgh, P.; Morrison, R.; Roulstone, V.; Twigger, K.; Coffey, M.; Mettinger, K.; Gill, G.; et al. A phase I study of the combination of intravenous reovirus type 3 Dearing and gemcitabine in patients with advanced cancer. *Clin. Cancer Res.* **2011**, *17*, 581–588. [CrossRef]

141. Roulstone, V.; Khan, K.; Pandha, H.S.; Rudman, S.; Coffey, M.; Gill, G.M.; Melcher, A.A.; Vile, R.; Harrington, K.J.; de Bono, J.; et al. Phase I trial of cyclophosphamide as an immune modulator for optimizing oncolytic reovirus delivery to solid tumors. *Clin. Cancer Res.* **2015**, *21*, 1305–1312. [CrossRef]

142. Galanis, E.; Markovic, S.N.; Suman, V.J.; Nuovo, G.J.; Vile, R.G.; Kottke, T.J.; Nevala, W.K.; Thompson, M.A.; Lewis, J.E.; Rumilla, K.M.; et al. Phase II trial of intravenous administration of Reolysin((R)) (Reovirus Serotype-3-dearing Strain) in patients with metastatic melanoma. *Mol. Ther. J. Am. Soc. Gene Ther.* **2012**, *20*, 1998–2003. [CrossRef] [PubMed]

143. Hughes, E.; Scurr, M.; Campbell, E.; Jones, E.; Godkin, A.; Gallimore, A. T-cell modulation by cyclophosphamide for tumour therapy. *Immunology* **2018**, *154*, 62–68. [CrossRef]

144. Huang, X.M.; Zhang, N.R.; Lin, X.T.; Zhu, C.Y.; Zou, Y.F.; Wu, X.J.; He, X.S.; He, X.W.; Wan, Y.L.; Lan, P. Antitumor immunity of low-dose cyclophosphamide: Changes in T cells and cytokines TGF-beta and IL-10 in mice with colon-cancer liver metastasis. *Gastroenterol. Rep. (Oxf.)* **2020**, *8*, 56–65. [CrossRef] [PubMed]

145. Ilett, E.J.; Prestwich, R.J.; Kottke, T.; Errington, F.; Thompson, J.M.; Harrington, K.J.; Pandha, H.S.; Coffey, M.; Selby, P.J.; Vile, R.G.; et al. Dendritic cells and T cells deliver oncolytic reovirus for tumour killing despite pre-existing anti-viral immunity. *Gene Ther.* **2009**, *16*, 689–699. [CrossRef]

146. Ilett, E.J.; Barcena, M.; Errington-Mais, F.; Griffin, S.; Harrington, K.J.; Pandha, H.S.; Coffey, M.; Selby, P.J.; Limpens, R.W.; Mommaas, M.; et al. Internalization of oncolytic reovirus by human dendritic cell carriers protects the virus from neutralization. *Clin. Cancer Res.* **2011**, *17*, 2767–2776. [CrossRef]

147. Ilett, E.; Kottke, T.; Donnelly, O.; Thompson, J.; Willmon, C.; Diaz, R.; Zaidi, S.; Coffey, M.; Selby, P.; Harrington, K.; et al. Cytokine conditioning enhances systemic delivery and therapy of an oncolytic virus. *Mol. Ther. J. Am. Soc. Gene Ther.* **2014**, *22*, 1851–1863. [CrossRef]

148. Berkeley, R.A.; Steele, L.P.; Mulder, A.A.; van den Wollenberg, D.J.M.; Kottke, T.J.; Thompson, J.; Coffey, M.; Hoeben, R.C.; Vile, R.G.; Melcher, A.; et al. Antibody-Neutralized Reovirus Is Effective in Oncolytic Virotherapy. *Cancer Immunol. Res.* **2018**, *6*, 1161–1173. [CrossRef] [PubMed]
149. Jennings, V.A.; Ilett, E.J.; Scott, K.J.; West, E.J.; Vile, R.; Pandha, H.; Harrington, K.; Young, A.; Hall, G.D.; Coffey, M.; et al. Lymphokine-activated killer and dendritic cell carriage enhances oncolytic reovirus therapy for ovarian cancer by overcoming antibody neutralization in ascites. *Int. J. Cancer* **2014**, *134*, 1091–1101. [CrossRef]
150. Twigger, K.; Vidal, L.; White, C.L.; De Bono, J.S.; Bhide, S.; Coffey, M.; Thompson, B.; Vile, R.G.; Heinemann, L.; Pandha, H.S.; et al. Enhanced in vitro and in vivo cytotoxicity of combined reovirus and radiotherapy. *Clin. Cancer Res.* **2008**, *14*, 912–923. [CrossRef]
151. Hingorani, P.; Zhang, W.; Lin, J.; Liu, L.; Guha, C.; Kolb, E.A. Systemic administration of reovirus (Reolysin) inhibits growth of human sarcoma xenografts. *Cancer* **2011**, *117*, 1764–1774. [CrossRef]
152. Heinemann, L.; Simpson, G.R.; Boxall, A.; Kottke, T.; Relph, K.L.; Vile, R.; Melcher, A.; Prestwich, R.; Harrington, K.J.; Morgan, R.; et al. Synergistic effects of oncolytic reovirus and docetaxel chemotherapy in prostate cancer. *BMC Cancer* **2011**, *11*, 221. [CrossRef] [PubMed]
153. Pandha, H.S.; Heinemann, L.; Simpson, G.R.; Melcher, A.; Prestwich, R.; Errington, F.; Coffey, M.; Harrington, K.J.; Morgan, R. Synergistic effects of oncolytic reovirus and cisplatin chemotherapy in murine malignant melanoma. *Clin. Cancer Res.* **2009**, *15*, 6158–6166. [CrossRef] [PubMed]
154. Roulstone, V.; Twigger, K.; Zaidi, S.; Pencavel, T.; Kyula, J.N.; White, C.; McLaughlin, M.; Seth, R.; Karapanagiotou, E.M.; Mansfield, D.; et al. Synergistic cytotoxicity of oncolytic reovirus in combination with cisplatin-paclitaxel doublet chemotherapy. *Gene Ther.* **2013**, *20*, 521–528. [CrossRef]
155. Gujar, S.A.; Clements, D.; Dielschneider, R.; Helson, E.; Marcato, P.; Lee, P.W. Gemcitabine enhances the efficacy of reovirus-based oncotherapy through anti-tumour immunological mechanisms. *Br. J. Cancer* **2014**, *110*, 83–93. [CrossRef] [PubMed]
156. Wang, T.H.; Wang, H.S.; Soong, Y.K. Paclitaxel-induced cell death: Where the cell cycle and apoptosis come together. *Cancer* **2000**, *88*, 2619–2628. [CrossRef]
157. Rouhimoghadam, M.; Safarian, S.; Carroll, J.S.; Sheibani, N.; Bidkhori, G. Tamoxifen-Induced Apoptosis of MCF-7 Cells via GPR30/PI3K/MAPKs Interactions: Verification by ODE Modeling and RNA Sequencing. *Front. Physiol.* **2018**, *9*, 907. [CrossRef] [PubMed]
158. Hill, R.; Rabb, M.; Madureira, P.A.; Clements, D.; Gujar, S.A.; Waisman, D.M.; Giacomantonio, C.A.; Lee, P.W. Gemcitabine-mediated tumour regression and p53-dependent gene expression: Implications for colon and pancreatic cancer therapy. *Cell Death Dis.* **2013**, *4*, e791. [CrossRef] [PubMed]
159. Morse, D.L.; Gray, H.; Payne, C.M.; Gillies, R.J. Docetaxel induces cell death through mitotic catastrophe in human breast cancer cells. *Mol. Cancer Ther.* **2005**, *4*, 1495–1504. [CrossRef]
160. Kelly, K.R.; Espitia, C.M.; Mahalingam, D.; Oyajobi, B.O.; Coffey, M.; Giles, F.J.; Carew, J.S.; Nawrocki, S.T. Reovirus therapy stimulates endoplasmic reticular stress, NOXA induction, and augments bortezomib-mediated apoptosis in multiple myeloma. *Oncogene* **2012**, *31*, 3023–3038. [CrossRef]
161. Roulstone, V.; Pedersen, M.; Kyula, J.; Mansfield, D.; Khan, A.A.; McEntee, G.; Wilkinson, M.; Karapanagiotou, E.; Coffey, M.; Marais, R.; et al. BRAF- and MEK-Targeted Small Molecule Inhibitors Exert Enhanced Antimelanoma Effects in Combination With Oncolytic Reovirus Through ER Stress. *Mol. Ther. J. Am. Soc. Gene Ther.* **2015**, *23*, 931–942. [CrossRef] [PubMed]
162. Kennedy, B.E.; Sadek, M.; Gujar, S.A. Targeted Metabolic Reprogramming to Improve the Efficacy of Oncolytic Virus Therapy. *Mol. Ther. J. Am. Soc. Gene Ther.* **2020**, *28*, 1417–1421. [CrossRef] [PubMed]
163. Kennedy, B.E.; Murphy, J.P.; Clements, D.R.; Konda, P.; Holay, N.; Kim, Y.; Pathak, G.P.; Giacomantonio, M.A.; Hiani, Y.E.; Gujar, S. Inhibition of Pyruvate Dehydrogenase Kinase Enhances the Antitumor Efficacy of Oncolytic Reovirus. *Cancer Res.* **2019**, *79*, 3824–3836. [CrossRef] [PubMed]
164. Kottke, T.; Hall, G.; Pulido, J.; Diaz, R.M.; Thompson, J.; Chong, H.; Selby, P.; Coffey, M.; Pandha, H.; Chester, J.; et al. Antiangiogenic cancer therapy combined with oncolytic virotherapy leads to regression of established tumors in mice. *J. Clin. Investig.* **2010**, *120*, 1551–1560. [CrossRef] [PubMed]
165. Kottke, T.; Chester, J.; Ilett, E.; Thompson, J.; Diaz, R.; Coffey, M.; Selby, P.; Nuovo, G.; Pulido, J.; Mukhopadhyay, D.; et al. Precise scheduling of chemotherapy primes VEGF-producing tumors for successful systemic oncolytic virotherapy. *Mol. Ther. J. Am. Soc. Gene Ther.* **2011**, *19*, 1802–1812. [CrossRef]

166. Ilett, E.; Kottke, T.; Thompson, J.; Rajani, K.; Zaidi, S.; Evgin, L.; Coffey, M.; Ralph, C.; Diaz, R.; Pandha, H.; et al. Prime-boost using separate oncolytic viruses in combination with checkpoint blockade improves anti-tumour therapy. *Gene Ther.* **2017**, *24*, 21–30. [CrossRef]

167. Mostafa, A.A.; Meyers, D.E.; Thirukkumaran, C.M.; Liu, P.J.; Gratton, K.; Spurrell, J.; Shi, Q.; Thakur, S.; Morris, D.G. Oncolytic Reovirus and Immune Checkpoint Inhibition as a Novel Immunotherapeutic Strategy for Breast Cancer. *Cancers (Basel)* **2018**, *10*, 205. [CrossRef]

168. Morris, D.G.; Feng, X.; DiFrancesco, L.M.; Fonseca, K.; Forsyth, P.A.; Paterson, A.H.; Coffey, M.C.; Thompson, B. REO-001: A phase I trial of percutaneous intralesional administration of reovirus type 3 dearing (Reolysin(R)) in patients with advanced solid tumors. *Investig. New Drugs* **2013**, *31*, 696–706. [CrossRef]

169. Gollamudi, R.; Ghalib, M.H.; Desai, K.K.; Chaudhary, I.; Wong, B.; Einstein, M.; Coffey, M.; Gill, G.M.; Mettinger, K.; Mariadason, J.M.; et al. Intravenous administration of Reolysin, a live replication competent RNA virus is safe in patients with advanced solid tumors. *Investig. New Drugs* **2010**, *28*, 641–649. [CrossRef]

170. Vidal, L.; Pandha, H.S.; Yap, T.A.; White, C.L.; Twigger, K.; Vile, R.G.; Melcher, A.; Coffey, M.; Harrington, K.J.; DeBono, J.S. A phase I study of intravenous oncolytic reovirus type 3 Dearing in patients with advanced cancer. *Clin. Cancer Res.* **2008**, *14*, 7127–7137. [CrossRef]

171. Forsyth, P.; Roldan, G.; George, D.; Wallace, C.; Palmer, C.A.; Morris, D.; Cairncross, G.; Matthews, M.V.; Markert, J.; Gillespie, Y.; et al. A phase I trial of intratumoral administration of reovirus in patients with histologically confirmed recurrent malignant gliomas. *Mol. Ther. J. Am. Soc. Gene Ther.* **2008**, *16*, 627–632. [CrossRef]

172. Sborov, D.W.; Nuovo, G.J.; Stiff, A.; Mace, T.; Lesinski, G.B.; Benson, D.M., Jr.; Efebera, Y.A.; Rosko, A.E.; Pichiorri, F.; Grever, M.R.; et al. A phase I trial of single-agent reolysin in patients with relapsed multiple myeloma. *Clin. Cancer Res.* **2014**, *20*, 5946–5955. [CrossRef]

173. Karnad, A.B.; Haigentz, M.; Miley, T.; Coffey, M.; Gill, G.; Mita, M. Abstract C22: A phase II study of intravenous wild-type reovirus (Reolysin®) in combination with paclitaxel plus carboplatin in patients with platinum refractory metastatic and/or recurrent squamous cell carcinoma of the head and neck. *Mol. Cancer Ther.* **2011**, *10*. [CrossRef]

174. Ocean, A.J.; Bekaii-Saab, T.S.; Chaudhary, I.; Palmer, R.; Christos, P.J.; Mercado, A.; Florendo, E.O.; Rosales, V.A.; Ruggiero, J.T.; Popa, E.C.; et al. A multicenter phase I study of intravenous administration of reolysin in combination with irinotecan/fluorouracil/leucovorin (FOLFIRI) in patients (pts) with oxaliplatin-refractory/intolerant KRAS-mutant metastatic colorectal cancer (mCRC). *J. Clin. Oncol.* **2013**, *31*, 450. [CrossRef]

175. Villalona-Calero, M.A.; Lam, E.; Otterson, G.A.; Zhao, W.; Timmons, M.; Subramaniam, D.; Hade, E.M.; Gill, G.M.; Coffey, M.; Selvaggi, G.; et al. Oncolytic reovirus in combination with chemotherapy in metastatic or recurrent non-small cell lung cancer patients with KRAS-activated tumors. *Cancer* **2016**, *122*, 875–883. [CrossRef] [PubMed]

176. Bernstein, V.E.; Dent, S.F.; Gelmon, K.A.; Dhesy-Thind, S.K.; Mates, M.; Salim, M.; Panasci, L.; Song, X.; Clemons, M.; Tu, D.; et al. A randomized (RCT) phase II study of oncolytic reovirus (pelareorep) plus standard weekly paclitaxel (P) as therapy for metastatic breast cancer (mBC) [Abstract nr CT131]. In Proceedings of the American Association for Cancer Research Annual Meeting 2017, Washington, DC, USA, 1–5 April 2017.

177. Gong, J.; Sachdev, E.; Mita, A.C.; Mita, M.M. Clinical development of reovirus for cancer therapy: An oncolytic virus with immune-mediated antitumor activity. *World J. Methodol.* **2016**, *6*, 25–42. [CrossRef]

178. Mita, A.; Sankhala, K.; Sarantopoulos, J. A phase II study of intravenous (IV) wild-type reovirus (Reolysin) in the treatment of patients with bone and soft tissue sarcomas metastatic to the lung. *J. Clin. Oncol.* **2009**, *27*, 10524. [CrossRef]

179. Mahalingam, D.; Fountzilas, C.; Moseley, J.L.; Noronha, N.; Cheetham, K.; Dzugalo, A.; Nuovo, G.; Gutierrez, A.; Arora, S.P. A study of REOLYSIN in combination with pembrolizumab and chemotherapy in patients (pts) with relapsed metastatic adenocarcinoma of the pancreas (MAP). *J. Clin. Oncol.* **2017**, *35*, e15753. [CrossRef]

180. Saunders, M.; Anthoney, A.; Coffey, M.; Mettinger, K.; Thompson, B.; Melcher, A.; Nutting, C.M.; Harrington, K. Results of a phase II study to evaluate the biological effects of intratumoral (ITu) reolysin in combination with low dose radiotherapy (RT) in patients (Pts) with advanced cancers. *J. Clin. Oncol.* **2009**, *27*. [CrossRef]

181. Jonker, D.J.; Tang, P.A.; Kennecke, H.; Welch, S.A.; Cripps, M.C.; Asmis, T.; Chalchal, H.; Tomiak, A.; Lim, H.; Ko, Y.J.; et al. A Randomized Phase II Study of FOLFOX6/Bevacizumab With or Without Pelareorep in Patients With Metastatic Colorectal Cancer: IND.210, a Canadian Cancer Trials Group Trial. *Clin. Colorectal. Cancer* **2018**, *17*, 231–239.e7. [CrossRef] [PubMed]

182. Mahalingam, D.; Fountzilas, C.; Moseley, J.; Noronha, N.; Tran, H.; Chakrabarty, R.; Selvaggi, G.; Coffey, M.; Thompson, B.; Sarantopoulos, J. A phase II study of REOLYSIN((R)) (pelareorep) in combination with carboplatin and paclitaxel for patients with advanced malignant melanoma. *Cancer Chemother. Pharmcol.* **2017**, *79*, 697–703. [CrossRef] [PubMed]

183. Mita, A.C.; Argiris, A.; Coffey, M.; Gill, G.; Mita, M. Abstract C70: A phase 2 study of intravenous administration of REOLYSIN®(reovirus type 3 dearing) in combination with paclitaxel (P) and carboplatin (C) in patients with squamous cell carcinoma of the lung. *Mol. Cancer Ther.* **2013**, *12*. [CrossRef]

184. Bradbury, P.A.; Morris, D.G.; Nicholas, G.; Tu, D.; Tehfe, M.; Goffin, J.R.; Shepherd, F.A.; Gregg, R.W.; Rothenstein, J.; Lee, C.; et al. Canadian Cancer Trials Group (CCTG) IND211: A randomized trial of pelareorep (Reolysin) in patients with previously treated advanced or metastatic non-small cell lung cancer receiving standard salvage therapy. *Lung Cancer (Amst. Neth.)* **2018**, *120*, 142–148. [CrossRef]

185. Eigl, B.J.; Chi, K.; Tu, D.; Hotte, S.J.; Winquist, E.; Booth, C.M.; Canil, C.; Potvin, K.; Gregg, R.; North, S.; et al. A randomized phase II study of pelareorep and docetaxel or docetaxel alone in men with metastatic castration resistant prostate cancer: CCTG study IND 209. *Oncotarget* **2018**, *9*, 8155–8164. [CrossRef] [PubMed]

186. Cohn, D.E.; Sill, M.W.; Walker, J.L.; O'Malley, D.; Nagel, C.I.; Rutledge, T.L.; Bradley, W.; Richardson, D.L.; Moxley, K.M.; Aghajanian, C. Randomized phase IIB evaluation of weekly paclitaxel versus weekly paclitaxel with oncolytic reovirus (Reolysin(R)) in recurrent ovarian, tubal, or peritoneal cancer: An NRG Oncology/Gynecologic Oncology Group study. *Gynecol. Oncol.* **2017**, *146*, 477–483. [CrossRef] [PubMed]

187. Harrington, K.J.; Karapanagiotou, E.M.; Roulstone, V.; Twigger, K.R.; White, C.L.; Vidal, L.; Beirne, D.; Prestwich, R.; Newbold, K.; Ahmed, M.; et al. Two-stage phase I dose-escalation study of intratumoral reovirus type 3 dearing and palliative radiotherapy in patients with advanced cancers. *Clin. Cancer Res.* **2010**, *16*, 3067–3077. [CrossRef]

188. Comins, C.; Spicer, J.; Protheroe, A.; Roulstone, V.; Twigger, K.; White, C.M.; Vile, R.; Melcher, A.; Coffey, M.C.; Mettinger, K.L.; et al. REO-10: A phase I study of intravenous reovirus and docetaxel in patients with advanced cancer. *Clin. Cancer Res.* **2010**, *16*, 5564–5572. [CrossRef]

189. van den Wollenberg, D.J.; Dautzenberg, I.J.; van den Hengel, S.K.; Cramer, S.J.; de Groot, R.J.; Hoeben, R.C. Isolation of reovirus T3D mutants capable of infecting human tumor cells independent of junction adhesion molecule-A. *PLoS ONE* **2012**, *7*, e48064. [CrossRef]

190. Mohamed, A.; Johnston, R.N.; Shmulevitz, M. Potential for Improving Potency and Specificity of Reovirus Oncolysis with Next-Generation Reovirus Variants. *Viruses* **2015**, *7*, 6251–6278. [CrossRef]

191. Kemp, V.; Hoeben, R.C.; van den Wollenberg, D.J. Exploring Reovirus Plasticity for Improving Its Use as Oncolytic Virus. *Viruses* **2016**, *8*, 4. [CrossRef]

192. van den Wollenberg, D.J.; Dautzenberg, I.J.; Ros, W.; Lipinska, A.D.; van den Hengel, S.K.; Hoeben, R.C. Replicating reoviruses with a transgene replacing the codons for the head domain of the viral spike. *Gene* **2015**, *22*, 267–279. [CrossRef]

193. Kemp, V.; van den Wollenberg, D.J.M. Arming oncolytic reovirus with GM-CSF gene to enhance immunity. *Cancer Gene Ther.* **2019**, *26*, 268–281. [CrossRef] [PubMed]

194. Rodríguez Stewart, R.M.; Berry, J.T.L.; Berger, A.K.; Yoon, S.B.; Hirsch, A.L.; Guberman, J.A.; Patel, N.B.; Tharp, G.K.; Bosinger, S.E.; Mainou, B.A. Enhanced Killing of Triple-Negative Breast Cancer Cells by Reassortant Reovirus and Topoisomerase Inhibitors. *J. Virol.* **2019**, *93*. [CrossRef]

Cancers **2020**, *12*, 3219

195. Fernandes, J.P.; Cristi, F.; Eaton, H.E.; Chen, P.; Haeflinger, S.; Bernard, I.; Hitt, M.M.; Shmulevitz, M. Breast Tumor-Associated Metalloproteases Restrict Reovirus Oncolysis by Cleaving the σ1 Cell Attachment Protein and Can Be Overcome by Mutation of σ1. *J. Virol.* **2019**, *93*. [CrossRef] [PubMed]

Publisher's Note: MDPI stays neutral with regard to jurisdictional claims in published maps and institutional affiliations.

Review

Newcastle Disease Virus at the Forefront of Cancer Immunotherapy

Bharat Burman [1,2]**, Giulio Pesci** [1,2] **and Dmitriy Zamarin** [1,2,3,4,*]

[1] Department of Medicine, Gynecologic Medical Oncology Service, Memorial Sloan Kettering Cancer Center, New York, NY 10065, USA; burmanb@mskcc.org (B.B.); pescig@mskcc.org (G.P.)

[2] Ludwig Collaborative Laboratory, Memorial Sloan Kettering Cancer Center, New York, NY 10065, USA

[3] Department of Medicine, Weill-Cornell Medical College, New York, NY 10065, USA

[4] Parker Institute for Cancer Immunotherapy, Memorial Sloan Kettering Cancer Center, New York, NY 10065, USA

* Correspondence: zamarind@mskcc.org

Received: 30 October 2020; Accepted: 24 November 2020; Published: 28 November 2020

Simple Summary: Newcastle disease virus (NDV) is an RNA virus belonging to the Paramyxoviridae family. In nature, NDV primarily infects birds, but poses no threat to human health. Multiple studies have demonstrated that NDV caries oncolytic potential due to its predilection for infection and replication in human cancer cells while sparing normal cells. In addition to its direct lytic effects, the virus triggers both innate and adaptive immune responses. In animal models, NDV injection into a tumor has been demonstrated to result in local inflammation and the recruitment of tumor-specific T cells, an effect that can be further potentiated through the use of viruses encoding immunomodulatory ligands and through combinations with immune checkpoint blockade. Initial clinical trials with naturally occurring NDV administered intravenously demonstrated durable responses across a number of cancer types. Clinical studies utilizing recombinant NDV in combination with immune checkpoint inhibitors are ongoing.

Abstract: Preclinical and clinical studies dating back to the 1950s have demonstrated that Newcastle disease virus (NDV) has oncolytic properties and can potently stimulate antitumor immune responses. NDV selectively infects, replicates within, and lyses cancer cells by exploiting defective antiviral defenses in cancer cells. Inflammation within the tumor microenvironment in response to NDV leads to the recruitment of innate and adaptive immune effector cells, presentation of tumor antigens, and induction of immune checkpoints. In animal models, intratumoral injection of NDV results in T cell infiltration of both local and distant non-injected tumors, demonstrating the potential of NDV to activate systemic adaptive antitumor immunity. The combination of intratumoral NDV with systemic immune checkpoint blockade leads to regression of both injected and distant tumors, an effect further potentiated by introduction of immunomodulatory transgenes into the viral genome. Clinical trials with naturally occurring NDV administered intravenously demonstrated durable responses across numerous cancer types. Based on these studies, further exploration of NDV is warranted, and clinical studies using recombinant NDV in combination with immune checkpoint blockade have been initiated.

Keywords: oncolytic virus; newcastle disease virus; NDV; cancer; immunotherapy; immune checkpoint inhibitor; PD-1; PD-L1; CTLA-4; type I interferon

---

1. Introduction

Observations that naturally occurring viral infections could cause spontaneous tumor regressions led to the search for viruses that could selectively lyse tumor cells with limited pathogenicity in humans [1]. In the 1950s, it was discovered that Newcastle disease virus (NDV), a highly virulent pathogen to over 240 species of birds, has oncolytic properties [2,3]. A decade later, NDV was injected intraperitoneally in mice with Ehrlich ascites, leading to tumor cell lysis and durable immunity upon tumor re-challenge [4,5]. Around the same time, NDV was tested clinically in a patient with acute myelogenous leukemia, who experienced transient anti-leukemic effect and clinical improvement with limited side effects [6].

NDV is an avian paramyxovirus type I virus belonging to the *Avulavirus* genus. NDV has a spherical morphology, formed by a lipid bilayer which surrounds the RNA genome. The genome consists of a 15,186-nucleotide negative single-strand RNA encoding six different genes: nucleocapsid protein (NP), phosphoprotein (P), matrix protein (M), fusion protein (F), haemagglutinin-neuraminidase (HN), and RNA-dependent RNA polymerase (L). NP, P, and L proteins form a ribonucleotide protein complex that embeds the genomic RNA. The lipid envelope surrounds the ribonucleotide protein complex [7–9]. NDV infection is initiated by binding of the viral surface HN and F glycoproteins to sialic acid-containing host cell surface proteins [10,11]. This triggers a conformational change in the F protein, which results in fusion of the viral envelope and the cell plasma membrane. Viral particles are internalized by endocytosis, and adjacent cells with attached particles may form syncytia due to the fusogenic F protein [8,11,12]. After viral entry, the M protein dissociates from the ribonucleotide protein complex in the cytoplasm, and the P and L proteins form a polymerase complex that initiates transcription of the viral RNA [10,13].

There are three main pathotypes of NDV, classified by the severity of disease caused in birds: lentogenic (avirulent), mesogenic (intermediate), and velogenic (highly virulent) [8]. Virulence is primarily determined by sequence variation in the F gene, which affects F protein cleavage efficiency [14,15]. Lentogenic viruses possess a monobasic F cleavage site and exhibit reduced capacity for multicycle replication and lysis. The mesogenic and velogenic NDV types possess a polybasic F cleavage site and have superior capacity for multicycle replication, syncytia formation, and tumor cell lysis. In birds, mesogenic strains cause mild respiratory and gastrointestinal disease, while velogenic strains cause severe respiratory and gastrointestinal disease as well as neurotoxicity [14–16]. In preclinical studies, the most commonly used strains are the mesogenic strains MTH-68/H, PV701, 73T, Italien, Beaudette C, and AF2240, and the lentogenic strains HUJ, Ulster, LaSota, Hitchner B1, and V40-UPM. Among these strains, the lentogenic NDV LaSota strain is a proven and safe vaccine vector that is commonly used as a live attenuated vaccine in the poultry industry [17]. Due to capacity for multicycle replication, mesogenic and velogenic exhibit superior capacity for direct virus-mediated lysis. It is incorrect, however, to classify the lentogenic NDV strains as completely nonlytic. In a number of studies using lentogenic NDV strains lacking the polybasic F cleavage site, the viruses still demonstrate capacity to infect and lyse cancer cells at multiplicity of infection as low as 0.001 [18].

Oncolytic properties of NDV derive primarily from deficient type I IFN signaling pathways and less sensitive type I IFN receptor-mediated signaling in tumor cells [19–21]. Mutations in genes related to the type I IFN pathway and the downstream Janus kinase (JAK)/signal transducer and activator of transcription (STAT) pathway are associated with NDV susceptibility and cytotoxicity [19,22,23]. Tumor cell susceptibility to NDV infection may also be based on the presence of sialic acid-containing cell surface proteins. It was proposed that the combination of altered type I IFN-related gene expression and sialic acid content could act as a clinical biomarker for determining susceptible tumor types [24]. Finally, defects in apoptotic pathways such as the Fas-FasL interaction or overexpression of antiapoptotic genes such as Livin and BcL-xL, which are documented in many tumor types, may increase susceptibility to NDV allowing for viral persistence, increased replication, and spread to surrounding cells [25–27].

NDV has been shown to cause cell death by apoptosis, necrosis, or autophagy mechanisms [26,28–30]. Viral HN protein can directly trigger the release of type I IFN and upregulates tumor necrosis factor (TNF)-related apoptosis inducing ligand (TRAIL) [31]. In human peripheral blood mononuclear cells (PBMCs), TRAIL signaling in turn upregulates apoptotic genes (FasL, Bax, caspase-8, caspase-9, and caspase-3) [32]. HN gene expression alone has been reported to induce apoptosis in human breast cancer MCF-7 cells [33]. NDV can also induce apoptosis through interferon-independent mechanisms such as the intrinsic mitochondrial death pathway [34]. Finally, the formation of syncytia by some NDV strains (termed "fusogenic" strains) ultimately leads syncytium disintegration either through necrosis or apoptosis [35].

2. Activation of the Innate Anti-Tumor Immune Response by NDV

The type I IFN pathway plays a central role in mediating antiviral immunity in mammals [36]. Type I IFNs have antiviral, proapoptotic, and immunomodulatory effects, all of which contribute in large part to the mechanism by which NDV induces antitumor response [36,37]. Type I IFN production in response to viral infection within the tumor microenvironment may have direct antiproliferative effects in some tumors [38]. More significantly, type I IFN signaling activates both innate and adaptive immunity through recruitment of innate cells including natural killer (NK) cells and antigen-presenting cells (APCs), upregulation of cell adhesion, major histocompatibility complex (MHC) and costimulatory molecules, and priming of antigen-specific T cells [37,39–42]. Thus, activation of type I IFN signaling is one of the key pathways being explored for cancer immunotherapy, and this is supported by the findings that tumors with high CD8+ T cell proliferation and responsiveness to immune checkpoint inhibitors are enriched for genes associated with type I IFN signaling [43].

Upon NDV infection, pathogen-associated molecular patterns (PAMPs) inherent to the virus and danger-associated molecular patterns (DAMPs) released by dying cells are recognized by pattern recognition receptors (PRRs) including extracellular Toll-like receptors (TLRs) 3, 7, 8, and 9; intracellular nucleotide-binding oligomerization domain (NOD) proteins; and intracellular RNA helicases such as RIG-1 or MDA5 [44,45] (Figure 1). Recognition of PAMPs and DAMPs by PRRs leads to the activation of transcription factors including IFN regulatory factor (IRF)3, IRF7, and nuclear factor kappa B via the adaptors interferon β stimulator-1 and stimulator of interferon genes (STING) [44]. This signaling cascade results in the transcription and expression of genes encoding proinflammatory cytokines and type I and type III IFN proteins [19,44]. In the case of NDV, cytosolic RNA generated by NDV infection is sensed by RIG-1, and reduction of RIG-1 protein levels has been shown to correlate with decreased intensity of type I IFN response to NDV in vitro [23,46] (Figure 1).

Tumor cells often have impaired type I IFN signaling, which is one of the principal mechanisms resulting in increased tumor cell sensitivity to NDV infection. Despite these deficiencies, the impairment in type I IFN production is typically not absolute, especially as NDV is capable of infecting normal cells in the tumor microenvironment, which have preserved type I IFN response [46–48]. Transcriptional profiling of mouse tumors injected with NDV reveals upregulation of type I IFN response-related genes and a range of cytokines and chemokines that mediate recruitment and proliferation of innate and adaptive immune cells [47,49]. Interestingly, this signature was shown to be independent of NDV-mediated replicative or lytic potential in a study utilizing the lentogenic NDV LaSota strain, indicating that type I IFN signaling activated to even a limited virus infection is sufficient to drive the inflammatory response [49].

While a strong type I IFN response to NDV results in a proinflammatory tumor microenvironment that contributes to the antitumor response, it may, on the other hand, limit therapeutic efficacy by suppressing NDV replication and virus-mediated lysis. Indeed, pretreatment with type I IFN has been shown to limit NDV replication in some tumor cell lines [20,23,46,48]. Therefore, a key unanswered question in the field concerns the timing of type I IFN induction, whereby a balance should be achieved between adequate virus replication and tumor lysis and induction of innate immune response to promote further adaptive immunity. A recombinant lentogenic NDV strain (Hitchner B1) expressing the influenza A virus IFN antagonist protein NS1, which suppresses RIG-1 receptor signaling, IRF3 dimerization, and expression of IFN-β, potently reduced IFN signaling across a panel of cancer cell lines and resulted in increased NDV replication and cytolysis [50]. In vivo, this virus was more effective in controlling tumor growth and prolonging survival in a syngeneic melanoma mouse model [50]. Similar results were demonstrated using the recombinant mesogenic Beaudette C NDV strain expressing an

IFN-antagonist protein which showed higher efficiency in tumor regression in a xenotransplanted fibrosarcoma mouse model [47]. Despite these findings, type I IFN has been shown to be essential for antitumor activity of NDV, and in mice lacking type I IFN receptor, the virus exhibited no ability to control tumor growth [51].

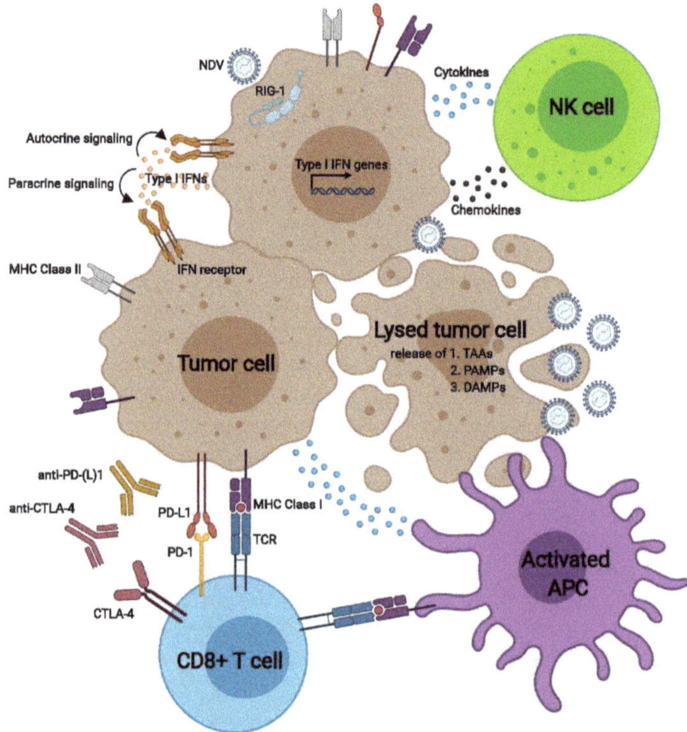

Figure 1. Newcastle disease virus (NDV) activates innate and adaptive anti-tumor immune responses. NDV selectively infects tumor cells that have defective anti-viral defenses. Extracellular and intracellular signaling mediated by sensors such as the RNA helicase RIG-1 leads to expression of type I IFN and related genes. Autocrine and paracrine IFN signaling upregulates MHC class I and II presentation, co-stimulatory molecules, and immune checkpoints on the cell surface. The release of cytokines and chemokines in addition results in the recruitment of innate effector cells such as NK cells and macrophages and antigen-presenting cells (APCs). Virus-mediated direct oncolysis leads to release of tumor antigens, PAMPs, and DAMPs that activate APCs including dendritic cells capable of antigen cross-presentation. Activated APCs prime T cells, resulting in generation of cytolytic T cells directed toward tumor and viral antigens; however, effector function of the activated T cells can be inhibited by upregulation of PD-L1 on tumor cells and APCs, and PD-1 and CTLA-4 on T cells. Upregulation of these negative feedback mechanisms provide the rationale for combining NDV with immune checkpoint inhibitors.

In addition to activation of tumor cell-inherent type I IFN signaling, the inflammatory environment generated by NDV results in the recruitment of innate effector cells and adaptive immune cells (discussed below) that contribute to antitumor immunity (Figure 1). In particular, intratumoral NDV injection leads to a significant tumor infiltration with natural killer (NK) cells [42,52,53]. Interestingly, depletion of NK cells prior to NDV treatment in a syngeneic mouse tumor model abrogated antitumor efficacy, while depletion of NK cells concomitantly with NDV treatment did not, suggesting that while NK cells are important early responders to NDV infection, their role appears to be essential only for the initial inflammatory response [53,54]. Last, NDV infection also results in the recruitment of myeloid cells, which have important roles in phagocytosis and antigen presentation [54,55].

3. Activation of the Adaptive Antitumor Immune Response by NDV

Activation of the innate immune system, largely mediated by type I IFN signaling in response to NDV infection, provides optimal conditions for stimulating adaptive antitumor immunity. Secretion of inflammatory mediators leads to the recruitment of both myeloid and lymphoid cells to the tumor microenvironment [41] (Figure 1). A key effector population is dendritic cells (DCs), a subset of which specialize in antigen cross-presentation (BATF3-dependent or CD8+ DCs) and priming of antigen-specific CD8+ T cells [43,56,57]. NDV infection can cause cell death by apoptosis, necrosis, or autophagy, all of which can lead to the release of viral and tumor-associated antigens and debris within the tumor microenvironment. Cross-presenting DCs become activated and mature in response to uptake of these antigens and in response to PAMPs and DAMPs [43,56,57]. Interleukin (IL)-12 produced by cross-presenting DCs, acting in concert with type I IFN signaling in the tumor microenvironment, leads to upregulation of MHC class I and II molecules, cell adhesion molecules, and co-stimulatory molecules, all of which promote priming of T cells by APCs [56,58]. In effect, tumor infection with NDV acts as an in situ vaccine by causing the release and presentation of tumor antigens in a setting of an inflammatory environment, eliminating the need for selection of antigens needed with other vaccine modalities [59] (Figure 1).

Evidence for NDV-induced tumor antigen-specific CD8+ T cell response comes from studies involving bilateral flank syngeneic tumor models, whereby lentogenic NDV LaSota strain is administered to a single flank tumor [49,51,53,54,60]. Due to restriction of virus replication to the injected tumor, such models allow for assessment of both local and distant immune effects. Interestingly, intratumoral therapy with NDV resulted in a marked increase in CD4+ and CD8+ T cell infiltration in both injected and non-injected tumors. Importantly, there was a greater increase in CD4+FoxP3− cells as compared to regulatory CD4+FoxP3+ cells [49,51,53,54,60]. Furthermore, tumor-infiltrating T cells isolated from both tumor sites expressed increased activation, proliferation, and lytic markers [51,54,60]. This was further supported by the finding of increased expression of other genes associated with T cell activation within the tumor microenvironment of both the injected and non-injected lesions [60]. Importantly, this expression profile was not observed in the spleen, suggesting that the activated T cell response was specific to tumors and not due to nonspecific inflammation [54]. Last, intratumoral NDV therapy resulted in tumor growth delay of both virus-injected and distant tumors and prolonged animal survival, implicating potential development of systemic tumor antigen-specific T cell responses [51]. Overall, these findings are consistent with clinical observations of intralesional administration of talimogene laherparepvec (T-VEC) in advanced melanoma leading to tumor immune infiltration and regression of both injected lesions and distant sites [61].

In the experiments discussed above, complete tumor regressions in the contralateral non-injected tumors were rare despite a marked increase in T cell infiltration, suggesting that compensatory immune inhibitory mechanisms may dampen the immune response. Indeed, upregulation of a number of immune checkpoints, including CTLA-4 and PD-1, was observed on tumor-infiltrating T cells in both virus-injected and distant tumors [51,54]. In addition, upregulation of PD-L1 was observed on tumor, myeloid, and stromal cells [54]. PD-L1 increase occurred early in the injected tumor and was found to be due to rapid upregulation of type I IFN in response to NDV injection. High levels of PD-L1 were also found in the distant non-injected lesion, albeit later in the treatment course, and were found to be upregulated in response to increase in tumor infiltrating lymphocytes. Interestingly, PD-L1 expression in distant tumors was more common in myeloid cells than in tumor cells [54]. Overall, these findings highlighted the rationale for combining NDV with immune checkpoint inhibitors as a means to alleviate the negative feedback mechanisms likely impacting therapeutic efficacy [41]. Indeed, combination of NDV with systemic anti-CTLA-4, anti-PD-1, or anti-PD-L1 resulted in enhanced rejection of bilateral tumors and prolonged animal survival compared to either treatment alone, an effect that was seen in multiple tumor types [51,54]. These findings highlight that intratumoral therapy with NDV can be an effective strategy to drive systemic efficacy of immune checkpoint inhibitors and have now been confirmed across a number of oncolytic viruses [61–73], including early clinical studies of immune checkpoint inhibitors in combination with T-VEC [61,74,75].

Despite these findings, the responses to oncolytic viruses in clinical trials have not been universal, and our understanding of the mechanisms by which oncolytic viruses activate antitumor immunity remains limited. For example, replicative capacity of oncolytic viruses is a subject of ongoing debate in the oncolytic virus field. As well-replicating viruses tend to exhibit superior lytic ability, many groups prefer well-replicating oncolytic viruses as a means to achieve a maximal tumor-debulking effect through direct virus-mediated lysis [76]. However, it is unclear how replicative capacity alters antitumor immunity. In human bladder cancer cell lines infected with lentogenic NDV LaSota strain, upregulation of innate immune response and antigen presentation machinery was not related to virus replication or tumor lysis [49]. Furthermore, intratumoral NDV therapy in the MB49 bladder cancer model, which is poorly susceptible to NDV-mediated lysis, resulted in complete regression of both virus-injected and distant tumors when used in combination with immune checkpoint inhibitors [49].

Related to the question of replicative capacity is the question of the impact of pre-existing anti-viral immunity. Adaptive immune responses towards an oncolytic virus can curtail anti-tumor efficacy by limiting virus persistence, replication and lysis [77,78]. While immunization of mice with NDV LaSota led to the development of neutralizing antibodies resulting in decreased NDV replication with subsequent challenge, antitumor efficacy was not compromised and, on the contrary, was superior in pre-immunized mice [53]. This was supported by increased T cell infiltration including T-helper cells and upregulation of immune-related gene expression in both treated and distant tumors [53]. Several potential mechanisms could contribute to enhanced antitumor efficacy observed with pre-existing immunity, including an antiviral memory response resulting in more rapid induction of tumor inflammatory response, bystander killing from virus-directed T cells, and epitope spreading [53,71,79]. A closer examination of antitumor versus antiviral immune responses elicited by NDV will be needed to answer these questions. In addition, further studies will be needed to understand if pre-existing antiviral immunity potentiates the antitumor response only within the setting of intratumoral therapy, although some patients who received systemically administered NDV in prior clinical trials experienced durable responses, the onset of which happened late in the treatment course [80,81].

4. Engineering NDV to Modulate Innate and Adaptive Immune Responses

With the development of reverse genetics, it has become possible to modify the NDV viral genome and introduce foreign sequences to potentially enhance oncolytic and immunostimulatory properties of these agents [82]. Several strategies to enhance innate and/or adaptive antitumor immunity by engineering NDV to express cytokines, antibodies, ligands, or tumor antigens have been explored, and a few are reviewed below (Figure 2). Given its ability to activate antigen-presenting cells, granulocyte–macrophage colony stimulating factor (GM-CSF) has been explored as a therapeutic transgene within the context of multiple oncolytic viruses, and T-VEC, an oncolytic herpes simplex virus expressing GM-CSF, was approved by the FDA for treatment of metastatic melanoma [74]. A recombinant strain based on the mesogenic NDV 73T strain currently in clinical development, MEDI5395, expressing human GM-CSF was recently shown to increase secretion of pro-inflammatory cytokines such as IFN-α, IL-6, IL-8, and TNF-α in PBMC samples from healthy volunteers, and stimulated PBMCs to exert antitumor effects in vitro [83]. In addition, infection of dendritic cells led to their maturation, and co-culture of dendritic cells with allogeneic T cells increased the levels of T cell effector cytokines IL-2 and IFN-γ [83]. In a separate study using NDV Hitchner B1 strains engineered to express either murine IL-2, IFN-γ, and GM-CSF in vivo, only NDV expressing IL-2 led to a significant increase in overall animal survival when compared to parental NDV [82]. Similar results were recently demonstrated with a lentogenic recombinant NDV strain expressing IL-24 [84].

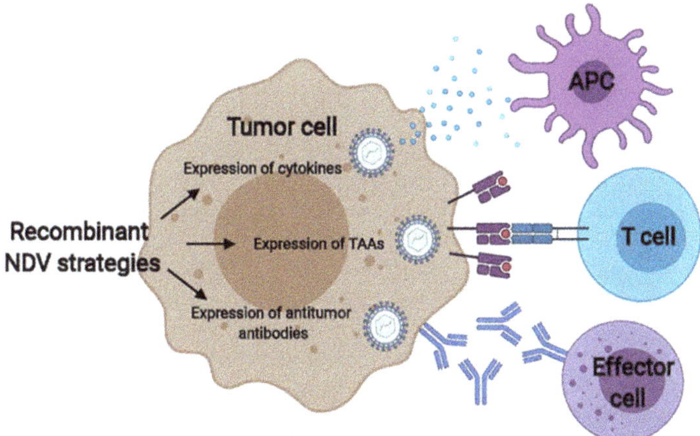

Figure 2. Strategies to enhance the NDV antitumor immune response by recombinant genetic engineering. Genetic engineering can be used to generate NDV strains with greater potential to stimulate antitumor immune response. First, NDV engineered to express cytokines such as GM-CSF or interleukins can increase recruitment of innate effector cells such as antigen-presenting cells (APCs). Second, NDV can be used as a therapeutic vaccine targeted to specific tumor antigens such as oncogenic viral antigens, frame shift mutations, or mutated self-antigens. Third, NDV can be engineered express single-chain variable fragments or full antitumor antibodies to induce antibody-dependent cellular cytotoxicity by effector cells.

Optimal immune mechanisms for intratumoral targeting with oncolytic virus are unknown. Gene expression profiling of tumors after NDV injection revealed the upregulation of T cell co-stimulatory receptors ICOS, 4-1BB, GITR, OX40, CD27, and CD40, all of which are currently being evaluated as therapeutic targets in clinic using monoclonal antibodies [60]. Targeting of ICOS within the context of tumor microenvironment using engineered cellular vaccines expressing ICOS ligand (ICOSL) has in particular been previously demonstrated to improve systemic efficacy of CTLA-4 blockade through potentiation of cytotoxic T cell function [85]. Intratumoral administration of engineered NDV LaSota expressing ICOSL resulted in enhanced infiltration of CD8+ and CD4+ T cells, tumor growth delay of both injected and non-injected tumors, and prolonged survival, as compared to wild type NDV, and this effect that was further enhanced when combined with anti-CTLA-4 blockade [60]. These findings highlight that stimulation of both innate and adaptive immune response pathways within the context of intratumoral NDV therapy may be required for optimal activation of antitumor immune response. Recently, recombinant NDV LaSota strains expressing soluble single-chain variable fragments for anti-CD28, anti-PD1, and anti-PDL1 were generated, as well as versions fused to IL-12 [86]. All of these strains showed improved tumor control and survival in a melanoma mouse model [86].

Engineering NDV to express a tumor-associated antigen represents another attractive strategy due to its potential to overcome immune tolerance within the context of NDV-induced inflammatory environment [59]. Such strategy was explored with NDV Hitchner B1 expressing an MHC class I restricted epitope of β-galactosidase (β-gal), a model antigen expressed by murine CT26 colorectal carcinoma cells [87]. Intratumoral therapy of CT26 tumor-bearing mice induced a β-gal-specific immune response and significant increase in the number of complete tumor regressions compared to parental NDV. This response was further boosted by co-administration of NDV expressing IL-2, with 90% tumor regression seen [87]. These findings warrant investigation of NDVs expressing other tumor-associated antigens, such as those caused by oncogenic viral antigens, frame shift mutations, and mutated self-antigens, but also highlight that combinatorial strategies using oncolytic viruses targeting different mechanisms (e.g., antigens and adaptive immunity) may be required to achieve optimal anti-tumor response.

5. Clinical Experience with NDV

The immunogenic properties of NDV were recognized early, and a number of studies have explored the virus for immunization of patients with virus-modified cancer cell vaccines [88–106]. Many of the early studies were performed by William Cassel and colleagues utilizing autologous or allogeneic NDV oncolysates for vaccination of patients with resected high risk melanoma, demonstrating improvement in overall survival when compared to historical controls [88,89,93,101,107]. A similar strategy was developed by Volker Schirrmacher and colleagues, where whole-cell autologous irradiated tumor cells were modified by infection with attenuated NDV [108]. The investigators evaluated vaccination with NDV-modified tumor cells in adjuvant or advanced disease setting across a number of cancers, demonstrating evidence of antitumor immunity (measured by delayed type hypersensitivity) and improvement in survival in some studies [91,92,94,95,98,109]. A similar approach was used by Liang and colleagues in a phase III trial in colorectal cancer, comparing adjuvant immunization with NDV-modified autologous cancer cells to resection alone [103]. The study reported improvement in overall survival in the vaccine group (7 vs. 4.5 years), which was statistically significant. Overall, these studies provide a proof of concept that infection of cancer cells by NDV can enhance cancer cell immunogenicity and has a potential to stimulate anti-tumor immunity. While the majority of the studies above are plagued by lack of control arms, prospective randomized studies are certainly warranted, especially in combination with modern immunotherapy agents such as immune checkpoint inhibitors.

As preparation of autologous virus-modified vaccines can be cumbersome, a number of studies have explored NDV for direct administration to cancer patients. In the first documented human use of NDV, administration of the mesogenic NDV Hickman strain to a patient with acute myelogenous leukemia resulted in reduction in leukemic blast count and transient improvement in symptoms [6]. In a case report, mesogenic NDV 73-T strain was used for intratumoral treatment of a patient with advanced cervical cancer, resulting in partial response [4]. Csatary and colleagues reported a case series of patients with various advanced cancers treated with mesogenic NDV strain MTH-68 using various routes of administration, with reported partial or even complete responses across a number of cancers [102,110]. In an additional series, fourteen patients with glioblastoma were treated intravenously with NDV MTH-68 on various schedules. Seven of the patients achieved response to therapy with four of the patients surviving between 5 and 9 years at the time of the publication in 2004 [111].

In the early 2000s, NDV strain PV701, derived from the mesogenic strain 73-T, was evaluated in three phase I trials in patients with advanced malignancies using intravenous administration [80,81,112,113]. In the initial study, in 79 patients there were two responses (one complete and one partial), with seven additional minor responses noted. In fourteen patients, a prolonged progression free survival that lasted from 4 to over 30 months was observed [112]. In a subsequent study of eighteen patients with various advanced cancers using slower infusion rate but higher therapeutic dosing, a higher response rate was observed, with demonstration of four major and two minor responses, with six patients surviving at least 2 years [80,81]. Despite the initial promising results, PV701 unfortunately was not evaluated in further studies, likely secondary to changes in regulatory guidelines surrounding the use of mesogenic and velogenic NDV strains. NDV strains that are highly virulent in birds are classified as USDA select agents, limiting their clinical applicability. Lentogenic NDV strain HUJ has been evaluated using an intravenous approach in 14 patients with recurrent glioblastoma, demonstrating a complete response in one patient. Across the studies, intravenous administration of NDV has in general been well tolerated, with flu-like symptoms being the most common reported adverse event.

While previous studies in humans have only explored naturally occurring NDV strains, genetically modified NDVs have recently entered therapeutic testing. As described above, recombinant NDV expressing GM-CSF (MEDI5395), also based on the 73-T strain, is being evaluated in patients with various advanced malignancies in combination with durvalumab using intravenous administration (NCT03889275). Additional recombinant NDVs are in various stages of development and are expected to enter clinic within the next year.

6. Conclusions

Over the past 60 years, NDV has repeatedly demonstrated its therapeutic potential, both as an oncolytic agent and an immunotherapeutic agent. With intravenous administration, NDV is one of the few viruses that has demonstrated an ability to result in partial and even complete responses as a single agent. Durability of these responses further highlights that the therapeutic effect of the virus is likely not solely dependent on direct oncolysis, but rather on the ability of the virus to induce durable immunity. While the use of mesogenic and velogenic (and thus most lytic) strains for antitumor therapy is limited due to their pathogenic potential in birds, data with fewer lytic strains nevertheless highlights their potential to incite antitumor immunity, with the recent data indicating their ability to potentiate the efficacy of systemic immune checkpoint inhibitors. Furthermore, with the advent of genetic engineering, it has become possible to modify NDV to further enhance its immunogenic potential, with introduction of transgenes targeting both innate and adaptive immune pathways. As with other oncolytic viruses, many questions surrounding therapy with NDV remain unanswered, including optimal route of administration, ideal strategies for genetic engineering, therapeutic sequencing with immune checkpoint inhibitors, and best combination partners. While preclinical syngeneic models have provided some answers to these questions, most, if not all, models fail to capture the heterogeneity of human cancers and are thus not sufficient for guiding therapy. It is thus imperative that within the context of clinical trials we collect as much information as possible, with translational endpoints being prioritized as essential elements of any study. Understanding of the evolution of immune response to the virus and the tumor, even in a trial with no clinical benefit, should be a key priority for any clinical trial utilizing oncolytic viruses, as it is the only way to guide the further development of these agents and move the field forward.

Author Contributions: B.B., G.P., and D.Z. all contributed to manuscript composition and editing. All authors have read and agreed to the published version of the manuscript.

Funding: This research was funded in part through the NIH/NCI Memorial Sloan Kettering Cancer Center Support Grant P30 CA008748. D.Z. is a member of the Parker Institute for Cancer Immunotherapy at MSKCC. D.Z. is supported by the Ovarian Cancer Research Foundation Liz Tilberis Award and the Department of Defense Ovarian Cancer Research Academy (OC150111). B.B. is supported by the NIH T32 Investigational Cancer Therapeutics Training Program Grant (T32-CA009207) and the ASCO Conquer Cancer Foundation Young Investigator Award.

Conflicts of Interest: D.Z. is an inventor on two patents related to use of NDV for cancer therapy. D.Z. reports personal/consultancy fees from Merck, Synlogic Therapeutics, Biomed Valley Discoveries, Trieza Therapeutics, Tesaro, and Agenus of the scope of this work. D.Z. reports institutional research support from Astra Zeneca, Plexxikon, and Genentech, outside of the scope of this work. B.B. and G.P. have no conflicts.

References

1. Kaufman, H.L.; Kohlhapp, F.J.; Zloza, A. Oncolytic viruses: A new class of immunotherapy drugs. *Nat. Rev. Drug Discov.* **2015**, *14*, 642–662. [CrossRef] [PubMed]
2. Flanagan, A.D.; Love, R.; Tesar, W. Propagation of Newcastle Disease Virus in Ehrlich Ascites Cells In Vitro and In vivo. *Exp. Biol. Med.* **1955**, *90*, 82–86. [CrossRef] [PubMed]
3. Sinkovics, J. Studies on the biological characteristics of the newcastle disease virus (NDV) adapted to the brain of newborne mice. *Arch. Virol.* **1957**, *7*, 403–411. [CrossRef] [PubMed]
4. Cassel, W.A.; Garrett, R.E. Newcastle disease virus as an antineoplastic agent. *Cancer* **1965**, *18*, 863–868. [CrossRef]
5. Cassel, W.A.; Garrett, R.E. Tumor immunity after viral oncolysis. *J. Bacteriol.* **1966**, *92*, 792. [CrossRef]
6. Wheelock, E.F.; Dingle, J.H. Observations on the Repeated Administration of Viruses to a Patient with Acute Leukemia. A Preliminary Report. *N. Engl. J. Med.* **1964**, *271*, 645–651. [CrossRef]
7. Lamb, R.; Parks, G. *Paramyxoviridae: The Viruses and Their Replication. Fields Virology*; Knipe, D.M., Howley, P.M., Eds.; Lippincott Williams & Wilkins: Philadelphia, PA, USA, 2007; pp. 1449–1496.
8. Sinkovics, J.G.; Horvath, J.C. Newcastle disease virus (NDV): Brief history of its oncolytic strains. *J. Clin. Virol.* **2000**, *16*, 1–15. [CrossRef]

9.   Leighton, F.A.; Heckert, R.A. Newcastle Disease and Related Avian Paramyxoviruses. *Infect. Dis. Wild Birds* **2008**, *1*, 1–16.

10. Dortmans, J.C.F.M.; Koch, G.; Rottier, P.J.M.; Peeters, B. Virulence of newcastle disease virus: What is known so far? *Vet. Res.* **2011**, *42*, 122. [CrossRef]

11. Connaris, H.; Takimoto, T.; Russell, R.; Crennell, S.; Moustafa, I.; Portner, A.; Taylor, G.L. Probing the Sialic Acid Binding Site of the Hemagglutinin-Neuraminidase of Newcastle Disease Virus: Identification of Key Amino Acids Involved in Cell Binding, Catalysis, and Fusion. *J. Virol.* **2002**, *76*, 1816–1824. [CrossRef]

12. Ito, Y.; Komada, H.; Kusagawa, S.; Tsurudome, M.; Matsumura, H.; Kawano, M.; Ohta, H.; Nishio, M. Fusion regulation proteins on the cell surface: Isolation and characterization of monoclonal antibodies which enhance giant polykaryocyte formation in Newcastle disease virus-infected cell lines of human origin. *J. Virol.* **1992**, *66*, 5999–6007. [CrossRef] [PubMed]

13. Lamb, R.A.; Jardetzky, T.S. Structural basis of viral invasion: Lessons from paramyxovirus F. *Curr. Opin. Struct. Biol.* **2007**, *17*, 427–436. [CrossRef] [PubMed]

14. De Leeuw, O.S.; Hartog, L.; Koch, G.; Peeters, B.P.H. Effect of fusion protein cleavage site mutations on virulence of Newcastle disease virus: Non-virulent cleavage site mutants revert to virulence after one passage in chicken brain. *J. Gen. Virol.* **2003**, *84*, 475–484. [CrossRef] [PubMed]

15. Panda, A.; Huang, Z.; Elankumaran, S.; Rockemann, D.D.; Samal, S.K. Role of fusion protein cleavage site in the virulence of Newcastle disease virus. *Microb. Pathog.* **2004**, *36*, 1–10. [CrossRef] [PubMed]

16. Seal, B.S.; King, D.J.; Bennett, J.D. Characterization of Newcastle disease virus isolates by reverse transcription PCR coupled to direct nucleotide sequencing and development of sequence database for pathotype prediction and molecular epidemiological analysis. *J. Clin. Microbiol.* **1995**, *33*, 2624–2630. [CrossRef] [PubMed]

17. Kim, S.-H.; Samal, S.K. Newcastle Disease Virus as a Vaccine Vector for Development of Human and Veterinary Vaccines. *Viruses* **2016**, *8*, 183. [CrossRef] [PubMed]

18. Ginting, T.E.; Suryatenggara, J.; Christian, S.; Mathew, G. Proinflammatory response induced by Newcastle disease virus in tumor and normal cells. *Oncolytic Virother.* **2017**, *6*, 21–30. [CrossRef]

19. Krishnamurthy, S.; Takimoto, T.; Scroggs, R.A.; Portner, A. Differentially Regulated Interferon Response Determines the Outcome of Newcastle Disease Virus Infection in Normal and Tumor Cell Lines. *J. Virol.* **2006**, *80*, 5145–5155. [CrossRef]

20. Fiola, C.; Peeters, B.; Fournier, P.; Arnold, A.; Bucur, M.; Schirrmacher, V. Tumor selective replication of Newcastle disease virus: Association with defects of tumor cells in antiviral defence. *Int. J. Cancer* **2006**, *119*, 328–338. [CrossRef]

21. Zamarin, D.; Palese, P. Oncolytic Newcastle disease virus for cancer therapy: Old challenges and new directions. *Future Microbiol.* **2012**, *7*, 347–367. [CrossRef]

22. Matveeva, O.V.; Chumakov, P.M. Defects in interferon pathways as potential biomarkers of sensitivity to oncolytic viruses. *Rev. Med. Virol.* **2018**, *28*, e2008. [CrossRef] [PubMed]

23. Wilden, H.; Fournier, P.; Zawatzky, R.; Schirrmacher, V. Expression of RIG-I, IRF3, IFN-beta and IRF7 determines resistance or susceptibility of cells to infection by Newcastle Disease Virus. *Int. J. Oncol.* **2009**, *34*, 971–982. [PubMed]

24. Liu, T.; Zhang, Y.; Cao, Y.; Jiang, S.; Sun, R.; Yin, J.; Gao, Z.; Ren, G.; Wang, Z.; Yu, Q.; et al. Optimization of oncolytic effect of Newcastle disease virus Clone30 by selecting sensitive tumor host and constructing more oncolytic viruses. *Gene Ther.* **2020**, 1–21. [CrossRef] [PubMed]

25. Mansour, M.; Palese, P.; Zamarin, D. Oncolytic Specificity of Newcastle Disease Virus Is Mediated by Selectivity for Apoptosis-Resistant Cells. *J. Virol.* **2011**, *85*, 6015–6023. [CrossRef] [PubMed]

26. Cuadrado-Castano, S.; Ayllon, J.; Mansour, M.; De La Iglesia-Vicente, J.; Jordan, S.; Tripathi, S.; García-Sastre, A.; Villar, E. Enhancement of the proapoptotic properties of newcastle disease virus promotes tumor remission in syngeneic murine cancer models. *Mol. Cancer Ther.* **2015**, *14*, 1247–1258. [CrossRef] [PubMed]

27. Lazar, I.; Yaacov, B.; Shiloach, T.; Eliahoo, E.; Kadouri, L.; Lotem, M.; Perlman, R.; Zakay-Rones, Z.; Panet, A.; Ben-Yehuda, D. The Oncolytic Activity of Newcastle Disease Virus NDV-HUJ on Chemoresistant Primary Melanoma Cells Is Dependent on the Proapoptotic Activity of the Inhibitor of Apoptosis Protein Livin. *J. Virol.* **2009**, *84*, 639–646. [CrossRef] [PubMed]

28. Cheng, J.-H.; Sun, Y.-J.; Zhang, F.-Q.; Zhang, X.-R.; Qiu, X.-S.; Yu, L.-P.; Wu, Y.; Ding, C. Newcastle disease virus NP and P proteins induce autophagy via the endoplasmic reticulum stress-related unfolded protein response. *Sci. Rep.* **2016**, *6*, 24721. [CrossRef] [PubMed]
29. Ye, T.; Jiang, K.; Wei, L.; Barr, M.P.; Xu, Q.; Zhang, G.; Ding, C.; Meng, S.; Piao, H. Oncolytic Newcastle disease virus induces autophagy-dependent immunogenic cell death in lung cancer cells. *Am. J. Cancer Res.* **2018**, *8*, 1514–1527.
30. Kroemer, G.; Galluzzi, L.; Kepp, O.; Zitvogel, L. Immunogenic Cell Death in Cancer Therapy. *Annu. Rev. Immunol.* **2013**, *31*, 51–72. [CrossRef]
31. Liao, Y.; Wang, H.-X.; Mao, X.; Fang, H.; Wang, H.; Li, Y.; Sun, Y.; Meng, C.; Tan, L.; Song, C.; et al. RIP1 is a central signaling protein in regulation of TNF-α/TRAIL mediated apoptosis and necroptosis during Newcastle disease virus infection. *Oncotarget* **2017**, *8*, 43201–43217. [CrossRef]
32. Zeng, J.; Fournier, P.; Schirrmacher, V. Induction of Interferon-α and Tumor Necrosis Factor-Related Apoptosis-Inducing Ligand in Human Blood Mononuclear Cells by Hemagglutinin-Neuraminidase but Not F Protein of Newcastle Disease Virus. *Virology* **2002**, *297*, 19–30. [CrossRef] [PubMed]
33. Ghrici, M.; El Zowalaty, M.E.; Omar, A.R.; Ideris, A. Induction of apoptosis in MCF-7 cells by the hemagglutinin-neuraminidase glycoprotein of Newcastle disease virus Malaysian strain AF2240. *Oncol. Rep.* **2013**, *30*, 1035–1044. [CrossRef]
34. Elankumaran, S.; Rockemann, D.; Samal, S.K. Newcastle Disease Virus Exerts Oncolysis by both Intrinsic and Extrinsic Caspase-Dependent Pathways of Cell Death. *J. Virol.* **2006**, *80*, 7522–7534. [CrossRef] [PubMed]
35. Zeng, J.; Fournier, P.; Schirrmacher, V. High cell surface expression of Newcastle disease virus proteins via replicon vectors demonstrates syncytia forming activity of F and fusion promotion activity of HN molecules. *Int. J. Oncol.* **2004**, *25*, 293–302. [CrossRef] [PubMed]
36. Samuel, C.E. Antiviral Actions of Interferons. *Clin. Microbiol. Rev.* **2001**, *14*, 778–809. [CrossRef]
37. Washburn, B.; Schirrmacher, V. Human tumor cell infection by Newcastle Disease Virus leads to upregulation of HLA and cell adhesion molecules and to induction of interferons, chemokines and finally apoptosis. *Int. J. Oncol.* **2002**, *21*, 85–93. [CrossRef]
38. Vitale, G.; van Eijck, C.H.; Koetsvelt, P.M.; Erdman, J.; Speel, E.J.M.; Mooji, D.M.; Colao, A.; Lombardi, G.; Croze, E.; Lamberts, S.W.J.; et al. Type I interferons in the treatment of pancreatic cancer: Mechanisms of action and role of related receptors. *Ann. Surg.* **2007**, *246*, 259–268. [CrossRef]
39. Ten, R.M.; Blank, V.; Le Bail, O.; Kourilsky, P.; Israël, A. Two factors, IRF1 and KBF1/NF-kappa B, cooperate during induction of MHC class I gene expression by interferon alpha beta or Newcastle disease virus. *Comptes Rendus Académie Sci. Ser. III* **1993**, *316*, 496–501.
40. Ni, J.; Galani, I.E.; Cerwenka, A.; Schirrmacher, V.; Fournier, P. Antitumor vaccination by Newcastle Disease Virus Hemagglutinin–Neuraminidase plasmid DNA application: Changes in tumor microenvironment and activation of innate anti-tumor immunity. *Vaccine* **2011**, *29*, 1185–1193. [CrossRef]
41. Zamarin, D.; Wolchok, J.D. Potentiation of immunomodulatory antibody therapy with oncolytic viruses for treatment of cancer. *Mol. Ther. Oncolytics* **2014**, *1*, 14004. [CrossRef]
42. Jarahian, M.; Watzl, C.; Fournier, P.; Arnold, A.; Djandji, D.; Zahedi, S.; Cerwenka, A.; Paschen, A.; Schirrmacher, V.; Momburg, F. Activation of Natural Killer Cells by Newcastle Disease Virus Hemagglutinin-Neuraminidase. *J. Virol.* **2009**, *83*, 8108–8121. [CrossRef] [PubMed]
43. Fuertes, M.B.; Kacha, A.K.; Kline, J.; Woo, S.-R.; Kranz, D.M.; Murphy, K.M.; Gajewski, T.F. Host type I IFN signals are required for antitumor CD8+ T cell responses through CD8α+ dendritic cells. *J. Exp. Med.* **2011**, *208*, 2005–2016. [CrossRef] [PubMed]
44. Takeuchi, O.; Akira, S. Pattern Recognition Receptors and Inflammation. *Cell* **2010**, *140*, 805–820. [CrossRef] [PubMed]
45. Kato, H.; Takeuchi, O.; Sato, S.; Yoneyama, M.; Yamamoto, M.; Matsui, K.; Uematsu, S.; Jung, A.; Kawai, T.; Ishii, K.J.; et al. Differential roles of MDA5 and RIG-I helicases in the recognition of RNA viruses. *Nat. Cell Biol.* **2006**, *441*, 101–105. [CrossRef] [PubMed]
46. Biswas, M.; Kumar, S.R.; Allen, A.; Yong, W.; Nimmanapalli, R.; Samal, S.K.; Elankumaran, S. Cell-Type-Specific Innate Immune Response to Oncolytic Newcastle Disease Virus. *Viral Immunol.* **2012**, *25*, 268–276. [CrossRef]

47. Elankumaran, S.; Chavan, V.; Qiao, D.; Shobana, R.; Moorkanat, G.; Biswas, M.; Samal, S.K. Type I Interferon-Sensitive Recombinant Newcastle Disease Virus for Oncolytic Virotherapy. *J. Virol.* **2010**, *84*, 3835–3844. [CrossRef]

48. Buijs, P.R.A.; Van Eijck, C.H.J.; Hofland, L.J.; Fouchier, R.A.M.; Hoogen, B.V.D. Different responses of human pancreatic adenocarcinoma cell lines to oncolytic Newcastle disease virus infection. *Cancer Gene Ther.* **2014**, *21*, 24–30. [CrossRef]

49. Oseledchyk, A.; Ricca, J.M.; Gigoux, M.; Ko, B.; Redelman-Sidi, G.; Walther, T.; Liu, C.; Iyer, G.; Merghoub, T.; Wolchok, J.D.; et al. Lysis-independent potentiation of immune checkpoint blockade by oncolytic virus. *Oncotarget* **2018**, *9*, 28702–28716. [CrossRef]

50. Zamarin, D.; Martínez-Sobrido, L.; Kelly, K.; Mansour, M.; Sheng, G.; Vigil, A.; García-Sastre, A.; Palese, P.; Fong, Y. Enhancement of Oncolytic Properties of Recombinant Newcastle Disease Virus Through Antagonism of Cellular Innate Immune Responses. *Mol. Ther.* **2009**, *17*, 697–706. [CrossRef]

51. Zamarin, D.; Holmgaard, R.B.; Subudhi, S.K.; Park, J.S.; Mansour, M.; Palese, P.; Merghoub, T.; Wolchok, J.D.; Allison, J.P. Localized Oncolytic Virotherapy Overcomes Systemic Tumor Resistance to Immune Checkpoint Blockade Immunotherapy. *Sci. Transl. Med.* **2014**, *6*, 226ra32. [CrossRef]

52. Schwaiger, T.; Knittler, M.R.; Grund, C.; Roemer-Oberdoerfer, A.; Kapp, J.-F.; Lerch, M.M.; Mettenleiter, T.C.; Mayerle, J.; Blohm, U. Newcastle disease virus mediates pancreatic tumor rejection via NK cell activation and prevents cancer relapse by prompting adaptive immunity. *Int. J. Cancer* **2017**, *141*, 2505–2516. [CrossRef] [PubMed]

53. Ricca, J.M.; Oseledchyk, A.; Walther, T.; Liu, C.; Mangarin, L.; Merghoub, T.; Wolchok, J.D.; Zamarin, D. Pre-existing Immunity to Oncolytic Virus Potentiates Its Immunotherapeutic Efficacy. *Mol. Ther.* **2018**, *26*, 1008–1019. [CrossRef] [PubMed]

54. Zamarin, D.; Ricca, J.M.; Sadekova, S.; Oseledchyk, A.; Yu, Y.; Blumenschein, W.M.; Wong, J.; Gigoux, M.; Merghoub, T.; Wolchok, J.D. PD-L1 in tumor microenvironment mediates resistance to oncolytic immunotherapy. *J. Clin. Investig.* **2018**, *128*, 1413–1428. [CrossRef] [PubMed]

55. Schirrmacher, V.; Bai, L.; Umansky, V.; Yu, L.; Xing, Y.; Qian, Z. Newcastle disease virus activates macrophages for anti-tumor activity. *Int. J. Oncol.* **2000**, *16*, 363–373. [CrossRef]

56. Shortman, K.; Heath, W.R. The CD8+ dendritic cell subset. *Immunol. Rev.* **2010**, *234*, 18–31. [CrossRef]

57. Hildner, K.; Edelson, B.T.; Purtha, W.E.; Diamond, M.S.; Matsushita, H.; Kohyama, M.; Calderon, B.; Schraml, B.U.; Unanue, E.R.; Schreiber, R.D.; et al. Batf3 Deficiency Reveals a Critical Role for CD8 + Dendritic Cells in Cytotoxic T Cell Immunity. *Science* **2008**, *322*, 1097–1100. [CrossRef]

58. Diamond, M.S.; Kinder, M.; Matsushita, H.; Mashayekhi, M.; Dunn, G.P.; Archambault, J.M.; Lee, H.; Arthur, C.D.; White, J.M.; Kalinke, U.; et al. Type I interferon is selectively required by dendritic cells for immune rejection of tumors. *J. Exp. Med.* **2011**, *208*, 1989–2003. [CrossRef]

59. Schirrmacher, V.; Van Gool, S.; Stuecker, W. Breaking Therapy Resistance: An Update on Oncolytic Newcastle Disease Virus for Improvements of Cancer Therapy. *Biomedicines* **2019**, *7*, 66. [CrossRef]

60. Zamarin, D.; Holmgaard, R.B.; Ricca, J.; Plitt, T.; Palese, P.; Sharma, P.; Merghoub, T.; Wolchok, J.D.; Allison, J.P. Intratumoral modulation of the inducible co-stimulator ICOS by recombinant oncolytic virus promotes systemic anti-tumour immunity. *Nat. Commun.* **2017**, *8*, 14340. [CrossRef]

61. Da Silva, E.S.; Flores, R.A.; Ribas, A.S.; Taschetto, A.P.D.; Faria, M.S.; Lima, L.B.; Metzger, M.; Donato, J.; Paschoalini, M.A. Injections of the of the α 1 -adrenoceptor antagonist prazosin into the median raphe nucleus increase food intake and Fos expression in orexin neurons of free-feeding rats. *Behav. Brain Res.* **2017**, *324*, 87–95. [CrossRef]

62. Porter, C.E.; Shaw, A.R.; Jung, Y.; Yip, T.; Castro, P.D.; Sandulache, V.C.; Sikora, A.; Gottschalk, S.; Ittman, M.M.; Brenner, M.K.; et al. Oncolytic Adenovirus Armed with BiTE, Cytokine, and Checkpoint Inhibitor Enables CAR T Cells to Control the Growth of Heterogeneous Tumors. *Mol. Ther.* **2020**, *28*, 1251–1262. [CrossRef] [PubMed]

63. Nakao, S.; Arai, Y.; Tasaki, M.; Yamashita, M.; Murakami, R.; Kawase, T.; Amino, N.; Nakatake, M.; Kurosaki, H.; Mori, M.; et al. Intratumoral expression of IL-7 and IL-12 using an oncolytic virus increases systemic sensitivity to immune checkpoint blockade. *Sci. Transl. Med.* **2020**, *12*, eaax7992. [CrossRef] [PubMed]

64. Kuryk, L.; Møller, A.-S.W.; Jaderberg, M. Combination of immunogenic oncolytic adenovirus ONCOS-102 with anti-PD-1 pembrolizumab exhibits synergistic antitumor effect in humanized A2058 melanoma huNOG mouse model. *OncoImmunology* **2019**, *8*, 1532763. [CrossRef] [PubMed]

65. Harrington, K.J.; Freeman, D.J.; Kelly, B.; Harper, J.; Soria, J.-C. Optimizing oncolytic virotherapy in cancer treatment. *Nat. Rev. Drug Discov.* **2019**, *18*, 689–706. [CrossRef] [PubMed]

66. Liu, Z.; Ge, Y.; Wang, H.; Ma, C.; Feist, M.; Ju, S.; Guo, Z.S.; Bartlett, D.L. Modifying the cancer-immune set point using vaccinia virus expressing re-designed interleukin-2. *Nat. Commun.* **2018**, *9*, 1–9. [CrossRef] [PubMed]

67. Liu, Z.; Ravindranathan, R.; Kalinski, P.; Guo, Z.S.; Bartlett, D.L. Rational combination of oncolytic vaccinia virus and PD-L1 blockade works synergistically to enhance therapeutic efficacy. *Nat. Commun.* **2017**, *8*, 14754. [CrossRef] [PubMed]

68. Ilett, E.; Kottke, T.; Thompson, J.; Rajani, K.; Zaidi, S.; Evgin, L.; Coffey, M.; Ralph, C.; Diaz, R.; Pandha, H.; et al. Prime-boost using separate oncolytic viruses in combination with checkpoint blockade improves anti-tumour therapy. *Gene Ther.* **2017**, *24*, 21–30. [CrossRef]

69. Chen, C.-Y.; Wang, P.-Y.; Hutzen, B.; Sprague, L.; Swain, H.M.; Love, J.K.; Stanek, J.R.; Boon, L.; Conner, J.; Cripe, T.P. Cooperation of Oncolytic Herpes Virotherapy and PD-1 Blockade in Murine Rhabdomyosarcoma Models. *Sci. Rep.* **2017**, *7*, 1–10. [CrossRef]

70. Hardcastle, J.; Mills, L.; Malo, C.S.; Jin, F.; Kurokawa, C.; Geekiyanage, H.; Schroeder, M.; Sarkaria, J.; Johnson, A.J.; Galanis, E. Immunovirotherapy with measles virus strains in combination with anti–PD-1 antibody blockade enhances antitumor activity in glioblastoma treatment. *Neuro-Oncology* **2016**, *19*, 493–502. [CrossRef]

71. Woller, N.; Gürlevik, E.; Fleischmann-Mundt, B.; Schumacher, A.; Knocke, S.; Kloos, A.M.; Saborowski, M.; Geffers, R.; Manns, M.P.; Wirth, T.C.; et al. Viral Infection of Tumors Overcomes Resistance to PD-1-immunotherapy by Broadening Neoantigenome-directed T-cell Responses. *Mol. Ther.* **2015**, *23*, 1630–1640. [CrossRef]

72. Shen, W.; Patnaik, M.M.; Ruiz, A.; Russell, S.J.; Peng, K.-W. Immunovirotherapy with vesicular stomatitis virus and PD-L1 blockade enhances therapeutic outcome in murine acute myeloid leukemia. *Blood* **2016**, *127*, 1449–1458. [CrossRef] [PubMed]

73. Rajani, K.; Parrish, C.; Kottke, T.; Thompson, J.; Zaidi, S.; Ilett, L.; Shim, K.G.; Diaz, R.-M.; Pandha, H.; Harrington, K.; et al. Combination Therapy With Reovirus and Anti-PD-1 Blockade Controls Tumor Growth Through Innate and Adaptive Immune Responses. *Mol. Ther.* **2016**, *24*, 166–174. [CrossRef] [PubMed]

74. Chesney, J.; Puzanov, I.; Collichio, F.; Singh, P.; Milhem, M.M.; Glaspy, J.; Hamid, O.; Ross, M.; Friedlander, P.; Garbe, C.; et al. Randomized, Open-Label Phase II Study Evaluating the Efficacy and Safety of Talimogene Laherparepvec in Combination With Ipilimumab Versus Ipilimumab Alone in Patients With Advanced, Unresectable Melanoma. *J. Clin. Oncol.* **2018**, *36*, 1658–1667. [CrossRef] [PubMed]

75. Kelly, C.M.; Antonescu, C.R.; Bowler, T.; Munhoz, R.; Chi, P.; Dickson, M.A.; Gounder, M.M.; Keohan, M.L.; Movva, S.; Dholakia, R.; et al. Objective Response Rate Among Patients With Locally Advanced or Metastatic Sarcoma Treated With Talimogene Laherparepvec in Combination With Pembrolizumab: A Phase 2 Clinical Trial. *JAMA Oncol.* **2020**, *6*, 402–408. [CrossRef] [PubMed]

76. Zamarin, D.; Pesonen, S. Replication-Competent Viruses as Cancer Immunotherapeutics: Emerging Clinical Data. *Hum. Gene Ther.* **2015**, *26*, 538–549. [CrossRef] [PubMed]

77. Wong, H.H.; Lemoine, N.R.; Wang, Y. Oncolytic Viruses for Cancer Therapy: Overcoming the Obstacles. *Viruses* **2010**, *2*, 78–106. [CrossRef] [PubMed]

78. Ferguson, M.S.; Lemoine, N.R.; Wang, Y. Systemic Delivery of Oncolytic Viruses: Hopes and Hurdles. *Adv. Virol.* **2012**, *2012*, 1–14. [CrossRef]

79. Bridle, B.W.; Stephenson, K.B.; Boudreau, J.E.; Koshy, S.; Kazdhan, N.; Pullenayegum, E.; Brunellière, J.; Bramson, J.L.; Lichty, B.D.; Wan, Y. Potentiating Cancer Immunotherapy Using an Oncolytic Virus. *Mol. Ther.* **2010**, *18*, 1430–1439. [CrossRef]

80. Hotte, S.J.; Lorence, R.M.; Hirte, H.; Polawski, S.R.; Bamat, M.K.; O'Neil, J.D.; Roberts, M.S.; Groene, W.S.; Major, P.P. An Optimized Clinical Regimen for the Oncolytic Virus PV701. *Clin. Cancer Res.* **2007**, *13*, 977–985. [CrossRef]

81. Lorence, R.M.; Roberts, M.S.; Neil, J.D.O.; Groene, W.S.; Miller, J.A.; Mueller, S.N.; Bamat, M.K. Phase 1 Clinical Experience Using Intravenous Administration of PV701, an Oncolytic Newcastle Disease Virus. *Curr. Cancer Drug Targets* **2007**, *7*, 157–167. [CrossRef]

82. Vigil, A.; Park, M.-S.; Martinez, O.; Chua, M.A.; Xiao, S.; Cros, J.; Martínez-Sobrido, L.; Woo, S.L.; García-Sastre, A. Use of Reverse Genetics to Enhance the Oncolytic Properties of Newcastle Disease Virus. *Cancer Res.* **2007**, *67*, 8285–8292. [CrossRef] [PubMed]

83. Burke, S.; Shergold, A.; Elder, M.J.; Whitworth, J.; Cheng, X.; Jin, H.; Wilkinson, R.W.; Harper, J.; Carroll, D.K. Oncolytic Newcastle disease virus activation of the innate immune response and priming of antitumor adaptive responses in vitro. *Cancer Immunol. Immunother.* **2020**, *69*, 1015–1027. [CrossRef] [PubMed]

84. Xu, X.; Yi, C.; Yang, X.; Xu, J.; Sun, Q.; Liu, Y.; Zhao, L. Tumor Cells Modified with Newcastle Disease Virus Expressing IL-24 as a Cancer Vaccine. *Mol. Ther. Oncolytics* **2019**, *14*, 213–221. [CrossRef] [PubMed]

85. Fan, X.; Quezada, S.A.; Sepulveda, M.A.; Sharma, P.; Allison, J.P. Engagement of the ICOS pathway markedly enhances efficacy of CTLA-4 blockade in cancer immunotherapy. *J. Exp. Med.* **2014**, *211*, 715–725. [CrossRef]

86. Vijayakumar, G.; McCroskery, S.; Palese, P. Engineering Newcastle Disease Virus as an Oncolytic Vector for Intratumoral Delivery of Immune Checkpoint Inhibitors and Immunocytokines. *J. Virol.* **2019**, *94*. [CrossRef]

87. Vigil, A.; Martinez, O.; Chua, M.A.; García-Sastre, A. Recombinant Newcastle Disease Virus as a Vaccine Vector for Cancer Therapy. *Mol. Ther.* **2008**, *16*, 1883–1890. [CrossRef]

88. Cassel, W.A.; Murray, D.R.; Torbin, A.H.; Olkowski, Z.L.; Moore, M.E. Viral oncolysate in the management of malignant melanoma. I. Preparation of the oncolysate and measurement of immunologic responses. *Cancer* **1977**, *40*, 672–679. [CrossRef]

89. Murray, D.R.; Cassel, W.A.; Torbin, A.H.; Olkowski, Z.L.; Moore, M.E. Viral oncolysate in the management of malignant melanoma. II. Clinical studies. *Cancer* **1977**, *40*, 680–686. [CrossRef]

90. Cassel, W.A.; Murray, D.R. Treatment of stage II malignant melanoma patients with a Newcastle disease virus oncolysate. *Nat. Immun. Cell Growth Regul.* **1988**, *7*, 351–352.

91. Lehner, B.; Schlag, P.; Liebrich, W.; Schirrmacher, V. Postoperative active specific immunization in curatively resected colorectal cancer patients with a virus-modified autologous tumor cell vaccine. *Cancer Immunol. Immunother.* **1990**, *32*, 173–178. [CrossRef]

92. Liebrich, W.; Schlag, P.; Manasterski, M.; Lehner, B.; Stohr, M.; Möller, P.; Schirrmacher, V. In vitro and clinical characterisation of a newcastle disease virus-modified autologous tumour cell vaccine for treatment of colorectal cancer patients. *Eur. J. Cancer Clin. Oncol.* **1991**, *27*, 703–710. [CrossRef]

93. Cassel, W.A.; Murray, D.R. A ten-year follow-up on stage II malignant melanoma patients treated postsurgically with Newcastle disease virus oncolysate. *Med. Oncol. Tumor Pharmacother.* **1992**, *9*, 169–171. [PubMed]

94. Schlag, P.; Manasterski, M.; Gerneth, T.; Hohenberger, P.; Dueck, M.; Herfarth, C.; Liebrich, W.; Schirrmacher, V. Active specific immunotherapy with Newcastle-disease-virus-modified autologous tumor cells following resection of liver metastases in colorectal cancer. First evaluation of clinical response of a phase II-trial. *Cancer Immunol. Immunother.* **1992**, *35*, 325–330. [CrossRef] [PubMed]

95. Schirrmacher, V.; Schlag, P.; Liebrich, W.; Patel, B.T.; Stoeck, M. Specific Immunotherapy of Colorectal Carcinoma with Newcastle-Disease Virus-Modified Autologous Tumor Cells Prepared from Resected Liver Metastasis. *Ann. N. Y. Acad. Sci.* **1993**, *690*, 364–366. [CrossRef]

96. Kirchner, H.; Anton, P.; Atzpodien, J. Adjuvant treatment of locally advanced renal cancer with autologous virus-modified tumor vaccines. *World J. Urol.* **1995**, *13*, 171–173. [CrossRef]

97. Pomer, S.; Schirrmacher, V.; Thiele, R.; Lohrke, H.; Brkovic, D.; Staehler, G. Tumor Response and 4 year survival-data of patients with advanced renal-cell carcinoma treated with autologous tumor vaccine AND subcutaneous R-IL-2 and Ifn-Alpha(2B). *Int. J. Oncol.* **1995**, *6*, 947–954. [CrossRef]

98. Ockert, D.; Schirrmacher, V.; Beck, N.; Stoelben, E.; Ahlert, T.; Flechtenmacher, J.; Hagmüller, E.; Buchcik, R.; Nagel, M.; Saeger, H.D. Newcastle disease virus-infected intact autologous tumor cell vaccine for adjuvant active specific immunotherapy of resected colorectal carcinoma. *Clin. Cancer Res.* **1996**, *2*, 21–28.

99. Ahlert, T.; Sauerbrei, W.; Bastert, G.; Ruhland, S.; Bartik, B.; Simiantonaki, N.; Schumacher, J.; Häcker, B.; Schirrmacher, V.; Schumacher, M. Tumor-cell number and viability as quality and efficacy parameters of autologous virus-modified cancer vaccines in patients with breast or ovarian cancer. *J. Clin. Oncol.* **1997**, *15*, 1354–1366. [CrossRef]

100. Zorn, U.; Duensing, S.; Langkopf, F.; Anastassiou, G.; Kirchner, H.; Hadam, M.; Knüver-Hopf, J.; Atzpodien, J. Active Specific Immunotherapy of Renal Cell Carcinoma: Cellular and Humoral Immune Responses. *Cancer Biother. Radiopharm.* **1997**, *12*, 157–165. [CrossRef]

101. Batliwalla, F.M.; Bateman, B.A.; Serrano, D.; Murray, D.; MacPhail, S.; Maino, V.C.; Ansel, J.C.; Gregersen, P.K.; Armstrong, C.A. A 15-year follow-up of AJCC stage III malignant melanoma patients treated postsurgically with Newcastle disease virus (NDV) oncolysate and determination of alterations in the CD8 T cell repertoire. *Mol. Med.* **1998**, *4*, 783–794. [CrossRef]

102. Csatary, L.K.; Moss, R.W.; Beuth, J.; Töröcsik, B.; Szeberenyi, J.; Bakacs, T. Beneficial treatment of patients with advanced cancer using a Newcastle disease virus vaccine (MTH-68/H). *Anticancer. Res.* **1999**, *19*, 635–638. [PubMed]

103. Liang, W.; Wang, H.; Sun, T.-M.; Yao, W.-Q.; Chen, L.; Jin, Y.; Li, C.-L.; Meng, F.-J. Application of autologous tumor cell vaccine and NDV vaccine in treatment of tumors of digestive tract. *World J. Gastroenterol.* **2003**, *9*, 495–498. [CrossRef] [PubMed]

104. Voit, C.A.; Kron, M.; Schwürzer-Voit, M.; Sterry, W. Intradermal injection of Newcastle disease virus-modified autologous melanoma cell lysate and interleukin-2 for adjuvant treatment of melanoma patients with resectable stage III disease. *J. Dtsch. Dermatol. Ges.* **2003**, *1*, 120–125. [CrossRef] [PubMed]

105. Steiner, H.-H.; Bonsanto, M.M.; Beckhove, P.; Brysch, M.; Geletneky, K.; Ahmadi, R.; Schuele-Freyer, R.; Kremer, P.; Ranaie, G.; Matejic, D.; et al. Antitumor Vaccination of Patients With Glioblastoma Multiforme: A Pilot Study to Assess Feasibility, Safety, and Clinical Benefit. *J. Clin. Oncol.* **2004**, *22*, 4272–4281. [CrossRef]

106. Schulze, T.; Kemmner, W.; Weitz, J.; Wernecke, K.-D.; Schirrmacher, V.; Schlag, P.M. Efficiency of adjuvant active specific immunization with Newcastle disease virus modified tumor cells in colorectal cancer patients following resection of liver metastases: Results of a prospective randomized trial. *Cancer Immunol. Immunother.* **2009**, *58*, 61–69. [CrossRef]

107. Cassel, W.A.; Murray, D.R.; Phillips, H.S. A phase II study on the postsurgical management of stage II malignant melanoma with a Newcastle disease virus oncolysate. *Cancer* **1983**, *52*, 856–860. [CrossRef]

108. Schirrmacher, V.; Fournier, P. Newcastle Disease Virus: A Promising Vector for Viral Therapy, Immune Therapy, and Gene Therapy of Cancer. *Methods Mol. Biol.* **2009**, *542*, 565–605. [CrossRef]

109. Bohle, W.; Schlag, P.; Liebrich, W.; Hohenberger, P.; Manasterski, M.; Möller, P.; Schirrmacher, V. Postoperative active specific immunization in colorectal cancer patients with virus-modified autologous tumor-cell vaccine. First clinical results with tumor-cell vaccines modified with live but avirulent newcastle disease virus. *Cancer* **1990**, *66*, 1517–1523. [CrossRef]

110. Csatary, L.K. Use of Newcastle Disease Virus Vaccine (MTH-68/H) in a Patient With High-grade Glioblastoma. *JAMA* **1999**, *281*, 1588–1589. [CrossRef]

111. Csatary, L.; Gosztonyi, G.; Szeberenyi, J.; Fabian, Z.; Liszka, V.; Bodey, B.; Csatary, C. MTH-68/H Oncolytic Viral Treatment in Human High-Grade Gliomas. *J. Neuro-Oncol.* **2004**, *67*, 83–93. [CrossRef]

112. Pecora, A.L.; Rizvi, N.; Cohen, G.I.; Meropol, N.J.; Sterman, D.; Marshall, J.L.; Goldberg, S.; Gross, P.; O'Neil, J.D.; Groene, W.S.; et al. Phase I Trial of Intravenous Administration of PV701, an Oncolytic Virus, in Patients with Advanced Solid Cancers. *J. Clin. Oncol.* **2002**, *20*, 2251–2266. [CrossRef] [PubMed]

113. Laurie, S.A.; Bell, J.C.; Atkins, H.L.; Roach, J.; Bamat, M.K.; O'Neil, J.D.; Roberts, M.S.; Groene, W.S.; Lorence, R.M. A Phase 1 Clinical Study of Intravenous Administration of PV701, an Oncolytic Virus, Using Two-Step Desensitization. *Clin. Cancer Res.* **2006**, *12*, 2555–2562. [CrossRef] [PubMed]

Publisher's Note: MDPI stays neutral with regard to jurisdictional claims in published maps and institutional affiliations.

Review

Measles Virus as an Oncolytic Immunotherapy

Christine E. Engeland [1,2,*] and Guy Ungerechts [1,*]

1   Clinical Cooperation Unit Virotherapy, German Cancer Research Center (DKFZ),
    National Center for Tumor Diseases (NCT) and Department of Medical Oncology,
    University Hospital Heidelberg, 69120 Heidelberg, Germany
2   Center for Biomedical Education and Research (ZBAF), Institute of Virology and Microbiology,
    Faculty of Health, School of Medicine, Witten/Herdecke University, 58453 Witten, Germany
*   Correspondence: christine.engeland@nct-heidelberg.de (C.E.E.); guy.ungerechts@nct-heidelberg.de (G.U.)

Simple Summary: Measles virus is currently under investigation as an innovative cancer treatment. The virus selectively replicates in and kills cancer cells. Furthermore, it can be genetically engineered to increase tumor specificity and therapeutic efficacy. Importantly, treatment with measles virus activates antitumor immune responses. A number of clinical trials using measles virus for cancer treatment have been completed or are ongoing. Future studies will further harness the possibilities of virus engineering and potential of combination immunotherapies to improve clinical outcome.

Abstract: Measles virus (MeV) preferentially replicates in malignant cells, leading to tumor lysis and priming of antitumor immunity. Live attenuated MeV vaccine strains are therefore under investigation as cancer therapeutics. The versatile MeV reverse genetics systems allows for engineering of advanced targeted, armed, and shielded oncolytic viral vectors. Therapeutic efficacy can further be enhanced by combination treatments. An emerging focus in this regard is combination immunotherapy, especially with immune checkpoint blockade. Despite challenges arising from antiviral immunity, availability of preclinical models, and GMP production, early clinical trials have demonstrated safety of oncolytic MeV and yielded promising efficacy data. Future clinical trials with engineered viruses, rational combination regimens, and comprehensive translational research programs will realize the potential of oncolytic immunotherapy.

Keywords: oncolytic virus; measles virus; cancer immunotherapy; vector engineering; vaccination; immune checkpoint blockade

Citation: Engeland, C.E.; Ungerechts, G. Measles Virus as an Oncolytic Immunotherapy. *Cancers* **2021**, *13*, 544. https://doi.org/10.3390/cancers13030544

Academic Editor: David Wong
Received: 31 December 2020
Accepted: 26 January 2021
Published: 1 February 2021

Publisher's Note: MDPI stays neutral with regard to jurisdictional claims in published maps and institutional affiliations.

1. Introduction—Measles Virus for Cancer Therapy

Measles virus (MeV) is a negative-strand RNA virus belonging to the family Paramyxoviridae, genus *Morbillivirus*. Its genome has a length of approximately 16 kb and encodes six structural and two non-structural proteins (Figure 1a,b). The viral glycoproteins hemagglutinin and fusion mediate receptor binding and fusion at the plasma membrane, respectively. While wild type MeV uses CD150/SLAM on lymphoid cells and epithelial nectin-4 as receptors, vaccine strains of MeV infect cells primarily via CD46 [1]. This is due to mutations in the receptor attachment protein hemagglutinin H in vaccine strain MeV, resulting in high affinity of H for CD46 [2–6]. MeV infection results in syncytia formation as typical cytopathic effect (Figure 1c).

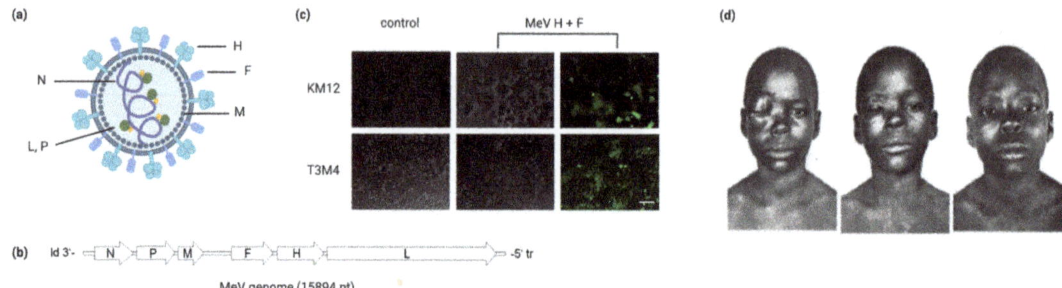

Figure 1. Measles as an oncolytic virus. (**a**) Schematic of the measles virus particle. The viral RNA genome is encapsulated by the nucleocapsid (N) protein and is associated with the viral polymerase (L, large protein) and its cofactor phosphoprotein (P), forming the ribonucleoprotein complex (RNP). The matrix (M) protein connects the RNP and the viral envelope. The surface glycoproteins hemagglutinin (H) and fusion (F) mediate receptor binding and cell fusion, respectively. (**b**) Schematic of the measles virus genome with open reading frames encoding the six structural proteins flanked by the 3′ leader (ld) and 5′ trailer (tr). (**c**) Syncytia formation as the typical cytopathic effect associated with measles virus infection. Human colorectal cancer (KM12, top) and pancreatic adenocarcinoma (T3M4, bottom) cells were transfected with plasmids encoding the MeV glycoproteins H and F as well as enhanced green fluorescent protein as reporter. Control cells were subjected to mock transfection. Phase contrast and fluorescence images were acquired with an Axiovert 200 microscope (Zeiss) at 36 h (KM12) and 12 h post-transfection (T3M4). Scale bar: 200 μm. (**d**) Lymphoma remission after measles infection. Left panel: The patient presented with orbital Burkitt's lymphoma. Middle panel: The patient was infected with measles; the typical skin rash is visible. Right panel: Without specific anti-lymphoma treatment, the orbital mass resolved. Reproduced from Lancet 10 July 1971; 2 (7715): 105–106, with permission.

Originally, the idea to treat cancer patients with MeV arose after case reports which linked measles infection to tumor remission [7]. One highly cited example relates to a boy suffering from Burkitt's lymphoma [8] (Figure 1d). These experiments of nature inspired the idea of using MeV in cancer treatment. However, measles is a severe infectious disease [9]. Thus, employing a pathogenic strain of MeV in cancer therapy is out of question. Live attenuated MeV strains for vaccination were licensed in the 1960s and have a proven safety record [10]. Several years later, testing of Edmonston B measles vaccine strain derivatives for cancer treatment began. In many early studies, hematological malignancies were chosen as target entities [11–14]. This was supported by the natural lymphotropism of MeV. However, other malignancies including ovarian cancer [15] and glioblastoma [16] were soon found to also be sensitive to MeV oncolysis, while normal cells are spared [15,17].

Meanwhile, preclinical efficacy of oncolytic MeV has been demonstrated against a broad range of cancer entities (reviewed in [18]). In addition to Edmonston B derivatives, also the vaccine strains Moraten-Schwarz [19], Edmonston-Zagreb and AIK-C [20], rMV-Hu191 [21], as well as Leningrad-16 [22] have been shown to exert oncolytic effects in preclinical studies.

Thus, MeV is one of several oncolytic platforms currently developed for cancer therapy. Advantages of MeV include the excellent safety profile of the oncolytic vaccine strains and lack of genotoxicity, its immunogenicity, and especially the plethora of engineering possibilities offered by the MeV reverse genetics system. Specific challenges related to MeV include pre-existing antiviral immunity, the choice of preclinical models and manufacturing. These assets and drawbacks are discussed in more detail within this review article.

2. Measles Virus Oncotropism

Measles vaccine strain oncotropism correlates with CD46 overexpression on malignantly transformed cells [23]. Although viral entry occurs in benign cells and at low CD46 receptor density, a certain threshold of expression is required for syncytia formation and cell death [24]. In myeloma, CD46 upregulation has been associated with abnormal p53 [25]. The epithelial receptor for MeV, nectin-4 [26,27], is also a tumor marker which may render carcinomas of pancreatic [28], colorectal [29], and mammary [30] origin susceptible to MeV oncolysis. Post-transcriptional regulation of nectin-4 levels by miR-31 and miR-128 has been demonstrated in breast cancer and glioblastoma [31]. In certain EBV-associated B cell lymphomas, viral latency may promote upregulation of the MeV receptor CD150/SLAM [32].

On the post-entry level, the cellular interferon (IFN) response has been identified as a key determinant of sensitivity to oncolytic MeV across several tumor entities, including the NCI60 panel of cancer cell lines [33]. In adult T cell leukemia/lymphoma, resistance to MeV oncolysis was associated with IFN-β production, while sensitive cells did not produce IFN [34]. In mesothelioma and melanoma, effects of treatment with oncolytic MeV were found not to correlate with CD46 expression, but rather with defects in the IFN response [35,36]. Consistently, expression of retinoic acid inducible gene I (RIG-I) and IFN-induced protein with tetratricopeptide repeats 1 (IFIT1) [37] and IFN-induced transmembrane protein 1 (IFITM1) [38] have been suggested as correlates of relative resistance to MeV oncolysis. Kurokawa et al. have devised a gene expression signature designating constitutive IFN pathway activation to predict outcome of oncolytic MeV treatment [39]. Further, RSAD2/viperin, encoded by an IFN-stimulated gene (ISG), has been shown to inhibit release of MeV progeny in ovarian cancer models [40].

Aside from the cellular antiviral response, several additional cellular factors have been associated with sensitivity to MeV oncolysis. For instance, apoptosis regulators appear to play a role. Caspase 3 has been implicated in MeV-induced cancer cell death [41,42] and overexpression of Bcl-2 reduces MeV-induced cell death in B cell lymphomas [43]. More broadly, basic cellular processes such as protein translation are necessary for efficient MeV replication and thus tumor cell killing. Stimulating cellular translation by insulin-like growth factor-I (IGF-I) or forced expression of eIF4E increases efficacy of oncolytic MeV, while inhibitors of cap-dependent translation reduce MeV oncolysis [44]. Furthermore, it has been reported that integrity of lipid rafts is a prerequisite for oncolysis with the MV-Hu191 strain [21]. Determinants of MeV oncotropism are summarized in Figure 2.

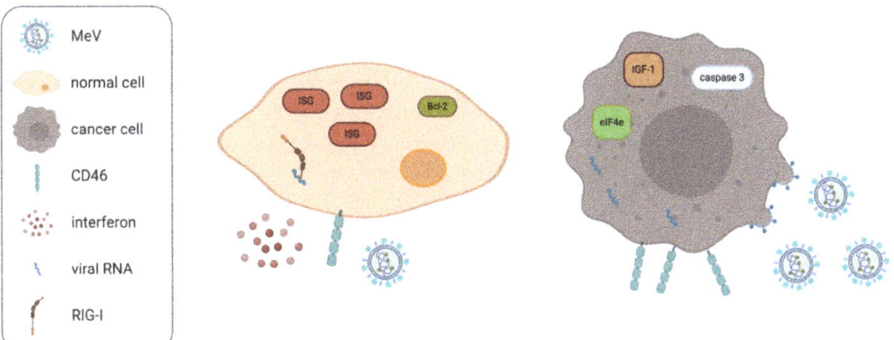

Figure 2. Measles virus oncotropism. Oncolytic measles virus (MeV) does not replicate productively in normal cells (**left**) in contrast to cancer cells (**right**). This oncotropism has been associated with differential expression of i.a. the depicted host cells factors. ISG: interferon-stimulated gene; RIG-I: retinoic acid-inducible gene I; IGF-1: insulin-like growth factor-I.

Overall, oncolytic MeV acts via mechanisms distinct from other established cancer treatments. Accordingly, gemcitabine-resistant pancreatic adenocarcinoma cells are susceptible to MeV oncolysis [45] and chemotherapy-induced senescence does not abrogate oncolysis [46].

3. Combination Therapies

Nevertheless, monotherapy with oncolytic MeV will often be insufficient to cure advanced stage malignancies. Modern medical oncology builds on effective combination therapies. Therefore, measles virotherapy has been combined with other established cancer therapies such as radiation and chemotherapy (recently reviewed in [47]). Synergistic effects of oncolytic MeV and radiotherapy against glioblastoma were observed in vitro and in a xenograft model [48]. In vitro studies have also demonstrated successful combination of oncolytic MeV with chemotherapies such as paclitaxel [49], camptothecin [50], and gemcitabine [51]. Combination with the anti-epidermal growth factor receptor (EGFR) monoclonal antibody nimotuzumab was reported to result in increased antitumor efficacy in laryngeal cancer models [52].

Several small molecules have also been shown to enhance MeV oncolysis by modulating host cell factors. MeV infection is associated with heat shock protein (Hsp) 70 up-regulation. Combination treatment with a Hsp90 inhibitor, resulting in increased Hsp70 expression [53,54], led to increased apoptosis [55]. Counteracting the IFN response, e.g., with janus-associated kinase (JAK) inhibitors such as ruxolitinib, enhances MeV replication in vitro [56]. Epigenetic modulation by histone deacetylase (HDAC) inhibition was also reported to increase efficacy of oncolytic MeV by preventing induction of ISGs in hepatocellular carcinoma [57], but by a different, so far unresolved mechanism in pancreatic adenocarcinoma [58]. As MeV spread and syncytia formation involves remodeling of the actin cytoskeleton, inhibition of Rho-associated coiled-coil forming kinase (ROCK) was tested during treatment of prostate, breast, and glioblastoma cancer cells with MeV, yielding increased viral replication, spread, and tumor cell killing [59]. Compounds which modulate cellular metabolism have also been tested in combination approaches. Blocking aerobic glycolysis with dichloroacetate was shown to increase cell death upon MeV treatment [60]. Furthermore, inducing autophagy has been suggested as a combination strategy to promote MeV oncolysis [61].

Even combination with other oncolytic viruses is conceivable. Along these lines, the combination of MeV with mumps virus showed increased efficacy in a human prostate cancer xenograft model [62].

4. Engineering Oncolytic MeV

Purposeful modification of oncolytic MeV vectors to enhance virotherapy was enabled by development of a reverse genetics system for rescue of MeV from cloned cDNA [63]. This system allows for insertion of transgenes via additional transcription units equipped with MeV polymerase regulatory sequences [64]. These genes are then expressed in infected cells, i.e., within the tumor. A plethora of genetic engineering approaches has been pursued which are summarized in the following, and in Figure 3 (for recent reviews, see [47,65]).

4.1. Tracking Viral Replication and Spread

Initially, reporter genes were inserted for tracking of MeV replication. Carcinoembryonic antigen (CEA) and β-human chorionic gonadotropin (HCG) were selected, which can be measured in routine clinical laboratory testing [12]. Encoding the sodium iodide symporter, NIS, yielding MV-NIS, allowed for γ-camera imaging of iodine-123 (^{123}I) or 99m-technetium uptake and also radiotherapy with ^{131}I [66]. In later studies, MV-NIS was used for advanced imaging techniques, such as pinhole micro-single photon emission computed tomography/computed tomography (SPECT/CT) [67] and contrast-enhanced CT [68]. Recently, a recombinant MeV variant encoding a fluorescent reporter gene was used for intravital imaging of viral spread at single-cell resolution by two-photon microscopy [69].

Data from preclinical studies with MV-NIS have also been used to develop mathematical models of oncolytic virotherapy and its combinations. This has been devised as a means to rationalize testing of distinct dosing and scheduling regimens [70–73].

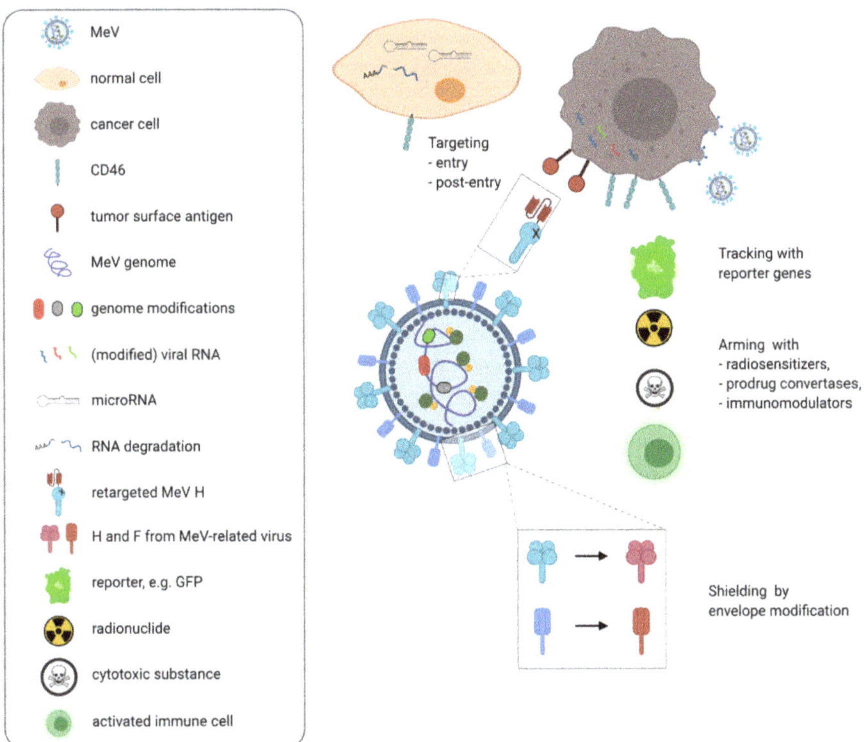

Figure 3. Engineering of oncolytic measles virus. Top: Targeting for increased tumor specificity can be achieved on the entry or post-entry level. For entry targeting, the viral attachment protein H can be mutated to ablate natural tropism and redirected by fusing targeting moieties such as antibody single-chain variable fragments to H (red). Post-entry targeting is achieved via target sites for microRNAs (gray) introduced into viral genes which are differentially expressed in malignant compared to healthy tissues, leading to degradation of the respective viral RNAs in normal cells. Middle: Viruses equipped with reporter genes encoding, e.g., fluorescent proteins (light green) can be used to track viral spread. To increase therapeutic efficacy, viruses can be armed with additional genes encoding radiosensitizers, prodrug convertases, or immunomodulators. Bottom: Shielding against neutralizing antibodies can be achieved by exchanging the viral envelope proteins for the counterparts from a related virus.

Valuable information was gained by employing viruses with reporter genes in clinical trials. After intraperitoneal administration of MV-CEA, dose-dependent increases in CEA levels were measured in peritoneal fluid and serum [74]. After intraperitoneal administration of MV-NIS, ^{123}I SPECT/CT scans were positive in three of 13 ovarian cancer patients, confirming viral gene expression at the tumor site. Scans were positive in eight of 31 multiple myeloma patients receiving MV-NIS i.v. [75]. In both studies, positive scans were associated with higher virus doses.

4.2. Retargeting MeV

Virus engineering has not only enabled tracking viral spread, but also modifying its tropism to increase tumor specificity. Retargeting of MeV was accomplished by mutating the intrinsic receptor binding sites and fusing antibody single-chain variable fragments (scFv) to the C-terminus of the viral hemagglutinin [76]. Using this strategy, oncolytic MeV were targeted to the myeloma surface antigen CD38 [13], to CD20 for targeting of B cell malignancies [77], folate receptor (FR)-α for treatment of ovarian cancer [78], and EGFRvIII expressed in glioblastoma [79], among others (reviewed in [18]). A range of different targeting moieties beyond scFv has been employed, such as the cytokine interleukin (IL)-13 [80] or the urokinase plasminogen activator [81] for direction of viral tropism to their respective receptors. Successful targeting has also been achieved using integrin-binding peptides [82], DARPins [83] and cystine knot proteins [84]. Viral tropism can be redirected to specific cell populations within the tumor, including tumor-initiating cells [85], the tumor stroma [86], and vasculature [81].

A sophisticated means of viral entry targeting employs proteases expressed within the tumor microenvironment. The MeV fusion protein encompasses a furin cleavage site and requires proteolytic processing for activity. Replacing the furin cleavage site with sequences recognized by matrix metalloproteinases or the urokinase-type plasminogen activator can increase tumor specificity [87,88].

Tumor targeting on the post-entry level was achieved using microRNA target sites inserted into the untranslated regions (UTRs) of viral genes [89]. This concept exploits downregulation of specific microRNAs in malignant vs. benign cells, leading to virus restriction in healthy tissue while spread within tumor tissue is unimpaired.

Proof-of-concept was also obtained for using riboswitches to control oncolytic MeV. Insertion of a ligand-activated ribozyme into the UTR of the MeV fusion gene enabled regulation of MeV infectivity and spread by addition of the cognate small molecule [90]. Recently, a photocontrollable MeV variant was reported which harbors a split L protein for control of viral replication by blue light illumination [91].

4.3. Arming with Additional Therapeutic Genes

While these means of targeting aim at enhancing specificity of virotherapy, a number of genetic engineering approaches have been developed to increase antitumor efficacy, often referred to as "arming". First arming strategies aimed at inducing bystander effects in combination radiotherapy and chemotherapy approaches. As mentioned above, MV-NIS allows for concentration of radioactive iodine in infected tumor cells [66].

MeV vectors encoding prodrug convertases were designed for local conversion of prodrugs into active chemotherapeutics. MeV encoding the purine nucleoside phosphorylase, which converts fludarabine into 2-fluoroadenine and 6-methylpurine-2′-deoxyriboside (MeP-dR) to 6-methylpurine, respectively, combined with prodrug administration improved outcome in lymphoma xenograft and immunocompetent murine colorectal cancer models [92,93]. Analogously, MeV was engineered to encode super cytosine deaminase (SCD), a fusion protein of yeast cytosine deaminase and yeast uracil phosphoribosyltransferase, which converts the prodrug 5-fluorocytosine (5-FC) to 5-fluorouracil (5-FU) [94–97].

Other engineering approaches to increase anti-tumor efficacy include insertion of a transgene encoding the proapoptotic protein BNiP3 [49] and the angiogenesis inhibitors endostatin and angiostatin to remodel the tumor microenvironment [98].

5. Immunovirotherapy

While early efforts in engineering oncolytic MeV mainly focused on maximizing direct tumor cell killing, there has been a recent shift from mainly oncolytic to mainly immunotherapeutic treatment strategies, spurred by the developments in cancer immunotherapy which have revolutionized medical oncology.

MeV oncolysis per se has pleiotropic effects on the anti-tumor immune response and supports all phases of the "cancer immunity cycle" (Figure 4; reviewed in [99]). MeV-induced cell death is immunogenic [100], induces a distinctive immunopeptidome [101], and promotes cross-priming of antitumor T cell responses by conventional and plasmacytoid dendritic cells [19,102]. MeV oncolysis has also been reported to increase tumor necrosis factor-related apoptosis-inducing ligand (TRAIL)-mediated cytotoxicity by myeloid and plasmacytoid DCs [103] as well as modulation of macrophages towards an antitumor phenotype [104]. Neutrophil activation also occurs, leading to secretion of IL-8, tumor necrosis factor (TNF)-α, monocyte chemoattractant protein (MCP)-1, and IFN-α, TRAIL expression, and degranulation [105], which may be beneficial or not depending on the tumor model [106].

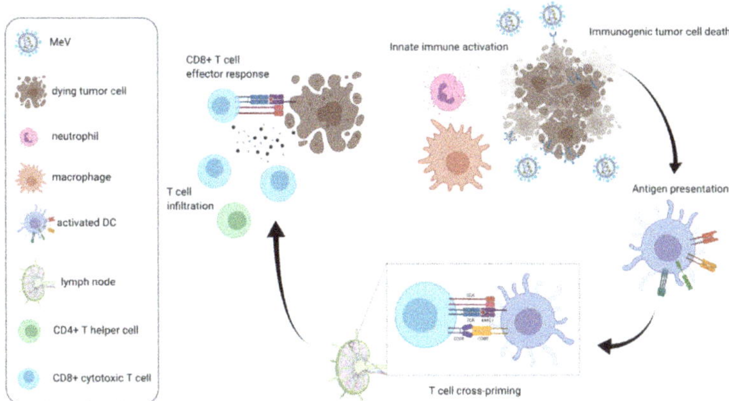

Figure 4. Measles virus as an oncolytic immunotherapy. Measles virus-mediated oncolysis has been shown to support different phases of the antitumor immune response: Oncolysis induces immunogenic cell death, which promotes dendritic cell activation, antigen presentation, and cross-priming of T cells. Measles virotherapy remodels the tumor microenvironment, thereby enhancing innate (macrophage repolarization and neutrophil degranulation) as well as adaptive antitumor immunity (T cell infiltration and CD8+ effector responses).

These immunotherapeutic effects can be enhanced by insertion of immunomodulatory transgenes into the MeV genome (Table 1). Further, MeV can serve as a vector to deliver immunomodulators to the tumor site which can be highly toxic when administered systemically. The first immunomodulatory transgene reported in the context of many oncolytic viruses and also MeV was the granulocyte macrophage colony stimulating factor, GM-CSF [14]. In a lymphoma xenograft model, MV GM-CSF led to increased neutrophil infiltration, which correlated with tumor regression. Further immunomodulators have been shown to increase innate immune activation in the context of MeV oncolysis. A MeV vector encoding IFN-β was reported to induce immune infiltration and remodeling of the tumor microenvironment in mesothelioma xenografts [107]. MeV encoding the immunomodulatory neutrophil-activating protein (NAP) of *H. pylori* prolonged survival and induced a beneficial cytokine response in breast cancer xenograft pleural effusion and lung colonization models [108].

Introduction of the first fully immunocompetent mouse model of MeV oncolysis, MC38cea [93], was the prerequisite to further study immunomodulatory MeV vectors and demonstrate induction of tumor-specific adaptive immune responses. This model consists of murine colorectal adenocarcinoma MC38, syngeneic to C57BL/6 mice and stably expressing the carcinoembryonic antigen (CEA), which are susceptible to CEA-targeted MeV [93]. In this model, treatment with MV GM-CSF led to prolonged survival compared to control MV. Forty percent of treated mice experienced complete tumor remission and were subsequently protected from tumor re-engraftment, indicating a tumor vaccination effect. Further, treatment with MV GM-CSF enhanced intratumoral T cell infiltration as well as tumor-specific T cell responses [109].

Table 1. Immunomodulatory oncolytic MeV. Overview of immunomodulatory transgenes that have been encoded in MeV, their anticipated immunological effects in the context of MeV oncolytic immunotherapy, and the outcome of the respective preclinical studies. GM-CSF: granulocyte–macrophage colony-stimulating factor; IFN: interferon; NAP: neutrophil activating protein; CTLA-4: cytotoxic T lymphocyte antigen-4; PD-L1: programmed cell death 1-ligand 1; Th: T helper cell; T_{eff}: effector T cell; T_{reg}: regulatory T cell; IL: interleukin; AICD: activation-induced cell death; BiTE: bispecific T cell engager; TAA: tumor-associated antigen; IFNAR: IFN-α receptor.

Immunomodulator	Anticipated Immunological Effects	Preclinical Data
GM-CSF	Dendritic cell activation and maturation; activation of monocytes, macrophages, neutrophils, NK cells	SCID model: increased antitumor efficacy, increased neutrophil infiltration [14] Immunocompetent model: increased antitumor efficacy, increased T cell infiltration, stronger tumor-specific T cell responses, rejection of tumor re-engraftment [109]
IFN-β	Enhanced antitumor response via innate and adaptive effector mechanisms	Athymic nude mouse model: increased CD68+ macrophage infiltration, reduced microvessel density; delayed tumor progression, prolonged survival [107]
H. pylori NAP	Inflammatory response, promotion of Th1-polarized immune responses	Athymic nude mouse model: prolonged survival, neutrophil infiltration, secretion of Th1-promoting cytokines [108]
Anti-CTLA-4, anti-PD-L1	Enhanced antitumor T cell response	Immunocompetent mouse model: delayed tumor progression, prolonged survival, increased T_{eff}/T_{reg} ratio, increased tumor-specific IFN-γ response [110]
IL-12	Activation and recruitment of T cells and NK cells	Immunocompetent mouse model: increased survival rates (CD8+-dependent), rejection of tumor re-engraftment, increased tumor-specific IFN-γ response, expression of effector cytokines, increased T cell infiltration, decrease in NK cells, increased proportion of activated CD8+ T cells and NK cells [111]
IL-15 superagonist	Activation of T cells and NK cells without induction of AICD	Immunocompetent mouse model: increased CD8+ T cell and NK cell infiltration and activation, antitumor efficacy inferior to MeV encoding IL-12 [112]
BiTEs	Recruitment of T cells, enhanced T cell antitumor cytotoxicity	Immunocompetent mouse model: increased T cell infiltration, prolonged survival, induction of tumor-specific immunity Patient-derived xenograft models: prolonged survival [113]
TAA	Priming and activation of TAA-specific T cells	IFNAR$^{-/-}$ CD46 transgenic mouse model: Induction of humoral and cellular responses against TAA, reduced tumor nodules and prolonged survival in lung colonization experiment [114] Ex vivo assays: Priming and activation of TAA-specific T cells [115]

Italic: bacterial taxa.

To develop a second immunocompetent model of MeV oncolysis in C57BL/6 mice, B16 melanoma cells were transduced for stable expression of the CD20 surface antigen for treatment with CD20-targeted MeV. In this model, MeV vectors encoding immune checkpoint antibodies against cytotoxic T lymphocyte-associated-4 (CTLA-4) and programmed cell death-ligand 1 (PD-L1) prolonged survival compared to MeV encoding the antibody constant region only [110]. Combination with systemically administered antibodies against CTLA-4, PD-1, and PD-L1 has also demonstrated the therapeutic value of this approach [110,116]. In the MC38cea model, systematic comparison of transgenes targeted at different phases of the cancer immunity cycle—GM-CSF, IFN-γ induced protein 10 (IP-10), membrane-bound CD80, anti-CTLA-4, IL-12, and anti-PD-L1 identified the latter two as the most potent [111]. MeV encoding IL-12 induced complete tumor remissions in 90% of treated mice, which were mediated by CD8+ effector T cell responses. Oncolytic MeV vectors encoding an IL-15 superagonist mediated T and NK cell activation, but were less effective than MeV encoding IL-12 [112]. Bispecific T cell engagers (BiTEs) simultaneously bind CD3 on T cells and a tumor surface antigen, thereby redirecting T cells to tumor cells to mediate antitumor T cell cytotoxicity. Oncolytic measles viruses encoding BiTEs were shown to promote T cell infiltration and activation in syngeneic and patient-derived tumor models [113].

For induction of T cell responses against specific antigens, MeV can also be employed as a heterologous, highly immunogenic vaccine vector (reviewed in [117]). This strategy has been used to develop vaccines against a range of pathogens, including emerging SARS CoV-2 [118]. This strategy has been adopted in oncolytic immunotherapy by encoding tumor-associated antigens in the MeV vector. MeV vectors encoding ovalbumin (OVA) as model antigen or the tumor antigen claudin-6 either in native form or in association with lentivirus-like particles were shown to induce antigen-specific humoral and cellular immune responses in IFN-α receptor (IFNAR)-deficient, CD46-transgenic mice and prolong survival in B16-derived tumor models [114]. Employing OVA and the melanoma antigen tyrosinase-related protein-2 (TRP-2), MeV vectors encoding the full-length antigens or their respective immunodominant CD8+ epitope or epitope variants were generated. The epitope variants are either secreted or targeted to the proteasome. Using these MeV vectors, activation and dendritic cell-mediated priming of cognate T cells was demonstrated ex vivo [115].

As another modality of immunovirotherapy, combination of oncolytic MeV with adoptive transfer of antitumor immune effector cells such as NK cells [119] or CD8+ NKG2D+ cells [120] has been reported.

Importantly, antitumor immune activation by MeV oncolysis has not only been demonstrated in preclinical models. Clinical data also suggest augmentation of antitumor immunity by oncolytic measles virotherapy. In cutaneous T cell lymphoma, a shift towards a Th1-biased T cell population in lymphoma lesions was noted after treatment [121]. In four ovarian cancer patients treated with MV-NIS, IFN-γ and IL-4 responses against the tumor antigens FRα and IGF binding protein 2 (IGFBP2) were detected by ELISPOT [122]. Increases in IFN-γ ELISPOT counts against cancer testis antigens were also observed in the majority of tested multiple myeloma patients treated with MV-NIS [123]. The myeloma patient with an exceptional response to MV-NIS had a high mutational load and high baseline T cell responses against several tumor antigens, which remained stable after virotherapy.

Of note, the clinical trials published thus far tested oncolytic MeV not encoding any additional immunotherapeutic payloads. Perhaps the fraction of patients showing immunological responses and overall therapeutic efficacy can be increased with novel immunomodulatory oncolytic MeV.

6. Antitumor vs. Antiviral Immunity

However, immune stimulation in the context of oncolytic virotherapy may hamper overall efficacy by premature viral clearance [124]. Though conferring a safety advantage, the antiviral immune response and specifically high measles seropositivity in the general population is one of the main reservations against using MeV for oncolytic virotherapy. Therefore, multiple strategies have been devised to circumvent anti-viral immunity. Substitution of the P/V/C and also N and L genes of attenuated oncolytic strains for their wild type counterparts has been shown to dampen the cellular IFN response and increase viral spread [125,126]. These variants resulted in higher progeny titers, increased viral gene expression, and cell killing in presence of interferon or in interferon-competent cells. Mutation of common antibody epitopes in the MeV envelope glycoproteins allows for evasion of virus neutralization in serum [127]. By exchanging the glycoproteins for their counterparts from a related morbillivirus, canine distemper virus, an envelope chimeric MeV was generated which showed similar replication kinetics and oncolytic properties as unmodified MeV, but was not neutralized by human MeV-immune sera [128]. However, these approaches may compromise safety. As alternatives, different "shielding" approaches have been developed to protect oncolytic MeV from antibody-mediated clearance.

One approach is to employ cell carriers to "deliver" oncolytic MeV to the tumor site. Successful tumor delivery by heterofusion of infected carrier cells and tumor cells was first demonstrated for infected monocytes, endothelial cells, and stimulated human peripheral blood cells. This allowed for effective oncolysis after i.v. or i.p. administration after passive immunization in xenograft models [129]. A range of different cell types have been employed as carriers, including T cells [130], cytokine-induced killer cells [131], mesenchymal stem cells [132], mesenchymal stromal cells [133], and also irradiated myeloma cells [134].

As an acellular shield, the scavenger receptor ligand polyinosinic acid can be used to prevent MeV sequestration by hepatic Kupffer cells after i.v. administration [135]. This was shown to enhance oncolytic efficacy in a nude mouse model. Multi-layer coating with ionic polymers and graphene oxide sheets [136] have also been reported as a means to protect MeV from premature clearance [137]. These modifications did not compromise infection of tumor cells and even enhanced oncolytic effects. Administration of UV-inactivated MeV as a decoy virus has been suggested as a means to sequester antiviral antibodies prior to treatment [138].

Instead of modifying the oncolytic agent, immune modulation in the patient has been envisaged to enable measles virotherapy. Clinically approved multidose cyclophosphamide regimens were shown to dampen both primary and secondary antibody responses to MeV [139]. Although pre-existing immunosuppression in advanced stage cancer and especially low antibody levels in myeloma patients were anticipated, cyclophosphamide was also tested in one cohort of the Phase I trial of MV-NIS for advanced multiple myeloma [75]. Clinical data in this regard are still limited, but so far no clear correlation between anti-measles immunity and therapeutic efficacy has been noted.

7. Preclinical Models

The conundrum of balancing antiviral immunity and antitumor immunity exemplifies the challenge to identify appropriate models for preclinical development of oncolytic MeV. Measles is a primate-adapted virus, thus rodents and other small animals commonly used in research are non-susceptible to the virus. CD46-transgenic, IFNAR-knockout (IFNAR$^{-/-}$ CD46Ge) mice which are supposed to mimic MeV replication and spread in humans are commonly used for study of MeV vaccines [117] and have also been used for testing of oncolytic MeV vaccines [114]. However, it remains unclear how the IFNAR$^{-/-}$ phenotype affects outcome of virotherapy. Syngeneic transplantable tumor models in fully immunocompetent mice have been widely adopted [99]. While these models have enabled proof-of-concept studies, they fail to recapitulate the genetic makeup, heterogeneity and evolution of human cancers. To address these issues, human precision cut liver slices [20], clinical samples [140], and patient-derived xenografts [113] have been used in preclinical testing of measles virotherapy. Successful targeting of cancer-initiating cells in patient-derived cultures, including glioma stem cells from neurospheres [141] and colorectal cancer tumor spheroids [85] have been reported. To address specific tumor niches, orthotopic models such as breast cancer pleural effusion [142] and intracranial glioblastoma models [116,141] have been studied, demonstrating efficacy of oncolytic MeV also in advanced preclinical models.

8. Pharmacokinetics and –Dynamics

In preparation of clinical trials, several preclinical toxicology and pharmacokinetic studies have been carried out in mice and non-human primates [143–147]. These studies confirmed safety of intravenous injection of up to 10^8 and 4×10^8 TCID$_{50}$/kg oncolytic

MeV in IFNAR$^{-/-}$ CD46Ge and squirrel monkeys, respectively. Further, intraventricular injection of oncolytic MeV into the cerebrospinal fluid of IFNAR$^{-/-}$ CD46Ge mice [147] and intrahepatic injection of prodrug convertase-armed oncolytic MeV in IFNAR$^{-/-}$ CD46Ge mice and rhesus macaques [146] were tolerated. Depending on the model, different pharmacokinetics and dose–response relationships were observed. Notably, despite detection of viral RNA, no significant shedding of infectious virus was reported.

This holds true in clinical settings. Saliva and urine samples were free of infectious virus after i.p. administration of MV-CEA in ovarian cancer patients [74]. Up to 10^9 TCID$_{50}$ i.p. and 10^{11} TCID$_{50}$ i.v. have been administered with manageable side effects [75,122]. The available clinical data also suggest a dose–response relationship, with higher doses associated with more favorable outcome.

9. Early Clinical Trials with MeV

Clinical trials in oncology typically enroll patients after failure of established therapies. In the first clinical trial with oncolytic MeV, patients with therapy-resistant or relapsed cutaneous T cell lymphomas received intralesional injections of Edmonston-Zagreb measles vaccine. As a safety measure, IFN-α was administered prior to treatment. Treatment was well tolerated and tumor regressions, also of non-injected lesions, were observed. Serial biopsies showed intralesional viral replication and favorable changes in the intralesional T cell populations [121].

Quite a high number of subsequent trials were conducted at Mayo Clinic in patients with very different cancer entities including ovarian cancer [74,122] (NCT02068794; NCT00390299), glioblastoma multiforme (NCT00390299), medulloblastoma (NCT02962167), mesothelioma (NCT01503177), breast cancer (NCT04521764), head and neck squamous cell carcinoma (NCT01846091), malignant peripheral nerve sheath tumors (NCT02700230), bladder cancer (NCT03171493), and multiple myeloma (NCT00450814; NCT02192775) using Edmonston B-derived attenuated MeV. These Phase I/II trials showed that MeV administration through all investigated routes including intraperitoneal, intracranial, intratumoral, intrapleural, and intravenous administration is safe, feasible, and may lead to a favorable outcome compared to expected median survival in the treated patient population [74,122]. In patients with multiple myeloma, treatment with oncolytic MeV led to transient drops in serum free light chains as myeloma marker in several patients. One patient experienced a durable complete remission which is still ongoing to date [75,148].

10. Translational Considerations, Perspectives, and Conclusions

As a consequence of the observed dose–response relationships, highest feasible doses are administered in current trials. However, large-scale manufacturing of the required high-titer, highly purified good manufacturing practice (GMP)-grade recombinant MeV remains challenging [149], despite development of processes including production in serum-free cell culture, tangential flow filtration, and diafiltration [150–152]. Nevertheless, these efforts seem worthwhile, given the versatility of MeV as an oncolytic vector platform [47], the excellent safety record of MeV vaccines [10], as well as the biosafety profile [153] and genetic stability [154] of recombinant MeV.

As outlined above, several rational combination approaches to cancer immunovirotherapy employing MeV and different immunomodulators will be under clinical investigation in the future. Other OVs have already been combined successfully with immune checkpoint inhibitors [155] in clinical trials. Moreover, clinical translation of second-generation MeV engineered to encode relevant immunomodulators as illustrated above will most likely further improve clinical outcomes.

Moving forward in this direction, it will be decisive to validate predictive markers of response and resistance in a clinical setting. These markers should not only incorporate tumor cell characteristics, but also signatures of antitumor immune activation. By defining criteria of successful immunovirotherapy, these results will also assist in prioritizing the most effective therapeutic payloads and combination therapies. Towards this end, even early stage clinical trials must encompass comprehensive correlative research programs to accelerate the advancement of effective immunovirotherapies.

Author Contributions: Conceptualization, C.E.E. and G.U.; writing—original draft preparation, C.E.E.; writing—review and editing, G.U. and C.E.E.; visualization, C.E.E. All authors have read and agreed to the published version of the manuscript.

Funding: This research received no external funding.

Institutional Review Board Statement: Not applicable.

Informed Consent Statement: Not applicable.

Acknowledgments: We acknowledge Gemma Pidelaserra Martí and Johannes Heidbuechel for comments on the draft manuscript. We thank Alessia Floerchinger for providing micrographs for Figure 1c. Graphical abstract, Figures 1a, 2–4 were created with Biorender.com.

Conflicts of Interest: C.E.E. and G.U. are listed as inventors on patent applications related to the use of measles viruses for cancer immunotherapy. G.U. acts as CMO, CSO, and COO of CanVirex.

References

1. Lin, L.T.; Richardson, C.D. The Host Cell Receptors for Measles Virus and Their Interaction with the Viral Hemagglutinin (H) Protein. *Viruses* **2016**, *8*, 250. [CrossRef] [PubMed]
2. Lecouturier, V.; Fayolle, J.; Caballero, M.; Carabaña, J.; Celma, M.L.; Fernandez-Munoz, R.; Wild, T.F.; Buckland, R. Identification of two amino acids in the hemagglutinin glycoprotein of measles virus (MV) that govern hemadsorption, HeLa cell fusion, and CD46 downregulation: Phenotypic markers that differentiate vaccine and wild-type MV strains. *J. Virol.* **1996**, *70*, 4200–4204. [CrossRef] [PubMed]
3. Hsu, E.C.; Sarangi, F.; Iorio, C.; Sidhu, M.S.; Udem, S.A.; Dillehay, D.L.; Xu, W.; Rota, P.A.; Bellini, W.J.; Richardson, C.D. A single amino acid change in the hemagglutinin protein of measles virus determines its ability to bind CD46 and reveals another receptor on marmoset B cells. *J. Virol.* **1998**, *72*, 2905–2916. [CrossRef]
4. Seki, F.; Takeda, M.; Minagawa, H.; Yanagi, Y. Recombinant wild-type measles virus containing a single N481Y substitution in its haemagglutinin cannot use receptor CD46 as efficiently as that having the haemagglutinin of the Edmonston laboratory strain. *J. Gen. Virol.* **2006**, *87*, 1643–1648. [CrossRef] [PubMed]
5. Tahara, M.; Takeda, M.; Seki, F.; Hashiguchi, T.; Yanagi, Y. Multiple amino acid substitutions in hemagglutinin are necessary for wild-type measles virus to acquire the ability to use receptor CD46 efficiently. *J. Virol.* **2007**, *81*, 2564–2572. [CrossRef]
6. Navaratnarajah, C.K.; Vongpunsawad, S.; Oezguen, N.; Stehle, T.; Braun, W.; Hashiguchi, T.; Maenaka, K.; Yanagi, Y.; Cattaneo, R. Dynamic interaction of the measles virus hemagglutinin with its receptor signaling lymphocytic activation molecule (SLAM, CD150). *J. Biol. Chem.* **2008**, *283*, 11763–11771. [CrossRef]
7. Kelly, E.; Russell, S.J. History of Oncolytic Viruses: Genesis to Genetic Engineering. *Mol. Ther.* **2007**, *15*, 651–659. [CrossRef]
8. Bluming, A.Z.; Ziegler, J.L. Regression of burkitt's lymphoma in association with measles infection. *Lancet* **1971**, *2*, 105–106. [CrossRef]
9. Moss, W.J. Measles. *Lancet* **2017**, *390*, 2490–2502. [CrossRef]
10. Di Pietrantonj, C.; Rivetti, A.; Marchione, P.; Debalini, M.G.; Demicheli, V. Vaccines for measles, mumps, rubella, and varicella in children. *Cochrane Database Syst. Rev.* **2020**, *4*, CD004407. [CrossRef]
11. Grote, D.; Russell, S.J.; Cornu, T.I.; Cattaneo, R.; Vile, R.; Poland, G.A.; Fielding, A.K. Live attenuated measles virus induces regression of human lymphoma xenografts in immunodeficient mice. *Blood* **2001**, *97*, 3746–3754. [CrossRef]
12. Peng, K.W.; Facteau, S.; Wegman, T.; O'Kane, D.; Russell, S.J. Non-invasive in vivo monitoring of trackable viruses expressing soluble marker peptides. *Nat. Med.* **2002**, *8*, 527–531. [CrossRef] [PubMed]
13. Peng, K.W.; Donovan, K.A.; Schneider, U.; Cattaneo, R.; Lust, J.A.; Russell, S.J. Oncolytic measles viruses displaying a single-chain antibody against CD38, a myeloma cell marker. *Blood* **2003**, *101*, 2557–2562. [CrossRef] [PubMed]
14. Grote, D.; Cattaneo, R.; Fielding, A.K. Neutrophils contribute to the measles virus-induced antitumor effect: Enhancement by granulocyte macrophage colony-stimulating factor expression. *Cancer Res.* **2003**, *63*, 6463–6468. [PubMed]
15. Peng, K.W.; TenEyck, C.J.; Galanis, E.; Kalli, K.R.; Hartmann, L.C.; Russell, S.J. Intraperitoneal therapy of ovarian cancer using an engineered measles virus. *Cancer Res.* **2002**, *62*, 4656–4662.
16. Phuong, L.K.; Allen, C.; Peng, K.-W.; Giannini, C.; Greiner, S.; TenEyck, C.J.; Mishra, P.; Macura, S.; Russell, S.J.; Galanis, E. Use of a vaccine strain of measles virus genetically engineered to produce carcinoembryonic antigen as a novel therapeutic agent against glioblastoma multiforme. *Cancer Res.* **2003**, *63*, 2462–2469.
17. Peng, K.W.; Ahmann, G.J.; Pham, L.; Greipp, P.R.; Cattaneo, R.; Russell, S.J. Systemic therapy of myeloma xenografts by an attenuated measles virus. *Blood* **2001**, *98*, 2002–2007. [CrossRef]
18. Mühlebach, M.D. Measles virus in cancer therapy. *Curr. Opin. Virol.* **2020**, *41*, 85–97. [CrossRef]

19. Gauvrit, A.; Brandler, S.; Sapede-Peroz, C.; Boisgerault, N.; Tangy, F.; Gregoire, M. Measles Virus Induces Oncolysis of Mesothelioma Cells and Allows Dendritic Cells to Cross-Prime Tumor-Specific CD8 Response. *Cancer Res.* **2008**, *68*, 4882–4892. [CrossRef]

20. Zimmermann, M.; Armeanu, S.; Smirnow, I.; Kupka, S.; Wagner, S.; Wehrmann, M.; Rots, M.G.; Groothuis, G.M.; Weiss, T.S.; Königsrainer, A.; et al. Human precision-cut liver tumor slices as a tumor patient-individual predictive test system for oncolytic measles vaccine viruses. *Int. J. Oncol.* **2009**, *34*, 1247–1256.

21. Lv, Y.; Zhou, D.; Hao, X.Q.; Zhu, M.Y.; Zhang, C.D.; Zhou, D.M.; Wang, J.H.; Liu, R.X.; Wang, Y.L.; Gu, W.Z.; et al. A recombinant measles virus vaccine strain rmv-hu191 has oncolytic effect against human gastric cancer by inducing apoptotic cell death requiring integrity of lipid raft microdomains. *Cancer Lett.* **2019**, *460*, 108–118. [CrossRef] [PubMed]

22. Ammour, Y.; Ryabaya, O.; Shchetinina, Y.; Prokofeva, E.; Gavrilova, M.; Khochenkov, D.A.; Vorobyev, D.; Faizuloev, E.; Shohin, I.E.; Zverev, V.V.; et al. The Susceptibility of Human Melanoma Cells to Infection with the Leningrad-16 Vaccine Strain of Measles Virus. *Viruses* **2020**, *12*, 173. [CrossRef] [PubMed]

23. Ong, H.T.; Timm, M.M.; Greipp, P.R.; Witzig, T.E.; Dispenzieri, A.; Russell, S.J.; Peng, K.-W. Oncolytic measles virus targets high CD46 expression on multiple myeloma cells. *Exp. Hematol.* **2006**, *34*, 713–720. [CrossRef] [PubMed]

24. Anderson, B.D.; Nakamura, T.; Russell, S.J.; Peng, K.-W. High CD46 Receptor Density Determines Preferential Killing of Tumor Cells by Oncolytic Measles Virus. *Cancer Res.* **2004**, *64*, 4919–4926. [CrossRef]

25. Lok, A.; Descamps, G.; Tessoulin, B.; Chiron, D.; Eveillard, M.; Godon, C.; Le Bris, Y.; Vabret, A.; Bellanger, C.; Maillet, L.; et al. p53 regulates CD46 expression and measles virus infection in myeloma cells. *Blood Adv.* **2018**, *2*, 3492–3505. [CrossRef]

26. Noyce, R.S.; Richardson, C.D. Nectin 4 is the epithelial cell receptor for measles virus. *Trends Microbiol.* **2012**, *20*, 429–439. [CrossRef]

27. Mühlebach, M.D.; Mateo, M.; Sinn, P.L.; Prüfer, S.; Uhlig, K.M.; Leonard, V.H.J.; Navaratnarajah, C.K.; Frenzke, M.; Wong, X.X.; Sawatsky, B.; et al. Adherens junction protein nectin-4 is the epithelial receptor for measles virus. *Nat. Cell Biol.* **2011**, *480*, 530–533. [CrossRef]

28. Awano, M.; Fujiyuki, T.; Shoji, K.; Amagai, Y.; Murakami, Y.; Furukawa, Y.; Sato, H.; Yoneda, M.; Kai, C. Measles virus selectively blind to signaling lymphocyte activity molecule has oncolytic efficacy against nectin-4-expressing pancreatic cancer cells. *Cancer Sci.* **2016**, *107*, 1647–1652. [CrossRef]

29. Amagai, Y.; Fujiyuki, T.; Yoneda, M.; Shoji, K.; Furukawa, Y.; Sato, H.; Kai, C. Oncolytic Activity of a Recombinant Measles Virus, Blind to Signaling Lymphocyte Activation Molecule, Against Colorectal Cancer Cells. *Sci. Rep.* **2016**, *6*, 24572. [CrossRef]

30. Delpeut, S.; Sisson, G.; Black, K.M.; Richardson, C.D. Measles Virus Enters Breast and Colon Cancer Cell Lines through a PVRL4-Mediated Macropinocytosis Pathway. *J. Virol.* **2017**, *91*, e02191-16. [CrossRef]

31. Geekiyanage, H.; Galanis, E. MiR-31 and miR-128 regulates poliovirus receptor-related 4 mediated measles virus infectivity in tumors. *Mol. Oncol.* **2016**, *10*, 1387–1403. [CrossRef] [PubMed]

32. Takeda, S.; Kanbayashi, D.; Kurata, T.; Yoshiyama, H.; Komano, J. Enhanced susceptibility ofBlymphoma cells to measles virus byEpstein–Barr virus typeIIIlatency that upregulatesCD150/signaling lymphocytic activation molecule. *Cancer Sci.* **2014**, *105*, 211–218. [CrossRef] [PubMed]

33. Noll, M.; Berchtold, S.; Lampe, J.; Malek, N.P.; Bitzer, M.; Lauer, U.M. Primary resistance phenomena to oncolytic measles vaccine viruses. *Int. J. Oncol.* **2013**, *43*, 103–112. [CrossRef] [PubMed]

34. Parrula, C.; Fernandez, S.A.; Zimmerman, B.; Lairmore, M.; Niewiesk, S. Measles virotherapy in a mouse model of adult T-cell leukaemia/lymphoma. *J. Gen. Virol.* **2011**, *92*, 1458–1466. [CrossRef] [PubMed]

35. Achard, C.; Boisgerault, N.; Delaunay, T.; Roulois, D.; Nedellec, S.; Royer, P.-J.; Pain, M.; Combredet, C.; Mesel-Lemoine, M.; Cellerin, L.; et al. Sensitivity of human pleural mesothelioma to oncolytic measles virus depends on defects of the type I interferon response. *Oncotarget* **2015**, *6*, 44892–44904. [CrossRef]

36. Allagui, F.; Achard, C.; Panterne, C.; Combredet, C.; Labarrière, N.; Dréno, B.; Elgaaied, A.B.; Pouliquen, D.; Tangy, F.; Fonteneau, J.F.; et al. Modulation of the type i interferon response defines the sensitivity of human melanoma cells to oncolytic measles virus. *Curr. Gene Ther.* **2017**, *16*, 419–428. [CrossRef]

37. Berchtold, S.; Lampe, J.; Weiland, T.; Smirnow, I.; Schleicher, S.; Handgretinger, R.; Kopp, H.-G.; Reiser, J.; Stubenrauch, F.; Mayer, N.; et al. Innate Immune Defense Defines Susceptibility of Sarcoma Cells to Measles Vaccine Virus-Based Oncolysis. *J. Virol.* **2013**, *87*, 3484–3501. [CrossRef]

38. Aref, S.; Castleton, A.Z.; Bailey, K.; Burt, R.; Dey, A.; Leongamornlert, D.; Mitchell, R.J.; Okasha, D.; Fielding, A.K. Type 1 Interferon Responses Underlie Tumor-Selective Replication of Oncolytic Measles Virus. *Mol. Ther.* **2020**, *28*, 1043–1055. [CrossRef]

39. Kurokawa, C.; Iankov, I.D.; Anderson, S.K.; Aderca, I.; Leontovich, A.A.; Maurer, M.J.; Oberg, A.L.; Schroeder, M.A.; Giannini, C.; Greiner, S.M.; et al. Constitutive Interferon Pathway Activation in Tumors as an Efficacy Determinant Following Oncolytic Virotherapy. *J. Natl. Cancer Inst.* **2018**, *110*, 1123–1132. [CrossRef]

40. Kurokawa, C.; Iankov, I.D.; Galanis, E. A key anti-viral protein, RSAD2/VIPERIN, restricts the release of measles virus from infected cells. *Virus Res.* **2019**, *263*, 145–150. [CrossRef]

41. Boisgerault, N.; Guillerme, J.-B.; Pouliquen, D.; Mesel-Lemoine, M.; Achard, C.; Combredet, C.; Fonteneau, J.-F.; Tangy, F.; Gregoire, M. Natural Oncolytic Activity of Live-Attenuated Measles Virus against Human Lung and Colorectal Adenocarcinomas. *BioMed. Res. Int.* **2013**, *2013*, 1–11. [CrossRef] [PubMed]

42. Wang, B.; Yan, X.; Guo, Q.; Li, Y.; Zhang, H.; Xie, J.S.; Meng, X. Deficiency of caspase 3 in tumor xenograft impairs therapeutic effect of measles virus Edmoston strain. *Oncotarget* **2015**, *6*, 16019–16030. [CrossRef] [PubMed]

43. Patel, B.; Dey, A.; Ghorani, E.; Kumar, S.; Malam, Y.; Rai, L.; Steele, A.J.; Thomson, J.; Wickremasinghe, R.G.; Zhang, Y.; et al. Differential Cytopathology and Kinetics of Measles Oncolysis in Two Primary B-cell Malignancies Provides Mechanistic Insights. *Mol. Ther.* **2011**, *19*, 1034–1040. [CrossRef] [PubMed]

44. Jacobson, B.A.; Sadiq, A.A.; Tang, S.; Jay-Dixon, J.; Patel, M.R.; Drees, J.; Sorenson, B.S.; Russell, S.J.; Kratzke, R.A. Cap-dependent translational control of oncolytic measles virus infection in malignant mesothelioma. *Oncotarget* **2017**, *8*, 63096–63109. [CrossRef]

45. Bossow, S.; Grossardt, C.; Temme, A.; Leber, M.F.; Sawall, S.; Rieber, E.P.; Cattaneo, R.; Von Kalle, C.; Ungerechts, G. Armed and targeted measles virus for chemovirotherapy of pancreatic cancer. *Cancer Gene Ther.* **2011**, *18*, 598–608. [CrossRef]

46. Weiland, T.; Lampe, J.; Essmann, F.; Venturelli, S.; Berger, A.; Bossow, S.; Berchtold, S.; Schulze-Osthoff, K.; Lauer, U.M.; Bitzer, M. Enhanced killing of therapy-induced senescent tumor cells by oncolytic measles vaccine viruses. *Int. J. Cancer* **2013**, *134*, 235–243. [CrossRef]

47. Leber, M.F.; Neault, S.; Jirovec, E.; Barkley, R.; Said, A.; Bell, J.C.; Ungerechts, G. Engineering and combining oncolytic measles virus for cancer therapy. *Cytokine Growth Factor Rev.* **2020**. [CrossRef]

48. Liu, C.; Sarkaria, J.N.; Petell, C.A.; Paraskevakou, G.; Zollman, P.J.; Schroeder, M.; Carlson, B.; Decker, P.A.; Wu, W.; James, C.D.; et al. Combination of measles virus virotherapy and radiation therapy has synergistic activity in the treatment of glioblastoma multiforme. *Clin. Cancer Res.* **2007**, *13*, 7155–7165. [CrossRef]

49. Lal, G.; Rajala, M.S. Combination of Oncolytic Measles Virus Armed With BNiP3, a Pro-apoptotic Gene and Paclitaxel Induces Breast Cancer Cell Death. *Front. Oncol.* **2018**, *8*, 676. [CrossRef]

50. Tai, C.J.; Liu, C.H.; Pan, Y.C.; Wong, S.H.; Tai, C.J.; Richardson, C.D.; Lin, L.T. Chemovirotherapeutic Treatment Using Camptothecin Enhances Oncolytic Measles Virus-Mediated Killing of Breast Cancer Cells. *Sci. Rep.* **2019**, *9*, 6767. [CrossRef]

51. May, V.; Berchtold, S.; Berger, A.; Venturelli, S.; Burkard, M.; Leischner, C.; Malek, N.P.; Lauer, U.M. Chemovirotherapy for pancreatic cancer: Gemcitabine plus oncolytic measles vaccine virus. *Oncol. Lett.* **2019**, *18*, 5534–5542. [CrossRef] [PubMed]

52. Toan, N.L.; Hang, N.T.; Luu, N.K.; Van Mao, C.; Van Ba, N.; Xuan, N.T.; Cam, T.D.; Yamamoto, N.; Van Tong, H.; Son, H.A. Combination of Vaccine Strain Measles Virus and Nimotuzumab in the Treatment of Laryngeal Cancer. *Anticancer Res.* **2019**, *39*, 3727–3737. [CrossRef] [PubMed]

53. Zou, J.; Guo, Y.; Guettouche, T.; Smith, D.F.; Voellmy, R. Repression of heat shock transcription factor HSF1 activation by HSP90 (HSP90 complex) that forms a stress-sensitive complex with HSF1. *Cell* **1998**, *94*, 471–480. [CrossRef]

54. Muchowski, P.J.; Wacker, J.L. Modulation of neurodegeneration by molecular chaperones. *Nat. Rev. Neurosci.* **2005**, *6*, 11–22. [CrossRef] [PubMed]

55. Liu, C.; Erlichman, C.; McDonald, C.J.; Ingle, J.N.; Zollman, P.; Iankov, I.; Russell, S.J.; Galanis, E. Heat shock protein inhibitors increase the efficacy of measles virotherapy. *Gene Ther.* **2008**, *15*, 1024–1034. [CrossRef] [PubMed]

56. Stewart, C.E.; Randall, R.E.; Adamson, C.S. Inhibitors of the interferon response enhance virus replication in vitro. *PLoS ONE* **2014**, *9*, e112014. [CrossRef]

57. Ruf, B.; Berchtold, S.; Venturelli, S.; Burkard, M.; Smirnow, I.; Prenzel, T.; Henning, S.W.; Lauer, U.M. Combination of the oral histone deacetylase inhibitor resminostat with oncolytic measles vaccine virus as a new option for epi-virotherapeutic treatment of hepatocellular carcinoma. *Mol. Ther. Oncolytics* **2015**, *2*, 15019. [CrossRef]

58. Ellerhoff, T.P.; Berchtold, S.; Venturelli, S.; Burkard, M.; Smirnow, I.; Wulff, T.; Lauer, U.M. Novel epi-virotherapeutic treatment of pancreatic cancer combining the oral histone deacetylase inhibitor resminostat with oncolytic measles vaccine virus. *Int. J. Oncol.* **2016**, *49*, 1931–1944. [CrossRef]

59. Opyrchal, M.; Allen, C.; Msaouel, P.; Iankov, I.; Galanis, E. Inhibition of Rho-associated coiled-coil-forming kinase increases efficacy of measles virotherapy. *Cancer Gene Ther.* **2013**, *20*, 630–637. [CrossRef]

60. Li, C.; Meng, G.; Su, L.; Chen, A.; Xia, M.; Xu, C.; Yu, D.; Jiang, A.; Wei, J. Dichloroacetate blocks aerobic glycolytic adaptation to attenuated measles virus and promotes viral replication leading to enhanced oncolysis in glioblastoma. *Oncotarget* **2015**, *6*, 1544–1555. [CrossRef]

61. Xia, M.; Gonzalez, P.; Li, C.; Meng, G.; Jiang, A.; Wang, H.; Gao, Q.; Debatin, K.M.; Beltinger, C.; Wei, J. Mitophagy enhances oncolytic measles virus replication by mitigating DDX58/RIG-I-like receptor signaling. *J. Virol.* **2014**, *88*, 5152–5164. [CrossRef] [PubMed]

62. Son, H.A.; Zhang, L.; Cuong, B.K.; Van Tong, H.; Cuong, L.D.; Hang, N.T.; Nhung, H.T.M.; Yamamoto, N.; Toan, N.L. Combination of Vaccine-Strain Measles and Mumps Viruses Enhances Oncolytic Activity against Human Solid Malignancies. *Cancer Invest.* **2018**, *36*, 106–117. [CrossRef] [PubMed]

63. Radecke, F.; Spielhofer, P.; Schneider, H.; Kaelin, K.; Huber, M.; Dötsch, C.; Christiansen, G.; Billeter, M.A. Rescue of measles viruses from cloned DNA. *Embo J.* **1995**, *14*, 5773–5784. [CrossRef] [PubMed]

64. Pfaller, C.K.; Cattaneo, R.; Schnell, M.J. Reverse genetics of Mononegavirales: How they work, new vaccines, and new cancer therapeutics. *Virology* **2015**, *479–480*, 331–344. [CrossRef] [PubMed]

65. Aref, S.; Bailey, K.; Fielding, A. Measles to the Rescue: A Review of Oncolytic Measles Virus. *Viruses* **2016**, *8*, 294. [CrossRef] [PubMed]

66. Dingli, D.; Peng, K.W.; Harvey, M.E.; Greipp, P.R.; O'Connor, M.K.; Cattaneo, R.; Morris, J.C.; Russell, S.J. Image-guided radiovirotherapy for multiple myeloma using a recombinant measles virus expressing the thyroidal sodium iodide symporter. *Blood* **2004**, *103*, 1641–1646. [CrossRef]

67. Penheiter, A.R.; Griesmann, G.E.; Federspiel, M.J.; Dingli, D.; Russell, S.J.; Carlson, S.K. Pinhole micro-SPECT/CT for noninvasive monitoring and quantitation of oncolytic virus dispersion and percent infection in solid tumors. *Gene Ther.* **2012**, *19*, 279–287. [CrossRef]

68. Penheiter, A.R.; Dingli, D.; Bender, C.E.; Russell, S.J.; Carlson, S.K. Monitoring the initial delivery of an oncolytic measles virus encoding the human sodium iodide symporter to solid tumors using contrast-enhanced computed tomography. *J. Gene Med.* **2012**, *14*, 590–597. [CrossRef]

69. Kemler, I.; Ennis, M.K.; Neuhauser, C.M.; Dingli, D. In Vivo Imaging of Oncolytic Measles Virus Propagation with Single-Cell Resolution. *Mol. Ther. Oncolytics* **2019**, *12*, 68–78. [CrossRef]

70. Dingli, D.; Cascino, M.D.; Josić, K.; Russell, S.J.; Bajzer, Z. Mathematical modeling of cancer radiovirotherapy. *Math Biosci.* **2006**, *199*, 55–78. [CrossRef]

71. Bajzer, Z.; Carr, T.; Josić, K.; Russell, S.J.; Dingli, D. Modeling of cancer virotherapy with recombinant measles viruses. *J Theor Biol* **2008**, *252*, 109–122. [CrossRef] [PubMed]

72. Berg, D.R.; Offord, C.P.; Kemler, I.; Ennis, M.K.; Chang, L.; Paulik, G.; Bajzer, Z.; Neuhauser, C.; Dingli, D. In vitro and in silico multidimensional modeling of oncolytic tumor virotherapy dynamics. *PLoS Comput. Biol.* **2019**, *15*, e1006773. [CrossRef]

73. Biesecker, M.; Kimn, J.H.; Lu, H.; Dingli, D.; Bajzer, Z. Optimization of virotherapy for cancer. *Bull Math Biol* **2010**, *72*, 469–489. [CrossRef] [PubMed]

74. Galanis, E.; Hartmann, L.C.; Cliby, W.A.; Long, H.J.; Peethambaram, P.P.; Barrette, B.A.; Kaur, J.S.; Haluska, P.J., Jr.; Aderca, I.; Zollman, P.J.; et al. Phase I trial of intraperitoneal administration of an oncolytic measles virus strain engineered to express carcinoembryonic antigen for recurrent ovarian cancer. *Cancer Res.* **2010**, *70*, 875–882. [CrossRef] [PubMed]

75. Dispenzieri, A.; Tong, C.; LaPlant, B.; Lacy, M.Q.; Laumann, K.; Dingli, D.; Zhou, Y.; Federspiel, M.J.; Gertz, M.A.; Hayman, S.; et al. Phase I trial of systemic administration of Edmonston strain of measles virus genetically engineered to express the sodium iodide symporter in patients with recurrent or refractory multiple myeloma. *Leukemia* **2017**, *31*, 2791–2798. [CrossRef] [PubMed]

76. Nakamura, T.; Peng, K.W.; Harvey, M.; Greiner, S.; Lorimer, I.A.; James, C.D.; Russell, S.J. Rescue and propagation of fully retargeted oncolytic measles viruses. *Nat. Biotechnol.* **2005**, *23*, 209–214. [CrossRef] [PubMed]

77. Bucheit, A.D.; Kumar, S.; Grote, D.M.; Lin, Y.; von Messling, V.; Cattaneo, R.B.; Fielding, A.K. An oncolytic measles virus engineered to enter cells through the CD20 antigen. *Mol. Ther.* **2003**, *7*, 62–72. [CrossRef]

78. Hasegawa, K.; Nakamura, T.; Harvey, M.; Ikeda, Y.; Oberg, A.; Figini, M.; Canevari, S.; Hartmann, L.C.; Peng, K.W. The use of a tropism-modified measles virus in folate receptor-targeted virotherapy of ovarian cancer. *Clin. Cancer Res.* **2006**, *12*, 6170–6178. [CrossRef] [PubMed]

79. Allen, C.; Vongpunsawad, S.; Nakamura, T.; James, C.D.; Schroeder, M.; Cattaneo, R.; Giannini, C.; Krempski, J.; Peng, K.W.; Goble, J.M.; et al. Retargeted oncolytic measles strains entering via the EGFRvIII receptor maintain significant antitumor activity against gliomas with increased tumor specificity. *Cancer Res.* **2006**, *66*, 11840–11850. [CrossRef]

80. Allen, C.; Paraskevakou, G.; Iankov, I.; Giannini, C.; Schroeder, M.; Sarkaria, J.; Puri, R.K.; Russell, S.J.; Galanis, E. Interleukin-13 Displaying Retargeted Oncolytic Measles Virus Strains Have Significant Activity Against Gliomas With Improved Specificity. *Mol. Ther.* **2008**, *16*, 1556–1564. [CrossRef]

81. Jing, Y.; Tong, C.; Zhang, J.; Nakamura, T.; Iankov, I.; Russell, S.J.; Merchan, J.R. Tumor and vascular targeting of a novel oncolytic measles virus retargeted against the urokinase receptor. *Cancer Res.* **2009**, *69*, 1459–1468. [CrossRef] [PubMed]

82. Ong, H.T.; Trejo, T.R.; Pham, L.D.; Oberg, A.L.; Russell, S.J.; Peng, K.W. Intravascularly administered RGD-displaying measles viruses bind to and infect neovessel endothelial cells in vivo. *Mol. Ther.* **2009**, *17*, 1012–1021. [CrossRef] [PubMed]

83. Friedrich, K.; Hanauer, J.R.; Prüfer, S.; Münch, R.C.; Völker, I.; Filippis, C.; Jost, C.; Hanschmann, K.M.; Cattaneo, R.; Peng, K.W.; et al. DARPin-targeting of measles virus: Unique bispecificity, effective oncolysis, and enhanced safety. *Mol. Ther.* **2013**, *21*, 849–859. [CrossRef]

84. Lal, S.; Raffel, C. Using Cystine Knot Proteins as a Novel Approach to Retarget Oncolytic Measles Virus. *Mol. Ther. Oncolytics* **2017**, *7*, 57–66. [CrossRef]

85. Bach, P.; Abel, T.; Hoffmann, C.; Gal, Z.; Braun, G.; Voelker, I.; Ball, C.R.; Johnston, I.C.; Lauer, U.M.; Herold-Mende, C.; et al. Specific elimination of CD133+ tumor cells with targeted oncolytic measles virus. *Cancer Res.* **2013**, *73*, 865–874. [CrossRef] [PubMed]

86. Jing, Y.; Chavez, V.; Ban, Y.; Acquavella, N.; El-Ashry, D.; Pronin, A.; Chen, X.; Merchan, J.R. Molecular Effects of Stromal-Selective Targeting by uPAR-Retargeted Oncolytic Virus in Breast Cancer. *Mol. Cancer Res.* **2017**, *15*, 1410–1420. [CrossRef] [PubMed]

87. Springfeld, C.; von Messling, V.; Frenzke, M.; Ungerechts, G.; Buchholz, C.J.; Cattaneo, R. Oncolytic efficacy and enhanced safety of measles virus activated by tumor-secreted matrix metalloproteinases. *Cancer Res.* **2006**, *66*, 7694–7700. [CrossRef]

88. Mühlebach, M.D.; Schaser, T.; Zimmermann, M.; Armeanu, S.; Hanschmann, K.M.; Cattaneo, R.; Bitzer, M.; Lauer, U.M.; Cichutek, K.; Buchholz, C.J. Liver cancer protease activity profiles support therapeutic options with matrix metalloproteinase-activatable oncolytic measles virus. *Cancer Res.* **2010**, *70*, 7620–7629. [CrossRef]

89. Leber, M.F.; Bossow, S.; Leonard, V.H.; Zaoui, K.; Grossardt, C.; Frenzke, M.; Miest, T.; Sawall, S.; Cattaneo, R.; von Kalle, C.; et al. MicroRNA-sensitive oncolytic measles viruses for cancer-specific vector tropism. *Mol. Ther.* **2011**, *19*, 1097–1106. [CrossRef]

90. Ketzer, P.; Kaufmann, J.K.; Engelhardt, S.; Bossow, S.; von Kalle, C.; Hartig, J.S.; Ungerechts, G.; Nettelbeck, D.M. Artificial riboswitches for gene expression and replication control of DNA and RNA viruses. *Proc. Natl. Acad. Sci. USA* **2014**, *111*, E554–E562. [CrossRef]

91. Tahara, M.; Takishima, Y.; Miyamoto, S.; Nakatsu, Y.; Someya, K.; Sato, M.; Tani, K.; Takeda, M. Photocontrollable mononegaviruses. *Proc. Natl. Acad. Sci. USA* **2019**, *116*, 11587–11589. [CrossRef]
92. Ungerechts, G.; Springfeld, C.; Frenzke, M.E.; Lampe, J.; Johnston, P.B.; Parker, W.B.; Sorscher, E.J.; Cattaneo, R. Lymphoma chemovirotherapy: CD20-targeted and convertase-armed measles virus can synergize with fludarabine. *Cancer Res.* **2007**, *67*, 10939–10947. [CrossRef] [PubMed]
93. Ungerechts, G.; Springfeld, C.; Frenzke, M.E.; Lampe, J.; Parker, W.B.; Sorscher, E.J.; Cattaneo, R. An immunocompetent murine model for oncolysis with an armed and targeted measles virus. *Mol. Ther.* **2007**, *15*, 1991–1997. [CrossRef] [PubMed]
94. Lange, S.; Lampe, J.; Bossow, S.; Zimmermann, M.; Neubert, W.; Bitzer, M.; Lauer, U.M. A novel armed oncolytic measles vaccine virus for the treatment of cholangiocarcinoma. *Hum. Gene Ther.* **2013**, *24*, 554–564. [CrossRef] [PubMed]
95. Hartkopf, A.D.; Bossow, S.; Lampe, J.; Zimmermann, M.; Taran, F.A.; Wallwiener, D.; Fehm, T.; Bitzer, M.; Lauer, U.M. Enhanced killing of ovarian carcinoma using oncolytic measles vaccine virus armed with a yeast cytosine deaminase and uracil phosphoribosyltransferase. *Gynecol. Oncol.* **2013**, *130*, 362–368. [CrossRef]
96. Kaufmann, J.K.; Bossow, S.; Grossardt, C.; Sawall, S.; Kupsch, J.; Erbs, P.; Hassel, J.C.; von Kalle, C.; Enk, A.H.; Nettelbeck, D.M.; et al. Chemovirotherapy of malignant melanoma with a targeted and armed oncolytic measles virus. *J. Invest. Dermatol.* **2013**, *133*, 1034–1042. [CrossRef]
97. Lampe, J.; Bossow, S.; Weiland, T.; Smirnow, I.; Lehmann, R.; Neubert, W.; Bitzer, M.; Lauer, U.M. An armed oncolytic measles vaccine virus eliminates human hepatoma cells independently of apoptosis. *Gene Ther.* **2013**, *20*, 1033–1041. [CrossRef]
98. Hutzen, B.; Bid, H.K.; Houghton, P.J.; Pierson, C.R.; Powell, K.; Bratasz, A.; Raffel, C.; Studebaker, A.W. Treatment of medulloblastoma with oncolytic measles viruses expressing the angiogenesis inhibitors endostatin and angiostatin. *BMC Cancer* **2014**, *14*, 206. [CrossRef]
99. Pidelaserra-Martí, G.; Engeland, C.E. Mechanisms of measles virus oncolytic immunotherapy. *Cytokine Growth Factor Rev.* **2020**. [CrossRef]
100. Donnelly, O.G.; Errington-Mais, F.; Steele, L.; Hadac, E.; Jennings, V.; Scott, K.; Peach, H.; Phillips, R.M.; Bond, J.; Pandha, H.; et al. Measles virus causes immunogenic cell death in human melanoma. *Gene Ther.* **2013**, *20*, 7–15. [CrossRef]
101. Rajaraman, S.; Canjuga, D.; Ghosh, M.; Codrea, M.C.; Sieger, R.; Wedekink, F.; Tatagiba, M.; Koch, M.; Lauer, U.M.; Nahnsen, S.; et al. Measles Virus-Based Treatments Trigger a Pro-inflammatory Cascade and a Distinctive Immunopeptidome in Glioblastoma. *Mol. Ther. Oncolytics* **2019**, *12*, 147–161. [CrossRef]
102. Guillerme, J.B.; Boisgerault, N.; Roulois, D.; Ménager, J.; Combredet, C.; Tangy, F.; Fonteneau, J.F.; Gregoire, M. Measles virus vaccine-infected tumor cells induce tumor antigen cross-presentation by human plasmacytoid dendritic cells. *Clin. Cancer Res.* **2013**, *19*, 1147–1158. [CrossRef]
103. Achard, C.; Guillerme, J.B.; Bruni, D.; Boisgerault, N.; Combredet, C.; Tangy, F.; Jouvenet, N.; Grégoire, M.; Fonteneau, J.F. Oncolytic measles virus induces tumor necrosis factor-related apoptosis-inducing ligand (TRAIL)-mediated cytotoxicity by human myeloid and plasmacytoid dendritic cells. *Oncoimmunology* **2017**, *6*, e1261240. [CrossRef]
104. Tan, D.Q.; Zhang, L.; Ohba, K.; Ye, M.; Ichiyama, K.; Yamamoto, N. Macrophage response to oncolytic paramyxoviruses potentiates virus-mediated tumor cell killing. *Eur. J. Immunol.* **2016**, *46*, 919–928. [CrossRef]
105. Zhang, Y.; Patel, B.; Dey, A.; Ghorani, E.; Rai, L.; Elham, M.; Castleton, A.Z.; Fielding, A.K. Attenuated, oncolytic, but not wild-type measles virus infection has pleiotropic effects on human neutrophil function. *J. Immunol.* **2012**, *188*, 1002–1010. [CrossRef]
106. Dey, A.; Zhang, Y.; Castleton, A.Z.; Bailey, K.; Beaton, B.; Patel, B.; Fielding, A.K. The Role of Neutrophils in Measles Virus-mediated Oncolysis Differs Between B-cell Malignancies and Is Not Always Enhanced by GCSF. *Mol. Ther.* **2016**, *24*, 184–192. [CrossRef]
107. Li, H.; Peng, K.W.; Dingli, D.; Kratzke, R.A.; Russell, S.J. Oncolytic measles viruses encoding interferon beta and the thyroidal sodium iodide symporter gene for mesothelioma virotherapy. *Cancer Gene Ther.* **2010**, *17*, 550–558. [CrossRef]
108. Iankov, I.D.; Allen, C.; Federspiel, M.J.; Myers, R.M.; Peng, K.W.; Ingle, J.N.; Russell, S.J.; Galanis, E. Expression of immunomodulatory neutrophil-activating protein of Helicobacter pylori enhances the antitumor activity of oncolytic measles virus. *Mol. Ther.* **2012**, *20*, 1139–1147. [CrossRef]
109. Grossardt, C.; Engeland, C.E.; Bossow, S.; Halama, N.; Zaoui, K.; Leber, M.F.; Springfeld, C.; Jaeger, D.; von Kalle, C.; Ungerechts, G. Granulocyte-macrophage colony-stimulating factor-armed oncolytic measles virus is an effective therapeutic cancer vaccine. *Hum. Gene Ther.* **2013**, *24*, 644–654. [CrossRef]
110. Engeland, C.E.; Grossardt, C.; Veinalde, R.; Bossow, S.; Lutz, D.; Kaufmann, J.K.; Shevchenko, I.; Umansky, V.; Nettelbeck, D.M.; Weichert, W.; et al. CTLA-4 and PD-L1 checkpoint blockade enhances oncolytic measles virus therapy. *Mol. Ther.* **2014**, *22*, 1949–1959. [CrossRef]
111. Veinalde, R.; Grossardt, C.; Hartmann, L.; Bourgeois-Daigneault, M.C.; Bell, J.C.; Jäger, D.; von Kalle, C.; Ungerechts, G.; Engeland, C.E. Oncolytic measles virus encoding interleukin-12 mediates potent antitumor effects through T cell activation. *Oncoimmunology* **2017**, *6*, e1285992. [CrossRef] [PubMed]
112. Backhaus, P.S.; Veinalde, R.; Hartmann, L.; Dunder, J.E.; Jeworowski, L.M.; Albert, J.; Hoyler, B.; Poth, T.; Jäger, D.; Ungerechts, G.; et al. Immunological Effects and Viral Gene Expression Determine the Efficacy of Oncolytic Measles Vaccines Encoding IL-12 or IL-15 Agonists. *Viruses* **2019**, *11*, 914. [CrossRef] [PubMed]

113. Speck, T.; Heidbuechel, J.P.W.; Veinalde, R.; Jaeger, D.; von Kalle, C.; Ball, C.R.; Ungerechts, G.; Engeland, C.E. Targeted BiTE Expression by an Oncolytic Vector Augments Therapeutic Efficacy Against Solid Tumors. *Clin. Cancer Res.* **2018**, *24*, 2128–2137. [CrossRef] [PubMed]

114. Hutzler, S.; Erbar, S.; Jabulowsky, R.A.; Hanauer, J.R.H.; Schnotz, J.H.; Beissert, T.; Bodmer, B.S.; Eberle, R.; Boller, K.; Klamp, T.; et al. Antigen-specific oncolytic MV-based tumor vaccines through presentation of selected tumor-associated antigens on infected cells or virus-like particles. *Sci. Rep.* **2017**, *7*, 16892. [CrossRef] [PubMed]

115. Busch, E.; Kubon, K.D.; Mayer, J.K.M.; Pidelaserra-Martí, G.; Albert, J.; Hoyler, B.; Heidbuechel, J.P.W.; Stephenson, K.B.; Lichty, B.D.; Osen, W.; et al. Measles Vaccines Designed for Enhanced CD8(+) T Cell Activation. *Viruses* **2020**, *12*, 242. [CrossRef]

116. Hardcastle, J.; Mills, L.; Malo, C.S.; Jin, F.; Kurokawa, C.; Geekiyanage, H.; Schroeder, M.; Sarkaria, J.; Johnson, A.J.; Galanis, E. Immunovirotherapy with measles virus strains in combination with anti-PD-1 antibody blockade enhances antitumor activity in glioblastoma treatment. *Neuro Oncol.* **2017**, *19*, 493–502. [CrossRef]

117. Mühlebach, M.D. Vaccine platform recombinant measles virus. *Virus Genes* **2017**, *53*, 733–740. [CrossRef]

118. Hörner, C.; Schürmann, C.; Auste, A.; Ebenig, A.; Muraleedharan, S.; Herrmann, M.; Schnierle, B.; Mühlebach, M.D. A Highly Immunogenic Measles Virus-based Th1-biased COVID-19 Vaccine. *bioRxiv* 2020.

119. Klose, C.; Berchtold, S.; Schmidt, M.; Beil, J.; Smirnow, I.; Venturelli, S.; Burkard, M.; Handgretinger, R.; Lauer, U.M. Biological treatment of pediatric sarcomas by combined virotherapy and NK cell therapy. *BMC Cancer* **2019**, *19*, 1172. [CrossRef]

120. Chen, A.; Zhang, Y.; Meng, G.; Jiang, D.; Zhang, H.; Zheng, M.; Xia, M.; Jiang, A.; Wu, J.; Beltinger, C.; et al. Oncolytic measles virus enhances antitumour responses of adoptive CD8(+)NKG2D(+) cells in hepatocellular carcinoma treatment. *Sci. Rep.* **2017**, *7*, 5170. [CrossRef]

121. Heinzerling, L.; Künzi, V.; Oberholzer, P.A.; Kündig, T.; Naim, H.; Dummer, R. Oncolytic measles virus in cutaneous T-cell lymphomas mounts antitumor immune responses in vivo and targets interferon-resistant tumor cells. *Blood* **2005**, *106*, 2287–2294. [CrossRef] [PubMed]

122. Galanis, E.; Atherton, P.J.; Maurer, M.J.; Knutson, K.L.; Dowdy, S.C.; Cliby, W.A.; Haluska, P., Jr.; Long, H.J.; Oberg, A.; Aderca, I.; et al. Oncolytic measles virus expressing the sodium iodide symporter to treat drug-resistant ovarian cancer. *Cancer Res.* **2015**, *75*, 22–30. [CrossRef] [PubMed]

123. Packiriswamy, N.; Upreti, D.; Zhou, Y.; Khan, R.; Miller, A.; Diaz, R.M.; Rooney, C.M.; Dispenzieri, A.; Peng, K.W.; Russell, S.J. Oncolytic measles virus therapy enhances tumor antigen-specific T-cell responses in patients with multiple myeloma. *Leukemia* **2020**. [CrossRef]

124. Dietz, L.; Engeland, C.E. Immunomodulation in Oncolytic Measles Virotherapy. *Methods Mol. Biol.* **2020**, *2058*, 111–126. [CrossRef]

125. Haralambieva, I.; Iankov, I.; Hasegawa, K.; Harvey, M.; Russell, S.J.; Peng, K.W. Engineering oncolytic measles virus to circumvent the intracellular innate immune response. *Mol. Ther.* **2007**, *15*, 588–597. [CrossRef]

126. Meng, X.; Nakamura, T.; Okazaki, T.; Inoue, H.; Takahashi, A.; Miyamoto, S.; Sakaguchi, G.; Eto, M.; Naito, S.; Takeda, M.; et al. Enhanced antitumor effects of an engineered measles virus Edmonston strain expressing the wild-type N, P, L genes on human renal cell carcinoma. *Mol. Ther.* **2010**, *18*, 544–551. [CrossRef]

127. Dyer, A.; Baugh, R.; Chia, S.L.; Frost, S.; Iris; Jacobus, E.J.; Khalique, H.; Pokrovska, T.D.; Scott, E.M.; Taverner, W.K.; et al. Turning cold tumours hot: Oncolytic virotherapy gets up close and personal with other therapeutics at the 11th Oncolytic Virus Conference. *Cancer Gene Ther.* **2019**, *26*, 59–73. [CrossRef]

128. Miest, T.S.; Yaiw, K.C.; Frenzke, M.; Lampe, J.; Hudacek, A.W.; Springfeld, C.; von Messling, V.; Ungerechts, G.; Cattaneo, R. Envelope-chimeric entry-targeted measles virus escapes neutralization and achieves oncolysis. *Mol. Ther.* **2011**, *19*, 1813–1820. [CrossRef]

129. Iankov, I.D.; Blechacz, B.; Liu, C.; Schmeckpeper, J.D.; Tarara, J.E.; Federspiel, M.J.; Caplice, N.; Russell, S.J. Infected cell carriers: A new strategy for systemic delivery of oncolytic measles viruses in cancer virotherapy. *Mol. Ther.* **2007**, *15*, 114–122. [CrossRef]

130. Ong, H.T.; Hasegawa, K.; Dietz, A.B.; Russell, S.J.; Peng, K.W. Evaluation of T cells as carriers for systemic measles virotherapy in the presence of antiviral antibodies. *Gene Ther.* **2007**, *14*, 324–333. [CrossRef]

131. Liu, C.; Suksanpaisan, L.; Chen, Y.W.; Russell, S.J.; Peng, K.W. Enhancing cytokine-induced killer cell therapy of multiple myeloma. *Exp. Hematol.* **2013**, *41*, 508–517. [CrossRef] [PubMed]

132. Mader, E.K.; Maeyama, Y.; Lin, Y.; Butler, G.W.; Russell, H.M.; Galanis, E.; Russell, S.J.; Dietz, A.B.; Peng, K.W. Mesenchymal stem cell carriers protect oncolytic measles viruses from antibody neutralization in an orthotopic ovarian cancer therapy model. *Clin. Cancer Res.* **2009**, *15*, 7246–7255. [CrossRef]

133. Castleton, A.; Dey, A.; Beaton, B.; Patel, B.; Aucher, A.; Davis, D.M.; Fielding, A.K. Human mesenchymal stromal cells deliver systemic oncolytic measles virus to treat acute lymphoblastic leukemia in the presence of humoral immunity. *Blood* **2014**, *123*, 1327–1335. [CrossRef] [PubMed]

134. Liu, C.; Russell, S.J.; Peng, K.W. Systemic therapy of disseminated myeloma in passively immunized mice using measles virus-infected cell carriers. *Mol. Ther.* **2010**, *18*, 1155–1164. [CrossRef] [PubMed]

135. Liu, Y.P.; Tong, C.; Dispenzieri, A.; Federspiel, M.J.; Russell, S.J.; Peng, K.W. Polyinosinic acid decreases sequestration and improves systemic therapy of measles virus. *Cancer Gene Ther.* **2012**, *19*, 202–211. [CrossRef] [PubMed]

136. Xia, M.; Luo, D.; Dong, J.; Zheng, M.; Meng, G.; Wu, J.; Wei, J. Graphene oxide arms oncolytic measles virus for improved effectiveness of cancer therapy. *J. Exp. Clin. Cancer Res.* **2019**, *38*, 408. [CrossRef]

137. Nosaki, K.; Hamada, K.; Takashima, Y.; Sagara, M.; Matsumura, Y.; Miyamoto, S.; Hijikata, Y.; Okazaki, T.; Nakanishi, Y.; Tani, K. A novel, polymer-coated oncolytic measles virus overcomes immune suppression and induces robust antitumor activity. *Mol. Ther. Oncolytics* **2016**, *3*, 16022. [CrossRef]

138. Xu, C.; Goß, A.V.; Dorneburg, C.; Debatin, K.M.; Wei, J.; Beltinger, C. Proof-of-principle that a decoy virus protects oncolytic measles virus against neutralizing antibodies. *Oncolytic. Virother.* **2018**, *7*, 37–41. [CrossRef]

139. Peng, K.W.; Myers, R.; Greenslade, A.; Mader, E.; Greiner, S.; Federspiel, M.J.; Dispenzieri, A.; Russell, S.J. Using clinically approved cyclophosphamide regimens to control the humoral immune response to oncolytic viruses. *Gene Ther.* **2013**, *20*, 255–261. [CrossRef]

140. Yaiw, K.C.; Miest, T.S.; Frenzke, M.; Timm, M.; Johnston, P.B.; Cattaneo, R. CD20-targeted measles virus shows high oncolytic specificity in clinical samples from lymphoma patients independent of prior rituximab therapy. *Gene Ther.* **2011**, *18*, 313–317. [CrossRef]

141. Allen, C.; Opyrchal, M.; Aderca, I.; Schroeder, M.A.; Sarkaria, J.N.; Domingo, E.; Federspiel, M.J.; Galanis, E. Oncolytic measles virus strains have significant antitumor activity against glioma stem cells. *Gene Ther.* **2013**, *20*, 444–449. [CrossRef] [PubMed]

142. Iankov, I.D.; Msaouel, P.; Allen, C.; Federspiel, M.J.; Bulur, P.A.; Dietz, A.B.; Gastineau, D.; Ikeda, Y.; Ingle, J.N.; Russell, S.J.; et al. Demonstration of anti-tumor activity of oncolytic measles virus strains in a malignant pleural effusion breast cancer model. *Breast Cancer Res. Treat.* **2010**, *122*, 745–754. [CrossRef] [PubMed]

143. Peng, K.W.; Hadac, E.M.; Anderson, B.D.; Myers, R.; Harvey, M.; Greiner, S.M.; Soeffker, D.; Federspiel, M.J.; Russell, S.J. Pharmacokinetics of oncolytic measles virotherapy: Eventual equilibrium between virus and tumor in an ovarian cancer xenograft model. *Cancer Gene Ther.* **2006**, *13*, 732–738. [CrossRef]

144. Myers, R.M.; Greiner, S.M.; Harvey, M.E.; Griesmann, G.; Kuffel, M.J.; Buhrow, S.A.; Reid, J.M.; Federspiel, M.; Ames, M.M.; Dingli, D.; et al. Preclinical pharmacology and toxicology of intravenous MV-NIS, an oncolytic measles virus administered with or without cyclophosphamide. *Clin. Pharmacol. Ther.* **2007**, *82*, 700–710. [CrossRef] [PubMed]

145. Myers, R.; Harvey, M.; Kaufmann, T.J.; Greiner, S.M.; Krempski, J.W.; Raffel, C.; Shelton, S.E.; Soeffker, D.; Zollman, P.; Federspiel, M.J.; et al. Toxicology study of repeat intracerebral administration of a measles virus derivative producing carcinoembryonic antigen in rhesus macaques in support of a phase I/II clinical trial for patients with recurrent gliomas. *Hum. Gene Ther.* **2008**, *19*, 690–698. [CrossRef]

146. Völker, I.; Bach, P.; Coulibaly, C.; Plesker, R.; Abel, T.; Seifried, J.; Heidmeier, S.; Mühlebach, M.D.; Lauer, U.M.; Buchholz, C.J. Intrahepatic application of suicide gene-armed measles virotherapeutics: A safety study in transgenic mice and rhesus macaques. *Hum. Gene Ther. Clin. Dev.* **2013**, *24*, 11–22. [CrossRef]

147. Lal, S.; Peng, K.W.; Steele, M.B.; Jenks, N.; Ma, H.; Kohanbash, G.; Phillips, J.J.; Raffel, C. Safety Study: Intraventricular Injection of a Modified Oncolytic Measles Virus into Measles-Immune, hCD46-Transgenic, IFNαRko Mice. *Hum. Gene Ther. Clin. Dev.* **2016**, *27*, 145–151. [CrossRef]

148. Russell, S.J.; Federspiel, M.J.; Peng, K.W.; Tong, C.; Dingli, D.; Morice, W.G.; Lowe, V.; O'Connor, M.K.; Kyle, R.A.; Leung, N.; et al. Remission of disseminated cancer after systemic oncolytic virotherapy. *Mayo Clin. Proc.* **2014**, *89*, 926–933. [CrossRef]

149. Ungerechts, G.; Bossow, S.; Leuchs, B.; Holm, P.S.; Rommelaere, J.; Coffey, M.; Coffin, R.; Bell, J.; Nettelbeck, D.M. Moving oncolytic viruses into the clinic: Clinical-grade production, purification, and characterization of diverse oncolytic viruses. *Mol. Ther. Methods Clin. Dev.* **2016**, *3*, 16018. [CrossRef]

150. Langfield, K.K.; Walker, H.J.; Gregory, L.C.; Federspiel, M.J. Manufacture of measles viruses. *Methods Mol. Biol.* **2011**, *737*, 345–366. [CrossRef]

151. Loewe, D.; Dieken, H.; Grein, T.A.; Weidner, T.; Salzig, D.; Czermak, P. Opportunities to debottleneck the downstream processing of the oncolytic measles virus. *Crit. Rev. Biotechnol.* **2020**, *40*, 247–264. [CrossRef]

152. Loewe, D.; Häussler, J.; Grein, T.A.; Dieken, H.; Weidner, T.; Salzig, D.; Czermak, P. Forced Degradation Studies to Identify Critical Process Parameters for the Purification of Infectious Measles Virus. *Viruses* **2019**, *11*, 725. [CrossRef]

153. Baldo, A.; Galanis, E.; Tangy, F.; Herman, P. Biosafety considerations for attenuated measles virus vectors used in virotherapy and vaccination. *Hum. Vaccin. Immunother.* **2016**, *12*, 1102–1116. [CrossRef]

154. Leber, M.F.; Hoyler, B.; Prien, S.; Neault, S.; Engeland, C.E.; Förster, J.M.; Bossow, S.; Springfeld, C.; von Kalle, C.; Jäger, D.; et al. Sequencing of serially passaged measles virus affirms its genomic stability and reveals a nonrandom distribution of consensus mutations. *J. Gen. Virol.* **2020**, *101*, 399–409. [CrossRef]

155. Ribas, A.; Dummer, R.; Puzanov, I.; VanderWalde, A.; Andtbacka, R.H.I.; Michielin, O.; Olszanski, A.J.; Malvehy, J.; Cebon, J.; Fernandez, E.; et al. Oncolytic Virotherapy Promotes Intratumoral T Cell Infiltration and Improves Anti-PD-1 Immunotherapy. *Cell* **2018**, *174*, 1031–1032. [CrossRef]

cancers

Review

Parvovirus-Based Combinatorial Immunotherapy: A Reinforced Therapeutic Strategy against Poor-Prognosis Solid Cancers

Assia Angelova [1,*] , **Tiago Ferreira** [2] , **Clemens Bretscher** [2] , **Jean Rommelaere** [1] and **Antonio Marchini** [2,3]

1   German Cancer Research Center (DKFZ), Research Program Infection, Inflammation and Cancer, Clinical Cooperation Unit Virotherapy, Im Neuenheimer Feld 242, 69120 Heidelberg, Germany; j.rommelaere@dkfz-heidelberg.de

2   German Cancer Research Center (DKFZ), Laboratory of Oncolytic-Virus-Immunotherapeutics (LOVIT), Im Neuenheimer Feld 242, 69120 Heidelberg, Germany; t.ferreira@dkfz-heidelberg.de (T.F.); c.bretscher@gmx.de (C.B.); antonio.marchini@lih.lu (A.M.)

3   Luxembourg Institute of Health (LIH), Laboratory of Oncolytic-Virus-Immunotherapeutics (LOVIT), 84 rue Val Fleuri, L-1526 Luxembourg, Luxembourg

*   Correspondence: a.angelova@dkfz-heidelberg.de; Tel.: +49-6221-42-4960

Simple Summary: Oncolytic virotherapy using oncolytic viruses with natural or engineered cancer-destroying capacities has emerged as a promising treatment concept in modern oncology. Rodent protoparvoviruses, in particular the rat H-1 parvovirus (H-1PV), have demonstrated their broad-range tumor-suppressive properties in both preclinical models and clinical studies. In addition to inducing selective tumor cell death, these viruses are also able to exert immunostimulating effects and reverse tumor-driven immune suppression. Parvovirotherapy holds therefore a potential for enhancing the efficacy of other cancer immunotherapies. The aim of this review is to provide an overview of all H-1PV-based combinatorial immunotherapeutic approaches against poor-prognosis human solid cancers that have been tested so far. Current challenges and future prospects of parvoviro-immunotherapy, notably parvovirus inclusion into various immunotherapeutic protocols against glioblastoma, pancreatic cancer, among other standard therapy-refractory solid malignancies, are also discussed in the light of H-1PV further clinical development.

Abstract: Resistance to anticancer treatments poses continuing challenges to oncology researchers and clinicians. The underlying mechanisms are complex and multifactorial. However, the immunologically "cold" tumor microenvironment (TME) has recently emerged as one of the critical players in cancer progression and therapeutic resistance. Therefore, TME modulation through induction of an immunological switch towards inflammation ("warming up") is among the leading approaches in modern oncology. Oncolytic viruses (OVs) are seen today not merely as tumor cell-killing (oncolytic) agents, but also as cancer therapeutics with multimodal antitumor action. Due to their intrinsic or engineered capacity for overcoming immune escape mechanisms, warming up the TME and promoting antitumor immune responses, OVs hold the potential for creating a proinflammatory background, which may in turn facilitate the action of other (immunomodulating) drugs. The latter provides the basis for the development of OV-based immunostimulatory anticancer combinations. This review deals with the smallest among all OVs, the H-1 parvovirus (H-1PV), and focuses on H-1PV-based combinatorial approaches, whose efficiency has been proven in preclinical and/or clinical settings. Special focus is given to cancer types with the most devastating impact on life expectancy that urgently call for novel therapies.

Keywords: parvovirus; oncolytic; tumor microenvironment; immunotherapy; combination therapy; glioblastoma; pancreatic cancer; colorectal cancer; melanoma

Citation: Angelova, A.; Ferreira, T.; Bretscher, C.; Rommelaere, J.; Marchini, A. Parvovirus-Based Combinatorial Immunotherapy: A Reinforced Therapeutic Strategy against Poor-Prognosis Solid Cancers. *Cancers* **2021**, *13*, 342. https://doi.org/10.3390/cancers13020342

Received: 17 December 2020
Accepted: 15 January 2021
Published: 19 January 2021

Publisher's Note: MDPI stays neutral with regard to jurisdictional claims in published maps and institutional affiliations.

1. Introduction

The rodent H-1 protoparvovirus (H-1PV) (for an overview of H-1PV classification and biology, we redirect the readers to a recent review by Bretscher and Marchini [1]) was first discovered as a contaminating agent in xeno-transplanted human tumor cell lines [2]. Originally identified as a pathogen, which lethally affects rat fetuses and newborn rats by causing cerebellar hypoplasia and hepatitis [3], H-1PV was later found to preferentially replicate in rat- and in human-transformed or tumor-derived cell cultures, while sparing their non-malignant counterparts [4,5]. H-1PV intrinsic oncotropism and oncoselectivity are a complex phenomenon based on multiple molecular determinants, which are underrepresented in normal cells, but characteristic of tumor cells [6]. Importantly, humans are not naturally infected with this virus, and no association between H-1PV and human disease has been observed [7]. Two early clinical studies of virus administration to cancer patients—dating back to the 1960s and 1990s of last century—demonstrated the lack of H-1PV pathogenic effects and the feasibility of the approach [8,9], thus laying the groundwork for the development of parvovirus (PV)-based oncolytic virotherapy. Three decades of laboratory efforts brought about extensive preclinical evidence of H-1PV broad tumor-suppressive potential [5,10]. Furthermore, it became increasingly apparent that in addition to directly inducing cancer cell death (oncolysis), H-1PV was also capable of exerting immuno-stimulating effects in various preclinical cancer models [11,12].

PV induced immune system stimulation results from multiple infection-associated immunogenic events. Depending on the tumor model, virus dose, route of administration and the immunological status of the host, one or another immunogenic stimulus may prevail [11]. Regardless of the particular mechanism involved, PV-mediated immunomodulation contributes to the "warming up" of the tumor microenvironment (TME) (Figure 1), increases tumor visibility and enhances immune cell reactivity [13]. H-1PV infection-associated immunogenic events and their impact on the immune system are reviewed in detail elsewhere [12,13], and briefly summarized below.

- Immunogenic cell death (ICD) of H-1PV-infected tumor cells (indirect immune cell stimulation): PVs are potent triggers of immunogenic stimuli through tumor cell ICD induction. Infected tumor cells release a spectrum of proinflammatory mediators, in particular chemo- and cyto-kines, and pathogen- and danger-associated molecular patterns (PAMPs, DAMPs), which are in turn capable of boosting the maturation and reactivity of distinct immune cell populations. This can be exemplified by H-1PV-infected human melanoma cells, which activate dendritic cell (DC) maturation through the release of heat shock protein 72 [14]. In line with this observation, H-1PV-infected pancreatic and colorectal carcinoma cells were shown to stimulate natural killer (NK) cell tumor-killing capacity through both the overexpression of ligands specific for NK cell activation receptors and the downregulation of MHC I on infected tumor cells [15,16]. Notably, productive infection of tumor cells is not required for immune stimulation. This was demonstrated by co-incubating H-1PV-infected semi-permissive pancreatic carcinoma cells with peripheral blood mononuclear cells (PBMC), under which conditions induction of Th1 signature and release of interferon-gamma (IFN-γ) and tumor necrosis factor-alpha (TNF-α) were detected in the PBMC population [17].
- H-1PV infection of immune cells (direct immune cell stimulation): H-1PV infection of human immune cell subpopulations has been documented in various preclinical settings. Virus entry may take place in T, B, NK, DC and monocytic populations; however, infection is aborted at subsequent virus intracellular replication steps [18]. Abortive infection can nevertheless exert multiple immuno-stimulating effects, such as expression of IFN-stimulated genes and proinflammatory cytokine production [17,18]. On the other hand, H-1PV is able to inhibit the immune suppressive activity of regulatory T (Treg) cells [18].
- H-1PV impact on tumor vasculature: It has been demonstrated that endothelial (precursor) cells may constitute direct targets for parvovirus-mediated toxicity. These cells sustain an abortive H-1PV infection in vitro. In animal models, virus treatment

inhibits the growth of lymphatic endothelium-derived tumors (Kaposi's sarcoma). Furthermore, recombinant propagation-deficient parvoviral vectors armed with angiostatic chemokines achieve significant reduction of vascular endothelial growth factor (VEGF) expression in Kaposi's sarcoma cells [19]. Given the control exerted by the vasculature of tumors over their infiltration with immune cells, these effects are likely to contribute to H-1PV immuno-stimulating activity, as further discussed below. Altogether, these data warrant validation of H-1PV as a tool against highly vascularized cancers, e.g., glioblastoma, one of the most angiogenic human tumors.

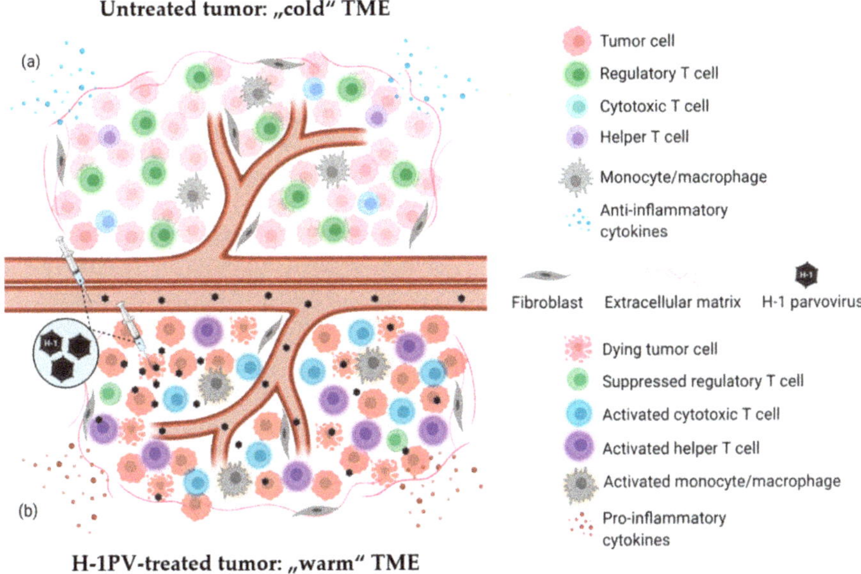

Figure 1. H-1PV-induced modulation of tumor microenvironment immune landscape. (**a**) Immunosuppressive ("cold") tumor microenvironment (TME) of a solid tumor. The tumor is often infiltrated with abundant immunosuppressive regulatory T cells (Treg)/myeloid-derived suppressor cells (MDSC). Tumor-infiltrating lymphocytes (TILs) (CD8+ CTLs, CD4+ Th cells) are scarce and/or anergic. Tumor and various TME cells produce anti-inflammatory cytokines to maintain immune suppression and facilitate tumor growth and dissemination. (**b**) Tumor infection with H-1PV results in immunogenic tumor cell death leading to the release proinflammatory cytokines, pathogen- and danger-associated molecular patterns (PAMPs and DAMPs), which alarm the immune system. The infection of tumor cells does not necessarily have to be productive for this immuno-stimulating effect to be achieved. Furthermore, abortive infection of immunocytes (CTLs, Th cells, monocytes/macrophages) with H-1PV can also lead to their activation. In contrast, H-1PV inhibits the immune suppressive functions of Treg cells. An immunological switch takes place and converts the "cold" TME into a "warmed up" (inflamed) one. Virus-mediated immuno-conversion of TME favors the mounting of enhanced antitumor immune responses.

The above-outlined H-1PV potential for creating a proinflammatory immune environment and alerting the immune system to the presence of a tumor opens prospects for combining the virus with various immunomodulators or other therapeutic agents endowed with immuno-stimulating properties. This combinatorial approach is in particular promising for the treatment of human tumors that remain presently incurable and pose continuing research and clinical challenges. Pancreatic ductal adenocarcinoma (PDAC), glioblastoma, colorectal cancer (CRC) and melanoma are among those cancers, which are urgently calling for novel therapeutic strategies. H-1PV-based immunotherapeutic combinations are reviewed below, which aim at targeting these devastating malignancies.

2. Parvovirus-Based Combinatorial Immunotherapy against Pancreatic Cancer

PDAC is the most common neoplasm of the pancreas and one of the most aggressive human cancers. It is characterized by quick progression, broad intraperitoneal dissemination (peritoneal carcinomatosis) and frequent resistance to conventional treatments. PDAC is usually diagnosed at advanced stages, when surgical resection is either not feasible or inefficient, as most patients eventually suffer from local recurrence and metachronous metastasis [20]. Current chemotherapy regimens achieve only minor improvements of PDAC dismal prognosis: the median survival time and overall 5-year survival remain as low as <12 months and approximately 5%, respectively [21]. Gemzar (gemcitabine) is the standard drug used to treat PDAC patients after surgery. Yet, gemcitabine only prolongs the survival of the majority (82%) of the patients by less than two-fold. On the same line, pathway-specific targeted therapies showed little efficacy against PDAC [22]. Therefore, new treatment paradigms need to be urgently explored in order to extend PDAC patient life expectancy and offer better quality of life.

H-1PV is among the oncolytic viruses (OVs), which have promising potential for efficiently targeting pancreatic cancer. PDAC sensitivity to H-1PV-induced oncolysis was demonstrated in various preclinical models [23,24]. Infection of human PDAC-derived cells leads to their killing, which is mediated at least in part by cathepsins [24]. Importantly, H-1PV sensitivity is preserved in gemcitabine-resistant cultures [23], thus opening up prospects to circumvent PDAC resistance to current standard death inducers.

2.1. H-1PV + Nucleoside Analogues (Gemcitabine)

As gemcitabine is currently considered the gold chemotherapeutic standard in PDAC clinical management, the therapeutic efficacy of gemcitabine in combination with H-1PV was tested in a rat syngeneic orthotopic PDAC model. H-1PV administration to gemcitabine-pretreated animals led to significant tumor suppression and survival prolongation in comparison with the mock-infected or gemcitabine-only treated groups [23]. These in vivo findings could not be straightforwardly ascribed to synergistic tumor cell death enhancement only. Indeed, in vitro studies showed that the cytotoxic effects of the combination, while allowing effective dose reduction for both agents, did not result in complete PDAC culture elimination. This prompted the investigation of the immunological effects exerted by the H-1PV + gemcitabine combination as an added value to direct tumor cell killing. Markers of ICD induction were analyzed in various PDAC cell lines, treated with either virus (or gemcitabine) alone or with H-1PV + gemcitabine. It was demonstrated that the release of high-mobility group box 1 protein (HMGB1) is a strikingly robust feature of H-1PV-infected PDAC cells [24]. Furthermore, H-1PV-triggered HMGB1 release did not require lytic infection, in line with the above-described PBMC activation by non-productively infected PDAC cells [17]. Gemcitabine alone was unable to induce HMGB1 secretion, yet H-1PV-induced HMGB1 release remained unaffected in gemcitabine-treated cells. Gemcitabine, on the other hand, was able to induce—albeit not in all cell lines tested—mature interleukin 1-beta (IL-1ß) accumulation in culture supernatants. Taken together, these data show that H-1PV and gemcitabine complement each other in the induction of immunogenic signals. The compatibility of H-1PV-induced alarmin (HMGB1) secretion with other (ICD-inducing) chemotherapeutic regimens warrants the consideration of PV inclusion into various multimodal anti-PDAC treatment protocols [24]. The therapeutic promise of H-1PV administration in gemcitabine-treated pancreatic cancer patients is further supported by several reports in the literature showing that, unlike most nucleoside analogues, gemcitabine is lacking immunosuppressive properties. On the contrary, gemcitabine may be beneficial not only to the cytocidal but also to the pro-immune outcome of H-1PV infection, as assumed from the findings below.

- One study conducted in gemcitabine-treated PDAC patients revealed the ability of the drug to enhance T cell-mediated and DC-dependent host immune responses [25].

- In keeping with the aforementioned data, it was documented that gemcitabine therapy may promote naïve T cell activation in PDAC patients and enhance their responsiveness to specific vaccination or to other forms of immunotherapy [26].
- The understanding of gemcitabine immunoregulating effects as a complementary constituent of tumor cell toxicity was extended by the demonstration that this drug alleviates pancreatic cancer immune escape through NK cell cytotoxicity enhancement [27].
- Studies conducted in murine orthotopic PDAC models provided yet another insight into gemcitabine-mediated immuno-stimulation, namely by indicating that low chemotherapeutic doses selectively deplete effector/memory Treg cell populations. The latter has a strong impact on PDAC microenvironment, as Tregs usually form large intra-tumoral infiltrates and trigger local immune suppression [28,29].
- Last but not least, in cancer models other than PDAC, gemcitabine enhances the efficacy of OV (e.g., reovirus) therapy. This complementation is achieved through gemcitabine-mediated inhibition of myeloid-derived suppressor cell (MDSC) recruitment to the TME and acceleration of reovirus-induced antitumor T cell immune responses [30].

Based on favorable preclinical data hinting at the potentiation of OV-induced antitumor effects in the presence of gemcitabine, a clinical trial, ParvOryx02 (NCT02653313), was designed and conducted with the aim to provide a clinical proof-of-principle of the safety (and efficacy) of H-1PV + gemcitabine co-treatment. Patients with inoperable metastatic (at least one hepatic metastasis) pancreatic cancer were treated with H-1PV. The virus was first administered intravenously (40% of the total virus dose on four consecutive days), and the remaining virus dose was then given intra-metastatically as single hepatic injection, followed by gemcitabine treatment [31]. Partial response and extended overall survival were observed in two out of seven trial patients, and immunological signatures most likely contributed to this improved outcome. The ParvOryx02 study therefore provided the first clinical indication that immune mechanisms underlie PV-mediated tumor suppression [32].

2.2. H-1PV + Histone Deacetylase Inhibitors (Valproic Acid)

Preclinical proof-of-concept was also obtained for another treatment combining H-1PV with the histone deacetylase (HDAC) inhibitor (HDACi) valproic acid (VPA) [33]. HDACis hold significant promise in cancer therapy, due to their ability to cause malignant cell growth inhibition, re-differentiation and death [34]. Most interestingly, HDAC inhibition was also found to potentiate the oncotoxicity of various OVs, including vesicular stomatitis [35], herpes- [36], adeno- [37] and parvo [33]-viruses (for a review, see Reference [38]). The synergism between HDACi and H-1PV was first demonstrated by Li et al., who conducted preclinical testing of this combination in cervical carcinoma and PDAC models [33]. VPA proved to synergize with H-1PV in inducing DNA damage, oxidative stress and death in PDAC-derived cell lines. This cooperation was traced back, at least in part, to the ability of VPA to stimulate the acetylation and, in consequence, the oncotoxic activity of the viral protein NS1. Interestingly, VPA-induced hyperacetylation of NS1 was also associated with enhanced H-1PV DNA replication and viral gene transcription, ultimately boosting virus multiplication in tumor cells. The VPA-dependent increase in both H-1PV intrinsic oncotoxicity and multiplication was reflected in the potentiation of tumor suppression in animal models. In order to establish a clinically relevant animal model of PDAC, patient-derived material was xeno-transplanted in non-obese diabetic/severe combined immunodeficiency disease (NOD/SCID) mice. Alternatively, the human AsPC-1 cell line was implanted into nude rats. Tumors were subjected to mono versus combinatorial treatment and tumor growth parameters were comparatively evaluated. In line with the in vitro observations, H-1PV + VPA administration resulted in enhanced NS1 and H-1PV intra-tumoral accumulation, correlating with an increase in oxidative stress and subsequent apoptosis in co-treated tumors. The combination achieved complete AsPC-1 tumor eradication. Patient-derived xenografts were also responsive, yet to a somewhat lesser extent, probably due to the characteristic PDAC intra-tumoral heterogeneity and prominent presence of stroma.

It is noteworthy that besides its effects on tumor cell growth and OV oncotoxicity, VPA was reported to modulate the immune system, providing an additional possible interface for cooperation with OVs at the level of their intrinsic immuno-stimulating activity. VPA was indeed shown to:

- Exert epigenetic regulation of various immune functions, e.g., attenuation of MDSC immunosuppressive effects [39].
- Induce the expression of MHC I-related chain A (MICA) and B (MICB) molecules, as well as of UL16-binding proteins (ULBPs) in human tumor cells, thereby triggering their enhanced recognition by NK cells [40], like H-1PV does (see above [15,16]).
- Mediate the inhibition of macrophage migration inhibitory factor (MIF) expression through local chromatin deacetylation-based transcription targeting [41].

As a whole, the above data speak for the high translational relevance of VPA to the future development of PV-based combinatorial (immuno) therapies.

In conclusion, two drugs that are available on the pharmaceutical market, i.e., gemcitabine (cytostatic) and VPA (antiepileptic), proved to be efficient in synergizing with H-1PV to suppress pancreatic cancer (Figure 2).

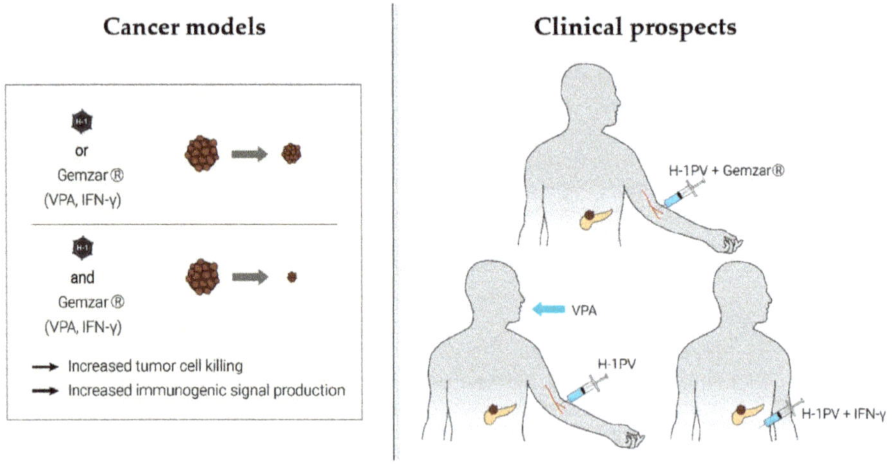

Figure 2. Parvovirus-based viro-immunotherapeutic combinations under development against pancreatic ducatl adenocarcinoma (PDAC). H-1PV-induced tumor cell lysis cooperates with gemcitabine-triggered programmed tumor cell death, valproic acid (VPA)-dependent epigenetic transcription regulation or interferon (IFN)-γ-induced immuno-stimulation to suppress PDAC. Preclinical data suggest that the immune system mediates, at least in part, this cooperation. H-1PV infection of tumor cells leads to the release of PAMPs/DAMPs, such as high-mobility group box 1 protein (HMGB1), which in turn alert the immune system to danger and mobilize an inflammatory antitumor immune response. Various aspects of H-1PV-, gemcitabine-, VPA- and IFN-γ-exerted immunomodulation may converge and synergize upon exposure of the host immune system to the respective combinations. The underlying mechanisms remain to be elucidated in detail by gathering extensive clinical experience. For details and references, see main text.

2.3. H-1PV + Proinflammatory Cytokines (Interferon-Gamma)

Another combination with substantial potential for clinical development relies on the mutual complementation of H-1PV- and IFN-γ-mediated immune stimulation. It was shown that IFN-γ improves the vaccination potential of the virus and diminishes the development of peritoneal carcinomatosis in preclinical PDAC models. Concomitant intraperitoneal administration of both H-1PV and IFN-γ in these models led to extended animal survival correlating with enhanced peritoneal macrophage and splenocyte responses against tumor cells [42].

3. Parvovirus-Based Combinatorial Immunotherapy against Glioblastoma

Glioblastoma multiforme (GBM) is the most common and aggressive human primary brain tumor. Similar to PDAC, GBM patients experience a very poor outcome. The 5-year overall survival rate is very low, around 5.1% [43]. GBM treatment faces a unique challenge: the presence of the blood–brain barrier (BBB), which largely prevents drugs, including small-molecule ones, from entering the central nervous system [44]. Current therapeutic approaches therefore include surgical resection of the tumor—to the largest extent feasible and safe—followed by radiotherapy and concomitant chemotherapy [45]. Unfortunately, despite all clinical efforts, tumor progression and recurrence typically occur, calling for alternative therapeutic solutions [46].

Based on the so far unmet need for novel, more efficient treatments, GBM was among the preclinical tumor models most extensively studied in our laboratory. H-1PV capacity for selectively killing glioma cells through cytosolic activation of lysosomal proteases was first demonstrated in vitro [47]. These results were validated in animal models, namely in immunocompetent rats bearing orthotopic autologous RG-2 tumors and in immunodeficient rats bearing xeno-transplanted human U87 gliomas. In these models, tumor regression after local, intravenous or intranasal virus administration was observed [47–49]. H-1PV treatment was not associated with any significant off-target toxicities; accordingly, virus transcription and NS1 protein accumulation could be detected in regressing tumor remnants and not in the surrounding normal tissues [48]. Interestingly, the therapeutic effect was potentiated in the presence of an intact host immune system. T cell depletion impaired H-1PV-induced glioma suppression; conversely, the presence of T cell only, in the absence of PV treatment, was not sufficient to inhibit tumor growth [11]. These preclinical observations provided the first hints of host T cell response involvement in PV-mediated glioma regression, hence the rationale for the development of PV-based immunotherapies against glioblastoma.

Pursuant to the above-described preclinical findings, the ParvOryx01 trial (NCT01301430) in recurrent glioblastoma patients delivered the first clinical proof-of-concept for tumor-infiltrating lymphocytes (TILs) playing substantial role in H-1PV-mediated immunomodulation of GBM TME. Although ParvOryx01 primary objectives were to determine virus safety, tolerability, pharmacokinetics, shedding and maximum tolerated dose, the analysis of post-virus-treatment resected tumor tissues revealed the presence of prominent immune cell infiltrates [50]. These infiltrates were comprised of CD45+CD3+CD4+ and CD45+CD3+CD8+ TILs. The latter contained both perforin and granzyme B-positive secretory granules, which is indicative of CTL cytolytic activity. TILs proved, in addition, to be CD25 (IL2 receptor alpha chain)-positive. Only a minor fraction of these cells expressed FOXP3, indicating the scarcity of Treg cells within the intra-tumoral immune infiltrates. Intra-tumoral production of proinflammatory cytokines (IFN-γ, IL-2) was also detected, together with inducible nitric oxide synthase (iNOS) expression in CD68+ tumor-associated microglia/macrophage cells [50,51]. Interestingly, tumor cells expressed the CD40 ligand (CD40L), a positive prognostic factor in glioblastoma [52]. Co-expression of CD40L and CD40, considered as a negative prognostic factor, was not seen [50,51]. Taken together, these first clinical findings indicated that H-1PV has the capacity to exert immuno-stimulating effects on glioblastoma TME. This makes the virus a worthwhile partner in therapeutic combinations, which aim at warming up the intrinsically immunosuppressive and immune-evasive environment of brain tumors.

3.1. H-1PV + Ionizing Radiation

We have previously shown that radiotherapy, one of the conventional first-line treatments in glioblastoma patients, sensitizes low-passage glioma cultures to H-1PV oncolysis. Pre-irradiation increases the susceptibility of these cells to virus infection. Interestingly,

H-1PV achieves killing both radiation-sensitive and resistant glioma cells [53]. Apart from triggering enhanced tumor cytolysis, the irradiation followed by H-1PV treatment holds, in addition, the potential—although not yet validated in animal models—of acting as combinatorial immunotherapy. Indeed, although irradiation was long regarded as a local anticancer therapy, the first reports on radiotherapy interactions with the host immune system can be traced back to the 1970s of the last century. In 1979, Slone et al. were the first to report that the radiation dose required to control 50% of mouse fibrosarcomas was twice as high in immunocompromised animals as in immunocompetent hosts [54]. Furthermore, tumor regression at sites distant to radiation fields, the so-called abscopal effect, has been systematically observed [55]. Radiation-triggered immunomodulation encompasses, among other effects, ICD induction, T and NK cell activation and MDSC suppression. These observations prompted the development of various combination therapy regimens based on radiation and other immunomodulating agents [56,57], including OVs (e.g., adeno-, herpes simplex-, measles- and vaccinia-viruses) against glioma [58].

3.2. H-1PV + Tumor Angiogenesis Inhibitors (Bevacizumab)

Another promising approach is the combination of H-1PV with bevacizumab (Avastin®). This co-treatment was evaluated in a series of compassionate virus uses in recurrent glioblastoma patients. Bevacizumab is an anti-VEGF-A monoclonal antibody available in Europe since 2005 for the treatment of breast, lung, kidney, colon, ovarian and endometrial carcinomas. In 2009, bevacizumab was approved by the Food and Drug Administration (FDA) for application in glioblastoma patients [59]. While achieving a steroid-sparing effect and alleviation of edema, bevacizumab monotherapy has, however, not demonstrated significant survival benefits [60]. On the other hand, scientists and clinicians have gathered an extensive—and yet to grow—knowledge of bevacizumab's mode of action. In particular, bevacizumab was found to exert immunomodulating activity by counteracting VEGF-induced negative effects on DC maturation, antigen presentation and lymphocytic trafficking [61]. These bevacizumab properties, together with the ParvOryx01 trial experience showing H-1PV treatment-associated immunogenic changes in glioblastoma TME, have opened up prospects for novel anti-glioma combinatorial immunotherapy development, i.e., H-1PV + bevacizumab (Figure 3). A compassionate use proof-of-concept program was conducted in five GBM patients, who developed a second or third recurrence after being treated in the ParvOryx01 trial. The patients underwent tumor resection, followed by local H-1PV administration and bevacizumab. The mean survival after treatment was extended to 15.4 months. Moreover, in three out of the five patients, striking remission of the recurrence was observed, providing first clinical hints of synergistic glioblastoma suppression through parvoviro-immunotherapy [62].

3.3. H-1PV + PD-1 Immune Checkpoint Inhibitors (Nivolumab)

Checkpoint blockade, a strategy which aims at overcoming immune system tolerance towards the tumor through the release from negative regulators of immune activation (immune checkpoints), is presently at the leading edge of cancer immunotherapy. Although efficient in controlling various other solid tumors, immune checkpoint inhibitors (ICIs) frequently fail to achieve a significant response in glioblastoma patients [63]. Several preclinical studies and clinical trials have therefore been initiated, in order to determine the optimal ICI-based combinations and redefine the future standards of care for this deadly disease [64].

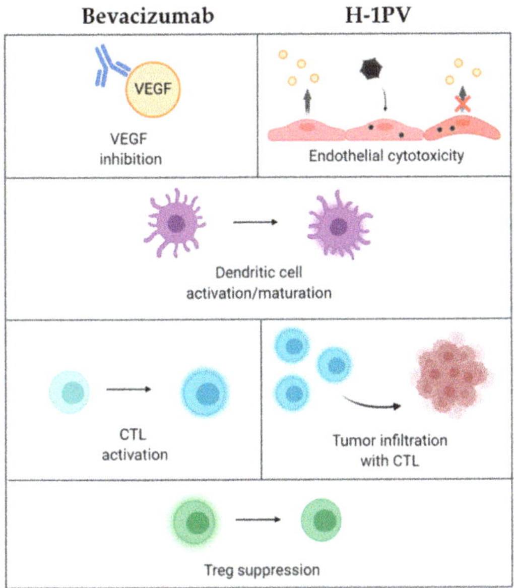

Figure 3. Rationale for combining H-1PV administration with bevacizumab treatment in patients with recurrent glioblastoma multiforme (GBM). Bevacizumab antibody and H-1PV infection share the capacity for inhibiting vascular endothelial growth factor (VEGF) (production) (upper row) and triggering distinct immuno-modulations (lower rows), raising hopes to improve antitumor immunity by combining both treatments. In support of this strategy, bevacizumab and H-1PV were found to jointly achieve significant clinical improvement in GBM patients at second or third recurrence, leading to remission of the recurrent tumor. The precise mechanisms of this therapeutic potentiation remain to be determined. However, the establishment by H-1PV of an immunologically "improved" proinflammatory background, which facilitates bevacizumab-mediated immuno-stimulating effects, is a likely scenario. For details and references, see the main text.

First clinical hints of improved antitumor effects of H-1PV virotherapy upon combination with checkpoint blockade were obtained through compassionate virus uses. A series of three patients with rapidly progressing recurrent glioblastoma were treated with H-1PV (two were irradiated prior to virus administration), followed by bevacizumab and the programmed cell death protein 1 (PD-1) inhibitor nivolumab. In addition, all patients received the HDACi VPA. This innovative PV-based multimodal strategy led to radiologically confirmed tumor regression accompanied by clinical improvement in all subjects 4 to 8 weeks after virus injection [65]. An objective tumor response was also seen in another group of primary or recurrent glioblastoma patients, who received H-1PV in combination with bevacizumab and checkpoint blockade. Complete to partial tumor remission was documented in 78% of the cases, which is a significantly higher response rate than the one reported in the literature for bevacizumab- and ICI-based monotherapies [66].

Altogether, the above data provide a strong impetus for further clinical development of H-1PV combinations with radiation and/or immunomodulators (in particular bevacizumab and ICIs) in the fight against glioblastoma.

4. Parvovirus-Based Combinatorial Immunotherapy against Colorectal Cancer

CRC is another major cause of cancer-related deaths worldwide. Although the implementation of early-detection screening programs has substantially improved the 5-year overall survival, prognosis for CRC patients with stage 4 metastatic disease remains poor [67]. Immunotherapy, in particular checkpoint blockade, has proved efficient against heavily mutated colorectal tumors. However, it fails to elicit sufficiently strong therapeutic responses in carcinomas, which are mismatch-repair-proficient (pMMR) and possess low levels of microsatellite instability (MSI-L). Low mutational burden, together with the lack of immune cell infiltration, contribute to pMMR-MSI-L immune resistance [68]. Novel approaches are therefore needed for the treatment of patients with advanced metastatic or low mutational burden CRC. One such approach, combinatorial immunotherapy, holds much potential for extending the scope of checkpoint blockade so as to bring benefit also to CRC patients with unfavorable prognosis.

H-1PV + CTLA-4 Immune Checkpoint Blockade (Tremelimumab)

Many tumor types, including CRC, overexpress the immune checkpoint cytotoxic T-lymphocyte-associated protein 4 (CTLA-4) and thus transmit inhibitory signals to T cells [69]. This immune evasion strategy creates an immunosuppressive environment, which allows the tumor to escape immune recognition and destruction. The anticancer effects of tremelimumab, a CTLA-4-specific human antibody, applied either alone or in combination with H-1PV, were studied by Heinrich et al. [70] n a human in vitro CRC model. H-1PV infection alone was found to reduce the viability of SW480 CRC cells and enhance extracellular CTLA-4 expression. SW480 cells and immature DCs (iDCs) co-culture experiments demonstrated that the expression of DC maturation and activation markers, namely CD83, CD80 and CD86, sharply increased when the tumor cells were infected with H-1PV. Notably, additional treatment of H-1PV-infected SW480 cells with tremelimumab resulted in IFN-γ enrichment of the co-culture supernatant [70].

5. Parvovirus-Based Combinatorial Immunotherapy against Melanoma

Cutaneous melanoma, also known as black skin cancer, is an aggressive tumor arising from the melanocytes. Over the past 10 years, melanoma has become a prototype for testing novel targeted therapies, first and foremost, immune checkpoint blockade. PD-1 inhibition has shown significant clinical success in controlling locoregional melanoma [71]. However, metastatic melanoma is a severe life-threatening condition for which reinforcement of current treatment tools and approaches is still needed.

H-1PV + CTLA-4 (Ipilimumab)/PD-1 (Nivolumab) Immune Checkpoint Blockade

In order to investigate the immunological effects of H-1PV in combination with ipilimumab and/or nivolumab, a human ex vivo melanoma model was used by Goepfert et al. Similar to the observations made in CRC-derived cells [70], upregulation of immune checkpoints, CTLA-4, PD-1 and PD-L1 in particular, was seen in H-1PV-infected melanoma cells. Yet, the virus potentiated the capacity of melanoma cells to induce iDC maturation in co-culture experiments. Nivolumab and ipilimumab, when added to the treatment scheme, triggered a further increase in the release into the co-culture supernatant of IFN-γ and TNF-α, respectively. Further to this, upon combination with H-1PV, the two ICIs induced stronger CTL activation, compared to virus alone [72]. Combining PV-induced immunogenic oncolysis with CTLA-4 and/or PD-1 blockade allows achieving a double goal, i.e., tumor cell killing and activation of the immune system against the tumor. This triggering of complementary events, centered on tumor destruction and immune-mediated elimination, renders the H-1PV + ICI approach promising for melanoma and other solid tumors' treatment.

6. Conclusions

Preclinical research and clinical experience have demonstrated the multimodal anticancer activity of the oncolytic parvovirus H-1PV. Two essential facets of H-1PV-induced tumor suppression consist of direct killing of malignant cells (oncolysis) and activation of cellular immune responses against the tumor. H-1PV infection, oncolysis and immune stimulation are interconnected, coordinated events, which cooperate towards multisided tumor elimination.

Glioblastoma and pancreatic adenocarcinoma are among the most devastating human malignancies, characterized by resistance to current therapies, tendency to recurrence and an overall poor outcome. H-1PV has undergone clinical testing in two recently conducted trials, ParvOryx01 in glioblastoma and ParvOryx02 in pancreatic carcinoma. Virus excellent safety and tolerability, together with the capacity for gentle TME immune landscape proinflammatory modulation, provide a strong impetus for further H-1PV clinical development. It should, however, be noted that in the clinical setting, various patient-dependent factors may result in suboptimal antitumor effects. Large intra-tumoral tissue heterogeneity, emergence of tumor cells resistant to virus infection/killing, dominance of the highly immunosuppressive TME, hampered virus spreading, off-target infection and virus neutralization by antiviral antibodies are among the major barriers to efficient H-1PV-induced tumor elimination. While various other approaches (capsid modification, chimera generation, fitness mutant selection, armed vector construction) to H-1PV treatment optimization are currently under investigation, PV-based combinatorial therapies are considered as a particularly promising avenue that holds the potential of enhancing both oncolysis and immune-mediated tumor destruction. Combinations of the virus with other anticancer approaches, namely irradiation, chemotherapy (gemcitabine), epigenetic modulation (HDACi), angiogenesis regulation (bevacizumab) or immunotherapy (immune checkpoint blockade), were evaluated in both preclinical models and in cancer patients. The combinatorial H-1PV-based viro(immuno)therapeutic strategy was proven to achieve greater anticancer effects compared to individual agents alone. The synergistic boost was particularly pronounced in combinations including the HDACi VPA, bevacizumab or the PD-1 inhibitor nivolumab. Glioblastoma patients treated with this combination showed striking tumor remission and extended survival, notably after second or even third recurrence. These early clinical observations speak in favor of considering H-1PV inclusion into various immunotherapeutic protocols against glioblastoma and other poor-prognosis solid tumors (Figure 4).

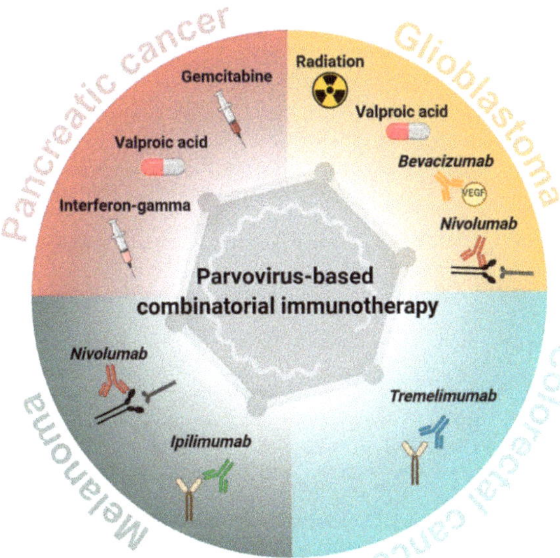

Figure 4. H-1PV inclusion into combinatorial anticancer immunotherapy regimens. The development of H-1PV combinations with ionizing radiation, chemotherapeutics, histone deacetylase inhibitors (HDACis), angiogenesis inhibitors and immunomodulators holds significant promise for the future of poor-prognosis solid cancer treatment.

Author Contributions: Writing—original draft preparation, A.A., T.F. and C.B.; writing—review and editing, A.A., J.R. and A.M. All authors have read and agreed to the published version of the manuscript.

Funding: The parvovirus research described in this review was partially funded by research grants from the Luxembourg Cancer Foundation and Télévie to A.M., and from ORYX GmbH & Co. KG to A.A., J.R. and A.M.

Acknowledgments: The authors wish to thank all members and collaborators of the former Tumor Virology Division at the German Cancer Research Center, for their contribution to the work described in this review. We are deeply indebted to the clinical trial and compassionate use program participants and their families, for their contribution to gathering clinical experience and knowledge. We are grateful to K. Geletneky for helpful discussions of the glioblastoma compassionate parvovirus use program.

Conflicts of Interest: A.A., T.F., J.R. and A.M. are holders of patents or patent applications related to H-1PV use for cancer therapeutic purposes. The parvovirus clinical trials (ParvOryx01, ParvOryx02) and compassionate H-1PV uses were financially supported by ORYX GmbH & Co. KG (Baldham, Germany). The funders had no role in the writing of the manuscript or in the decision to submit it for publication.

References

1. Bretscher, C.; Marchini, A. H-1 parvovirus as a cancer-killing agent: Past, present, and future. *Viruses* **2019**, *11*, 562. [CrossRef] [PubMed]
2. Toolan, H.W.; Dalldore, G.; Barclay, M.; Chandra, S.; Moore, A.E. An unidentified, filtrable agent isolated from transplanted human tumors. *Proc. Natl. Acad. Sci. USA* **1960**, *46*, 1256–1258. [CrossRef] [PubMed]
3. Besselsen, D.G.; Besch-Williford, C.L.; Pintel, D.J.; Franklin, C.L.; Hook, R.R.; Riley, L.K. Detection of H-1 parvovirus and Kilham rat virus by PCR. *J. Clin. Microbiol.* **1995**, *33*, 1699–1703. [CrossRef] [PubMed]
4. Rommelaere, J.; Giese, N.; Cziepluch, C.; Cornelis, J.J. Parvoviruses as anticancer agents. In *Viral Therapy of Human Cancers*; Sinkovics, J.G., Horvath, J.C., Eds.; Marcel Dekker: New York, NY, USA, 2005; pp. 627–675.
5. Rommelaere, J.; Geletneky, K.; Angelova, A.L.; Daeffler, L.; Dinsart, C.; Kiprianova, I.; Schlehofer, J.R.; Raykov, Z. Oncolytic parvoviruses as cancer therapeutics. *Cytokine Growth Factor Rev.* **2010**, *21*, 185–195. [CrossRef]
6. Angelova, A.L.; Geletneky, K.; Nüesch, J.P.F.; Rommelaere, J. Tumor selectivity of oncolytic parvoviruses: From in vitro and animal models to cancer patients. *Front. Bioeng. Biotechnol.* **2015**, *3*, 55. [CrossRef]
7. Newman, S.J.; McCallin, P.F.; Sever, J.L. Attempts to isolate H-1 virus from spontaneous human abortions: A negative report. *Teratology* **1970**, *3*, 279–281. [CrossRef]
8. Toolan, H.W.; Saunders, E.L.; Southam, C.M.; Moore, A.E.; Levin, A.G. H-1 virus viremia in the human. *Exp. Biol. Med.* **1965**, *119*, 711–715. [CrossRef]
9. Le Cesne, A.; Dupressoir, T.; Janin, N.; Spielmann, M.; Le Chevalier, T.; Sancho-Garnier, H.; Paoletti, C.; Rommelaere, J.; Stehelin, D.; Tursz, T.; et al. Intralesional administration of a live virus, parvovirus H1 (PVH1) in cancer patients: A feasibility study. *Proc. Am. Soc. Clin. Oncol.* **1993**, *12*, 297.
10. Hartley, A.; Kavishwar, G.; Salvato, I.; Marchini, A. A roadmap for the success of oncolytic parvovirus-based anticancer therapies. *Annu. Rev. Virol.* **2020**, *7*, 537–557. [CrossRef]
11. Geletneky, K.; Nüesch, J.P.; Angelova, A.L.; Kiprianova, I.; Rommelaere, J. Double-faceted mechanism of parvoviral oncosuppression. *Curr. Opin. Virol.* **2015**, *13*, 17–24. [CrossRef]
12. Angelova, A.L.; Rommelaere, J. Immune System Stimulation by Oncolytic Rodent Protoparvoviruses. *Viruses* **2019**, *11*, 415. [CrossRef] [PubMed]
13. Marchini, A.; Daeffler, L.; Pozdeev, V.I.; Angelova, A.; Rommelaere, J. Immune conversion of tumor microenvironment by oncolytic viruses: The protoparvovirus H-1PV case study. *Front. Immunol.* **2019**, *10*, 1848. [CrossRef] [PubMed]
14. Moehler, M.; Zeidler, M.; Schede, J.; Rommelaere, J.; Galle, P.R.; Cornelis, J.J.; Heike, M. Oncolytic parvovirus H1 induces release of heat-shock protein HSP72 in susceptible human tumor cells but may not affect primary immune cells. *Cancer Gene Ther.* **2003**, *10*, 477–480. [CrossRef] [PubMed]
15. Bhat, R.; Dempe, S.; Dinsart, C.; Rommelaere, J. Enhancement of NK cell antitumor responses using an oncolytic parvovirus. *Int. J. Cancer* **2010**, *128*, 908–919. [CrossRef]
16. Bhat, R.; Rommelaere, J. NK-cell-dependent killing of colon carcinoma cells is mediated by natural cytotoxicity receptors (NCRs) and stimulated by parvovirus infection of target cells. *BMC Cancer* **2013**, *13*, 367. [CrossRef]
17. Grekova, S.P.; Aprahamian, M.; Giese, N.; Schmitt, S.; Giese, T.; Falk, C.S.; Daeffler, L.; Cziepluch, C.; Rommelaere, J.; Raykov, Z. Immune cells participate in the oncosuppressive activity of parvovirus H-1PV and are activated as a result of their abortive infection with this agent. *Cancer Biol. Ther.* **2010**, *10*, 1280–1289. [CrossRef]

18. Moralès, O.; Richard, A.; Martin, N.; Mrizak, D.; Sénéchal, M.; Miroux, C.; Pancré, V.; Rommelaere, J.; Caillet-Fauquet, P.; De Launoit, Y.; et al. Activation of a Helper and Not Regulatory Human CD4+ T Cell Response by Oncolytic H-1 Parvovirus. *PLoS ONE* **2012**, *7*, e32197. [CrossRef]

19. Lavie, M.; Struyf, S.; Stroh-Dege, A.; Rommelaere, J.; Van Damme, J.; Dinsart, C. Capacity of wild-type and chemokine-armed parvovirus H-1PV for inhibiting neo-angiogenesis. *Virology* **2013**, *447*, 221–232. [CrossRef]

20. Felsenstein, M.; Hruban, R.H.; Wood, L.D. New developments in the molecular mechanisms of pancreatic tumorigenesis. *Adv. Anat. Pathol.* **2018**, *25*, 131–142. [CrossRef]

21. Siegel, R.L.; Mph, K.D.M.; Jemal, A. Cancer statistics, 2017. *CA A Cancer J. Clin.* **2017**, *67*, 7–30. [CrossRef]

22. Neoptolemos, J.; Kleeff, J.; Michl, P.; Costello, E.; Greenhalf, W.; Palmer, D.H. Therapeutic developments in pancreatic cancer: Current and future perspectives. *Nat. Rev. Gastroenterol. Hepatol.* **2018**, *15*, 333–348. [CrossRef] [PubMed]

23. Angelova, A.L.; Aprahamian, M.; Grekova, S.P.; Hajri, A.; Leuchs, B.; Giese, N.A.; Dinsart, C.; Herrmann, A.; Balboni, G.; Rommelaere, J.; et al. Improvement of Gemcitabine-Based Therapy of Pancreatic Carcinoma by Means of Oncolytic Parvovirus H-1PV. *Clin. Cancer Res.* **2009**, *15*, 511–519. [CrossRef] [PubMed]

24. Angelova, A.L.; Grekova, S.P.; Heller, A.; Kuhlmann, O.; Soyka, E.; Giese, T.; Aprahamian, M.; Bour, G.; Rüffer, S.; Cziepluch, C.; et al. Complementary Induction of Immunogenic Cell Death by Oncolytic Parvovirus H-1PV and Gemcitabine in Pancreatic Cancer. *J. Virol.* **2014**, *88*, 5263–5276. [CrossRef] [PubMed]

25. Plate, J.M.D.; Harris, J.E. Effects of gemcitabine treatment on immune cells and functions in pancreatic cancer patients. *Cancer Res.* **2004**, *64* (Suppl. 7), 500.

26. Plate, J.M.D.; Plate, A.E.; Shott, S.; Bograd, S.; Harris, J.E. Effect of gemcitabine on immune cells in subjects with adenocarcinoma of the pancreas. *Cancer Immunol. Immunother.* **2005**, *54*, 915–925. [CrossRef] [PubMed]

27. Lin, X.; Huang, M.; Xie, F.; Zhou, H.; Yang, J.; Huang, Q. Gemcitabine inhibits immune escape of pancreatic cancer by down regulating the soluble ULBP2 protein. *Oncotarget* **2016**, *7*, 70092–70099. [CrossRef]

28. Shevchenko, I.; Karakhanova, S.; Soltek, S.; Link, J.; Bayry, J.; Werner, J.; Umansky, V.; Bazhin, A.V. Low-dose gemcitabine depletes regulatory T cells and improves survival in the orthotopic Panc02 model of pancreatic cancer. *Int. J. Cancer* **2013**, *133*, 98–107. [CrossRef] [PubMed]

29. Homma, Y.; Taniguchi, K.; Nakazawa, M.; Matsuyama, R.; Mori, R.; Takeda, K.; Ichikawa, Y.; Tanaka, K.; Endo, I. Changes in the immune cell population and cell proliferation in peripheral blood after gemcitabine-based chemotherapy for pancreatic cancer. *Clin. Transl. Oncol.* **2014**, *16*, 330–335. [CrossRef]

30. Gujar, S.A.; Clements, D.A.; Dielschneider, R.; Helson, E.; Marcato, P.S.; Lee, P.W. Gemcitabine enhances the efficacy of reovirus-based oncotherapy through anti-tumour immunological mechanisms. *Br. J. Cancer* **2014**, *110*, 83–93. [CrossRef]

31. Hajda, J.; Lehmann, M.; Krebs, O.; Kieser, M.; Geletneky, K.; Jäger, D.; Dahm, M.; Huber, B.; Schöning, T.; Sedlaczek, O.; et al. A non-controlled, single arm, open label, phase II study of intravenous and intratumoral administration of ParvOryx in patients with metastatic, inoperable pancreatic cancer: ParvOryx02 protocol. *BMC Cancer* **2017**, *17*, 1–11. [CrossRef]

32. ParvOryx02: A Phase II Trial of Intravenous and Intratumoral Administration of H-1 Parvovirus in Patients with Metastatic Pancreatic Cancer. Available online: http://oryx-medicine.com/fileadmin/user_upload/uploads/News/Publications/201910_IOVC_Ungerechts_ParvOryx02.pdf (accessed on 16 November 2020).

33. Li, J.; Bonifati, S.; Hristov, G.; Marttila, T.; Valmary-Degano, S.; Stanzel, S.; Schnölzer, M.; Mougin, C.; Aprahamian, M.; Grekova, S.P.; et al. Synergistic combination of valproic acid and oncolytic parvovirus H-1 PV as a potential therapy against cervical and pancreatic carcinomas. *EMBO Mol. Med.* **2013**, *5*, 1537–1555. [CrossRef] [PubMed]

34. Minucci, S.; Pelicci, P.G. Histone deacetylase inhibitors and the promise of epigenetic (and more) treatments for cancer. *Nat. Rev. Cancer* **2006**, *6*, 38–51. [CrossRef] [PubMed]

35. Alvarez-Breckenridge, C.A.; Yu, J.; Price, R.L.; Wei, M.; Wang, Y.; Nowicki, M.O.; Ha, Y.P.; Bergin, S.M.; Hwang, C.; Fernandez, S.A.; et al. The histone deacetylase inhibitor valproic acid lessens NK cell action against oncolytic virus-infected glioblastoma cells by inhibition of STAT5/T-BET signaling and generation of gamma interferon. *J. Virol.* **2012**, *86*, 4566–4577. [CrossRef] [PubMed]

36. Otsuki, A.; Patel, A.; Kasai, K.; Suzuki, M.; Kurozumi, K.; Chiocca, E.A.; Saeki, Y. Histone Deacetylase Inhibitors Augment Antitumor Efficacy of Herpes-based Oncolytic Viruses. *Mol. Ther.* **2008**, *16*, 1546–1555. [CrossRef]

37. VanOosten, R.L.; Earel, J.K., Jr.; Griffith, T.S. Histone deacetylase inhibitors enhance Ad5-TRAIL killing of TRAIL-resistant prostate tumor cells through increased caspase-2 activity. *Apoptosis* **2006**, *12*, 561–571. [CrossRef]

38. Marchini, A.; Scott, E.M.; Rommelaere, J. Overcoming barriers in oncolytic virotherapy with HDAC inhibitors and immune checkpoint blockade. *Viruses* **2016**, *8*, 9. [CrossRef]

39. Xie, Z.; Ago, Y.; Okada, N.; Tachibana, M. Valproic acid attenuates immunosuppressive function of myeloid-derived suppressor cells. *J. Pharmacol. Sci.* **2018**, *137*, 359–365. [CrossRef]

40. Armeanu, S.; Bitzer, M.; Lauer, U.M.; Venturelli, S.; Pathil, A.; Krusch, M.; Kaiser, S.; Jobst, J.; Smirnow, I.; Wagner, A.; et al. Natural Killer Cell–Mediated Lysis of Hepatoma Cells via Specific Induction of NKG2D Ligands by the Histone Deacetylase Inhibitor Sodium Valproate. *Cancer Res.* **2005**, *65*, 6321–6329. [CrossRef]

41. Lugrin, J.; Ding, X.C.; Le Roy, D.; Chanson, A.-L.; Sweep, F.C.; Calandra, T.; Roger, T. Histone deacetylase inhibitors repress macrophage migration inhibitory factor (MIF) expression by targeting MIF gene transcription through a local chromatin deacetylation. *Biochim. Biophys. Acta BBA Bioenerg.* **2009**, *1793*, 1749–1758. [CrossRef]

42. Grekova, S.P.; Aprahamian, M.; Daeffler, L.; Leuchs, B.; Angelova, A.; Giese, T.; Galabov, A.; Heller, A.; Giese, N.A.; Rommelaere, J.; et al. Interferon γ improves the vaccination potential of oncolytic parvovirus H-1PV for the treatment of peritoneal carcinomatosis in pancreatic cancer. *Cancer Biol. Ther.* **2011**, *12*, 888–895. [CrossRef]

43. Ostrom, Q.T.; Gittleman, H.; Truitt, G.; Boscia, A.; Kruchko, C.; Barnholtz-Sloan, J.S. CBTRUS statistical report: Primary brain and other central nervous system tumors diagnosed in the United States in 2011–2015. *Neuro Oncol.* **2018**, *20*, iv1–iv86. [CrossRef] [PubMed]

44. Xu, Y.-Y.; Gao, P.; Sun, Y.; Duan, Y. Development of targeted therapies in treatment of glioblastoma. *Cancer Biol. Med.* **2015**, *12*, 223–237. [PubMed]

45. Davis, M.E. Glioblastoma: Overview of Disease and Treatment. *Clin. J. Oncol. Nurs.* **2016**, *20*, S2–S8. [CrossRef] [PubMed]

46. Gallego, O. Nonsurgical treatment of recurrent glioblastoma. *Curr. Oncol.* **2015**, *22*, 273–281. [CrossRef] [PubMed]

47. Di Piazza, M.; Mader, C.; Geletneky, K.; Calle, M.H.Y.; Weber, E.; Schlehofer, J.; Deleu, L.; Rommelaere, J. Cytosolic Activation of cathepsins mediates parvovirus H-1-induced killing of cisplatin and trail-resistant glioma cells. *J. Virol.* **2007**, *81*, 4186–4198. [CrossRef]

48. Geletneky, K.; Kiprianova, I.; Ayache, A.; Koch, R.; Herrero y Calle, M.; Deleu, L.; Sommer, C.; Thomas, N.; Rommelaere, J.; Schlehofer, J.R. Regression of advanced rat and human gliomas by local or systemic treatment with oncolytic parvovirus H-1 in rat models. *Neuro Oncol.* **2010**, *12*, 804–814. [CrossRef]

49. Kiprianova, I.; Thomas, N.; Ayache, A.; Fischer, M.; Leuchs, B.; Klein, M.; Rommelaere, J.; Schlehofer, J.R. Regression of glioma in rat models by intranasal application of parvovirus H-1. *Clin. Cancer Res.* **2011**, *17*, 5333–5342. [CrossRef]

50. Geletneky, K.; Hajda, J.; Angelova, A.L.; Leuchs, B.; Capper, D.; Bartsch, A.J.; Neumann, J.-O.; Schöning, T.; Hüsing, J.; Beelte, B.; et al. Oncolytic H-1 parvovirus shows safety and signs of immunogenic activity in a first phase I/IIa glioblastoma trial. *Mol. Ther.* **2017**, *25*, 2620–2634. [CrossRef]

51. Angelova, A.L.; Barf, M.; Geletneky, K.; Unterberg, A.; Rommelaere, J. Immunotherapeutic potential of oncolytic H-1 parvovirus: Hints of glioblastoma microenvironment conversion towards immunogenicity. *Viruses* **2017**, *9*, 382. [CrossRef]

52. Chonan, M.; Saito, R.; Shoji, T.; Shibahara, I.; Kanamori, M.; Sonoda, Y.; Watanabe, M.; Kikuchi, T.; Ishii, N.; Tominaga, T. CD40/CD40L expression correlates with the survival of patients with glioblastomas and an augmentation in CD40 signaling enhances the efficacy of vaccinations against glioma models. *Neuro Oncol.* **2015**, *17*, 1453–1462. [CrossRef]

53. Geletneky, K.; Hartkopf, A.D.; Krempien, R.; Rommelaere, J.; Schlehofer, J.R. Improved killing of human high-grade glioma cells by combining ionizing radiation with oncolytic parvovirus H-1 infection. *J. Biomed. Biotechnol.* **2010**, *2010*, 1–9. [CrossRef] [PubMed]

54. Slone, H.B.; Peters, L.J.; Milas, L. Effect of host immune capability on radiocurability and subsequent transplantability of a murine fibrosarcoma. *J. Natl. Cancer Inst.* **1979**, *63*, 1229–1235. [CrossRef]

55. Reynders, K.; Illidge, T.; Siva, S.; Chang, J.Y.; De Ruysscher, D. The abscopal effect of local radiotherapy: Using immunotherapy to make a rare event clinically relevant. *Cancer Treat. Rev.* **2015**, *41*, 503–510. [CrossRef] [PubMed]

56. Van Limbergen, E.J.; De Ruysscher, D.K.; Pimentel, V.O.; Marcus, D.; Berbee, M.; Hoeben, A.; Rekers, N.H.; Theys, J.; Yaromina, A.; Dubois, L.J.; et al. Combining radiotherapy with immunotherapy: The past, the present and the future. *Br. J. Radiol.* **2017**, *90*, 20170157. [CrossRef]

57. Zhao, X.; Shao, C. Radiotherapy-Mediated Immunomodulation and Anti-Tumor Abscopal Effect Combining Immune Checkpoint Blockade. *Cancers* **2020**, *12*, 2762. [CrossRef]

58. Touchefeu, Y.; Vassaux, G.; Harrington, K.J. Oncolytic viruses in radiation oncology. *Radiother. Oncol.* **2011**, *99*, 262–270. [CrossRef]

59. Garcia, J.; Hurwitz, H.I.; Sandler, A.B.; Miles, D.; Coleman, R.L.; Deurloo, R.; Chinot, O.L. Bevacizumab (Avastin®) in cancer treatment: A review of 15 years of clinical experience and future outlook. *Cancer Treat. Rev.* **2020**, *86*, 102017. [CrossRef]

60. Gilbert, M.R.; Dignam, J.J.; Armstrong, T.S.; Wefel, J.S.; Blumenthal, D.T.; Vogelbaum, M.A.; Colman, H.; Chakravarti, A.; Pugh, S.; Won, M.; et al. A randomized trial of bevacizumab for newly diagnosed glioblastoma. *N. Engl. J. Med.* **2014**, *370*, 699–708. [CrossRef]

61. Brown, N.F.; Carter, T.J.; Ottaviani, D.; Mulholland, P. Harnessing the immune system in glioblastoma. *Br. J. Cancer* **2018**, *119*, 1171–1181. [CrossRef]

62. Geletneky, K.; Angelova, A.; Leuchs, B.; Bartsch, A.; Capper, D.; Hajda, J.; Rommelaere, J. ATNT-07. Favorable response of patients with glioblastoma at second or third recurrence to repeated injection of oncolytic parvovirus H-1 in combination with bevacicumab. *Neuro Oncol.* **2015**, *17*, v11. [CrossRef]

63. Reardon, D.A.; Omuro, A.; Brandes, A.A.; Rieger, J.; Wick, A.; Sepulveda, J.; Phuphanich, S.; De Souza, P.; Ahluwalia, M.S.; Lim, M.; et al. OS10.3 randomized phase 3 study evaluating the efficacy and safety of nivolumab vs bevacizumab in patients with recurrent glioblastoma: CheckMate 143. *Neuro Oncol.* **2017**, *19*, iii21. [CrossRef]

64. Desai, K.; Hubben, A.; Ahluwalia, M. The role of checkpoint inhibitors in glioblastoma. *Target. Oncol.* **2019**, *14*, 375–394. [CrossRef] [PubMed]

65. Geletneky, K.; Weiss, C.; Bernhard, H.; Capper, D.; Leuchs, B.; Marchini, A.; Rommelaere, J. ATIM-29. First clinical observation of improved anti-tumor effects of viro-immunotherapy with oncolytic parvovirus H-1 in combination with PD-1 checkpoint blockade and bevacicumab in patients with recurrent glioblastoma. *Neuro Oncol.* **2016**, *18*, vi24. [CrossRef]

66. Geletneky, K.; Bartsch, A.; Weiss, C.; Bernhard, H.; Marchini, A.; Rommelaere, J. ATIM-40. High rate of objective anti-tumor response in 9 patients with glioblastoma after viro-immunotherapy with oncolytic parvovirus H-1 in combination with bevacicumab and PD-1 checkpoint blockade. *Neuro Oncol.* **2018**, *20*, vi10. [CrossRef]

67. Siegel, R.L.; DeSantis, C.; Jemal, A. Colorectal cancer statistics, 2014. *CA A Cancer J. Clin.* **2014**, *64*, 104–117. [CrossRef]

68. Le, D.T.; Durham, J.N.; Smith, K.N.; Wang, H.; Bartlett, B.R.; Aulakh, L.K.; Lu, S.; Kemberling, H.; Wilt, C.; Luber, B.S.; et al. Mismatch repair deficiency predicts response of solid tumors to PD-1 blockade. *Science* **2017**, *357*, 409–413. [CrossRef]

69. Mazzolini, G. Immunotherapy and immunoescape in colorectal cancer. *World J. Gastroenterol.* **2007**, *13*, 5822–5831. [CrossRef]

70. Heinrich, B.; Goepfert, K.; Delic, M.; Galle, P.R.; Moehler, M. Influence of the oncolytic parvovirus H-1, CTLA-4 antibody tremelimumab and cytostatic drugs on the human immune system in a human in vitro model of colorectal cancer cells. *OncoTargets Ther.* **2013**, *6*, 1119–1127. [CrossRef]

71. Huang, A.C.; Orlowski, R.J.; Xu, X.; Mick, R.; George, S.M.; Yan, P.K.; Manne, S.; Kraya, A.A.; Wubbenhorst, B.; Dorfman, L.; et al. A single dose of neoadjuvant PD-1 blockade predicts clinical outcomes in resectable melanoma. *Nat. Med.* **2019**, *25*, 454–461. [CrossRef]

72. Goepfert, K.; Dinsart, C.; Rommelaere, J.; Foerster, F.; Moehler, M. Rational combination of parvovirus h1 with ctla-4 and pd-1 checkpoint inhibitors dampens the tumor induced immune silencing. *Front. Oncol.* **2019**, *9*, 425. [CrossRef]

Review

Oncolytic Adenovirus in Cancer Immunotherapy

Malin Peter and Florian Kühnel *

Department of Gastroenterology, Hepatology, and Endocrinology, Hannover Medical School, 30625 Hannover, Germany; peter.malin@mh-hannover.de
* Correspondence: kuehnel.florian@mh-hannover.de; Tel.: +49-511-532-3995

Received: 9 October 2020; Accepted: 11 November 2020; Published: 13 November 2020

Simple Summary: Oncolytic adenoviruses are engineered to selectively replicate in and destroy cancer tissue. Moreover, these viruses are promising tools to restore antitumor immune response in cancer patients due to their high immunogenicity and the ability to interfere with the immunosuppressive tumor microenvironment. Due to these characteristics, oncolytic adenoviruses can activate tumors for already existing, systemic immunotherapies. The goal of this review is to provide an introduction into the common concepts of oncolytic adenoviruses, and to present their current status in clinical development. We also want to report in detail on strategies to optimize the immunoactivating properties of these agents for future application in multistage cancer immunotherapies.

Abstract: Tumor-selective replicating "oncolytic" viruses are novel and promising tools for immunotherapy of cancer. However, despite their first success in clinical trials, previous experience suggests that currently used oncolytic virus monotherapies will not be effective enough to achieve complete tumor responses and long-term cure in a broad spectrum of cancers. Nevertheless, there are reasonable arguments that suggest advanced oncolytic viruses will play an essential role as enablers of multi-stage immunotherapies including established systemic immunotherapies. Oncolytic adenoviruses (oAds) display several features to meet this therapeutic need. oAds potently lyse infected tumor cells and induce a strong immunogenic cell death associated with tumor inflammation and induction of antitumor immune responses. Furthermore, established and versatile platforms of oAds exist, which are well suited for the incorporation of heterologous genes to optimally exploit and amplify the immunostimulatory effect of viral oncolysis. A considerable spectrum of functional genes has already been integrated in oAds to optimize particular aspects of immune stimulation including antigen presentation, T cell priming, engagement of additional effector functions, and interference with immunosuppression. These advanced concepts have the potential to play a promising future role as enablers of multi-stage immunotherapies involving adoptive cell transfer and systemic immunotherapies.

Keywords: oncolytic adenovirus; cancer immunotherapy; multi-stage; immunostimulatory; arming

---

1. Introduction

Oncolytic viruses (OV) preferably replicate in and lyse tumor cells, and thus leave, healthy tissue unharmed. This common feature is either intrinsic or a consequence of genetic engineering [1]. OVs were initially designed to enable effective tumor cell lysis and virus spreading thereby ensuring a reliable control of viral replication in normal cells. However, it has been recognized that OVs exert multiple antitumor functions including the induction of innate and adaptive immune responses against the tumor. Oncolytic adenoviruses (oAds) have been among the earliest OVs to enter clinical trials. Onyx-015, an E1B55k mutant adenovirus for selective replication in p53 dysfunctional tumor cells, has been intensively investigated [2]. Though tumor responses had been observed in patients, particularly in combination with chemotherapy, the therapeutic efficacy did not meet the high expectations [3]. Nevertheless, these pioneering studies demonstrated that administration of oAds is well tolerated and safe. Additionally, lessons have been learned for the development of the next generation of viruses. Several advances in the fields of tumor immunotherapy have stimulated the interest in adenoviruses as oncolytic agents. First, there was the successful phase III study of the herpesvirus T-Vec and its subsequent approval by the U.S. food and drug administration (FDA), which delivered the final proof that OVs provide a clinical benefit for cancer patients [4]. There was also the growing perception that adaptive, tumor-directed immune responses are the essential therapeutic outcome of virotherapy [5]. A striking advance was the success of checkpoint inhibitors demonstrating that tumor immunotherapy even facilitates long-term cure [6]. However, the observation that the vast majority of cancer patients do not respond to these therapies opened up new future perspectives for the clinical application of OVs. The ability of oncolysis to induce tumor inflammation and to interfere with impaired immune functions in tumors suggests that OVs are promising agents to sensitize tumors for checkpoint inhibitors. Corresponding clinical studies are ongoing and are expected to deliver results soon [7]. Regarding the immunogenic properties of adenoviruses, oAds are presumably well qualified to meet this therapeutic need. Furthermore, oAds can be easily equipped with immunostimulatory transgenes to modulate the tumor microenvironment and to engage specific immune effector mechanisms. In this review, we want to give an overview on the currently existing platforms of oncolytic adenoviruses as well as the current state of their clinical development. We will report on current concepts of arming oAds with cytokines or alternative immune activators and will finally discuss the future prospects of oAds as an integrative part of multi-stage immunotherapies.

2. Adenovirus Cell Entry, Replication, and Immunogenicity

With more than 55 different serotypes, adenoviruses are ubiquitous pathogens that cause infections of the eyes, the respiratory or gastric tract with rather mild clinical manifestations in immunocompetent individuals [8]. Adenoviruses are non-enveloped, icosahedral viruses approximately 90 nm in size with a linear, non-integrating dsDNA genome ranging from 30–38 kb depending on the serotype. The proteins, which are expressed early during the viral replication cycle, exert regulatory functions including cell cycle induction, prevention of premature apoptosis, interference with pathogen defense, and escape from immune recognition. Typical late proteins are major capsid components such as hexon, penton, and fiber. Viral DNA replication and capsid assembly take place in the nucleus and infected cells undergo a lytic process to release the virus progeny. Since adenoviruses are able to infect a large spectrum of epithelial cells, they have been preferably adopted for gene transfer purposes and as oncolytic agents [9]. The commonly used serotype 5 infects cells by recognizing the coxsackievirus adenovirus receptor. After association with this primary receptor on the surface of a target cell, subsequent recognition of integrins by an RGD-motif, located in the capsid protein penton, initiates the endocytotic uptake of the viral particle.

Adenoviruses are highly immunogenic. Components of the virus capsid, the viral DNA, and specific intermediates expressed during the replication cycle are strong pathogen-associated molecular patterns (PAMPs) that are detected on all levels of cell entry. Their recognition by cellular pattern recognition receptors (PRR) triggers an inflammatory response comprising the release of numerous cytokines and chemokines (for review see [10]). In infected cells, oAds induce immunogenic cells death (ICD), an essential process for triggering adaptive antitumor immune responses and antitumoral memory. oAds kill infected tumor cells with features of necrosis/necroptosis, and autophagy [11–13] accompanied by release of high mobility group box-1 (HMGB-1), calreticulin, extracellular ATP, and heat shock protein 70 (Hsp70) [14–16]. Adenovirus-mediated ICD has been associated with induction of antitumor immune responses [17]. Consistently, it has been demonstrated by depletion of T cells in a Syrian hamster model that therapeutic efficacy of oAds was largely T-cell mediated [18]. In summary, these observations indicate that immunogenic cell death is an important prerequisite for therapeutic efficacy of oAds.

3. Current Concepts of Tumor-Selective Replicating Adenoviruses

The adenoviral E1 proteins drive infected cells into the S-Phase of the cell cycle and prevent premature apoptosis. These proteins have therefore been preferred targets of genetic manipulation to generate tumor-selective replicating adenoviruses. The adenovirus mutant dl-1520 (Onyx-015) lacks E1B-55k, a potent inhibitor of p53. As Onyx-015 is unable to degrade p53, it was originally assumed to productively replicate in p53 mutant tumor cells but not in cells with functional p53 [2]. However, subsequent studies have shown that replication in tumor cells was rather associated with other cellular functions such as late mRNA export [19,20]. A closely related oAd, the E1B-55k-deleted H101, has been approved for the treatment of head and neck cancer in China. Alternatively, p53-dysfunction has been addressed with oAds that exploit the virus-induced upregulation of p53 to activate mechanisms to suppress the expression of E1A and to inhibit the onset of adenovirus replication in normal cells [21,22]. Further mutant adenoviruses contain deletions in E1A, an essential protein for the onset of replication [23,24]. E1A binds to complexes containing E2F and the retinoblastoma protein (Rb) promoting the release of free E2F, which drives the cell to enter the cell cycle. Disruption of the Rb-binding site disables E1A to support adenoviral replication in resting cells without free E2F. Such E1A mutants are the basic modification of several oAds (referred to as dl922-947 or Δ24) and variants thereof are currently under clinical investigation [25]. Deletions in the N-terminus of E1A, responsible for binding to p300, further improve tumor-selective replication of the virus [26]. It is not well understood how these frequently used genetic modifications affect immunogenicity and immunogenic cell death. The sequence that is missing in Δ24-E1A is important for blocking the pathway of cyclic GMP-AMP synthase/stimulator of interferon genes (cGAS/STING) [27]. Proteins of the E3 region, also frequently deleted in oAds, are involved in regulation of adenoviral immune escape and nuclear factor kappa B (NFκB) activity levels [28,29]. It has been pursued to increase intrinsic immunogenicity, e.g., by inserting immunostimulatory CpG islands into the adenoviral backbone to enhance toll-like receptor (TLR)-9 signaling upon intracellular detection of the virus [30]. However, systematic studies on adenovirus immunogenicity in vivo are difficult, since syngeneic mouse models do not support productive replication of human adenovirus and may not reflect immunogenicity in the human system [31,32].

Targeted transcriptional control of E1A by tumor-specific promoters has also been a favored tool to generate oAds. Promoters for established tumor markers such as α-fetoprotein (AFP) or mucin-1 (MUC1) have been used for this purpose [33,34]. To broaden the spectrum of target cancers, promoters have been used that are activated by pan-cancer molecular alterations. Targeting tumor cells with a defective Rb-pathway, the E2F-1 promoter has been employed for control of E1A to facilitate tumor-selective replication [35,36]. As the hTert subunit of the human telomerase is expressed in 90% of human tumors, oAds have been generated containing E1A under control of the hTert-promoter [37,38]. Some of these viruses are subject of current clinical trials (see below). Additionally, artificial promoters have been developed to improve tumor selective E1A expression and virus replication. The oAd ICOVIR-7 has been provided with an insulated E2F promoter harboring additional E2F-responsive sites. The increased E2F-depency reduced systemic toxicity in immunocompetent mice [39]. The original hTert-promoter has been modified to contain further Sp1 and c-myc binding sites or a TATA-Box to increase effective replication in tumor cells [40,41]. Alternative strategies depend on aberrant expression of oncoproteins such as Y-box-binding protein 1 (YB-1) [42] or exploit hypoxic conditions in the tumor core by using a hypoxia-inducible factor 1α (HIF1α)-dependent promoter [43]. Recently, hybrid promoters containing hypoxia response elements (HRE) linked to either the E2F or hTert-promoter have been established to achieve potent viral replication in both hypoxic and normoxic regions of the tumor [44].

A further important aspect of tumor-selectivity is the ability to preferably recognize and infect cancer cells. Tumors have a tendency to downregulate the Ad5 primary receptor, which may impair adenoviral transduction of tumor tissue [45]. The capsid protein fiber, responsible for recognition of the coxsackievirus adenovirus receptor by adenoviruses, has therefore been a frequent target of genetic manipulations to improve tumor infection. One approach has been the introduction of an RGD-motif into the knob domain of fiber [46]. Circumventing the need of binding to the known primary receptors, this allows direct binding to cell surface integrins. More complex alterations comprise the exchange of the knob domain by the equivalent structures from alternative serotypes, such as Ad 3 or 35 [47,48]. These fibers bind to CD46 or desmoglein-2, which are not downregulated on the surface of tumor cells. Additionally, various strategies exist to redirect oncolytic adenoviruses to defined molecular targets on tumor cells (see review in [49]).

4. Translational Efforts and Clinical Development of Oncolytic Adenoviruses

Numerous clinical trials with oAds have been reported and several studies are currently ongoing. Without the claim of being exhaustive, Table 1 gives an overview of currently running clinical trials listed on https://clinicaltrials.gov. In the following section, some examples of oAds that have already entered clinical trials are described in more detail.

In preclinical studies, the telomerase-dependent OBP-301/telomelysin showed growth suppression in a panel of tumor cells and in xenograft models of lung cancer [38]. It was also demonstrated that oAds spread to the lymph nodes yielding an antimetastatic effect [50]. Safety of OBP-301 has been confirmed in phase I studies in various advanced solid tumors [51] and is currently being tested in phase II studies in metastatic melanoma, and in esophagogastric cancers in combination with pembrolizumab. It has been observed in bilateral, syngeneic models of colorectal and pancreatic cancer that a variant of OBP-301 in combination with an antibody targeting programmed cell death protein-1 (PD-1) yielded an abscopal effect on non-treated tumors confirming that oAds are promising agents to immunize tumors for checkpoint inhibitors [52].

DNX2401 (tasadenoturev), a delta24-RGD adenovirus, contains delta24-E1A to facilitate selective replication in Rb-dysfunctional cells. The ability to infect tumor cells is enhanced through the integration of an RGD-motif in the fiber. DNX-2401 treatment was effective in preclinical glioma models and showed immunoactivating properties in syngeneic pancreas tumors in mice [53,54]. Regarding therapeutic efficacy, Lang et al. reported that after a single intratumoral injection of DNX-2401 in glioma, 20% of patients survived more than three years, and also almost complete responses could be observed, resulting in a progression-free survival of more than three years [25]. Investigations on tumor specimens from patients receiving a neoadjuvant treatment suggest that DNX-2401 replicates and spreads within the tumor. Signs of effective immune activation, such as infiltration of CD8 T cells and T-bet+ cells, have been reported. DNX-2401 is now under clinical investigation with pembrolizumab in brain cancers. A further variant expressing OX40L (DNX-2440) has already been generated and subjected to clinical testing.

VCN-01 (Ad-E2F-Δ24RGD-PH20) is also an oAd for use in Rb-dysfunctional tumors. In this virus, Δ24-E1A is transcriptionally controlled by a promoter harboring E2F-1 responsive elements to boost replication in tumor cells through a positive feedback loop. The virus additionally expresses a hyaluronidase for improved virus spreading. VCN-01 has shown tumor selectivity in vitro, antitumoral effects in murine xenograft models, and increased spreading of virus infection [55,56]. VCN-01 effectively killed patient-derived retinoblastoma in vitro. Intravitreous administration in retinoblastoma xenografts led to tumor necrosis, improved ocular survival, and prevented dissemination. Data from a phase I trial showed the feasibility of vitreous administration and antitumor activity in vitreous seeds. Local inflammation of the retina has been observed, but no systemic complications occurred [57]. VCN-01 is currently involved in clinical studies in pancreatic cancer in combination with gemcitabine and nab-paclitaxel and also in head and neck cancer in combination with pembrolizumab.

Table 1. Currently running clinical trials on oncolytic adenoviruses according to https://clinicaltrials.gov.

Agent/Virus Name	Virus Type	Trial No.	Status/Start Date	Indication	Admin.	Phase	Co-Therapy	Arming
Phase I								
BM-hMSCs-DNX-2401	Ad5-delta24-RGD (MSCs as carriers)	NCT03896568	Recr. 02/2019	Recurrent glioma	i.a.	I	Surgery	none
DNX-2401	Ad5-delta24-RGD	NCT03178032	Active, not recr. 05/2017	Brainstem glioma DIPG	local	I	Radiotherapy, chemotherapy	none
DNX-2440	Ad5-delta24-RGD-OX40L	NCT03714334	Recr. 10/2018	Glioblastoma	i.t.	I	-	OX40L
CAdVec	Binary oAd: Onc.Ad + helper-dependent (HD)-Ad	NCT03740256	Active, not yet recr. 09/2020	Diverse HER2 positive solid tumors	i.t.	I	HER2-specific autol. CAR T cells	not disclosed
Enadenotucirev/ Colo-Ad1	Ad3/11 Chimera	NCT03916510	Recr. 07/2019	Locally adv. rectal cancer		I	Chemoradiation	none
NG-641	Ad3/11 Chimera	NCT04053283	Recr. 01/2020	Adv./Metastatic epithelial tumors	i.t., i.v.	I	Chemotherapy, checkpoint inhibitors	FAP/CD3, CXCL9, CXCL10, IFNα
NG-350A	Ad3/11 Chimera	NCT03852511	Recr. 02/2019	Adv./Metastatic epithelial tumors	i.t., i.v.	I	-	Anti-CD40 Ab
VCN-01	Ad-DM-E2F-K-Δ24RGD-PH20	NCT03284268	Recr. 09/2017	Refractory retinoblastoma	intravitreal	I	-	Hyaluronidase
VCN-01	Ad-DM-E2F-K-Δ24RGD-PH20	NCT03799744	Recr. 05/2019	Squamous cell carcinoma of head and neck	i.v.	I	Durvalumab	Hyaluronidase
VCN-01	Ad-DM-E2F-K-Δ24RGD-PH20	NCT02045602	Active, not recr. 01/2014	Adv. solid tumors PDAC	i.v.	I	Gemcitabine, Abraxane	Hyaluronidase
Ad5-yCD/ mutTKSR39rep-hIL12	Ad5-yCD/mutTKSR39rep-hIL12	NCT02555397	Unknown 08/2015	Prostate cancer	intra prostatic	I	-	yCD/mutTk/IL-12
Ad5-yCD/ mutTKSR39rep-hIL12	Ad5-yCD/mutTKSR39rep-hIL12	NCT03281382	Recr. 07/2017	Metastatic PDAC	i.t.	I	5-FC, chemotherapy	yCD/mutTk/IL-12
OBP-301	Ad5-hTert-E1A-IRES-E1B	NCT02293850	Recr. 10/2014	HCC	i.t.	I	-	none
ONCOS-102	Ad5/3-D24-GMCSF	NCT03003676	Active, not recr. 12/2016	Advanced or unresectable melanoma	i.t.	I	Cyclophosphamide, pembrolizumab	GM-CSF
TILT 123	Ad5/3-D24-TNFα-IRES-IL2	NCT04217473	Recr. 02/2020	Metastatic melanoma		I	TIL	TNFα, IL2

Table 1. *Cont.*

Agent/Virus Name	Virus Type	Trial No.	Status/Start Date	Indication	Admin.	Phase	Co-Therapy	Arming
Phase I/II								
ONCOS-102	Ad5/3-D24-GMCSF	NCT02963831	Recr. 09/2017	Colorectal, chemoresistant ovarian, appendiceal cancer	i.p.	I/II	Durvalumab	GM-CSF
ONCOS-102	Ad5/3-D24-GMCSF	NCT03514836	Recr. 05/2018	Castration-resistant advanced metastatic prostate cancer	i.t.	I/II	DCVAC/PCa	GM-CSF
ONCOS-102	Ad5/3-D24-GMCSF	NCT02879669	Active, not recr. 06/2016	Unresectable malignant pleural mesothelioma		I/II	Carboplatin, cyclophosphamide	GM-CSF
LOAd-703	Ad5/35	NCT04123470	Recr. 01/2020	Malignant melanoma	i.t.	I/II	Atezolizumab	CD40L, 4-1BBL
LOAd-703	Ad5/35	NCT03225989	Recr. 03/2018	PDAC/ovarian, biliary, colorectal cancer	i.t.	I/II	Standard chemotherapy or Gemcitabine	CD40L, 4-1BBL
LOAd-703	Ad5/35	NCT02705196	Recr. 11/2016	PDAC	i.t.	I/II	Gemcitabine, Nab-Paclitaxel, atezolizumab	CD40L, 4-1BBL
AdVince	Ad5(PTD)CgA-E1AmiR122	NCT02749331	Recr. 03/2016	Neuroendocrine tumors	i.a.	I/II	-	none
ORCA-010	Ad5-Δ24RGD; T1-mut.	NCT04097002	Recr. 11/2019	Prostate cancer	i.t.	I/II	-	none

Table 1. *Cont.*

Agent/Virus Name	Virus Type	Trial No.	Status/Start Date	Indication	Admin.	Phase	Co-Therapy	Arming
Phase II								
ADV/HSV-tk	Ad5	NCT03004183	Recr. 07/2017	Metastatic NSCLC TNBC	i.t.	II	Valacyclovir, SBRT radiation, pembrolizumab	HSV-tk
DNX-2401	Ad5-delta24-RGD	NCT02798406	Active, not recr. 06/2016	Brain cancer	i.t.	II	Pembrolizumab	none
OBP-301	Ad5-hTert-E1A-IRES-E1B	NCT03190824	Active, not recr. 12/2016	Melanoma stage III|stage IV	i.t.	II	-	none
OBP-301	Ad5-hTert-E1A-IRES-E1B	NCT03921021	Recr. 05/2019	Esophagogastric adenocarcinoma	i.t.	II	Pembrolizumab	none
CG0070	Ad-E2F-E1A-E3-GM-CSF	NCT04387461	Not yet recr. 08/2015	NMIBC	intra vesical	II	Pembrolizumab	GM-CSF
Phase III								
CG0070	Ad-E2F-E1A-E3-GM-CSF	NCT04452591	Not yet recr. 09/2020	NMIBC	intra vesical	III	N-dodecyl-B-D-maltoside	GM-CSF
H101	Ad5	NCT03780049	Recr. 10/2018	Non-resectable HCC	i.a.	III	HAIC 5-FU, leucovorin	none

Abbreviations: i.a., intraarterial; i.v., intravenous; i.t., intratumoral; i.p., intraperitoneal; HCC, hepatocellular carcinoma; NMIBC, non-muscular invasive bladder cancer; NSCLC, non-small cell lung cancer; TNBC, triple negative breast cancer, PDAC, pancreatic adenocarcinoma; DIPG, diffuse intrinsic pontine glioma; GM-CSF, granulocyte-macrophage stimulating factor; 5-FU, 5-fluorouracil; 5-FC, 5-fluorocytosine; HAIC, hepatic artery infusion chemotherapy; SBRT, stereotactic body radiation therapy; hMSC, human mesenchymal stem cells; DCVAC/PCa, autologous dendritic cells pulsed with killed LNCaP prostate cancer cells; TIL, tumor-infiltrating lymphocytes; TNFα, tumor necrosis factor alpha; IL2, interleukin 2; IL-12, interleukin 12; HSV-tk, herpes simplex thymidine kinase; CXCL, chemokine ligand; FAP, fibroblast activation protein; IFNα, interferon alpha; CAR, chimeric antigen receptor.

CG0070 and ONCOS-102 are both selective for Rb-dysfunctional tumor cells and express granulocyte-macrophage colony-stimulating factor (GM-CSF). CG0070, an oAd5 that uses an E2F-responsive promoter for control of E1A, has been developed for application in non-muscle invasive bladder cancer (NMIBC) [58]. A phase II trial has shown an overall complete response rate of 47% at 6 months in patients with Bacillus Calmette–Guerin (BCG)-unresponsive NMIBC with acceptable toxicity [59]. ONCOS-102 is an Ad5/3 fiber chimeric oAd with favorable toxicity data in a phase I study [47]. Clinical studies of ONCOS-102 combined with chemotherapy in mesothelioma, with pembrolizumab in advanced melanoma, and together with a dendritic cell (DC)-vaccine for treatment of prostate cancer are ongoing. LoAd703 is an oAd5 containing chimeric fibers with an Ad35 knob and is additionally armed with the costimulatory factors CD40L and 4-1BBL. Intratumoral injection of LoAd703 inhibited tumor growth in a syngeneic pancreatic tumor model in mice, which could be further enhanced with gemcitabine [48]. LoAd infection promoted lymphocyte migration and stimulated DCs resulting in the activation of natural killer (NK) cells and the triggering of tumor-directed T cell responses. Currently, the safety and viroimmunotherapeutic activity of LoAd703 in combination with atezolizumab are being clinically investigated in several cancer entities including pancreatic cancer.

Enadenotucirev (or EnAd, formerly Colo-Ad1) has been generated by in vitro chimerization using adenovirus serotypes 3 and 11, and subsequent coevolution by serial passaging in colon cancer cells [60]. Safety of intravenous injections has been demonstrated in a phase I study [61]. An additional phase I study is being carried out in colorectal cancer patients prior to surgical removal and in combination with chemoradiotherapy. A further advancement is the EnAd-variant NG-641, expressing a fibroblast activation protein (FAP)-CD3 bispecific T-cell engager (BiTE), the chemokines chemokine ligand 9 (CXCL9) and CXCL10, and interferon alpha (IFNα), in an approach to attract and stimulate T cells to attack both tumor cells and cancer-associated fibroblasts (CAFs). This variant and also the variant NG-350A, which expresses a costimulatory, agonistic CD40 antibody, are in clinical testing.

5. Armed Oncolytic Adenoviruses

5.1. Arming with Transgenes to Amplify Tumor Lysis

Once the limitations of using first-generation oncolytic viruses as monotherapy became apparent, transgenes were introduced into existing oAd platforms with the intention to amplify tumor lysis and viral spread. When expressed in cells, the herpes simplex virus thymidine kinase (HSV-tk) gene converts a non-toxic prodrug ganciclovir (GCV) into a toxic agent, which is also distributed to neighboring cells to cause bystander cytotoxicity. Application of a E1B-55k-deleted oAd expressing HSV-tk and GCV improved survival of human colon carcinoma xenografts in mice [62]. Furthermore, oAds expressing HSV-tk and cytosine deaminase have been generated for the treatment of prostate cancer [63]. These oAds have also been used to deliver further cytotoxic or immunostimulatory payloads such as adenovirus death protein (ADP) or interleukin 12 (IL-12), respectively [64,65]. Ad5-yCD/mutTKSR39rep-mIL12, which expresses murine IL-12, improved local and metastatic tumor control in a preclinical prostate adenocarcinoma model accompanied by only mild local inflammation. The corresponding oAd expressing human IL-12 is being investigated in clinical trials.

The tumor necrosis factor- (TNF)-related apoptosis-inducing ligand (TRAIL) has been used in oAds with serotype 5/35 chimeric fibers. In vitro, Ad5/35.IR-E1A/TRAIL showed efficient virus spread and induction of apoptosis. Systemic administration eliminated preestablished liver metastasis in mice [66]. Fernández-Ulibarri et al. developed an oAd expressing a soluble RNase onconase fused to a tumor ligand (ONC$_{EGFR}$). Upon internalization, the molecule induces tumor cell death through RNA degradation [67].

A critical aspect of arming with cytotoxic transgenes is the non-virus mediated cell killing, which may affect the productivity of viral infection [68]. It is also unclear how non-viral cell death affects the induction of tumor-directed immune responses.

Some monoclonal antibodies (mAb) against growth factor receptors are approved anticancer drugs. When systemically applied, these immunotherapies cannot exploit their full potential because of poor tumor penetration and side effects through normal tissue exposure. Taking advantage of a clinically established antibody against human epidermal growth factor receptor 2 (HER2), a full-length trastuzumab-expressing oAd has been constructed [69]. Ad5/3-Δ24-tras showed improved cytotoxicity in a panel of HER2 + cell lines and enhanced antitumor efficacy in a xenograft model of gastric cancer. Viral oncolysis by Ad5/3-Δ24-tras activated CD11c + DCs in lymph nodes in a NK cell-dependent manner. The Fc-terminus of the antibody also labels target cells for recognition by innate immune cells, which may induce antibody-dependent cell-mediated cytotoxicity (ADCC). This approach therefore combines direct antitumor activity and the engagement of additional immune effector mechanisms.

5.2. Arming with Matrix-Modifying Genes to Enhance Intratumoral Virus Spreading

Tumor cells are embedded in a dense network of extracellular matrix (ECM) and infection-resistant stroma cells, which impair effective distribution of the virus. To address this issue, oAds have been provided with matrix modifying genes such as TIMP2, TIMP3, MMP8, and relaxin. Expression of these matrix modifiers enhanced intratumoral viral spread and effectively inhibited tumor growth in cancer xenograft models in mice [70–73].

VCN-01, a clinically investigated oAd (see above), is armed with a soluble human sperm hyaluronidase (PH-20), which effectively degrades hyaluronan. Degradation of the ECM by PH-20 results in enhanced virus spreading in xenografted tumors [56]. Mutants of the proteoglycan decorin have been used to improve viral distribution and tumor penetration by oAds [74]. In the future, it will be interesting to see how degradation of the ECM can promote leukocyte infiltration and immune activation of the tumor microenvironment. Recently, it has been shown with a relaxin-expressing oAd that the ECM degradation enhanced tumor penetration by a systemically administered therapeutic antibody. When oAds additionally expressed IL-12 and granulocyte-macrophage colony-stimulating factor (GM-CSF), tumors were effectively converted into an immunoactivated state responsive to PD-1 checkpoint inhibition [75].

5.3. Arming with Antiangiogenic Transgenes

Angiogenesis is an important target of immunotherapies in clinical oncology. oAds have been armed with antiangiogenic mechanisms to enhance the antitumor effect of oncolysis. In human hepatocellular carcinoma (HCC) cells and in xenografts in mice, Li et al. showed anti-angiogenesis and antitumoral effects when endostatin was expressed by the E1B-55k deleted oAd CNHK200-mE [76]. Xiao and colleagues generated ZD55-VEGI-251, also an E1B55k-deleted oAd, armed with a secreted isoform of vascular endothelial cell growth inhibitor [77]. VEGI-251 inhibited angiogenesis in chick chorioallantoic membranes and suppressed tumor growth in xenograft models. Decorin, which is able to suppress multiple tyrosine kinase receptors including c-Met and the Wnt/β-catenin pathway, has also been employed. In a nude mice model of human prostate cancer, the decorin-expressing Ad.dcn reduced tumor burden, significantly inhibited skeletal metastases and improved survival [78]. The group of Chae-Ok Yun suppressed vascular endothelial growth factor (VEGF) by expressing VEGF-specific short-hairpin RNA (shRNA) or by expression of an artificial zinc-finger protein (F435-KOX) targeting the VEGF promoter [79,80].

5.4. Arming with Immunostimulatory Cytokines and Chemokines

Corresponding to the diversity of immune mechanisms that can be dysfunctional in tumors, various immunostimulatory transgenes have been integrated into oAds to stimulate effective antitumor immune responses. Since systemic administration of potent immunostimulatory factors, such as type I Interferons, tumor necrosis factor alpha (TNFα), or interleukin 12 (IL-12), may have considerable side effects, delivery by oAds provides an attractive option to focus cytokine activity on the target tumor.

In tumor cells, IFNs exert pleiotropic effects including the activation of the immune proteasome, the upregulation of major histocompatibility complex (MHC) class I and II, and potent activation of NK cells and cytotoxic T lymphocytes (CTLs). Shashkova et al. integrated IFNα into an oAd (KD3-IFN), which should render replication more sensitive to the IFNα response in normal cells [81]. The authors were able to confirm a decreased off-target toxicity in HCC xenografts in nude mice and in an immunocompetent model of kidney cancer in Syrian hamsters. A cyclooxygenase (Cox) 2-dependent oAd expressing IFNα was capable of inhibiting tumor growth in a Syrian hamster model of pancreatic cancer [82]. In an immunocompetent mouse model of Lewis lung carcinoma, co-application of an oAd in combination with a non-replicating Ad-IFNβ has been investigated [83]. This binary strategy prolonged interferon expression and improved antitumoral immune responses. Efficient delivery of a non-replicating transgenic adenovirus by coinfection with an oAd has been initially shown in an approach of cancer gene therapy [84]. Regarding armed virotherapeutic vectors, this binary approach is particularly promising for immunostimulatory transgenes. Assuming that enough events with single virus transduction will occur, the binary method holds promise to maintain cytokine expression beyond clearance of the oAd.

The potent antitumor functions of TNFα have been well known for decades. Loco-regional delivery of TNFα by oAds promises potent antitumor activity with limited side effects. Hirvinen et al. showed that the TNFα-armed Ad5/3-E2F-delta24 vector led to increased tumor destruction due to TNFα-mediated apoptosis, immunogenic cell death, and induction of antitumor immune responses, including tumor-antigen-specific T cells [85]. A corresponding virus with additional expression of interleukin 2 (IL-2) (Ad5/3-E2F-D24-hTNFα-IRES-hIL2 or TILT-123) is currently under clinical investigations. IL-2 is a central cytokine for survival and proliferation of T cells qualifying TILT-123 to augment the transfer of tumor infiltrating lymphocytes (TIL). In an immunocompetent Syrian hamster tumor model, concomitant transfer of TILs and virus application resulted in a 100% cure of treated animals [86]. The virus has also been used to support tumor infiltration with chimeric antigen receptor (CAR) transgenic T cells [87]. By using an ex vivo ovarian cancer (OVCA) model derived from patient samples, enhanced levels of proinflammatory signals (IFNγ, CXCL10, TNFα and IL-2) associated with a concomitant activation of CD4 and CD8 TILs could be observed when tumor cells were infected with TILT-123 [88]. In response to autologous, T cell-depleted OVCA cultures, which had been infected with TILT-123, TILs secreted high levels of IFNγ. These observations confirmed the use of TILT-123 in adoptive cell transfer.

Several oAds have been armed with IL-12, an essential cytokine involved in inflammation and proliferation of effector T cells and NK cells. Using the hypoxia-dependent Ad-DhscIL12 in a Syrian hamster model of pancreatic cancer, Bortolanza et al. showed active viral replication and enhanced transgene expression in vivo resulting in potent antitumor effects and less toxicity due to shorter systemic exposure [89]. Lee et al. investigated the oAd YKL-IL12/B7 expressing IL-12 and B/7-1 (CD80), a ligand of the costimulatory CD28 receptor, on T cells. In a syngeneic murine B16-F10 melanoma, the virus showed effective tumor growth inhibition including complete tumor regressions and improved survival [90]. Using the oncolytic Ad-TD-nsIL12, which expresses a non-secreted version of IL-12, Wang et al. were able to reduce off-target toxicity of IL-12 [91].

IL-24 is an immunomodulatory cytokine with profound antitumor effects through immune activation, induction of tumor cell apoptosis and inhibition of angiogenesis. IL-24-expressing oAds have shown antitumor efficacy in vitro and in xenografts in mice [92,93]. IL-4 has been used to promote intratumoral leukocyte infiltration [94]. The cytokine IL-18 induces IFNγ production through T cells and NK cells. Using the IL-18-armed ZD55 in xenograft models, Zheng et al. could observe stronger antitumor responses and inhibition of tumor angiogenesis [95]. Choi et al. generated oAd expressing IL-12 in combination with IL-18, or IL-23, respectively, and demonstrated enhanced antitumor efficacy in B16-F10 melanoma associated with an improved Th1/Th2 cytokine ratio and infiltration of NK and T cells [96,97]. Cytokines have also been combined with the chemokine CCL21, which binds to CCR7 on naïve T cells and DCs and promotes their attraction to the tumor [98,99].

Alternative options for immune arming are factors that directly target immunosuppression in the tumor microenvironment. Seth et al. have targeted transforming growth factor β (TGFβ) with a soluble TGFβ-receptor II protein fused to a human immunoglobulin (IgG) Fc fragment [100]. By using an oAd expressing sTGFβRII-Fc (rAd.sT), the authors showed in a xenograft mouse model tumor regression in 85% of treated animals. rAd.sT enhanced the efficacy of concomitant anti-PD-1 and anti-cytotoxic T-lymphocyte-associated protein 4 (CTLA-4) treatment in an immunocompetent 4T1 breast cancer model [101].

5.5. Immunological Arming to Improve Antigen Presentation

To enable successful tumor-directed T-cell immunity, effective presentation of tumor antigen by DCs needs to be restored. The chemokine GM-CSF promotes maturation and activation of antigen presenting DCs from myeloid precursors. oAds armed with GM-CSF have been used to elicit T-cell mediated antitumoral responses [58,102]. CG0070 is a GM-CSF-armed oncolytic Ad5 involved in clinical investigation as described above. Using Ad5-Δ24-GMCSF, Cerullo et al. showed tumor-specific immunity in an immunocompetent syngeneic hamster model [102]. In 20 patients with advanced solid tumors, responses could be observed including two complete tumor responses. The administration of Ad5/3-Δ24-GMCSF has been investigated in tumor patients showing a clinical benefit according to RECIST criteria in 8/12 radiologically evaluated individuals. The data revealed that oAd treatment affected immune responses specific for the tumor antigen survivin [103]. A correlation of antitumoral and antiviral immune responses has been confirmed by Kanerva et al. [104]. An oAd expressing both GM-CSF and IL-12 has been used to support the administration of a DC vaccine. Tumor infection with Ad-ΔB7/IL12/GMCSF promoted migration of DCs to tumor-draining lymph nodes [105]. However, GM-CSF also has protumorigenic and immunosuppressive functions by recruiting myeloid suppressor cells and impairing immune responses [106,107]. In pancreatic cancer, tumor-released GM-CSF supports the development of an immunosuppressive subset of DCs, which promotes metastasis [108]. Alternative options to improve intratumoral antigen presentation by oAds include the co-expression of Fms-like tyrosine kinase-3 ligand (Flt3L) and GM-CSF [109], or a combination of Flt3L with macrophage inflammatory protein 1α (MIP-1α, CCL3) [110].

5.6. Arming with Transgenes Addressing T Cell Costimulation or Immune Checkpoints

Pharmacological blockade of inhibitory immune checkpoints or activation of costimulatory receptors are potent strategies to activate antitumor T cells. Dias et al. used a full length anti-CTLA4 monoclonal antibody expressed by Ad5/3-Δ24-CTLA to combine oncolysis and checkpoint inhibition [111]. Intratumoral expression allowed high local levels of the checkpoint inhibitor. In patient-derived peripheral blood mononuclear cells (PBMCs), the authors observed T cell activation and αCTLA4-mediated apoptosis. The PD-1/PD-L1 is an inhibitory checkpoint regulating the activity of peripheral T cells. In prostate cancer models, Tanoue et al. showed that an oAd combined with a helper-Ad expressing a PD-L1-blocking mini-antibody supported the intratumoral activity of adoptively transferred CAR T cells [112]. The specific benefit of viral delivery was confirmed by the demonstration that local expression of the PD-L1-blocking minibody was superior compared with systemic infusion of αPD-L1 IgG. An equivalent approach using helper-dependent adenoviruses for expression of the PD-L1 blocking mini-antibody and IL-12-p70 for immune stimulation augmented the activity and persistence of CAR T cells in murine models of head and neck squamous cell carcinoma (HNSCC) [113].

CD40 is a costimulatory receptor expressed on antigen-presenting cells (APCs), mostly B cells, macrophages and DCs. Interaction of CD40 with its ligand CD40L induces cytokine production, increases MHC class II-dependent antigen presentation and thus supports the priming and expansion of T cells. Tumor infection with the CD40L-armed AdEHCD40L reduced the growth of xenografted human myeloma [114]. The oncolytic Ad5/3-hTERT-E1A-hCD40L (CGTG-401) induced multiple antitumor effects including reduced tumor growth via apoptosis, increased number of cytotoxic CD8 T cells in the tumor, and upregulation of T_H1 associated cytokines [115]. Administration of CGTG-401 in nine patients with advanced solid tumors demonstrated that the treatment was well tolerated, and immunological responses could be confirmed [116]. The Hemminki group also recently showed that a CD40L-expressing oAd enabled effective antitumoral DC-therapies in humanized mice [117].

APCs express the co-stimulatory molecule 4-1BB ligand (4-1BBL), and 4-1BB antibodies are known to stimulate potent antitumor immune responses. The oncolytic adenovirus LoAd703, armed with 4-1BBL together with a trimerized CD40L, is currently being tested in clinical trials as described above [48]. A further approach studying the combined expression of 4-1BBL and IL-12 (Ad-ΔB7-IL12/4-1BBL) demonstrated a synergistic enhancement of IFNγ levels compared to single cytokine viruses and supported the administration of DCs through an enhanced T_H1-mediated antitumor immune response [118].

The costimulatory OX40 ligand (OX40L) binds to OX40 on T cells and promotes T cell activation. Application of the OX40L-expressing oAd Delta-24-RGDOX showed intratumoral activation of lymphocytes and the development of a tumor-specific CD8 T-cell immune memory in syngeneic mouse models of glioma [119]. Delta24-RGDOX is currently being tested in clinical trials (see above).

Another co-stimulatory receptor is glucocorticoid-induced TNFR family-related gene (GITR). Stimulation with GITR-ligand (GITRL) leads to activation and proliferation of antigen-primed CD4 and CD8 T cells. Glioma treatment with the GITRL-armed oAd Delta24-GREAT resulted in expansion and activation of T cells with a high frequency of central memory CD8 T cells [120].

5.7. Arming with T Cell Engager Proteins

A promising strategy to redirect T cells to cancer cells are bispecific T-cell engagers (BiTEs) [121]. BiTEs are composed of a tumor ligand and a single-chain antibody fragment, which facilitates binding to CD3. BiTE-mediated clustering of tumor cells and T cells leads to T-cell activation and antitumor cytotoxicity thereby circumventing T-cell receptor (TCR)-mediated antigen recognition. However, the side effects of BiTEs can be considerable and therapeutic success in solid tumors has been rather limited. OVs armed with BiTEs can warrant high BiTE levels in tumor tissue and thus optimizes the ratio of on-target/off-tumor toxicity. Furthermore, viral tumor inflammation promotes intratumoral T-cell infiltration and thus provides an appropriate T-cell pool for BiTE-mediated T-cell retargeting. Fajardo et al. engineered an oAd to express an EGFR-BiTE (ICOVIR-15K-cBiTE) [122]. After expression in cancer cells, the BiTEs activated T cells in PBMCs in vitro. In murine xenograft models, ICOVIR-15K-cBiTE supported tumor infiltration and persistence of adoptively transferred T cells. ICOVIR-15K-cBiTE was also employed to overcome the limits of a CAR T-cell therapy because the EGFR-BiTE was able to redirect T cells against tumor cells that had lost the recognition antigen of the CAR [123]. Freedman et al. constructed a variant of enadenotucirev (EnAd) expressing a BiTE which targets epithelial cell adhesion molecule (EpCAM) [124]. The authors showed that crosslinking of EpCAM-expressing target cells and PBMC-derived T cells activated both CD8 and CD4 T cells. Furthermore, T cells in ascites fluid from cancer patients were activated by the virus-encoded BiTE and EpCAM-positive tumor cells were successfully depleted.

The use of BiTEs targeting components of the tumor microenvironment is a promising approach to reverse tumor immunosuppression. Freedman et al. modified EnAd to express a fibroblast activation protein (FAP)-targeting BiTE to redirect T cells to cancer-associated fibroblasts (CAFs) [125]. Treatment of biopsies of ascites or solid prostate cancer tissue samples with FAP-BiTE- expressing variant of EnAd was capable of activating tumor-infiltrating $PD1^+$ T cells to kill CAFs. This in turn interfered with CAF-associated immunosuppression and resulted in an upregulation of proinflammatory cytokines, increased the presentation of tumor antigen, and finally led to improved T cell function. A comparable strategy has been pursued by generating the oAd ICO15K-FBiTE [126]. In a xenograft model, the expression of FBiTE led to an increased intratumoral T-cell accumulation and decreased the intratumor levels of FAP.

Tumor-associated macrophages (TAMs), and particularly the M2-polarized subset, contribute to immunosuppression. To deplete cancer-promoting TAMs and to reverse immunosuppression, Scott et al. recently developed EnAd-variants equipped with bi- and trivalent T-cell engagers targeting CD206 or folate receptor β on M2-like macrophages [127]. By detecting selective T-cell cytotoxicity against M2-TAMs in cancer patient biopsies, they could demonstrate that these BiTEs allow selective depletion of tumor-promoting TAMs whilst sparing those with potential antitumor features.

6. Oncolytic Adenoviruses in Multi-Stage Immunotherapies

During carcinogenesis, tumors use a wide variety of mechanisms to escape immunosurveillance, which may explain the heterogenous responses to checkpoint inhibitors. To make more tumors sensitive for this therapy, multi-stage immunotherapies are required in the future that address T-cell paucity and immunosuppression in the tumor from several sides. These multi-stage immunotherapies may include a virotherapy part for initial immunoactivation, an external support with tumor-directed T cells, and a systemic checkpoint intervention to maintain T cell activity. First steps towards multi-stage therapies are the current investigations on the synergy of OVs with checkpoint inhibitors. T-Vec and pembrolizumab have yielded encouraging interim results and data on long-term survival are eagerly awaited [7]. Corresponding studies with oAds are ongoing. Results from experimental models support this perspective. It has been shown after infection of B16 melanoma with Newcastle disease virus (NDV) that localized virotherapy and systemic CTLA-4 blockade led to rejection of infected and distant/non-infected tumors [128]. Using an oncolytic adenovirus, we demonstrated that intratumoral application sensitized CMT64 tumors for a systemic PD-1 antibody resulting in epitope spreading of neoantigen-specific T cells [129].

As an initiation step in multi-stage immunotherapies, virotherapeutic vectors must provide a solid basis for follow-up interventions. An important aim is to increase the immunogenicity of the used oAds e.g., by including additional danger signals such as CpG motifs [30]. Furthermore, oAds need to stimulate immune cells that augment direct cytotoxicity of oncolytic viruses or which support the shaping of optimal antitumor immune responses. It has been shown that contact dependent stimulation of NK cells can augment the therapeutic potential of oncolytic adenoviruses [130]. Based on the paradigm that effective CD8 T cell responses require the help of CD4 T cells, the Cerullo group has recently reengaged a pathogen-related CD4 T cell response to support an antitumor vaccination using peptide-loaded oAds (PeptiCRAds). In mice that had been preimmunized with tetanus, CD8 dependent immune responses, elicited with the oncolytic vaccine, were more effective when the used oAds were additionally loaded with a CD4-restricted tetanus peptide [131].

Based on early experiences, repetitive dosage of oncolytic viruses has been regarded as mandatory to achieve a sufficient extent of tumor lysis. However, such a procedure may not necessarily yield the most effective anti-tumor immune responses. Robust anti-adenovirus responses may interfere with the activity as an in-situ vaccine. In the STEP-trial, an adenovirus-based vaccine against human immunodeficiency virus (HIV) was not capable of preventing HIV-infection. Instead, vaccine-treated men showed an even increased infection risk compared to control patients [132]. Studies with adenoviral vaccines in mice have confirmed that strong adenovirus epitopes may cause unresponsiveness to the vaccine [133]. Considering the prime/boost characteristics of repetitive oAd application, strong virus-derived antigens may outcompete the supposedly weaker tumor-associated antigens. Heterologous use of OVs is a promising approach to prevent the dominance of virus-specific immune responses. Application of an oAd followed by an oncolytic vaccinia virus eradicated established tumors in Syrian hamsters predominantly via strong tumor-specific T-cell immune response [134]. Interestingly, Tysome et al. found that this specific sequence was superior compared with the reverse combination, suggesting that viruses are differentially qualified for prime or boost, respectively. As an example for such a coordinated virus choice, it has been demonstrated in a murine B16 model that reovirus for triggering a CD8 Th1-dominated immune response can be combined with a subsequent CD4 Th17 helper response by vesicular stomatitis virus to achieve a potent T-cell pool for PD-1 inhibition [135].

Adenoviruses induce strong CD8 effector memory responses with a rather moderate potential for further expansion [136,137] and are therefore probably better suited to amplify an immune response initiated by alternative OVs. Heterologous administration also provides an option to select vectors with immunostimulatory arming adapted to specific needs of tumor immune activation. Whereas initial OV applications need to optimize antigen presentation and T-cell priming, subsequent applications need to promote T-cell migration and tumor infiltration (Figure 1).

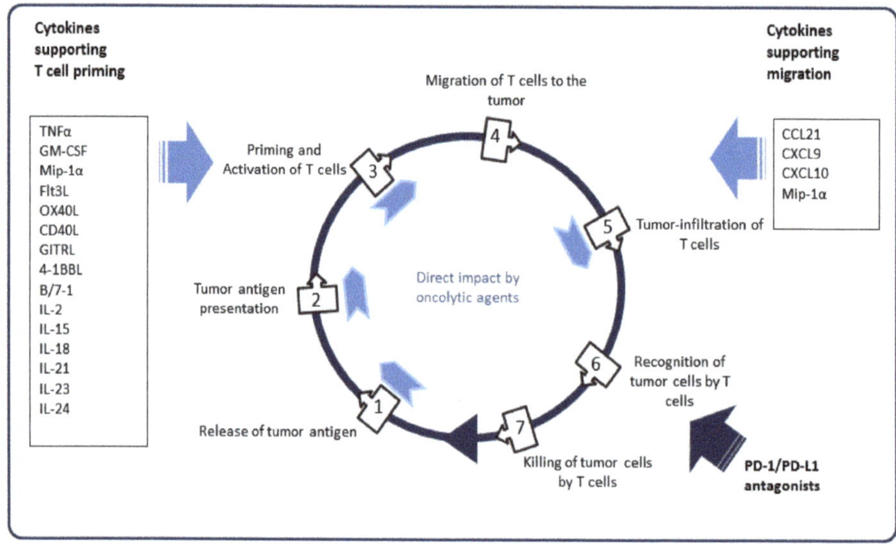

Figure 1. Immunostimulatory transgenes currently used in oncolytic adenoviruses in the context of the cancer immunity cycle.

OAds cause strong humoral immune responses. These neutralizing antibodies have been mostly regarded as an undesired adverse event in virotherapy that severely reduces virus efficacy and applicability. However, it has been demonstrated that fully neutralized OVs can still exert antitumoral effects through delivery in monocytes [138]. Moreover, it has been recently shown that preexisting immune responses can even improve the immune effect of oncolytic viruses [139], suggesting that neutralizing antibodies represent a so far unharnessed immune potential. We have recently described a strategy using bispecific adapter molecules to retarget adenovirus-neutralizing antibodies against tumor cells. This approach led to NK-cell dependent triggering of antitumor CD8 T cells and thus converted a limiting factor of virotherapy into an immunotherapeutic tool [140]. Tumor retargeting of antibodies could be a further option to fully exploit virotherapy-mediated tumor immune activation in multi-stage immunotherapies.

Regarding future multi-stage immunotherapies, oAds are promising tools for immunoactivation of solid tumors to facilitate adoptive cell therapy including CAR T cells. Tähtinen et al. have shown that tumor infection by oAds attracted leukocytes to the tumor, which promoted the intratumoral activity of adoptively transferred OT-1 T cells [141]. Watanabe and colleagues have demonstrated that the oAd TILT-123 improved the outcome of mesothelin-directed CAR-T cell therapy in models of pancreatic adenocarcinoma (PDAC) [87]. Intratumoral virotherapy with TILT-123 supported tumor infiltration with T cells expressing a mesothelin-targeted CAR and significant tumor regression could be confirmed. The authors have shown increased M1-polarization of macrophages and dendritic cell maturation, indicating that this combined therapy is able to overcome the highly immunosuppressive tumor microenvironment of this tumor entity.

When oAds are used for expression of immunostimulatory transgenes, immune modulation ends with the termination of oAd infection. Porter et al. developed an interesting strategy to uncouple the cytokine activity from the limitations of oAd infection by using helper-dependent adenoviruses (hdAd) [142]. The oAd, added to hdAds at a low ratio, provides the factors in trans that are required for efficient virus replication and spreading of helper-dependent vectors. This in turn warrants a prolonged expression of immunomodulators beyond the elimination of the oAd through the host's immune response. A further advantage is the huge capacity of the helper-dependent viruses and the option to incorporate various effector genes to realize a multi-stage immunotherapy. The authors have generated a hdAd expressing an antitumor BiTE (against CD44v6), IL-12, and an anti-PD-1L minibody. This immunostimulatory array allowed additionally transferred CAR-T cells to control tumor growth in xenograft models including an orthotopic model of HNSCC. In summary, combinations of oncolytic Ads, hdAds, checkpoint inhibition, and autologous CAR T cells are a strategy with significant regulatory and technical challenges but with unparalleled clinical potential for cancer immunotherapy.

7. Conclusions

Oncolytic viruses have shown a tolerable safety profile in cancer immunotherapy. Current oncolytic viruses have demonstrated therapeutic efficacy but also limitations when applied as a monotherapy. Nevertheless, oncolytic viruses have an outstanding potential to immunoactivate tumors that are unresponsive to systemic immunotherapies. To convert these tumors into an immunoactivated state that is more likely to respond to systemic checkpoint inhibitor application, oAds are well suited. oAds are established and highly versatile platforms for the local delivery of immunoactivating factors to modulate intratumoral immune cell contexture and to break immune suppression. oAd-based strategies that address tumor-specific immune dysfunction by employing variable immune modifiers or by delivering complex arrays of immune stimulatory factors show great promise as an essential part of multi-stage immunotherapies.

Funding: This research was funded by the Deutsche Forschungsgemeinschaft (DFG) SFB/TRR209-314905040 "Liver cancer", and Deutsche Krebshilfe e.V. (project No. 70113873).

Conflicts of Interest: The authors declare no conflict of interest.

References

1. Russell, S.J.; Peng, K.W.; Bell, J.C. Oncolytic virotherapy. *Nat. Biotechnol.* **2012**, *30*, 658–670. [CrossRef] [PubMed]
2. Bischoff, J.R.; Kirn, D.H.; Williams, A.; Heise, C.; Horn, S.; Muna, M.; Ng, L.; Nye, J.A.; Sampson-Johannes, A.; Fattaey, A.; et al. An Adenovirus Mutant That Replicates Selectively in p53-Deficient Human Tumor Cells. *Science* **1996**, *274*, 373–376. [CrossRef] [PubMed]
3. Liu, T.C.; Galanis, E.; Kirn, D. Clinical trial results with oncolytic virotherapy: A century of promise, a decade of progress. *Nat. Clin. Pract. Oncol.* **2007**, *4*, 101–117. [CrossRef] [PubMed]
4. Andtbacka, R.H.I.; Kaufman, H.L.; Collichio, F.; Amatruda, T.; Senzer, N.; Chesney, J.; Delman, K.A.; Spitler, L.E.; Puzanov, I.; Agarwala, S.S.; et al. Talimogene Laherparepvec Improves Durable Response Rate in Patients with Advanced Melanoma. *J. Clin. Oncol.* **2015**, *33*, 2780–2788. [CrossRef] [PubMed]
5. Prestwich, R.J.; Harrington, K.J.; Pandha, H.S.; Vile, R.G.; Melcher, A.A.; Errington, F. Oncolytic viruses: A novel form of immunotherapy. *Expert Rev. Anticancer Ther.* **2008**, *8*, 1581–1588. [CrossRef] [PubMed]
6. Topalian, S.L.; Hodi, F.S.; Brahmer, J.R.; Gettinger, S.N.; Smith, D.C.; McDermott, D.F.; Powderly, J.D.; Carvajal, R.D.; Sosman, J.A.; Atkins, M.B.; et al. Safety, activity, and immune correlates of anti-PD-1 antibody in cancer. *N. Engl. J. Med.* **2012**, *366*, 2443–2454. [CrossRef] [PubMed]
7. Ribas, A.; Dummer, R.; Puzanov, I.; VanderWalde, A.; Andtbacka, R.H.I.; Michielin, O.; Olszanski, A.J.; Malvehy, J.; Cebon, J.; Fernandez, E.; et al. Oncolytic Virotherapy Promotes Intratumoral T Cell Infiltration and Improves Anti-PD-1 Immunotherapy. *Cell* **2017**, *170*, 1109–1119. [CrossRef]
8. Mennechet, F.J.D.; Paris, O.; Ouoba, A.R.; Salazar Arenas, S.; Sirima, S.B.; Takoudjou Dzomo, G.R.; Diarra, A.; Traore, I.T.; Kania, D.; Eichholz, K.; et al. A review of 65 years of human adenovirus seroprevalence. *Expert Rev. Vaccines* **2019**, *18*, 597–613. [CrossRef]

9. Wold, W.; Toth, K. Adenovirus Vectors for Gene Therapy, Vaccination and Cancer Gene Therapy. *Curr. Gene Ther.* **2014**, *13*, 421–433. [CrossRef]

10. Shaw, A.R.; Suzuki, M. Immunology of Adenoviral Vectors in Cancer Therapy. *Mol. Ther. Methods Clin. Dev.* **2019**, *15*, 418–429. [CrossRef]

11. Abou El Hassan, M.A.I.; van der Meulen-Muileman, I.; Abbas, S.; Kruyt, F.A.E. Conditionally Replicating Adenoviruses Kill Tumor Cells via a Basic Apoptotic Machinery-Independent Mechanism That Resembles Necrosis-Like Programmed Cell Death. *J. Virol.* **2004**, *78*, 12243–12251. [CrossRef] [PubMed]

12. Ito, H.; Aoki, H.; Kühnel, F.; Kondo, Y.; Kubicka, S.; Wirth, T.; Iwado, E.; Iwamaru, A.; Fujiwara, K.; Hess, K.R.; et al. Autophagic cell death of malignant glioma cells induced by a conditionally replicating adenovirus. *J. Natl. Cancer Inst.* **2006**, *98*, 625–636. [CrossRef] [PubMed]

13. Baird, S.K.; Aerts, J.L.; Eddaoudi, A.; Lockley, M.; Lemoine, N.R.; McNeish, I.A. Oncolytic adenoviral mutants induce a novel mode of programmed cell death in ovarian cancer. *Oncogene* **2008**, *27*, 3081–3090. [CrossRef] [PubMed]

14. Di Somma, S.; Iannuzzi, C.A.; Passaro, C.; Forte, I.M.; Iannone, R.; Gigantino, V.; Indovina, P.; Botti, G.; Giordano, A.; Formisano, P.; et al. The Oncolytic Virus dl922-947 Triggers Immunogenic Cell Death in Mesothelioma and Reduces Xenograft Growth. *Front. Oncol.* **2019**, *9*, 564. [CrossRef] [PubMed]

15. Liikanen, I.; Ahtiainen, L.; Hirvinen, M.L.; Bramante, S.; Cerullo, V.; Nokisalmi, P.; Hemminki, O.; Diaconu, I.; Pesonen, S.; Koski, A.; et al. Oncolytic Adenovirus With Temozolomide Induces Autophagy and Antitumor Immune Responses in Cancer Patients. *Mol. Ther.* **2013**, *21*, 1212–1223. [CrossRef] [PubMed]

16. Ma, J.; Ramachandran, M.; Jin, C.; Quijano-Rubio, C.; Martikainen, M.; Yu, D.; Essand, M. Characterization of virus-mediated immunogenic cancer cell death and the consequences for oncolytic virus-based immunotherapy of cancer. *Cell Death Dis.* **2020**, *11*, 48. [CrossRef]

17. Boozari, B.; Mundt, B.; Woller, N.; Struver, N.; Gurlevik, E.; Schache, P.; Kloos, A.; Knocke, S.; Manns, M.P.; Wirth, T.C.; et al. Antitumoural immunity by virus-mediated immunogenic apoptosis inhibits metastatic growth of hepatocellular carcinoma. *Gut* **2010**, *59*, 1416–1426. [CrossRef]

18. Li, X.; Wang, P.; Li, H.; Du, X.; Liu, M.; Huang, Q.; Wang, Y.; Wang, S. The Efficacy of Oncolytic Adenovirus Is Mediated by T-cell Responses against Virus and Tumor in Syrian Hamster Model. *Clin. Cancer Res.* **2017**, *23*, 239–249. [CrossRef]

19. Koch, P.; Gatfield, J.; Löber, C.; Hobom, U.; Lenz-Stöppler, C.; Roth, J.; Dobbelstein, M. Efficient replication of adenovirus despite the overexpression of active and nondegradable p53. *Cancer Res.* **2001**, *61*, 5941–5947.

20. O'Shea, C.C.; Johnson, L.; Bagus, B.; Choi, S.; Nicholas, C.; Shen, A.; Boyle, L.; Pandey, K.; Soria, C.; Kunich, J.; et al. Late viral RNA export, rather than p53 inactivation, determines ONYX-015 tumor selectivity. *Cancer Cell* **2004**, *6*, 611–623. [CrossRef]

21. Ramachandra, M.; Rahman, A.; Zou, A.; Vaillancourt, M.; Howe, J.A.; Antelman, D.; Sugarman, B.; Demers, G.W.; Engler, H.; Johnson, D.; et al. Re-engineering adenovirus regulatory pathways to enhance oncolytic specificity and efficacy. *Nat. Biotechnol.* **2001**, *19*, 1035–1041. [CrossRef] [PubMed]

22. Gurlevik, E.; Woller, N.; Schache, P.; Malek, N.P.; Wirth, T.C.; Zender, L.; Manns, M.P.; Kubicka, S.; Kuhnel, F. p53-dependent antiviral RNA-interference facilitates tumor-selective viral replication. *Nucleic Acids Res.* **2009**, *37*, e84. [CrossRef] [PubMed]

23. Heise, C.; Hermiston, T.; Johnson, L.; Brooks, G.; Sampson-Johannes, A.; Williams, A.; Hawkins, L.; Kirn, D. An adenovirus E1A mutant that demonstrates potent and selective systemic anti-tumoral efficacy. *Nat. Med.* **2000**, *6*, 1134–1139. [CrossRef] [PubMed]

24. Fueyo, J.; Gomez-Manzano, C.; Alemany, R.; Lee, P.S.; McDonnell, T.J.; Mitlianga, P.; Shi, Y.X.; Levin, V.A.; Yung, W.K.A.; Kyritsis, A.P. A mutant oncolytic adenovirus targeting the Rb pathway produces anti-glioma effect in vivo. *Oncogene* **2000**, *19*, 2–12. [CrossRef] [PubMed]

25. Lang, F.F.; Conrad, C.; Gomez-Manzano, C.; Yung, W.K.A.; Sawaya, R.; Weinberg, J.S.; Prabhu, S.S.; Rao, G.; Fuller, G.N.; Aldape, K.D.; et al. Phase I Study of DNX-2401 (Delta-24-RGD) Oncolytic Adenovirus: Replication and Immunotherapeutic Effects in Recurrent Malignant Glioma. *J. Clin. Oncol.* **2018**, *36*, 1419–1427. [CrossRef]

26. Ulasov, I.V.; Tyler, M.A.; Rivera, A.A.; Nettlebeck, D.M.; Douglas, J.T.; Lesniak, M.S. Evaluation of E1A double mutant oncolytic adenovectors in anti-glioma gene therapy. *J. Med. Virol.* **2008**, *80*, 1595–1603. [CrossRef]

27. Lau, L.; Gray, E.E.; Brunette, R.L.; Stetson, D.B. DNA tumor virus oncogenes antagonize the cGAS-STING DNA-sensing pathway. *Science* **2015**, *350*, 568–571. [CrossRef]

28. Spurrell, E.; Gangeswaran, R.; Wang, P.; Cao, F.; Gao, D.; Feng, B.; Wold, W.; Tollefson, A.; Lemoine, N.R.; Wang, Y. STAT1 Interaction with E3-14.7K in Monocytes Affects the Efficacy of Oncolytic Adenovirus. *J. Virol.* **2014**, *88*, 2291–2300. [CrossRef]

29. Zeng, X.; Carlin, C.R. Adenovirus early region 3 RIDα protein limits NFκB signaling through stress-activated EGF receptors. *PLoS Pathog.* **2019**, *15*, e1008017. [CrossRef]

30. Cerullo, V.; Diaconu, I.; Romano, V.; Hirvinen, M.; Ugolini, M.; Escutenaire, S.; Holm, S.L.; Kipar, A.; Kanerva, A.; Hemminki, A. An Oncolytic Adenovirus Enhanced for Toll-like Receptor 9 Stimulation Increases Antitumor Immune Responses and Tumor Clearance. *Mol. Ther.* **2012**, *20*, 2076–2086. [CrossRef]

31. Young, A.M.; Archibald, K.M.; Tookman, L.A.; Pool, A.; Dudek, K.; Jones, C.; Williams, S.L.; Pirlo, K.J.; Willis, A.E.; Lockley, M.; et al. Failure of Translation of Human Adenovirus mRNA in Murine Cancer Cells Can be Partially Overcome by L4-100K Expression In Vitro and In Vivo. *Mol. Ther.* **2012**, *20*, 1676–1688. [CrossRef] [PubMed]

32. Lei, J.; Jacobus, E.J.; Taverner, W.K.; Fisher, K.D.; Hemmi, S.; West, K.; Slater, L.; Lilley, F.; Brown, A.; Champion, B.; et al. Expression of human CD46 and trans-complementation by murine adenovirus 1 fails to allow productive infection by a group B oncolytic adenovirus in murine cancer cells. *J. Immunother. Cancer* **2018**, *6*, 55. [CrossRef] [PubMed]

33. Hallenbeck, P.L.; Chang, Y.N.; Hay, C.; Golightly, D.; Stewart, D.; Lin, J.; Phipps, S.; Chiang, Y.L. A novel tumor-specific replication-restricted adenoviral vector for gene therapy of hepatocellular carcinoma. *Hum. Gene Ther.* **1999**, *10*, 1721–1733. [CrossRef] [PubMed]

34. Kurihara, T.; Brough, D.E.; Kovesdi, I.; Kufe, D.W. Selectivity of a replication-competent adenovirus for human breast carcinoma cells expressing the MUC1 antigen. *J. Clin. Investig.* **2000**, *106*, 763–771. [CrossRef]

35. Jakubczak, J.L.; Ryan, P.; Gorziglia, M.; Clarke, L.; Hawkins, L.K.; Hay, C.; Huang, Y.; Kaloss, M.; Marinov, A.; Phipps, S.; et al. An oncolytic adenovirus selective for retinoblastoma tumor suppressor protein pathway-defective tumors: Dependence on E1A, the E2F-1 promoter, and viral replication for selectivity and efficacy. *Cancer Res.* **2003**, *63*, 1490–1499.

36. Alonso, M.M.; Cascallo, M.; Gomez-Manzano, C.; Jiang, H.; Bekele, B.N.; Perez-Gimenez, A.; Lang, F.F.; Piao, Y.; Alemany, R.; Fueyo, J. ICOVIR-5 shows E2F1 addiction and potent antiglioma effect in vivo. *Cancer Res.* **2007**, *67*, 8255–8263. [CrossRef]

37. Lanson, N.A.; Friedlander, P.L.; Schwarzenberger, P.; Kolls, J.K.; Wang, G. 1114. Replication of an Adenoviral Vector Controlled by the Human Telomerase Reverse Transcriptase Promoter Causes Tumor-Selective Cell Lysis. *Mol. Ther.* **2003**, *7*, S429. [CrossRef]

38. Kawashima, T.; Kagawa, S.; Kobayashi, N.; Shirakiya, Y.; Umeoka, T.; Teraishi, F.; Taki, M.; Kyo, S.; Tanaka, N.; Fujiwara, T. Telomerase-specific replication-selective virotherapy for human cancer. *Clin. Cancer Res.* **2004**, *10*, 285–292. [CrossRef]

39. Rojas, J.J.; Cascallo, M.; Guedan, S.; Gros, A.; Martinez-Quintanilla, J.; Hemminki, A.; Alemany, R. A modified E2F-1 promoter improves the efficacy to toxicity ratio of oncolytic adenoviruses. *Gene Ther.* **2009**, *16*, 1441–1451. [CrossRef]

40. Kim, E.; Kim, J.H.; Shin, H.Y.; Lee, H.; Yang, J.M.; Kim, J.; Sohn, J.H.; Kim, H.; Yun, C.O. Ad-mTERT-delta19, a conditional replication-competent adenovirus driven by the human telomerase promoter, selectively replicates in and elicits cytopathic effect in a cancer cell-specific manner. *Hum. Gene Ther.* **2003**, *14*, 1415–1428. [CrossRef]

41. Wirth, T.; Zender, L.; Schulte, B.; Mundt, B.; Plentz, R.; Rudolph, K.L.; Manns, M.; Kubicka, S.; Kühnel, F. A telomerase-dependent conditionally replicating adenovirus for selective treatment of cancer. *Cancer Res.* **2003**, *63*, 3181–3188.

42. Mantwill, K.; Köhler-Vargas, N.; Bernshausen, A.; Bieler, A.; Lage, H.; Kaszubiak, A.; Surowiak, P.; Dravits, T.; Treiber, U.; Hartung, R.; et al. Inhibition of the Multidrug-Resistant Phenotype by Targeting YB-1 with a Conditionally Oncolytic Adenovirus: Implications for Combinatorial Treatment Regimen with Chemotherapeutic Agents. *Cancer Res.* **2006**, *66*, 7195–7202. [CrossRef] [PubMed]

43. Cuevas, Y.; Hernández-Alcoceba, R.; Aragones, J.; Naranjo-Suárez, S.; Castellanos, M.C.; Esteban, M.A.; Martín-Puig, S.; Landazuri, M.O.; del Peso, L. Specific oncolytic effect of a new hypoxia-inducible factor-dependent replicative adenovirus on von Hippel-Lindau-defective renal cell carcinomas. *Cancer Res.* **2003**, *63*, 6877–6884. [PubMed]

44. Li, Y.; Hong, J.; Oh, J.E.; Yoon, A.R.; Yun, C.O. Potent antitumor effect of tumor microenvironment-targeted oncolytic adenovirus against desmoplastic pancreatic cancer. *Int. J. Cancer* **2018**, *142*, 392–413. [CrossRef] [PubMed]

45. Douglas, J.T.; Kim, M.; Sumerel, L.A.; Carey, D.E.; Curiel, D.T. Efficient oncolysis by a replicating adenovirus (ad) in vivo is critically dependent on tumor expression of primary ad receptors. *Cancer Res.* **2001**, *61*, 813–817. [PubMed]

46. Suzuki, K.; Fueyo, J.; Krasnykh, V.; Reynolds, P.N.; Curiel, D.T.; Alemany, R. A conditionally replicative adenovirus with enhanced infectivity shows improved oncolytic potency. *Clin. Cancer Res.* **2001**, *7*, 120–126. [PubMed]

47. Ranki, T.; Pesonen, S.; Hemminki, A.; Partanen, K.; Kairemo, K.; Alanko, T.; Lundin, J.; Linder, N.; Turkki, R.; Ristimäki, A.; et al. Phase I study with ONCOS-102 for the treatment of solid tumors—An evaluation of clinical response and exploratory analyses of immune markers. *J. Immunother. Cancer* **2016**, *4*, 17. [CrossRef] [PubMed]

48. Eriksson, E.; Milenova, I.; Wenthe, J.; Ståhle, M.; Leja-Jarblad, J.; Ullenhag, G.; Dimberg, A.; Moreno, R.; Alemany, R.; Loskog, A. Shaping the Tumor Stroma and Sparking Immune Activation by CD40 and 4-1BB Signaling Induced by an Armed Oncolytic Virus. *Clin. Cancer Res.* **2017**, *23*, 5846–5857. [CrossRef]

49. Beatty, M.S.; Curiel, D.T. Adenovirus Strategies for Tissue-Specific Targeting. *Adv. Cancer Res.* **2012**, *115*, 39–67.

50. Kishimoto, H.; Kojima, T.; Watanabe, Y.; Kagawa, S.; Fujiwara, T.; Uno, F.; Teraishi, F.; Kyo, S.; Mizuguchi, H.; Hashimoto, Y.; et al. In vivo imaging of lymph node metastasis with telomerase-specific replication-selective adenovirus. *Nat. Med.* **2006**, *12*, 1213–1219. [CrossRef]

51. Nemunaitis, J.; Tong, A.W.; Nemunaitis, M.; Senzer, N.; Phadke, A.P.; Bedell, C.; Adams, N.; Zhang, Y.A.; Maples, P.B.; Chen, S.; et al. A Phase I Study of Telomerase-specific Replication Competent Oncolytic Adenovirus (Telomelysin) for Various Solid Tumors. *Mol. Ther.* **2010**, *18*, 429–434. [CrossRef] [PubMed]

52. Kanaya, N.; Kuroda, S.; Kakiuchi, Y.; Kumon, K.; Tsumura, T.; Hashimoto, M.; Morihiro, T.; Kubota, T.; Aoyama, K.; Kikuchi, S.; et al. Immune Modulation by Telomerase-Specific Oncolytic Adenovirus Synergistically Enhances Antitumor Efficacy with Anti-PD1 Antibody. *Mol. Ther.* **2020**, *28*, 794–804. [CrossRef] [PubMed]

53. Fueyo, J.; Alemany, R.; Gomez-Manzano, C.; Fuller, G.N.; Khan, A.; Conrad, C.A.; Liu, T.J.; Jiang, H.; Lemoine, M.G.; Suzuki, K.; et al. Preclinical characterization of the antiglioma activity of a tropism-enhanced adenovirus targeted to the retinoblastoma pathway. *J. Natl. Cancer Inst.* **2003**, *95*, 652–660. [CrossRef] [PubMed]

54. Dai, B.; Roife, D.; Kang, Y.; Gumin, J.; Rios Perez, M.V.; Li, X.; Pratt, M.; Brekken, R.A.; Fueyo-Margareto, J.; Lang, F.F.; et al. Preclinical Evaluation of Sequential Combination of Oncolytic Adenovirus Delta-24-RGD and Phosphatidylserine-Targeting Antibody in Pancreatic Ductal Adenocarcinoma. *Mol. Cancer Ther.* **2017**, *16*, 662–670. [CrossRef] [PubMed]

55. Rodriguez-Garcia, A.; Gimenez-Alejandre, M.; Rojas, J.J.; Moreno, R.; Bazan-Peregrino, M.; Cascallo, M.; Alemany, R. Safety and Efficacy of VCN-01, an Oncolytic Adenovirus Combining Fiber HSG-Binding Domain Replacement with RGD and Hyaluronidase Expression. *Clin. Cancer Res.* **2015**, *21*, 1406–1418. [CrossRef]

56. Guedan, S.; Rojas, J.J.; Gros, A.; Mercade, E.; Cascallo, M.; Alemany, R. Hyaluronidase expression by an oncolytic adenovirus enhances its intratumoral spread and suppresses tumor growth. *Mol. Ther.* **2010**. [CrossRef]

57. Pascual-Pasto, G.; Bazan-Peregrino, M.; Olaciregui, N.G.; Restrepo-Perdomo, C.A.; Mato-Berciano, A.; Ottaviani, D.; Weber, K.; Correa, G.; Paco, S.; Vila-Ubach, M.; et al. Therapeutic targeting of the RB1 pathway in retinoblastoma with the oncolytic adenovirus VCN-01. *Sci. Transl. Med.* **2019**, *11*, eaat9321. [CrossRef]

58. Ramesh, N. CG0070, a Conditionally Replicating Granulocyte Macrophage Colony-Stimulating Factor-Armed Oncolytic Adenovirus for the Treatment of Bladder Cancer. *Clin. Cancer Res.* **2006**, *12*, 305–313. [CrossRef]

59. Packiam, V.T.; Lamm, D.L.; Barocas, D.A.; Trainer, A.; Fand, B.; Davis, R.L.; Clark, W.; Kroeger, M.; Dumbadze, I.; Chamie, K.; et al. An open label, single-arm, phase II multicenter study of the safety and efficacy of CG0070 oncolytic vector regimen in patients with BCG-unresponsive non–muscle-invasive bladder cancer: Interim results. *Urol. Oncol. Semin. Orig. Investig.* **2018**, *36*, 440–447. [CrossRef]

60. Kuhn, I.; Harden, P.; Bauzon, M.; Chartier, C.; Nye, J.; Thorne, S.; Reid, T.; Ni, S.; Lieber, A.; Fisher, K.; et al. Directed Evolution Generates a Novel Oncolytic Virus for the Treatment of Colon Cancer. *PLoS ONE* **2008**, *3*, e2409. [CrossRef]

61. Machiels, J.P.; Salazar, R.; Rottey, S.; Duran, I.; Dirix, L.; Geboes, K.; Wilkinson-Blanc, C.; Pover, G.; Alvis, S.; Champion, B.; et al. A phase 1 dose escalation study of the oncolytic adenovirus enadenotucirev, administered intravenously to patients with epithelial solid tumors (EVOLVE). *J. Immunother. Cancer* **2019**, *7*, 20. [CrossRef] [PubMed]

62. Wildner, O.; Blaese, R.M.; Morris, J.C. Therapy of colon cancer with oncolytic adenovirus is enhanced by the addition of herpes simplex virus-thymidine kinase. *Cancer Res.* **1999**, *59*, 410–413. [PubMed]

63. Freytag, S.O.; Khil, M.; Stricker, H.; Peabody, J.; Menon, M.; DePeralta-Venturina, M.; Nafziger, D.; Pegg, J.; Paielli, D.; Brown, S.; et al. Phase I study of replication-competent adenovirus-mediated double suicide gene therapy for the treatment of locally recurrent prostate cancer. *Cancer Res.* **2002**, *62*, 4968–4976. [PubMed]

64. Barton, K.N.; Paielli, D.; Zhang, Y.; Koul, S.; Brown, S.L.; Lu, M.; Seely, J.; Kim, J.H.; Freytag, S.O. Second-generation replication-competent oncolytic adenovirus armed with improved suicide genes and ADP gene demonstrates greater efficacy without increased toxicity. *Mol. Ther.* **2006**, *13*, 347–356. [CrossRef]

65. Freytag, S.O.; Barton, K.N.; Zhang, Y. Efficacy of oncolytic adenovirus expressing suicide genes and interleukin-12 in preclinical model of prostate cancer. *Gene Ther.* **2013**, *20*, 1131–1139. [CrossRef]

66. Sova, P.; Ren, X.W.; Ni, S.; Bernt, K.M.; Mi, J.; Kiviat, N.; Lieber, A. A Tumor-Targeted and Conditionally Replicating Oncolytic Adenovirus Vector Expressing TRAIL for Treatment of Liver Metastases. *Mol. Ther.* **2004**, *9*, 496–509. [CrossRef]

67. Fernández-Ulibarri, I.; Hammer, K.; Arndt, M.A.E.; Kaufmann, J.K.; Dorer, D.; Engelhardt, S.; Kontermann, R.E.; Hess, J.; Allgayer, H.; Krauss, J.; et al. Genetic delivery of an immunoRNase by an oncolytic adenovirus enhances anticancer activity. *Int. J. Cancer* **2015**, *136*, 2228–2240. [CrossRef]

68. Wildner, O.; Hoffmann, D.; Jogler, C.; Überla, K. Comparison of HSV-1 thymidine kinase-dependent and -independent inhibition of replication-competent adenoviral vectors by a panel of drugs. *Cancer Gene Ther.* **2003**, *10*, 791–802. [CrossRef]

69. Liikanen, I.; Tahtinen, S.; Guse, K.; Gutmann, T.; Savola, P.; Oksanen, M.; Kanerva, A.; Hemminki, A. Oncolytic Adenovirus Expressing Monoclonal Antibody Trastuzumab for Treatment of HER2-Positive Cancer. *Mol. Cancer Ther.* **2016**, *15*, 2259–2269. [CrossRef]

70. Yang, S.W.; Cody, J.J.; Rivera, A.A.; Waehler, R.; Wang, M.; Kimball, K.J.; Alvarez, R.A.; Siegal, G.P.; Douglas, J.T.; Ponnazhagan, S. Conditionally replicating adenovirus expressing TIMP2 for ovarian cancer therapy. *Clin. Cancer Res.* **2011**, *17*, 538–549. [CrossRef]

71. Lamfers, M.L.M.; Gianni, D.; Tung, C.H.; Idema, S.; Schagen, F.H.E.; Carette, J.E.; Quax, P.H.A.; Van Beusechem, V.W.; Vandertop, W.P.; Dirven, C.M.F.; et al. Tissue inhibitor of metalloproteinase-3 expression from an oncolytic adenovirus inhibits matrix metalloproteinase activity in vivo without affecting antitumor efficacy in malignant glioma. *Cancer Res.* **2005**, *65*, 9398–9405. [CrossRef] [PubMed]

72. Cheng, J.; Sauthoff, H.; Huang, Y.; Kutler, D.I.; Bajwa, S.; Rom, W.N.; Hay, J.G. Human matrix metalloproteinase-8 gene delivery increases the oncolytic activity of a replicating adenovirus. *Mol. Ther.* **2007**, *15*, 1982–1990. [CrossRef] [PubMed]

73. Kim, J.H.; Lee, Y.S.; Kim, H.; Huang, J.H.; Yoon, A.R.; Yun, C.O. Relaxin Expression From Tumor-Targeting Adenoviruses and Its Intratumoral Spread, Apoptosis Induction, and Efficacy. *JNCI J. Natl. Cancer Inst.* **2006**, *98*, 1482–1493. [CrossRef] [PubMed]

74. Choi, I.K.; Lee, Y.S.; Yoo, J.Y.; Yoon, A.R.; Kim, H.; Kim, D.S.; Seidler, D.G.; Kim, J.H.; Yun, C.O. Effect of decorin on overcoming the extracellular matrix barrier for oncolytic virotherapy. *Gene Ther.* **2010**, *17*, 190–201. [CrossRef] [PubMed]

75. Jung, B.K.; Ko, H.Y.; Kang, H.; Hong, J.; Ahn, H.M.; Na, Y.; Kim, H.; Kim, J.S.; Yun, C.O. Relaxin-expressing oncolytic adenovirus induces remodeling of physical and immunological aspects of cold tumor to potentiate PD-1 blockade. *J. Immunother. Cancer* **2020**, *8*, e000763. [CrossRef]

76. Li, G.; Sham, J.; Yang, J.; Su, C.; Xue, H.; Chua, D.; Sun, L.; Zhang, Q.; Cui, Z.; Wu, M.; et al. Potent antitumor efficacy of an E1B 55kDa-deficient adenovirus carrying murineendostatin in hepatocellular carcinoma. *Int. J. Cancer* **2005**, *113*, 640–648. [CrossRef]

77. Xiao, T.; Fan, J.K.; Huang, H.L.; Gu, J.F.; Li, L.Y.; Liu, X.Y. VEGI-armed oncolytic adenovirus inhibits tumor neovascularization and directly induces mitochondria-mediated cancer cell apoptosis. *Cell Res.* **2010**, *20*, 367–378. [CrossRef]

78. Xu, W.; Neill, T.; Yang, Y.; Hu, Z.; Cleveland, E.; Wu, Y.; Hutten, R.; Xiao, X.; Stock, S.R.; Shevrin, D.; et al. The systemic delivery of an oncolytic adenovirus expressing decorin inhibits bone metastasis in a mouse model of human prostate cancer. *Gene Ther.* **2015**, *22*, 247–256. [CrossRef]

79. Yoo, J.Y.; Kim, J.H.; Kwon, Y.G.; Kim, E.C.; Kim, N.K.; Choi, H.J.; Yun, C.O. VEGF-specific Short Hairpin RNA–expressing Oncolytic Adenovirus Elicits Potent Inhibition of Angiogenesis and Tumor Growth. *Mol. Ther.* **2007**, *15*, 295–302. [CrossRef]

80. Kang, Y.A.; Shin, H.C.; Yoo, J.Y.; Kim, J.H.; Kim, J.S.; Yun, C.O. Novel Cancer Antiangiotherapy Using the VEGF Promoter-targeted Artificial Zinc-finger Protein and Oncolytic Adenovirus. *Mol. Ther.* **2008**, *16*, 1033–1040. [CrossRef]

81. Shashkova, E.V.; Spencer, J.F.; Wold, W.S.M.; Doronin, K. Targeting Interferon-α Increases Antitumor Efficacy and Reduces Hepatotoxicity of E1A-mutated Spread-enhanced Oncolytic Adenovirus. *Mol. Ther.* **2007**, *15*, 598–607. [CrossRef] [PubMed]

82. LaRocca, C.J.; Han, J.; Gavrikova, T.; Armstrong, L.; Oliveira, A.R.; Shanley, R.; Vickers, S.M.; Yamamoto, M.; Davydova, J. Oncolytic adenovirus expressing interferon alpha in a syngeneic Syrian hamster model for the treatment of pancreatic cancer. *Surgery* **2015**, *157*, 888–898. [CrossRef]

83. Park, M.Y.; Kim, D.R.; Jung, H.W.; Yoon, H.I.; Lee, J.H.; Lee, C.T. Genetic immunotherapy of lung cancer using conditionally replicating adenovirus and adenovirus-interferon-β. *Cancer Gene Ther.* **2010**, *17*, 356–364. [CrossRef] [PubMed]

84. Lee, C.T.; Park, K.H.; Yanagisawa, K.; Adachi, Y.; Ohm, J.E.; Nadaf, S.; Dikov, M.M.; Curiel, D.T.; Carbone, D.P. Combination Therapy with Conditionally Replicating Adenovirus and Replication Defective Adenovirus. *Cancer Res.* **2004**, *64*, 6660–6665. [CrossRef] [PubMed]

85. Hirvinen, M.; Rajecki, M.; Kapanen, M.; Parviainen, S.; Rouvinen-Lagerström, N.; Diaconu, I.; Nokisalmi, P.; Tenhunen, M.; Hemminki, A.; Cerullo, V. Immunological Effects of a Tumor Necrosis Factor Alpha–Armed Oncolytic Adenovirus. *Hum. Gene Ther.* **2015**, *26*, 134–144. [CrossRef]

86. Havunen, R.; Siurala, M.; Sorsa, S.; Grönberg-Vähä-Koskela, S.; Behr, M.; Tähtinen, S.; Santos, J.M.; Karell, P.; Rusanen, J.; Nettelbeck, D.M.; et al. Oncolytic Adenoviruses Armed with Tumor Necrosis Factor Alpha and Interleukin-2 Enable Successful Adoptive Cell Therapy. *Mol. Ther. Oncolytics* **2017**, *4*, 77–86. [CrossRef]

87. Watanabe, K.; Luo, Y.; Da, T.; Guedan, S.; Ruella, M.; Scholler, J.; Keith, B.; Young, R.M.; Engels, B.; Sorsa, S.; et al. Pancreatic cancer therapy with combined mesothelin-redirected chimeric antigen receptor T cells and cytokine-armed oncolytic adenoviruses. *JCI Insight* **2018**, *3*. [CrossRef]

88. Santos, J.M.; Heiniö, C.; Cervera-Carrascon, V.; Quixabeira, D.C.A.; Siurala, M.; Havunen, R.; Butzow, R.; Zafar, S.; de Gruijl, T.; Lassus, H.; et al. Oncolytic adenovirus shapes the ovarian tumor microenvironment for potent tumor-infiltrating lymphocyte tumor reactivity. *J. Immunother. Cancer* **2020**, *8*, e000188. [CrossRef]

89. Bortolanza, S.; Bunuales, M.; Otano, I.; Gonzalez-Aseguinolaza, G.; Ortiz-de-Solorzano, C.; Perez, D.; Prieto, J.; Hernandez-Alcoceba, R. Treatment of pancreatic cancer with an oncolytic adenovirus expressing interleukin-12 in Syrian hamsters. *Mol. Ther.* **2009**, *17*, 614–622. [CrossRef]

90. Lee, Y.S.; Kim, J.H.; Choi, K.J.; Choi, I.K.; Kim, H.; Cho, S.; Cho, B.C.; Yun, C.O. Enhanced Antitumor Effect of Oncolytic Adenovirus Expressing Interleukin-12 and B7-1 in an Immunocompetent Murine Model. *Clin. Cancer Res.* **2006**, *12*, 5859–5868. [CrossRef]

91. Wang, P.; Li, X.; Wang, J.; Gao, D.; Li, Y.; Li, H.; Chu, Y.; Zhang, Z.; Liu, H.; Jiang, G.; et al. Re-designing Interleukin-12 to enhance its safety and potential as an anti-tumor immunotherapeutic agent. *Nat. Commun.* **2017**, *8*, 1395. [CrossRef] [PubMed]

92. Zhao, L.; Gu, J.; Dong, A.; Zhang, Y.; Zhong, L.; He, L.; Wang, Y.; Zhang, J.; Zhang, Z.; Huiwang, J.; et al. Potent Antitumor Activity of Oncolytic Adenovirus Expressing mda-7/IL-24 for Colorectal Cancer. *Hum. Gene Ther.* **2005**, *16*, 845–858. [CrossRef] [PubMed]

93. Luo, J.; Xia, Q.; Zhang, R.; Lv, C.; Zhang, W.; Wang, Y.; Cui, Q.; Liu, L.; Cai, R.; Qian, C. Treatment of Cancer with a Novel Dual-Targeted Conditionally Replicative Adenovirus Armed with mda-7/IL-24 Gene. *Clin. Cancer Res.* **2008**, *14*, 2450–2457. [CrossRef] [PubMed]

94. Post, D.E.; Sandberg, E.M.; Kyle, M.M.; Devi, N.S.; Brat, D.J.; Xu, Z.; Tighiouart, M.; Van Meir, E.G. Targeted Cancer Gene Therapy Using a Hypoxia Inducible Factor–Dependent Oncolytic Adenovirus Armed with Interleukin-4. *Cancer Res.* **2007**, *67*, 6872–6881. [CrossRef] [PubMed]

95. Zheng, J.N.; Pei, D.S.; Mao, L.J.; Liu, X.Y.; Sun, F.H.; Zhang, B.F.; Liu, Y.Q.; Liu, J.J.; Li, W.; Han, D. Oncolytic adenovirus expressing interleukin-18 induces significant antitumor effects against melanoma in mice through inhibition of angiogenesis. *Cancer Gene Ther.* **2010**, *17*, 28–36. [CrossRef]

96. Choi, I.K.; Lee, J.S.; Zhang, S.N.; Park, J.; Lee, K.M.; Sonn, C.H.; Yun, C.O. Oncolytic adenovirus co-expressing IL-12 and IL-18 improves tumor-specific immunity via differentiation of T cells expressing IL-12Rβ2 or IL-18Rα. *Gene Ther.* **2011**, *18*, 898–909. [CrossRef]

97. Choi, I.K.; Li, Y.; Oh, E.; Kim, J.; Yun, C.O. Oncolytic Adenovirus Expressing IL-23 and p35 Elicits IFN-γ- and TNF-α-Co-Producing T Cell-Mediated Antitumor Immunity. *PLoS ONE* **2013**, *8*, e67512. [CrossRef]

98. Li, Y.; Li, Y.; Si, C.; Zhu, Y.; Jin, Y.; Zhu, T.; Liu, M.; Liu, G. CCL21/IL21-armed oncolytic adenovirus enhances antitumor activity against TERT-positive tumor cells. *Virus Res.* **2016**, *220*, 172–178. [CrossRef]

99. Ye, J.F.; Lin, Y.Q.; Yu, X.H.; Liu, M.Y.; Li, Y. Immunotherapeutic effects of cytokine-induced killer cells combined with CCL21/IL15 armed oncolytic adenovirus in TERT-positive tumor cells. *Int. Immunopharmacol.* **2016**, *38*, 460–467. [CrossRef]

100. Hu, Z.; Gerseny, H.; Zhang, Z.; Chen, Y.J.; Berg, A.; Zhang, Z.; Stock, S.; Seth, P. Oncolytic Adenovirus Expressing Soluble TGFβ Receptor II-Fc-mediated Inhibition of Established Bone Metastases: A Safe and Effective Systemic Therapeutic Approach for Breast Cancer. *Mol. Ther.* **2011**, *19*, 1609–1618. [CrossRef]

101. Yang, Y.; Xu, W.; Peng, D.; Wang, H.; Zhang, X.; Wang, H.; Xiao, F.; Zhu, Y.; Ji, Y.; Gulukota, K.; et al. An Oncolytic Adenovirus Targeting Transforming Growth Factor β Inhibits Protumorigenic Signals and Produces Immune Activation: A Novel Approach to Enhance Anti-PD-1 and Anti-CTLA-4 Therapy. *Hum. Gene Ther.* **2019**, *30*, 1117–1132. [CrossRef] [PubMed]

102. Cerullo, V.; Pesonen, S.; Diaconu, I.; Escutenaire, S.; Arstila, P.T.; Ugolini, M.; Nokisalmi, P.; Raki, M.; Laasonen, L.; Särkioja, M.; et al. Oncolytic adenovirus coding for granulocyte macrophage colony-stimulating factor induces antitumoral immunity in cancer patients. *Cancer Res.* **2010**. [CrossRef] [PubMed]

103. Koski, A.; Kangasniemi, L.; Escutenaire, S.; Pesonen, S.; Cerullo, V.; Diaconu, I.; Nokisalmi, P.; Raki, M.; Rajecki, M.; Guse, K.; et al. Treatment of Cancer Patients With a Serotype 5/3 Chimeric Oncolytic Adenovirus Expressing GMCSF. *Mol. Ther.* **2010**, *18*, 1874–1884. [CrossRef] [PubMed]

104. Kanerva, A.; Nokisalmi, P.; Diaconu, I.; Koski, A.; Cerullo, V.; Liikanen, I.; Tahtinen, S.; Oksanen, M.; Heiskanen, R.; Pesonen, S.; et al. Antiviral and Antitumor T-cell Immunity in Patients Treated with GM-CSF-Coding Oncolytic Adenovirus. *Clin. Cancer Res.* **2013**, *19*, 2734–2744. [CrossRef]

105. Zhang, S.N.; Choi, I.K.; Huang, J.H.; Yoo, J.Y.; Choi, K.J.; Yun, C.O. Optimizing DC Vaccination by Combination With Oncolytic Adenovirus Coexpressing IL-12 and GM-CSF. *Mol. Ther.* **2011**, *19*, 1558–1568. [CrossRef]

106. Serafini, P.; Carbley, R.; Noonan, K.A.; Tan, G.; Bronte, V.; Borrello, I. High-dose granulocyte-macrophage colony-stimulating factor-producing vaccines impair the immune response through the recruitment of myeloid suppressor cells. *Cancer Res.* **2004**. [CrossRef]

107. Marigo, I.; Bosio, E.; Solito, S.; Mesa, C.; Fernandez, A.; Dolcetti, L.; Ugel, S.; Sonda, N.; Bicciato, S.; Falisi, E.; et al. Tumor-Induced Tolerance and Immune Suppression Depend on the C/EBPβ Transcription Factor. *Immunity* **2010**, *32*, 790–802. [CrossRef]

108. Kenkel, J.A.; Tseng, W.W.; Davidson, M.G.; Tolentino, L.L.; Choi, O.; Bhattacharya, N.; Seeley, E.S.; Winer, D.A.; Reticker-Flynn, N.E.; Engleman, E.G. An immunosuppressive dendritic cell subset accumulates at secondary sites and promotes metastasis in pancreatic cancer. *Cancer Res.* **2017**. [CrossRef]

109. Bernt, K.M.; Ni, S.; Tieu, A.T.; Lieber, A. Assessment of a combined, adenovirus-mediated oncolytic and immunostimulatory tumor therapy. *Cancer Res.* **2005**. [CrossRef]

110. Ramakrishna, E.; Woller, N.; Mundt, B.; Knocke, S.; Gürlevik, E.; Saborowski, M.; Malek, N.; Manns, M.P.; Wirth, T.; Kühnel, F.; et al. Antitumoral Immune Response by Recruitment and Expansion of Dendritic Cells in Tumors Infected with Telomerase-Dependent Oncolytic Viruses. *Cancer Res.* **2009**, *69*, 1448–1458. [CrossRef]

111. Dias, J.D.; Hemminki, O.; Diaconu, I.; Hirvinen, M.; Bonetti, A.; Guse, K.; Escutenaire, S.; Kanerva, A.; Pesonen, S.; Löskog, A.; et al. Targeted cancer immunotherapy with oncolytic adenovirus coding for a fully human monoclonal antibody specific for CTLA-4. *Gene Ther.* **2012**, *19*, 988–998. [CrossRef] [PubMed]

112. Tanoue, K.; Rosewell Shaw, A.; Watanabe, N.; Porter, C.; Rana, B.; Gottschalk, S.; Brenner, M.; Suzuki, M. Armed Oncolytic Adenovirus–Expressing PD-L1 Mini-Body Enhances Antitumor Effects of Chimeric Antigen Receptor T Cells in Solid Tumors. *Cancer Res.* **2017**, *77*, 2040–2051. [CrossRef] [PubMed]

113. Rosewell Shaw, A.; Porter, C.E.; Watanabe, N.; Tanoue, K.; Sikora, A.; Gottschalk, S.; Brenner, M.K.; Suzuki, M. Adenovirotherapy Delivering Cytokine and Checkpoint Inhibitor Augments CAR T Cells against Metastatic Head and Neck Cancer. *Mol. Ther.* **2017**, *25*, 2440–2451. [CrossRef] [PubMed]

114. Fernandes, M.S.; Gomes, E.M.; Butcher, L.D.; Hernandez-Alcoceba, R.; Chang, D.; Kansopon, J.; Newman, J.; Stone, M.J.; Tong, A.W. Growth Inhibition of Human Multiple Myeloma Cells by an Oncolytic Adenovirus Carrying the CD40 Ligand Transgene. *Clin. Cancer Res.* **2009**, *15*, 4847–4856. [CrossRef]

115. Diaconu, I.; Cerullo, V.; Hirvinen, M.L.M.; Escutenaire, S.; Ugolini, M.; Pesonen, S.K.; Bramante, S.; Parviainen, S.; Kanerva, A.; Loskog, A.S.I.; et al. Immune response is an important aspect of the antitumor effect produced by a CD40L-encoding oncolytic adenovirus. *Cancer Res.* **2012**, *72*, 2327–2338. [CrossRef]

116. Pesonen, S.; Diaconu, I.; Kangasniemi, L.; Ranki, T.; Kanerva, A.; Pesonen, S.K.; Gerdemann, U.; Leen, A.M.; Kairemo, K.; Oksanen, M.; et al. Oncolytic Immunotherapy of Advanced Solid Tumors with a CD40L-Expressing Replicating Adenovirus: Assessment of Safety and Immunologic Responses in Patients. *Cancer Res.* **2012**, *72*, 1621–1631. [CrossRef]

117. Zafar, S.; Sorsa, S.; Siurala, M.; Hemminki, O.; Havunen, R.; Cervera-Carrascon, V.; Santos, J.M.; Wang, H.; Lieber, A.; De Gruijl, T.; et al. CD40L coding oncolytic adenovirus allows long-term survival of humanized mice receiving dendritic cell therapy. *Oncoimmunology* **2018**, *7*, e1490856. [CrossRef]

118. Huang, J.H.; Zhang, S.N.; Choi, K.J.; Choi, I.K.; Kim, J.H.; Lee, M.; Kim, H.; Yun, C.O. Therapeutic and Tumor-specific Immunity Induced by Combination of Dendritic Cells and Oncolytic Adenovirus Expressing IL-12 and 4-1BBL. *Mol. Ther.* **2010**, *18*, 264–274. [CrossRef]

119. Jiang, H.; Rivera-Molina, Y.; Gomez-Manzano, C.; Clise-Dwyer, K.; Bover, L.; Vence, L.M.; Yuan, Y.; Lang, F.F.; Toniatti, C.; Hossain, M.B.; et al. Oncolytic Adenovirus and Tumor-Targeting Immune Modulatory Therapy Improve Autologous Cancer Vaccination. *Cancer Res.* **2017**, *77*, 3894–3907. [CrossRef]

120. Rivera-Molina, Y.; Jiang, H.; Fueyo, J.; Nguyen, T.; Shin, D.H.; Youssef, G.; Fan, X.; Gumin, J.; Alonso, M.M.; Phadnis, S.; et al. GITRL-armed Delta-24-RGD oncolytic adenovirus prolongs survival and induces anti-glioma immune memory. *Neuro Oncol. Adv.* **2019**, *1*. [CrossRef]

121. Runcie, K.; Budman, D.R.; John, V.; Seetharamu, N. Bi-specific and tri-specific antibodies- the next big thing in solid tumor therapeutics. *Mol. Med.* **2018**, *24*, 50. [CrossRef] [PubMed]

122. Fajardo, C.A.; Guedan, S.; Rojas, L.A.; Moreno, R.; Arias-Badia, M.; de Sostoa, J.; June, C.H.; Alemany, R. Oncolytic Adenoviral Delivery of an EGFR-Targeting T-cell Engager Improves Antitumor Efficacy. *Cancer Res.* **2017**, *77*, 2052–2063. [CrossRef] [PubMed]

123. Wing, A.; Fajardo, C.A.; Posey, A.D.; Shaw, C.; Da, T.; Young, R.M.; Alemany, R.; June, C.H.; Guedan, S. Improving CART-Cell Therapy of Solid Tumors with Oncolytic Virus–Driven Production of a Bispecific T-cell Engager. *Cancer Immunol. Res.* **2018**, *6*, 605–616. [CrossRef] [PubMed]

124. Freedman, J.D.; Hagel, J.; Scott, E.M.; Psallidas, I.; Gupta, A.; Spiers, L.; Miller, P.; Kanellakis, N.; Ashfield, R.; Fisher, K.D.; et al. Oncolytic adenovirus expressing bispecific antibody targets T-cell cytotoxicity in cancer biopsies. *EMBO Mol. Med.* **2017**, *9*, 1067–1087. [CrossRef] [PubMed]

125. Freedman, J.D.; Duffy, M.R.; Lei-Rossmann, J.; Muntzer, A.; Scott, E.M.; Hagel, J.; Campo, L.; Bryant, R.J.; Verrill, C.; Lambert, A.; et al. An Oncolytic Virus Expressing a T-cell Engager Simultaneously Targets Cancer and Immunosuppressive Stromal Cells. *Cancer Res.* **2018**, *78*, 6852–6865. [CrossRef] [PubMed]

126. de Sostoa, J.; Fajardo, C.A.; Moreno, R.; Ramos, M.D.; Farrera-Sal, M.; Alemany, R. Targeting the tumor stroma with an oncolytic adenovirus secreting a fibroblast activation protein-targeted bispecific T-cell engager. *J. Immunother. Cancer* **2019**, *7*, 19. [CrossRef]

127. Scott, E.M.; Jacobus, E.J.; Lyons, B.; Frost, S.; Freedman, J.D.; Dyer, A.; Khalique, H.; Taverner, W.K.; Carr, A.; Champion, B.R.; et al. Bi- And tri-valent T cell engagers deplete tumour-associated macrophages in cancer patient samples. *J. Immunother. Cancer* **2019**. [CrossRef]

128. Zamarin, D.; Holmgaard, R.B.; Subudhi, S.K.; Park, J.S.; Mansour, M.; Palese, P.; Merghoub, T.; Wolchok, J.D.; Allison, J.P. Localized Oncolytic Virotherapy Overcomes Systemic Tumor Resistance to Immune Checkpoint Blockade Immunotherapy. *Sci. Transl. Med.* **2014**, *6*, 226ra32. [CrossRef]

129. Woller, N.; Gürlevik, E.; Fleischmann-Mundt, B.; Schumacher, A.; Knocke, S.; Kloos, A.M.; Saborowski, M.; Geffers, R.; Manns, M.P.; Wirth, T.C.; et al. Viral Infection of Tumors Overcomes Resistance to PD-1-immunotherapy by Broadening Neoantigenome-directed T-cell Responses. *Mol. Ther.* **2015**, *23*, 1630–1640. [CrossRef]

130. Leung, E.Y.L.; Ennis, D.P.; Kennedy, P.R.; Hansell, C.; Dowson, S.; Farquharson, M.; Spiliopoulou, P.; Nautiyal, J.; McNamara, S.; Carlin, L.M.; et al. NK Cells Augment Oncolytic Adenovirus Cytotoxicity in Ovarian Cancer. *Mol. Ther. Oncolytics* **2020**, *16*, 289–301. [CrossRef]

131. Tähtinen, S.; Feola, S.; Capasso, C.; Laustio, N.; Groeneveldt, C.; Ylösmäki, E.O.; Ylösmäki, L.; Martins, B.; Fusciello, M.; Medeot, M.; et al. Exploiting Preexisting Immunity to Enhance Oncolytic Cancer Immunotherapy. *Cancer Res.* **2020**, *80*, 2575–2585. [CrossRef] [PubMed]

132. Buchbinder, S.P.; Mehrotra, D.V.; Duerr, A.; Fitzgerald, D.W.; Mogg, R.; Li, D.; Gilbert, P.B.; Lama, J.R.; Marmor, M.; Del Rio, C.; et al. Efficacy assessment of a cell-mediated immunity HIV-1 vaccine (the Step Study): A double-blind, randomised, placebo-controlled, test-of-concept trial. *Lancet* **2008**, *372*, 1881–1893. [CrossRef]

133. Schöne, D.; Hrycak, C.P.; Windmann, S.; Lapuente, D.; Dittmer, U.; Tenbusch, M.; Bayer, W. Immunodominance of Adenovirus-Derived CD8+ T Cell Epitopes Interferes with the Induction of Transgene-Specific Immunity in Adenovirus-Based Immunization. *J. Virol.* **2017**, *91*. [CrossRef] [PubMed]

134. Tysome, J.R.; Li, X.; Wang, S.; Wang, P.; Gao, D.; Du, P.; Chen, D.; Gangeswaran, R.; Chard, L.S.; Yuan, M.; et al. A Novel Therapeutic Regimen to Eradicate Established Solid Tumors with an Effective Induction of Tumor-Specific Immunity. *Clin. Cancer Res.* **2012**, *18*, 6679–6689. [CrossRef]

135. Ilett, E.; Kottke, T.; Thompson, J.; Rajani, K.; Zaidi, S.; Evgin, L.; Coffey, M.; Ralph, C.; Diaz, R.; Pandha, H.; et al. Prime-boost using separate oncolytic viruses in combination with checkpoint blockade improves anti-tumour therapy. *Gene Ther.* **2017**, *24*. [CrossRef]

136. Bassett, J.D.; Yang, T.C.; Bernard, D.; Millar, J.B.; Swift, S.L.; McGray, A.J.R.; VanSeggelen, H.; Boudreau, J.E.; Finn, J.D.; Parsons, R.; et al. CD8+ T-cell expansion and maintenance after recombinant adenovirus immunization rely upon cooperation between hematopoietic and nonhematopoietic antigen-presenting cells. *Blood* **2011**, *117*, 1146–1155. [CrossRef]

137. Lee, J.; Hashimoto, M.; Im, S.J.; Araki, K.; Jin, H.T.; Davis, C.W.; Konieczny, B.T.; Spies, G.A.; McElrath, M.J.; Ahmed, R. Adenovirus Serotype 5 Vaccination Results in Suboptimal CD4 T Helper 1 Responses in Mice. *J. Virol.* **2017**, *91*. [CrossRef]

138. Berkeley, R.A.; Steele, L.P.; Mulder, A.A.; van den Wollenberg, D.J.M.; Kottke, T.J.; Thompson, J.; Coffey, M.; Hoeben, R.C.; Vile, R.G.; Melcher, A.; et al. Antibody-Neutralized Reovirus Is Effective in Oncolytic Virotherapy. *Cancer Immunol. Res.* **2018**, *6*, 1161–1173. [CrossRef]

139. Ricca, J.M.; Oseledchyk, A.; Walther, T.; Liu, C.; Mangarin, L.; Merghoub, T.; Wolchok, J.D.; Zamarin, D. Pre-existing Immunity to Oncolytic Virus Potentiates Its Immunotherapeutic Efficacy. *Mol. Ther.* **2018**, *26*, 1008–1019. [CrossRef]

140. Niemann, J.; Woller, N.; Brooks, J.; Fleischmann-Mundt, B.; Martin, N.T.; Kloos, A.; Knocke, S.; Ernst, A.M.; Manns, M.P.; Kubicka, S.; et al. Molecular retargeting of antibodies converts immune defense against oncolytic viruses into cancer immunotherapy. *Nat. Commun.* **2019**, *10*, 3236. [CrossRef]

141. Tahtinen, S.; Gronberg-Vaha-Koskela, S.; Lumen, D.; Merisalo-Soikkeli, M.; Siurala, M.; Airaksinen, A.J.; Vaha-Koskela, M.; Hemminki, A. Adenovirus Improves the Efficacy of Adoptive T-cell Therapy by Recruiting Immune Cells to and Promoting Their Activity at the Tumor. *Cancer Immunol. Res.* **2015**, *3*, 915–925. [CrossRef] [PubMed]

142. Porter, C.E.; Rosewell Shaw, A.; Jung, Y.; Yip, T.; Castro, P.D.; Sandulache, V.C.; Sikora, A.; Gottschalk, S.; Ittman, M.M.; Brenner, M.K.; et al. Oncolytic Adenovirus Armed with BiTE, Cytokine, and Checkpoint Inhibitor Enables CAR T Cells to Control the Growth of Heterogeneous Tumors. *Mol. Ther.* **2020**, *28*, 1251–1262. [CrossRef] [PubMed]

Publisher's Note: MDPI stays neutral with regard to jurisdictional claims in published maps and institutional affiliations.

Review

Hitting the Target but Missing the Point: Recent Progress towards Adenovirus-Based Precision Virotherapies

Tabitha G. Cunliffe [†], Emily A. Bates [†] and Alan L. Parker *

Division of Cancer and Genetics, School of Medicine, Cardiff University, Cardiff CF14 4XN, UK;
cunliffetg@cardiff.ac.uk (T.G.C.); batese@cardiff.ac.uk (E.A.B.)
* Correspondence: parkeral@cardiff.ac.uk
† Authors contributed equally to this work.

Received: 4 October 2020; Accepted: 9 November 2020; Published: 11 November 2020

Simple Summary: If harnessed appropriately, oncolytic viruses offer significant potential as anti-cancer agents. Such virotherapies can be engineered to replicate inside cancerous cells, stimulating the immune system, spreading daughter virions to surrounding cells and producing additional anticancer agents as a by-product of infection. To achieve this necessitates deep understanding of the biology of the virus and tumour cell, to tailor viruses from naturally pathogenic agents into refined, tumour selective "precision virotherapies" suitable for clinical translation. Here, we focus on the adenovirus, which in its pathogenic form causes transient and mild ocular, respiratory or gastrointestinal tract infections, depending on the serotype. We highlight advances that have been made in refining adenovirus to ablate natural means of infection and the strategies that have been employed to engineer viral tropism and selectivity for tumour cells. Further advances in these strategies will be required to deliver fully bespoke and efficacious precision virotherapies to the clinic.

Abstract: More people are surviving longer with cancer. Whilst this can be partially attributed to advances in early detection of cancers, there is little doubt that the improvement in survival statistics is also due to the expansion in the spectrum of treatments available for efficacious treatment. Transformative amongst those are immunotherapies, which have proven effective agents for treating immunogenic forms of cancer, although immunologically "cold" tumour types remain refractive. Oncolytic viruses, such as those based on adenovirus, have great potential as anti-cancer agents and have seen a resurgence of interest in recent years. Amongst their many advantages is their ability to induce immunogenic cell death (ICD) of infected tumour cells, thus providing the alluring potential to synergise with immunotherapies by turning immunologically "cold" tumours "hot". Additionally, enhanced immune mediated cell killing can be promoted through the local overexpression of immunological transgenes, encoded from within the engineered viral genome. To achieve this full potential requires the development of refined, tumour selective "precision virotherapies" that are extensively engineered to prevent off-target up take via native routes of infection and targeted to infect and replicate uniquely within malignantly transformed cells. Here, we review the latest advances towards this holy grail within the adenoviral field.

Keywords: adenovirus; oncolytic; virotherapy; targeting; immunotherapy; immunogenic cell death; $\alpha v \beta 6$ integrin

---

1. Introduction

Cancer treatment has come a long way in recent years, with 10-year survival rates increasing to around 50%, double that of 40 years ago [1]. While some of these improvements are credited to better and earlier diagnoses, a proportion of the advances in survival rates are attributable to the better understanding of cancer genetics, and thus how a patient may respond to a particular treatment. These advances have allowed clinicians to design and implement more efficacious and safer personalised treatment plans. Despite these advances, more progress remains to be made until fully personalised medicines are available for all patients. The need for such specific knowledge with a range of treatment options can lead onto the emerging era of targeted cancer medicines, in which the therapies act on a specific molecular target associated with the patient's cancer [2]. Some targeted therapeutics are already being used as the first line treatment for patients in the clinic, such as the monoclonal antibodies Herceptin and the newer pertuzumab that target the receptor HER2, which is overexpressed in cancers such as metastatic breast cancer [3–5]. Herceptin treatment in patients with HER2 overexpressing tumours results in significantly better survival rates [6]. Despite these advances in targeted therapies, continued progress into the understanding of cancer-specific markers, such as upregulated HER2, and the development of targeted treatments are required to improve patient survival further.

1.1. Use of Targeted Therapies

Targeted therapies can come in many forms, from antibody treatments, such as Herceptin, or treatment involving inhibitors to important enzymes, including upregulated kinases, such as mitogen-activated protein kinase 3 (MAPK3), which has abnormal expression in many forms of cancer [2]. Small molecular weight cancer drugs are also important since they readily enter cells and affect changes compared to large molecular weight drugs such as antibodies. These inhibitors can be used to target and block many enzymatic pathways which support cancer progression. For example, vascular endothelial growth factor receptor (VEGFR) inhibitors which aim to reduce angiogenesis are currently being developed, with just under 20 different molecules being tested in in vitro and clinical trials [7].

1.2. Oncolytic Viruses

One rejuvenated area of research for targeted therapy is oncolytic viruses (OVs). These cancer-killing viruses exploit the natural ability of a virus to infect, replicate and lyse cells. An additional benefit of OVs is that the lytic nature of cell killing induces immunogenic cell death (ICD), increasing recruitment of immune cells to the tumour site. This ability of OVs to enhance host anti-tumour immune responses through ICD has the potential to turn immunologically "cold" tumours "hot", thus sensitising otherwise refractory tumours to subsequent immunotherapies. This exciting potential of OVs to synergise with immunotherapies is the subject of significant ongoing clinical investigation, with compelling preliminary data obtained in several early phase clinical trials in brain tumours [8] and breast cancer [9], with other trials continuing in sarcoma, melanoma and breast cancer [10–12]. Efficacy may be further enhanced, with correspondingly reduced dose-limiting toxicities, through the development of rationally and effectively tumour targeted OVs suitable for intravenous applications.

The overarching aim in the development of OVs is to engineer viruses that can selectively target and/or replicate within cancer cells, leaving normal cells and tissues uninfected. The pro-immunogenic environment induced by OVs through the release of tumour antigens during cell lysis can be further enhanced by engineering the viral genome to over-express therapeutic transgenes, thus adding an extra layer to their therapeutic power. These multiple levels of activity are overviewed in Figure 1.

To limit "off target" uptake of OVs and subsequent killing of healthy cells, genetic engineering approaches have been undertaken. These have broadly focused on two approaches—firstly the active targeting of virions to tumour cells through the manipulation of viral capsid proteins to enhance uptake of virus into tumour cells and limit uptake via healthy cells. This has been largely achieved through the rational engineering of the viral genome and thus capsid to engineer tumour tropism via tumour associated antigens and receptors [13–15]. A second approach is to engineer in selectivity "post entry", such that viral replication is blocked in non-transformed cells, but subtle modifications within the viral early genes permit replication to proceed within malignantly transformed cells. Using combinations of these approaches, it is possible to achieve tightly controlled tumour cell killing, and thus creating a more effective cancer treatment.

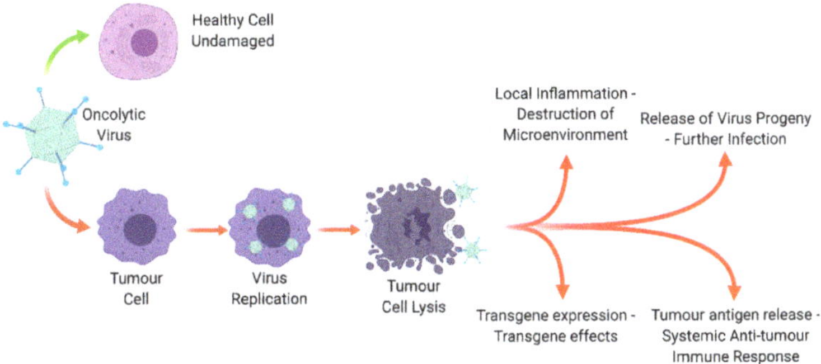

Figure 1. The mode of action of an oncolytic virus. Oncolytic viruses leave healthy cells undamaged, whilst leading to a range of effects in tumour cells which lead to lysis, further infection and an immunological response. Created with https://biorender.com.

The first OV licensed in the western world by both the FDA and EMA is talimogene laherparepvec (T-Vec, Imlygic™), which exemplifies the early therapeutic potential of oncolytic viruses. T-Vec is a herpes simplex virus 1, with several genetic modifications. Deletions of the ICP34.5 gene reduces neurovirulence whilst also suppressing its ability to reactivate, which prevents it producing cold sores in the patient. A deletion in the ICP47 gene leads to enhanced antigen loading on MHC class I molecules. The MHC loading results in infected cells presenting tumour antigens, leading to enhanced immune recognition. These deletions improve the safety of this treatment as it limits the ability of the virus to completely evade immune responses. In addition, T-Vec is engineered to overexpress the cytokine GM-CSF (granulocyte-macrophage colony-stimulating factor) to further enhance immunogenicity through T-cell priming [16]. A truncated version of the US11 gene is also included to enhance the lytic activity of the virus by partial de-attenuation [17]. The virus takes advantage of disrupted anti-viral pathways in cancer cells to enable selective virus replication. One such pathway is the PKR (protein kinase R) pathway which is key in regulating cell proliferation. Normally, this pathway is activated by the dsRNA produced when the virus replicates, thus triggering protein synthesis inhibition. However, in cancer cells with aberrant regulation, the dsRNA production warning sign is ignored, allowing the virus to replicate, leading to viral propagation and cell lysis. Disruption in the PKR pathway has been found in approximately 70% of melanoma cells, and thus T-Vec is a good replication selective therapy for melanoma cancers [18,19].

T-Vec was licensed for localised treatment of recurrent melanoma [20,21]. In phase III trials using T-Vec to treat unresectable late stage melanoma, durable response rates of 16% vs. 2% were shown, compared to control group, increasing the survival rates of these patients [22,23]. Moreover, recent studies have demonstrated that virotherapies such as T-Vec can be a powerful tool in adjuvant therapy and have been used to sensitise triple negative breast cancers to follow up treatment with immunotherapies [9,24].

1.3. Adenovirus as an Oncolytic Virotherapy

Despite the success of T-Vec, the therapy is only licensed for local intra-tumoural applications. Therefore, its use for treatment of metastases, which kills the majority of cancer patients, is limited. Metastasis treatment would require blood system disseminated therapy to individually target each lesion through intravenous (IV) delivery, or utilising the abscopal effect from treatment of one or more lesions, which could lead to a systemic anti-tumour immune response extending to the milieu of micro metastatic deposits in the body [25–27]. Therefore, significant research is ongoing into other viral vectors that may be better suited to IV delivery, including adenoviruses (Ads) [28], reovirus [29] and vaccinia virus [30]. Ads have been shown to be the most durable of these options and are the most studied clinically and experimentally. Ads also have the advantage of being naturally lytic, immunogenic and can be produced to high titres and purity, all important considerations when developing such therapeutics for widespread clinical application.

1.3.1. Adenovirus Cell Entry and Trafficking

Human adenoviruses can be classified into seven species (termed A–G) comprised of over 100 serotypes [31]. Ads commonly infect the respiratory system however different serotypes can also infect the gastrointestinal, cardiac, neurological, ophthalmological and genitourinary tissues which can result in an array of clinical pathologies [32]. Ads bind to receptors on the membrane surface. Ad5 has been well described as recognising the coxsackievirus and adenovirus receptor (hCAR) [33]. Other receptors involved in adenoviral attachment are CD46 [34], desmoglein 2 (DSG2) [35] and sialic acid [36]. Primary receptor binding is dependent on the Ad serotype but generally species A, C, E and F interact with hCAR while species B and D utilise other receptors [37]. Species B is reported as using primarily CD46 and DSG2, and, although the individual receptors for many species D serotypes are unknown, it is thought there is sialic acid involvement [38,39]. Initial receptor attachment is followed by internalisation mediated by $\alpha v \beta 3/5$ integrin binding through a conserved Arg-Gly-Asp (RGD) motif [40]. Upon cell entry, the adenovirus is partially disassembled via endosome acidification and the uncoated is released and transports viral DNA into the nucleus [41]. Viral replication takes place in two phases: early phase and late phase. An overview of the adenoviral replication cycle is illustrated in Figure 2.

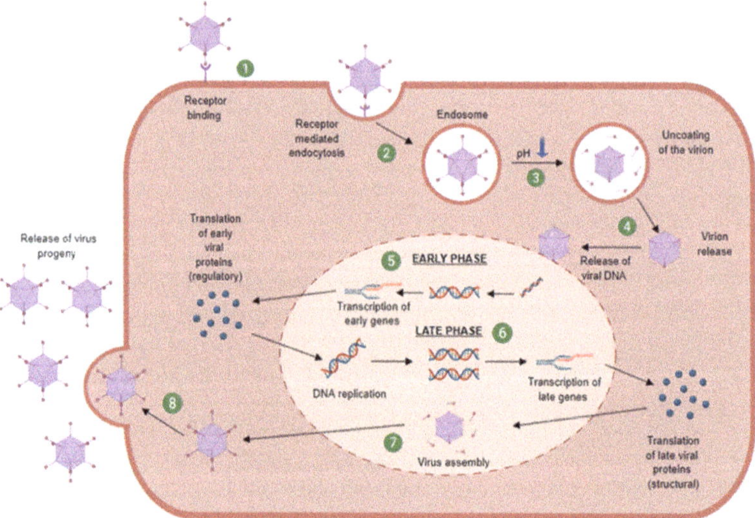

Figure 2. Adenovirus replication cycle. (**1**) Virus attachment to receptors on the host cell surface. (**2**) Internalisation of the virus by endocytosis. (**3**) Low pH results in endosomal acidification and partial disassembly of the virion. (**4**) Virion released from endosome and trafficked to the nuclear pore complex where it releases viral DNA into the nucleus. (**5**) Early phase: Transcription and subsequent translation of early genes to the regulatory early proteins. (**6**) Late phase: Transcription and subsequent translation of late genes to the late structural proteins. (**7**) Assembly of progeny virion. (**8**) Cell lysis resulting in release of mature virus. Created with https://biorender.com. Figure adapted from [41–44].

1.3.2. Oncolytic Adenovirus

Many Ad-based therapies in the clinic currently use replication-based control as a mechanism for cancer targeting. They are reliant on cancer selectivity at the point of cellular replication at a post cell entry stage rather than bona-fide tumour tropism. The concept that adenovirus infection mirrors several key hallmarks of cancer [45] underpins the mechanism that restricts the replication of many early oncolytic adenoviruses to tumour cells [46]. An oncolytic adenovirus that preferentially replicates in a tumour cell environment often takes advantage of genes that are frequently overexpressed in cancer. Deletion of viral replication genes to render the virus replication-incompetent in healthy cells, but replication proficient in tumour cells with dysregulated or inactivated tumour suppressor gene function [47,48]. These are termed replication selective oncolytic or conditionally replicating adenovirus [49,50].

Adenovirus have many benefits for use as oncolytics. These include their relative ease of manipulation, being double stranded DNA viruses. Their capacity for transgene incorporation, being around 6 kb for non-replicating vectors and around 2 kb for oncolytic vectors is more than sufficient to encode therapeutic transgenes (or combinations of transgenes) to enhance the pro-immunogenic tumour microenvironment. These transgenes could include immune checkpoint inhibitors targeting PD-L1 or CTLA4 [51,52], cytokines such as TNFα and IL-2 [53] or chemokines including CCL5 [54]. Alternatively, genes encoding proteins that are directly cytotoxic to the tumour cells such as REIC/DKK-3 can also be incorporated [55].

Although early clinical data for many oncolytic viruses are encouraging, the exact mechanism of cell killing often remains unclear [56]. It is evident the viral and host cell interactions are complex, particularly in the context of systemic delivery and within the tumour microenvironment, and an understanding of the tumour and virus biology will provide insight and enhance future oncolytic virotherapies [57]. The popularity of adenovirus is evidenced by the sheer number of clinical trials, standing at 237 at the time of writing, that use adenovirus for cancer treatment in some form [58]. These trials have demonstrated safety and feasibility; however, delivery and efficacy must be improved if oncolytic adenovirus is to achieve its full promise as an effective cancer therapy [59,60].

Despite their immense potential, adenoviruses, especially those based on the species C serotype Ad5, have several pitfalls which need to be carefully addressed to tailor the OV in to an effective therapeutic. These disadvantages include the high rates of pre-existing immunity against Ad5 in the populations where Ad5 is a common pathogen. These levels of pre-existing immunity vary geographically from >90% in sub-Saharan Africa [61] to ~30% in the UK population [62]. High levels of pre-existing immunity will promote the rapid removal and destruction of the therapeutic by the reticuloendothelial system, resulting in limited bioavailability for active tumour targeting [63]. A further limitation stems from the native infectious routes via the capsid proteins of Ad5 that can also cause dose-limiting interactions and toxicity (Figure 3). The widespread anatomical expression of the primary receptor, Coxsackie and adenovirus receptor (hCAR) [33], means vectors based on Ad5 will be sequestered and infect a wide range of off target (non-cancerous) tissues in the body [64] or may become irreversibly trapped in the blood [65]. One way this can be overcome is by genetic modification of the amino acids 408 and 409 within the AB loop of the fibre knob protein (Fkn) to remove binding to hCAR (called the KO1 mutation) [66]. Although Ad5 predominantly uses hCAR there are alternative receptors utilised by other species including CD46 and desmoglein-2 (DSG-2) which are the primary entry route for Species B adenovirus [34,35]. The species B Ad3 pseudotype is a prominent oncolytic virus which uses both CD46 and DSG-2 for cell entry. CD46 is expressed on almost all nucleated cells, and DSG-2 is a cardiomyocyte [67] and tight junctions restricted receptor [68], and therefore present additional considerations for ablation of native binding tropisms. Species D does not appear to bind these three known adenoviral receptors with any significant affinity. There is evidence to suggest these viruses may be more likely to use sialic acid as their mode of entry [36,37,69].

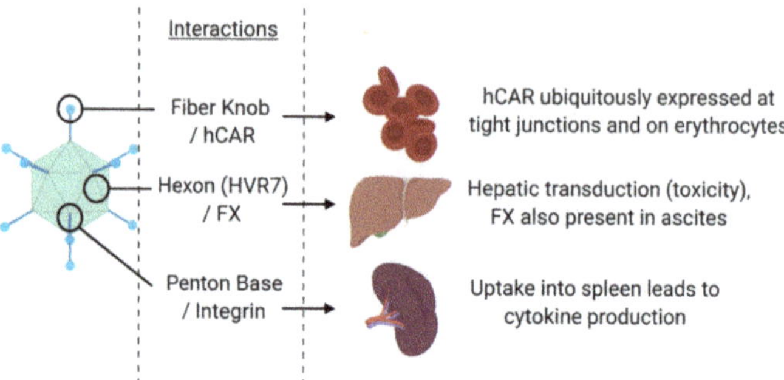

Figure 3. Dose-limiting Ad5 interactions in vivo. The fibre knob protein binds to hCAR expressed at tight junctions and on erythrocytes, the hexon binds to Factor X (FX) in the blood and the penton base binds to $\alpha v \beta 3/5$ integrins. These binding interactions would lead to off-target effects. Created with https://biorender.com.

Other capsid proteins can also cause off-target binding and sequestration issues. The hexon protein of Ad5 binds with high affinity to the blood clotting factor X (FX), which results in rapid and efficient transduction of hepatocytes, with consequent potential hepatotoxicity resulting from of Ad5 vectors [70,71]. The penton base protein on the capsid also has implications for off-target effects. The RGD domain in the pentameric protein group binds to integrins $\alpha v \beta 3/\alpha v \beta 5$ leading to downstream signalling for internalisation [72]. These interactions are also thought to lead to uptake in the spleen inducing consequent pro-inflammatory responses against the Ad [73,74]. Therefore, mutation within the RGD binding region in the penton may be important in limiting these off-target effects [75].

In this review, we discuss the current approaches and significant refinements to the Ad5 capsid necessary to prevent off target interactions. We also consider alternative approaches to circumvent the Ad5 associated limitations and generate precisely guided cancer therapeutics.

2. Genetic Engineering of Oncolytic Adenovirus

The adenoviral genome is organised into early (E) and late (L) genes (Figure 4). The early phase genes encode proteins that regulate the host and viral proteins, avoid premature cell lysis and prepare components for DNA replication. Late phase produces structural proteins that are required for the assembly of mature virions [44].

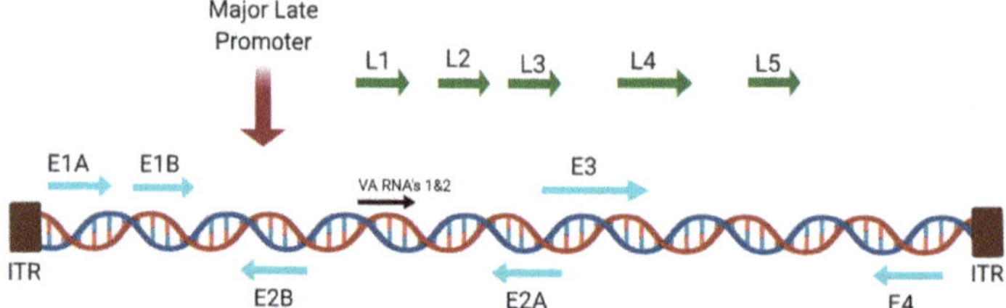

Figure 4. Adenoviral genome, highlighting key genes which are often modified or deleted in oncolytic therapeutics. Created with https://biorender.com.

The standard approach in the design of novel oncolytic virotherapies involves making modifications viral genes to improve cancer cell selectivity and oncolytic potency. The ability to engineer the double stranded DNA genome of adenovirus with relative ease has been proven for clinical applications from vectors for gene therapy and vaccines to oncolytic viruses [76]. A key feature in development of adenoviral vectors are the modifications to reduce the immunogenicity and bypass innate anti-viral immune responses.

First generation adenoviral vectors harbour deletions in the E1 and E3 regions [77,78]. These deletions not only improved the vector safety profile but also create significant space necessary for the insertion of transgenes [79]. The E1 genes encode proteins necessary for viral replication, therefore E1 deletion results in a replication deficient virus [44]. Consequently, vectors with this deletion must be propagated in cell lines expressing E1 products in trans, such as 293 or PER.C6 cells [80–82]. E3 encoded viral proteins are involved in evading host antiviral immunity but are not essential for viral replication, deletion of this region allows insertion of larger genes but may reduce the oncolytic potency [83]. Second generation adenoviral vectors may also have the E2 and E4 regions deleted which eliminates expression of most Ad genes and allows more room for transgene insertion [84]. However, these vectors must be propagated in cell lines that express E1, E2 and E4 gene products. The late genes are involved in structure and therefore are required for production of mature virions. A final generation of Ad vectors that are lacking all viral coding regions have been developed. These are termed gutless or helper-dependent Ads as they require co-infection of a wild-type adenovirus or helper vectors [85]. These have promising therapeutic advantages but are difficult to manufacture in high quantities.

Conditionally replicating adenoviruses (CRAds) encompass several oncolytic adenovirus therapies in the clinic. They can be classified into two types, however both approaches involve modifications in the E1 region of the adenoviral genome. Adenovirus E1 is comprised of two genes: early region 1A (E1A) and early region 1B (E1B). E1A is the first transcribed gene post infection and promotes progression into S-phase of the cell cycle. E1B encodes genes that protect the cell from undergoing apoptosis as a result of E1A induction of S phase and enables the virus to undergo productive replication in the host cell. The first strategy employed when designing CRAds is to replace the E1 promotor with a tumour specific promotor, therefore preventing induction of E1A mediated viral replication in the absence of the appropriate promoter [86–88]. This approach can be used to restrict replication and to start the expression of the treatment transgenes within tumour cells. One example is the promotor survivin, which has been used to this effect, regulating the expression of the heat shock protein 70 (Hsp70) that inhibited tumour growth in gastric cancer and adult T-cell leukaemia (ATL) [86,89]. Another promotor of note is human telomerase reverse transcriptase promotor (hTERT). This promotor can enhance cell lysis, leads to increased release of viral progeny for further infection and shows reduced hepatocyte effects compared to ONYX-015 in solid tumour in-vivo models [90].

The second strategy relies on modifications within the E1 region preventing the virus from restricting host cell defences (for example, pRb mediated apoptosis) and therefore the virus is only able to replicate in tumour cells defective in these pathways. One of the most effective mutations described to date is the dl24 (Δ24) mutation. This mutation is a 24-base pair deletion in the constant region of E1A gene. This deletion is in the region that is responsible for binding the Rb protein and so targets replication to cells with abnormal Rb control that can bypass this pathway. This leads to selective replication in cells that are defective in the Rb/p16 pathway, which has been identified in the majority of cancers, including gliomas and ovarian cancers [91–94].

Another mutation used is the T1 mutation, which has a truncating insertion in the E3/19K protein. The T1 mutation means that this protein is relocated to the plasma membrane and enhances the release of virus from infected cells [95]. Therefore, this mutation may be a useful addition to a tumour-selective Ad-based therapy, such as in the oncolytic ORCA-010 [96].

3. Current Clinical Applications of Oncolytic Adenoviral Therapies

ONYX-015 (also referred to as Δ1520) was one of the first replication selective oncolytic adenoviruses tested in a clinical setting for the treatment of head and neck cancer [97,98]. ONYX-015 harbours an E1B55K deletion, which was originally thought to be essential for viral replication as it sequesters p53 and promotes cell cycle transition [99]. p53 is commonly lost or downregulated through mutations in multiple cancer cells [100], and it is therefore considered that ONYX-015 would replicate almost exclusively in cancer cells lacking p53. However, more recent research suggests that this mutation is more likely to work through the loss of late viral RNA export, rather than through p53 status alone. The mechanism of action may be more complex than originally thought, likely due to multi-modal action of the p53 pathway [49,101,102]. Whilst ONYX-015 and a variant H101 (E1B55K and E3 deletion) demonstrated the safety of oncolytic adenoviruses, the efficacy was limited by attenuated viral replication and spread [96,103]. Subsequent generations of tumour selective oncolytic adenoviruses contain mutations in the E1A gene that functions through binding the retinoblastoma protein (pRb) [91]. Several oncolytic adeno-virotherapies have since entered clinical trials (Table 1) and have demonstrated safety and feasibility. However, delivery and efficacy must be improved if oncolytic adenovirus is to be used as an effective cancer therapy, especially as a systemically administered agent capable of effectively targeting tumours and metastases [59,60].

The vast majority of adenovirus research to date has focused on the species C, Ad5. Several oncolytic adenoviruses (OAds) have demonstrated limited efficacy in clinical trials as a result of poor viral persistence [104,105]. Although this is, in part, related to the early design of these viruses, it may also result from high levels of pre-existing immunity [106]. A substantial proportion of the population will have experienced an acute adenovirus infection, and many will have developed neutralising antibodies against the most common Ad serotypes [32]. Activation of anti-tumour immunity whilst dampening the innate host anti-viral immune response is essential to the success of OAds. An alternative approach, therefore, may be through the development of alternative, low seroprevalence adenoviral species, such as those from Species B or D, which tend to have naturally low levels of pre-existing immunity in the population [107]. Such serotypes may also exhibit naturally lower levels of off target uptake due to reduced interactions with components of the blood. Ad5 is known to bind to FX in serum which mediates sequestration by the liver and can impact virus delivery to the tumour [108]. FX binds the hexon protein on Ad5 capsid and can result in off target uptake [109]. It was observed that alternative species, for example species D Ad26, does not bind FX in the same manner as Ad5 [109]. This knowledge was used to identify key residues in FX binding through sequence alignment and has fed into the production of retargeted Ad vectors however the use of alternative species as an alternative to Ad5 has not been fully explored [71]. The use of novel oncolytics developed from rarely isolated serotypes from species D may represent an exciting and alluring possibility, where the diverse nature of this species represents a significant and largely untapped repository for investigation. Recent significant progress has been made in this regard by the Ehrhardt laboratory, who have investigated a larger spectrum of adenoviral vectors and begun to evaluate their potential for oncology applications [110,111].

Alternative adenoviral species offer many advantages over the commonly used Ad5-based therapies however, despite their increasing popularity as platforms for vaccine applications, they are poorly represented in the clinical oncology setting. Ad5/kn3 pseudotype has demonstrated some limited efficacy in ovarian cancer [112,113]. This targets via species B receptors which are not cancer specific and maintain the previously outlined issues associated with Ad5-based vectors.

Table 1. A summary table of Ads in clinical trials, with their base genome, modifications and their target. Data obtained from https://www.cancer.gov/ (Accessed on: 30 October 2020).

Biologic	Synonyms	Adenovirus genome	Modifications	Targeting	NCI Identifier
GM-CSF-encoding Oncolytic Adenovirus CGTG-102	ONCOS-102	Adenovirus serotype 5/3 (capsid-modified)	Ad5 capsid protein replaced with Ad3 knob domain. Granulocyte-macrophage colony stimulating factor (GM-CSF)	Selective replication in Rb/p16 defective cells. Ad3 receptors.	C98287
OX40L-expressing Oncolytic Adenovirus DNX-2440	Oncolytic Adenovirus Armed with OX40L DNX-2440	Adenovirus serotype 5	Expresses OX40 ligand (OX40L). Δ24 mutation	Selective replication in Rb/p16 defective cells	C160192
Oncolytic Adenovirus ORCA-010	Modified Ad5 ORCA-010 / Oncolytic Adenovirus ORCA-010	Ad5/3	Δ24 mutation. RGD-4C motif. T1 mutation in E3/19K gene	Selective replication in Rb/p16 defective cells. T1 mutation enhances Ad5 release, Ad3 receptors	C168607
Oncolytic Adenovirus Encoding GM-CSF	CG0070	Adenovirus serotype 5	E2F-1 promotor. Granulocyte-macrophage colony stimulating factor (GM-CSF) in E3 region	Selective replication in Rb/p16 defective cells	C48412
Delolimogene Mupadenorepvec	Double-armed TMZ-CD40L/4-1BBL Oncolytic Ad5/35 Adenovirus LOAd703	Adenovirus serotype 5 with L5 segment of fiber replaced with Ad35 fiber	Expresses trimerized CD40 ligand. Δ24 mutation in E1A	Targets CD46. Selective replication in Rb/p16 defective cells	C148462
Oncolytic Adenovirus ICOVIR5-infected Autologous Mesenchymal Stem Cells	LOAd 703	Wildtype human adenovirus 5	RGD-4C motif allows integrin binding. Δ24 in E1A prevents Rb complex and transition into S phase	Bone marrow-derived MSCs target and deliver adenovirus to tumour	C107160
Tasadenoturev	DNX-2401 / (Oncolytic Adenovirus) Ad5-Δ24RGD / Oncolytic Adenovirus Ad5-DNX-2401	Adenovirus serotype 5	RGD-4C motif allows integrin binding. Δ24 in E1A prevents Rb complex and transition into S phase	CAR independent. Selective replication in Rb/p16 defective cells	C74067
Tasadenoturev-infected Allogeneic Bone Marrow-derived Mesenchymal Stem Cells	Ad5-DNX-2401-infected Allogeneic Bone Marrow Mesenchymal Stem Cells / (Allogeneic) BM-hMSC-Δ24 / (Allogeneic) BM-hMSC-Δ24-RGD	Ad5-DNX-2401	RGD-4C motif, Δ24 in E1A prevents Rb complex and transition into S phase	Bone marrow-derived MSCs target and deliver adenovirus to tumour	C159798
Ad5-yCD/mutTKSR39rep-hIL12	Oncolytic Adenovirus Ad5-yCD/mutTKSR39rep-hIL12	Adenovirus serotype 5	Encodes murine interleukin-12 (IL-12) gene in E3 region and a suicide fusion gene (yCD/HSV-1 TKSR39) in E1 region	E1B55K deletion	C123930
Enadenotucirev	ColoAd-1 / EnAd	Chimeric Oncolytic Adenovirus Ad3/Ad11p	Deletions in E3 Region (2444 bp) and E4 Region (24 bp) and 197 Non-homologous nucleotides in the E2B Region	Not fully understood	C113786

Species B adenovirus have demonstrated the potential to play an important role in the field of adenoviral oncolytics. An alternative strategy to the rational development of novel Ad serotypes for oncology applications has been to develop a novel chimeric OV through a process of natural selection for recombinants with enhanced cell killing activity in cancer cells such as with enadenotucirev (EnAd, formally known as ColoAd1; PsiOxus Therapeutics Ltd., Abingdon, UK). Several adenovirus serotype recombinants were selected for on colon cancer cells (HT29) by this method of "directed evolution" (Figure 5) [114]. Despite the initial pool representing diverse serotypes from species B–F, the resultant chimera is a fully species B recombinant, derived from Ad11p and Ad3 [114]. This virus demonstrated potency and selectivity greater than ONYX-015 and Ad5 [115]. The tumour selectivity mechanism for EnAd is not fully understood however initial clinical trials have demonstrated durability and tolerability, not only for colorectal cancers but with other solid cancers [115,116]. EnAd is thought to act through a non-apoptotic mechanism termed ischemic cell death and possesses pro-inflammatory properties [117]. Species B therefore represents an exciting avenue for development of oncolytics, however the lack of intrinsic tumour specificity and high prevalence of species B receptors, CD46 and DSG-2, on healthy cells is an important consideration that may result in the depletion of virotherapy available for active targeting of tumours through off-target sequestration. Ablation of these native interactions or use of Ads that bind receptors with weak affinity may result in improved novel retargeted oncolytics.

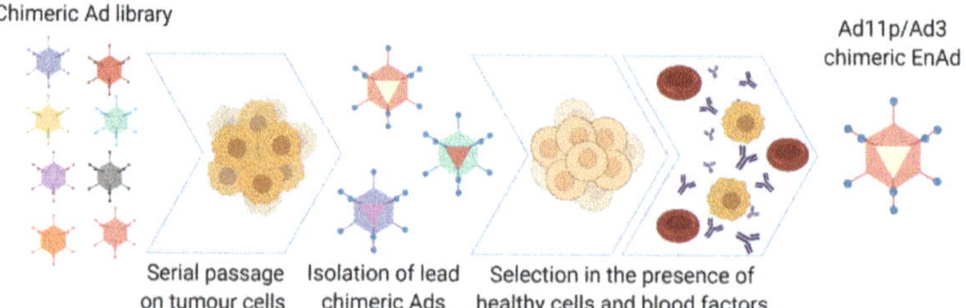

Figure 5. Method of production for oncolytic virus, EnAd (PsiOxus Therapeutics Ltd., Abingdon, UK). Serial passages of an Adenoviral library in tumour cells lead to chimeric Ads development. Selection with healthy cells and blood factors removes those that bind off-target receptors, lead to Ad11p/Ad3 chimeric EnAd being selected for. Created with https://biorender.com.

4. Detargeting Ads

Adenovirus has not evolved as a cancer selective pathogen, and therefore requires engineering in order to effectively target cancer cells. In Ad-based therapies, the efficiency of the treatment can be greatly affected by their native binding interactions. The most studied Ad, Ad5, binds hCAR which localises to the tight junctions between cells and is expressed ubiquitously throughout the body [33,118]. hCAR has also been reported as downregulated in certain cancers [119–121]. Therefore, reliance upon hCAR as an entry receptor for any Ad-based cancer therapy would greatly limit its uses, as transduction would be limited to cancer cells with high-hCAR expression, and off-target transduction could lead to tissue toxicity. Moreover, more aggressive cancer growth correlates with loss of hCAR expression, and so non-targeted Ad therapy is unlikely to treat aggressive cancers through hCAR transduction alone [118].

hCAR binding can be ablated by mutating the key amino acids in the fibre knob AB loop (L5 gene), using the KO1 mutations, S408E and P409A [66,122]. hCAR mediated cell entry is a two-step mechanism. First the virus attaches to hCAR and, secondly the virion internalised through binding of an Arg-Gly-Asp (RGD) motif in the penton base to $\alpha v \beta 3/\alpha v \beta 5$ transmembrane integrins on the cell surface [123]. This secondary interaction has been exploited to further detarget Ad5 through the RGD to RGE modification of the penton base and improve cancer targeting. The Ad5$_{NULL}$ vector encompasses the KO1 mutation, the RGD to RGE modification as well as a modification within the hexon hypervariable region (HVR7) in order to ablate binding to coagulation FX and prevent sequestration by and transduction of liver hepatocytes [13,124].

Larger modifications can also be made to detarget the virus particle, such as replacing the fibre shaft of Ad5 with shorter variations found in other serotypes such as Ad40, or Ad41. This approach has shown reduced binding to cells [125–127].

Chemical modifications can also be used as a means of detargeting through polymer coating of the Ad particle. Polyethylene glycol (PEG) is commonly used for this purpose, due to its cationic properties [128]. The main advantage of coating the Ad is to prevent neutralisation by pre-existing antibodies which reduces the efficiency in patients previously exposed to the Ad. The use of chemical modifications was well reviewed elsewhere previously by Kreppel and Kochanek [129] and Kim et al. [130]. A major disadvantage of non-genetic means of targeting is that daughter virions produced through replication will not harbour the modifications necessary to target tumour cells, and therefore genetic strategies which are heritable are therefore more commonly preferred.

5. Retargeting Strategies

Here, a range of retargeting strategies is discussed, and both CRAds and oncolytic virus results are considered in this section together. This is due to the abilities of the targeting strategies to be applied in either context.

5.1. Pseudotyping

One relatively common method to introduce new tropism to an Ad-based therapeutic is to use chimeric fibre knob/shaft proteins through pseudotyping. Pseudotypes are recombinant adenoviruses that combine different aspects and structural proteins from differing serotypes into a chimera, and are often generated to cherry pick optimal features associated with different serotypes [131]. This genetic strategy uses Fkn (fibre knob) proteins from less-commonly used Ad species that do not use hCAR as a primary transduction mechanism and substituting these onto an OV based on Ad5. This confers new binding abilities without having to move away from the large knowledge base of commonly used Ad5. This has shown some success in colon cancer [132] and ovarian cancer [133] amongst others.

The use of chimeric Fkns (cFkns) has been extended by sequences from non-human Ad species [134] and through chimeric fibres created from several sequences such as bacteriophage T4 fibritin and human CD40 ligand (CD40L) in conjunction with the Ad5 Fkn [15].

Pseudotyping can also be used for the whole fibre protein comprising both the shaft and knob domain into the Ad5 capsid. However, the "tail domain" is an important consideration in this strategy, as the maintenance of a portion of the N-terminus of the parent fibre shaft is required for the translation of binding to the penton base of the parental capsid [135]. Shayakhmetov and Lieber demonstrated that pseudotyping the fibre proteins can also alter the binding capabilities of the virus. Pseudotyping the fibre shaft and the Fkn proteins from Ad5, Ad9 and Ad35 results in different receptor usage and intracellular trafficking, partly down to the length and geometry of the fibre shaft [136].

Though this method has shown promising results, it has its limitations. Creating a chimera that can correctly fold once it is attached to the fibre shaft can be a limitation in itself. The cFkn formed and its binding capabilities are likely dependent on its ability to form trimers, which is required for native Ad binding [137]. If such a protein is found that can correctly fold once attached to the fibre shaft, and form the trimers required for binding, it then must confer a novel binding tropism. The natural array of binding tropisms already understood from the less-commonly studied Ad species that can be pseudotyped is limited. Therefore, although this method has its place as a tool for investigating the tropism of rarely isolated adenoviral species, other mechanisms may prove more useful in the context of developing tumour tropism.

5.2. Peptide Retargeting

One mechanism employed experimentally to enhance hCAR independent uptake of Ads into cells is to enhance targeting to upregulated αvβ3/5 integrins on tumour cells. The most successfully deployed has been the RGD-4C motif, incorporated into the HI loop of the fibre knob, which has demonstrated improved uptake in cancers expressing high levels of integrins such as ovarian cancer and glioma [138,139]. This modification enables suitable presentation of the integrin interacting RGD motif, held in position by the pair of disulphide bonds between the cysteine residues, to successfully engage with cellular integrins and stimulate uptake via endosomes.

Other retargeting methods have also proven effective, such as insertion of peptide sequences which confer a known binding ability within the Fkn protein, though this comes with its own limitations. The sites within the Ad5 fibre knob domain in which peptide sequences can be inserted successfully have been narrowed down through structural studies, demonstrating the HI loop and the C-terminus of the protein as the most promising sites [123,140]. Within other serotypes, hypervariable nature of the loops within the fibre knob protein have demonstrated that other loops are more surface exposed and thus better suited to genetic insertions, for example the DG loop in Ad48 has been shown to be the region best suited to genetic manipulation [141,142]. Insertions in these sites have shown promising results in targeting Ads, such as targeting towards ovarian, breast and prostate cancer cell lines by insertion of Her2/neu-reactive Affibody into the HI loop of a native-binding ablated Ad5 vector [143,144].

Peptide sequences from other viruses had also been shown to target an Ad-based therapy towards cancer cells. Insertion of a 20-aa peptide, NAVPNLRGDLQVLAQKART, native to foot and mouth disease virus (FMDV) was identified as a binding peptide sequence to αvβ6 integrin [145], an integrin that is reported to be upregulated in certain epithelial cancers, including breast, ovarian, pancreatic and colorectal [146–148]. Figure 6 illustrates the Ad5 fibre knob protein engineered to present the A20 peptide within the HI loop in complex with αvβ6 integrin. This peptide has previously been studied in terms of cancer research, used for non-invasive radiolabelled peptide for cancer imagery, whilst antibodies and CAR-T cells directed to αvβ6 integrin are also being investigated, underpinning the potential of this biomarker for cancer selective targeting [145,149,150]. The A20 peptide has been successfully incorporated into the Ad5$_{NULL}$ to create the Ad5$_{NULL}$-A20 OV, which is a highly selective virotherapy targeting ovarian cancer [13], and with significant promise to target other epithelial cancers expressing high levels of αvβ6 integrin.

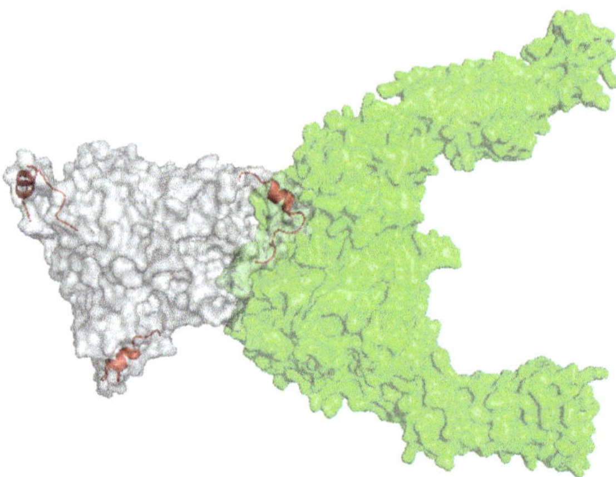

Figure 6. Ad5 knob (white) with the A20 peptide insertion (red) in complex with αvβ6 (green). Image created using PyMol.

Other sites in the capsid that are amenable to insertions have also been explored, such as the hexon gene. Insertion into the hexon has good potential for targeting due to its abundance in the capsid. If all copies displayed the insertion, it could lead to a coated Ad capsid with 720 copies of the targeting peptide. This has shown some recent success, through insertion of muscle binding peptides into the hypervariable region 4 (HVR4) in the hexon protein [14], as well as the RGD-4C peptide [151,152].

When modifying viral capsid proteins, additional considerations should be made. The size, structure and charge of the insert can affect its success due to steric hindrances. The rate-limiting step is the lack of efficacious tumour targeting peptides that can be incorporated into the viral capsid efficiently to retarget towards cancer cells with limited off-site effects.

Targeting peptides which have proven effective when presented within the context of the three-dimensional fibre knob protein often appear to be those which have a degree of secondary structure. The secondary structure is important to consider as strategies using linear peptides have proven to be less successful. For example, the RGD-4C peptide is designed to present the RGD integrin interacting tripeptide at the apex of the loop, held in place by disulphide linkages. Similarly, the A20 peptide forms an alpha helical confirmation both in its native context within the vp1 protein of FMDV, which it retains when transferred into the Ad5 fibre knob protein, thus retaining the geometry required for receptor engagement. For future targeting strategies to be successful, it will likely require the development of sophisticated molecular technologies, capable of high throughput evolution, screening and selection of knob variants with increased binding affinity for tumour associated antigens of interest.

5.3. Techniques for Targeting Peptide Discovery

Other methods for retargeting Ads have also had some success. Despite the success of Ad5$_{NULL}$-A20, there has been limited continued success using the Ad5$_{NULL}$ platform to target other tumour-associated antigens (TAAs) and receptors. The rate-limiting step is the lack of efficacious tumour targeting peptides that can be incorporated into the viral capsid efficiently to retarget the Ad5$_{NULL}$ platform towards tumour cells. Previously, we and others have utilised methods such as bacteriophage (phage) biopanning to identify peptides that can bind TAAs [153–155].

Biopanning is an approach that uses affinity-based selection. Random peptide libraries can be created and displayed on the phage, often in the pVII or pIII gene of filamentous phage M13 (Figure 7). M13 has around 5 copies of each pVII/pIII gene products in the capsid, located at the end of the cylindrical phage. The resultant library is incubated with a target protein (or cell line), allowing binding to occur. Unbound phage, or those with low affinity, are then washed away. Finally, those random peptides still bound strongly to the target are eluted, either by changing the pH or by competitive inhibition. The process is repeated to identify peptides with the highest affinity for the given target, which can be sequenced for further use [156]. Phage display allows for high throughput analysis of peptide libraries for targeting a specific receptor protein. In fact, this method has been used frequently, with peptides targeting EGFR [157] and HER2 [158]. Promising peptides can be inserted into the permissive regions of Ad5 Fkn [155]. This has been tried with several different peptides targeting cancer-specific markers, such as folate receptor α (FRα) commonly upregulated in ovarian cancers [159]. However, after binding, the FRα mediated cell entry mechanism does not allow for correct intracellular trafficking, showing retention of targeted virotherapies in late endosomes in FRα positive ovarian cancer cells, with limited successful transduction [153]. This highlights another potential consideration when developing targeting approaches for adenoviral-based oncolytics—not all TAA receptor pathways will be compatible with clathrin-mediated endocytosis pathways and thus represent viable entry routes for adenoviral-based virotherapies.

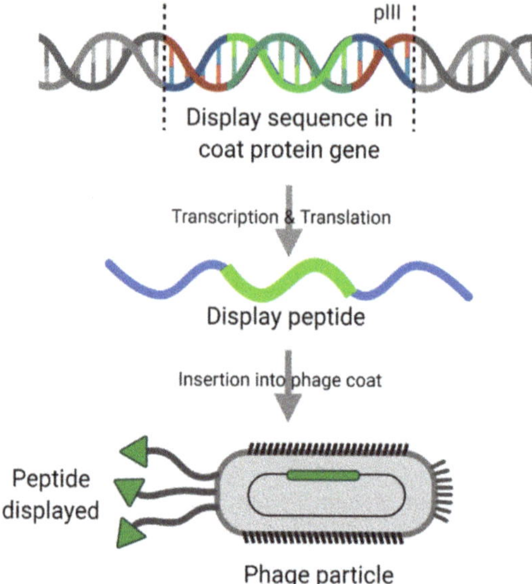

Figure 7. M13-based phage display mechanism. Created with https://biorender.com.

Additional limitations to this approach revolve around the linear orientation of the peptides being selected and displayed. Although promising peptides can be found through this approach, once they are incorporated back in the Fkn, the peptide can change confirmation due to the three-dimensional nature of the viral capsid protein into which the peptide is engineered. It could be assumed that the difference, and therefore the success, of the A20-peptide is due to the constrained orientation was maintained. A20 was identified in FMDV and transposed into Ad, and thus the orientation was conferred. Conversely, in phage-display technology, the selection is based around incorporation into the capsid coat proteins in a linear orientation.

A secondary issue of note for the future of this technology is that insertions can only be made in one linear string of DNA, creating one addition to the coat proteins in the phage particle to be used in selection. For this technology to be the answer to extending the use of OAds to many different cancer types, with very different surface protein expression profiles, there needs to be a way of incorporating, and thus selecting for, multiple regions that confer binding but are not next to each other in linear DNA sequence. This would create additional problems such as the requirement for multiple incorporation sites in the Ad Fkn protein and the complexity of maintaining the correct confirmation of these multiple sites in protein space (for example, distance apart on the Fkn, interactions with polar amino acids nearby limiting availability for binding and flexibility of the insert). However, if these were to be achieved, it would allow for quick, efficient and effective selection of cancer-specific binding peptide regions for incorporation into an Ad-based vector.

Lupold et al. also developed a useful technique for retargeting, using the Ad particle itself. The pTex system uses a similar approach to pseudotyping mixed with peptide insertions, but allowing isolated and randomised mutation of the fibre knob for later incorporation into the capsid. This system overcomes some of the issues, such as linear display, although it may be limited for targeting towards a specific receptors of interest without the issue of previously understood peptide-ligand interactions [160].

The detargeting and retargeting methods highlighted in this review are overviewed in Figure 8.

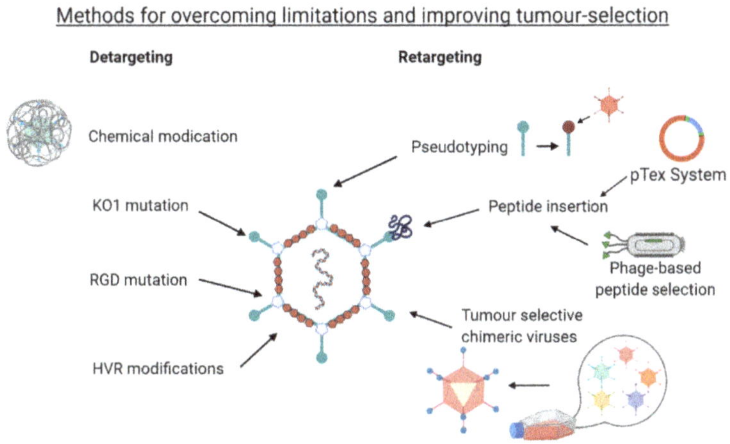

Figure 8. Overview of putative adenoviral detargeting and retargeting approaches. Created with https://biorender.com.

6. Conclusions

Oncolytic adenoviruses are powerful therapeutic agents with great potential in the clinical arena, combining multiple cell-killing effects on the tumour microenvironment. Firstly, the life cycle of adenovirus induces immunogenic cell death. Moreover, the process of replication and lysis results in the production of many tens of thousands of additional daughter virions, which when released, infect surrounding cells, thus repeating and amplifying the process. Additional engineering of the viral genome to encode therapeutic transgenes, such as immunotherapies, cytokines or pro-apoptotic proteins can further enhance the immunogenicity of the tumour microenvironment effectively turning the tumour into a factory producing protein to promote its own destruction.

Despite some early evidence of efficacy as a combination therapy in the clinic, the development of fully refined oncolytic adenoviruses has failed to reach its full potential. There are numerous obstacles to sequentially consider when developing novel adenovirus based oncolytic virotherapies, including infection of non-cancerous cells, activation of the anti-viral immune response and a limit in the number of viruses with native cancer tropism. The ability to modify the adenoviral genome and overcome these limitations makes them attractive candidates for targeted oncolytic virotherapies. Furthermore, there is a vast repository of alternative adenoviral serotypes, possessing known advantages over Ad5-based therapies, that are yet to be explored in an oncology setting. Employing techniques such as peptide insertion has had promising pre-clinical results and if combined with additional modifications to further detarget and arm with therapeutic transgenes, the result would be highly potent targeted oncolytic virotherapies. Whilst significant progress has been made in developing such systems (e.g., the Ad5$_{NULL}$ platform), step changing technologies will be required to develop optimally targeted "precision virotherapies" to tumour specific molecular addresses, and thus to deliver truly personalised virotherapies moving forwards.

Therefore, the remaining limitations for targeted oncolytic applications using this approach are the identification of ligands that are cancer-specific [161], coupled with the poor ability to transfer linearly selected peptides from phage libraries into the three-dimensional Ad structure. In a sense, whilst technologies exist to elucidate peptides or antibody fragments that allow us to "hit" tumour targets of interest, the success of such targeting technologies when translated into oncolytic virotherapies will require smarter systems, designed to engineer tropisms directly into the viral capsid protein of relevance to be successful, or else they will continue to "miss the point". Developing technologies deigned to overcome these limitations will be key to the future success and efficacy in the clinic.

Cancers **2020**, 12, 3327

Funding: TGC is funded by Knowledge Economy Skills Scholarships (KESS 2) PhD studentship to ALP: 515374, EAB is funded by a Cardiff University School of Medicine PhD award to ALP (AC1170AP02). ALP is funded by HEFCW.

Conflicts of Interest: The authors declare no conflict of interest.

References

1. Cancer Research UK. Cancer Survival Statistics|Cancer Research UK. Available online: https://www.cancerresearchuk.org/health-professional/cancer-statistics/survival (accessed on 23 April 2019).

2. Yan, L.; Rosen, N.; Arteaga, C. Targeted cancer therapies. *Chin. J. Cancer* **2011**, 30, 1–4. [CrossRef] [PubMed]

3. Arteaga, C.L. Trastuzumab, an appropriate first-line single-agent therapy for HER2-overexpressing metastatic breast cancer. *Breast Cancer Res.* **2003**, 5, 96–100. [CrossRef] [PubMed]

4. Eiger, D.; Pondé, N.F.; De Azambuja, E. Pertuzumab in HER2-positive early breast cancer: Current use and perspectives. *Futur. Oncol.* **2019**, 15, 1823–1843. [CrossRef] [PubMed]

5. Vogel, C.L.; Cobleigh, M.A.; Tripathy, D.; Gutheil, J.C.; Harris, L.N.; Fehrenbacher, L.; Slamon, D.J.; Murphy, M.; Novotny, W.F.; Burchmore, M.; et al. Efficacy and Safety of Trastuzumab as a Single Agent in First-Line Treatment of HER2 -Overexpressing Metastatic Breast Cancer. *J. Clin. Oncol.* **2002**, 20, 719–726. [CrossRef]

6. Lv, S.; Wang, Y.; Sun, T.; Wan, D.; Sheng, L.; Li, W.; Zhu, H.; Li, Y.; Lu, J. Overall Survival Benefit from Trastuzumab-Based Treatment in HER2-Positive Metastatic Breast Cancer: A Retrospective Analysis. *Oncol. Res. Treat.* **2018**, 41, 450–455. [CrossRef]

7. Ivy, S.P.; Wick, J.Y.; Kaufman, B.M. an overview of small-molecule inhibitors of VegFr signaling. *Nat. Rev. Clin. Oncol* **2009**, 6, 569–579. [CrossRef]

8. Samson, A.; Scott, K.J.; Taggart, D.; West, E.J.; Wilson, E.; Nuovo, G.J.; Thomson, S.; Corns, R.; Mathew, R.K.; Fuller, M.J.; et al. Intravenous delivery of oncolytic reovirus to brain tumor patients immunologically primes for subsequent checkpoint blockade. *Sci. Transl. Med.* **2018**, 10, eaam7577. [CrossRef]

9. Bourgeois-Daigneault, M.-C.; Roy, D.G.; Aitken, A.S.; El Sayes, N.; Martin, N.T.; Varette, O.; Falls, T.; St-Germain, L.E.; Pelin, A.; Lichty, B.D.; et al. Neoadjuvant oncolytic virotherapy before surgery sensitizes triple-negative breast cancer to immune checkpoint therapy. *Sci. Transl. Med.* **2018**, 10, eaao1641. [CrossRef]

10. Chawla, S.P.; Chua, V.S.; Kim, K.; Dy, P.S.; Paz, M.K.; Angel, N.; Wang, K.; Moradkhani, A.; Quon, D.; Wong, S.; et al. The TNT protocol: A phase II study using talimogene laherparepvec (TVEC), nivolumab (N) and trabectedin (T) as first, second/third line therapy for advanced sarcoma, including desmoid tumor and chordoma. *J. Clin. Oncol.* **2020**, 38, TPS11572. [CrossRef]

11. ClinicalTrials.gov. Talimogene Laherparepvec and Pembrolizumab in Treating Patients With Stage III-IV Melanoma-Full Text View-ClinicalTrials.gov. Available online: https://clinicaltrials.gov/ct2/show/NCT02965716 (accessed on 28 October 2020).

12. ClinicalTrials.gov. Ipilimumab, Nivolumab, and Talimogene Laherparepvec Before Surgery in Treating Participants With Localized, Triple-Negative or Estrogen Receptor Positive, HER2 Negative Breast Cancer-deleted-Full Text View-ClinicalTrials.gov. Available online: https://clinicaltrials.gov/ct2/show/NCT04185311 (accessed on 28 October 2020).

13. Uusi-Kerttula, H.; Davies, J.A.; Thompson, J.M.; Wongthida, P.; Evgin, L.; Shim, K.G.; Bradshaw, A.; Baker, A.T.; Rizkallah, P.J.; Jones, R.; et al. Ad5NULL-A20: A Tropism-Modified, αvβ6 Integrin-Selective Oncolytic Adenovirus for Epithelial Ovarian Cancer Therapies. *Clin. Cancer Res.* **2018**, 24, 4215–4224. [CrossRef]

14. Nguyen, T.V.; Anguiano-Zarate, S.S.; Matchett, W.E.; Barry, M.E.; Barry, M.A. Retargeted and detargeted adenovirus for gene delivery to the muscle. *Virology* **2018**, 514, 118–123. [CrossRef]

15. Belousova, N.; Korokhov, N.; Krendelshchikova, V.; Simonenko, V.; Mikheeva, G.; Triozzi, P.L.; Aldrich, W.A.; Banerjee, P.T.; Gillies, S.D.; Curiel, D.T.; et al. Genetically Targeted Adenovirus Vector Directed to CD40-Expressing Cells. *J. Virol.* **2003**, 77, 11367–11377. [CrossRef]

16. Kaufman, H.L.; Ruby, C.E.; Hughes, T.; Slingluff, C.L. Current status of granulocyte-macrophage colony-stimulating factor in the immunotherapy of melanoma. *J. Immunother. Cancer* **2014**, 2, 11. [CrossRef]

17. Kohlhapp, F.J.; Kaufman, H.L. Molecular pathways: Mechanism of action for talimogene laherparepvec, a new oncolytic virus immunotherapy. *Clin. Cancer Res.* **2016**, 22, 1048–1054. [CrossRef]

18. Farassati, F.; Yang, A.D.; Lee, P.W.K. Oncogenes in Ras signalling pathway dictate host-cell permissiveness to herpes simplex virus 1. *Nat. Cell Biol.* **2001**, *3*, 745–750. [CrossRef]

19. Conry, R.M.; Westbrook, B.; McKee, S.; Norwood, T.G. Talimogene laherparepvec: First in class oncolytic virotherapy. *Hum. Vaccin. Immunother.* **2018**, *14*, 839–846. [CrossRef]

20. Shen, Y.; Nemunaitis, J. Herpes simplex virus 1 (HSV-1) for cancer treatment. *Cancer Gene Ther.* **2006**, *13*, 975–992. [CrossRef]

21. Rehman, H.; Silk, A.W.; Kane, M.P.; Kaufman, H.L. Into the clinic: Talimogene laherparepvec (T-VEC), a first-in-class intratumoral oncolytic viral therapy. *J. Immunother. Cancer* **2016**, *4*, 53. [CrossRef]

22. Andtbacka, R.H.I.; Kaufman, H.L.; Collichio, F.A.; Amatruda, T.; Menunaitis, J.J.; Chesney, J.; Puzanov, I.; Harrington, K.; Zhang, Y.; Chen, L.; et al. Durable complete responses (CR) in patients (pts) with stage IIIB-IV melanoma treated with talimogene laherparepvec (T-VEC) in OPTiM. In Proceedings of the Society for Melanoma Research 2015 Congress, San Francisco, CA USA, 18–21 November 2015; Volume 28, pp. 753–826. [CrossRef]

23. Greig, S.L. Talimogene Laherparepvec: First Global Approval. *Drugs* **2016**, *76*, 147–154. [CrossRef]

24. Sun, L.; Funchain, P.; Song, J.M.; Rayman, P.; Tannenbaum, C.; Ko, J.; Mcnamara, M.; Marcela Diaz-Montero, C.; Gastman, B. Talimogene Laherparepvec combined with anti-PD-1 based immunotherapy for unresectable stage III-IV melanoma: A case series. *J. Immunother. Cancer* **2018**, *6*, 36. [CrossRef]

25. Seyfried, T.N.; Huysentruyt, L.C. On the Origin of Cancer Metastasis. *Crit. Rev. Oncog.* **2013**, *18*, 43–73. [PubMed]

26. Russell, L.; Peng, K.W. The emerging role of oncolytic virus therapy against cancer. *Chin. Clin. Oncol.* **2018**, *7*, 16. [PubMed]

27. Kuryk, L.; Møller, A.S.W.; Jaderberg, M. Abscopal effect when combining oncolytic adenovirus and checkpoint inhibitor in a humanized NOG mouse model of melanoma. *J. Med. Virol.* **2019**, *91*, 1702–1706. [CrossRef] [PubMed]

28. Wold, W.; Toth, K. Adenovirus Vectors for Gene Therapy, Vaccination and Cancer Gene Therapy. *Curr. Gene Ther.* **2014**, *13*, 421–433.

29. Phillips, M.; Stuart, J.; Rodríguez Stewart, R.; Berry, J.; Mainou, B.; Boehme, K. Current understanding of reovirus oncolysis mechanisms. *Oncolytic Virother.* **2018**, *7*, 53–63. [CrossRef] [PubMed]

30. Guse, K.; Cerullo, V.; Hemminki, A. Oncolytic vaccinia virus for the treatment of cancer. *Expert Opin. Biol. Ther.* **2011**, *11*, 595–608. [CrossRef]

31. HAdV Working Group. Available online: http://hadvwg.gmu.edu/ (accessed on 30 October 2020).

32. Lynch, J.P.; Kajon, A.E. Adenovirus: Epidemiology, Global Spread of Novel Serotypes, and Advances in Treatment and Prevention. *Semin. Respir. Crit. Care Med.* **2016**, *37*, 586–602. [CrossRef]

33. Bergelson, J.M.; Cunningham, J.A.; Droguett, G.; Kurt-Jones, E.A.; Krithivas, A.; Hong, J.S.; Horwitz, M.S.; Crowell, R.L.; Finberg, R.W. Isolation of a common receptor for coxsackie B viruses and adenoviruses 2 and 5. *Science (80-)* **1997**, *275*, 1320–1323.

34. Gaggar, A.; Shayakhmetov, D.M.; Lieber, A. CD46 is a cellular receptor for group B adenoviruses. *Nat. Med.* **2003**, *9*, 1408–1412. [CrossRef]

35. Wang, H.; Li, Z.Y.; Liu, Y.; Persson, J.; Beyer, I.; Möller, T.; Koyuncu, D.; Drescher, M.R.; Strauss, R.; Zhang, X.B.; et al. Desmoglein 2 is a receptor for adenovirus serotypes 3, 7, 11 and 14. *Nat. Med.* **2011**, *17*, 96–104. [CrossRef]

36. Arnberg, N.; Edlund, K.; Kidd, A.H.; Wadell, G. Adenovirus Type 37 Uses Sialic Acid as a Cellular Receptor. *J. Virol.* **2000**, *74*, 42–48. [CrossRef]

37. Cupelli, K.; Stehle, T. Viral attachment strategies: The many faces of adenoviruses. *Curr. Opin. Virol.* **2011**, *1*, 84–91.

38. Burmeister, W.P.; Guilligay, D.; Cusack, S.; Wadell, G.; Arnberg, N. Crystal Structure of Species D Adenovirus Fiber Knobs and Their Sialic Acid Binding Sites. *J. Virol.* **2004**, *78*, 7727–7736. [CrossRef]

39. Marttila, M.; Persson, D.; Gustafsson, D.; Liszewski, M.K.; Atkinson, J.P.; Wadell, G.; Arnberg, N. CD46 Is a Cellular Receptor for All Species B Adenoviruses except Types 3 and 7. *J. Virol.* **2005**, *79*, 14429–14436.

40. Stewart, P.; Chiu, C.; Huang, S.; Muir, T.; Zhao, Y.; Chait, B.; Mathias, P.; Nemerow, G. Cryo-EM visualization of an exposed RGD epitope on adenovirus that escapes antibody neutralization cell membrane permeabilization. *EMBO J.* **1997**, *16*, 1189–1198.

41. Meier, O.; Greber, U.F. Adenovirus endocytosis. *J. Gene Med.* **2003**, *5*, 451–462. [CrossRef]

42. Russell, W.C. Update on adenovirus and its vectors. *J. Gen. Virol.* **2000**, *81*, 2573–2604.
43. Russell, W.C. Adenoviruses: Update on structure and function. *J. Gen. Virol.* **2009**, *90*, 1–20.
44. Hoeben, R.C.; Uil, T.G. Adenovirus DNA replication. *Cold Spring Harb. Perspect. Biol.* **2013**, *5*, a013003. [CrossRef]
45. Hanahan, D.; Weinberg, R.A. Hallmarks of cancer: The next generation. *Cell* **2011**, *144*, 646–674.
46. Seymour, L.W.; Fisher, K.D. Oncolytic viruses: Finally delivering. *Br. J. Cancer* **2016**, *114*, 357–361. [CrossRef]
47. Cheng, P.H.; Wechman, S.L.; McMasters, K.M.; Zhou, H.S. Oncolytic replication of E1b-deleted adenoviruses. *Viruses* **2015**, *7*, 5767–5779. [CrossRef]
48. Everts, B.; Van Der Poel, H.G. Replication-selective oncolytic viruses in the treatment of cancer. *Cancer Gene Ther.* **2005**, *12*, 141–161. [CrossRef]
49. Larson, C.; Oronsky, B.; Scicinski, J.; Fanger, G.R.; Stirn, M.; Oronsky, A.; Reid, T.R. Going viral: A review of replication-selective oncolytic adenoviruses. *Oncotarget* **2015**, *6*, 19976–19989. [CrossRef]
50. Rodriguez, R.; Schuur, E.R.; Lim, H.Y.; Henderson, G.A.; Simons, J.W.; Henderson, D.R. Prostate Attenuated Replication Competent Adenovirus (ARCA) CN706: A Selective Cytotoxic for Prostate-specific Antigen-positive Prostate Cancer Cells. *Cancer Res.* **1997**, *57*, 2559–2563.
51. Tanoue, K.; Shaw, A.R.; Watanabe, N.; Porter, C.; Rana, B.; Gottschalk, S.; Brenner, M.; Suzuki, M. Armed oncolytic adenovirus-expressing PD-L1 mini-body enhances antitumor effects of Chimeric antigen receptor t cells in solid tumors. *Cancer Res.* **2017**, *77*, 2040–2051. [CrossRef]
52. Dias, J.D.; Hemminki, O.; Diaconu, I.; Hirvinen, M.; Bonetti, A.; Guse, K.; Escutenaire, S.; Kanerva, A.; Pesonen, S.; Löskog, A.; et al. Targeted cancer immunotherapy with oncolytic adenovirus coding for a fully human monoclonal antibody specific for CTLA-4. *Gene Ther.* **2012**, *19*, 988–998. [CrossRef]
53. Cervera-Carrascon, V.; Siurala, M.; Santos, J.M.; Havunen, R.; Tähtinen, S.; Karell, P.; Sorsa, S.; Kanerva, A.; Hemminki, A. TNFa and IL-2 armed adenoviruses enable complete responses by anti-PD-1 checkpoint blockade. *Oncoimmunology* **2018**, *7*, e1412902. [CrossRef]
54. Lapteva, N.; Aldrich, M.; Weksberg, D.; Rollins, L.; Goltsova, T.; Chen, S.Y.; Huang, X.F. Targeting the intratumoral dendritic cells by the oncolytic adenoviral vaccine expressing RANTES elicits potent antitumor immunity. *J. Immunother.* **2009**, *32*, 145–156. [CrossRef]
55. Kumon, H.; Ariyoshi, Y.; Sasaki, K.; Sadahira, T.; Araki, M.; Ebara, S.; Yanai, H.; Watanabe, M.; Nasu, Y. Adenovirus vector carrying REIC/DKK-3 gene: Neoadjuvant intraprostatic injection for high-risk localized prostate cancer undergoing radical prostatectomy. *Cancer Gene Ther.* **2016**, *23*, 400. [CrossRef]
56. Singh, G.; Robinson, C.M.; Dehghan, S.; Schmidt, T.; Seto, D.; Jones, M.S.; Dyer, D.W.; Chodosh, J. Overreliance on the Hexon Gene, Leading to Misclassification of Human Adenoviruses. *J. Virol.* **2012**, *86*, 4693–4695. [CrossRef]
57. Pearl, T.M.; Markert, J.M.; Cassady, K.A.; Ghonime, M.G. Oncolytic Virus-Based Cytokine Expression to Improve Immune Activity in Brain and Solid Tumors. *Mol. Ther. Oncolytics* **2019**, *13*, 14–21. [CrossRef]
58. ClinicalTrials.gov. Search of: Adenovirus|Cancer-List Results-ClinicalTrials.gov. Available online: https://clinicaltrials.gov/ct2/results?cond=Cancer&term=Adenovirus&cntry=&state=&city=&dist= &Search=Search (accessed on 30 April 2019).
59. Jiang, H.; Gomez-Manzano, C.; Lang, F.; Alemany, R.; Fueyo, J. Oncolytic Adenovirus: Preclinical and Clinical Studies in Patients with Human Malignant Gliomas. *Curr. Gene Ther.* **2009**, *9*, 422–427. [CrossRef]
60. Freytag, S.O.; Movsas, B.; Stricker, H. Clinical Trials of Oncolytic Adenovirus-Mediated Gene Therapy. *Mol. Ther.* **2016**, *24*, S205. [CrossRef]
61. Abbink, P.; Lemckert, A.A.C.; Ewald, B.A.; Lynch, D.M.; Denholtz, M.; Smits, S.; Holterman, L.; Damen, I.; Vogels, R.; Thorner, A.R.; et al. Comparative Seroprevalence and Immunogenicity of Six Rare Serotype Recombinant Adenovirus Vaccine Vectors from Subgroups B and D. *J. Virol.* **2007**, *81*, 4654–4663. [CrossRef]
62. Dakin, R.S.; Parker, A.L.; Delles, C.; Nicklin, S.A.; Baker, A.H. Efficient Transduction of Primary Vascular Cells by the Rare Adenovirus Serotype 49 Vector. *Hum. Gene Ther.* **2015**, *26*, 312–319. [CrossRef]
63. Parker, A.L.; Waddington, S.N.; Buckley, S.M.K.; Custers, J.; Havenga, M.J.E.; van Rooijen, N.; Goudsmit, J.; McVey, J.H.; Nicklin, S.A.; Baker, A.H. Effect of Neutralizing Sera on Factor X-Mediated Adenovirus Serotype 5 Gene Transfer. *J. Virol.* **2009**, *83*, 479–483. [CrossRef]
64. Alemany, R. Oncolytic Adenoviruses in Cancer Treatment. *Biomedicines* **2014**, *2*, 36–49. [CrossRef]

65. Carlisle, R.C.; Di, Y.; Cerny, A.M.; Sonnen, A.F.P.; Sim, R.B.; Green, N.K.; Subr, V.; Ulbrich, K.; Gilbert, R.J.C.; Fisher, K.D.; et al. Human erythrocytes bind and inactivate type 5 adenovirus by presenting Coxsackie virus-adenovirus receptor and complement receptor 1. *Blood* **2009**, *113*, 1909–1918. [CrossRef]

66. Smith, T.; Idamakanti, N.; Kylefjord, H.; Rollence, M.; King, L.; Kaloss, M.; Kaleko, M.; Stevenson, S.C. In Vivo Hepatic Adenoviral Gene Delivery Occurs Independently of the Coxsackievirus–Adenovirus Receptor. *Mol. Ther.* **2002**, *5*, 770–779. [CrossRef]

67. Schlipp, A.; Schinner, C.; Spindler, V.; Vielmuth, F.; Gehmlich, K.; Syrris, P.; McKenna, W.J.; Dendorfer, A.; Hartlieb, E.; Waschke, J. Desmoglein-2 interaction is crucial for cardiomyocyte cohesion and function. *Cardiovasc. Res.* **2014**, *104*, 245–257. [CrossRef]

68. Nava, P.; Hopkins, A.N.; Laukoetter, M.G.; Green, K.J.; Parkos, C.A.; Nusrat, A. A role of Desmoglein 2 in intestinal epithelial apoptosis. *FASEB J.* **2007**, *21*, A192. [CrossRef]

69. Baker, A.T.; Mundy, R.M.; Davies, J.A.; Rizkallah, P.J.; Parker, A.L. Human adenovirus type 26 uses sialic acid-bearing glycans as a primary cell entry receptor. *Sci. Adv.* **2019**, *5*, eaax3567. [CrossRef] [PubMed]

70. Parker, A.L.; Waddington, S.N.; Nicol, C.G.; Shayakhmetov, D.M.; Buckley, S.M.; Denby, L.; Kemball-Cook, G.; Ni, S.; Lieber, A.; McVey, J.H.; et al. Multiple vitamin K-dependent coagulation zymogens promote adenovirus-mediated gene delivery to hepatocytes. *Blood* **2006**, *108*, 2554–2561. [CrossRef] [PubMed]

71. Waddington, S.N.; McVey, J.H.; Bhella, D.; Parker, A.L.; Barker, K.; Atoda, H.; Pink, R.; Buckley, S.M.K.; Greig, J.A.; Denby, L.; et al. Adenovirus Serotype 5 Hexon Mediates Liver Gene Transfer. *Cell* **2008**, *132*, 397–409. [CrossRef]

72. Stewart, P.L.; Nemerow, G.R. Cell integrins: Commonly used receptors for diverse viral pathogens. *Trends Microbiol.* **2007**, *15*, 500–507. [CrossRef]

73. Nemerow, G.R. A new link between virus cell entry and inflammation: Adenovirus interaction with integrins induces specific proinflammatory responses. *Mol. Ther.* **2009**, *17*, 1490–1491. [CrossRef]

74. Di Paolo, N.C.; Miao, E.A.; Iwakura, Y.; Murali-Krishna, K.; Aderem, A.; Flavell, R.A.; Papayannopoulou, T.; Shayakhmetov, D.M. Virus Binding to a Plasma Membrane Receptor Triggers Interleukin-1α-Mediated Proinflammatory Macrophage Response In Vivo. *Immunity* **2009**, *31*, 110–121. [CrossRef]

75. Bradshaw, A.C.; Coughlan, L.; Miller, A.M.; Alba, R.; Van Rooijen, N.; Nicklin, S.A.; Baker, A.H. Biodistribution and inflammatory profiles of novel penton and hexon double-mutant serotype 5 adenoviruses. *J. Control. Release* **2012**, *164*, 394–402. [CrossRef]

76. Fukuhara, H.; Ino, Y.; Todo, T. Oncolytic virus therapy: A new era of cancer treatment at dawn. *Cancer Sci.* **2016**, *107*, 1373–1379. [CrossRef]

77. Armendáriz-Borunda, J.; Bastidas-Ramírez, B.E.; Sandoval-Rodríguez, A.; González-Cuevas, J.; Gómez-Meda, B.; García-Bañuelos, J. Production of first generation adenoviral vectors for preclinical protocols: Amplification, purification and functional titration. *JBIOSC* **2011**, *112*, 415–421. [CrossRef]

78. Rowe, D.T.; Branton, P.E.; Graham, F.L. The kinetics of synthesis of early viral proteins in KB cells infected with wild-type and transformation-defective host-range mutants of human adenovirus type 5. *J. Gen. Virol.* **1984**, *65*, 585–597. [CrossRef]

79. Saha, B.; Parks, R.J. Human adenovirus type 5 vectors deleted of early region 1 (E1) undergo limited expression of early replicative E2 proteins and DNA replication in non-permissive cells. *PLoS ONE* **2017**, *12*, e0181012. [CrossRef]

80. Kovesdi, I.; Hedley, S.J. Adenoviral producer cells. *Viruses* **2010**, *2*, 1681. [CrossRef]

81. Graham, F.L.; Smiley, J.; Russell, W.C.; Nairn, R. Characteristics of a human cell line transformed by DNA from human adenovirus type 5. *J. Gen. Virol.* **1977**, *36*, 59–72. [CrossRef]

82. Fallaux, F.J.; Bout, A.; Van Der Velde, I.; Van Den Wollenberg, D.J.M.; Hehir, K.M.; Keegan, J.; Auger, C.; Cramer, S.J.; Van Ormondt, H.; Van Der Eb, A.J.; et al. New helper cells and matched early region 1-deleted adenovirus vectors prevent generation of replication-competent adenoviruses. *Hum. Gene Ther.* **1998**, *9*, 1909–1917. [CrossRef]

83. Suzuki, K.; Alemany, R.; Yamamoto, M.; Curiel, D.T.; Program, G.T. The Presence of the Adenovirus E3 Region Improves the Oncolytic Potency of Conditionally Replicative Adenoviruses 1. *Clin. Cancer Res.* **2002**, *8*, 3348–3359.

84. Gorziglia, M.I.; Lapcevich, C.; Roy, S.; Kang, Q.; Kadan, M.; Wu, V.; Pechan, P.; Kaleko, M. Generation of an Adenovirus Vector Lacking E1, E2a, E3, and All of E4 except Open Reading Frame 3. *J. Virol.* **1999**, *73*, 6048–6055. [CrossRef]

85. Alba, R.; Bosch, A.; Chillon, M. Gutless adenovirus: Last-generation adenovirus for gene therapy. *Gene Ther.* **2005**, *12*, S18–S27. [CrossRef]

86. Wang, W.; Ji, W.; Hu, H.; Ma, J.; Li, X.; Mei, W.; Xu, Y.; Hu, H.; Yan, Y.; Song, Q.; et al. Survivin promoter-regulated oncolytic adenovirus with Hsp70 gene exerts effective antitumor efficacy in gastric cancer immunotherapy. *Oncotarget* **2014**, *5*, 150–160. [CrossRef]

87. Wirth, T.; Zender, L.; Schulte, B.; Mundt, B.; Plentz, R.; Rudolph, K.L.; Manns, M.; Kubicka, S.; Kuhnel, F. A telomerase-dependent conditionally replicating adenovirus for selective treatment of cancer (Cancer Research (2003) 63 (3181-3188)). *Cancer Res.* **2018**, *78*, 6027.

88. Berns, K.I.; Giraud, C. Adenovirus and Adeno-Associated Virus as Vectors for Gene Therapy. *Ann. N. Y. Acad. Sci.* **1995**, *772*, 95–104. [PubMed]

89. Suzuki, S.; Kofune, H.; Uozumi, K.; Yoshimitsu, M.; Arima, N.; Ishitsuka, K.; Ueno, S.I.; Kosai, K.I. A survivin-responsive, conditionally replicating adenovirus induces potent cytocidal effects in adult T-cell leukemia/lymphoma. *BMC Cancer* **2019**, *19*, 516.

90. Doloff, J.C.; Waxman, D.J.; Jounaidi, Y. Human Telomerase Reverse Transcriptase Promoter-Driven Oncolytic Adenovirus with E1B-19 kDa and E1B-55 kDa Gene Deletions. *Hum. Gene Ther.* **2008**, *19*, 1383–1399. [CrossRef] [PubMed]

91. Fueyo, J.; Gomez-Manzano, C.; Alemany, R.; Lee, P.S.Y.; McDonnell, T.J.; Mitlianga, P.; Shi, Y.X.; Levin, V.A.; Yung, W.K.A.; Kyritsis, A.P. A mutant oncolytic adenovirus targeting the Rb pathway produces anti-glioma effect in vivo. *Oncogene* **2000**, *19*, 2–12. [CrossRef] [PubMed]

92. Hulin-Curtis, S.L.; Davies, J.A.; Jones, R.; Hudson, E.; Hanna, L.; Chester, J.D.; Parker, A.L. Histone deacetylase inhibitor trichostatin A sensitises cisplatinresistant ovarian cancer cells to oncolytic adenovirus. *Oncotarget* **2018**, *9*, 26328–26341. [CrossRef] [PubMed]

93. Konecny, G.E.; Winterhoff, B.; Kolarova, T.; Qi, J.; Manivong, K.; Dering, J.; Yang, G.; Chalukya, M.; Wang, H.J.; Anderson, L.; et al. Expression of p16 and retinoblastoma determines response to CDK4/6 inhibition in ovarian cancer. *Clin. Cancer Res.* **2011**, *17*, 1591–1602.

94. Sellers, W.R.; Kaelin, W.G. Role of the retinoblastoma protein in the pathogenesis of human cancer. *J. Clin. Oncol.* **1997**, *15*, 3301–3312. [CrossRef]

95. Gros, A.; Martínez-Quintanilla, J.; Puig, C.; Guedan, S.; Molleví, D.G.; Alemany, R.; Cascallo, M. Bioselection of a gain of function mutation that enhances adenovirus 5 release and improves its antitumoral potency. *Cancer Res.* **2008**, *68*, 8928–8937.

96. Dong, W.; Van Ginkel, J.W.H.; Au, K.Y.; Alemany, R.; Meulenberg, J.J.M.; Van Beusechem, V.W. ORCA-010, a novel potency-enhanced oncolytic adenovirus, exerts strong antitumor activity in preclinical models. *Hum. Gene Ther.* **2014**, *25*, 897–904.

97. Ganly, I.; Kirn, D.; Eckhardt, S.G.; Rodriguez, G.I.; Soutar, D.S.; Otto, R.; Robertson, A.G.; Park, O.; Gulley, M.L.; Heise, C.; et al. A Phase I Study of Onyx-015, an E1B Attenuated Adenovirus, Administered Intratumorally to Patients with Recurrent Head and Neck Cancer. *Clin. Cancer Res.* **2000**, *6*, 798. [PubMed]

98. Garber, K. China approves world's first oncolytic virus therapy for cancer treatment. *J. Natl. Cancer Inst.* **2006**, *98*, 298–300. [CrossRef] [PubMed]

99. Ries, S.; Korn, W.M. ONYX-015: Mechanisms of action and clinical potential of a replication-selective adenovirus. *Br. J. Cancer* **2002**, *86*, 5–11. [CrossRef] [PubMed]

100. Muller, P.A.J.; Vousden, K.H. P53 mutations in cancer. *Nat. Cell Biol.* **2013**, *15*, 2–8. [CrossRef] [PubMed]

101. O'Shea, C.C.; Johnson, L.; Bagus, B.; Choi, S.; Nicholas, C.; Shen, A.; Boyle, L.; Pandey, K.; Soria, C.; Kunich, J.; et al. Late viral RNA export, rather than p53 inactivation, determines ONYX-015 tumor selectivity. *Cancer Cell* **2004**, *6*, 611–623. [CrossRef] [PubMed]

102. Raman, S.S.; Hecht, J.R.; Chan, E. Talimogene laherparepvec: Review of its mechanism of action and clinical efficacy and safety. *Immunotherapy* **2019**, *11*, 705–723. [CrossRef] [PubMed]

103. Kirn, D. Oncolytic virotherapy for cancer with the adenovirus dl1520 (Onyx-015): Results of Phase I and II trials. *Expert Opin. Biol. Ther.* **2001**, *1*, 525–538. [CrossRef]

104. Ram, Z.; Culver, K.W.; Oshiro, E.M.; Viola, J.J.; Devroom, H.L.; Otto, E.; Long, Z.; Chiang, Y.; Mcgarrity, G.J.; Muul, L.M.; et al. Therapy of malignant brain tumors by intratumoral implantation of retroviral vector-producing cells. *Nat. Med.* **1997**, *3*, 1354–1361. [CrossRef]

105. Sterman, D.H.; Treat, J.; Litzky, L.A.; Amin, K.M.; Coonrod, L.; Molnar-Kimber, K.; Recio, A.; Knox, L.; Wilson, J.M.; Albelda, S.M.; et al. Adenovirus-mediated herpes simplex virus thymidine kinase/ganciclovir gene therapy in patients with localized malignancy: Results of a phase I clinical trial in malignant mesothelioma. *Hum. Gene Ther.* **1998**, *9*, 1083–1092. [CrossRef]

106. Dhar, D.; Spencer, J.F.; Toth, K.; Wold, W.S.M. Pre-existing immunity and passive immunity to adenovirus 5 prevents toxicity caused by an oncolytic adenovirus vector in the syrian hamster model. *Mol. Ther.* **2009**, *17*, 1724–1732. [CrossRef]

107. Belousova, N.; Mikheeva, G.; Xiong, C.; Stagg, L.J.; Gagea, M.; Fox, P.S.; Bassett, R.L.; Ladbury, J.E.; Braun, M.B.; Stehle, T.; et al. Native and engineered tropism of vectors derived from a rare species D adenovirus serotype 43. *Oncotarget* **2016**, *7*, 53414–53429. [CrossRef] [PubMed]

108. Greig, J.A.; Buckley, S.M.K.; Waddington, S.N.; Parker, A.L.; Bhella, D.; Pink, R.; Rahim, A.A.; Morita, T.; Nicklin, S.A.; McVey, J.H.; et al. Influence of coagulation factor X on in vitro and in vivo gene delivery by adenovirus (Ad) 5, Ad35, and chimeric Ad5/Ad35 vectors. *Mol. Ther.* **2009**, *17*, 1683–1691. [CrossRef] [PubMed]

109. Alba, R.; Bradshaw, A.C.; Parker, A.L.; Bhella, D.; Waddington, S.N.; Nicklin, S.A.; Van Rooijen, N.; Custers, J.; Goudsmit, J.; Barouch, D.H.; et al. Identification of coagulation factor (F)X binding sites on the adenovirus serotype 5 hexon: Effect of mutagenesis on FX interactions and gene transfer. *Blood* **2009**, *114*, 965–971. [CrossRef]

110. Gao, J.; Zhang, W.; Ehrhardt, A. Expanding the spectrum of adenoviral vectors for cancer therapy. *Cancers (Basel)* **2020**, *12*, 1139. [CrossRef] [PubMed]

111. Mach, N.; Gao, J.; Schaffarczyk, L.; Janz, S.; Ehrke-Schulz, E.; Dittmar, T.; Ehrhardt, A.; Zhang, W. Spectrum-wide exploration of human adenoviruses for breast cancer therapy. *Cancers (Basel)* **2020**, *12*, 1403. [CrossRef]

112. Kanerva, A.; Wang, M.; Bauerschmitz, G.J.; Lam, J.T.; Desmond, R.A.; Bhoola, S.M.; Barnes, M.N.; Alvarez, R.D.; Siegal, G.P.; Curiel, D.T.; et al. Gene transfer to ovarian cancer versus normal tissues with fiber-modified adenoviruses. *Mol. Ther.* **2002**, *5*, 695–704. [CrossRef]

113. Krasnykh, V.N.; Mikheeva, G.V.; Douglas, J.T.; Curiel, D.T. Generation of recombinant adenovirus vectors with modified fibers for altering viral tropism. *J. Virol.* **1996**, *70*, 6839–6846. [CrossRef]

114. Kuhn, I.; Harden, P.; Bauzon, M.; Chartier, C.; Nye, J.; Thorne, S.; Reid, T.; Ni, S.; Lieber, A.; Fisher, K.; et al. Directed evolution generates a novel oncolytic virus for the treatment of colon cancer. *PLoS ONE* **2008**, *3*, e2409. [CrossRef]

115. Garcia-Carbonero, R.; Salazar, R.; Duran, I.; Osman-Garcia, I.; Paz-Ares, L.; Bozada, J.M.; Boni, V.; Blanc, C.; Seymour, L.; Beadle, J.; et al. Phase 1 study of intravenous administration of the chimeric adenovirus enadenotucirev in patients undergoing primary tumor resection. *J. Immunother. Cancer* **2017**, *5*, 71. [CrossRef]

116. McNeish, I.A.; Oza, A.M.; Coleman, R.L.; Scott, C.L.; Konecny, G.E.; Tinker, A.; O'Malley, D.M.; Brenton, J.; Kristeleit, R.S.; Bell-McGuinn, K.; et al. Results of ARIEL2: A Phase 2 trial to prospectively identify ovarian cancer patients likely to respond to rucaparib using tumor genetic analysis. *J. Clin. Oncol.* **2015**, *33*, 5508. [CrossRef]

117. Dyer, A.; Di, Y.; Calderon, H.; Illingworth, S.; Kueberuwa, G.; Tedcastle, A.; Jakeman, P.; Chia, S.L.; Brown, A.; Silva, M.A.; et al. Oncolytic Group B Adenovirus Enadenotucirev Mediates Non-apoptotic Cell Death with Membrane Disruption and Release of Inflammatory Mediators. *Mol. Ther. Oncolytics* **2017**, *4*, 18–30. [CrossRef] [PubMed]

118. Cohen, C.J.; Shieh, J.T.C.; Pickles, R.J.; Okegawa, T.; Hsieh, J.T.; Bergelson, J.M. The coxsackievirus and adenovirus receptor is a transmembrane component of the tight junction. *Proc. Natl. Acad. Sci. USA* **2001**, *98*, 15191–15196. [CrossRef] [PubMed]

119. Korn, W.M.; Macal, M.; Christian, C.; Lacher, M.D.; McMillan, A.; Rauen, K.A.; Warren, R.S.; Ferrell, L. Expression of the coxsackievirus- and adenovirus receptor in gastrointestinal cancer correlates with tumor differentiation. *Cancer Gene Ther.* **2006**, *13*, 792–797. [CrossRef] [PubMed]

120. Sachs, M.D.; Rauen, K.A.; Ramamurthy, M.; Dodson, J.L.; De Marzo, A.M.; Putzi, M.J.; Schoenberg, M.P.; Rodriguez, R. Integrin αv and coxsackie adenovirus receptor expression in clinical bladder cancer. *Urology* **2002**, *60*, 531–536. [CrossRef]

121. Rauen, K.A.; Sudilovsky, D.; Le, J.L.; Chew, K.L.; Hann, B.; Weinberg, V.; Schmitt, L.D.; Mccormick, F. Expression of the Coxsackie Adenovirus Receptor in Normal Prostate and in Primary and Metastatic Prostate Carcinoma: Potential Relevance to Gene Therapy 1. *Cancer Res.* **2002**, *62*, 3812–3818. [PubMed]

122. Nicklin, S.A.; Von Seggern, D.J.; Work, L.M.; Pek, D.C.K.; Dominiczak, A.F.; Nemerow, G.R.; Baker, A.H. Ablating adenovirus type 5 fiber-CAR binding and HI loop insertion of the SIGYPLP peptide generate an endothelial cell-selective adenovirus. *Mol. Ther.* **2001**, *4*, 534–542. [CrossRef] [PubMed]

123. Wickham, T.J.; Tzeng, E.; Shears Ii, L.L.; Roelvink, P.W.; Li, Y.; Lee, G.M.; Brough, D.E.; Lizonova, A.; Kovesdi, I. Increased In Vitro and In Vivo Gene Transfer by Adenovirus Vectors Containing Chimeric Fiber Proteins. *J. Virol.* **1997**, *71*, 8221–8229. [CrossRef] [PubMed]

124. Parker, A.; Davies, J.; Marlow, G.; Uusi-Kerttula, H.; Seaton, G.; Piggott, L.; Clarkson, R.; Chester, J. Ad5 (NULL)-A20-a precision virotherapy that efficiently and selectively targets αvβ6 positive cancers following intravenous administration. *Br. J. Cancer* **2019**, *121*, 16.

125. Nakamura, T.; Sato, K.; Hamada, H. Reduction of Natural Adenovirus Tropism to the Liver by both Ablation of Fiber-Coxsackievirus and Adenovirus Receptor Interaction and Use of Replaceable Short Fiber. *J. Virol.* **2003**, *77*, 2512–2521. [CrossRef] [PubMed]

126. Schoggins, J.W.; Gall, J.G.D.; Falck-Pedersen, E. Subgroup B and F Fiber Chimeras Eliminate Normal Adenovirus Type 5 Vector Transduction In Vitro and In Vivo. *J. Virol.* **2003**, *77*, 1039–1048. [CrossRef]

127. Kashentseva, E.A.; Douglas, J.T.; Zinn, K.R.; Curiel, D.T.; Dmitriev, I.P. Targeting of Adenovirus Serotype 5 Pseudotyped with Short Fiber from Serotype 41 to c-erbB2-Positive Cells using Bispecific Single-Chain Diabody. *J. Mol. Biol.* **2009**, *388*, 443–461. [CrossRef] [PubMed]

128. Hofherr, S.E.; Shashkova, E.V.; Weaver, E.A.; Khare, R.; Barry, M.A. Modification of adenoviral vectors with polyethylene glycol modulates in vivo tissue tropism and gene expression. *Mol. Ther.* **2008**, *16*, 1276–1282. [CrossRef] [PubMed]

129. Kreppel, F.; Kochanek, S. Modification of adenovirus gene transfer vectors with synthetic polymers: A scientific review and technical guide. *Mol. Ther.* **2008**, *16*, 16–29. [CrossRef] [PubMed]

130. Kim, J.; Kim, P.H.; Kim, S.W.; Yun, C.O. Enhancing the therapeutic efficacy of adenovirus in combination with biomaterials. *Biomaterials* **2012**, *33*, 1838–1850. [CrossRef] [PubMed]

131. Kaufmann, J.K.; Nettelbeck, D.M. Virus chimeras for gene therapy, vaccination, and oncolysis: Adenoviruses and beyond. *Trends Mol. Med.* **2012**, *18*, 365–376. [CrossRef] [PubMed]

132. Silver, J.; Mei, Y.-F. Transduction and Oncolytic Profile of a Potent Replication-Competent Adenovirus 11p Vector (RCAd11pGFP) in Colon Carcinoma Cells. *PLoS ONE* **2011**, *6*, e17532. [CrossRef]

133. Rein, D.T.; Volkmer, A.; Beyer, I.M.; Curiel, D.T.; Janni, W.; Dragoi, A.; Hess, A.P.; Maass, N.; Baldus, S.E.; Bauerschmitz, G.; et al. Treatment of chemotherapy resistant ovarian cancer with a MDR1 targeted oncolytic adenovirus. *Gynecol. Oncol.* **2011**, *123*, 138–146. [CrossRef]

134. Nakayama, M.; Both, G.W.; Banizs, B.; Tsuruta, Y.; Yamamoto, S.; Kawakami, Y.; Douglas, J.T.; Tani, K.; Curiel, D.T.; Glasgow, J.N. An adenovirus serotype 5 vector with fibers derived from ovine atadenovirus demonstrates CAR-independent tropism and unique biodistribution in mice. *Virology* **2006**, *350*, 103–115. [CrossRef]

135. Nicklin, S.A.; Wu, E.; Nemerow, G.R.; Baker, A.H. The influence of adenovirus fiber structure and function on vector development for gene therapy. *Mol. Ther.* **2005**, *12*, 384–393. [CrossRef]

136. Shayakhmetov, D.M.; Lieber, A. Dependence of Adenovirus Infectivity on Length of the Fiber Shaft Domain. *J. Virol.* **2000**, *74*, 10274–10286. [CrossRef]

137. Xial, D.; Henry2, L.J.; Gerard2, R.D.; Eisenhoferl, J. Crystal structure of the receptor-binding domain of adenovirus type 5 fiber protein at 1.7 A resolution. *Structure* **1994**, *2*, 1259–1270.

138. Hemminki, A.; Wang, M.; Desmond, R.A.; Strong, T.V.; Alvarez, R.D.; Curiel, D.T. Serum and ascites neutralizing antibodies in ovarian cancer patients treated with intraperitoneal adenoviral gene therapy. *Hum. Gene Ther.* **2002**, *13*, 1505–1514. [CrossRef] [PubMed]

139. Kiyokawa, J.; Wakimoto, H. Preclinical And Clinical Development Of Oncolytic Adenovirus For The Treatment Of Malignant Glioma. *Oncolytic Virotherapy* **2019**, *8*, 27–37. [CrossRef] [PubMed]

140. Krasnykh, V.; Dmitriev, I.; Mikheeva, G.; Miller, C.R.; Belousova, N.; Curiel, D.T. Characterization of an Adenovirus Vector Containing a Heterologous Peptide Epitope in the HI Loop of the Fiber Knob. *J. Virol.* **1998**, *72*, 1844–1852. [CrossRef] [PubMed]

141. Coughlan, L.; Uusi-Kerttula, H.; Ma, J.; Degg, B.P.; Parker, A.L.; Baker, A.H. Retargeting adenovirus serotype 48 fiber knob domain by peptide incorporation. *Hum. Gene Ther.* **2014**, *25*, 385–394. [CrossRef]

142. Uusi-Kerttula, H.; Davies, J.; Coughlan, L.; Hulin-Curtis, S.; Jones, R.; Hanna, L.; Chester, J.D.; Parker, A.L. Pseudotyped $\alpha v \beta 6$ integrin-targeted adenovirus vectors for ovarian cancer therapies. *Oncotarget* **2016**, *7*, 27926–27937. [CrossRef]

143. Magnusson, M.K.; Henning, P.; Myhre, S.; Wikman, M.; Uil, T.G.; Friedman, M.; Andersson, K.M.E.; Hong, S.S.; Hoeben, R.C.; Habib, N.A.; et al. Adenovirus 5 vector genetically re-targeted by an Affibody molecule with specificity for tumor antigen HER2/neu. *Cancer Gene Ther.* **2007**, *14*, 468–479. [CrossRef]

144. Magnusson, M.K.; Kraaij, R.; Leadley, R.M.; De Ridder, C.M.A.; Van Weerden, W.M.; Van Schie, K.A.J.; Van Der Kroeg, M.; Hoeben, R.C.; Maitland, N.J.; Lindholm, L. A transductionally retargeted adenoviral vector for virotherapy of her2/neu-expressing prostate cancer. *Hum. Gene Ther.* **2012**, *23*, 70–82. [CrossRef]

145. Whilding, L.M.; Parente-Pereira, A.C.; Zabinski, T.; Davies, D.M.; Petrovic, R.M.G.; Kao, Y.V.; Saxena, S.A.; Romain, A.; Costa-Guerra, J.A.; Violette, S.; et al. Targeting of Aberrant $\alpha v \beta 6$ Integrin Expression in Solid Tumors Using Chimeric Antigen Receptor-Engineered T Cells. *Mol. Ther.* **2017**, *25*, 2427. [CrossRef]

146. Ahmed, N.; Riley, C.; Rice, G.E.; Quinn, M.A.; Baker, M.S. $\alpha v \beta 6$ integrin-A marker for the malignant potential of epithelial ovarian cancer. *J. Histochem. Cytochem.* **2002**, *50*, 1371–1379. [CrossRef]

147. Reader, C.S.; Vallath, S.; Steele, C.W.; Haider, S.; Brentnall, A.; Desai, A.; Moore, K.M.; Jamieson, N.B.; Chang, D.; Bailey, P.; et al. The integrin $\alpha v \beta 6$ drives pancreatic cancer through diverse mechanisms and represents an effective target for therapy. *J. Pathol.* **2019**, *249*, 332–342. [CrossRef]

148. Sun, Q.; Sun, F.; Wang, B.; Liu, S.; Niu, W.; Liu, E.; Peng, C.; Wang, J.; Gao, H.; Liang, B.; et al. Interleukin-8 promotes cell migration through integrin $\alpha v \beta 6$ upregulation in colorectal cancer. *Cancer Lett.* **2014**, *354*, 245–253. [CrossRef]

149. Hausner, S.H.; DiCara, D.; Marik, J.; Marshall, J.F.; Sutcliffe, J.L. Use of a peptide derived from foot-and-mouth disease virus for the noninvasive imaging of human cancer: Generation and evaluation of 4-[^{18}F]fluorobenzoyl A20FMDV2 for in vivo imaging of integrin $\alpha v \beta 6$ expression with positron emission tomography. *Cancer Res.* **2007**, *67*, 7833–7840. [CrossRef] [PubMed]

150. Moore, K.M.; Desai, A.; de Luxán Delgado, B.; Trabulo, S.M.D.; Reader, C.; Brown, N.F.; Murray, E.R.; Brentnall, A.; Howard, P.; Masterson, L.; et al. Integrin $\alpha v \beta 6$-specific therapy for pancreatic cancer developed from foot-and-mouth-disease virus. *Theranostics* **2020**, *10*, 2930–2942. [CrossRef] [PubMed]

151. Robertson, S.; Parker, A.L.; Clarke, C.; Duffy, M.R.; Alba, R.; Nicklin, S.A.; Bakerk, A.H. Retargeting FX-binding-ablated HAdV-5 to vascular cells by inclusion of the RGD-4C peptide in hexon hypervariable region 7 and the HI loop. *J. Gen. Virol.* **2016**, *97*, 1911–1916. [CrossRef] [PubMed]

152. Vigne, E.; Mahfouz, I.; Dedieu, J.-F.; Brie, A.; Perricaudet, M.; Yeh, P. RGD Inclusion in the Hexon Monomer Provides Adenovirus Type 5-Based Vectors with a Fiber Knob-Independent Pathway for Infection. *J. Virol.* **1999**, *73*, 5156–5161. [CrossRef]

153. Hulin-Curtis, S.L.; Davies, J.A.; Nestić, D.; Bates, E.A.; Baker, A.T.; Cunliffe, T.G.; Majhen, D.; Chester, J.D.; Parker, A.L. Identification of folate receptor α (FRα) binding oligopeptides and their evaluation for targeted virotherapy applications. *Cancer Gene Ther.* **2020**, 1–14. [CrossRef] [PubMed]

154. Maruta, F.; Parker, A.L.; Fisher, K.D.; Hallissey, M.T.; Ismail, T.; Rowlands, D.C.; Chandler, L.A.; Kerr, D.J.; Seymour, L.W. Identification of FGF receptor-binding peptides for cancer gene therapy. *Cancer Gene Ther.* **2002**, *9*, 543–552. [CrossRef]

155. Uusi-Kerttula, H.; Legut, M.; Davies, J.; Jones, R.; Hudson, E.; Hanna, L.; Stanton, R.J.; Chester, J.D.; Parker, A.L. Incorporation of Peptides Targeting EGFR and FGFR1 into the Adenoviral Fiber Knob Domain and Their Evaluation as Targeted Cancer Therapies. *Hum. Gene Ther.* **2015**, *26*, 320–329. [CrossRef]

156. Ehrlich, G.K.; Berthold, W.; Bailon, P. Phage Display Technology: Affinity Selection by Biopanning. In *Affinity Chromatography*; Humana Press: Totowa, NJ, USA, 2000; Volume 147, pp. 195–208.

157. Li, Z.; Zhao, R.; Wu, X.; Sun, Y.; Yao, M.; Li, J.; Xu, Y.; Gu, J. Identification and characterization of a novel peptide ligand of epidermal growth factor receptor for targeted delivery of therapeutics. *FASEB J.* **2005**, *19*, 1978–1985. [CrossRef]

158. Urbanelli, L.; Ronchini, C.; Fontana, L.; Menard, S.; Orlandi, R.; Monaci, P. Targeted gene transduction of mammalian cells expressing the HER2/neu receptor by filamentous phage 1 1Edited by J. Karn. *J. Mol. Biol.* **2001**, *313*, 965–976. [CrossRef]

159. Senol, S.; Ceyran, A.B.; Aydin, A.; Zemheri, E.; Ozkanli, S.; Kösemetin, D.; Sehitoglu, I.; Akalin, I. Folate receptor α expression and significance in endometrioid endometrium carcinoma and endometrial hyperplasia. *Int. J. Clin. Exp. Pathol.* **2015**, *8*, 5633–5641.
160. Lupold, S.E.; Kudrolli, T.A.; Chowdhury, W.H.; Wu, P.; Rodriguez, R. A novel method for generating and screening peptides and libraries displayed on adenovirus fiber. *Nucleic Acids Res.* **2007**, *35*, 138. [CrossRef]
161. Brown, K.C. Peptidic tumor targeting agents: The road from phage display peptide selections to clinical applications. *Curr. Pharm. Des.* **2010**, *16*, 1040–1054. [CrossRef]

Publisher's Note: MDPI stays neutral with regard to jurisdictional claims in published maps and institutional affiliations.

cancers

Review

Immunotherapeutic Efficacy of Retargeted oHSVs Designed for Propagation in an Ad Hoc Cell Line

Andrea Vannini [1,*], Valerio Leoni [1], Mara Sanapo [1], Tatiana Gianni [1], Giorgia Giordani [2], Valentina Gatta [1], Catia Barboni [3], Anna Zaghini [3] and Gabriella Campadelli-Fiume [1,*]

1   Department of Experimental, Diagnostic and Specialty Medicine, University of Bologna, 40126 Bologna, Italy; valerio.leoni2@unibo.it (V.L.); mara.sanapo4@unibo.it (M.S.); tatiana.gianni3@unibo.it (T.G.); valentina.gatta6@unibo.it (V.G.)
2   Department of Pharmacy and Biotechnology, University of Bologna, 40126 Bologna, Italy; giorgia.giordani3@unibo.it
3   Department of Veterinary Medical Sciences, University of Bologna, 40126 Bologna, Italy; catia.barboni@unibo.it (C.B.); anna.zaghini@unibo.it (A.Z.)
*   Correspondence: andrea.vannini5@unibo.it (A.V.); gabriella.campadelli@unibo.it (G.C.-F.); Tel.: +39-051-209-4741 (A.V.); +39-051-209-4733 (G.C.-F.)

Simple Summary: The onco-immunotherapeutic viruses, among which stand the onco-immunotherapeutic herpes simplex viruses, have gained renewed interest due to their ability to unlock the potential of checkpoint inhibitors in preclinical and clinical settings. In prior decades, safety concerns led to the generation of overall safe, partially or highly attenuated oncolytic viruses. Current focus is on more efficacious onco-immunotherapeutic viruses with limited ability to cause off-tumor and off-target infections and the capability to subvert the tumor microenvironment immunosuppression—hence to potentiate checkpoint inhibitors. These viruses might serve as potential partners of T-cell therapies.

Abstract: Our laboratory has pursued the generation of cancer-specific oncolytic herpes simplex viruses (oHSVs) which ensure high efficacy while maintaining a high safety profile. Their blueprint included retargeting to a Tumor-Associated Antigen, e.g., HER2, coupled to detargeting from natural receptors to avoid off-target and off-tumor infections and preservation of the full complement of unmodified viral genes. These oHSVs are "fully virulent in their target cancer cells". The 3rd generation retargeted oHSVs carry two distinct retargeting moieties, which enable infection of a producer cell line and of the target cancer cells, respectively. They can be propagated in an ad hoc Vero cell derivative at about tenfold higher yields than 1st generation recombinants, and more effectively replicate in human cancer cell lines. The R-335 and R-337 prototypes were armed with murine IL-12. Intratumorally-administered R-337 conferred almost complete protection from LLC-1-HER2 primary tumors, unleashed the tumor microenvironment immunosuppression, synergized with the checkpoint blockade and conferred long-term vaccination against distant challenge tumors. In summary, the problem intrinsic to the propagation of retargeted oHSVs—which strictly require cells positive for targeted receptors—was solved in 3rd generation viruses. They are effective as immunotherapeutic agents against primary tumors and as antigen-agnostic vaccines.

Keywords: oncolytic virus; herpes simplex virus; retargeted virus; tropism retargeting; tumor; immunotherapy; checkpoint inhibitor; vaccination; antigen-agnostic vaccination; HER2

Citation: Vannini, A.; Leoni, V.; Sanapo, M.; Gianni, T.; Giordani, G.; Gatta, V.; Barboni, C.; Zaghini, A.; Campadelli-Fiume, G. Immunotherapeutic Efficacy of Retargeted oHSVs Designed for Propagation in an Ad Hoc Cell Line. *Cancers* **2021**, *13*, 266. https://doi.org/10.3390/cancers13020266

Received: 30 October 2020
Accepted: 8 January 2021
Published: 12 January 2021

Publisher's Note: MDPI stays neutral with regard to jurisdictional claims in published maps and institutional affiliations.

1. Cancer-Selective Oncolytic Herpes Simplex Viruses and Synergism with Check Point Blockade

Herpes simplex viruses (HSVs) were among the first viruses taken into consideration as candidate oncolytic viruses (OVs) [1–4] and still rank high in the list of OVs in clinical trials [5]. The early approaches to generate OVs, including oHSVs, were rather conservative.

Safety was a major concern, probably because scientists wanted to avoid the problems which characterized the initial approaches to gene therapy, and because of the frailty of patients. Most of the OVs that entered the clinical trials were attenuated, or over-attenuated. In reality, safety proved not to be a major clinical issue. oHSVs, and OVs in general, are well tolerated in clinical applications with very limited description of serious adverse effects [6,7]. However, the efficacy in humans did not keep up to the expectations raised by animal experimentation. For most OVs, and particularly for oHSVs, attenuation has been the prerequisite for cancer selectivity, and hence for safety. Cancer cells exhibit varying defects in innate responses and weakly contrast viral replication. Attenuated viruses exploit such defects to target cancers cells, which sustain the replication of both wt and attenuated viruses, and to spare non-cancer cells, in which the replication of attenuated viruses is contrasted, but not fully abolished, by the antiviral innate response. One such example is the attenuation conferred by the deletion of the γ34.5 virulence gene [2–4]. Since attenuation also weakens virus replication in cancer cells, additional modifications were introduced in the Δγ34.5 oHSVs to rescue replication and virulence. The α47 open reading frame was deleted to augment antigen presentation. The deletion modified the expression profile of the late US11 gene. Immunomodulatory genes were engineered in the viral genome. This is essentially the genotype of the approved OncovexGM-CSF, also known as T-VEC or Imlygic® [2,8].

The interest in OVs, including oHSVs, was boosted by check point inhibitors (CPIs) [9]. These molecules disable the breaks imposed by some tumors on the anti-cancer immune machinery and unleash the T-cell response against tumors. In humans, CPIs are limited by the fact that they target a restricted range of tumors—typically those with high mutational load, high tumor-specific T-cell infiltration and low T-cell activity due to the checkpoint brakes—and are effective only against a fraction of patients. The rationale for combining oncolytic viruses, like OncovexGM-CSF, with checkpoint inhibitors is compelling [10–13]. OVs inflame tumors, overcome the tumor microenvironment (TME) immunosuppression, and unlock the potential of checkpoint inhibitors across many cancer types. The underlying mechanism rests on the direct oncolysis induced by virus replication and consequent increase in both total and cancer-specific antigenic load in the TME, on the ability to recruit the immune cells to TME and to induce the expression of pro-inflammatory molecules. In turn, the latter activate immune effector cells and cause tumor inflammation. Given the heterogeneity in cancer genotypes and phenotypes, it is foreseen that even more complex combinations of immunomodulatory agents may be required to obtain consistent and durable therapeutic responses against a broad spectrum of cancers [14,15]. oHSVs are well equipped to do this job, since their large genome has space for extra genes. GM-CSF is a potent pro-inflammatory cytokine, most frequently employed as a payload in oHSVs [5]; it targets mainly the myeloid lineage, activates the dendritic cells, and enhances anti-cancer effects; it is present in OncovexGM-CSF. IL-12 (interleukin-12) [16] is another highly effective pro-inflammatory cytokine transgenically expressed by OVs, since it orchestrates the innate and the adaptive immune response against cancer and pathogens [17]. It is present in the oHSV named M032, currently in clinical trial [18,19]. Additional payloads recently engineered in oHSVs include the ligands to co-stimulatory immune receptors, CPI or combinations thereof [20,21].

2. Strategies towards Cancer-Specific and Efficacious oHSVs

In recent years efforts were made to generate cancer-specific rather than cancer-selective oHSVs. Such viruses gain safety from specificity, contain the entire set of viral genes so that they counteract the antiviral innate response, replicate robustly in the tumor cells, and are effective in releasing the immune suppression typical of cancers and in reactivating tumor recognition by the immune system.

2.1. Tropism Retargeting

The notion of tropism retargeting was pioneered by Glorioso and co-workers, and by Roizman and Zhou, and has been reviewed [22–25]. It entails the genetic engineering of a ligand to a selected cancer receptor into one of the viral glycoproteins that mediate HSV entry into the cells, most frequently gD [24,26,27]. Crucial to this development has been the elucidation of the molecular events that govern HSV entry. It requires four essential glycoproteins, gD, gH, gL and gB, which are activated in a cascade fashion, and two major alternative receptors HVEM (herpesvirus entry mediator) and nectin1 which interact with gD and activate it; they serve as major tropism determinants. Thereafter, the receptor-activated gD and additional integrin receptors activate the heterodimer gH/gL, and finally gB. The latter executes the fusion between the virion envelope and the cell membranes—plasma membranes or endocytic membranes [28–32]. The presence of cell surface receptors is a requirement for HSV entry [33].

The strategy developed in our laboratory to generate cancer-specific oHSVs and increase their efficacy entails (a) the retargeting of the virus tropism to Tumor-Associated Antigens (TAAs), i.e., to molecules that are selectively present at cancer cell surfaces; (b) the detargeting of the virus tropism from the natural receptors HVEM and nectin, to avoid off-target and off-tumor infections. When combined, retargeting and detargeting provide specificity and safety; (c) preservation of the full spectrum of viral genes, to enable a robust anti-tumor response [26,27]. Since the retargeted oHSVs do not carry any genetic modification in virulence or other genes, they are "fully virulent viruses in their target cancer cells" (Figure 1). Clearly, the extent of cancer-specificity reflects the specificity of the target to which the oHSVs are addressed. Some TAAs—such as EGFRVIII (epidermal growth factor receptor Variant III), IL-13 Receptor 2-α and HER2 (human epidermal growth factor receptor 2) are more specific than others. A further improvement was introduced by Glorioso and coworkers through point mutations in gD that impair the binging of neutralizing antibodies, and thus make the oHSV stealth to anti-HSV antibodies present in the human population [34].

Bench and preclinical studies foresee the following advantages for the tropism re-targeted oHSVs. In contrast to the $\Delta\gamma34.5$ oHSVs that can replicate solely in cancer cells that carry defects in certain pathways of the innate response and may potentially replicate in non-cancerous cells defective in innate responses, the retargeted oHSVs infect cancer cells irrespective of the status of their innate response and are designed to prevent off-target infections. The extent of infections in tissues with low level expression of the targeted receptors remains to be verified in humans. Moreover, once the retargeted oHSVs infect the tumor cells, they are essentially wt-viruses, they blunt the cell innate response and promote high viral replication [35–38]. Lastly, in the tumor bed, the retargeted oHSVs exclusively infect the cancer cells, whereas the $\Delta\gamma34.5$ and the attenuated viruses additionally infect immune cells, with unclear effects on the global immune response.

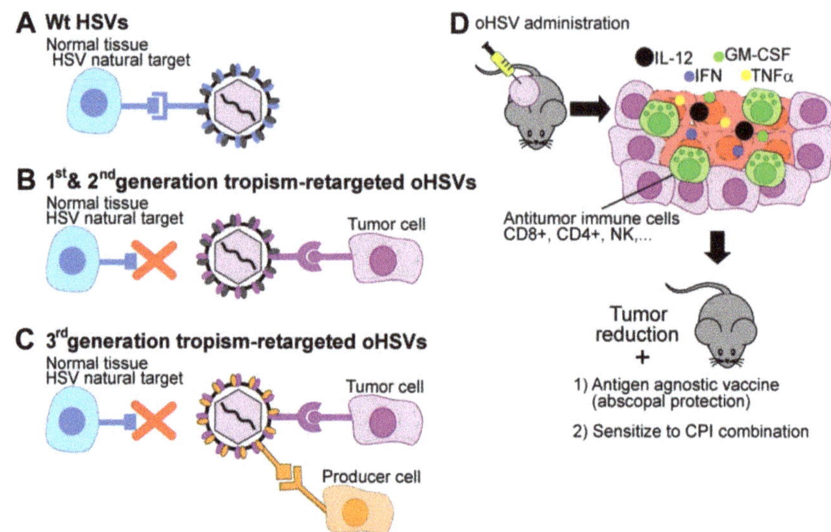

Figure 1. Schematic view of HSV tropism retargeting and immunotherapy induced by armed oHSVs. (**A–D**). (**A**) wt-HSV infects normal tissues, i.e., its natural target cells through natural major receptors HVEM and nectin1 (blue). (**B**) 1st and 2nd generation tropism retargeted oHSVs infect cancer cells expressing the target Tumor Associated Antigen (TAA, violet) and fails to infect its natural targets. The scFv to TAA which mediated the virus retargeting to cancer cells is engineered in the following virion glycoproteins: gD (1st generation), either gB or gH (2nd generation). (**C**) 3rd generation tropism retargeted oHSV, can infect both the cancer cells that express the target TAA (violet) and the producer cells that express an artificial receptor (orange), and fails to infect its natural targets. (**D**) The IL-12-armed retargeted oHSV specifically infects tumors cells, causes immunogenic cell death of cancer cells and elicits immuno-therapeutic response, that result in inhibition of tumor growth, sensitization to checkpoint blockade and antigen-agnostic vaccination.

2.2. Transcriptional and Post-Transcriptional Retargeting Strategies

Additional strategies to attain cancer specificity and preserve viral virulence include transcriptional retargeting—i.e., placing a critical viral gene under the control of a cancer-specific promoter, post-transcriptional retargeting and combinations thereof [39]. Examples include the control of the key γ34.5 virulence gene by a hybrid nestin enhancer-HSP68 minimal promoter which ensures expression of the HSV genome specifically in the nestin-positive glioblastoma cells [40] and the insertion of miRNA target sequences specific for selected tissues (e.g., brain, heart, or liver) to avoid off-tumor expression of key herpesviral proteins like ICP4 (infected cell protein 4), ICP27, UL8, and γ34.5 [20]. Such approaches have been elegantly reviewed recently [23] and are beyond the scope of current review.

3. Properties of the Tropism-Retargeted oHSVs Generated in Our Laboratory

3.1. Three Generations of Tropism Retargeted oHSVs

For heuristic reasons we divide the retargeted oHSVs generated in our laboratory into three groups (Table 1). In all, the retargeting was achieved by insertion of a single chain antibody (scFv) to the receptor of choice, while detargeting was achieved by deletion of appropriate portions in gD [27,41] [WO2009144755] (Table 1).

Table 1. 1st, 2nd and 3rd generation retargeted oHSV, genotypic modifications for retargeting and detargeting purposes.

Generation of Recombinant	Name of Recombinant	scFv for Retargeting to Tumor Cells Inserted	GCN4 Peptide for Retargeting to Producer Cells Inserted in	Detargeting Strategy, Deletions @ gD	Ref
1st	R-LM113 R-115 R-123	HER2 @ gD	Absent	Δ aa 6–38	[21,27,42]
	R-LM249	HER2 @ gD	Absent	Δ aa 61–218	[41]
	R-611	EGFR @ gD	Absent	Δ aa 6–38	[42]
	R-613	EGFRVIII @ gD	Absent	Δ aa 6–38	[42]
	R-593	PSMA @ gD	Absent	Δ aa 6–38	[42]
2nd	R-803 R-809	HER2 @gH	Absent	No deletion, or Δ aa 6–38	[43]
	R-903 R-909	HER2 @ gB	Absent	No deletion, or Δ aa 6–38	[44]
3rd	R-313, R-315 R-317 R-319	HER2 @ gD	@ gB	Δ aa 6–38	[46]
	R-213	HER2 @ gD	@ gH	Δ aa 6–38	[45]
	R-87 R-89 R-97 R-99 R-99-2	HER2 @ gD	@ gD	Deletions, various	[47]
	R-321, R-335 R-337	HER2 @ gD	@ gB	Δ aa30 and aa38	[46] this paper

The 1st generation recombinants carry the scFv in gD, in place of either aa 6-38 or aa 61–218. Such deletions eliminate the portions in gD responsible for interactions with HVEM and nectin1, and confer full detargeting. In different recombinants, the scFvs were addressed alternatively to HER2 (human epithelial growth factor receptor 2), EGFR (epithelial growth factor receptor), EGFRVIII (EGFR variant III) or PSMA (prostate specific membrane antigen) [27,42] and WO2009144755.

The 2nd generation recombinants carried the scFv to HER2 or to EGFR in either gH or gB. This recombinant group explored the possibility that glycoproteins essential for HSV entry, other than gD, serve as vector for the scFv. They carry the Δ6–38 in gD [43,44] and WO201612849.

The 3rd generation recombinants simultaneously carry two retargeting moieties. The rationale is detailed below (see, paragraph 3.3). One moiety is the anti-HER2 scFv inserted in gD for cancer cell retargeting. The other moiety is the GCN4 peptide engineered alternatively in gD, gH or gB for retargeting to an ad hoc producer cell line. For detargeting purposes, the 3rd generation recombinants contained one of the following deletions in gD: aa 6–38, two single amino acids—ΔD30 and ΔY38, or deletions in the nectin binding site encompassing aa 214–223 [44–47] [WO2017211941, WO2017211944, WO2017211945].

3.2. The Retargeted oHSVs are a Platform

TAAs constitutes a family of molecules, with varying degrees of cancer specificity. Very often, the encoding genes are genetically amplified in cancer cells, such that the TAAs are overexpressed in cancer cells, and poorly or not expressed in non-cancerous cells. Many are located on the cell surface. Since each member of the family is expressed across several cancer types [48], a single retargeted oHSV can potentially be employed against a number of different cancers. In most of our studies we selected HER2, expressed and amplified in a number of cancers, including breast, ovary, stomach, lung and pancreas cancers and glioblastoma, and is a relevant target in cancer immunotherapy. Thus, a HER2-retargeted oHSV can potentially be employed against a variety of indications. Glorioso laboratory, as well as our additionally generated oHSVs retargeted to EGFR, EpCAM, EGFRVIII specific for glioblastoma puntiforme, and PSMA, present in prostate cancers [23,26,49–51]. The EGFRVIII recombinant was further improved by insertion of a matrix metalloproteinase which enhanced intratumoral vector distribution and efficacy in a glioblastoma model [52]. Essentially, the retargeted oHSVs are a platform and can potentially be addressed to different TAAs. It is envisioned that the intensive molecular profiling programs carried out worldwide may lead to the discovery of novel TAAs, even more specific than the ones currently known.

3.3. Cultivation of Tropism-Retargeted oHSVs in Non-Cancerous Producer Cells

A critical feature of the retargeted oHSVs generated in our laboratory is that they are strictly dependent on the targeted cancer receptor for infection, including infection of the producer cells. Often, the target receptor is an oncogene, i.e., it contributes to the oncogenic potential of the cancer cells. While the retargeted oHSVs can be readily cultivated in human cancer cells lines positive for the targeted receptor, approval of clinical grade virus production in cancer cells by competent authorities might likely imply specific motivations. A goal in the design of the 3rd generation retargeted oHSVs was to generate a non-cancerous producer cell line for the in vitro growth of the retargeted oHSVs. To this end, as mentioned above, we designed recombinants that simultaneously carry two retargeting moieties (Figures 1 and 2A). The scFv retargets HSV to the cancer receptor (HER2, in our case). The second one consists of a small high affinity ligand (GCN4 peptide) engineered in one of the entry glycoproteins—gD, gH or gB (Figure 2A) [45–47]. The producer cell line is a Vero cell derivative, named Vero-GCN4R-HER2, which expresses an artificial receptor for the GCN4 peptide, along with human HER2.

In subsequent sections of this review, we shall focus on the two most advanced recombinants from the 3rd generation group, R-335 and R-337. They carry (i) the insertion of the GCN4 peptide in gB between residues 81 and 82; (ii) the deletion of only two amino acids in gD—D30 and Y38—for HVEM and nectin1 detargeting; (iii) the insertion in gD of scFv to HER2 in place of Y38. R-335 and R-337 carry mIL-12 in the US1/US2 intergenic region, a site that enables a high expression level. While R-335 carried the natural form of mIL-12, made of p40 and p35 subunits, R-337 carried the fusion form, in which the two subunits are held together by a linker to form a single peptide. In cultured cells, the fusion form was produced at 50 to 100-fold higher amounts than the dimeric form and possibly was more stable. R-335 and R-337 genotypes are depicted in Figure 2A. The properties of OVs, and particularly oHSVs expressing IL-12 are reviewed in [53].

Preliminarily, we quantified the ability of the two recombinants to grow in Vero-GCN4R-HER2 and in the human HER2-positive cancer cell line SK-OV-3. The yields of R-335 and R-337 in Vero-GCN4R-HER2 are shown in Figure 2B, which also shows the yields of R-LM5 (an essentially wt HSV carrying EGFP [27]) and of the 1st generation recombinant R-115. R-335 and R-337 infect these cells through both the GCN4R and HER2 receptors, whereas R-115 infects only through HER2. Figure 2C shows the fold-increase of the yields relative to that of R-115. Two features emerged. R-335 and R-337 replicated to seven to eight-fold higher yields than R-115. As expected, R-335 and R-337 replicated to fivefold lower yields than R-LM5; this is a common feature for recombinant viruses and accounts for different receptor usage, in that R-LM5 infection occurs through the simian orthologs of the natural receptors nectin1 and HVEM, which ensure the best possible interaction for HSV entry into the cells.

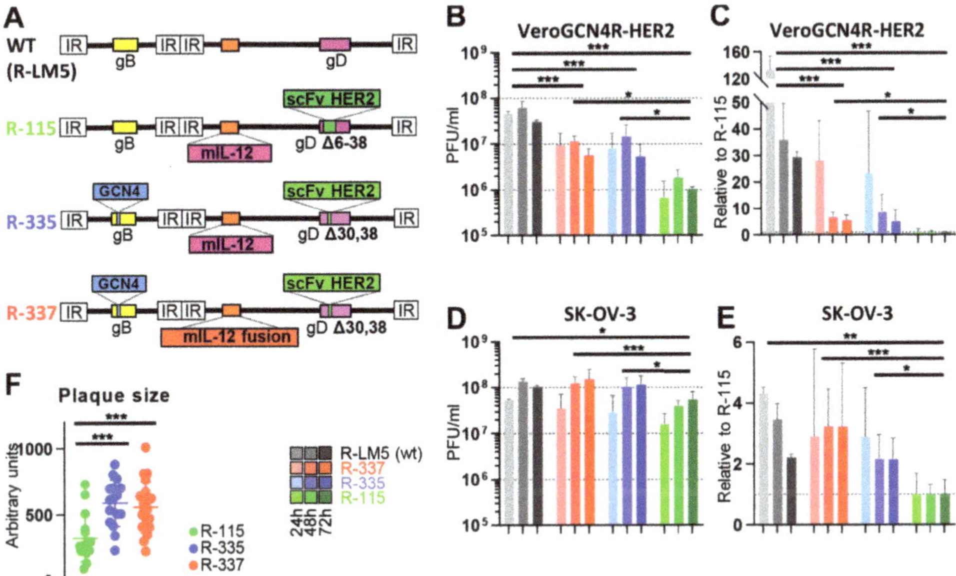

Figure 2. Growth kinetics of 1st and 3rd generation HER2-retargeted oHSVs in cancerous and non-cancerous producer cells. (**A**) Schematic representation of genomes of wt HSV named R-LM5 (carrying GFP), R-115 (1st generation), R-335 and R-337 (3rd generation) retargeted oHSVs. Indicated are the genetic loci of gB, gD, the insertion site of mIL-12 in the US1 and US2 intergenic locus. 1st and 3rd generation recombinant viruses carry the insertion of scFv anti-HER2 for the retargeting to HER2-positive cells, and the deletion of indicated portions of gD for the detargeting from HSV-1 natural receptors HVEM and nectin1. The 3rd generation R-335 and R-337 viruses carry the GCN4 peptide in gB between aa 81 and 82, and were engineered as detailed in [17,54]. (**B–E**) Yields of the wt HSV named R-LM5 (carrying GFP), R-115, R-335 and R-337 in VERO-GCN4R-HER2 (**B,C**) and in SK-OV-3 (**D,E**) cells at 24, 48 and 72 h after infection. Replicate cultures of each cell line were infected with the indicated viruses at 0.1 PFU/cell as titrated in SK-OV-3 cell line. Progeny virus was titrated in SK-OV-3 cells. In panels C and E, yields are expressed relative to that of R-115. The data represent the average of at least five independent experiments ± SD. (**F**) Plaque size of the indicated viruses five days after infection. For each virus-infected culture 20 plaque pictures were taken, expressed as arbitrary units and plotted ± SD. (**B–F**) Statistical significance was calculated by One Way ANOVA test and expressed as * = *p*-value < 0.05; ** = *p*-value < 0.01; *** = *p*-value < 0.001.

In our experience the highest yields for both wt-HSV and HER2-reatargeted oHSVs are obtained in SK-OV-3. Figure 2D,E show that the wt R-LM5 grew somewhat better in SK-OV-3 than in Vero-GCN4R-HER2 cells, as expected. The growth of R-335 and R-337 could not be differentiated from that of R-LM5 and was about three-fold higher than that of R-115. In addition, R-335 and R-337 plaques in SK-OV-3 cells were doubled in size relative to those from R-115, in agreement with the virus yields results (Figure 2F). The higher replication of R-335 and R-337 relative to that of R-115 was surprising in that SK-OV-3 cells lack the receptor for GCN4 peptide. We interpret these results to indicate that the smaller deletion in gD improved the glycoprotein performance, and that the GCN-4 insertion in gB somehow activated gB or a combination of these effects.

3.4. Safety Profile of Retargeted oHSVs

Safety of retargeted oHSVs rests on their specificity for cancer cells and on genetic stability and is documented by the following lines of evidence. In vitro, both 1st and 3rd generation retargeted oHSVs infected almost exclusively the cancer cells positive for the targeted receptor and failed to infect or infected very poorly receptor-negative cancer cells and non-cancerous cells, unless they transgenically expressed the targeted receptor [27, 41]. In no cases did the infection of few cells in a culture of receptor-negative cells result in a virus that could be serially passaged. The viruses exhibit genetic stability in that they have been passaged in cultures for several months (3rd generation) or years (1st generation), without any change in retargeting/detargeting properties. In vivo, upon intratumoral (i.t.) administration, the 1st generation R-115 was detectable only in the tumors, and not in serum or other organs (Figure 3A) [55]. When administered intraperitoneally (i.p.), the 1st generation oHSVs did not cause any pathological signs, including brain infections, even at the highest amounts (2×10^9 PFU). Under the same conditions, the wt-HSV killed all mice (Figure 3B) [41,56]. In vivo, upon intracranial administration, the 1st generation R-LM113 virus did not infect the brains, whereas the wt-HSV readily did (Figure 3C) [57]. Altogether, the results support the notion that (i) that in vitro infection of human cells only occurs at high level HER2 expression, and (ii) the HER2-retargeted oHSVs do not cause detectable off-target infections in mice. A detailed analysis on bio-distribution to human tissues, especially in tissues with low level HER2 expression, remains to be performed.

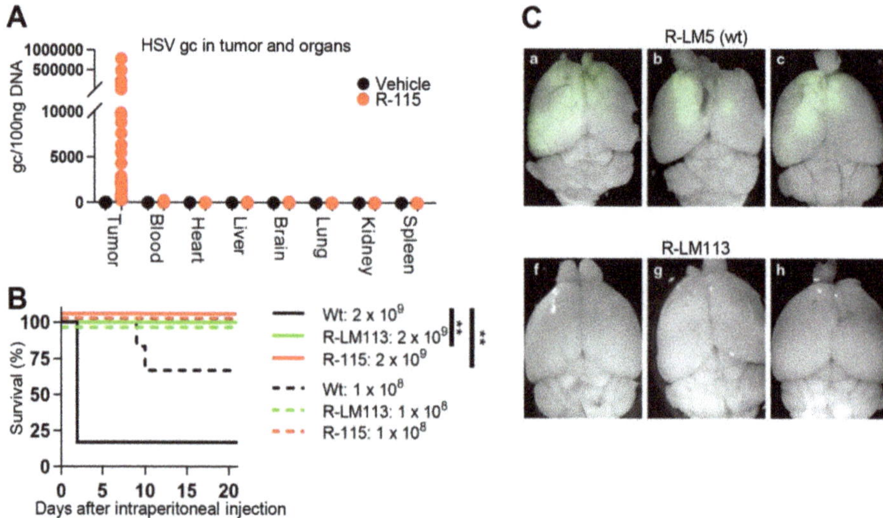

Figure 3. The retargeted oHSVs are safe in mice upon intraperitoneal, intratumoral or intracranial routes of administration. (**A**) R-115 biodistribution to the indicated organs following four intratumoral injections (1×10^8 PFU/dose or vehicle), started at d 10 after tumor implantation. Organs and tumors were explanted at d 26, and, after homogenization, the total DNA was extracted. R-115 genome copy numbers were determined by qPCR in comparison with a standard curve obtained with purified HSV DNA, and expressed as gc/100 ng of DNA or gc/100 µl blood. (**B**) Kaplan Meier survival curves of the C57BL/6 mice intraperitoneally injected with 1×10^8 or 2×10^9 PFU of R-LM5 (wt HSV), R-LM113 and R-115 (1st generation) oHSVs. (**C**) Merged fluorescence and bright-field images of adult nonobese diabetic/severe combined immunodeficient (NOD/SCID) mouse brains after injection with R-LM5 (1×10^5 PFU) or R-LM113 (3×10^5 PFU) viruses. Viral spread is visualized by enhanced green fluorescent protein fluorescence. (**A**) Statistical significance was calculated using the Log-rank (Mantel-Cox) test. Panels (**A–C**), reproduced with permission. ** = *p*-value < 0.01.

4. In Vivo Efficacy of Retargeted oHSVs in Immunocompetent Mouse Models

4.1. Efficacy against LLC-1-HER2 Primary Tumors

Early studies from our laboratory indicated that retargeted oHSVs are highly effective in nude mice, a property which only accounted for direct oncolytic effects [41,56]. The key question arose as to how effective the retargeted oHSVs are in immunocompetent mice, in particular in eliciting the innate response to the virus, the innate and the adaptive long-term immunity to the tumor. To address this question here we provide the first description of the efficacy of R-335 and R-337 and review previously described efficacy data on 1st generation recombinants R-LM113, R-115 and R-123. A list of the most significant preclinical studies carried out in our laboratory is reported in Table 2. As discussed above, the HER2-retargeted oHSVs are strictly dependent on (human) HER2 to carry out infection, a feature that required an ad hoc immunocompetent murine model. Preliminarily, we screened a number of murine tumor cell lines and found that LLC-1 (Lewis lung carcinoma-1) cells enabled the highest HSV replication [55]. The ad hoc model consists of LLC-1 cells made transgenic for HER2 (LLC-1-HER2) and of the syngeneic C57Bl/6 mice transgenic and tolerant (TG) to HER2. Mouse tolerance to HER2 was critical to prevent that the immune response to the tumor was mainly driven by the allogeneic HER2 [55]. With respect to antitumor activity, the 1st generation R-115 protected 60% of mice, of which 16% exhibited a complete response (CR) and 44% a partial response (PR) [55]. The experimental design in anti-tumor efficacy experiments is illustrated in Figure 4, panel A. Briefly, mice were implanted with subcutaneous tumors; the recombinants were administered intratumorally (i.t.) to well-developed tumors. The mice that survived the primary tumor received a second challenge tumor, which was untreated. The R-335 and R-337 recombinants were administered i.t. to well-developed tumors as five injections of 1×10^8 PFU each, every other day (Figure 4A). The antitumor activity of R-335 was similar to that of R-115, while that of R-337 was higher. In particular, R-335 protected 60% of mice, 30% of which exhibited a complete response (CR) (Figure 4B–E). R-337 protected 100% of the mice, 80% of which exhibited CR (Figure 4B–E). The Kaplan Meier survival curve shows highly statically significant differences between each virus and the control, and between the two viruses (Figure 4F). The superior efficacy R-337 relative to R-335 should be interpreted in light of the fact the only genotypic difference between the two viruses resides in mIL-12, which is a heterodimer in R-335 and a fusion form in R-337. The results clearly indicate that a significant contribution to the control of primary tumor growth is immune mediated.

Table 2. 1st and 3rd generation retargeted oHSVs. Summary of preclinical studies.

	Name of Recombinant	Payload	Preclinical Assays and Main Results							Ref
			Tumor	Mice	Efficacy against Primary Tumor	Efficacy against Distant Challenge Tumor	Modifications to TME	Combo with CPI	Route of Administration	
	R-LM249	None	Human Ovary SK-OV-3	Immune-deficient	Very high protection. Protection from disseminated metastases.	NT	NT	NT	I.T. I.P. Carrier cells	[41,56,58]
	R-LM113	None	Mouse lung (LLC-1-HER2)	Immune-competent, HER2-tolerant	Complete response against primary tumor in 30% mice in early treatment.	High protection	Low modifications	NT	I.T.	[55]
	R-LM113	None	Human high grade glioma AND Murine high grade glioma	Immune-deficient AND Immune-competent	Doubling in survival time AND Doubling in survival time	NT	NT	NT	I.C.	[57,59]
1st generation	R-115	mIL-12	LLC-1-HER2	Immune-competent, HER2-tolerant	Complete response against primary tumor in 70% mice in early treatment.	High protection	Infiltration by immune cells. Increased cytokines		I.T.	[55]
	R-115	mIL-12	Murine high grade glioma (HGG-HER2)	Immune-competent	Complete eradication in 30% mice in late treatment	High protection	Increased infiltration by CD4 CD8		I.C.	[60]
	R-123	mIL-12 + mGM-CSF	LLC-1-HER2	Immune-competent, HER2-tolerant	Complete response against primary tumor in 40% mice in late treatment.	NT		Full protection	I.T. I.V.	[21]

Table 2. *Cont.*

	Name of Recombinant	Payload	Preclinical Assays and Main Results							Ref
			Tumor	Mice	Efficacy against Primary Tumor	Efficacy against Distant Challenge Tumor	Modifications to TME	Combo with CPI	Route of Administration	
3rd generation	R-335	mIL-12	LLC-1-HER2	Immune-competent, HER2-tolerant	Complete response in 30% mice in late treatment	High protection	NT	Low poten-tiation	I.T.	This paper
	R-337	mIL-12 fusion protein	LLC-1-HER2	Immune-competent, HER2-tolerant	Complete response in 80% mice in late treatment	High protection	Increased infiltration by immune cells; increased expression of cytokines	Potentiation	I.T.	This paper

NT, not tested.

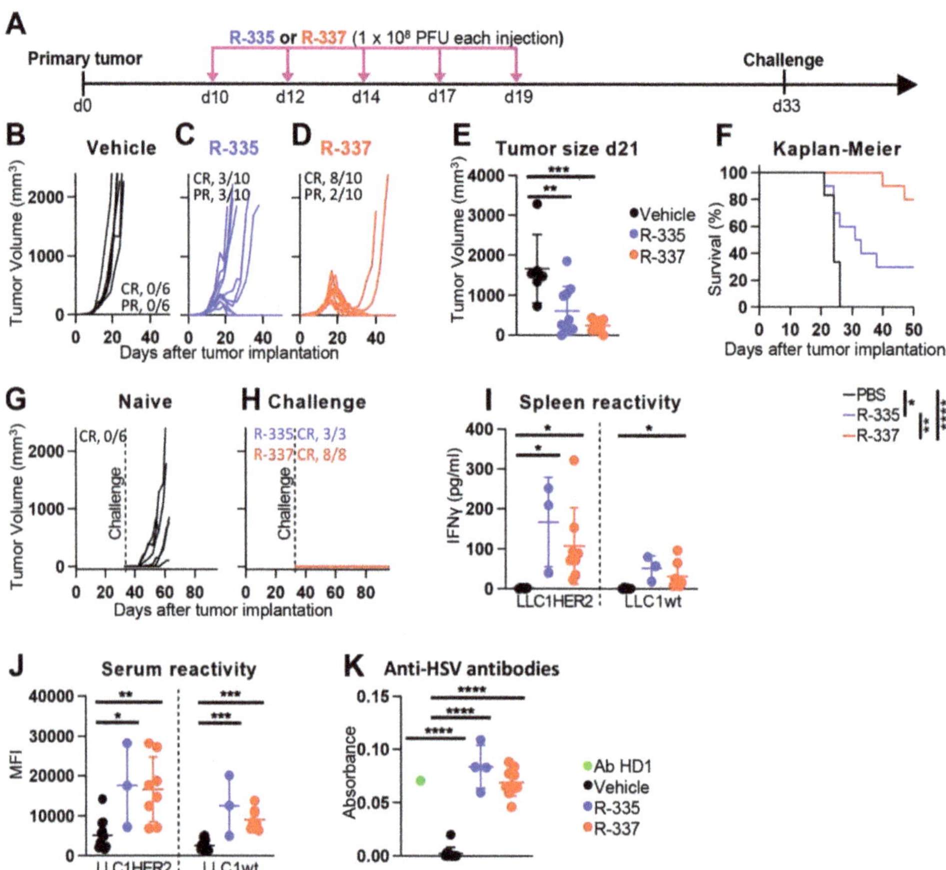

Figure 4. Efficacy of R-335 and R-337 monotherapy on the growth of LLC-1-HER2 tumors. (**A**) Schedule of treatments. The six-to-eight weeks old HER2-transgenic/tolerant (HER2-TG) C57BL/6 mice were subcutaneously implanted in the left flank with 5×10^5 LLC-1-HER2 cells in 100 μL of PBS [55]. 10 d later, when the tumor volumes averaged 70–100 mm^3, mice received 5 intratumoral injections of R-335, R-337 (1×10^8 PFU per injection, diluted in 50 μL PBS) or vehicle (50 μL PBS), at 2–3 day intervals. At d 33, the mice which survived the primary tumor received a contralateral challenge LLC-1-HER2 tumor of 5 $\times 10^5$ cell per mouse. Tumor volume was calculated using the formula: largest diameter x (smallest diameter) 2 × 0.5. Mice were sacrificed when tumor volumes exceed 1000–2000 mm^3, ulceration occurred, or animals exhibited distress or pain. (**B–D**) Kinetics of tumor growth in mice treated with vehicle (**B**), R-335 (**C**) or R-337 (**D**). The numbers reported in each panel indicate the numbers of mice which were completely cured from tumors (complete response, CR), or which showed a delay/reduction in tumor growth (partial response, PR). The mice were scored PR when the tumor volume was <50% smaller than the mean size of the tumors in the vehicle group. (**E**) Volumes of the primary tumors at d 21 after implantation. Black (vehicle), blue (R-335) and red (R-337) circles. (**F**) Kaplan-Meier survival curves of the three groups of mice. (**G,H**) Kinetics of growth of contralateral challenge tumor in naïve mice (**G**), and in the R-335 or R-337 (**H**) arms. (**I**) Immune response in splenocytes harvested at sacrifice. To isolate splenocytes, spleens were smashed through a 70 μm cell strainer in PBS, red blood cells were lysed with ACK buffer, and samples were resuspended in medium (RPMI 1640 containing 10% heat inactivated FBS, 1% penicillin/streptomycin). Splenocytes (1×10^6 cell/well) were incubated with 1×10^5 LLC-1-HER2 or LLC-1 cells in 0.5 mL medium, and cocultured for 48 h. The amount of secreted IFNγ (quantified by ELISA) was a measure of the splenic anti-LLC-1 and anti- LLC-1-HER2 immune response [55]. (**J,K**) Antibody reactivity in sera harvested at sacrifice to LLC-1-HER2 or LLC-1 cells (**J**), and to HSV-1-infected cells (**K**), as determined by cell enzyme-linked immunosorbent assay (CELISA).

Wt-LLC-1 and LLC-1-HER2 single cell preparations were reacted with mouse serum, diluted 1:150 in flow cytometry buffer (PBS + 2% FBS), in ice for 1 h, washed with flow cytometry buffer and incubated with anti-mouse PE (1:400). Data were acquired on BD C6 Accuri. For CELISA assay, RS cells were infected with HSV-1 at 3 PFU/cell for 24 h, then they were fixed with paraformaldehyde, reacted with mouse serum diluted 1:60, or with the anti-gD monoclonal antibody HD1 (green) diluted 1:400 (positive control), followed by anti-mouse peroxidase. Peroxidase substrate o-phenylenediamine dihydrochloride was added and plates were read at 490 nm as detailed [55]. (F) Statistical significance was calculated by the Log-rank (Mantel-Cox) test. (**E,I–K**) Statistical significance was calculated by means of the One Way ANOVA test and expressed as * = p-value < 0.05; ** = p-value < 0.01; *** = p-value < 0.001; **** = p-value < 0.0001. Color code: mice which received Vehicle, R-335 or R-337 are indicated in black, blue or red, respectively.

4.2. Retargeted oHSVs Promote Antigen Agnostic Vaccination Effect against Distant Tumors

A notable property of R-115 is the long-term abscopal efficacy [55]. Mice which survived the primary tumor were fully protected from a distal tumor implanted at later times. Essentially, R-115 vaccinated mice against a subsequent challenge tumor. Even the 3rd generation R-335 and R-337 proved to be particularly effective. Of the mice described in Figure 4C,D, those which survived the primary tumor received a challenge LLC-1-HER2 tumor 33 days later. All these mice were fully protected (Figure 4G,H). Inasmuch as the mice did not receive any treatment after the implantation of the distant challenge tumor, any protection seen against such tumors was immune-mediated.

The mice protected from distant tumors exhibited a T-cell immune response documented as splenocyte reactivity to tumor cells (Figure 4I), in agreement with similar finding with R-115 [55]. In particular, the splenocytes from both R-335- and R-337-treated mice, harvested at sacrifice, reacted strongly to LLC-1-HER2 cells, and weakly to LLC-1 cells (Figure 4I). The antibody response reflected at large the T-cell response, in that the sera from R-335- and R-337-treated mice carried antibodies to LLC-1-HER2 and, to a lesser extent, to LLC-1 cells (Figure 4J). The extent of protection against distant wt-LLC-1 tumors will be evaluated in detail in future studies. Current results argue that the intratumoral treatment of LLC-1-HER2 tumors with R-335 or R-337 can elicit a protective response also to wt-LLC-1 cell neoantigens. The mice sera showed seroconversion also to HSV-1 (Figure 4K), as expected.

4.3. Retargeted oHSVs Subvert TME Immunosuppression

The purpose of this series of experiments was to provide evidence that the long-term distant protection was mediated by an immune response, documented as dramatic changes to the immunosuppressive TME. In these experiments, mice were treated i.t. with the R-337, and sacrificed a few days after the end of treatment, at a time when tumors were decreasing in size (Figure 5A–C). Analyses were carried out on tumor infiltrating lymphocytes and cytokines, on the reactivity of splenocytes and of serum antibodies to tumors cells, with the aim to detect local and systemic modifications. In R-115-treatred mice, the major modifications consisted in the tumor infiltration by CD4+, CD8+ and activated CD8+, NK (natural killer) and activated NK, Tregs (T-regulatory), along with the reduction in intratumoral CD11b+ leucocytes [55]. The immune landscape of LLC-1-HER2 TME is that of an immunologically desert tumor, characterized by low infiltration from anti-tumor immune subpopulations and low levels of immune activation markers, co-stimulatory molecules and pro-inflammatory cytokines [61]. In essence, the host immune system is unable to recognize and react against LLC-1 tumors. Figure 5D–K documents the modifications detected in R-337-treated mice. Worth noting are the increase in tumor infiltrating leucocytes, specifically CD4+, CD8+ and activated CD8+, DCs, and NK and activated NK cells. The CD11b-positive population, which includes the immunosuppressive myeloid derived suppressor cells, was decreased (Figure 5L). FoxP3+ cells, which include the T-regulatory cells, were also increased (Figure 5H), in agreement with previous reports [55]. Transcriptional analysis of the tumor specimens revealed an increase in IFNγ, IL-12 (most of which likely expressed from the viral genome), CXCL11 chemokine and t-bet transcription factor (Figure 5M–Q), hallmarks of inflamed TME and polarization to activated Th1 cells. Analysis of the systemic effect was carried out on spleen samples. The modifications were essentially similar to those detected in the tumor samples (Figure 5R–W), except that the increase in NK cells was non statistically significant. The splenocyte reactivity and the antibody response to LLC-1-HER2 cells were essentially similar to those detected in mice sacrificed at about 100 days after primary tumor implantation (Figure 5X,Y). Altogether, i.t.- administered R-337 elicited a strong systemic and intratumoral immune response, and the inflammation of the LLC-1-HER2 TME.

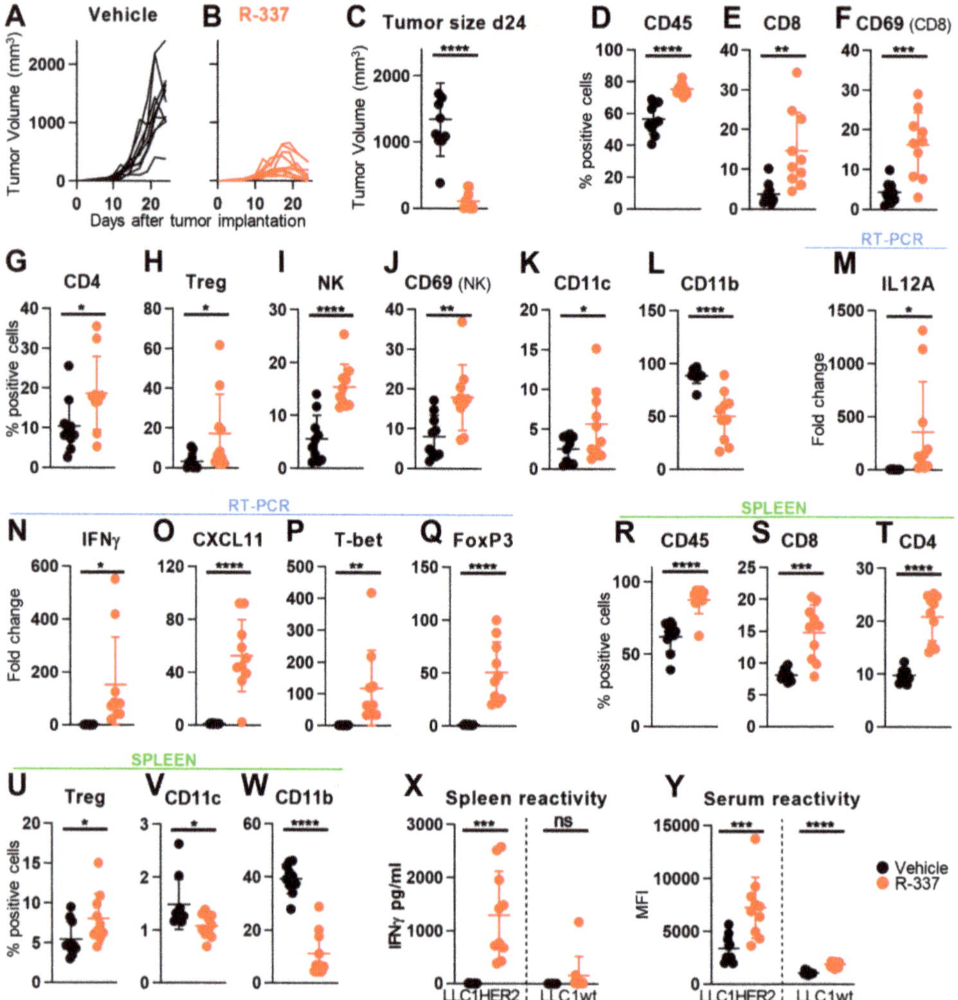

Figure 5. Immune heating of TME and spleen modifications induced by intratumoral R-337 monotherapy. (**A,B**) Kinetics of tumor growth in HER2-TG C57BL/6 mice treated with vehicle (**A**) or R-337 (**B**), according to the schedule reported in Figure 4A. (**C**) Tumor volumes at d 24. Black (vehicle) and red (R-337) circles. (**D–L**) Immune cell populations in tumors. Single cell suspensions were prepared from freshly isolated LLC1-HER2 tumors at sacrifice. Tumors were minced in small pieces, digested with collagenase, passed through 70 µm cell strainer and rinsed with FACS buffer. For each sample, 2×10^6 cells were blocked with α-CD16/32 Ab (clone 93), and then reacted with the antibodies CD4-FITC (clone GK1.5), CD8a-PE (clone 53-6.7), CD45-Percp-Cy7 (clone 30-F11), CD335-APC (clone 29A1.4), FoxP3-PE (clone 150d/e4), CD11b-FITC (clone M1/70), CD11c-PE (clone N418) and CD69-PercP (clone H1-2F3). Data were acquired on BD C6 Accuri. CD4 (CD4+ cells), CD8 (CD8+ cells), NK (CD335+ cells) and myeloid cells (CD11b+ cells) were gated on CD45+ subpopulation.

Activated (CD69+) CD8 and NK cells were gated on CD8+ and CD335+ subpopulations, respectively. DC cells (CD11c+CD11b+) were gated on CD11b+ population. Tregs (FoxP3+CD4+) were gated on CD4+ population. (**M–Q**). Expression profile of cytokines, immune related transcription factor and immune markers. Tumor homogenates (a few mgs) were employed for total RNA purification and 1.2 µg of RNA was employed for the cDNA synthesis. Diluted cDNAs (1:4) were assayed by real-time PCR with Taqman probes. The levels of expression were determined using the ΔΔCt method, normalized on the Rpl13a housekeeping gene and on the mean of the vehicle-treated group. (**R–W**) Immune cell populations in spleens. Sample preparation and staining as described for tumors. (**X**) Immune response in splenocytes to LLC-1-HER2 and LLC-1 cells was quantified as IFNγ secretion in the culture medium. For the details, see Figure 4. (**Y**) Serum antibody reactivity to LLC-1-HER2 and LLC-1 cells. For the details, see Figure 4C–Y Statistical significance was calculated by the t-test and expressed as * = *p*-value < 0.05; ** = *p*-value < 0.01; *** = *p*-value < 0.001; **** = *p*-value < 0.0001, ns = not significant. Color code, mice treated with vehicle or R-337 are indicated in black or red, respectively.

4.4. The "Immune Heating" of the Tumor Predisposes to Combination Therapy

The distant long-term protection, along with the dramatic changes to TME induced by R-337, suggested that the recombinant could render immunologically cold and CPI-resistant tumors immunologically hot and possibly CPI-sensitive. LLC-1-HER2 tumors recapitulate tumors that are completely insensitive to anti-PD-1 (compare Figure 6B with Figure 4B), in agreement with the low immunogenicity of these tumors [61]. The experiment documented in Figure 6 was designed to ascertain whether R-335 and R-337 synergize with anti-PD-1. Mice were treated as in Figure 4, and additionally received anti-PD-1, administered i.p. (see Figure 6, panel A). It can be seen that, when combined with anti-PD-1 in a simultaneous regimen [62–65], R-335 displayed a tendency to increase efficacy (Figure 6B–E). Thus, in mice treated with R-335 alone, CR and PR occurred in 31 and 25% of mice, respectively, in agreement with data shown in Figure 4. In mice which received the combination therapy, CR and PR occurred in 41 and 35% of mice, i.e., 76% mice were protected, completely or partially. The Kaplan Meier survival curve is reported in Figure 6F. The mice which survived the primary tumor were fully protected from a challenge distant tumor (Figure 6G,H). The long-term protection was most likely based on the systemic immune response, documented as splenocyte and antibody reactivities to LLC-1-HER2 and wt-LLC-1 cells (Figure 6I,J).

To evaluate the efficacy of R-337 in combination with anti-PD-1, we decreased the overall amount of virus from five injections of 1×10^8 PFU each to three injections of 0.3×10^8 PFU each (in total, 0.9×10^8 vs 5×10^8) (Figure 6K). At this lower dosage, R-337 monotherapy induced CR in 36% mice and PR in 18%, with an overall response rate of about 55%. In the combination arm, 80% of mice exhibited CR, and 10% exhibited PR (Figure 6L–N). The difference between monotherapy and combination therapy was statistically significant with respect to tumor size (Figure 6O) and Kaplan Meier survival curve (Figure 6P). The surviving mice were fully protected from a distant challenge made of LLC-1-HER2 cells (Figure 6Q,R). At sacrifice, 80 days after primary tumor implantation, the mice treated with the combination therapy showed a tendency to increased splenocyte response (Figure 6S), and an increase in antibody response (Figure 6T). The results show that the R-337 and anti-PD-1 combination therapy was highly effective and are consistent with the view that the distant protection was immune-based.

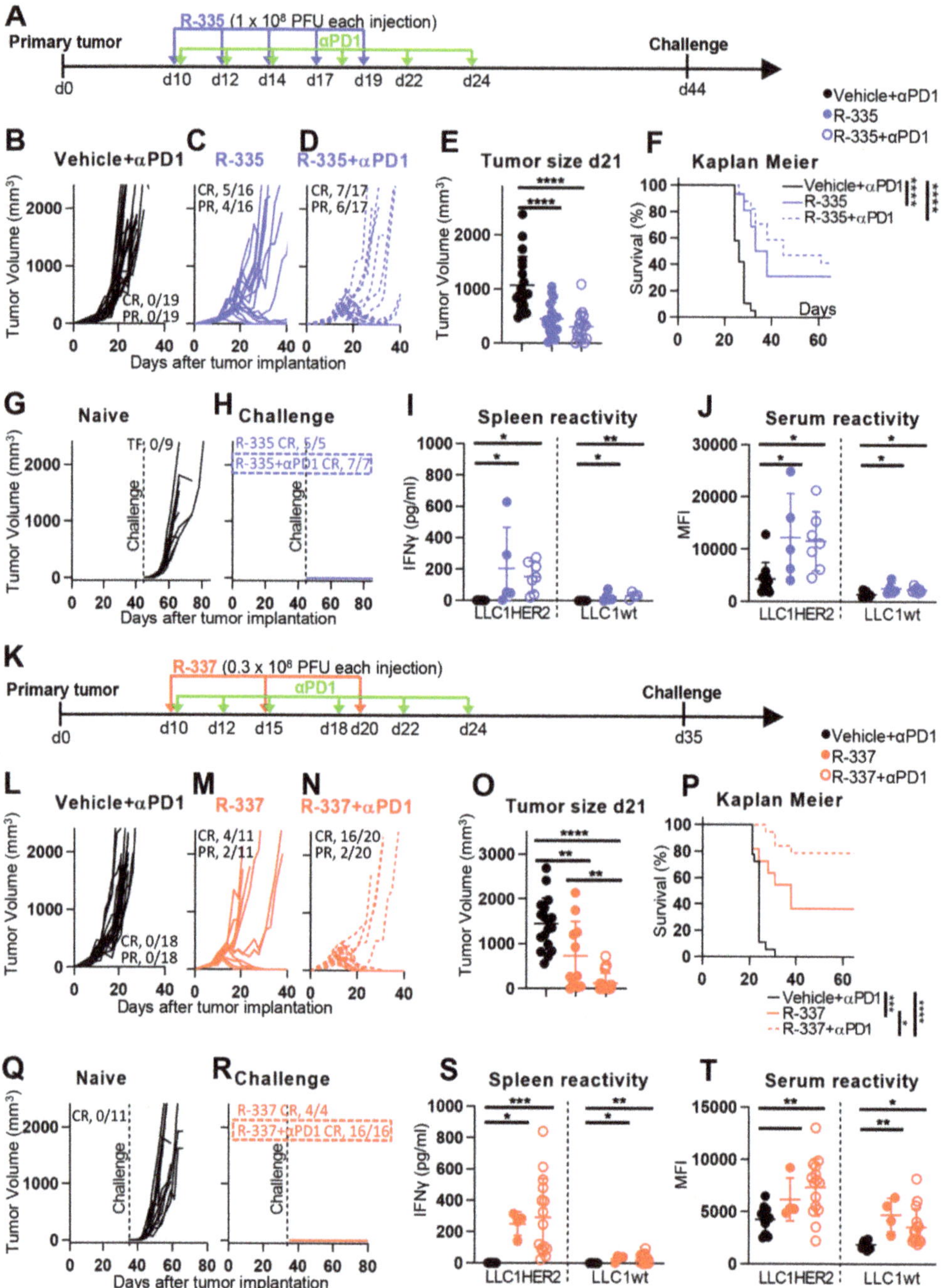

Figure 6. Efficacy of R-335 of R-337 in combination with anti-PD1 antibodies on the growth of LLC-1-HER2 tumors. (**A**) Schedule of the treatments. The HER2-TG C57BL/6 mice were implanted with LLC-1-HER2 cells. At d 10 after implantation, when tumors reached the average volume of 70–100 mm^3, mice received 5 i.t. injections of R-335, or R-335 plus i.p. injections of anti-PD-1, at 2–3 days intervals. The administration schedule of oHSV and anti-PD-1 treatments was according to [62–65]. At d 44, the mice which survived the primary tumor received a contralateral challenge LLC-1-HER2 tumor.

For the details, see Figure 4B–D Kinetics of tumor growth in mice treated with vehicle (**B**), R-335 alone (**C**), or R-335 plus anti-PD-1 combination therapy (**D**). Figures in panels indicate the number of mice exhibiting complete response (CR) or partial response (PR). (**E**) Volumes of the primary tumors at d 21 after implantation. Black (vehicle), blue (R-335) and open blue (combination) circles. (**F**) Kaplan-Meier survival curves of the three groups of mice. (**G–H**) Growth kinetics of contralateral challenge tumors in naïve mice (**G**), and in R-335 or combination arms (**H**). (**I**) Immune response in splenocytes harvested at sacrifice. Splenocytes were incubated with LLC-1-HER2 or LLC-1 cells. Activation was quantified as IFNγ secretion in the culture medium. (**J**) Serum antibody reactivity to LLC-1-HER2 or LLC-1 cells. (**K**) Schedule of the treatment with R-337 with or without combination with anti-PD-1 antibodies. The HER2-TG C57BL/6 mice were implanted with LLC-1-HER2 cells. At 10 d after tumor implantation, mice received 3 i.t injections of R-337 at 5 days interval, and, where indicated, i.p. injections of anti-PD-1 antibodies, as detailed in the drawing. At d 35, the mice which survived the primary tumor received a contralateral challenge LLC-1-HER2 tumor. (**L–T**) Kinetics of tumor growth (**L–N**), tumor size at d 21 (**O**), Kaplan Meier survival curves (**P**), growth curves of challenge tumors (**Q–R**), immune response in splenocytes (**S**), antibody reactivity to LLC-1-HER2 or LLC-1 cells (**T**). (**F,P**) Statistical significance was calculated by the Log-rank (Mantel-Cox) test. (E, I, J, O, S, T) Statistical significance was calculated by means of the ANOVA test and expressed as * = p-value < 0.05; ** = p-value < 0.01; *** = p-value < 0.001; **** = p-value < 0.0001. Color codes: mice treated with vehicle, R-335, R-337 are indicated in black, blue or red, respectively. Full circles and continuous lines, monotherapies. Open circles and dotted lines, combination therapies.

4.5. Retargeted oHSVs Eradicate High Grade Gliomas (HGG) in Preclinical Models

Glioblastomas (GBMs) are among the tumors with highest resistance to surgery, chemo- and radiotherapy, and highest mortality rate. Essentially, the natural history of these tumors has not changed in the last 50 years. GBMs have been the subject of intense interest as targets of OV-based therapy, and especially of oHSVs, in part because of the natural tropism of HSV for the nervous system. Human GBMs express TAAs, such as HER-2, EGFRvIII, IL-13R2α, EGFR and others.

We provided proof of principle that GBM can potentially be treated with retargeted oHSVs. In initial studies, Malatesta and his group developed a high-grade glioma (HGG) model, consisting of human GBM cells genetically modified to express HER2 and ortho-topically implanted in the brains of immunodeficient mice [57]. When the 1st generation unarmed R-LM113 was administered i.t. as single dose, it more than doubled the survival time, and fully protected about 20% of the mice [57]. These findings confirmed and extended similar finding on EGFRvIII-expressing GBM cells [49]. Subsequently, the Malatesta group developed a genetically engineered HGG preclinical model in immunocompetent syngeneic BALB/c mice. The tumors cells were derived upon overexpression of platelet-derived growth factor B (PDGF-B). This model exhibits a gene expression profile typical of oligodendrocyte precursor cells and histopathological features typical of GBM, thus reca-pitulating GBM [66]. The cancer cells were made transgenic for HER2, and orthotopically implanted in the brains of BALB/c mice. A single dose of R-115 administered intracranially (i.c.) fully protected 30% of mice. At sacrifice, the protected mice did not harbor any remnant of tumor. Interestingly, as we observed with the LLC-1-HER2-bearing mice, all the R-115-treated mice which survived the primary tumor exhibited a long-term distant protection and developed immune response to the tumor. This consisted of a systemic IgG response, as well as of a local response, whereby tumors became infiltrated with CD4+ and CD8+ lymphocytes [60].

4.6. Efficacy of Systemically Administered Retargeted oHSVs

A major aim in the OV field has been the development of agents suitable for systemic treatment of metastatic cancer. Given the natural history of the viruses from which the different OVs are derived, some OVs are better suited than others for systemic administration. In addition to unspecific uptake by parenchymal organs—a barrier for all OVs, the OVs based on some human viruses, like HSV, human Adenoviruses, measles viruses need to contrast the prior immunity that exists in humans. The question arose whether retargeted oHSVs are sufficiently robust for systemic delivery. In early studies we found that a 1st generation retargeted oHSV, named R-LM249, administered by weekly i.p. injections, significantly reduced brain and ovarian metastases by HER2-positive human tumors in immunodeficient mice carrying multiorgan tumors [56]. R-LM249 could also be delivered by means of mesenchymal carrier cells and decreased metastatic lung burden [58]. Most recently, in a immunocompetent mouse model that takes advantage of the immunotherapeutic effect, a 1st generation oHSV payloaded with IL-12 and GM-CSF, and administered in combination with anti-PD1, was capable of strongly decreasing the burden of lung tumor nodules induced by intravenous administration of LLC-1-HER2 cells, a model of metastatic lung disease [21]. Remarkably, the tumor growth inhibition occurred also in HSV-preimmunized mice [21]. The finding provides evidence that appropriately armed retargeted oHSVs in combination with checkpoint blockade can be a suitable agent for systemic administration.

4.7. Patents

Herpes Simplex Virus (HSV) with modified tropism, uses and process of preparation thereof—WO2009144755 (divisional patent EP2700405)

Retargeted herpesvirus with a glycoprotein H fusion—WO201612849

Herpesvirus with Modified Glycoprotein B—WO2017211941

Herpesvirus with Modified Glycoprotein D—WO2017211944

Herpesvirus with Modified Glycoprotein H For Propagation In A Cell—WO2017211945

5. Conclusions

The main features to emerge from current review on retargeted oHSVs are as follows.

A major argument that may be raised against tropism retargeted oHSVs is that they are not anti-pan-tumor agents. This argument does not take into account that any selected TAA is expressed across a number of cancer types, and therefore any retargeted oHSV can potentially be employed against a variety of indications. Moreover, clinical practice indicates that, due to cancer heterogeneity, the therapeutic effects of potentially pan-tumor OVs is not uniformly exerted on any type of tumor, or of cancer patient. The current trend in anti-cancer therapy is to develop therapeutic agents tailored on the patient characteristic, since it is futile and even counterproductive to administer a therapy to a patient who will not benefit of it (see reference [67] in this issue). The retargeted oHSVs meet this need. It is interesting to note that the restriction to predefined targets is shared with CAR-Ts (chimeric antigen receptor T cells) to solid tumors, which are essentially addressed to the same targets as the retargeted oHSVs [68,69]. The retargeted oHSVs and CAR-Ts could potentially be combined, and the combination could overcome some of the limits encountered by CAR-Ts, e.g., the difficulty to populate the tumors and to overcome the TME immunosuppression, essentially hostile to lymphocyte proliferation and activation [70].

A second Con is that the retargeted oHSVs strictly depend on the presence of the targeted receptor for infection, also for infection of producer cells. This property greatly limits the repertoire of cells which can be employed for production. Inasmuch as it advisable to make use of non-cancerous cells for clinical grade virus production, we set-up a strategy for virus growth, based on a double retargeting. Finally, it has long been debated that a disadvantage of employing human viruses as OVs is the prior immunity, which can block virions, particularly if they are administered by systemic routes. This notion should be reassessed in light of the interesting discovery that prior immunity, or an unrelated immune response, can actually contribute to unleashing the TME immunosuppression [71]. Indeed, the high efficacy deployed by the retargeted oHSVs enabled a systemic delivery, even in mice carrying a prior immunity to HSV [21]. In the field of vaccines to infectious agents, an analogous potentiation of the innate response by immunologically unrelated priming was also reported [72]. Whether such effects apply to humans treated with HSV-based OVs remains to be determined.

On the Pro side, the IL-12-positive R-337 was highly effective in eradicating the primary tumor, modifying the TME, and eliciting an antigen-agnostic long-term anti-tumor vaccination. How do we explain the high abscopal efficacy and antigen-agnostic vaccination? It is well known that wt-HSVs cause immunogenic cell death and are inducers of innate response, which they blunt later in infection [35–38]. Typical features of the innate response to HSV are the secretion of IFN-α, -β, and -γ—through STING and other sensors—and of additional pro-inflammatory cytokines, infiltration by and activation of NK cells, of CD4+ and CD8+ lymphocytes, and of dendritic cells (DCs). The high levels of IFN-γ and the recruitment and activation of DCs, T and NK cells are hallmarks of the early phases of innate and adaptive responses to the tumor. Thus, the innate antiviral response may well serve as the switch that turns on the anti-tumor innate response; the latter then evolves into the adaptive response [73]. The LLC-1-HER2 tumor employed in current studies exemplifies a tumor insensitive to anti-PD-1. Treatment with R-337 efficiently unleashed the resistance, such that the efficacy of both R-337 and anti-PD-1 were dramatically increased in a combination regimen and resulted in high protection. Thus, the retargeted oHSVs appear to serve well the function of augmenting the tumor sensitivity to CPIs.

In mice there was no detectable off-tumor and off-target infections, even though the potential of spread to tissues with low level expression of the targeted receptor (HER-2, in our case) remains to be verified in humans for any given retargeted oHSV.

In conclusion, major difficulties facing the clinical translation of retargeted oHSVs to the clinic have been tackled and solutions have been found. The stage is ready for further developments and for verification at the bedside of how well this class of therapeutic agents will hold promise.

Author Contributions: Conceptualization and writing original draft G.C.-F. and A.V.; review and editing, all; Data curation, Methodology, A.V., V.L., M.S., T.G., C.B., G.G., V.G.; Supervision, G.C.-F. and A.Z.; Animal experimentation responsibility A.Z. All authors have read and agreed to the published version of the manuscript.

Funding: The work was funded by European Research Council (ERC) Advanced Grant number 340060 to G.C.-F., by RFO (University of Bologna) and Pallotti Legacy to T.G. The authors received funds for covering the costs to publish in open access (ERC grant).

Institutional Review Board Statement: Animal experiments were performed according to European directive 2010/63/UE, Italian law 26/2014. The experimental protocols were reviewed and approved by the University of Bologna Animal Care and Use Committee ("Comitato per il Benessere degli Animali", COBA), and approved by the Italian Ministry of Health. C57BL/6 mice transgenic for and tolerant to HER2 (B6.Cg-Pds5bTg(Wap-ERBB2)229Wzw/J) were obtained from Wayne State University through The Jackson Laboratories, and bred in the facility of the Department of Veterinary Medical Sciences, University of Bologna.

Informed Consent Statement: Not applicable.

Data Availability Statement: Data is contained within current article or previous publications.

Conflicts of Interest: G.C.-F. is a minor shareholder in Nouscom.

References

1. Martuza, R.L.; Malick, A.; Markert, J.M.; Ruffner, K.L.; Coen, D.M. Experimental therapy of human glioma by means of a genetically engineered virus mutant. *Science* **1991**, *252*, 854–856. [CrossRef]
2. Liu, B.L.; Robinson, M.; Han, Z.Q.; Branston, R.H.; English, C.; Reay, P.; McGrath, Y.; Thomas, S.K.; Thornton, M.; Bullock, P.; et al. ICP34.5 deleted herpes simplex virus with enhanced oncolytic, immune stimulating, and anti-tumour properties. *Gene Ther.* **2003**, *10*, 292–303. [CrossRef]
3. Chou, J.; Kern, E.R.; Whitley, R.J.; Roizman, B. Mapping of herpes simplex virus-1 neurovirulence to gamma 134.5, a gene nonessential for growth in culture. *Science* **1990**, *250*, 1262–1266. [CrossRef]
4. Chambers, R.; Gillespie, G.Y.; Soroceanu, L.; Andreansky, S.; Chatterjee, S.; Chou, J.; Roizman, B.; Whitley, R.J. Comparison of genetically engineered herpes simplex viruses for the treatment of brain tumors in a scid mouse model of human malignant glioma. *Proc. Natl. Acad. Sci. USA* **1995**, *92*, 1411–1415. [CrossRef]
5. Macedo, N.; Miller, D.M.; Haq, R.; Kaufman, H.L. Clinical landscape of oncolytic virus research in 2020. *J. Immunol. Ther. Cancer* **2020**, *8*, e001486. [CrossRef]

6.   Patel, M.R.; Kratzke, R.A. Oncolytic virus therapy for cancer: The first wave of translational clinical trials. *Transl. Res.* **2013**, *161*, 355–364. [CrossRef]

7.   Buijs, P.R.; Verhagen, J.H.; van Eijck, C.H.; van den Hoogen, B.G. Oncolytic viruses: From bench to bedside with a focus on safety. *Hum. Vaccines Immunother.* **2015**, *11*, 1573–1584. [CrossRef] [PubMed]

8.   Hu, J.C.; Coffin, R.S.; Davis, C.J.; Graham, N.J.; Groves, N.; Guest, P.J.; Harrington, K.J.; James, N.D.; Love, C.A.; McNeish, I.; et al. A phase I study of OncoVEXGM-CSF, a second-generation oncolytic herpes simplex virus expressing granulocyte macrophage colony-stimulating factor. *Clin. Cancer Res.* **2006**, *12*, 6737–6747. [CrossRef] [PubMed]

9.   Mellman, I.; Coukos, G.; Dranoff, G. Cancer immunotherapy comes of age. *Nature* **2011**, *480*, 480–489. [CrossRef]

10. Sivanandam, V.; LaRocca, C.J.; Chen, N.G.; Fong, Y.; Warner, S.G. Oncolytic viruses and immune checkpoint inhibition: The best of both worlds. *Mol. Ther. Oncol.* **2019**, *13*, 93–106. [CrossRef] [PubMed]

11. Keller, B.A.; Bell, J.C. Oncolytic viruses-immunotherapeutics on the rise. *J. Mol. Med.* **2016**, *94*, 979–991. [CrossRef]

12. Gujar, S.; Bell, J.; Diallo, J.-S. SnapShot: Cancer immunotherapy with oncolytic viruses. *Cell* **2019**, *176*, 1240.e1. [CrossRef] [PubMed]

13. Ribas, A.; Dummer, R.; Puzanov, I.; VanderWalde, A.; Andtbacka, R.H.I.; Michielin, O.; Olszanski, A.J.; Malvehy, J.; Cebon, J.; Fernandez, E.; et al. Oncolytic Virotherapy Promotes Intratumoral T Cell Infiltration and Improves Anti-PD-1 Immunotherapy. *Cell* **2017**, *170*, 1109–1119. [CrossRef] [PubMed]

14. Bommareddy, P.K.; Shettigar, M.; Kaufman, H.L. Integrating oncolytic viruses in combination cancer immunotherapy. *Nat. Rev. Immunol.* **2018**, *18*, 498. [CrossRef] [PubMed]

15. Markert, J.M.; Razdan, S.N.; Kuo, H.-C.; Cantor, A.; Knoll, A.; Karrasch, M.; Nabors, L.B.; Markiewicz, M.; Agee, B.S.; Coleman, J.M. A phase 1 trial of oncolytic HSV-1, G207, given in combination with radiation for recurrent GBM demonstrates safety and radiographic responses. *Mol. Ther.* **2014**, *22*, 1048–1055. [CrossRef] [PubMed]

16. Trinchieri, G. Interleukin-12 and the regulation of innate resistance and adaptive immunity. *Nat. Rev. Immunol.* **2003**, *3*, 133–146. [CrossRef] [PubMed]

17. Vannini, A.; Petrovic, B.; Gatta, V.; Leoni, V.; Pepe, S.; Menotti, L.; Campadelli-Fiume, G.; Gianni, T. Rescue, Purification, and Characterization of a Recombinant HSV Expressing a Transgenic Protein. In *Herpes Simplex Virus*; Springer: Cham, Switzerland, 2020; pp. 153–168.

18. Parker, J.N.; Gillespie, G.Y.; Love, C.E.; Randall, S.; Whitley, R.J.; Markert, J.M. Engineered herpes simplex virus expressing IL-12 in the treatment of experimental murine brain tumors. *Proc. Natl. Acad. Sci. USA* **2000**, *97*, 2208–2213. [CrossRef] [PubMed]

19. Patel, D.M.; Foreman, P.M.; Nabors, L.B.; Riley, K.O.; Gillespie, G.Y.; Markert, J.M. Design of a phase I clinical trial to evaluate M032, a genetically engineered HSV-1 expressing IL-12, in patients with recurrent/progressive glioblastoma multiforme, anaplastic astrocytoma, or gliosarcoma. *Hum. Gene Ther. Clin. Dev.* **2016**, *27*, 69–78. [CrossRef]

20. Kennedy, E.M.; Farkaly, T.; Grzesik, P.; Lee, J.; Denslow, A.; Hewett, J.; Bryant, J.; Behara, P.; Goshert, C.; Wambua, D. Design of an interferon-resistant oncolytic HSV-1 incorporating redundant safety modalities for improved tolerability. *Mol. Ther. Oncol.* **2020**, *18*, 476–490. [CrossRef]

21. De Lucia, M.; Cotugno, G.; Bignone, V.; Garzia, I.; Nocchi, L.; Langone, F.; Petrovic, B.; Sasso, E.; Pepe, S.; Froechlich, G.; et al. Retargeted and multi-cytokine armed herpes virus is a potent cancer endovaccine for local and systemic anti-tumor treatment in combination with anti-PD1. *Mol. Ther. Oncol.* **2020**, *19*, 253–264. [CrossRef]

22. Laquerre, S.; Anderson, D.B.; Stolz, D.B.; Glorioso, J.C. Recombinant herpes simplex virus type 1 engineered for targeted binding to erythropoietin receptor-bearing cells. *J. Virol.* **1998**, *72*, 9683–9697. [CrossRef] [PubMed]

23. Glorioso, J.C.; Cohen, J.B.; Goins, W.F.; Hall, B.; Jackson, J.W.; Kohanbash, G.; Amankulor, N.; Kaur, B.; Caligiuri, M.A.; Chiocca, E.A.; et al. Oncolytic HSV Vectors and Anti-Tumor Immunity. *Curr. Issues Mol. Biol.* **2020**, *41*, 381–468. [PubMed]

24. Zhou, G.; Roizman, B. Construction and properties of a herpes simplex virus 1 designed to enter cells solely via the IL-13alpha2 receptor. *Proc. Natl. Acad. Sci. USA* **2006**, *103*, 5508–5513. [CrossRef] [PubMed]

25. Campadelli-Fiume, G.; Petrovic, B.; Leoni, V.; Gianni, T.; Avitabile, E.; Casiraghi, C.; Gatta, V. Retargeting Strategies for Oncolytic Herpes Simplex Viruses. *Viruses* **2016**, *8*, 63. [CrossRef] [PubMed]

26. Menotti, L.; Cerretani, A.; Campadelli-Fiume, G. A herpes simplex virus recombinant that exhibits a single-chain antibody to HER2/neu enters cells through the mammary tumor receptor, independently of the gD receptors. *J. Virol.* **2006**, *80*, 5531–5539. [CrossRef]

27. Menotti, L.; Cerretani, A.; Hengel, H.; Campadelli-Fiume, G. Construction of a fully retargeted herpes simplex virus 1 recombinant capable of entering cells solely via human epidermal growth factor receptor 2. *J. Virol.* **2008**, *20*, 10153–10161. [CrossRef]

28. Campadelli-Fiume, G.; Amasio, M.; Avitabile, E.; Cerretani, A.; Forghieri, C.; Gianni, T.; Menotti, L. The multipartite system that mediates entry of herpes simplex virus into the cell. *Rev. Med. Virol.* **2007**, *17*, 313–326. [CrossRef]

29. Campadelli-Fiume, G.; Collins-McMillen, D.; Gianni, T.; Yurochko, A.D. Integrins as Herpesvirus Receptors and Mediators of the Host Signalosome. *Annu Rev. Virol.* **2016**, *3*, 215–236. [CrossRef]

30. Heldwein, E.E.; Lou, H.; Bender, F.C.; Cohen, G.H.; Eisenberg, R.J.; Harrison, S.C. Crystal structure of glycoprotein B from herpes simplex virus 1. *Science* **2006**, *313*, 217–220. [CrossRef]

31. Gianni, T.; Salvioli, S.; Chesnokova, L.S.; Hutt-Fletcher, L.M.; Campadelli-Fiume, G. alphavbeta6- and alphavbeta8-integrins serve as interchangeable receptors for HSV gH/gL to promote endocytosis and activation of membrane fusion. *PLoS Pathog.* **2013**, *9*, e1003806. [CrossRef]

32. Rey, F.A. Molecular gymnastics at the herpesvirus surface. *EMBO Rep.* **2006**, *7*, 1000–1005. [CrossRef] [PubMed]

33. Cocchi, F.; Menotti, L.; Mirandola, P.; Lopez, M.; Campadelli-Fiume, G. The ectodomain of a novel member of the immunoglobulin subfamily related to the poliovirus receptor has the attributes of a bona fide receptor for herpes simplex virus types 1 and 2 in human cells. *J. Virol.* **1998**, *72*, 9992–10002. [CrossRef] [PubMed]

34. Tuzmen, C.; Cairns, T.M.; Atanasiu, D.; Lou, H.; Saw, W.T.; Hall, B.L.; Cohen, J.B.; Cohen, G.H.; Glorioso, J.C. Point mutations in retargeted gD eliminate the sensitivity of EGFR/EGFRvIII-targeted HSV to key neutralizing antibodies. *Mol. Ther. Methods Clin. Dev.* **2020**, *16*, 145–154. [CrossRef] [PubMed]

35. Paludan, S.R.; Bowie, A.G.; Horan, K.A.; Fitzgerald, K.A. Recognition of herpesviruses by the innate immune system. *Nat. Rev. Immunol.* **2011**, *11*, 143–154. [CrossRef]

36. Paludan, S.R.; Bowie, A.G. Immune sensing of DNA. *Immunity* **2013**, *38*, 870–880. [CrossRef]

37. Paladino, P.; Collins, S.E.; Mossman, K.L. Cellular localization of the herpes simplex virus ICP0 protein dictates its ability to block IRF3-mediated innate immune responses. *PLoS ONE* **2010**, *5*, e10428. [CrossRef]

38. Kalamvoki, M.; Roizman, B. HSV-1 degrades, stabilizes, requires, or is stung by STING depending on ICP0, the US3 protein kinase, and cell derivation. *Proc. Natl. Acad. Sci. USA* **2014**, *111*, E611–E617. [CrossRef]

39. Sasso, E.; Froechlich, G.; Cotugno, G.; D'Alise, A.M.; Gentile, C.; Bignone, V.; De Lucia, M.; Petrovic, B.; Campadelli-Fiume, G.; Scarselli, E. Replicative conditioning of Herpes simplex type 1 virus by Survivin promoter, combined to ERBB2 retargeting, improves tumour cell-restricted oncolysis. *Sci. Rep.* **2020**, *10*, 1–12. [CrossRef]

40. Kambara, H.; Okano, H.; Chiocca, E.A.; Saeki, Y. An oncolytic HSV-1 mutant expressing ICP34.5 under control of a nestin promoter increases survival of animals even when symptomatic from a brain tumor. *Cancer Res.* **2005**, *65*, 2832–2839. [CrossRef]

41. Menotti, L.; Nicoletti, G.; Gatta, V.; Croci, S.; Landuzzi, L.; De Giovanni, C.; Nanni, P.; Lollini, P.L.; Campadelli-Fiume, G. Inhibition of human tumor growth in mice by an oncolytic herpes simplex virus designed to target solely HER-2-positive cells. *Proc. Natl. Acad. Sci. USA* **2009**, *106*, 9039–9044. [CrossRef]

42. Menotti, L.; Avitabile, E.; Gatta, V.; Malatesta, P.; Petrovic, B.; Campadelli-Fiume, G. HSV as A Platform for the Generation of Retargeted, Armed, and Reporter-Expressing Oncolytic Viruses. *Viruses* **2018**, *10*, 352. [CrossRef] [PubMed]

43. Gatta, V.; Petrovic, B.; Campadelli-Fiume, G. The Engineering of a Novel Ligand in gH Confers to HSV an Expanded Tropism Independent of gD Activation by Its Receptors. *PLoS Pathog.* **2015**, *11*, e1004907. [CrossRef] [PubMed]

44. Petrovic, B.; Gianni, T.; Gatta, V.; Campadelli-Fiume, G. Insertion of a ligand to HER2 in gB retargets HSV tropism and obviates the need for activation of the other entry glycoproteins. *PLoS Pathog.* **2017**, *13*, e1006352. [CrossRef] [PubMed]

45. Leoni, V.; Gatta, V.; Casiraghi, C.; Nicosia, A.; Petrovic, B.; Campadelli-Fiume, G. A Strategy for Cultivation of Retargeted Oncolytic Herpes Simplex Viruses in Non-cancer Cells. *J. Virol.* **2017**, *91*. [CrossRef] [PubMed]

46. Petrovic, B.; Leoni, V.; Gatta, V.; Zaghini, A.; Vannini, A.; Campadelli-Fiume, G. Dual Ligand Insertion in gB and gD of Oncolytic Herpes Simplex Viruses for Retargeting to a Producer Vero Cell Line and to Cancer Cells. *J. Virol.* **2018**, *92*. [CrossRef] [PubMed]

47. Leoni, V.; Petrovic, B.; Gianni, T.; Gatta, V.; Campadelli-Fiume, G. Simultaneous Insertion of Two Ligands in gD for Cultivation of Oncolytic Herpes Simplex Viruses in Noncancer Cells and Retargeting to Cancer Receptors. *J. Virol.* **2018**, *92*. [CrossRef] [PubMed]

48. Cheever, M.A.; Allison, J.P.; Ferris, A.S.; Finn, O.J.; Hastings, B.M.; Hecht, T.T.; Mellman, I.; Prindiville, S.A.; Viner, J.L.; Weiner, L.M.; et al. The prioritization of cancer antigens: A national cancer institute pilot project for the acceleration of translational research. *Clin. Cancer Res.* **2009**, *15*, 5323–5337. [CrossRef]

49. Uchida, H.; Marzulli, M.; Nakano, K.; Goins, W.F.; Chan, J.; Hong, C.S.; Mazzacurati, L.; Yoo, J.Y.; Haseley, A.; Nakashima, H.; et al. Effective treatment of an orthotopic xenograft model of human glioblastoma using an EGFR-retargeted oncolytic herpes simplex virus. *Mol. Ther.* **2013**, *21*, 561–569. [CrossRef]

50. Shibata, T.; Uchida, H.; Shiroyama, T.; Okubo, Y.; Suzuki, T.; Ikeda, H.; Yamaguchi, M.; Miyagawa, Y.; Fukuhara, T.; Cohen, J. Development of an oncolytic HSV vector fully retargeted specifically to cellular EpCAM for virus entry and cell-to-cell spread. *Gene Ther.* **2016**, *23*, 479–488. [CrossRef]

51. Hall, B.L.; Leronni, D.; Miyagawa, Y.; Goins, W.F.; Glorioso, J.C.; Cohen, J.B. Generation of an Oncolytic Herpes Simplex Viral Vector Completely Retargeted to the GDNF Receptor GFRα1 for Specific Infection of Breast Cancer Cells. *Int. J. Mol. Sci.* **2020**, *21*, 8815. [CrossRef]

52. Sette, P.; Amankulor, N.; Li, A.; Marzulli, M.; Leronni, D.; Zhang, M.; Goins, W.F.; Kaur, B.; Bolyard, C.; Cripe, T.P. GBM-targeted oHSV armed with matrix metalloproteinase 9 enhances anti-tumor activity and animal survival. *Mol. Ther. Oncol.* **2019**, *15*, 214–222. [CrossRef] [PubMed]

53. Vannini, A.; Leoni, V.; Campadelli-Fiume, G. Targeted Delivery of IL-12 adjuvants immunotherapy by oncolytic viruses. In *Tumor Microenvironment*; Birbrair, A., Ed.; Springer: Cham, Switzerland, 2021; Volume 1290.

54. Menotti, L.; Leoni, V.; Gatta, V.; Petrovic, B.; Vannini, A.; Pepe, S.; Gianni, T.; Campadelli-Fiume, G. oHSV Genome Editing by Means of galK Recombineering. In *Herpes Simplex Virus*; Springer: Berlin, Germany, 2020; pp. 131–151.

55. Leoni, V.; Vannini, A.; Gatta, V.; Rambaldi, J.; Sanapo, M.; Barboni, C.; Zaghini, A.; Nanni, P.; Lollini, P.L.; Casiraghi, C.; et al. A fully-virulent retargeted oncolytic HSV armed with IL-12 elicits local immunity and vaccine therapy towards distant tumors. *PLoS Pathog.* **2018**, *14*, e1007209. [CrossRef] [PubMed]

56. Nanni, P.; Gatta, V.; Menotti, L.; De Giovanni, C.; Ianzano, M.; Palladini, A.; Grosso, V.; Dall'ora, M.; Croci, S.; Nicoletti, G.; et al. Preclinical Therapy of Disseminated HER-2(+) Ovarian and Breast Carcinomas with a HER-2-Retargeted Oncolytic Herpesvirus. *PLoS Pathog.* **2013**, *9*, e1003155. [CrossRef] [PubMed]

57. Gambini, E.; Reisoli, E.; Appolloni, I.; Gatta, V.; Campadelli-Fiume, G.; Menotti, L.; Malatesta, P. Replication-competent herpes simplex virus retargeted to HER2 as therapy for high-grade glioma. *Mol. Ther.* **2012**, *20*, 994–1001. [CrossRef]

58. Leoni, V.; Gatta, V.; Palladini, A.; Nicoletti, G.; Ranieri, D.; Dall'Ora, M.; Grosso, V.; Rossi, M.; Alviano, F.; Bonsi, L.; et al. Systemic delivery of HER2-retargeted oncolytic-HSV by mesenchymal stromal cells protects from lung and brain metastases. *Oncotarget* **2015**, *6*, 34774. [CrossRef] [PubMed]

59. Reisoli, E.; Gambini, E.; Appolloni, I.; Gatta, V.; Barilari, M.; Menotti, L.; Malatesta, P. Efficacy of HER2 retargeted herpes simplex virus as therapy for high-grade glioma in immunocompetent mice. *Cancer Gene Ther.* **2012**, *19*, 788–795. [CrossRef]

60. Alessandrini, F.; Menotti, L.; Avitabile, E.; Appolloni, I.; Ceresa, D.; Marubbi, D.; Campadelli-Fiume, G.; Malatesta, P. Eradication of glioblastoma by immuno-virotherapy with a retargeted oncolytic HSV in a preclinical model. *Oncogene* **2019**, *38*, 4467–4479. [CrossRef]

61. Lechner, M.G.; Karimi, S.S.; Barry-Holson, K.; Angell, T.E.; Murphy, K.A.; Church, C.H.; Ohlfest, J.R.; Hu, P.; Epstein, A.L. Immunogenicity of murine solid tumor models as a defining feature of in vivo behavior and response to immunotherapy. *J. Immunother.* **2013**, *36*, 477. [CrossRef]

62. Yamada, T.; Tateishi, R.; Iwai, M.; Koike, K.; Todo, T. Neoadjuvant use of oncolytic herpes virus G47Δ enhances the antitumor efficacy of radiofrequency ablation. *Mol. Ther. Oncol.* **2020**, *18*, 535–545. [CrossRef]

63. Zhu, Y.; Hu, X.; Feng, L.; Yang, Z.; Zhou, L.; Duan, X.; Cheng, S.; Zhang, W.; Liu, B.; Zhang, K. Enhanced therapeutic efficacy of a novel oncolytic herpes simplex virus Type 2 encoding an antibody against programmed cell death 1. *Mol. Ther. Oncol.* **2019**, *15*, 201–213. [CrossRef]

64. Saha, D.; Martuza, R.L.; Rabkin, S.D. Macrophage Polarization Contributes to Glioblastoma Eradication by Combination Immunovirotherapy and Immune Checkpoint Blockade. *Cancer Cell* **2017**, *32*, 253–267. [CrossRef] [PubMed]

65. Russell, L.; Peng, K.W.; Russell, S.J.; Diaz, R.M. Oncolytic viruses: Priming time for cancer immunotherapy. *BioDrugs* **2019**, *33*, 485–501. [CrossRef] [PubMed]

66. Alessandrini, F.; Menotti, L.; Apolloni, I.; Avitabile, E.; Ceresa, D.; Marubbi, D.; Campadelli-Fiume, G.; Malatesta, P. Il-12 secreting herpes simplex virus retargeted to HER2 as therapy for high grade glioma. *Viruses* **2018**, *10*, 352.

67. Matveeva, O.V.; Shabalina, S.A. Prospects for Using Expression Patterns of Paramyxovirus Receptors as Biomarkers for Oncolytic Virotherapy. *Cancers* **2020**, *12*, 3659. [CrossRef]

68. Newick, K.; O'Brien, S.; Moon, E.; Albelda, S.M. CAR T cell therapy for solid tumors. *Ann. Rev. Med.* **2017**, *68*, 139–152. [CrossRef]

69. Schmidts, A.; Maus, M.V. Making CAR T cells a solid option for solid tumors. *Front. Immunol.* **2018**, *9*, 2593. [CrossRef]

70. Nishio, N.; Diaconu, I.; Liu, H.; Cerullo, V.; Caruana, I.; Hoyos, V.; Bouchier-Hayes, L.; Savoldo, B.; Dotti, G. Armed oncolytic virus enhances immune functions of chimeric antigen receptor–modified T cells in solid tumors. *Cancer Res.* **2014**, *74*, 5195–5205. [CrossRef]

71. Ricca, J.M.; Oseledchyk, A.; Walther, T.; Liu, C.; Mangarin, L.; Merghoub, T.; Wolchok, J.D.; Zamarin, D. Pre-existing Immunity to Oncolytic Virus Potentiates Its Immunotherapeutic Efficacy. *Mol. Ther.* **2018**, *26*, 1008–1019. [CrossRef]

72. Mantovani, A.; Netea, M.G. Trained Innate Immunity, Epigenetics, and Covid-19. *N. Engl. J. Med.* **2020**, *383*, 1078–1080. [CrossRef]

73. Gujar, S.; Pol, J.G.; Kim, Y.; Lee, P.W.; Kroemer, G. Antitumor Benefits of Antiviral Immunity: An Underappreciated Aspect of Oncolytic Virotherapies. *Trends Immunol.* **2017**. [CrossRef]

Review

Oncolytic Virotherapy in Solid Tumors: The Challenges and Achievements

Ke-Tao Jin [1,†], Wen-Lin Du [2,3,†], Yu-Yao Liu [1], Huan-Rong Lan [4], Jing-Xing Si [3,*] and Xiao-Zhou Mou [3,*]

1   Department of Colorectal Surgery, Affiliated Jinhua Hospital, Zhejiang University School of Medicine, Jinhua 321000, China; jinketao2001@zju.edu.cn (K.-T.J.); lyy21918555@zju.edu.cn (Y.-Y.L.)
2   Key Laboratory of Gastroenterology of Zhejiang Province, Zhejiang Provincial People's Hospital, People's Hospital of Hangzhou Medical College, Hangzhou 310014, China; dwl1209@163.com
3   Clinical Research Institute, Zhejiang Provincial People's Hospital, People's Hospital of Hangzhou Medical College, Hangzhou 310014, China
4   Department of Breast and Thyroid Surgery, Affiliated Jinhua Hospital, Zhejiang University School of Medicine, Jinhua 321000, China; lanhr2018@163.com
*   Correspondence: sijingxing@hmc.edu.cn (J.-X.S.); mouxz@zju.edu.cn (X.-Z.M.);
    Tel./Fax: +86-571-85893781 (J.-X.S.); +86-571-85893985 (X.-Z.M.)
†   These authors contributed equally to this work.

Simple Summary: Oncolytic virotherapy (OVT) is a promising approach in cancer immunotherapy. Oncolytic viruses (OVs) could be applied in cancer immunotherapy without in-depth knowledge of tumor antigens. Improving efficacy, employing immunostimulatory elements, changing the immunosuppressive tumor microenvironment (TME) to inflammatory TME, optimizing their delivery system, and increasing the safety are the main areas of OVs manipulations. Recently, the reciprocal interaction of OVs and TME has become a hot topic for investigators to enhance the efficacy of OVT with less off-target adverse events. Current investigations suggest that the main application of OVT is to provoke the antitumor immune response in the TME, which synergize the effects of other immunotherapies such as immune-checkpoint blockers and adoptive cell therapy. In this review, we focused on the effects of OVs on the TME and antitumor immune responses. Furthermore, OVT challenges, including its moderate efficiency, safety concerns, and delivery strategies, along with recent achievements to overcome challenges, are thoroughly discussed.

Abstract: Oncolytic virotherapy (OVT) is a promising approach in cancer immunotherapy. Oncolytic viruses (OVs) could be applied in cancer immunotherapy without in-depth knowledge of tumor antigens. The capability of genetic modification makes OVs exciting therapeutic tools with a high potential for manipulation. Improving efficacy, employing immunostimulatory elements, changing the immunosuppressive tumor microenvironment (TME) to inflammatory TME, optimizing their delivery system, and increasing the safety are the main areas of OVs manipulations. Recently, the reciprocal interaction of OVs and TME has become a hot topic for investigators to enhance the efficacy of OVT with less off-target adverse events. Current investigations suggest that the main application of OVT is to provoke the antitumor immune response in the TME, which synergize the effects of other immunotherapies such as immune-checkpoint blockers and adoptive cell therapy. In this review, we focused on the effects of OVs on the TME and antitumor immune responses. Furthermore, OVT challenges, including its moderate efficiency, safety concerns, and delivery strategies, along with recent achievements to overcome challenges, are thoroughly discussed.

Keywords: oncolytic virus; tumor microenvironment; antitumor immune response; delivery; genetic modification

Citation: Jin, K.-T.; Du, W.-L.; Liu, Y.-Y.; Lan, H.-R.; Si, J.-X.; Mou, X.-Z. Oncolytic Virotherapy in Solid Tumors: The Challenges and Achievements. *Cancers* **2021**, *13*, 588. https://doi.org/10.3390/cancers13040588

Academic Editors: Antonio Marchini, Carolina S. Ilkow and Alan Melcher
Received: 31 December 2020
Accepted: 30 January 2021
Published: 3 February 2021

Publisher's Note: MDPI stays neutral with regard to jurisdictional claims in published maps and institutional affiliations.

1. Introduction

The first hints of the possible anticancer effects of viruses occurred during the early 20th century, with evidence of tumor regression in patients with simultaneous viral infections [1]. Such reports persisted until the 1950s, when the primary clinical studies on the tumor-killing ability of viruses that form the cornerstone of today's achievements were carried-out [2]. Since then, various preclinical and clinical studies have attempted to optimize the viruses for increasing specificity, efficiency, and reducing adverse events (AEs), which led to the introduction of oncolytic virotherapy (OVT) as emerging immunotherapy of cancers [3]. Oncolytic virus (OVs) or cancer-killing viruses are defined as natural or genetically modified viruses that are able to selectively proliferate in tumor cells without damaging normal cells [4]. This natural tropism of some viruses to tumors is due to an increase in some receptors (such as CD54) on the surfaces of tumor cells or defects of tumor cells to induce innate immunity against viruses [5]. So far, various DNA and RNA OVs have been used to treat cancer [6]. The majority of DNA viruses are double-stranded, while RNA viruses are predominantly single-stranded. The advantages of double-stranded DNA viruses are their large genomes which enable them to carry large eukaryotic transgenes and high fidelity DNA polymerase, maintaining the virus genome integrity during replication [7]. Regarding their relatively small size, RNA viruses cannot encode large transgenes. However, they are better candidates in the delivery system due to less induction of immune responses [8]. Several RNA viruses and DNA viruses, including reovirus (RV), Seneca Valley virus (SVV), poliovirus (PoV), parvovirus (PV), vaccinia virus (VACV), and herpes simplex virus (HSV) have the ability to cross the blood-brain barrier (BBB) enabling their use in brain tumors [9–14]. OVT started with wild-type viruses such as Newcastle disease virus (NDV), myxoma virus (MYXV), SVV, PV, coxsackievirus (CV), and RV [3]. However, genetic modification was a revolutionary achievement in the OVT providing greater specificity and efficacy against tumors with higher safety for healthy cells [15]. Genetically modified OVs (GMOVs) mainly include PoV, measles virus (MeV), adenovirus (AdV), VACV, HSV, and vesicular stomatitis virus (VSV) [3]. The first GMOV was HSV-1, introduced in 1991 [16]. So far, three OV-based drugs have been approved for cancer treatment, the first of which was an unmodified ECHO-7 virus called Rigavirus which was approved in 2004 in Lativa under the brand name Rigvir for melanoma [17]. However, the approval was withdrawn in 2019 due to its low efficacy. The two other approved OVs are GMOVs include Oncorine (H101 adenovirus), which obtained approval for head and neck cancer in China in 2005 [3], and T-VEC or Imlygic (HSV-1), which was approved in 2015 in the United States and Europe for non-surgical melanoma [18]. The efficacy of OVs on many cancers, such as melanoma, glioblastoma, triple-negative breast cancer (TNBC), head and neck cancers, and colorectal cancers has been elucidated [19–23], and a large number of clinical trials are currently evaluating the wild-type and GMOVs efficiency and safety in various cancers which are listed in Table 1. Along with the therapeutic approaches, GMOVs expressing reporter genes can be applied in the diagnosis of various cancers by positron emission tomography or single-photon emission computed tomography [24].

OVs can kill the tumor cells in the following main ways: 1. OVs infect and replicate specifically in tumor cells leading to direct lysis of tumor cells. Malignant cells have defects in antiviral responses allowing OVs to replicate and lyse malignant cells [7]; 2. OVs can induce different types of immunogenic cell death (ICD), including necrosis, necroptosis, immunologic apoptosis, pyroptosis, and autophagy. Tumor cell death or lysis causes the release of tumor-associated antigens (TAA) and neoantigens (TAN) and damage-associated molecular patterns (DAMPs), which increase inflammation and improve the efficacy of immunotherapy [25,26]; 3. OVs, especially GMOVs, can enhance tumor antigen presentation and prime the immune response in the tumor microenvironment (TME) by induction of antiviral responses, inflammation, cytokine production, and expression of costimulatory molecules [26,27]; 4. The infection of vascular endothelial cells (vECs) by OVs destroys tumor vasculature, resulting in tumor necrosis and the infiltration of immune cells into the TME [28].

Table 1. Oncolytic viruses that reached the clinical phase.

Oncolytic Virus	Modification	Combination Therapy	Cancer Type (Clinical Trial Phase)	Ref.
HSV-1	Virulence gene ICP34.5 and ICP47 are deleted and human GM-CSF gene is inserted	ICIs (anti-PD1, anti-CTLA4	Melanoma (I, II), Sarcoma (I, II)	[29–31]
		-	Breast cancer (I), Head and neck cancer (I, I/II), Gastrointestinal cancers (I), Melanoma (I, II, III)	[32–36]
	Virulence gene ICP34.5 is deleted	-	Oral SCC (I), Pediatric extracranial cancers (I)	[37,38]
		Chemotherapy	Chemo-resistant metastatic colon cancer (I, I/II)	[39,40]
	Virulence gene ICP34.5 is deleted and ICP6 gene is inactivated	Radiotherapy	Glioblastoma (I)	[41]
	Naturally mutated	-	Pancreatic cancer (I)	[42]
NDV	Autologous tumor lysate and IL-2 is added	-	Stage III of Melanoma (I)	[43]
	Naturally attenuated	-	Advanced solid tumors (I)	[44]
	One-cycle replicating cytopathogenic NDV	-	Glioblastoma (I/II)	[45]
CVA21	-	ICIs (anti-PD1)	NSCLC (Ib), Bladder cancer (Ib)	[46]
		-	Bladder cancer (II), Advanced melanoma (II)	[47,48]
RV	-	-	Advanced solid tumors (I), Recurrent glioma (I), Extracranial solid tumors (I), Melanoma (II), Pancreatic adenocarcinoma (II)	[49–53]
		Chemotherapy	Advanced solid tumors (I), Ovarian cancer (IIb), Peritoneal cancer (IIb), Melanoma (II), Metastatic breast cancer (II), Advanced head and neck cancer (I/II) Pancreatic adenocarcinoma (II)	[54–58]
		Radiotherapy	Advanced solid tumors (I)	[59]
PoV	Recombinant oral PoV Sabin-1: the internal ribosome entry site (IRES) is replaced with the IRES from human rhinovirus-2: nonpathogenic	-	Recurrent glioblastoma (I)	[60]

154

Table 1. *Cont.*

Oncolytic Virus	Modification	Combination Therapy	Cancer Type (Clinical Trial Phase)	Ref.
AdV	AdV3 fiber knob is inserted into the backbone of AdV5, A 24-base pair in the E1 gene is deleted: CRAd	-	Ovarian Cancer (I), Gynecologic malignancies (I), Advanced solid tumors (I)	[61–63]
	GM-CSF gene is inserted	Chemotherapy	Chemo-resistant advanced solid tumors (I)	[61]
	RGD motif is inserted into the AdV5 fiber knob: Integrin targeted instead of CAR dependence GM-CSF gene is inserted	-	Chemo-resistant advanced solid tumors (I)	[64]
	Prostate-specific antigen (PSA)-selective	Radiotherapy	Metastatic prostate cancer (I)	[65]
	Conditionally replicating GM-CSF expressing AdV	-	Bladder Cancer (I, II), Head and neck cancers (I)	[66,67]
	Human telomerase reverse transcriptase (hTERT) is inserted: tumor selective replication	-	Advanced solid tumors (I)	[68]
	E1B-deleted AdV: selective replication in P53-deficient cells	Chemotherapy	Advanced solid tumors (I), Malignant glioma (I), Recurrent head and neck cancer (I, II), Gastrointestinal cancers (II), Colorectal cancer (II), Advanced sarcoma (I/II),	[69–72]
	Chimeric AdV: Ad11p/Ad3, AdV5- cytosine deaminase/HSV-1 thymidine kinase: suicide gene for safety	-	RCC (I), NSCLC (I), Colorectal cancer (I), Urothelial cancer (I),Prostate cancer (I, II), Glioma (II)	[73–76]
VACV	GM-CSF gene is inserted Thymidine kinase gene is deleted	Chemotherapy	Metastatic melanoma (I), HCC (I, II), Colorectal cancer (I), Ewing sarcoma (I), neuroblastoma (I),	[77–80]
	FCU1 transgene is inserted: metabolize 5-FC to 5-FU-monophosphate	Chemotherapy	Chemo-resistant liver tumors (I)	[81]
	Thymidine kinase gene and hemagglutinin gene and F14.5 gene are deleted Luciferase gene, beta-galactosidase, and beta-glucuronidase are inserted	Chemotherapy and radiotherapy	Head and neck cancer (I), Colorectal cancer (I) Advanced solid tumors (I)	[82–84]

Table 1. *Cont.*

Oncolytic Virus	Modification	Combination Therapy	Cancer Type (Clinical Trial Phase)	Ref.
MeV	Genetically modified to express carcinoembryonic antigen	-	Ovarian cancer (I)	[85]
SVV	-	-	Neuroblstoma (I), rhabdomyosarcoma (I), Neuroendocrine malignancies (I)	[86,87]
Poxvirus	Genetically modified expressing costimulatory and adhesion molecules such as B7-1, LFA-3, ICAM-1	-	Colorectal cancer (I), Melanoma (I)	[88]
PV	-	-	Glioblastoma (I/II)	[89]

HSV-1. Herpes simplex virus-1; ICP. Infected cell protein; GM-CSF. Granulocyte-macrophage colony-stimulating factor; ICI. Immune-checkpoint inhibitor; PD1. Programmed cell death protein 1; CTLA4. cytotoxic T-lymphocyte-associated protein 4; SCC. Squamous cell carcinoma; NDV. Newcastle disease virus; CVA21. Coxsackievirus A21; NSCLC. Nonsmall-cell lung carcinoma; RV. Reovirus; PoV. Poliovirus; AdV. Adenovirus; CRAd. Conditionally replicative adenoviruses; RGD. Arginine-Glycine-Aspartate; CAR. Coxsackievirus and adenovirus receptor; RCC. Renal cell carcinoma; VACV. Vaccinia virus; HCC. Hepatocellular carcinoma; FCU1. Fusion suicide gene; 5-FC. 5-fluorocytosine; 5-FU.5-Fluorouracil; MeV. Measles virus; SVV. Seneca Valley virus; LFA-3. Lymphocyte function-associated antigen-3; ICAM-1. Intercellular adhesion molecule-1; PV. Parvovirus.

Accordingly, a considerable part of OVT effects on tumors is achieved by changing the TME from an immunosuppressive to the immunostimulatory microenvironment and affecting the tumor vasculature and matrix. Moreover, the success of OVT in solid tumors largely depends on the OV access to the tumor. Here, we review the effects of OVs on the TME and antitumor immune responses. Furthermore, OVT challenges, including its moderate efficiency and safety concerns, along with recent achievements to overcome challenges, are thoroughly discussed. Regarding the critical role of OV delivery strategy in the efficacy of OVT, recent approaches enhancing OV delivery into the TME are also provided.

2. Oncolytic Virus Effects on TME

The long-term effects of immunotherapy in solid tumors are mostly unsatisfactory, partly due to the immunosuppressive condition of TME and low infiltration of immune cells. TME consists of tumor cells, tumor-associated fibroblasts (TAF), vEC, mesenchymal cells, myeloid-derived suppressor cells (MDSCs), and tumor-infiltrating leukocytes (TILs), such as T cells, B cells, dendritic cells (DCs), natural killer (NK) cells, macrophages, and neutrophils [90]. The presence of exhausted cytotoxic T lymphocytes (CTLs), helper T-cells (THs), and NK cells, as well as a large number of regulatory T-cells (Tregs), tolerogenic DCs, MDSC, and M2-macrophages, induce immunosuppressive milieu in the TME through inhibitory ligands and secretion of inhibitory cytokines such as interleukin (IL)-10, tumor growth factor (TGF)-β, IL-35, and IL-27 [91]. OVs can change the paradigm in the TME and convert cold tumors to hot ones by various mechanisms.

2.1. OV-Mediated Lysis of Tumor

Direct oncolysis activity of OVs is the first stimulus of the immune response in the TME [92]. Overexpression of surface receptors such as CD46, CD54, CD155, CD55, and integrins enhances OVs' preferable entry to tumor cells [93–97]. In normal cells, viral components known as pathogen-associated molecular patterns (PAMPs) are sensed by pattern recognition receptors (PRRs) and induce the production of interferon (IFN)-I through the Janus kinase signal transducer and activator of transcription (JAK-STAT) and Nuclear Factor (NF)-kB signaling pathways. IFN-I activates the protein kinase RNA-activated (PKR) signaling pathway leading to protein synthesis blockade and viral clearance [98]. Tumor cells have defects in antiviral pathways such as IFN-I, PKR, and JAK-STAT, resulting in the survival and proliferation of OVs, specifically in tumor cells [99–101]. Lysis of OV-infected cells releases a very diverse TAAs that prime immune cells to induce a local and systemic vaccination against the released TAAs [92]. While many cancer immunotherapies depend on identifying and targeting TAAs (one or several limited TAAs), OVT can vaccinate patients against the entire TAA and TAN treasure of cancer through a phenomenon called antigen/epitope spreading. Hence, OVT could be considered a kind of personalized immunotherapy. Interestingly enough, recent studies have reported the increase of TAA- and TAN-specific T cells in the blood of patients with melanoma and ovarian cancer treated with OVs, suggesting that the in situ OV injection might enhance the systemic antitumor response [102–104]. This finding raises hopes for the anti-metastatic effects of OVT. TANs are assumed to be derived from high mutational burden of tumor cells [105,106]. These immunogenic TANs are capable of eliciting tumor-specific immune responses and serve as ideal targets in immunotherapy [105–107]. However, TAN-specific T cells are not activated enough in cancer patients due to the poor presentation of TANs, lack of costimulatory signals, and abundance of inhibitory immune checkpoints in the TME [107]. OVs, especially armed OVs, have been shown to activate the TANs-specific T cells by increasing the access of APCs to the TANs (epitope spreading), enhancing the TANs processing and presentation by APCs, and providing costimulatory signals [107–109]. Accordingly, Wang et al. demonstrated that VACV armed with PD-L1 inhibitor and GM-CSF enhanced TANs presentation and activated systemic T cell responses against dominant and subdominant (cryptic) neoantigens [107], so OVT could potentiate the antitumor immune responses by activating the TANs-specific T cells.

2.2. Induction of Immunologic Cell Death

Apart from the direct lysis of cancer cells, OVs can induce various ICDs in virus-infected cells through induction of endoplasmic reticulum (ER) stress [110]. Infection of tumor cells with AdV, CV-B3, MeV, VACV, HSV, and H1-PV has been shown to induce ICD and autophagy in cancer cells [111,112]. ICD is characterized by the expression and release of DAMPs such as ATP, uric acid, heat shock proteins, ecto-calreticulin, and HMGB1, as well as extracellular proinflammatory cytokines [113]. Extracellular ATP acts as a danger signal which attracts and activates DCs [114]. HMGB1 and calreticulin can activate DCs via toll-like receptor (TLR)-4 signaling [115]. In addition, calreticulin neutralizes CD47 receptors on the tumor cell surface, and thereby, increases the tumor cell engulfment by macrophages [116]. OV-mediated ICD, along with other ICD-inducing methods such as chemotherapy and radiotherapy, break immune tolerance against the tumor and increase lymphocyte and neutrophil infiltration, leading to antitumor response and more survival in preclinical models [111].

2.3. Stimulation of Antitumor Immune Response

Besides the release of DAMPs, cancer cell death also causes the release of viral PAMPs in the TME. These PAMPs mainly include DNA, ssRNA, dsRNA, proteins, and capsid contents that activate innate immune cells through stimulating PRRs such as retinoic acid-inducible gene (RIG)-1, cyclic GMP-AMP synthase (cGAS), and stimulator of interferon genes (STING) [113]. DCs, as a bridge between the innate and adaptive immune systems, play a critical role in generating the antitumor response. DCs elicit a specific response against TAA-expressing tumor cells by engulfing OV-infected cells and cross-presentation of TAAs to CD8+ T and CD4+ T cells [117]. On the other hand, the OVs-derived PAMPs cause maturation of myeloid and plasmacytoid DCs, leading to the production of proinflammatory cytokines such as IFN-α, IFN-γ, IL-12, IL-1β, IL-6, IL-8, and tumor necrosis factor (TNF)-α [90,118,119]. These functional DCs, mainly CD103+ and BATF3+, prime CD8+ T cells against tumors [120]. Innate immune signaling, such as the cGAS-STING pathway, plays a pivotal role in the recruitment of lymphocytes to the TME through the expression of CXCL9 and CXCL10 [121]. Parallel to DCs, innate lymphoid cells (ILCs) also respond to the released PAMPs leading to higher inflammation and antitumor responses [18]. As an example, arenavirus-infected melanoma cells produce a high level of CCL5, leading to recruitment of NK cells and melanoma regression [122]. Interestingly, in situ antitumor responses following OVT are mainly mediated by IFN-I, whereas OVT-mediated systemic antitumor responses appear to be mediated by IFN-II excreted from TILs [123]. In general, the innate immune response to OVs increases lymphocyte infiltration, antigen presentation, and activation of the antitumor adaptive immune response through an IFN-mediated mechanism [18]. T cell activation requires at least three consecutive signals (peptide-MHC, CD28-B7, and stimulatory cytokines), all of which are defected in TME to escape adaptive immune responses. OVs, as potent immunogens, induce all three signals needed to activate T cells [18]. OVT increases the expression of B7-1/2 and CD40 on the surface of DCs and induces the expression of MHC-peptide on the surface of tumor cells leading to optimal activation of T cells [124]. Conversion of the TME phenotype from immunologically inert to immunologically active status can augment the effectiveness of the immunotherapeutic modalities.

2.4. Effect of OV on Tumor Vasculature

Some OVs, such as HSVs and VACVs, can target tumor stromal cells, such as TAFs, vECs, and pericytes, thereby destroy the tumor's complex structure [26]. TGF-β secreted by tumor cells makes TAFs susceptible to OV infection [125]. OVs also reduce the fibrosis in the TME. VSV has been shown to infect hepatic stellate cells (HSCs), leading to tumor fibrosis reduction [126]. OVs affect the tumors vasculature by replicating in the tumor vECs. Vascular endothelial growth factor (VEGF) secreted from tumor vECs suppresses the antiviral response and allows the replication of OVs in endothelial cells through ERK1/2 and STAT3 pathways [127]. Following infection and replication, the OVs reduce VEGF production from the infected cell resulting in angiogenesis prevention in the tumor. OVs' antiangiogenic properties further limit tumor growth by decreasing the oxygen and nutrition supplies [6]. VACV is shown to replicate in the tumor vEC and cause vascular destruction and ischemia [28]. Neutrophil infiltration into the TME seems essential for OVT-mediated ischemia through the induction of thrombosis in small tumor vessels [28]. It has been shown that the administration of JX-594 in hepatocellular carcinoma destroyed tumor vasculature without affecting patients' normal vessels [28]. Thus, targeting of stromal cells by OVs increases the infiltration of immune cells into the TME, and converts immuno-deserted or immune-excluded tumors (with low TILs) into immune-infiltrated tumors [18]. OVT-mediated changes in the TME, including lymphocyte infiltration into the tumor, enhancement of TAAs/TANs presentation, and heating the TME can improve other immunotherapies such as adoptive cell therapy (ACT) and immune checkpoint inhibitors (ICIs) [90].

3. OVT Challenges and Achievements

3.1. Tumor Targeting

Although OVs have tumor tropism based on some overexpressed receptors and adhesion molecules on the tumor cells, the tumor tropism of wild OVs is not enough. GMOVs can express receptors with a high affinity for TAAs. For instance, insertion of single-chain antibodies (scAb) against human epidermal growth factor receptor (HER)-2, epithelial cell adhesion molecule (EpCAM), and carcinoembryonic antigen (CEA) increases the specificity of OVs to tumors [128–130]. Insertion of sequences such as the arginine-glycine-aspartate (RGD) motif or specific domain from AdV3 and AdV35 to AdV5, makes AdV5 specific for integrins, desmoglein-2, and CD46, which are overexpressed in tumors [131,132]. VSV expressing HIV-derived glycoprotein (gp)-160 is a specific VSV against leukemia and T lymphomas [133].

Defects in the IFN-I antiviral response, lack of tumor suppressor genes such as the retinoblastoma (Rb), and increased Ras signaling in tumor cells lead to the specific proliferation of OVs in tumor cells [134]. Insertion of tumor-specific promoters such as prostate-specific antigen (PSA) and human telomerase reverse transcriptase (hTERT) promoters, which are highly expressed in tumor cells, causes specific expression of viral genes in tumor cells [135,136]. Some micro-RNAs (miRNAs) are overexpressed in healthy cells while they are at negligible levels in tumor cells. Hence, targeting these miRNAs by miRNA-targeting sequences (miRNA-TS) destroys viral RNA in normal cells. Low expression of miRNA-TS targets in tumor cells causes viral RNAs to remain and replicate in tumor cells [137].

3.2. Improving Antitumor Efficacy

Genetic modifications of OVs to increase the expression of cytokines, chemokines, costimulatory molecules, tumor extracellular matrix (ECM)-degrading enzymes, and antiangiogenic molecules can enhance their antitumor effects (Figure 1). Granulocyte-macrophage colony-stimulating factor (GM-CSF) gene-bearing OVs such as T-VEC, Pexa-Vec, and CG0070 recruit antigen-presenting cells (APCs) and CTLs, resulting in a better TAA presentation with minimal antiviral response induction [6]. GMOVs expressing proinflammatory cytokines showed enhanced antitumor efficacy. Despite the considerable antitumor response, IL-2-secreting OVs cause systemic toxicity. The design of VACV expressing membranous IL-2 rather than secretory form increases local antitumor response with significantly reduced toxicity [138]. The use of IL-12, IL-15, IL-18, TNF-α, IL-24, and IFN-γ genes in OVs also enhances antitumor effects with much lower toxicity than IL-2 [6,139–141]. Interestingly, the application of the non-secretory form of these cytokines causes local effects rather than systemic AEs [142]. Expression of specific chemokines such as CCL5, CCL19, CCL20, CCL21 by engineered OVs (mainly VACV) increases the infiltration of naïve and memory T lymphocytes and DCs into the TME [143–146]. Simultaneously, employment of one or multiple costimulatory ligands, including CD40L, 4-1BBL, OX40L, and B7-1 in OVs such as LOAd703 (the combination of CD40L and 4-1BBL) increases antigen presentation and T cell priming [6,26,96]. Besides, insertion of TLR ligands such as CpG-rich regions in the OVs genome stimulates TLRs and further activates innate and acquired immunity [138].

Another way to enhance the immune responses in the TME is the elimination of immunosuppressive cells. GMOVs that express the hydroxyprostaglandin dehydrogenase (HPGD) enzyme inactivate PGE2 and reduce the presence of MDSCs in the TME [147]. Soluble CXCR4 expressed by GMOVs binds to CXCL12 secreted by tumor cells as a decoy receptor and inhibits the effects of CXCL12 on angiogenesis, metastasis, and recruitment of MDSCs [148].

Although OVT can release TAAs through various mechanisms, the expression of TAAs by GMOVs or coating the TAA-derived peptides on the surface of OVs increases T cell response and improves OVT. A large number of TAAs and peptides have been studied so far [26]. The advantage of peptide coating over peptide expression is the convenience, speed, lower cost, and the possibility of personalization for each patient in the peptide coating method [26].

OVs can be engineered to express proapoptotic proteins such as TNF-related apoptosis-inducing ligand (TRAIL) and apoptin that can induce specific apoptosis in tumor cells [149,150]. Insertion of the oncogene suppressor small interfering RNAs (siRNAs) in OVs could also suppress oncogene expression and inhibit tumor growth [151,152].

The host antiviral response ensures that OVs disappear after a while and prevents the AEs of their long presence. However, the host antiviral response might cause rapid clearance of OVs before fulfilling their antitumor activity [153]. Expression of IFN-I antagonists by OVs or some non-pathogenic bacteria reduces the innate immune response against OV and delays their clearance [154]. Also, the use of stem cells, polymers, and liposomes as OV carriers reduces the immunogenicity of OVs, shields them from neutralizing antibodies (nAbs), and improves their transmission to the TME, which is listed in Table 2. An interesting way to optimize cytokine production with minimal antiviral responses is to insert inducible promoters or regulatory genes so that the cytokine expression is exogenously induced after sufficient replication of OVs in tumor cells [155].

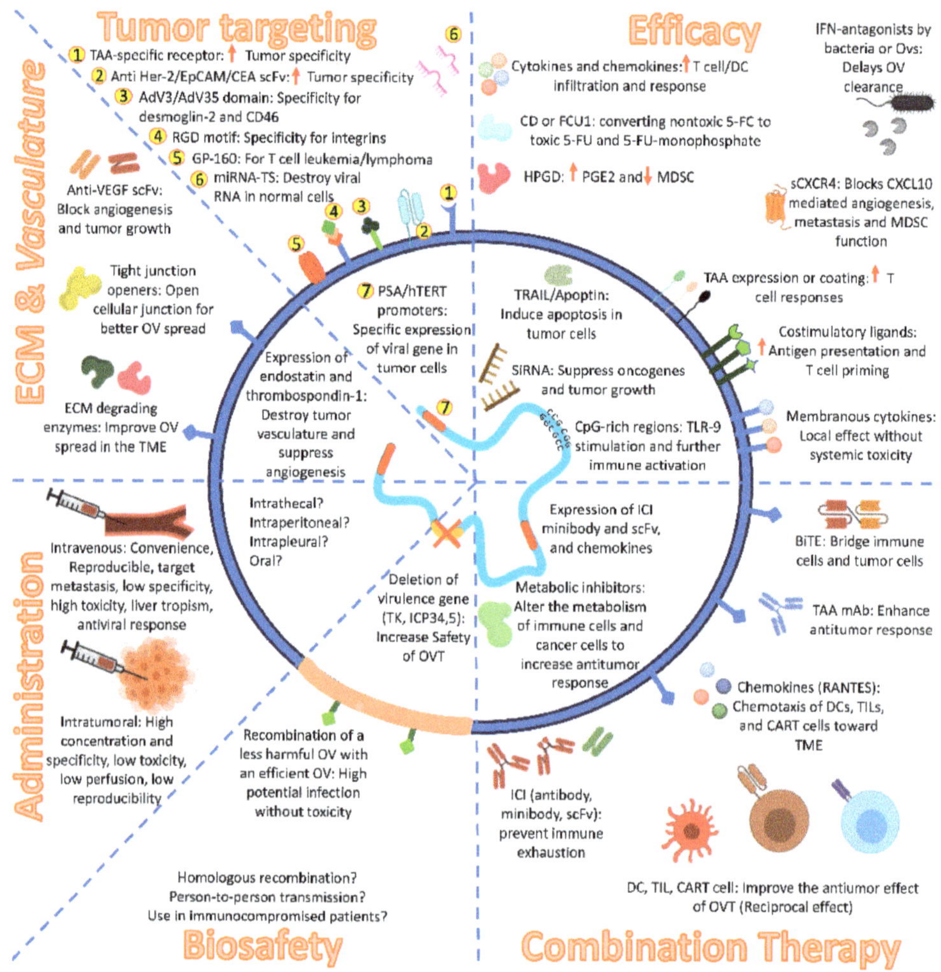

Figure 1. Recent approaches in oncolytic virotherapy. Expression of TAA-receptors and scFvs, recombination of specific domains and motifs, using tumor-specific promoters, and application of miRNA-TS could enhance the tumor targeting. GMOVs expressed inflammatory cytokines, enzymes, chemokine receptors, costimulatory molecules, and proapoptotic proteins achieve high antitumor potency. ECM and vasculature degradation by enzymes and molecules result in a higher spread of OVs. Administration routs are a critical factor in achieving better results with lower adverse effects. Besides, deleting virulence genes and recombination of OVs together could diminish the concerns of adverse events. However, several biosafety concerns still remained unmet. The combination of OVT with other immunotherapy, such as ICIs, TIL therapy, CART cell therapy, DC vaccines, mAbs, BiTEs, and metabolic inhibitors could potentiate the immunotherapy against tumors. OV. Oncolytic virus; OVT. OV therapy; TAA. Tumor-associated antigen; scFv. Single-chain variable fragment; Her-2. Human epidermal growth factor receptor 2; EpCAM. Epithelial cell adhesion molecule; CEA. Carcinoembryonic antigen; AdV. Adenovirus; RGD. Arginine-glycine-aspartate; GP-160. Glycoprotein-160; miRNA-TS. microRNA targeting sequence; PSA. Prostate-specific antigen; hTERT. Human telomerase reverse transcriptase; DC. Dendritic cell; CD. Cytosine deaminase; 5-FC. 5-fuorocytosine; 5-FU. 5-fluorouracil; HPGD. Hydroxyprostaglandin dehydrogenase; PGE2. Prostagalndin-E2; MDSC. Myeloid-derived suppressor cell; IFN. Interferon; TRAIL. TNF-related apoptosis-inducing ligand; siRNA. Small interfering RNA; TLR-9. Toll-like receptor-9; ICI. Immune checkpoint inhibitor; BiTE. Bispecific T cell engager; mAb. Monoclonal antibody; TIL. Tumor0infiltraring lymphocyte; CART cell. Chimeric-antigen receptor T cell; TME. Tumor microenvironment; ECM. Extracellular matrix; VEGF. Vascular-endothelial growth factor.

Table 2. Delivery approaches to enhance tumor access by oncolytic viruses.

Strategy	Approach	Method	Outcome	References
	Stem cell carrier	Mesenchymal Stem cells (Bone marrow, adipose, umbilical cord-menstrual blood)	Off-the-shelf; both systemic and local application; OV shielding; better replication; deliver more viral copies; enhanced tumor tropism; delivery of Ovs to hard-to-access metastatic foci; antiviral immune response evasion; increased persistence of OVs; enhance tumor cell apoptosis; probable toxicity due to trapping mesenchymal stem cells in the lung	[156–158]
		Neural Stem cells	Off-the-shelf; improved OV delivery to brain tumors; better response in chemo-resistant ovarian cancer	[159]
Organic carriers	Immune and blood cell carrier	Granulocytes, neutrophils	Delivery to cancer cells located in the bone marrow or spleen; circumvent the problems of systemic delivery; OV shielding	[160,161]
		Dendritic cells	Protect OVs from systemic neutralization, long-term tumor regression; decrease pleural exudation in breast cancer	[162,163]
		T cells	Facilitate systemic OVT in the presence of antiviral nAbs; delivery to cancer cells located in the bone marrow or spleen; Increased efficacy in intratumoral injection; prolonged survival; enhance the efficacy of adoptive cell therapy and OVT; increase selectivity for metastatic tumors; viral concentration in tumor;	[163–165]
		Macrophage	Migration to hypoxic tumors; enhanced OV proliferation and antitumor effect in hypoxia; inhibited tumor growth and metastasis; more resistant to antibody neutralization	[166,167]
		Natural killer cell (NK-92 cells transduced with Ad5/37 chimeric fiber)	Strong antitumor effects	[168]
		Cytokine-induced killer cells	Improved tumor trafficking; enhanced antitumor effects	[169]
		Peripheral-blood mononuclear cells	OV shielding from nAb; retained proliferation and selective cytotoxicity for tumor cells, enhanced OV delivery to treat minimal residual disease	[160,170,171]

Table 2. *Cont.*

Strategy	Approach	Method	Outcome	References
		Myeloid-derived suppressor cells (MDSC)	Avoid of antiviral responses; preferential migration into tumors; less toxicity following multiple administration; induction of MDSC differentiation towards the M1-like macrophage	[172]
		Platelets	OV shielding from nAb, retained proliferation and cytotoxicity	[160]
		Monocytes	OVs shielding; possibility of multiple administration; more resistant to antibody neutralization	[160,173]
		Erythrocyte, Sickle cell	Improved transfection, high absorption and infection despite nAbs presence	[174,175]
	Carrier Cell lines	HS 578T HeLa A549 MCF-7 CT26 SF268 U937 UR-D7 MC38 MH3924A	Better in vitro manipulation, Trapped in small vessels and decrease circulation, more resistant to antibody neutralization, Iproved viral delivery, replication and intratumoral spread, reduce OV spreading to peripheral organs	[168,176–178]
	Other Cells	Blood outgrowth endothelial cells	Shield OVs from nAbs; reduced tumor burden; superior antitumor activity;	[179]
	Extracellular vesicles (EVs)	infected cell-derived EVs, A549-derived EVs, LL/2-derived EVs	increased the transduction and infectious titer; reduced tumor growth; specifically target the tumor; immunological cell death; immune cell infiltration; localized inflammatory effect; provide alternative entry pathways into tumor cells	[180,181]

163

Table 2. *Cont.*

Strategy	Approach	Method	Outcome	References
	Tumor cell membrane	ExtraCRAd (Extra conditionallyreplicating adenoviruses): Membrane of B16.OVA, B16.F10, LL/2, CMT64.OVA, MB49, A549, and SKOV-3 cell lines	OVs wrapped with cancer cell membranes carrying TAA, increased in vitro and in vivo infectivity; control tumor growth with preventive and therapeutic applications; high specific antitumor immune response	[182]
	Implant	3D-engineered conformal implant	Constant release of OVs; apoptosis induction; delays tumor recurrence; eradicating post-surgery residual tumors	[183]
Biomaterials Polymeric carriers		Silica, Biosilicification	reduced viral clearance in the liver; evaded nAbs; efficacious anticancer effect; biocompatibility	[184]
		Polylysine-encoded fiber, poly-L-lysine polymer	Better infection capacity	[185,186]
		Lactic-co-glycolic acid nanofiber	Enhanced delivery and therapeutic efficacy; reduced antiviral response	[187]
		multilayer ionic polymer	enhanced oncolytic activity; complement-dependent cytotoxicity; prolonged antitumor activities	[188]
	Polymers	Alginate	Reduced antiviral response	[189]
		Poly-2-dibutylamino-ethylamine-L-glutamate	High safety and efficacy	[190]
		PolycationsPolybrene	Shielding OVs; bridge virion and cell surface; efficient gene transduction and viral progeny	[191]
		core-cross-linked polyethyleneimine	low immunogenicity and toxicity; higher transduction; stability; improved anticancer cytotoxicity	[192]
		Polyethylene glycol (PEG)ylation, PH-sensitive pegylation	shield virions from nAbs; possibility of dose reduction; increased half-life in circulation	[178,193]

Table 2. *Cont.*

Strategy	Approach	Method	Outcome	References
		Poly hydroxypropyl methacrylamide	OVs shielding, increased half-life	[194]
		Polysaccharide	Failed to evade nAbs	[195]
		Silk-elastin-like polymer	OV shielding; better delivery and transduction; higher expression of viral genes; cause acute toxicity	[196]
		Chitosan	OV shielding; enhanced infectivity; induce cell fusion; delay in tumor growth	[197]
		Fibrin and collagen	Sustained release of viral particles	[198]
	Dendrimers	EGFR-targeted dendrimer, Poly-amidoamine dendrimers	selective internalization into EGFR-positive cells; low immunogenicity, toxicity and liver sequestration; Better transduction; OV shielding from nAbs	[199]
	Hydrogel	gelatin-based hydrogel	Decrease antiviral phagocyte response; better DC migration and activation; induction of tumor-specific IFN-γ+ immune cells	[200]
	Scaffolds	Microporous scaffolds	Prevent phagocytosis	[201]
Lipid-based carriers	Liposomes	Anionic liposome, Cationic liposome, Clondrosome (clodronate-loaded liposomes)	Shielding OVs; promoted OV delivery to the cytosol; enhance the tumor cell killing; macrophage depletion and better OV replication; induced expression of antitumorigenic genes	[202–204]
	Micelles	Micelles	higher transduction; efficient cellular internalization; improved cancer cell killing; attenuated the host antiviral response; minimal hepatotoxicity; good safety	[205]

Table 2. *Cont.*

Strategy	Approach	Method	Outcome	References
Metal-based carriers	Magnetosome	alternating magnetic field (iron oxide) Magnetic nanoparticles magnetically label OV-loaded macrophages and cells OVs labeled with magnetic particle	ECM degradation; enhanced OV uptake; prevention of tumor growth and metastasis; improved targeted therapy; increased tumor macrophage infiltrations; Protection against nAbs,	[206–208]
	Metal nanoparticles	Gold nanoparticles	Protected OVs; efficient transduction; enhanced viral cytopathic effect; safe vector for OVs	[209]
Ultrasound		Ultrasound-induced cavitation, Ultrasound + polymers, Ultrasound mediated microbubbles, Ultrasound contrast agents	improve OV extravasation and distribution; kill tumor cells within the ultrasound focal area; retardation of tumor growth; enhanced cell-based OVT	[210–212]
Photodynamics	Infrared	Near infrared light (plus gold nanoparticle)	ECM degradation	[213]
	Blue light	Photoactivatable OVs + blue light irradiation	induced replication; no off-tumor toxicity; inhibition subcutaneous tumor growth; therapeutic effect on cancer stem cells	[214]
Pre-treatment	Preconditioning	Granulocyte-macrophage colony-stimulating factor (GM-CSF)	Provide a pool of Potential OV carriers in the circulation: monocyte, macrophage, granulocytes, MDSCs, and CD11b+ cells; enhanced viral delivery; protected OVs from nAb	[215]

Table 2. *Cont.*

Strategy	Approach	Method	Outcome	References
Targeting ligands	Nanoparticle natural and engineered ligands	BiTEs, Trispecific Abs, Arginine-glycine-aspartic acid motif (RGD), Glycoprotein B/C, Neurotensin, Folic acid, Trastuzumab, Cetuximab, VEGF/bFGF, Biotin-EGF, CD71 and CD62E/P-immunovirosomes	Bridge tumor cells and OVs/immune cells; OV release in hypoxic/acidic TME; better cell entry; enhanced tumor tropism, nAb evasion; prolonged blood retention time; improved transduction;	[15,157,178,216–218]
	Viral particles	extracellular enveloped viral particle	Rapid OV spread within the TME; prevent removal by immune response; well adapted for systemic infusion;	[219]
Pharmacologic manipulation	Systemic and local pharmacotherapy	Angiotensin receptor blocker, paclitaxel, nitric oxide, nitroglycerin, bradykinin, Histamine TH-302 and PR-104 IC87114 or idelalisib (PI3Kδ-inhibitor)	activated local matrix metalloproteinases to disrupt the ECM; temporal vasodilation and better perfusion; OV activation in hypoxia; potentiateintravenous delivery of OV	[178,211,220,221]
		Cobra venom factor (CP40)	Complement inhibition; increase in OV titer in the blood; Prolonged OV existence	[222]
		Cyclophosphamide, Rapamycin	enhanced OV replication and activity; Avoid antiviral immune response	[223]
		Polyinosinic acide	Saturate scavenger receptors; prevent OV sequestration by Kupffer cells (liver macrophages); requirement of low dose OVs; lower toxicity; improve transduction	[224]

Table 2. *Cont.*

Strategy	Approach	Method	Outcome	References
Intratumoral spread of OVs	ECM-degradation	Hyaloronidase, Decorin, Relaxin, Chondroitinase, Matrix metalloproteinases, Collagenase, Bromelain, TAF depletion, LOX inhibition antibodies	Enhanced intratumoral spread of OVs; decrease matrix crosslinking and deposition	[6,211,225,226]
	Cellular junction opener	penton-dodecahedra, Junction Opener-1	Enhanced intratumoral spread of OVs	[226]
	Fusogenic proteins	Natural or engineered fusogenic OVs: MeV, NDV, RV, SeV, MuV, RSV, GALV, PoxV, VACV, VSV, HSV, and AdV	Improved infection; Enhanced tumor killing capability	[226–232]
	Vasculature degradation agents	Trombospondin-1 (TSP-1) TSP-1 peptide 3TSR Endostatin Anti-VEGF scAb	Better perfusion and delivery; enhanced intratumoral spread of OVs; Tumor necrosis; reduced hypoxia	[6,233]

OV. Oncolytic virus; OVT. OV therapy; ECM. Extracellular matrix; BiTE. Bispecific T cell engager; VEGF. Vascular-endothelial growth factor; bFGF. basic fibroblast GF; PI3K. Phosphoinositide 3-kinase; TAF. Tumor-associated fibroblast; Lox. Lipoxygenase.

3.3. Tumor ECM and Vasculature Degradation

Tumor ECM is a barrier to access tumor cells. Co-administration of ECM-degrading enzymes such as relaxin [234], matrix metalloproteinase (MMP)-1, -8, -9 [131,226], chondroitinase [235], and hyaluronidase [226] with OVT, or induction of their genes expression in GLV-1h255 (VACV) and VCN-01 (OAdV) can increase OV spread into the TME and improved OVT efficiency in cancers such as retinoblastoma and pancreatic carcinoma [236,237]. Cellular tight junctions are also accounted as barriers for OV distribution. GMOVs can trigger the production of proteins such as penton-dodecahedra and junction opener-1, which open the cellular junction through binding to desmoglein-2 [226]. However, there are concerns about increasing the likelihood of metastasis in this method that needs further investigation.

On the other hand, the insertion of endostatin and thrombospondin-1 genes in HSV-Endo and T-TSP-1 (both are HSV) destroys tumor vasculature. It suppresses angiogenesis in lung and gastric cancer by inhibiting migration and enhancing apoptosis in vECs [238,239]. Also, the expression of anti-VEGF sc-Ab by VACV increases antiangiogenic and antitumor properties [240].

3.4. Biosafety of OVT

Besides tumor cells, some OVs might replicate in normal cells and cause damage. For instance, T-VEC might remain a latent infection and cause long-term neurological AEs [153]. Using OVs with low pathogenicity in humans, such as parvovirus and reovirus, weakening OVs through repeated passages or deleting virulence genes, can increase the safety of OVT [241,242]. Thymidine kinase (TK) and infected cell protein (ICP)34.5 genes play a vital role in VACV and HSV-1 replication. The products of such genes are abundant in tumor cells, so the GMOVs lacking these genes can replicate in tumor cells, while the virus replication is impaired in healthy cells due to the low expression of such products [243,244]. The GL-ONC1 and Pexa-Vec (JX-594) are TK-free VACVs, and the T-VEC, HSV-1716, and G207 are ICP34.5-free HSVs showing acceptable safety in clinical trials [6,245–247]. Wild ZIKA virus has oncolytic potential in glioblastoma but also infects normal nerves with severe complications. Removal of 10 nucleotides from 3' of its genome can increase safety without reducing oncolytic activity [248]. Mutation or deletion of the E1 gene in AdV, and deletion of TK, vaccinia growth factor (VGF), hemagglutinin, and B18R genes in poxvirus reduce the virulence of OVs in normal cells [153,249]. However, deleting virulence genes to increase safety sometimes reduces OVs' antitumor activity [250].

Recombination of a safe OV such as NDV with an efficient OV like VSV is another way to increase the safety of OVs. Recombinant VSV-NDV (rVSV-NDV) comprises the envelope contents from NDV and the original backbone of VSV. Recombination of AdV with less harmful coxsackievirus or parvovirus constitutes OVs with high potency in tumor cell infection without damage to normal cells [250–252]. Using Ebolavirus (EBOV) glycoproteins also reduces the neurotoxicity of VSV in rVSV-EBOV [253]. Nevertheless, naturally occurring homologous recombination of GMOVs and wild-type OVs might result in a transgenic and pathogenic virus [153]. Transmission of OVs through body fluids to other people is rare but still a concern [254]. Also, the safety of OVT in immunocompromised individuals receiving radiotherapy and chemotherapy, as well as in pregnant women is still debated [153]. In general, due to the emergence of OVT with GMOVs, its long-term AEs are still unknown and require caution and further investigations.

3.5. Administration Routs

One of the factors influencing the response to OVT is the way of administration. Intratumoral injection results in precise control of the OV concentration in the TME, resulting in better therapeutic outcomes [255,256]. However, the complexity of intratumoral injection limits dosing repetition [257]. Besides, low perfusion of OVs into dense tumors requires ECM-degradation strategies [226]. Intravenous injection is popular due to its convenience, reproducibility, and possibility to target metastatic foci [258,259]. However, it requires tumor- specific delivery systems and is more likely to cause systemic toxicity [257]. Liver tropism, physical barriers such as BBB, complement activation, and the immune system response to OV before accessing the TME are the other disadvantages of intravenous injection [226,257]. Intraperitoneal, intrathecal/intracranial, and intrapleural injections are suitable for targeting intra-abdominal organs, central nervous system (CNS), and lung tumors, respectively, but are limited to use in laboratory animals [153]. The best route of administration is still a matter of debate with no specific guidelines. It seems that the less aggressive administration routs such as oral/mucosal and nasal administration, at least for gastrointestinal and cerebral malignancies, could increase the acceptability for patients and should be considered in future studies.

4. Combination Therapy

4.1. Immune-Checkpoint and Cell Therapy

Despite all the benefits, OVT as monotherapy cannot have a dramatic effect on tumor suppression and, like other immunotherapy methods, is used as combination therapy. A common complementary treatment strategy for OVT is ICI [260]. The overexpression of various immune checkpoints in the TME suppresses the response of immune cells. OVT and ICI seem to have synergistic effects [114,260]. OVT facilitates the infiltration of immune cells into the TME, and ICIs prevent the suppression of infiltrated immune cells activity. OVT also improves ICI access to the TME by destroying ECM and tumor vessels [6]. Recently, the use of OVs expressing mini-antibody (minibody) and single-chain variable fragment (scFv) against checkpoints has been able to block checkpoints locally in the TME, with fewer AEs [261,262]. Many clinical trials are currently examining the combination of ICI and OVT, the results of which primarily suggest that in order to achieve a better outcome, ICI should be prescribed after the onset of OV responses [6,263]. OVT increases the effectiveness of TIL and CAR-T cell therapy. OVT can increase the access of TILs and CAR-T cells to the tumor by altering the tumor matrix and increasing the chemokines such as CCL5 [264]. The secretion of IL-15, TNF-α and IL-2 from OVs in the TME increase the in situ proliferation and activation of TILs and enhances tumor response to CAR-T cell therapy [265,266]. Bispecific T-cell engagers (BiTEs) are fusion proteins containing two scAbs against tumor antigens and T cell surface CD3 [267]. The use of BiTE-expressing OVs can bridge T/CAR-T cells to TAA-expressing cells in the TME [267]. Furthermore, concomitant use of TAA-specific mAbs with OVT can enhance the antitumor response. However, the small size of OVs genome has made it difficult to encode whole antibodies [268] Combination of OVT with DC vaccines also improves the efficacy of DC vaccines by altering the TME immunosuppressive conditions [269]. OVs could be utilized as tumor vaccines in order to enhance the immune responses against established tumors or even prevent tumor recurrence. The main function of such OV-based tumor vaccines is the recruitment of APCs, facilitating the phagocytosis of tumor cells by APCs, and promoting the APC maturation to induce appropriate antitumor immune responses [270–272].

4.2. Metabolic Inhibitors as an Emerging Combination Therapy

Given the OV dependence on host cell metabolism for replication, the metabolic pathways can be considered effective modalities in OVT. For example, due to the role of glycolysis in the antiviral response, blocking this pathway increases the sensitivity of cells to OV infection [273,274]. On the other hand, increasing pyruvate flux into the tricarboxylic acid cycle, the increment of oxidative phosphorylation, and reactive oxygen species production lead to enhance OV replication and oncolytic activity [275–277]. However, there are contradictions in the enhancing or dampening roles of these metabolic pathways in the replication and function of OVs [278]. These discrepancies indicate that the metabolic pathway targeting should be based on the type of cancer and employed OV. Tumor cells deplete the glucose, tryptophan, and glutamine required by immune cells and produce lactate, kynurenine, and adenosine [279]. These changes cause induction of exhausted CTLs, M2-macrophages, and Tregs, creating an immunosuppressive TME [275–278]. The combination of metabolic inhibitors with OVT and the application of GMOVs to express metabolic inhibitors can alter the metabolism of cancer cells and immune cells to increase antitumor responses [278,280].

4.3. Other Combination Therapies

Along with the growing interest in OVT in the field of cancer treatment, many preclinical and clinical studies have suggested the use of OVs in combination with other common cancer therapies. OVT has been shown to potentiate the response to chemotherapy and radiotherapy, so that it could re-sensitize the chemo-/radio-resistant cells. Therefore, the combination of OVT with chemotherapy and radiotherapy is currently being evaluated in several clinical trials for chemo-resistant patients (Table 1). One of the shared mechanisms of OVT and chemo-/radiotherapy is ICD, in which a plethora of DAMPs is released, resulting in maximum induction of innate and adaptive immune responses. Hence, using OVT along with chemo-/radiotherapy could decrease the required doses of toxic agents and consequently lessen the adverse events of high dose treatments. Recombinant OVs can express enzymes such as cytosine deaminase, which converts the non-toxic prodrug 5-fuorocytosine (5-FC) into a toxic drug 5-fluorouracil (5-FU) in the tumor milieu [152]. Such local production of chemotherapeutic agents would decrease the systemic adverse events. GMOVs encoding the FCU1 gene can produce two enzymes, FCY1 and FUR1, that convert 5-FC to 5-FU and consequently 5-FU-monophosphate to target 5-FU-resistant tumors [152]. The tumor ECM prevents the access of therapeutic agents to the tumor cells, making the tumor resistant to chemotherapy [281]. Combination of ECM-degrading GMOVs with chemotherapy overrides the ECM-induced chemo-resistance observed in solid tumors [281]. Combination therapy of OVs and chemotherapy has been shown to exert synergistic antitumor activities via enhancing tumor cell killing capacity of chemotherapeutic agents, increasing virus proliferation in tumor cells, and invigorating oncolytic activities of OVs [282,283].

Besides conventional chemotherapy and radiotherapy, OVs could be administered in combination with targeted therapies [284]. Histone deacetylase inhibitors (HDIs) are recently entered the clinic as a promising treatment for cancers [285]. The companion of HDIs with OVT increases viral replication, upregulates the transgene expression (such as GM-CSF in T-VEC), enhances virus spread through the tumor cells, and augments oncolytic activities [286,287]. Moreover, HDIs induce antitumor immunity by inducing the expression of NK cell activating ligands and expression of TAAs, resulting NK cells and CTLs priming [286]. Co-administration of OVs with some protein kinase inhibitors such as MEK-1/2 and BRAF, and also inhibitors of some transcription factors like STAT-1 and NK-κB has been shown to enhance the oncolytic activities of OVs [288,289]. MEK/BRAF inhibitors do not affect viral replication. Instead, they enhance ER stress-induced apoptosis following OVT [288]. Interferon-stimulated genes (ISGs) are associated with resistance of tumors to chemotherapy, radiotherapy, and OVT. STAT-1 and NK-κB inhibitors diminish the expression of ISGs and thereby increase the cytotoxicity of OVs [289].

5. Conclusions

Although the OVT is not a new concept in cancer, the concerns of possible adverse events and unspecific infection hamper enough development in this era. The emerging genetic manipulations of OVs facilitate clinical studies with much lower concerns and reintroduce OVT as a promising immunotherapeutic approach. However, many questions should still be addressed. Finding the appropriate OV for each tumor, the best combination therapy, higher OVT efficacy and safety, and optimal delivery system require further knowledge about the cellular and molecular interaction between the OVs and the cells present in the TME. The results of current clinical trials could pave the way for OVT in the clinic.

Author Contributions: X.-Z.M. and J.-X.S.: Conception, design and inviting co-authors to participate. K.-T.J., W.-L.D., Y.-Y.L. and H.-R.L.: Writing original manuscript draft. X.-Z.M. and J.-X.S.: Review and editing of manuscript critically for important intellectual content and provided comments and feedback for the scientific contents of the manuscript. All authors have read and agreed to the published version of the manuscript.

Funding: This research received no external funding.

Acknowledgments: This work was supported by Zhejiang Provincial Science and Technology Projects (No. LGD19H160001 to K.-T.J.), Zhejiang Provincial Natural Science Foundation of China (No. LY21H160048 to J.-X.S.), National Natural Science Foundation of China (No. 81672430 to X.-Z.M.).

Conflicts of Interest: The authors declare no conflict of interest.

References

1. Dock, G. The influence of complicating diseases upon leukaemia. *Am. J. Med. Sci. (1827–1924)* **1904**, *127*, 563. [CrossRef]
2. Kelly, E.; Russell, S.J. History of oncolytic viruses: Genesis to genetic engineering. *Mol. Ther.* **2007**, *15*, 651–659. [CrossRef] [PubMed]
3. Russell, S.J.; Peng, K.-W.; Bell, J.C. Oncolytic virotherapy. *Nat. Biotechnol.* **2012**, *30*, 658. [CrossRef] [PubMed]
4. Chiocca, E.A.; Rabkin, S.D. Oncolytic viruses and their application to cancer immunotherapy. *Cancer Immunol. Res.* **2014**, *2*, 295–300. [CrossRef] [PubMed]
5. Annels, N.E.; Mansfield, D.; Arif, M.; Ballesteros-Merino, C.; Simpson, G.R.; Denyer, M.; Sandhu, S.S.; Melcher, A.A.; Harrington, K.J.; Davies, B. Phase I trial of an ICAM-1-targeted immunotherapeutic-coxsackievirus A21 (CVA21) as an oncolytic agent against non muscle-invasive bladder cancer. *Clin. Cancer Res.* **2019**, *25*, 5818–5831. [CrossRef]
6. Lan, Q.; Xia, S.; Wang, Q.; Xu, W.; Huang, H.; Jiang, S.; Lu, L. Development of oncolytic virotherapy: From genetic modification to combination therapy. *Front. Med.* **2020**, *14*, 1–25. [CrossRef]
7. Kaufman, H.L.; Kohlhapp, F.J.; Zloza, A. Oncolytic viruses: A new class of immunotherapy drugs. *Nat. Rev. Drug Discov.* **2015**, *14*, 642–662. [CrossRef]
8. Bommareddy, P.K.; Patel, A.; Hossain, S.; Kaufman, H.L. Talimogene laherparepvec (T-VEC) and other oncolytic viruses for the treatment of melanoma. *Am. J. Clin. Dermatol.* **2017**, *18*, 1–15. [CrossRef]
9. Samson, A.; Scott, K.J.; Taggart, D.; West, E.J.; Wilson, E.; Nuovo, G.J.; Thomson, S.; Corns, R.; Mathew, R.K.; Fuller, M.J. Intravenous delivery of oncolytic reovirus to brain tumor patients immunologically primes for subsequent checkpoint blockade. *Sci. Transl. Med.* **2018**, *10*, eaam7577. [CrossRef]
10. Geletneky, K.; Hajda, J.; Angelova, A.L.; Leuchs, B.; Capper, D.; Bartsch, A.J.; Neumann, J.-O.; Schöning, T.; Hüsing, J.; Beelte, B. Oncolytic H-1 parvovirus shows safety and signs of immunogenic activity in a first phase I/IIa glioblastoma trial. *Mol. Ther.* **2017**, *25*, 2620–2634. [CrossRef]
11. Yu, L.; Baxter, P.A.; Zhao, X.; Liu, Z.; Wadhwa, L.; Zhang, Y.; Su, J.M.; Tan, X.; Yang, J.; Adesina, A. A single intravenous injection of oncolytic picornavirus SVV-001 eliminates medulloblastomas in primary tumor-based orthotopic xenograft mouse models. *Neuro Oncol.* **2010**, *13*, 14–27. [CrossRef] [PubMed]
12. Ohka, S.; Nihei, C.-I.; Yamazaki, M.; Nomoto, A. Poliovirus trafficking toward central nervous system via human poliovirus receptor-dependent and-independent pathway. *Front. Microbiol.* **2012**, *3*, 147. [CrossRef] [PubMed]
13. Garcel, A.; Fauquette, W.; Dehouck, M.-P.; Crance, J.-M.; Favier, A.-L. Vaccinia virus-induced smallpox postvaccinal encephalitis in case of blood–brain barrier damage. *Vaccine* **2012**, *30*, 1397–1405. [CrossRef] [PubMed]
14. Liu, H.; Qiu, K.; He, Q.; Lei, Q.; Lu, W. Mechanisms of blood-brain barrier disruption in herpes simplex encephalitis. *J. Neuroimmune Pharmacol.* **2019**, *14*, 157–172. [CrossRef] [PubMed]
15. Yu, F.; Wang, X.; Guo, Z.S.; Bartlett, D.L.; Gottschalk, S.M.; Song, X.-T. T-cell engager-armed oncolytic vaccinia virus significantly enhances antitumor therapy. *Mol. Ther.* **2014**, *22*, 102–111. [CrossRef]
16. Martuza, R.L.; Malick, A.; Markert, J.M.; Ruffner, K.L.; Coen, D.M. Experimental therapy of human glioma by means of a genetically engineered virus mutant. *Science* **1991**, *252*, 854–856. [CrossRef]
17. Alberts, P.; Tilgase, A.; Rasa, A.; Bandere, K.; Venskus, D. The advent of oncolytic virotherapy in oncology: The Rigvir®story. *Eur. J. Pharmacol.* **2018**, *837*, 117–126. [CrossRef]
18. Bommareddy, P.K.; Shettigar, M.; Kaufman, H.L. Integrating oncolytic viruses in combination cancer immunotherapy. *Nat. Rev. Immunol.* **2018**, *18*, 498. [CrossRef]
19. Bourgeois-Daigneault, M.-C.; Roy, D.G.; Aitken, A.S.; El Sayes, N.; Martin, N.T.; Varette, O.; Falls, T.; St-Germain, L.E.; Pelin, A.; Lichty, B.D. Neoadjuvant oncolytic virotherapy before surgery sensitizes triple-negative breast cancer to immune checkpoint therapy. *Sci. Transl. Med.* **2018**, *10*, 1641–1650. [CrossRef]
20. Delwar, Z.M.; Kuo, Y.; Wen, Y.H.; Rennie, P.S.; Jia, W. Oncolytic virotherapy blockade by microglia and macrophages requires STAT1/3. *Cancer Res.* **2018**, *78*, 718–730. [CrossRef]

21. Garcia-Carbonero, R.; Salazar, R.; Duran, I.; Osman-Garcia, I.; Paz-Ares, L.; Bozada, J.M.; Boni, V.; Blanc, C.; Seymour, L.; Beadle, J. Phase 1 study of intravenous administration of the chimeric adenovirus enadenotucirev in patients undergoing primary tumor resection. *J. Immunother. Cancer* **2017**, *5*, 1–13. [CrossRef] [PubMed]

22. Lawler, S.E.; Speranza, M.-C.; Cho, C.-F.; Chiocca, E.A. Oncolytic viruses in cancer treatment: A review. *JAMA Oncol.* **2017**, *3*, 841–849. [CrossRef] [PubMed]

23. Martikainen, M.; Essand, M. Virus-based immunotherapy of glioblastoma. *Cancers* **2019**, *11*, 186. [CrossRef] [PubMed]

24. Haddad, D. Genetically engineered vaccinia viruses as agents for cancer treatment, imaging, and transgene delivery. *Front. Oncol.* **2017**, *7*, 96. [CrossRef] [PubMed]

25. Bai, Y.; Hui, P.; Du, X.; Su, X. Updates to the antitumor mechanism of oncolytic virus. *Thorac. Cancer* **2019**, *10*, 1031–1035. [CrossRef] [PubMed]

26. Ylösmäki, E.; Cerullo, V. Design and application of oncolytic viruses for cancer immunotherapy. *Curr. Opin. Biotechnol.* **2020**, *65*, 25–36. [CrossRef]

27. De Graaf, J.F.; de Vor, L.; Fouchier, R.A.M.; Van Den Hoogen, B.G. Armed oncolytic viruses: A kick-start for anti-tumor immunity. *Cytokine Growth Factor Rev.* **2018**, *41*, 28–39. [CrossRef]

28. Breitbach, C.J.; Arulanandam, R.; De Silva, N.; Thorne, S.H.; Patt, R.; Daneshmand, M.; Moon, A.; Ilkow, C.; Burke, J.; Hwang, T.-H. Oncolytic vaccinia virus disrupts tumor-associated vasculature in humans. *Cancer Res.* **2013**, *73*, 1265–1275. [CrossRef]

29. Chesney, J.; Puzanov, I.; Collichio, F.; Singh, P.; Milhem, M.M.; Glaspy, J.; Hamid, O.; Ross, M.; Friedlander, P.; Garbe, C. Randomized, open-label phase II study evaluating the efficacy and safety of talimogene laherparepvec in combination with ipilimumab versus ipilimumab alone in patients with advanced, unresectable melanoma. *J. Clin. Oncol.* **2018**, *36*, 1658. [CrossRef]

30. Kelly, C.M.; Antonescu, C.R.; Bowler, T.; Munhoz, R.; Chi, P.; Dickson, M.A.; Gounder, M.M.; Keohan, M.L.; Movva, S.; Dholakia, R. Objective response rate among patients with locally advanced or metastatic sarcoma treated with talimogene laherparepvec in combination with pembrolizumab: A phase 2 clinical trial. *JAMA Oncol.* **2020**, *6*, 402–408. [CrossRef]

31. Ribas, A.; Dummer, R.; Puzanov, I.; VanderWalde, A.; Andtbacka, R.H.; Michielin, O.; Olszanski, A.J.; Malvehy, J.; Cebon, J.; Fernandez, E. Oncolytic virotherapy promotes intratumoral T cell infiltration and improves anti-PD-1 immunotherapy. *Cell* **2017**, *170*, 1109–1119. e1110. [CrossRef] [PubMed]

32. Chesney, J.; Awasthi, S.; Curti, B.; Hutchins, L.; Linette, G.; Triozzi, P.; Tan, M.C.; Brown, R.E.; Nemunaitis, J.; Whitman, E. Phase IIIb safety results from an expanded-access protocol of talimogene laherparepvec for patients with unresected, stage IIIB–IVM1c melanoma. *Melanoma Res.* **2018**, *28*, 44–51. [CrossRef] [PubMed]

33. Andtbacka, R.; Kaufman, H.L.; Collichio, F.; Amatruda, T.; Senzer, N.; Chesney, J.; Delman, K.A.; Spitler, L.E.; Puzanov, I.; Agarwala, S.S. Talimogene laherparepvec improves durable response rate in patients with advanced melanoma. *J. Clin. Oncol.* **2015**, *33*, 2780–2788. [CrossRef] [PubMed]

34. Harrington, K.J.; Hingorani, M.; Tanay, M.A.; Hickey, J.; Bhide, S.A.; Clarke, P.M.; Renouf, L.C.; Thway, K.; Sibtain, A.; McNeish, I.A. Phase I/II study of oncolytic HSVGM-CSF in combination with radiotherapy and cisplatin in untreated stage III/IV squamous cell cancer of the head and neck. *Clin. Cancer Res.* **2010**, *16*, 4005–4015. [CrossRef]

35. Senzer, N.N.; Kaufman, H.L.; Amatruda, T.; Nemunaitis, M.; Reid, T.; Daniels, G.; Gonzalez, R.; Glaspy, J.; Whitman, E.; Harrington, K. Phase II clinical trial of a granulocyte-macrophage colony-stimulating factor-encoding, second-generation oncolytic herpesvirus in patients with unresectable metastatic melanoma. *J. Clin. Oncol.* **2009**, *27*, 5763. [CrossRef] [PubMed]

36. Hu, J.C.; Coffin, R.S.; Davis, C.J.; Graham, N.J.; Groves, N.; Guest, P.J.; Harrington, K.J.; James, N.D.; Love, C.A.; McNeish, I. A phase I study of OncoVEXGM-CSF, a second-generation oncolytic herpes simplex virus expressing granulocyte macrophage colony-stimulating factor. *Clin. Cancer Res.* **2006**, *12*, 6737–6747. [CrossRef] [PubMed]

37. Streby, K.A.; Geller, J.I.; Currier, M.A.; Warren, P.S.; Racadio, J.M.; Towbin, A.J.; Vaughan, M.R.; Triplet, M.; Ott-Napier, K.; Dishman, D.J. Intratumoral injection of HSV1716, an oncolytic herpes virus, is safe and shows evidence of immune response and viral replication in young cancer patients. *Clin. Cancer Res.* **2017**, *23*, 3566–3574. [CrossRef] [PubMed]

38. Mace, A.T.; Ganly, I.; Soutar, D.S.; Brown, S.M. Potential for efficacy of the oncolytic Herpes simplex virus 1716 in patients with oral squamous cell carcinoma. *Head Neck J. Sci. Spec. Head Neck* **2008**, *30*, 1045–1051. [CrossRef]

39. Geevarghese, S.K.; Geller, D.A.; de Haan, H.A.; Hörer, M.; Knoll, A.E.; Mescheder, A.; Nemunaitis, J.; Reid, T.R.; Sze, D.Y.; Tanabe, K.K. Phase I/II study of oncolytic herpes simplex virus NV1020 in patients with extensively pretreated refractory colorectal cancer metastatic to the liver. *Hum. Gene Ther.* **2010**, *21*, 1119–1128. [CrossRef]

40. Fong, Y.; Kim, T.; Bhargava, A.; Schwartz, L.; Brown, K.; Brody, L.; Covey, A.; Karrasch, M.; Getrajdman, G.; Mescheder, A. A herpes oncolytic virus can be delivered via the vasculature to produce biologic changes in human colorectal cancer. *Mol. Ther.* **2009**, *17*, 389–394. [CrossRef]

41. Markert, J.M.; Razdan, S.N.; Kuo, H.-C.; Cantor, A.; Knoll, A.; Karrasch, M.; Nabors, L.B.; Markiewicz, M.; Agee, B.S.; Coleman, J.M. A phase 1 trial of oncolytic HSV-1, G207, given in combination with radiation for recurrent GBM demonstrates safety and radiographic responses. *Mol. Ther.* **2014**, *22*, 1048–1055. [CrossRef] [PubMed]

42. Hirooka, Y.; Kasuya, H.; Ishikawa, T.; Kawashima, H.; Ohno, E.; Villalobos, I.B.; Naoe, Y.; Ichinose, T.; Koyama, N.; Tanaka, M. A Phase I clinical trial of EUS-guided intratumoral injection of the oncolytic virus, HF10 for unresectable locally advanced pancreatic cancer. *BMC Cancer* **2018**, *18*, 1–9. [CrossRef] [PubMed]

43. Voit, C.; Kron, M.; Schwürzer-Voit, M.; Sterry, W. Intradermal injection of Newcastle disease virus-modified autologous melanoma cell lysate and interleukin-2 for adjuvant treatment of melanoma patients with resectable stage III disease: Adjuvante Therapie von Melanompatienten im Stadium III mit einer Kombination aus Newcastle-disease-Virus-modifiziertem Tumorzelllysat und Interleukin-2. *JDDG J. Der Dtsch. Dermatol. Ges.* **2003**, *1*, 120–125.

44. Pecora, A.L.; Rizvi, N.; Cohen, G.I.; Meropol, N.J.; Sterman, D.; Marshall, J.L.; Goldberg, S.; Gross, P.; O'Neil, J.D.; Groene, W.S. Phase I trial of intravenous administration of PV701, an oncolytic virus, in patients with advanced solid cancers. *J. Clin. Oncol.* **2002**, *20*, 2251–2266. [CrossRef] [PubMed]

45. Freeman, A.I.; Zakay-Rones, Z.; Gomori, J.M.; Linetsky, E.; Rasooly, L.; Greenbaum, E.; Rozenman-Yair, S.; Panet, A.; Libson, E.; Irving, C.S. Phase I/II trial of intravenous NDV-HUJ oncolytic virus in recurrent glioblastoma multiforme. *Mol. Ther.* **2006**, *13*, 221–228. [CrossRef]

46. Rudin, C.; Pandha, H.; Gupta, S.; Zibelman, M.; Akerley, W.; Day, D.; Hill, A.; Sanborn, R.; O'Day, S.; Clay, T. Phase Ib KEYNOTE-200: A study of an intravenously delivered oncolytic virus, coxsackievirus A21 in combination with pembrolizumab in advanced NSCLC and bladder cancer patients. *Ann. Oncol.* **2018**, *29*, viii732. [CrossRef]

47. Cook, M.; Chauhan, A. Clinical Application of Oncolytic Viruses: A Systematic Review. *Int. J. Mol. Sci.* **2020**, *21*, 7505. [CrossRef] [PubMed]

48. A Study of Intratumoral CAVATAK in Patients with Stage IIIc and Stage IV Malignant Melanoma (VLA-007 CALM) (CALM). Available online: https://clinicaltrials.gov/ct2/show/results/NCT01227551 (accessed on 23 January 2021).

49. Morris, D.G.; Feng, X.; DiFrancesco, L.M.; Fonseca, K.; Forsyth, P.A.; Paterson, A.H.; Coffey, M.C.; Thompson, B. REO-001: A phase I trial of percutaneous intralesional administration of reovirus type 3 dearing (Reolysin®) in patients with advanced solid tumors. *Investig. New Drugs* **2013**, *31*, 696–706. [CrossRef] [PubMed]

50. Kicielinski, K.P.; Chiocca, E.A.; John, S.Y.; Gill, G.M.; Coffey, M.; Markert, J.M. Phase 1 clinical trial of intratumoral reovirus infusion for the treatment of recurrent malignant gliomas in adults. *Mol. Ther.* **2014**, *22*, 1056–1062. [CrossRef]

51. Kolb, E.A.; Sampson, V.; Stabley, D.; Walter, A.; Sol-Church, K.; Cripe, T.; Hingorani, P.; Ahern, C.H.; Weigel, B.J.; Zwiebel, J. A phase I trial and viral clearance study of reovirus (Reolysin) in children with relapsed or refractory extra-cranial solid tumors: A Children's Oncology Group Phase I Consortium report. *Pediatr. Blood Cancer* **2015**, *62*, 751–758. [CrossRef]

52. Noonan, A.M.; Farren, M.R.; Geyer, S.M.; Huang, Y.; Tahiri, S.; Ahn, D.; Mikhail, S.; Ciombor, K.K.; Pant, S.; Aparo, S. Randomized phase 2 trial of the oncolytic virus pelareorep (Reolysin) in upfront treatment of metastatic pancreatic adenocarcinoma. *Mol. Ther.* **2016**, *24*, 1150–1158. [CrossRef] [PubMed]

53. Galanis, E.; Markovic, S.N.; Suman, V.J.; Nuovo, G.J.; Vile, R.G.; Kottke, T.J.; Nevala, W.K.; Thompson, M.A.; Lewis, J.E.; Rumilla, K.M. Phase II trial of intravenous administration of Reolysin®(Reovirus Serotype-3-dearing Strain) in patients with metastatic melanoma. *Mol. Ther.* **2012**, *20*, 1998–2003. [CrossRef] [PubMed]

54. Bernstein, V.; Ellard, S.; Dent, S.; Tu, D.; Mates, M.; Dhesy-Thind, S.; Panasci, L.; Gelmon, K.; Salim, M.; Song, X. A randomized phase II study of weekly paclitaxel with or without pelareorep in patients with metastatic breast cancer: Final analysis of Canadian Cancer Trials Group IND. 213. *Breast Cancer Res. Treat.* **2018**, *167*, 485–493. [CrossRef]

55. Mahalingam, D.; Goel, S.; Aparo, S.; Patel Arora, S.; Noronha, N.; Tran, H.; Chakrabarty, R.; Selvaggi, G.; Gutierrez, A.; Coffey, M. A phase II study of pelareorep (REOLYSIN®) in combination with gemcitabine for patients with advanced pancreatic adenocarcinoma. *Cancers* **2018**, *10*, 160. [CrossRef]

56. Mahalingam, D.; Fountzilas, C.; Moseley, J.; Noronha, N.; Tran, H.; Chakrabarty, R.; Selvaggi, G.; Coffey, M.; Thompson, B.; Sarantopoulos, J. A phase II study of REOLYSIN®(pelareorep) in combination with carboplatin and paclitaxel for patients with advanced malignant melanoma. *Cancer Chemother. Pharmacol.* **2017**, *79*, 697–703. [CrossRef]

57. Cohn, D.E.; Sill, M.W.; Walker, J.L.; O'Malley, D.; Nagel, C.I.; Rutledge, T.L.; Bradley, W.; Richardson, D.L.; Moxley, K.M.; Aghajanian, C. Randomized phase IIB evaluation of weekly paclitaxel versus weekly paclitaxel with oncolytic reovirus (Reolysin®) in recurrent ovarian, tubal, or peritoneal cancer: An nrg oncology/gynecologic oncology group study. *Gynecol. Oncol.* **2017**, *146*, 477–483. [CrossRef]

58. Roulstone, V.; Khan, K.; Pandha, H.S.; Rudman, S.; Coffey, M.; Gill, G.M.; Melcher, A.A.; Vile, R.; Harrington, K.J.; De Bono, J. Phase I trial of cyclophosphamide as an immune modulator for optimizing oncolytic reovirus delivery to solid tumors. *Clin. Cancer Res.* **2015**, *21*, 1305–1312. [CrossRef]

59. Harrington, K.J.; Karapanagiotou, E.M.; Roulstone, V.; Twigger, K.R.; White, C.L.; Vidal, L.; Beirne, D.; Prestwich, R.; Newbold, K.; Ahmed, M. Two-stage phase I dose-escalation study of intratumoral reovirus type 3 dearing and palliative radiotherapy in patients with advanced cancers. *Clin. Cancer Res.* **2010**, *16*, 3067–3077. [CrossRef]

60. Desjardins, A.; Gromeier, M.; Herndon, J.E.; Beaubier, N.; Bolognesi, D.P.; Friedman, A.H.; Friedman, H.S.; McSherry, F.; Muscat, A.M.; Nair, S. Recurrent glioblastoma treated with recombinant poliovirus. *N. Engl. J. Med.* **2018**, *379*, 150–161. [CrossRef] [PubMed]

61. Nokisalmi, P.; Pesonen, S.; Escutenaire, S.; Särkioja, M.; Raki, M.; Cerullo, V.; Laasonen, L.; Alemany, R.; Rojas, J.; Cascallo, M. Oncolytic adenovirus ICOVIR-7 in patients with advanced and refractory solid tumors. *Clin. Cancer Res.* **2010**, *16*, 3035–3043. [CrossRef]

62. Kimball, K.J.; Preuss, M.A.; Barnes, M.N.; Wang, M.; Siegal, G.P.; Wan, W.; Kuo, H.; Saddekni, S.; Stockard, C.R.; Grizzle, W.E. A phase I study of a tropism-modified conditionally replicative adenovirus for recurrent malignant gynecologic diseases. *Clin. Cancer Res.* **2010**, *16*, 5277–5287. [CrossRef] [PubMed]

63. Kim, K.H.; Dmitriev, I.P.; Saddekni, S.; Kashentseva, E.A.; Harris, R.D.; Aurigemma, R.; Bae, S.; Singh, K.P.; Siegal, G.P.; Curiel, D.T. A phase I clinical trial of Ad5/3-Δ24, a novel serotype-chimeric, infectivity-enhanced, conditionally-replicative adenovirus (CRAd), in patients with recurrent ovarian cancer. *Gynecol. Oncol.* **2013**, *130*, 518–524. [CrossRef] [PubMed]

64. Pesonen, S.; Diaconu, I.; Cerullo, V.; Escutenaire, S.; Raki, M.; Kangasniemi, L.; Nokisalmi, P.; Dotti, G.; Guse, K.; Laasonen, L. Integrin targeted oncolytic adenoviruses Ad5-D24-RGD and Ad5-RGD-D24-GMCSF for treatment of patients with advanced chemotherapy refractory solid tumors. *Int. J. Cancer* **2012**, *130*, 1937–1947. [CrossRef] [PubMed]

65. DeWeese, T.L.; van der Poel, H.; Li, S.; Mikhak, B.; Drew, R.; Goemann, M.; Hamper, U.; DeJong, R.; Detorie, N.; Rodriguez, R. A phase I trial of CV706, a replication-competent, PSA selective oncolytic adenovirus, for the treatment of locally recurrent prostate cancer following radiation therapy. *Cancer Res.* **2001**, *61*, 7464–7472. [PubMed]

66. Chang, J.; Zhao, X.; Wu, X.; Guo, Y.; Guo, H.; Cao, J.; Guo, Y.; Lou, D.; Yu, D.; Li, J. A Phase I study of KH901, a conditionally replicating granulocyte-macrophage colony-stimulating factor: Armed oncolytic adenovirus for the treatment of head and neck cancers. *Cancer Biol. Ther.* **2009**, *8*, 676–682. [CrossRef] [PubMed]

67. Packiam, V.T.; Lamm, D.L.; Barocas, D.A.; Trainer, A.; Fand, B.; Davis III, R.L.; Clark, W.; Kroeger, M.; Dumbadze, I.; Chamie, K. An open label, single-arm, phase II multicenter study of the safety and efficacy of CG0070 oncolytic vector regimen in patients with BCG-unresponsive non–muscle-invasive bladder cancer: Interim results. *Urol. Oncol. Semin. Ori.* **2018**, *36*, 440–447. [CrossRef] [PubMed]

68. Nemunaitis, J.; Tong, A.W.; Nemunaitis, M.; Senzer, N.; Phadke, A.P.; Bedell, C.; Adams, N.; Zhang, Y.-A.; Maples, P.B.; Chen, S. A phase I study of telomerase-specific replication competent oncolytic adenovirus (telomelysin) for various solid tumors. *Mol. Ther.* **2010**, *18*, 429–434. [CrossRef]

69. Nemunaitis, J.; Ganly, I.; Khuri, F.; Arseneau, J.; Kuhn, J.; McCarty, T.; Landers, S.; Maples, P.; Romel, L.; Randlev, B. Selective replication and oncolysis in p53 mutant tumors with ONYX-015, an E1B-55kD gene-deleted adenovirus, in patients with advanced head and neck cancer: A phase II trial. *Cancer Res.* **2000**, *60*, 6359–6366.

70. Galanis, E.; Okuno, S.H.; Nascimento, A.; Lewis, B.; Lee, R.; Oliveira, A.; Sloan, J.A.; Atherton, P.; Edmonson, J.; Erlichman, C. Phase I–II trial of ONYX-015 in combination with MAP chemotherapy in patients with advanced sarcomas. *Gene Ther.* **2005**, *12*, 437–445. [CrossRef]

71. Reid, T.R.; Freeman, S.; Post, L.; McCormick, F.; Sze, D.Y. Effects of Onyx-015 among metastatic colorectal cancer patients that have failed prior treatment with 5-FU/leucovorin. *Cancer Gene Ther.* **2005**, *12*, 673–681. [CrossRef]

72. Nemunaitis, J.; Senzer, N.; Sarmiento, S.; Zhang, Y.; Arzaga, R.; Sands, B.; Maples, P.; Tong, A. A phase I trial of intravenous infusion of ONYX-015 and enbrel in solid tumor patients. *Cancer Gene Ther.* **2007**, *14*, 885–893. [CrossRef] [PubMed]

73. Wheeler, L.A.; Manzanera, A.G.; Bell, S.D.; Cavaliere, R.; McGregor, J.M.; Grecula, J.C.; Newton, H.B.; Lo, S.S.; Badie, B.; Portnow, J. Phase II multicenter study of gene-mediated cytotoxic immunotherapy as adjuvant to surgical resection for newly diagnosed malignant glioma. *Neuro Oncol.* **2016**, *18*, 1137–1145. [CrossRef] [PubMed]

74. Freytag, S.O.; Stricker, H.; Lu, M.; Elshaikh, M.; Aref, I.; Pradhan, D.; Levin, K.; Kim, J.H.; Peabody, J.; Siddiqui, F. Prospective randomized phase 2 trial of intensity modulated radiation therapy with or without oncolytic adenovirus-mediated cytotoxic gene therapy in intermediate-risk prostate cancer. *Int. J. Radiat. Oncol. Biol. Phys.* **2014**, *89*, 268–276. [CrossRef] [PubMed]

75. Freytag, S.O.; Movsas, B.; Aref, I.; Stricker, H.; Peabody, J.; Pegg, J.; Zhang, Y.; Barton, K.N.; Brown, S.L.; Lu, M. Phase I trial of replication-competent adenovirus-mediated suicide gene therapy combined with IMRT for prostate cancer. *Mol. Ther.* **2007**, *15*, 1016–1023. [CrossRef]

76. Machiels, J.-P.; Salazar, R.; Rottey, S.; Duran, I.; Dirix, L.; Geboes, K.; Wilkinson-Blanc, C.; Pover, G.; Alvis, S.; Champion, B. A phase 1 dose escalation study of the oncolytic adenovirus enadenotucirev, administered intravenously to patients with epithelial solid tumors (EVOLVE). *J. Immunother. Cancer* **2019**, *7*, 1–15. [CrossRef]

77. Cripe, T.P.; Ngo, M.C.; Geller, J.I.; Louis, C.U.; Currier, M.A.; Racadio, J.M.; Towbin, A.J.; Rooney, C.M.; Pelusio, A.; Moon, A. Phase 1 study of intratumoral Pexa-Vec (JX-594), an oncolytic and immunotherapeutic vaccinia virus, in pediatric cancer patients. *Mol. Ther.* **2015**, *23*, 602–608. [CrossRef]

78. Heo, J.; Breitbach, C.; Cho, M.; Hwang, T.-H.; Kim, C.W.; Jeon, U.B.; Woo, H.Y.; Yoon, K.T.; Lee, J.W.; Burke, J. *Phase II Trial of Pexa-Vec (Pexastimogene Devacirepvec; JX-594), an Oncolytic and Immunotherapeutic Vaccinia Virus, Followed by Sorafenib in Patients with Advanced Hepatocellular Carcinoma (HCC)*; American Society of Clinical Oncology: Alexandria, VA, USA, 2013.

79. Park, S.H.; Breitbach, C.J.; Lee, J.; Park, J.O.; Lim, H.Y.; Kang, W.K.; Moon, A.; Mun, J.-H.; Sommermann, E.M.; Avidal, L.M. Phase 1b trial of biweekly intravenous Pexa-Vec (JX-594), an oncolytic and immunotherapeutic vaccinia virus in colorectal cancer. *Mol. Ther.* **2015**, *23*, 1532–1540. [CrossRef]

80. Hwang, T.-H.; Moon, A.; Burke, J.; Ribas, A.; Stephenson, J.; Breitbach, C.J.; Daneshmand, M.; De Silva, N.; Parato, K.; Diallo, J.-S. A mechanistic proof-of-concept clinical trial with JX-594, a targeted multi-mechanistic oncolytic poxvirus, in patients with metastatic melanoma. *Mol. Ther.* **2011**, *19*, 1913–1922. [CrossRef]

81. Husseini, F.; Delord, J.-P.; Fournel-Federico, C.; Guitton, J.; Erbs, P.; Homerin, M.; Halluard, C.; Jemming, C.; Orange, C.; Limacher, J.-M. Vectorized gene therapy of liver tumors: Proof-of-concept of TG4023 (MVA-FCU1) in combination with flucytosine. *Ann. Oncol.* **2017**, *28*, 169–174. [CrossRef]

82. Downs-Canner, S.; Guo, Z.S.; Ravindranathan, R.; Breitbach, C.J.; O'malley, M.E.; Jones, H.L.; Moon, A.; McCart, J.A.; Shuai, Y.; Zeh, H.J. Phase 1 study of intravenous oncolytic poxvirus (vvDD) in patients with advanced solid cancers. *Mol. Ther.* **2016**, *24*, 1492–1501. [CrossRef]

83. Zeh, H.J.; Downs-Canner, S.; McCart, J.A.; Guo, Z.S.; Rao, U.N.; Ramalingam, L.; Thorne, S.H.; Jones, H.L.; Kalinski, P.; Wieckowski, E. First-in-man study of western reserve strain oncolytic vaccinia virus: Safety, systemic spread, and antitumor activity. *Mol. Ther.* **2015**, *23*, 202–214. [CrossRef] [PubMed]

84. Mell, L.K.; Brumund, K.T.; Daniels, G.A.; Advani, S.J.; Zakeri, K.; Wright, M.E.; Onyeama, S.-J.; Weisman, R.A.; Sanghvi, P.R.; Martin, P.J. Phase I trial of intravenous oncolytic vaccinia virus (GL-ONC1) with cisplatin and radiotherapy in patients with locoregionally advanced head and neck carcinoma. *Clin. Cancer Res.* **2017**, *23*, 5696–5702. [CrossRef] [PubMed]

85. Galanis, E.; Hartmann, L.C.; Cliby, W.A.; Long, H.J.; Peethambaram, P.P.; Barrette, B.A.; Kaur, J.S.; Haluska, P.J.; Aderca, I.; Zollman, P.J. Phase I trial of intraperitoneal administration of an oncolytic measles virus strain engineered to express carcinoembryonic antigen for recurrent ovarian cancer. *Cancer Res.* **2010**, *70*, 875–882. [CrossRef] [PubMed]

86. Rudin, C.M.; Poirier, J.T.; Senzer, N.N.; Stephenson, J.; Loesch, D.; Burroughs, K.D.; Reddy, P.S.; Hann, C.L.; Hallenbeck, P.L. Phase I clinical study of Seneca Valley Virus (SVV-001), a replication-competent picornavirus, in advanced solid tumors with neuroendocrine features. *Clin. Cancer Res.* **2011**, *17*, 888–895. [CrossRef] [PubMed]

87. Burke, M.J.; Ahern, C.; Weigel, B.J.; Poirier, J.T.; Rudin, C.M.; Chen, Y.; Cripe, T.P.; Bernhardt, M.B.; Blaney, S.M. Phase I trial of Seneca Valley Virus (NTX-010) in children with relapsed/refractory solid tumors: A report of the Children's Oncology Group. *Pediatr. Blood Cancer* **2015**, *62*, 743–750. [CrossRef] [PubMed]

88. Kaufman, H.L.; Kim, D.W.; Kim-Schulze, S.; DeRaffele, G.; Jagoda, M.C.; Broucek, J.R.; Zloza, A. Results of a randomized phase I gene therapy clinical trial of nononcolytic fowlpox viruses encoding T cell costimulatory molecules. *Hum. Gene Ther.* **2014**, *25*, 452–460. [CrossRef]

89. Angelova, A.L.; Barf, M.; Geletneky, K.; Unterberg, A.; Rommelaere, J. Immunotherapeutic potential of oncolytic H-1 parvovirus: Hints of glioblastoma microenvironment conversion towards immunogenicity. *Viruses* **2017**, *9*, 382. [CrossRef]

90. Achard, C.; Surendran, A.; Wedge, M.-E.; Ungerechts, G.; Bell, J.; Ilkow, C.S. Lighting a fire in the tumor microenvironment using oncolytic immunotherapy. *EBioMedicine* **2018**, *31*, 17–24. [CrossRef]

91. Allegrezza, M.J.; Conejo-Garcia, J.R. Targeted therapy and immunosuppression in the tumor microenvironment. *Trends Cancer* **2017**, *3*, 19–27. [CrossRef]

92. Pol, J.G.; Bridle, B.W.; Lichty, B.D. Detection of Tumor Antigen-Specific T-Cell Responses after Oncolytic Vaccination. In *Oncolytic Viruses*; Springer: Humana, NY, USA, 2020; pp. 191–211. [CrossRef]

93. Bakhshaei, P.; Kazemi, M.H.; Golara, M.; Abdolmaleki, S.; Khosravi-Eghbal, R.; Khoshnoodi, J.; Judaki, M.A.; Salimi, V.; Douraghi, M.; Jeddi-Tehrani, M.; et al. Investigation of the Cellular Immune Response to Recombinant Fragments of Filamentous Hemagglutinin and Pertactin of Bordetella pertussis in BALB/c Mice. *J. Interferon Cytokine Res.* **2018**, *38*. [CrossRef]

94. Kuryk, L.; Møller, A.S.W. Chimeric oncolytic Ad5/3 virus replicates and lyses ovarian cancer cells through desmoglein-2 cell entry receptor. *J. Med. Virol.* **2020**, *92*, 1309–1315. [CrossRef] [PubMed]

95. Wang, M.; Liu, W.; Zhang, Y.; Dang, M.; Zhang, Y.; Tao, J.; Chen, K.; Peng, X.; Teng, Z. Intercellular adhesion molecule 1 antibody-mediated mesoporous drug delivery system for targeted treatment of triple-negative breast cancer. *J. Colloid Interface Sci.* **2019**, *538*, 630–637. [CrossRef] [PubMed]

96. Wenthe, J.; Naseri, S.; Hellström, A.-C.; Wiklund, H.J.; Eriksson, E.; Loskog, A. Immunostimulatory oncolytic virotherapy for multiple myeloma targeting 4-1BB and/or CD40. *Cancer Gene Ther.* **2020**, *27*, 948–959. [CrossRef] [PubMed]

97. Zhang, Y.; Ye, M.; Huang, F.; Wang, S.; Wang, H.; Mou, X.; Wang, Y. Oncolytic Adenovirus Expressing ST13 Increases Antitumor Effect of Tumor Necrosis Factor-Related Apoptosis-Inducing Ligand Against Pancreatic Ductal Adenocarcinoma. *Hum. Gene Ther.* **2020**, *31*, 15–16. [CrossRef] [PubMed]

98. Heiniö, C.; Havunen, R.; Santos, J.; Lint, K.D.; Cervera-Carrascon, V.; Kanerva, A.; Hemminki, A. TNFa and IL2 encoding oncolytic adenovirus activates pathogen and danger-associated immunological signaling. *Cells* **2020**, *9*, 798. [CrossRef]

99. Delaunay, T.; Achard, C.; Boisgerault, N.; Grard, M.; Petithomme, T.; Chatelain, C.; Dutoit, S.; Blanquart, C.; Royer, P.-J.; Minvielle, S. Frequent homozygous deletions of type I interferon genes in pleural mesothelioma confer sensitivity to oncolytic measles virus. *J. Thorac. Oncol.* **2020**, *15*, 827–842. [CrossRef]

100. Hindupur, S.V.; Schmid, S.C.; Koch, J.A.; Youssef, A.; Baur, E.-M.; Wang, D.; Horn, T.; Slotta-Huspenina, J.; Gschwend, J.E.; Holm, P.S. STAT3/5 inhibitors suppress proliferation in bladder cancer and enhance oncolytic adenovirus therapy. *Int. J. Mol. Sci.* **2020**, *21*, 1106. [CrossRef]

101. McLaughlin, M.; Pedersen, M.; Roulstone, V.; Bergerhoff, K.F.; Smith, H.G.; Whittock, H.; Kyula, J.N.; Dillon, M.T.; Pandha, H.S.; Vile, R. The PERK Inhibitor GSK2606414 Enhances Reovirus Infection in Head and Neck Squamous Cell Carcinoma via an ATF4-Dependent Mechanism. *Mol. Ther. Oncolytics* **2020**, *16*, 238–249. [CrossRef]

102. Kaufman, H.L.; Kim, D.W.; DeRaffele, G.; Mitcham, J.; Coffin, R.S.; Kim-Schulze, S. Local and distant immunity induced by intralesional vaccination with an oncolytic herpes virus encoding GM-CSF in patients with stage IIIc and IV melanoma. *Ann. Surg. Oncol.* **2010**, *17*, 718–730. [CrossRef]

103. Moesta, A.K.; Cooke, K.; Piasecki, J.; Mitchell, P.; Rottman, J.B.; Fitzgerald, K.; Zhan, J.; Yang, B.; Le, T.; Belmontes, B. Local Delivery of OncoVEXmGM-CSF Generates Systemic Antitumor Immune Responses Enhanced by Cytotoxic T-Lymphocyte–Associated Protein Blockade. *Clin. Cancer Res.* **2017**, *23*, 6190–6202. [CrossRef]

104. Galanis, E.; Atherton, P.J.; Maurer, M.J.; Knutson, K.L.; Dowdy, S.C.; Cliby, W.A.; Haluska, P.; Long, H.J.; Oberg, A.; Aderca, I. Oncolytic measles virus expressing the sodium iodide symporter to treat drug-resistant ovarian cancer. *Cancer Res.* **2015**, *75*, 22–30. [CrossRef] [PubMed]

105. DuPage, M.; Mazumdar, C.; Schmidt, L.M.; Cheung, A.F.; Jacks, T. Expression of tumour-specific antigens underlies cancer immunoediting. *Nature* **2012**, *482*, 405–409. [CrossRef] [PubMed]

106. Segal, N.H.; Parsons, D.W.; Peggs, K.S.; Velculescu, V.; Kinzler, K.W.; Vogelstein, B.; Allison, J.P. Epitope landscape in breast and colorectal cancer. *Cancer Res.* **2008**, *68*, 889–892. [CrossRef] [PubMed]

107. Wang, G.; Kang, X.; Chen, K.S.; Jehng, T.; Jones, L.; Chen, J.; Huang, X.F.; Chen, S.-Y. An engineered oncolytic virus expressing PD-L1 inhibitors activates tumor neoantigen-specific T cell responses. *Nat. Commun.* **2020**, *11*, 1–14. [CrossRef]

108. Kanerva, A.; Nokisalmi, P.; Diaconu, I.; Koski, A.; Cerullo, V.; Liikanen, I.; Tähtinen, S.; Oksanen, M.; Heiskanen, R.; Pesonen, S. Antiviral and antitumor T-cell immunity in patients treated with GM-CSF–coding oncolytic adenovirus. *Clin. Cancer Res.* **2013**, *19*, 2734–2744. [CrossRef] [PubMed]

109. Woller, N.; Gürlevik, E.; Fleischmann-Mundt, B.; Schumacher, A.; Knocke, S.; Kloos, A.M.; Saborowski, M.; Geffers, R.; Manns, M.P.; Wirth, T.C. Viral infection of tumors overcomes resistance to PD-1-immunotherapy by broadening neoantigenome-directed T-cell responses. *Mol. Ther.* **2015**, *23*, 1630–1640. [CrossRef]

110. Wang, X.; Shao, X.; Gu, L.; Jiang, K.; Wang, S.; Chen, J.; Fang, J.; Guo, X.; Yuan, M.; Shi, J. Targeting STAT3 enhances NDV-induced immunogenic cell death in prostate cancer cells. *J. Cell. Mol. Med.* **2020**, *24*, 4286–4297. [CrossRef]

111. Ma, J.; Ramachandran, M.; Jin, C.; Quijano-Rubio, C.; Martikainen, M.; Yu, D.; Essand, M. Characterization of virus-mediated immunogenic cancer cell death and the consequences for oncolytic virus-based immunotherapy of cancer. *Cell Death Dis.* **2020**, *11*, 1–15. [CrossRef]

112. van Vloten, J.P.; Workenhe, S.T.; Wootton, S.K.; Mossman, K.L.; Bridle, B.W. Critical interactions between immunogenic cancer cell death, oncolytic viruses, and the immune system define the rational design of combination immunotherapies. *J. Immunol.* **2018**, *200*, 450–458. [CrossRef]

113. Guo, Z.S.; Liu, Z.; Bartlett, D.L. Oncolytic immunotherapy: Dying the right way is a key to eliciting potent antitumor immunity. *Front. Oncol.* **2014**, *4*, 74. [CrossRef]

114. Hajifathali, A.; Parkhideh, S.; Kazemi, M.H.; Chegeni, R.; Roshandel, E.; Gholizadeh, M. Immune checkpoints in hematologic malignancies: What made the immune cells and clinicians exhausted! *J. Cell. Physiol.* **2020**, *235*, 9080–9097. [CrossRef]

115. Kepp, O.; Senovilla, L.; Vitale, I.; Vacchelli, E.; Adjemian, S.; Agostinis, P.; Apetoh, L.; Aranda, F.; Barnaba, V.; Bloy, N. Consensus guidelines for the detection of immunogenic cell death. *Oncoimmunology* **2014**, *3*, e955691. [CrossRef] [PubMed]

116. Lin, H.D.; Fong, C.-Y.; Biswas, A.; Bongso, A. Hypoxic Wharton's Jelly Stem Cell Conditioned Medium Induces Immunogenic Cell Death in Lymphoma Cells. *Stem Cells Int.* **2020**, *20*, 1–14. [CrossRef] [PubMed]

117. Burke, S.; Shergold, A.; Elder, M.J.; Whitworth, J.; Cheng, X.; Jin, H.; Wilkinson, R.W.; Harper, J.; Carroll, D.K. Oncolytic Newcastle disease virus activation of the innate immune response and priming of antitumor adaptive responses in vitro. *Cancer Immunol. Immunother.* **2020**, *69*, 1015–1027. [CrossRef] [PubMed]

118. Ghasemi, K.; Parkhideh, S.; Kazemi, M.H.; Salimi, M.; Salari, S.; Nalini, R.; Hajifathali, A. The role of serum uric acid in the prediction of graft-versus-host disease in allogeneic hematopoietic stem cell transplantation. *J. Clin. Lab. Anal.* **2020**, *34*, e23271. [CrossRef] [PubMed]

119. Schuster, P.; Lindner, G.; Thomann, S.; Haferkamp, S.; Schmidt, B. Prospect of plasmacytoid dendritic cells in enhancing anti-tumor immunity of oncolytic herpes viruses. *Cancers* **2019**, *11*, 651. [CrossRef] [PubMed]

120. Dai, P.; Wang, W.; Yang, N.; Serna-Tamayo, C.; Ricca, J.M.; Zamarin, D.; Shuman, S.; Merghoub, T.; Wolchok, J.D.; Deng, L. Intratumoral delivery of inactivated modified vaccinia virus Ankara (iMVA) induces systemic antitumor immunity via STING and Batf3-dependent dendritic cells. *Sci. Immunol.* **2017**, *2*, 1713–1725. [CrossRef] [PubMed]

121. Woo, S.-R.; Fuertes, M.B.; Corrales, L.; Spranger, S.; Furdyna, M.J.; Leung, M.Y.K.; Duggan, R.; Wang, Y.; Barber, G.N.; Fitzgerald, K.A. STING-dependent cytosolic DNA sensing mediates innate immune recognition of immunogenic tumors. *Immunity* **2014**, *41*, 830–842. [CrossRef]

122. Bhat, H.; Zaun, G.; Hamdan, T.A.; Lang, J.; Adomati, T.; Schmitz, R.; Friedrich, S.-K.; Bergerhausen, M.; Cham, L.B.; Li, F. Arenavirus Induced CCL5 Expression Causes NK Cell-Mediated Melanoma Regression. *Front. Immunol.* **2020**, *11*, 1849. [CrossRef]

123. Zamarin, D.; Ricca, J.M.; Sadekova, S.; Oseledchyk, A.; Yu, Y.; Blumenschein, W.M.; Wong, J.; Gigoux, M.; Merghoub, T.; Wolchok, J.D. PD-L1 in tumor microenvironment mediates resistance to oncolytic immunotherapy. *J. Clin. Investig.* **2018**, *128*, 1413–1428. [CrossRef]

124. Gujar, S.A.; Lee, P.W.K. Oncolytic virus-mediated reversal of impaired tumor antigen presentation. *Front. Oncol.* **2014**, *4*, 77. [CrossRef] [PubMed]

125. Ilkow, C.S.; Marguerie, M.; Batenchuk, C.; Mayer, J.; Neriah, D.B.; Cousineau, S.; Falls, T.; Jennings, V.A.; Boileau, M.; Bellamy, D. Reciprocal cellular cross-talk within the tumor microenvironment promotes oncolytic virus activity. *Nat. Med.* **2015**, *21*, 530. [CrossRef] [PubMed]

126. Altomonte, J.; Marozin, S.; De Toni, E.N.; Rizzani, A.; Esposito, I.; Steiger, K.; Feuchtinger, A.; Hellerbrand, C.; Schmid, R.M.; Ebert, O. Antifibrotic properties of transarterial oncolytic VSV therapy for hepatocellular carcinoma in rats with thioacetamide-induced liver fibrosis. *Mol. Ther.* **2013**, *21*, 2032–2042. [CrossRef] [PubMed]

127. Arulanandam, R.; Batenchuk, C.; Angarita, F.A.; Ottolino-Perry, K.; Cousineau, S.; Mottashed, A.; Burgess, E.; Falls, T.J.; De Silva, N.; Tsang, J. VEGF-mediated induction of PRD1-BF1/Blimp1 expression sensitizes tumor vasculature to oncolytic virus infection. *Cancer Cell* **2015**, *28*, 210–224. [CrossRef]

128. Alessandrini, F.; Menotti, L.; Avitabile, E.; Appolloni, I.; Ceresa, D.; Marubbi, D.; Campadelli-Fiume, G.; Malatesta, P. Eradication of glioblastoma by immuno-virotherapy with a retargeted oncolytic HSV in a preclinical model. *Oncogene* **2019**, *38*, 4467–4479. [CrossRef]

129. Shibata, T.; Uchida, H.; Shiroyama, T.; Okubo, Y.; Suzuki, T.; Ikeda, H.; Yamaguchi, M.; Miyagawa, Y.; Fukuhara, T.; Cohen, J.B. Development of an oncolytic HSV vector fully retargeted specifically to cellular EpCAM for virus entry and cell-to-cell spread. *Gene Ther.* **2016**, *23*, 479–488. [CrossRef]

130. Uchida, H.; Marzulli, M.; Nakano, K.; Goins, W.F.; Chan, J.; Hong, C.-S.; Mazzacurati, L.; Yoo, J.Y.; Haseley, A.; Nakashima, H. Effective treatment of an orthotopic xenograft model of human glioblastoma using an EGFR-retargeted oncolytic herpes simplex virus. *Mol. Ther.* **2013**, *21*, 561–569. [CrossRef]

131. Foreman, P.M.; Friedman, G.K.; Cassady, K.A.; Markert, J.M. Oncolytic virotherapy for the treatment of malignant glioma. *Neurotherapeutics* **2017**, *14*, 333–344. [CrossRef]

132. Stepanenko, A.A.; Chekhonin, V.P. Tropism and transduction of oncolytic adenovirus 5 vectors in cancer therapy: Focus on fiber chimerism and mosaicism, hexon and pIX. *Virus Res.* **2018**, *257*, 40–51. [CrossRef]

133. Betancourt, D.; Ramos, J.C.; Barber, G.N. Retargeting oncolytic vesicular stomatitis virus to human T-cell lymphotropic virus type 1-associated adult T-cell leukemia. *J. Virol.* **2015**, *89*, 11786–11800. [CrossRef]

134. Garant, K.A.; Shmulevitz, M.; Pan, L.; Daigle, R.M.; Ahn, D.G.; Gujar, S.A.; Lee, P.W.K. Oncolytic reovirus induces intracellular redistribution of Ras to promote apoptosis and progeny virus release. *Oncogene* **2016**, *35*, 771–782. [CrossRef] [PubMed]

135. Lu, Y.; Zhang, Y.; Chang, G.; Zhang, J. Comparison of prostate-specific promoters and the use of PSP-driven virotherapy for prostate cancer. *BioMed Res. Int.* **2013**, *13*, 1–15. [CrossRef]

136. Zhang, W.; Ge, K.; Zhao, Q.; Zhuang, X.; Deng, Z.; Liu, L.; Li, J.; Zhang, Y.; Dong, Y.; Zhang, Y. A novel oHSV-1 targeting telomerase reverse transcriptase-positive cancer cells via tumor-specific promoters regulating the expression of ICP4. *Oncotarget* **2015**, *6*, 20345. [CrossRef] [PubMed]

137. Leber, M.F.; Baertsch, M.-A.; Anker, S.C.; Henkel, L.; Singh, H.M.; Bossow, S.; Engeland, C.E.; Barkley, R.; Hoyler, B.; Albert, J. Enhanced control of oncolytic measles virus using microrna target sites. *Mol. Ther. Oncolytics* **2018**, *9*, 30–40. [CrossRef]

138. Liu, W.; Dai, E.; Liu, Z.; Ma, C.; Guo, Z.S.; Bartlett, D.L. In Situ Therapeutic Cancer Vaccination with an Oncolytic Virus Expressing Membrane-Tethered IL-2. *Mol. Ther. Oncolytics* **2020**, *17*, 350–360. [CrossRef] [PubMed]

139. Ardakani, M.T.; Mehrpooya, M.; Mehdizadeh, M.; Beiraghi, N.; Hajifathali, A.; Kazemi, M.H. Sertraline treatment decreased the serum levels of interleukin-6 and high-sensitivity C-reactive protein in hematopoietic stem cell transplantation patients with depression; a randomized double-blind, placebo-controlled clinical trial. *Bone Marrow Transpl.* **2020**, *55*, 830–832. [CrossRef]

140. Hock, K.; Laengle, J.; Kuznetsova, I.; Egorov, A.; Hegedus, B.; Dome, B.; Wekerle, T.; Sachet, M.; Bergmann, M. Oncolytic influenza A virus expressing interleukin-15 decreases tumor growth in vivo. *Surgery* **2017**, *161*, 735–746. [CrossRef]

141. Patel, D.M.; Foreman, P.M.; Nabors, L.B.; Riley, K.O.; Gillespie, G.Y.; Markert, J.M. Design of a phase I clinical trial to evaluate M032, a genetically engineered HSV-1 expressing IL-12, in patients with recurrent/progressive glioblastoma multiforme, anaplastic astrocytoma, or gliosarcoma. *Hum. Gene Ther. Clin. Dev.* **2016**, *27*, 69–78. [CrossRef]

142. Pearl, T.M.; Markert, J.M.; Cassady, K.A.; Ghonime, M.G. Oncolytic virus-based cytokine expression to improve immune activity in brain and solid tumors. *Mol. Ther. Oncolytics* **2019**, *13*, 14–21. [CrossRef]

143. Li, J.; O'Malley, M.; Sampath, P.; Kalinski, P.; Bartlett, D.L.; Thorne, S.H. Expression of CCL19 from oncolytic vaccinia enhances immunotherapeutic potential while maintaining oncolytic activity. *Neoplasia* **2012**, *14*, 1115–1121. [CrossRef]

144. Li, Y.; Li, Y.-F.; Si, C.-Z.; Zhu, Y.-H.; Jin, Y.; Zhu, T.-T.; Liu, M.-Y.; Liu, G.-Y. CCL21/IL21-armed oncolytic adenovirus enhances antitumor activity against TERT-positive tumor cells. *Virus Res.* **2016**, *220*, 172–178. [CrossRef] [PubMed]

145. Ye, J.-F.; Qi, W.-X.; Liu, M.-Y.; Li, Y. The combination of NK and CD8+ T cells with CCL20/IL15-armed oncolytic adenoviruses enhances the growth suppression of TERT-positive tumor cells. *Cell. Immunol.* **2017**, *318*, 35–41. [CrossRef] [PubMed]

146. Kazemi, M.H.; Malakootikhah, F.; Momeni-Varposhti, Z.; Falak, R.; Delbandi, A.-A.; Tajik, N. Human platelet antigen 1-6, 9 and 15 in the Iranian population: An anthropological genetic analysis. *Sci. Rep.* **2020**, *10*, 1–10. [CrossRef]

147. Hou, W.; Sampath, P.; Rojas, J.J.; Thorne, S.H. Oncolytic virus-mediated targeting of PGE2 in the tumor alters the immune status and sensitizes established and resistant tumors to immunotherapy. *Cancer Cell* **2016**, *30*, 108–119. [CrossRef] [PubMed]

148. Gil, M.; Komorowski, M.P.; Seshadri, M.; Rokita, H.; McGray, A.J.R.; Opyrchal, M.; Odunsi, K.O.; Kozbor, D. CXCL12/CXCR4 blockade by oncolytic virotherapy inhibits ovarian cancer growth by decreasing immunosuppression and targeting cancer-initiating cells. *J. Immunol.* **2014**, *193*, 5327–5337. [CrossRef]

149. Cui, C.-X.; Li, Y.-Q.; Sun, Y.-J.; Zhu, Y.-L.; Fang, J.-B.; Bai, B.; Li, W.-J.; Li, S.-Z.; Ma, Y.-Z.; Li, X. Antitumor effect of a dual cancer-specific oncolytic adenovirus on prostate cancer PC-3 cells. *Urol. Oncol. Semin. Orig.* **2019**, *37*, 351–352. [CrossRef]

150. Jeong, S.-N.; Yoo, S.Y. Novel Oncolytic Virus Armed with Cancer Suicide Gene and Normal Vasculogenic Gene for Improved Anti-Tumor Activity. *Cancers* **2020**, *12*, 1070. [CrossRef]

151. Hajeri, P.B.; Sharma, N.S.; Yamamoto, M. Oncolytic Adenoviruses: Strategies for Improved Targeting and Specificity. *Cancers* **2020**, *12*, 1504. [CrossRef]

152. Ricordel, M.; Foloppe, J.; Antoine, D.; Findeli, A.; Kempf, J.; Cordier, P.; Gerbaud, A.; Grellier, B.; Lusky, M.; Quemeneur, E. Vaccinia virus shuffling: deVV5, a novel chimeric poxvirus with improved oncolytic potency. *Cancers* **2018**, *10*, 231. [CrossRef]

153. Li, L.; Liu, S.; Han, D.; Tang, B.; Ma, J. Delivery and Biosafety of Oncolytic Virotherapy. *Front. Oncol.* **2020**, *10*, 475. [CrossRef]

154. Cronin, M.; Le Boeuf, F.; Murphy, C.; Roy, D.G.; Falls, T.; Bell, J.C.; Tangney, M. Bacterial-mediated knockdown of tumor resistance to an oncolytic virus enhances therapy. *Mol. Ther.* **2014**, *22*, 1188–1197. [CrossRef] [PubMed]

155. Chen, H.; Sampath, P.; Hou, W.; Thorne, S.H. Regulating cytokine function enhances safety and activity of genetic cancer therapies. *Mol. Ther.* **2013**, *21*, 167–174. [CrossRef] [PubMed]

156. Draganov, D.D.; Santidrian, A.F.; Minev, I.; Nguyen, D.; Kilinc, M.O.; Petrov, I.; Vyalkova, A.; Lander, E.; Berman, M.; Minev, B. Delivery of oncolytic vaccinia virus by matched allogeneic stem cells overcomes critical innate and adaptive immune barriers. *J. Transl. Med.* **2019**, *17*, 1–22. [CrossRef] [PubMed]

157. Hadryś, A.; Sochanik, A.; McFadden, G.; Jazowiecka-Rakus, J. Mesenchymal stem cells as carriers for systemic delivery of oncolytic viruses. *Eur. J. Pharmacol.* **2020**, *874*, 172991. [CrossRef]

158. Josiah, D.T.; Zhu, D.; Dreher, F.; Olson, J.; McFadden, G.; Caldas, H. Adipose-derived stem cells as therapeutic delivery vehicles of an oncolytic virus for glioblastoma. *Mol. Ther.* **2010**, *18*, 377–385. [CrossRef]

159. Hammad, M.; Cornejo, Y.R.; Batalla-Covello, J.; Majid, A.A.; Burke, C.; Liu, Z.; Yuan, Y.-C.; Li, M.; Dellinger, T.H.; Lu, J. Neural Stem Cells Improve the Delivery of Oncolytic Chimeric Orthopoxvirus in a Metastatic Ovarian Cancer Model. *Mol. Ther. Oncolytics* **2020**, *18*, 326–334. [CrossRef]

160. Adair, R.A.; Roulstone, V.; Scott, K.J.; Morgan, R.; Nuovo, G.J.; Fuller, M.; Beirne, D.; West, E.J.; Jennings, V.A.; Rose, A. Cell carriage, delivery, and selective replication of an oncolytic virus in tumor in patients. *Sci. Transl. Med.* **2012**, *4*, 138ra77. [CrossRef]

161. Lilly, C.L.; Villa, N.Y.; de Matos, A.L.; Ali, H.M.; Dhillon, J.-K.S.; Hofland, T.; Rahman, M.M.; Chan, W.; Bogen, B.; Cogle, C. Ex vivo oncolytic virotherapy with myxoma virus arms multiple allogeneic bone marrow transplant leukocytes to enhance graft versus tumor. *Mol. Ther. Oncolytics* **2017**, *4*, 31–40. [CrossRef]

162. Iankov, I.D.; Msaouel, P.; Allen, C.; Federspiel, M.J.; Bulur, P.A.; Dietz, A.B.; Gastineau, D.; Ikeda, Y.; Ingle, J.N.; Russell, S.J. Demonstration of anti-tumor activity of oncolytic measles virus strains in a malignant pleural effusion breast cancer model. *Breast Cancer Res. Treat.* **2010**, *122*, 745–754. [CrossRef]

163. Ilett, E.J.; Prestwich, R.J.; Kottke, T.; Errington, F.; Thompson, J.M.; Harrington, K.J.; Pandha, H.S.; Coffey, M.; Selby, P.J.; Vile, R.G. Dendritic cells and T cells deliver oncolytic reovirus for tumour killing despite pre-existing anti-viral immunity. *Gene Ther.* **2009**, *16*, 689–699. [CrossRef]

164. Balvers, R.K.; Belcaid, Z.; Van den Hengel, S.K.; Kloezeman, J.; De Vrij, J.; Wakimoto, H.; Hoeben, R.C.; Debets, R.; Leenstra, S.; Dirven, C. Locally-delivered T-cell-derived cellular vehicles efficiently track and deliver adenovirus delta24-RGD to infiltrating glioma. *Viruses* **2014**, *6*, 3080–3096. [CrossRef] [PubMed]

165. Kanzaki, A.; Kasuya, H.; Yamamura, K.; Sahin, T.T.; Nomura, N.; Shikano, T.; Shirota, T.; Tan, G.; Fukuda, S.; Misawa, M. Antitumor efficacy of oncolytic herpes simplex virus adsorbed onto antigen-specific lymphocytes. *Cancer Gene Ther.* **2012**, *19*, 292–298. [CrossRef]

166. Muthana, M.; Giannoudis, A.; Scott, S.D.; Fang, H.-Y.; Coffelt, S.B.; Morrow, F.J.; Murdoch, C.; Burton, J.; Cross, N.; Burke, B. Use of macrophages to target therapeutic adenovirus to human prostate tumors. *Cancer Res.* **2011**, *71*, 1805–1815. [CrossRef] [PubMed]

167. Muthana, M.; Rodrigues, S.; Chen, Y.-Y.; Welford, A.; Hughes, R.; Tazzyman, S.; Essand, M.; Morrow, F.; Lewis, C.E. Macrophage delivery of an oncolytic virus abolishes tumor regrowth and metastasis after chemotherapy or irradiation. *Cancer Res.* **2013**, *73*, 490–495. [CrossRef] [PubMed]

168. Gao, J.; Zhang, W.; Mese, K.; Bunz, O.; Lu, F.; Ehrhardt, A. Transient chimeric Ad5/37 fiber enhances NK-92 carrier cell mediated delivery of oncolytic adenovirus type 5 to tumor cells. *Mol. Ther. Methods Clin. Dev.* **2020**, *18*, 376–389. [CrossRef] [PubMed]

169. Liu, C.; Suksanpaisan, L.; Chen, Y.-W.; Russell, S.J.; Peng, K.-W. Enhancing cytokine-induced killer cell therapy of multiple myeloma. *Exp. Hematol.* **2013**, *41*, 508–517. [CrossRef]

170. Adair, R.A.; Scott, K.J.; Fraser, S.; Errington-Mais, F.; Pandha, H.; Coffey, M.; Selby, P.; Cook, G.P.; Vile, R.; Harrington, K.J. Cytotoxic and immune-mediated killing of human colorectal cancer by reovirus-loaded blood and liver mononuclear cells. *Int. J. Cancer* **2013**, *132*, 2327–2338. [CrossRef] [PubMed]

171. Villa, N.Y.; Rahman, M.M.; Mamola, J.; D'Isabella, J.; Goras, E.; Kilbourne, J.; Lowe, K.; Daggett-Vondras, J.; Torres, L.; Christie, J. Autologous Transplantation Using Donor Leukocytes Loaded Ex Vivo with Oncolytic Myxoma Virus Can Eliminate Residual Multiple Myeloma. *Mol. Ther. Oncolytics* **2020**, *18*, 171–188. [CrossRef]

172. Eisenstein, S.; Coakley, B.A.; Briley-Saebo, K.; Ma, G.; Chen, H.-M.; Meseck, M.; Ward, S.; Divino, C.; Woo, S.; Chen, S.-H. Myeloid-derived suppressor cells as a vehicle for tumor-specific oncolytic viral therapy. *Cancer Res.* **2013**, *73*, 5003–5015. [CrossRef]

173. Marelli, G.; Howells, A.; Lemoine, N.R.; Wang, Y. Oncolytic viral therapy and the immune system: A double-edged sword against cancer. *Front. Immunol.* **2018**, *9*, 866. [CrossRef]

174. Liu, X.; Li, Y.-P.; Zhong, Z.-M.; Tan, H.-Q.; Lin, H.-P.; Chen, S.-J.; Fu, Y.-C.; Xu, W.-C.; Wei, C.-J. Incorporation of viral glycoprotein VSV-G improves the delivery of DNA by erythrocyte ghost into cells refractory to conventional transfection. *Appl. Biochem. Biotechnol.* **2017**, *181*, 748–761. [CrossRef] [PubMed]

175. Sun, C.W.; Willmon, C.; Wu, L.-C.; Knopick, P.; Thoerner, J.; Vile, R.; Townes, T.M.; Terman, D.S. sickle cells abolish Melanoma Tumorigenesis in hemoglobin ss Knockin Mice and augment the Tumoricidal effect of Oncolytic Virus In Vivo. *Front. Oncol.* **2016**, *6*, 166. [CrossRef] [PubMed]

176. Guo, Z.S.; Parimi, V.; O'Malley, M.E.; Thirunavukarasu, P.; Sathaiah, M.; Austin, F.; Bartlett, D.L. The combination of immunosuppression and carrier cells significantly enhances the efficacy of oncolytic poxvirus in the pre-immunized host. *Gene Ther.* **2010**, *17*, 1465–1475. [CrossRef]

177. Iankov, I.D.; Blechacz, B.; Liu, C.; Schmeckpeper, J.D.; Tarara, J.E.; Federspiel, M.J.; Caplice, N.; Russell, S.J. Infected cell carriers: A new strategy for systemic delivery of oncolytic measles viruses in cancer virotherapy. *Mol. Ther.* **2007**, *15*, 114–122. [CrossRef] [PubMed]

178. Yokoda, R.; Nagalo, B.M.; Vernon, B.; Oklu, R.; Albadawi, H.; DeLeon, T.T.; Zhou, Y.; Egan, J.B.; Duda, D.G.; Borad, M.J. Oncolytic virus delivery: From nano-pharmacodynamics to enhanced oncolytic effect. *Oncolytic Virother.* **2017**, *6*, 39. [CrossRef]

179. Patel, M.R.; Jacobson, B.A.; Ji, Y.; Hebbel, R.P.; Kratzke, R.A. Blood Outgrowth Endothelial Cells as a Cellular Carrier for Oncolytic Vesicular Stomatitis Virus Expressing Interferon-β in Preclinical Models of Non-Small Cell Lung Cancer. *Transl. Oncol.* **2020**, *13*, 100782. [CrossRef]

180. Garofalo, M.; Villa, A.; Rizzi, N.; Kuryk, L.; Rinner, B.; Cerullo, V.; Yliperttula, M.; Mazzaferro, V.; Ciana, P. Extracellular vesicles enhance the targeted delivery of immunogenic oncolytic adenovirus and paclitaxel in immunocompetent mice. *J. Control. Release* **2019**, *294*, 165–175. [CrossRef]

181. Saari, H.; Turunen, T.; Lõhmus, A.; Turunen, M.; Jalasvuori, M.; Butcher, S.J.; Ylä-Herttuala, S.; Viitala, T.; Cerullo, V.; Siljander, P.R.M. Extracellular vesicles provide a capsid-free vector for oncolytic adenoviral DNA delivery. *J. Extracell. Vesicles* **2020**, *9*, 1747206. [CrossRef]

182. Fusciello, M.; Fontana, F.; Tähtinen, S.; Capasso, C.; Feola, S.; Martins, B.; Chiaro, J.; Peltonen, K.; Ylösmäki, L.; Ylösmäki, E. Artificially cloaked viral nanovaccine for cancer immunotherapy. *Nat. Commun.* **2019**, *10*, 1–13. [CrossRef]

183. Yang, Y.; Du, T.; Zhang, J.; Kang, T.; Luo, L.; Tao, J.; Gou, Z.; Chen, S.; Du, Y.; He, J. A 3D-Engineered Conformal Implant Releases DNA Nanocomplexs for Eradicating the Postsurgery Residual Glioblastoma. *Adv. Sci.* **2017**, *4*, 1600491. [CrossRef]

184. Kong, H.; Zhao, R.; Zhang, Q.; Iqbal, M.Z.; Lu, J.; Zhao, Q.; Luo, D.; Feng, C.; Zhang, K.; Liu, X. Biosilicified Oncolytic Adenovirus for Cancer Viral Gene Therapy. *Biomater. Sci.* **2020**, *8*, 5317–5328. [CrossRef] [PubMed]

185. Kiyokawa, J.; Wakimoto, H. Preclinical and clinical development of oncolytic adenovirus for the treatment of malignant glioma. *Oncolytic Virother.* **2019**, *8*, 27. [CrossRef] [PubMed]

186. Park, J.W.; Mok, H.; Park, T.G. Physical adsorption of PEG grafted and blocked poly-L-lysine copolymers on adenovirus surface for enhanced gene transduction. *J. Control. Release* **2010**, *142*, 238–244. [CrossRef] [PubMed]

187. Badrinath, N.; Jeong, Y.I.; Woo, H.Y.; Bang, S.Y.; Kim, C.; Heo, J.; Kang, D.H.; Yoo, S.Y. Local delivery of a cancer-favoring oncolytic vaccinia virus via poly (lactic-co-glycolic acid) nanofiber for theranostic purposes. *Int. J. Pharm.* **2018**, *552*, 437–442. [CrossRef]

188. Nosaki, K.; Hamada, K.; Takashima, Y.; Sagara, M.; Matsumura, Y.; Miyamoto, S.; Hijikata, Y.; Okazaki, T.; Nakanishi, Y.; Tani, K. A novel, polymer-coated oncolytic measles virus overcomes immune suppression and induces robust antitumor activity. *Mol. Ther. Oncolytics* **2016**, *3*, 16022. [CrossRef]

189. Mok, H.; Park, J.W.; Park, T.G. Microencapsulation of PEGylated adenovirus within PLGA microspheres for enhanced stability and gene transfection efficiency. *Pharm. Res.* **2007**, *24*, 2263–2269. [CrossRef]

190. Choi, J.-W.; Kim, J.; Bui, Q.N.; Li, Y.; Yun, C.-O.; Lee, D.S.; Kim, S.W. Tuning surface charge and PEGylation of biocompatible polymers for efficient delivery of nucleic acid or adenoviral vector. *Bioconjug. Chem.* **2015**, *26*, 1818–1829. [CrossRef]

191. Kim, J.; Kim, P.-H.; Kim, S.W.; Yun, C.-O. Enhancing the therapeutic efficacy of adenovirus in combination with biomaterials. *Biomaterials* **2012**, *33*, 1838–1850. [CrossRef]

192. Choi, J.-W.; Nam, J.-P.; Nam, K.; Lee, Y.S.; Yun, C.-O.; Kim, S.W. Oncolytic adenovirus coated with multidegradable bioreducible core-cross-linked polyethylenimine for cancer gene therapy. *Biomacromolecules* **2015**, *16*, 2132–2143. [CrossRef]

193. Oh, I.K.; Mok, H.; Park, T.G. Folate immobilized and PEGylated adenovirus for retargeting to tumor cells. *Bioconjugate Chem.* **2006**, *17*, 721–727. [CrossRef]

194. Green, N.K.; Herbert, C.W.; Hale, S.J.; Hale, A.B.; Mautner, V.; Harkins, R.; Hermiston, T.; Ulbrich, K.; Fisher, K.D.; Seymour, L.W. Extended plasma circulation time and decreased toxicity of polymer-coated adenovirus. *Gene Ther.* **2004**, *11*, 1256–1263. [CrossRef] [PubMed]

195. Espenlaub, S.; Wortmann, A.; Engler, T.; Corjon, S.; Kochanek, S.; Kreppel, F. Reductive amination as a strategy to reduce adenovirus vector promiscuity by chemical capsid modification with large polysaccharides. *J. Gene Med. Cross Discip. J. Res. Sci. Gene Transf. Clin. Appl.* **2008**, *10*, 1303–1314. [CrossRef] [PubMed]

196. Price, D.L.; Li, P.; Chen, C.H.; Wong, D.; Yu, Z.; Chen, N.G.; Yu, Y.A.; Szalay, A.A.; Cappello, J.; Fong, Y. Silk-elastin-like protein polymer matrix for intraoperative delivery of an oncolytic vaccinia virus. *Head Neck* **2016**, *38*, 237–246. [CrossRef] [PubMed]

197. Robles-Planells, C.; Sánchez-Guerrero, G.; Barrera-Avalos, C.; Matiacevich, S.; Rojo, L.E.; Pavez, J.; Salas-Huenuleo, E.; Kogan, M.J.; Escobar, A.; Milla, L.A. Chitosan-based nanoparticles for intracellular delivery of ISAV fusion protein cDNA into melanoma cells: A path to develop oncolytic anticancer therapies. *Mediat. Inflamm.* **2020**, *20*, 1–13. [CrossRef] [PubMed]

198. Breen, A.; Strappe, P.; Kumar, A.; O'Brien, T.; Pandit, A. Optimization of a fibrin scaffold for sustained release of an adenoviral gene vector. *J. Biomed. Mater. Res. Part A Off. J. Soc. Biomater. Jpn. Soc. Biomater. Aust. Soc. Biomater. Korean Soc. Biomater.* **2006**, *78*, 702–708. [CrossRef]

199. Yoon, A.-R.; Kasala, D.; Li, Y.; Hong, J.; Lee, W.; Jung, S.-J.; Yun, C.-O. Antitumor effect and safety profile of systemically delivered oncolytic adenovirus complexed with EGFR-targeted PAMAM-based dendrimer in orthotopic lung tumor model. *J. Control. Release* **2016**, *231*, 2–16. [CrossRef]

200. Oh, E.; Oh, J.-E.; Hong, J.; Chung, Y.; Lee, Y.; Park, K.D.; Kim, S.; Yun, C.-O. Optimized biodegradable polymeric reservoir-mediated local and sustained co-delivery of dendritic cells and oncolytic adenovirus co-expressing IL-12 and GM-CSF for cancer immunotherapy. *J. Control. Release* **2017**, *259*, 115–127. [CrossRef]

201. Liao, I.-C.; Chen, S.; Liu, J.B.; Leong, K.W. Sustained viral gene delivery through core-shell fibers. *J. Control. Release* **2009**, *139*, 48–55. [CrossRef]

202. Denton, N.L.; Chen, C.-Y.; Hutzen, B.; Currier, M.A.; Scott, T.; Nartker, B.; Leddon, J.L.; Wang, P.-Y.; Srinivas, R.; Cassady, K.A. Myelolytic treatments enhance oncolytic herpes virotherapy in models of Ewing sarcoma by modulating the immune microenvironment. *Mol. Ther. Oncolytics* **2018**, *11*, 62–74. [CrossRef]

203. Sakurai, F.; Inoue, S.; Kaminade, T.; Hotani, T.; Katayama, Y.; Hosoyamada, E.; Terasawa, Y.; Tachibana, M.; Mizuguchi, H. Cationic liposome-mediated delivery of reovirus enhances the tumor cell-killing efficiencies of reovirus in reovirus-resistant tumor cells. *Int. J. Pharm.* **2017**, *524*, 238–247. [CrossRef]

204. Zhong, Z.; Shi, S.; Han, J.; Zhang, Z.; Sun, X. Anionic liposomes increase the efficiency of adenovirus-mediated gene transfer to coxsackie-adenovirus receptor deficient cells. *Mol. Pharm.* **2010**, *7*, 105–115. [CrossRef] [PubMed]

205. Kasala, D.; Lee, S.-H.; Hong, J.W.; Choi, J.-W.; Nam, K.; Chung, Y.H.; Kim, S.W.; Yun, C.-O. Synergistic antitumor effect mediated by a paclitaxel-conjugated polymeric micelle-coated oncolytic adenovirus. *Biomaterials* **2017**, *145*, 207–222. [CrossRef] [PubMed]

206. Almstätter, I.; Mykhaylyk, O.; Settles, M.; Altomonte, J.; Aichler, M.; Walch, A.; Rummeny, E.J.; Ebert, O.; Plank, C.; Braren, R. Characterization of magnetic viral complexes for targeted delivery in oncology. *Theranostics* **2015**, *5*, 667. [CrossRef] [PubMed]

207. Muthana, M.; Kennerley, A.J.; Hughes, R.; Fagnano, E.; Richardson, J.; Paul, M.; Murdoch, C.; Wright, F.; Payne, C.; Lythgoe, M.F. Directing cell therapy to anatomic target sites in vivo with magnetic resonance targeting. *Nat. Commun.* **2015**, *6*, 1–11. [CrossRef] [PubMed]

208. Roy, D.G.; Bell, J.C.; Bourgeois-Daigneault, M.-C. Magnetic targeting of oncolytic VSV-based therapies improves infection of tumor cells in the presence of virus-specific neutralizing antibodies in vitro. *Biochem. Biophys. Res. Commun.* **2020**, *526*, 641–646. [CrossRef] [PubMed]

209. Sendra, L.; Miguel, A.; Navarro-Plaza, M.C.; Herrero, M.J.; de la Higuera, J.; Cháfer-Pericás, C.; Aznar, E.; Marcos, M.D.; Martínez-Máñez, R.; Rojas, L.A. Gold Nanoparticle-Assisted Virus Formation by Means of the Delivery of an Oncolytic Adenovirus Genome. *Nanomaterials* **2020**, *10*, 1183. [CrossRef]

210. Bazan-Peregrino, M.; Arvanitis, C.D.; Rifai, B.; Seymour, L.W.; Coussios, C.-C. Ultrasound-induced cavitation enhances the delivery and therapeutic efficacy of an oncolytic virus in an in vitro model. *J. Control. Release* **2012**, *157*, 235–242. [CrossRef]

211. He, X.; Yang, Y.; Li, L.; Zhang, P.; Guo, H.; Liu, N.; Yang, X.; Xu, F. Engineering extracellular matrix to improve drug delivery for cancer therapy. *Drug Discov. Today* **2020**, *25*, 1727–1734. [CrossRef]

212. Nande, R.; Howard, C.M.; Claudio, P.P. Ultrasound-mediated oncolytic virus delivery and uptake for increased therapeutic efficacy: State of art. *Oncolytic Virother.* **2015**, *4*, 193.

213. Raeesi, V.; Chan, W.C.W. Improving nanoparticle diffusion through tumor collagen matrix by photo-thermal gold nanorods. *Nanoscale* **2016**, *8*, 12524–12530. [CrossRef]

214. Hagihara, Y.; Sakamoto, A.; Tokuda, T.; Yamashita, T.; Ikemoto, S.; Kimura, A.; Haruta, M.; Sasagawa, K.; Ohta, J.; Takayama, K. Photoactivatable oncolytic adenovirus for optogenetic cancer therapy. *Cell Death Dis.* **2020**, *11*, 1–9. [CrossRef] [PubMed]

215. Ilett, E.; Kottke, T.; Donnelly, O.; Thompson, J.; Willmon, C.; Diaz, R.; Zaidi, S.; Coffey, M.; Selby, P.; Harrington, K. Cytokine conditioning enhances systemic delivery and therapy of an oncolytic virus. *Mol. Ther.* **2014**, *22*, 1851–1863. [CrossRef] [PubMed]

216. Howard, F.; Muthana, M. Designer nanocarriers for navigating the systemic delivery of oncolytic viruses. *Nanomedicine* **2020**, *15*, 93–110. [CrossRef] [PubMed]

217. Na, Y.; Choi, J.-W.; Kasala, D.; Hong, J.; Oh, E.; Li, Y.; Jung, S.-J.; Kim, S.W.; Yun, C.-O. Potent antitumor effect of neurotensin receptor-targeted oncolytic adenovirus co-expressing decorin and Wnt antagonist in an orthotopic pancreatic tumor model. *J. Control. Release* **2015**, *220*, 766–782. [CrossRef]

218. Choi, J.-W.; Kim, H.A.; Nam, K.; Na, Y.; Yun, C.-O.; Kim, S. Hepatoma targeting peptide conjugated bio-reducible polymer complexed with oncolytic adenovirus for cancer gene therapy. *J. Control. Release* **2015**, *220*, 691–703. [CrossRef] [PubMed]

219. Kirn, D.H.; Wang, Y.; Liang, W.; Contag, C.H.; Thorne, S.H. Enhancing poxvirus oncolytic effects through increased spread and immune evasion. *Cancer Res.* **2008**, *68*, 2071–2075. [CrossRef]

220. Baran, N.; Konopleva, M. Molecular pathways: Hypoxia-activated prodrugs in cancer therapy. *Clin. Cancer Res.* **2017**, *23*, 2382–2390. [CrossRef]

221. Ferguson, M.; Dunmall, L.S.C.; Gangeswaran, R.; Marelli, G.; Tysome, J.R.; Burns, E.; Whitehead, M.A.; Aksoy, E.; Alusi, G.; Hiley, C. Transient inhibition of phosphoinositide 3-kinase δ enhances the therapeutic effect of intravenous delivery of oncolytic Vaccinia virus. *Mol. Ther.* **2020**, *28*, 1263–1275. [CrossRef]

222. Evgin, L.; Acuna, S.A.; De Souza, C.T.; Marguerie, M.; Lemay, C.G.; Ilkow, C.S.; Findlay, C.S.; Falls, T.; Parato, K.A.; Hanwell, D. Complement inhibition prevents oncolytic vaccinia virus neutralization in immune humans and cynomolgus macaques. *Mol. Ther.* **2015**, *23*, 1066–1076. [CrossRef]

223. Lun, X.Q.; Jang, J.-H.; Tang, N.; Deng, H.; Head, R.; Bell, J.C.; Stojdl, D.F.; Nutt, C.L.; Senger, D.L.; Forsyth, P.A. Efficacy of systemically administered oncolytic vaccinia virotherapy for malignant gliomas is enhanced by combination therapy with rapamycin or cyclophosphamide. *Clin. Cancer Res.* **2009**, *15*, 2777–2788. [CrossRef]

224. Liu, Y.-P.; Tong, C.; Dispenzieri, A.; Federspiel, M.J.; Russell, S.J.; Peng, K.-W. Polyinosinic acid decreases sequestration and improves systemic therapy of measles virus. *Cancer Gene Ther.* **2012**, *19*, 202–211. [CrossRef] [PubMed]
225. Berkey, S.E.; Thorne, S.H.; Bartlett, D.L. Oncolytic virotherapy and the tumor microenvironment. In *Tumor Immune Microenvironment in Cancer Progression and Cancer Therapy*; Springer: Berlin, Germany, 2017; pp. 157–172.
226. Goradel, N.H.; Negahdari, B.; Ghorghanlu, S.; Jahangiri, S.; Arashkia, A. Strategies for enhancing intratumoral spread of oncolytic adenoviruses. *Pharmacol. Ther.* **2020**, *213*, 75–86. [CrossRef]
227. Ravindra, P.; Tiwari, A.K.; Ratta, B.; Chaturvedi, U.; Palia, S.K.; Chauhan, R. Newcastle disease virus-induced cytopathic effect in infected cells is caused by apoptosis. *Virus Res.* **2009**, *141*, 13–20. [CrossRef] [PubMed]
228. Duncan, R. Fusogenic reoviruses and their fusion-associated small transmembrane (FAST) proteins. *Annu. Rev. Virol.* **2019**, *6*, 341–363. [CrossRef] [PubMed]
229. Moss, B. Membrane fusion during poxvirus entry. Semin. *Cell Dev. Biol.* **2016**, *60*, 89–96.
230. Nakatake, M.; Kuwano, N.; Kaitsurumaru, E.; Kurosaki, H.; Nakamura, T. Fusogenic oncolytic vaccinia virus enhances systemic antitumor immune response by modulating the tumor microenvironment. *Mol. Ther.* **2020**, *12*, 24–38.
231. Burton, C.; Bartee, E. Syncytia formation in oncolytic virotherapy. *Mol. Ther. Oncolytics* **2019**, *15*, 131–139. [CrossRef] [PubMed]
232. Krabbe, T.; Altomonte, J. Fusogenic viruses in oncolytic immunotherapy. *Cancers* **2018**, *10*, 216. [CrossRef] [PubMed]
233. Matuszewska, K.; Santry, L.A.; Van Vloten, J.P.; AuYeung, A.W.K.; Major, P.P.; Lawler, J.; Wootton, S.K.; Bridle, B.W.; Petrik, J. Combining vascular normalization with an oncolytic virus enhances immunotherapy in a preclinical model of advanced-stage ovarian cancer. *Clin. Cancer Res.* **2019**, *25*, 1624–1638. [CrossRef]
234. Na, Y.; Nam, J.-P.; Hong, J.; Oh, E.; Shin, H.C.; Kim, H.S.; Kim, S.W.; Yun, C.-O. Systemic administration of human mesenchymal stromal cells infected with polymer-coated oncolytic adenovirus induces efficient pancreatic tumor homing and infiltration. *J. Control. Release* **2019**, *305*, 75–88. [CrossRef]
235. Jaime-Ramirez, A.C.; Dmitrieva, N.; Yoo, J.Y.; Banasavadi-Siddegowda, Y.; Zhang, J.; Relation, T.; Bolyard, C.; Wojton, J.; Kaur, B. Humanized chondroitinase ABC sensitizes glioblastoma cells to temozolomide. *J. Gene Med.* **2017**, *19*, e2942. [CrossRef] [PubMed]
236. Pascual-Pasto, G.; Bazan-Peregrino, M.; Olaciregui, N.G.; Restrepo-Perdomo, C.A.; Mato-Berciano, A.; Ottaviani, D.; Weber, K.; Correa, G.; Paco, S.; Vila-Ubach, M. Therapeutic targeting of the RB1 pathway in retinoblastoma with the oncolytic adenovirus VCN-01. *Sci. Transl. Med.* **2019**, *11*, 1–11. [CrossRef] [PubMed]
237. Schäfer, S.; Weibel, S.; Donat, U.; Zhang, Q.; Aguilar, R.J.; Chen, N.G.; Szalay, A.A. Vaccinia virus-mediated intra-tumoral expression of matrix metalloproteinase 9 enhances oncolysis of PC-3 xenograft tumors. *BMC Cancer* **2012**, *12*, 366. [CrossRef] [PubMed]
238. Goodwin, J.M.; Schmitt, A.D.; McGinn, C.M.; Fuchs, B.C.; Kuruppu, D.; Tanabe, K.K.; Lanuti, M. Angiogenesis inhibition using an oncolytic herpes simplex virus expressing endostatin in a murine lung cancer model. *Cancer Investig.* **2012**, *30*, 243–250. [CrossRef]
239. Tsuji, T.; Nakamori, M.; Iwahashi, M.; Nakamura, M.; Ojima, T.; Iida, T.; Katsuda, M.; Hayata, K.; Ino, Y.; Todo, T. An armed oncolytic herpes simplex virus expressing thrombospondin-1 has an enhanced in vivo antitumor effect against human gastric cancer. *Int. J. Cancer* **2013**, *132*, 485–494. [CrossRef]
240. Adelfinger, M.; Bessler, S.; Frentzen, A.; Cecil, A.; Langbein-Laugwitz, J.; Gentschev, I.; Szalay, A.A. Preclinical testing oncolytic vaccinia virus strain GLV-5b451 expressing an anti-VEGF single-chain antibody for canine cancer therapy. *Viruses* **2015**, *7*, 4075–4092. [CrossRef]
241. Angelova, A.L.; Geletneky, K.; Nüesch, J.P.F.; Rommelaere, J. Tumor selectivity of oncolytic parvoviruses: From in vitro and animal models to cancer patients. *Front. Bioeng. Biotechnol.* **2015**, *3*, 55. [CrossRef]
242. Yamaguchi, T.; Uchida, E. Oncolytic virus: Regulatory aspects from quality control to clinical studies. *Curr. Cancer Drug Targets* **2018**, *18*, 202–208. [CrossRef]
243. Chiocca, E.A.; Nakashima, H.; Kasai, K.; Fernandez, S.A.; Oglesbee, M. Preclinical toxicology of rQNestin34. 5v. 2, an oncolytic herpes virus with transcriptional regulation of the ICP34. 5 neurovirulence gene. *Mol. Ther. Methods Clin. Dev.* **2020**, *17*, 871–893. [CrossRef]
244. Islam, S.M.; Lee, B.; Jiang, F.; Kim, E.-K.; Ahn, S.C.; Hwang, T.-H. Engineering and Characterization of Oncolytic Vaccinia Virus Expressing Truncated Herpes Simplex Virus Thymidine Kinase. *Cancers* **2020**, *12*, 228. [CrossRef]
245. Bernstock, J.D.; Bag, A.; Fiveash, J.; Kachurak, K.; Elsayed, G.; Chagoya, G.; Gessler, F.; Valdes, P.A.; Madan-Swain, A.; Whitley, R. Design and Rationale for First-In-Human Phase 1 Immunovirotherapy Clinical Trial of Oncolytic HSV G207 to Treat Malignant Pediatric Cerebellar Brain Tumors. *Hum. Gene Ther.* **2020**, *31*, 19–20. [CrossRef] [PubMed]
246. Mori, K.M.; Giuliano, P.D.; Lopez, K.L.; King, M.M.; Bohart, R.; Goldstein, B.H. Pronounced clinical response following the oncolytic vaccinia virus GL-ONC1 and chemotherapy in a heavily pretreated ovarian cancer patient. *Anti Cancer Drugs* **2019**, *30*, 1064–1066. [CrossRef] [PubMed]
247. Tenneti, P.; Borad, M.J.; Babiker, H.M. Exploring the role of oncolytic viruses in hepatobiliary cancers. *Immunotherapy* **2018**, *10*, 971–986. [CrossRef]
248. Chen, Q.; Wu, J.; Ye, Q.; Ma, F.; Zhu, Q.; Wu, Y.; Shan, C.; Xie, X.; Li, D.; Zhan, X. Treatment of human glioblastoma with a live attenuated Zika virus vaccine candidate. *MBio* **2018**, *9*, e01683-18. [CrossRef] [PubMed]
249. Gao, J.; Zhang, W.; Ehrhardt, A. Expanding the Spectrum of Adenoviral Vectors for Cancer Therapy. *Cancers* **2020**, *12*, 1139. [CrossRef]

250. Abdullahi, S.; Jäkel, M.; Behrend, S.J.; Steiger, K.; Topping, G.; Krabbe, T.; Colombo, A.; Sandig, V.; Schiergens, T.S.; Thasler, W.E. A novel chimeric oncolytic virus vector for improved safety and efficacy as a platform for the treatment of hepatocellular carcinoma. *J. Virol.* **2018**, *92*, 1386–1398. [CrossRef]

251. Coughlan, L.; Vallath, S.; Gros, A.; Giménez-Alejandre, M.; Van Rooijen, N.; Thomas, G.J.; Baker, A.H.; Cascalló, M.; Alemany, R.; Hart, I.R. Combined fiber modifications both to target αvβ6 and detarget the coxsackievirus–adenovirus receptor improve virus toxicity profiles in vivo but fail to improve antitumoral efficacy relative to adenovirus serotype 5. *Hum. Gene Ther.* **2012**, *23*, 960–979. [CrossRef]

252. Marchini, A.; Bonifati, S.; Scott, E.M.; Angelova, A.L.; Rommelaere, J. Oncolytic parvoviruses: From basic virology to clinical applications. *Virol. J.* **2015**, *12*, 1–16. [CrossRef]

253. Zhang, X.; Zhang, T.; Davis, J.N.; Marzi, A.; Marchese, A.M.; Robek, M.D.; van den Pol, A.N. Mucin-like domain of Ebola virus glycoprotein enhances selective oncolytic actions against brain tumors. *J. Virol.* **2020**, *94*, 967–979. [CrossRef]

254. Harrington, K.J.; Michielin, O.; Malvehy, J.; Grüter, I.P.; Grove, L.; Frauchiger, A.L.; Dummer, R. A practical guide to the handling and administration of talimogene laherparepvec in Europe. *Oncotargets Ther.* **2017**, *10*, 3867. [CrossRef]

255. Selman, M.; Ou, P.; Rousso, C.; Bergeron, A.; Krishnan, R.; Pikor, L.; Chen, A.; Keller, B.A.; Ilkow, C.; Bell, J.C. Dimethyl fumarate potentiates oncolytic virotherapy through NF-κB inhibition. *Sci. Transl. Med.* **2018**, *10*, eaao1613. [CrossRef] [PubMed]

256. Xiao, X.; Liang, J.; Huang, C.; Li, K.; Xing, F.; Zhu, W.; Lin, Z.; Xu, W.; Wu, G.; Zhang, J. DNA-PK inhibition synergizes with oncolytic virus M1 by inhibiting antiviral response and potentiating DNA damage. *Nat. Commun.* **2018**, *9*, 1–15. [CrossRef] [PubMed]

257. Tang, B.; Guo, Z.S.; Bartlett, D.L.; Liu, J.; McFadden, G.; Shisler, J.L.; Roy, E.J. A cautionary note on the selectivity of oncolytic poxviruses. *Oncolytic Virother.* **2019**, *8*, 3. [CrossRef] [PubMed]

258. Liu, Y.; Cai, J.; Liu, W.; Lin, Y.; Guo, L.; Liu, X.; Qin, Z.; Xu, C.; Zhang, Y.; Su, X. Intravenous injection of the oncolytic virus M1 awakens antitumor T cells and overcomes resistance to checkpoint blockade. *Cell Death Dis.* **2020**, *11*, 1–13. [CrossRef] [PubMed]

259. Puzanov, I.; Milhem, M.M.; Minor, D.; Hamid, O.; Li, A.; Chen, L.; Chastain, M.; Gorski, K.S.; Anderson, A.; Chou, J. Talimogene laherparepvec in combination with ipilimumab in previously untreated, unresectable stage IIIB-IV melanoma. *J. Clin. Oncol.* **2016**, *34*, 2619. [CrossRef] [PubMed]

260. Sivanandam, V.; LaRocca, C.J.; Chen, N.G.; Fong, Y.; Warner, S.G. Oncolytic viruses and immune checkpoint inhibition: The best of both worlds. *Mol. Ther. Oncolytics* **2019**, *13*, 93–106. [CrossRef] [PubMed]

261. Lin, C.; Ren, W.; Luo, Y.; Li, S.; Chang, Y.; Li, L.; Xiong, D.; Huang, X.; Xu, Z.; Yu, Z. Intratumoral Delivery of a PD-1–Blocking scFv Encoded in Oncolytic HSV-1 Promotes Antitumor Immunity and Synergizes with TIGIT Blockade. *Cancer Immunol. Res.* **2020**, *8*, 632–647. [CrossRef] [PubMed]

262. Postow, M.A.; Sidlow, R.; Hellmann, M.D. Immune-related adverse events associated with immune checkpoint blockade. *N. Engl. J. Med.* **2018**, *378*, 158–168. [CrossRef] [PubMed]

263. Rojas, J.J.; Sampath, P.; Hou, W.; Thorne, S.H. Defining effective combinations of immune checkpoint blockade and oncolytic virotherapy. *Clin. Cancer Res.* **2015**, *21*, 5543–5551. [CrossRef]

264. Santos, J.M.; Havunen, R.; Hemminki, A. Modulation of the tumor microenvironment with an oncolytic adenovirus for effective T-cell therapy and checkpoint inhibition. *Meth. Enzymol.* **2020**, *635*, 205–230.

265. Nishio, N.; Diaconu, I.; Liu, H.; Cerullo, V.; Caruana, I.; Hoyos, V.; Bouchier-Hayes, L.; Savoldo, B.; Dotti, G. Armed oncolytic virus enhances immune functions of chimeric antigen receptor–modified T cells in solid tumors. *Cancer Res.* **2014**, *74*, 5195–5205. [CrossRef] [PubMed]

266. Watanabe, K.; Luo, Y.; Da, T.; Guedan, S.; Ruella, M.; Scholler, J.; Keith, B.; Young, R.M.; Engels, B.; Sorsa, S. Pancreatic cancer therapy with combined mesothelin-redirected chimeric antigen receptor T cells and cytokine-armed oncolytic adenoviruses. *JCI Insight* **2018**, *3*, 1–17. [CrossRef] [PubMed]

267. Scott, E.M.; Duffy, M.R.; Freedman, J.D.; Fisher, K.D.; Seymour, L.W. Solid Tumor Immunotherapy with T Cell Engager-Armed Oncolytic Viruses. *Macromol. Biosci.* **2018**, *18*, 1700187. [CrossRef] [PubMed]

268. Hamano, S.; Mori, Y.; Aoyama, M.; Kataoka, H.; Tanaka, M.; Ebi, M.; Kubota, E.; Mizoshita, T.; Tanida, S.; Johnston, R.N. Oncolytic reovirus combined with trastuzumab enhances antitumor efficacy through TRAIL signaling in human HER2-positive gastric cancer cells. *Cancer Lett.* **2015**, *356*, 846–854. [CrossRef] [PubMed]

269. Koske, I.; Rössler, A.; Pipperger, L.; Petersson, M.; Barnstorf, I.; Kimpel, J.; Tripp, C.H.; Stoitzner, P.; Bánki, Z.; von Laer, D. Oncolytic virotherapy enhances the efficacy of a cancer vaccine by modulating the tumor microenvironment. *Int. J. Cancer* **2019**, *145*, 1958–1969. [CrossRef]

270. Schirrmacher, V. Cancer vaccines and oncolytic viruses exert profoundly lower side effects in cancer patients than other systemic therapies: A comparative analysis. *Biomedicines* **2020**, *8*, 61. [CrossRef]

271. Lemay, C.G.; Rintoul, J.L.; Kus, A.; Paterson, J.M.; Garcia, V.; Falls, T.J.; Ferreira, L.; Bridle, B.W.; Conrad, D.P.; Tang, V.A. Harnessing oncolytic virus-mediated antitumor immunity in an infected cell vaccine. *Mol. Ther.* **2012**, *20*, 1791–1799. [CrossRef]

272. Xu, X.; Sun, Q.; Mei, Y.; Liu, Y.; Zhao, L. Newcastle disease virus co-expressing interleukin 7 and interleukin 15 modified tumor cells as a vaccine for cancer immunotherapy. *Cancer Sci.* **2018**, *109*, 279–288. [CrossRef]

273. Burke, J.D.; Platanias, L.C.; Fish, E.N. Beta interferon regulation of glucose metabolism is PI3K/Akt dependent and important for antiviral activity against coxsackievirus B3. *J. Virol.* **2014**, *88*, 3485–3495. [CrossRef]

274. Jiang, H.; Shi, H.; Sun, M.; Wang, Y.; Meng, Q.; Guo, P.; Cao, Y.; Chen, J.; Gao, X.; Li, E. PFKFB3-driven macrophage glycolytic metabolism is a crucial component of innate antiviral defense. *J. Immunol.* **2016**, *197*, 2880–2890. [CrossRef]

275. da Costa, L.S.; da Silva, A.P.P.; Da Poian, A.T.; El-Bacha, T. Mitochondrial bioenergetic alterations in mouse neuroblastoma cells infected with Sindbis virus: Implications to viral replication and neuronal death. *PloS ONE* **2012**, *7*, e33871. [CrossRef] [PubMed]

276. Kennedy, B.E.; Murphy, J.P.; Clements, D.R.; Konda, P.; Holay, N.; Kim, Y.; Pathak, G.P.; Giacomantonio, M.A.; El Hiani, Y.; Gujar, S. Inhibition of pyruvate dehydrogenase kinase enhances the antitumor efficacy of oncolytic reovirus. *Cancer Res.* **2019**, *79*, 3824–3836. [CrossRef] [PubMed]

277. Li, C.; Meng, G.; Su, L.; Chen, A.; Xia, M.; Xu, C.; Yu, D.; Jiang, A.; Wei, J. Dichloroacetate blocks aerobic glycolytic adaptation to attenuated measles virus and promotes viral replication leading to enhanced oncolysis in glioblastoma. *Oncotarget* **2015**, *6*, 1544. [CrossRef] [PubMed]

278. Kennedy, B.E.; Sadek, M.; Gujar, S.A. Targeted metabolic reprogramming to improve the efficacy of oncolytic virus therapy. *Mol. Ther.* **2020**, *28*, 1417–1421. [CrossRef] [PubMed]

279. Roy, D.G.; Kaymak, I.; Williams, K.S.; Ma, E.H.; Jones, R.G. Immunometabolism in the Tumor Microenvironment. *Annu. Rev. Cancer Biol.* **2020**, *4*, 17–40.

280. Chang, C.-H.; Qiu, J.; O'Sullivan, D.; Buck, M.D.; Noguchi, T.; Curtis, J.D.; Chen, Q.; Gindin, M.; Gubin, M.M.; Van Der Windt, G.J.W. Metabolic competition in the tumor microenvironment is a driver of cancer progression. *Cell* **2015**, *162*, 1229–1241. [CrossRef]

281. Jung, K.H.; Choi, I.-K.; Lee, H.-S.; Yan, H.H.; Son, M.K.; Ahn, H.M.; Hong, J.; Yun, C.-O.; Hong, S.-S. Oncolytic adenovirus expressing relaxin (YDC002) enhances therapeutic efficacy of gemcitabine against pancreatic cancer. *Cancer Lett.* **2017**, *396*, 155–166. [CrossRef]

282. Habiba, U.; Hossain, E.; Yanagawa-Matsuda, A.; Chowdhury, A.F.M.A.; Tsuda, M.; Zaman, A.-U.; Tanaka, S.; Higashino, F. Cisplatin relocalizes RNA binding protein HuR and enhances the oncolytic activity of E4orf6 deleted adenovirus. *Cancers* **2020**, *12*, 809. [CrossRef]

283. Gomez-Gutierrez, J.G.; Nitz, J.; Sharma, R.; Wechman, S.L.; Riedinger, E.; Martinez-Jaramillo, E.; Zhou, H.S.; McMasters, K.M. Combined therapy of oncolytic adenovirus and temozolomide enhances lung cancer virotherapy in vitro and in vivo. *Virology* **2016**, *487*, 249–259. [CrossRef]

284. Malfitano, A.M.; Di Somma, S.; Iannuzzi, C.A.; Pentimalli, F.; Portella, G. Virotherapy: From single agents to combinatorial treatments. *Biochem. Pharmacol.* **2020**, 113986. [CrossRef]

285. Ganesan, A. HDAC inhibitors in cancer therapy. *Histone Modif. Ther.* **2020**, *2*, 19–49.

286. Jennings, V.A.; Scott, G.B.; Rose, A.M.; Scott, K.J.; Migneco, G.; Keller, B.; Reilly, K.; Donnelly, O.; Peach, H.; Dewar, D. Potentiating oncolytic virus-induced immune-mediated tumor cell killing using histone deacetylase inhibition. *Mol. Ther.* **2019**, *27*, 1139–1152. [CrossRef] [PubMed]

287. Fox, C.R.; Parks, G.D. Histone Deacetylase Inhibitors Enhance Cell Killing and Block Interferon-Beta Synthesis Elicited by Infection with an Oncolytic Parainfluenza Virus. *Viruses* **2019**, *11*, 431. [CrossRef] [PubMed]

288. Roulstone, V.; Pedersen, M.; Kyula, J.; Mansfield, D.; Khan, A.A.; McEntee, G.; Wilkinson, M.; Karapanagiotou, E.; Coffey, M.; Marais, R. BRAF-and MEK-targeted small molecule inhibitors exert enhanced antimelanoma effects in combination with oncolytic reovirus through ER stress. *Mol. Ther.* **2015**, *23*, 931–942. [CrossRef]

289. Jackson, J.D.; Markert, J.M.; Li, L.; Carroll, S.L.; Cassady, K.A. STAT1 and NF-κB inhibitors diminish basal interferon-stimulated gene expression and improve the productive infection of oncolytic HSV in MPNST cells. *Mol. Cancer Res.* **2016**, *14*, 482–492. [CrossRef]

Review

Parking CAR T Cells in Tumours: Oncolytic Viruses as Valets or Vandals?

Laura Evgin [1,2] and Richard G. Vile [3,4,*]

[1] Canada's Michael Smith Genome Sciences Centre, BC Cancer, Vancouver, BC V5Z 1L3, Canada; levgin@bcgsc.ca

[2] Department of Medical Genetics, University of British Columbia, Vancouver, BC V6H 3N1, Canada

[3] Department of Molecular Medicine, Mayo Clinic, Rochester, MN 55905, USA

[4] Department of Immunology, Mayo Clinic, Rochester, MN 55905, USA

* Correspondence: vile.richard@mayo.edu; Tel.: +1-507-284-9941

Simple Summary: Chimeric Antigen Receptor (CAR) modified T cell therapy has revolutionized the treatment of B cell malignancies, however transposition of the technology to the solid tumour setting has been met with more therapeutic resistance. Oncolytic Viruses (OVs) are multi-modal agents, possessing tumour cell cytolytic capabilities as well as strong immune stimulatory properties. Although combination therapy poses great promise, great care must be employed so as to maximize the output of each modality and minimize interference.

Abstract: Oncolytic viruses (OVs) and adoptive T cell therapy (ACT) each possess direct tumour cytolytic capabilities, and their combination potentially seems like a match made in heaven to complement the strengths and weakness of each modality. While providing strong innate immune stimulation that can mobilize adaptive responses, the magnitude of anti-tumour T cell priming induced by OVs is often modest. Chimeric antigen receptor (CAR) modified T cells bypass conventional T cell education through introduction of a synthetic receptor; however, realization of their full therapeutic properties can be stunted by the heavily immune-suppressive nature of the tumour microenvironment (TME). Oncolytic viruses have thus been seen as a natural ally to overcome immunosuppressive mechanisms in the TME which limit CAR T cell infiltration and functionality. Engineering has further endowed viruses with the ability to express transgenes in situ to relieve T cell tumour-intrinsic resistance mechanisms and decorate the tumour with antigen to overcome antigen heterogeneity or loss. Despite this helpful remodeling of the tumour microenvironment, it has simultaneously become clear that not all virus induced effects are favourable for CAR T, begging the question whether viruses act as valets ushering CAR T into their active site, or vandals which cause chaos leading to both tumour and T cell death. Herein, we summarize recent studies combining these two therapeutic modalities and seek to place them within the broader context of viral T cell immunology which will help to overcome the current limitations of effective CAR T therapy to make the most of combinatorial strategies.

Keywords: oncolytic virus; adoptive T cell therapy; CAR T cell; immunotherapy

Citation: Evgin, L.; Vile, R.G. Parking CAR T Cells in Tumours: Oncolytic Viruses as Valets or Vandals? *Cancers* **2021**, *13*, 1106. https://doi.org/10.3390/cancers13051106

Academic Editors: Antonio Marchini, Carolina S. Ilkow and Alan Melcher

Received: 29 January 2021
Accepted: 3 March 2021
Published: 5 March 2021

Publisher's Note: MDPI stays neutral with regard to jurisdictional claims in published maps and institutional affiliations.

1. CARs: The Ultimate Tumour Killing Machines?

An insufficiency of the breadth, or functionality, of the tumour reactive T cell repertoire can be overcome through the use of adoptive T cell therapy (ACT) in which T cells specific to the antigenic constituency of the tumour are generated ex vivo and re-introduced to the patient. These cells may be directly expanded from the tumour [1] or derived from peripheral blood in which novel specificity is conferred by expression of an ectopic T cell receptor (TCR) or a chimeric antigen receptor (CAR) [2]. This allows for T cells to be cultured to possess desirable phenotypes in vitro, and patients to be treated with preconditioning regimes to promote T cell engraftment and minimize suppression. T cells generated through these means have led to a large subset of complete responses in otherwise treatment refractory disease, with CAR T cells in particular experiencing unprecedented success against B lymphoid cancers [3].

While TCRs recognize intracellular antigens presented in the context of the major histocompatibility complex (MHC), the synthetic CAR confers specificity in an MHC unrestricted manner to cell surface, and now recently, soluble antigens [4]. The CAR design is modular with each domain contributing to the resulting functional outcome. In its most basic configuration, the CAR molecule is composed of an extracellular antigen binding domain (most commonly an scFv), an extracellular hinge region, a transmembrane domain, and an intracellular signaling domain (including the CD3ζ and costimulatory domains) [5].

The signaling region of the earliest iterations of CARs comprised only the CD3ζ endodomain and triggered effector function but had limited therapeutic potential [6,7]. A multi-step activation model is required to mount effective T cell responses, with signal one being derived from the TCR, signal two from co-stimulatory ligation, and signal three from cytokine exposure. Co-stimulatory domains have thus been included proximal to CD3ζ in second generation constructs to promote persistence and anti-tumour activity [8–10]. Third and fourth generation CARs include two or more co-stimulatory domains, or other transgenes including cytokines [11–14]. CD28 and CD137 (4-1BB) have been most rigorously explored preclinically and are included in the clinically approved constructs axicabtagene ciloleucel (Yescarta), and tisagenlecleucel (Kymriah), respectively [15–17]. Several other costimulatory domains have however been successfully evaluated including CD27, OX40, CD40L, and ICOS, and the incorporation of distinct costimulatory domains has been shown to have profound effects on phenotype, expansion kinetics, metabolism, and persistence [18–22]. Sophisticated synthetic biology circuits have been designed to incorporate additional regulatory tuning and recognition capacity [23]. The functional properties of each domain have been explored and are extensively reviewed elsewhere [24–27].

Retro- or lentiviral vectors are the primary means of introducing stable expression of the CAR into T cells, but random integration into the genome has the potential to lead to insertional mutagenesis and variegated expression of the CAR, thus prompting use of targeted means to introduce the CAR into genomic safe harbours such as the T-cell receptor α constant (TRAC) locus using CRISPR/Cas9 [28,29]. Although the CAR has primarily been introduced into autologous T cells to generate a bespoke patient product, disruption of the TRAC simultaneously enables the use of allogeneic T cells by preventing the development of graft vs. host disease (GVHD) [30].

2. Switching on the Ignition with Oncolytic Viruses

Through natural tropism, or genetic engineering, oncolytic viruses are a class of viruses which share a preference for replication in malignant cells over normal tissue. A broad diversity of viruses has been defined as having oncolytic properties and possess distinct genomes (RNA and DNA), entry specificities, replication mechanics, immune-evasion machinery, and genetic modifications which collectively confer tumour specificity. Infection of tumour cells can be facilitated by the overexpression of viral binding and entry receptors, such as CD46 for measles [31] or ICAM-1 for Coxsackievirus A21 [32]. Deletion of viral genes, complemented by high level expression in tumour cells, such as those involved in nucleotide metabolism (thymidine kinase, ribonucleotide reductase, uracil DNA glycosylase) has been employed with herpes simplex virus (HSV) and vaccinia virus (VACV) strains [33–36]. The adenovirus E1A protein binds the cellular retinoblastoma protein to drive S-phase entry, allowing it to access the cellular DNA replication and protein synthesis machinery, and a 24 amino acid deletion restricts the virus to rapidly proliferating cells [37].

The primary innate antiviral mechanism, type I interferon (IFN), is known to be anti-angiogenic, and to promote growth arrest and apoptosis [38]. While many tumours harbour mutations in key IFN genes or epigenetically silence them, the activation of oncogenic pathways or loss of tumour suppressors such as EGFR, Wnt B catenin, or Pten, all have links to IFN production or responsiveness [39,40]. The net effect is that an estimated 65–70% of cancer cell lines are thought to have defects in their ability to produce, or respond to, type I IFN [41]. Compromised IFN signaling thus underlies the tumour selectivity of OVs and is particularly relevant to Vesicular stomatitis virus (VSV) and Newcastle disease virus (NDV) [41,42]. The safety and specificity of oncolytic VSV is further enhanced through deletion or mutation at position 51 in the matrix protein whose normal function is to block nucleocytoplasmic trafficking of mRNA, thus preventing the translation of IFN and interferon stimulated genes (ISGs) [41,43]. An analogous approach encodes IFNβ in the viral genome, with the added benefit that the cytokine promotes dendritic cell (DC) activation and acts as a signal 3 cytokine for T cell priming [44–46].

3. Combination OV and T Cell Therapies: Driving CAR T to the Tumour

Inter and intra-patient tumour heterogeneity, the plastic nature of cancer genomes, and the dynamic state of the tumour microenvironment all contribute to the likely failure of monotherapy approaches to cancer treatment. In this respect, it would seem on the surface that a partnership between OVs and CAR T cells offers a perfect opportunity to orchestrate a multi-pronged approach against often rapidly evolving targets on multiple fronts. Although combination strategies using multiple biologic agents may face more regulatory hurdles, significant clinical development of both platforms individually may pave a way forward. Significant toxicities have been well described for CAR T therapy, including cytokine release syndrome (CRS) and neurotoxicity, and it will be paramount to establish a robust safety profile of any combination strategy. A multiplexed approach has now made the jump from the bench [47] to clinic as the investigation of HER2 CAR T cells and oncolytic and helper dependent adenovirus expressing IL12 and anti-PDL1 is now underway (NCT03740256).

Herein we review how intrinsic and engineered properties of oncolytic viral vectors may be exploited to enable CAR T to overcome barriers to effective therapy in the solid tumour setting, including restricted infiltration, interaction with immunosuppressive soluble mediators and cellular players, and antigen heterogeneity and escape (Figure 1). However, combination with OVs does not automatically guarantee a superior therapeutic outcome as they can lead to both helpful and deleterious consequences for CAR T cells, and thus act both as valets and vandals. The studies highlighted herein illustrate the complex biology of each living drug and the importance of highly tailored therapeutic strategies.

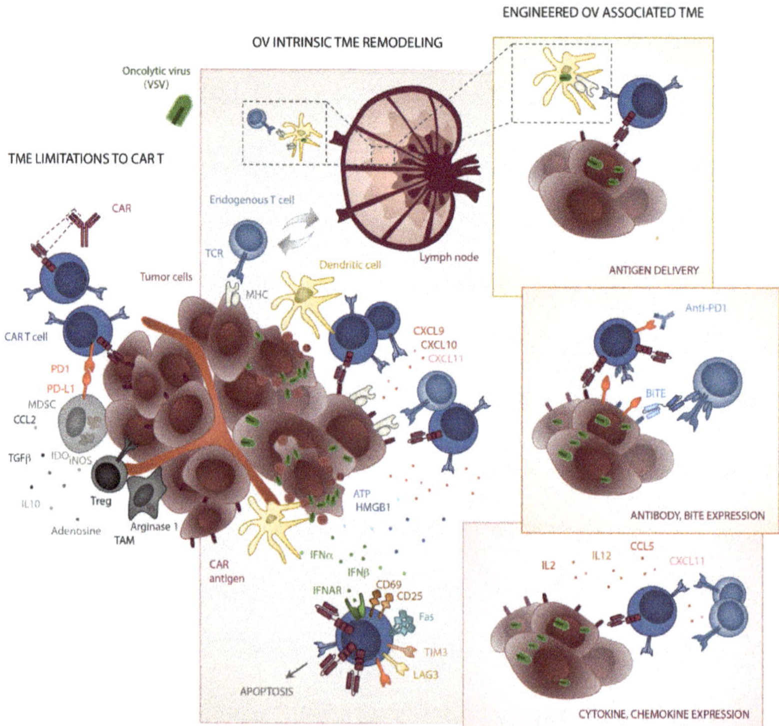

Figure 1. Strategic combination of oncolytic viruses with CAR T cells. The TME presents many immunosuppressive barriers to CAR T trafficking (high levels of CCL2, low levels of T cell chemotactic chemokines), as well as functionality through cytokines (TGFβ and IL10), metabolic dysregulation (arginase 1, inducible NO synthase (iNOS), indoleamine 2,3-dioxygenase (IDO), and CD39 and CD73 production of adenosine), and inhibitory ligands (PDL1 etc.). Many of these factors are expressed by tumor associated macrophages (TAMs), myeloid derived suppressor cells (MDSCs), regulatory T cells (Tregs) or the tumor cells themselves. Viral infection and oncolysis of tumor cells lead to the production of type I interferons (IFNs), danger-associated molecular pattern molecules such as HMGB1 and ATP, and CXCL9, 10, and 11 which in turn recruit additional T cells and dendritic cells. Exposure to high level type I IFN can also have inadvertent negative consequences for CAR T cells leading the upregulation of various inhibitory receptors including PD1, TIM-3 and LAG-3, as well Fas, leading to apoptosis. In contrast to some of these intrinsic properties, OVs can be armed with transgenes such as cytokines (IL2, IL12), chemokines (CCL5, CXCL11), checkpoint blocking antibodies (anti-PD1, etc.), BiTEs (EGFR, etc.) or the CAR antigen itself (CD19, etc.).

4. The TME—A Breaker's Yard for CAR T Cells

The tumour microenvironment is composed not only of cancer cells, but heterogenous levels of a variety of immune cell types whose location and density can profoundly affect prognosis and therapeutic response [48]. Although effector T cells, NK cells and B cells can be present to variable degrees, suppressive immune cell types, including regulatory T cells (T_{reg}) and aberrantly matured myeloid cells such as myeloid derived suppressor cells (MDSCs) and tumour-associated macrophages (TAMs) are often found within the tumour core and the invasive front [48,49]. This constellation of stromal cells coordinates a network of overlapping regulatory mechanisms which mask the tumour from immune destruction, beginning with the expression of chemokines which disfavour the recruitment of effector T cells. Expression of the counter-ligands for CXCR3 on activated lymphocytes, CXCL9, 10, and 11, is associated with good prognosis and is required for T-cell trafficking across tumour vascular checkpoints [50–52]. However, tumours may reduce levels of these ligands through epigenetic silencing or the co-expression of chemokine- cleaving proteases [53,54], and instead express CCL2 which recruits immature myeloid cells and TAMS [55–57]. In turn, MDSCs and TAMs lead to metabolic and cytokine mediated dysfunction of T cells through the production of arginase 1, inducible NO synthase (iNOS), indoleamine 2,3-dioxygenase (IDO), transforming growth factor beta (TGFβ) and IL10, respectively [58,59]. T$_{regs}$ exploit cytokine-mediated and contact-dependent mechanisms to limit effector T function, including competition for IL2, expression of CD39 and CD73 leading to the production of adenosine, secretion of TGFβ and IL10, and expression of checkpoint ligands such as CTLA-4 and PDL-1 [60]. Finally, tumour cells themselves, immune cells, and exosomes can all express ligands or release soluble factors which engage checkpoint receptors on CAR T cells, including PD-1, TIM-3, LAG-3 and TIGIT, leading to dysfunction and apoptosis [61]. Overall, these factors contribute to making the tumour more akin to a scrap yard than the exclusive valet parked ramp where you would, ideally, want to leave your meticulously engineered CAR.

5. TME Make over by Oncolytic Viruses

OV infection leads to a cascade of inflammatory events, stimulating innate and adaptive immune responses, and thus changing the cytokine, chemokine and cellular composition of tumours. Viral nucleic acids serve as pathogen-associated molecular patterns (PAMPs) to activate cytoplasmic RNA and DNA sensors and Toll-like receptors (TLRs), converging on TRIF and MyD88 to activate type I IFN signaling [62,63]. OV infection upregulates calreticulin (CRT) on the cell surface, and oncolysis releases adenosine triphosphate (ATP) and high-mobility group box 1 (HMGB1) into the extracellular environment, all members of the danger-associated molecular pattern (DAMP) family [64,65]. Together, in concert with type I IFN, these signals promote the recruitment and maturation of DCs which take up virus and tumour debris, traffic antigen back to lymph nodes, and prime naïve T cells. Although all CXCR3 ligands are induced by IFNγ, CXCL10 and 11 are directly agonized by type I IFN [66].

Thus, infection leads to a global change in the cellular composition of the tumour and the corresponding derived soluble mediators. Mouse models have demonstrated that OVs promote the infiltration and activation of CD8 T cells, CD11c+ DC, NK cells, M1-like macrophages, and concomitantly reduce the proportion of Tregs and MDSCs [67–74]. Although a large fraction of infiltrating T cells is likely to be specific to viral antigens, oncolysis can act as a tumour antigen agnostic vaccine, priming T cells against public and private neoantigens [73,75]. Oncolytic infection and type I IFN concomitantly induce the upregulation of checkpoint receptor ligands such as PDL1, and thus combination therapy with pharmacologic or viral expression of checkpoint blocking antibodies with vaccinia [67], VSV [68,76], reovirus [77], measles [70,78], HSV [72] and NDV [71] have provided superior tumour outcomes. Immune correlative studies in the clinical setting have corroborated preclinical findings, showing that talimogene laherparepvec (HSV) and reovirus treatment promotes an increase in CD8 T cell density in post treatment biopsies [79,80], and measles treatment facilitates T cell priming against tumour antigens [81]. The representation of virus specific or tumour antigen specific T cells which infiltrate into a tumour is not well characterized, however is likely to be skewed toward viral specificities due to high level expression of viral epitopes which are not subject to tolerance mechanisms.

6. Fiddling while CARs Burn

On the face of it, these pro-inflammatory effects of OV infection of tumours would be predicted to provide an excellent make over to convert the tumour from a parking site that is essentially 'closed for business' to T cells to one that now is replete with special offers for both long and short term parking deals -suggesting that OVs should significantly improve CAR T efficacy in the solid tumour setting. However, woven throughout the potential benefits of type I IFN responses on T cell recruitment/activation are negative feedback mechanisms which have evolved to promote de novo anti-viral T cell responses, and subsequently restrain inflammation to prevent autoimmunity. While the upregulation of checkpoint ligands can be blocked by the simultaneous use of checkpoint inhibitors, the more insidious role of type I IFN is its direct effect on CD8 T cell biology. Depending on the timing, memory status and concentration, type I IFN exerts pleiotropic effects on T cells. Type I IFN supports the expansion and differentiation of naïve T cells, thus playing a key role in cell fate decisions as a signal 3 cytokine [46,82]. However, via a mechanism that is thought to make space for T cells specific to incoming pathogens, type I IFNs also promote acute apoptosis of memory T cells [83–86]. Indeed, priming the tumor with oncolytic VSV expressing IFNβ simultaneously promoted significant CAR T attrition in a type I IFN dependent manner [87]. Although the effect was largely T cell intrinsic as adoptively transferred transgenic Pmel T cells underwent the same IFN associated fate, additional CAR specific effects were observed. Virus derived IFN upregulated the expression of the CAR, promoting downstream effects of tonic signaling, including high level expression of inhibitory receptors. Although apoptosis was averted through the use of transgenic or CRISPR edited interferon alpha receptor (IFNAR1) deficient T cells, and thus allowed for enhanced combination therapy in lymphodepleted animals, this engineering strategy inadvertently sensitized the CAR T cells to NK cell attack [87–89]. These effects are thought to be broadly relevant to other OVs. While the underlying biology was enhanced by the expression of IFNβ from the VSV vector, oncolytic reovirus also induced CAR T cell attrition, albeit to a more moderate extent [87].

High levels of VEGF in the tumor have been shown to attenuate type I IFN signaling in tumour-associated endothelial cells through Blimp-1, thus sensitizing them to OV infection [90]. Together with the neutrophil-dependent induction of microclots, several OVs, including reovirus, VSV, vaccinia virus, and NDV, have been shown to induce vascular shutdown in tumours [90–95]. While vascular collapse may starve tumour cells, it may simultaneously limit the access of CAR T cells to their targets. Vascular normalization using 3TSR prior to NDV therapy has been shown to increase immune cell trafficking into the tumour [95], and may thus represent an important third-party consideration for combination therapy.

7. Viruses as Micro-Pharmacies for T Cells

Although the virus-intrinsic effects of infection on the tumour composition are potentially overwhelmingly favourable, an additional therapeutic strength may be in the ability of OVs to deliver desirable transgenes locoregionally. The magnitude and timing of chemokine induction varies depending on virus biology and several OVs have been engineered to express chemokines to enhance recruitment of CAR T to the tumour. Oncolytic adenovirus armed with the chemokine RANTES (CCL5) to promote infiltration, as well as the cytokine IL15 to support T cell survival once in the tumour conferred enhanced therapeutic benefit when used in combination with GD2 CAR T [96]. A similar strategy incorporating the CXCL11 transgene into vaccinia virus enhanced CD8 T cell infiltration and enhanced mesothelin specific CAR T therapy of murine TC1 tumours [97].

In order to sidestep any reduction in replication and oncolytic capacity, Shaw et al. used a gutted (helper-dependent) adenovirus to deliver various cytokine payloads in combination with replication competent oncolytic adenovirus and HER2 specific CAR T. Among the candidate cytokines IL2, IL7, IL-12p70, IL15, and IL2, expression of IL-12p70 was found to potentiate CAR T efficacy in a xenograft model of head and neck squamous cell carcinoma [47]. Further incorporation of a PD-L1 blocking antibody in the helper dependent adenovirus increased anti-tumour efficacy [47,98] and provides the rationale for clinical evaluation. Notably, local production of the anti-PDL1 antibody from the virus was superior to systemic administration of anti-PD-L1 IgG, thus highlighting the benefit of in situ transgene production [98]. Encoding the checkpoint blockade molecule and IL12 in the virus is a particularly attractive strategy to produce locally high concentrations at bioactive sites, where systemic delivery is associated with adverse events [99].

A similar strategy by Wanatabe et al. employed oncolytic adenovirus armed with TNFα and IL2 to enhance both human and mouse mesothelin specific CAR T. Treatment with Ad5/3-OAd-TNFα-IL2 induced more robust and persistent localization of human CAR T in the tumour and correspondingly induced sustained regression. Adenovirus encoding the murine cytokines increased CD80 and CD86 expression on tumour resident macrophages and dendritic cells, upregulated CXCL10 production in the tumour, and was associated with higher levels of infiltrating CD4 CAR T, and CD4 and CD8 endogenous T cells; all of which contributed to increased tumour control in the combination arm [100].

8. Graffitiing Antigenic Specificity onto Tumours

Oncolytic viruses exhibit a range of infectivity within tumours suggesting that highly susceptible cells would be killed, whilst cells refractory to oncolysis, but in which viral genes are expressed, could be targeted by CAR T. In solid tumours where no specific CAR targets have been identified, OVs could be used to deliver ectopic antigens to tumour cells [101,102]. This strategy could also be applied to re-target CAR T cells to antigen negative tumour cells, a common mechanism of treatment failure [103,104]. Proof of concept studies from Aalipour et al. and Park et al. used oncolytic vaccinia viruses to decorate cancer cells with CD19 and demonstrated targeting of CD19 CAR T to previously unrecognized tumour cells both in vitro and in vivo. Although CD19 delivery leads to unnecessary induction of B cell aplasia, the modular nature of the approach suggests that it could be extended to other antigens. One significant limitation would be heterogenous expression of the CAR target antigens and incomplete targeting by either modality. However, both OVs and CAR T have been demonstrated to elicit endogenous T and B cell responses to tumour associated antigens (TAA) through epitope spreading [80,105–107], and indeed Park et al. show that mice which were cured by combination therapy were partially protected against re-challenge with parental tumours which did not express the CAR antigen.

The immune adjuvanticity and transgene expression capabilities of OVs allow them to act as strong vaccines. TAAs can be expressed to a high degree in tumour cells, and upon cell lysis, taken up by dendritic cells or other phagocytic cells for presentation to T cells. Viruses may also initiate abortive non-lytic infections in non-transformed cells, such as APCs, which lead to the expression of viral genes and the subsequent priming of T cells. Thus, oncolytic vaccination against single TAAs, or even a library of antigens, provides stronger anti-tumour therapy than parental strains [108–110]. The size of the T cell pool is further magnified through the use of heterologous vectors for priming and boosting which encode the same antigen [111–113], or by adoptive transfer of transgenic antigen specific T cells [114–116]. Although CAR modified T cells acquire a novel specificity, they also retain the specificity conferred by their native TCR, and parallel work has sought to boost CAR T cells through the TCR. Transgenic T cells expressing both a CAR specific to HER2 and bearing a TCR specific for either the gp100 or OVA antigens are expanded by treatment with vaccinia expressing the cognate epitope, leading to accumulation of T cells in the tumour and eradication of large established tumours [117]. Clinical evaluation of a mixed infusion product containing Epstein-Barr Virus (EBV) TCR specific and open TCR repertoire GD2 CAR T demonstrated that virus specific CAR T circulate at a higher frequency than the counterpart control T cells [7]. Viral reactivation or vaccination further supports the expansion of virus specific CAR T cells [118,119].

Bispecific T-cell engager (BiTE) technology links an anti-CD3 scFv to a tumour antigen specific scFv, and thus bypasses both the TCR-MHC interaction and the CAR to engage effector function. In this way, OVs and CAR T cells engineered to express BiTEs can redirect endogenous T cells, or CAR T themselves, against a second tumour antigen specificity [120–123]. Putting these platforms together, Wing et al. demonstrated that oncolytic adenovirus expressing an EGFR-targeting BiTE improved the activation, proliferation, and cytokine production of CART cells targeting the folate receptor alpha (FR-α) and can help to overcome antigen heterogeneity [124]. Use of the virally expressed BiTE redirected both CAR T and CAR negative nonspecific T cells and provided superior anti-tumour efficacy compared to each monotherapy. A multiplex strategy has also combined BiTE expression (specific to CD44v6) with cytokine (IL12) and checkpoint (anti-PDL1) delivery using the oncolytic and helper dependent adenovirus system in conjunction with HER2 CAR T to combat several mechanisms of tumour escape simultaneously [125].

9. Virus CAR-Pooling to Tumours

Although we have discussed primarily the use of viruses to improve various aspects of T cell anti-tumour function, so T cells can also be exploited to improve the efficiency of oncolytic virotherapy—most notably perhaps through helping to deliver the viruses to their active site. In vitro pre-loaded antigen specific T cells, and cytokine-induced killer cells, have been reported to traffic OVs, allowing replication and oncolysis within the tumour. This smuggling of viruses to tumours has been shown to be possible even in pre-vaccinated hosts, thus bypassing circulating anti-viral antibodies which have often proved to be the Achilles heel of systemic OV therapy [126–131]. Similarly, murine and human HER2 CAR T cells loaded with low doses of oncolytic VSV or vaccinia virus have been shown to deposit their cargo without compromising the function of the CAR T cells [132]. As discussed above, the TME represents a very unwelcoming parking place for (CAR) T cells. The same is true for highly immunogenic viruses trying to passage through and then exit selectively from a circulatory highway system heavily patrolled by neutralising antibodies, complement, and other anti-viral effectors. However, CAR-pooling of precisely engineered tumour-targeting viruses may overcome several of the barriers to effective combination CAR/OV therapy.

10. Conclusions: OV Enhanced CAR T Cell Therapy—And Vice Versa

In the quest for a systemic, potent anti-tumour therapy, both adoptive (CAR) T cell transfer and oncolytic viruses have enormous curative potential. Both intrinsic and engineered capabilities endow OVs with a unique potential to serve as a platform to enhance adoptive T cell therapy. However, in order to reach their destination, both (CAR) T cells and viruses have to navigate a circulatory highway fraught with diversions, patrols and obstacles. Even if/when they successfully reach the tumour, the TME represents a highly immune-suppressive, neutralizing destination. This neighbourhood is unlikely to appeal to any owner of such highly sophisticated and engineered anti-tumour killing machines (cells or viruses) as a safe and effective parking place. However, tumour infection by OVs has the potential to effect a dramatic make over and convert this hostile, T cell repellent TME into a highly attractive haven, open for business for an influx of CAR T cells. In this respect, the inflammatory profile induced by OV infection, as well as OV-triggered transgene expression, needs to be carefully crafted. Moving forward, models in which the safety and efficacy of combination strategies that intricately engage innate and adaptive immunity are evaluated must account for factors which both recruit and support activated CAR T cell therapies, as well as those compensatory mechanisms which restrain and inhibit the (T cell) immune system. Importantly, a variety of models should be used which, combinatorially, analyze the plethora of factors which may be absent from specific model systems, or which may be present but which are non-reactive in the specific model being tested—such as type I IFN, which binds to species specific IFNAR [133]. Thus, it will be critical to prevent OV infection from simply converting a T cell freezing TME into an incendiary, CAR T-vandalizing, TME. With appropriate design of the levels, nature and timing of inflammatory cytokine expression from OV infection, it will be possible to generate an optimal, climate-controlled environment that nurtures the gentle, valet parking of CAR T cells inside the tumour—where the T cells can go on to do their worst against solid tumours. Therefore, we believe that by generating novel designer combinations of paired viruses and engineered T cells it will be possible to create a powerful synergy between adoptive T cell therapies and OV infection, whereby each one enhances the tumour trafficking, selectivity and potency of the other.

Funding: Funding provided by the National Institutes of Health (P50 CA210964 and R01CA108961 to R.G.V.), The Richard M. Schulze Family Foundation, Mayo Foundation, The Minnesota Partnership for Biotechnology and Medical Genomics, Shannon O'Hara Foundation, Hyundai Hope on Wheels as well as BC Cancer Research and BC Cancer Foundation (L.E.).

Acknowledgments: The authors thank Toni L. Woltman for expert secretarial assistance.

Conflicts of Interest: L.E. and R.G.V. are inventors on patent applications related to the use of oncolytic viruses and CAR T cells for the treatment of cancer.

References

1. Hinrichs, C.S.; Rosenberg, S.A. Exploiting the curative potential of adoptive T-cell therapy for cancer. *Immunol. Rev.* **2013**, *257*, 56–71. [CrossRef]
2. Weber, E.W.; Maus, M.V.; Mackall, C.L. The Emerging Landscape of Immune Cell Therapies. *Cell* **2020**, *181*, 46–62. [CrossRef] [PubMed]
3. June, C.H.; Sadelain, M. Chimeric Antigen Receptor Therapy. *N. Engl. J. Med.* **2018**, *379*, 64–73. [CrossRef]
4. Chang, Z.L.; Lorenzini, M.H.; Chen, X.; Tran, U.; Bangayan, N.J.; Chen, Y.Y. Rewiring T-cell responses to soluble factors with chimeric antigen receptors. *Nat. Chem. Biol.* **2018**, *14*, 317–324. [CrossRef] [PubMed]
5. Sadelain, M.; Brentjens, R.; Rivière, I. The Basic Principles of Chimeric Antigen Receptor Design. *Cancer Discov.* **2013**, *3*, 388–398. [CrossRef]
6. Eshhar, Z.; Waks, T.; Gross, G.; Schindler, D.G. Specific activation and targeting of cytotoxic lymphocytes through chimeric single chains consisting of antibody-binding domains and the gamma or zeta subunits of the immunoglobulin and T-cell receptors. *Proc. Natl. Acad. Sci. USA* **1993**, *90*, 720–724. [CrossRef]
7. Pule, M.A.; Savoldo, B.; Myers, G.D.; Rossig, C.; Russell, H.V.; Dotti, G.; Huls, M.H.; Liu, E.; Gee, A.P.; Mei, Z.; et al. Virus-specific T cells engineered to coexpress tumor-specific receptors: Persistence and antitumor activity in individuals with neuroblastoma. *Nat. Med.* **2008**, *14*, 1264–1270. [CrossRef]
8. Maher, J.; Brentjens, R.J.; Gunset, G.; Rivière, I.; Sadelain, M. Human T-lymphocyte cytotoxicity and proliferation directed by a single chimeric TCRζ /CD28 receptor. *Nat. Biotechnol.* **2002**, *20*, 70–75. [CrossRef] [PubMed]
9. Brentjens, R.J.; Santos, E.; Nikhamin, Y.; Yeh, R.; Matsushita, M.; La Perle, K.; Quintás-Cardama, A.; Larson, S.M.; Sadelain, M. Genetically Targeted T Cells Eradicate Systemic Acute Lymphoblastic Leukemia Xenografts. *Clin. Cancer Res.* **2007**, *13*, 5426–5435. [CrossRef]
10. Imai, C.; Mihara, K.; Andreansky, M.; Nicholson, I.C.; Pui, C.-H.; Geiger, T.L.; Campana, D. Chimeric receptors with 4-1BB signaling capacity provoke potent cytotoxicity against acute lymphoblastic leukemia. *Leukemia* **2004**, *18*, 676–684. [CrossRef]
11. Ramos, C.A.; Rouce, R.; Robertson, C.S.; Reyna, A.; Narala, N.; Vyas, G.; Mehta, B.; Zhang, H.; Dakhova, O.; Carrum, G.; et al. In Vivo Fate and Activity of Second- versus Third-Generation CD19-Specific CAR-T Cells in B Cell Non-Hodgkin's Lymphomas. *Mol. Ther.* **2018**, *26*, 2727–2737. [CrossRef]
12. Zhong, X.-S.; Matsushita, M.; Plotkin, J.; Riviere, I.; Sadelain, M. Chimeric Antigen Receptors Combining 4-1BB and CD28 Signaling Domains Augment PI3kinase/AKT/Bcl-XL Activation and CD8+ T Cell–mediated Tumor Eradication. *Mol. Ther.* **2010**, *18*, 413–420. [CrossRef]
13. Karlsson, H.; Svensson, E.; Gigg, C.; Jarvius, M.; Olsson-Strömberg, U.; Savoldo, B.; Dotti, G.; Loskog, A. Evaluation of Intracellular Signaling Downstream Chimeric Antigen Receptors. *PLoS ONE* **2015**, *10*, e0144787. [CrossRef] [PubMed]
14. Chmielewski, M.; Hombach, A.A.; Abken, H. Of CARs and TRUCKs: Chimeric antigen receptor (CAR) T cells engineered with an inducible cytokine to modulate the tumor stroma. *Immunol. Rev.* **2013**, *257*, 83–90. [CrossRef] [PubMed]
15. Maude, S.L.; Laetsch, T.W.; Buechner, J.; Rives, S.; Boyer, M.; Bittencourt, H.; Bader, P.; Verneris, M.R.; Stefanski, H.E.; Myers, G.D.; et al. Tisagenlecleucel in Children and Young Adults with B-Cell Lymphoblastic Leukemia. *N. Engl. J. Med.* **2018**, *378*, 439–448. [CrossRef] [PubMed]
16. Neelapu, S.S.; Locke, F.L.; Bartlett, N.L.; Lekakis, L.J.; Miklos, D.B.; Jacobson, C.A.; Braunschweig, I.; Oluwole, O.O.; Siddiqi, T.; Lin, Y.; et al. Axicabtagene Ciloleucel CAR T-Cell Therapy in Refractory Large B-Cell Lymphoma. *N. Engl. J. Med.* **2017**, *377*, 2531–2544. [CrossRef]
17. Schuster, S.J.; Bishop, M.R.; Tam, C.S.; Waller, E.K.; Borchmann, P.; McGuirk, J.P.; Jäger, U.; Jaglowski, S.; Andreadis, C.; Westin, J.R.; et al. Tisagenlecleucel in Adult Relapsed or Refractory Diffuse Large B-Cell Lymphoma. *N. Engl. J. Med.* **2019**, *380*, 45–56. [CrossRef]
18. Philipson, B.I.; O'Connor, R.S.; May, M.J.; June, C.H.; Albelda, S.M.; Milone, M.C. 4-1BB costimulation promotes CAR T cell survival through noncanonical NF-κB signaling. *Sci. Signal.* **2020**, *13*, eaay8248. [CrossRef] [PubMed]
19. Song, D.-G.; Ye, Q.; Poussin, M.; Harms, G.M.; Figini, M.; Powell, D.J. CD27 costimulation augments the survival and antitumor activity of redirected human T cells in vivo. *Blood* **2012**, *119*, 696–706. [CrossRef]
20. Hombach, A.A.; Heiders, J.; Foppe, M.; Chmielewski, M.; Abken, H. OX40 costimulation by a chimeric antigen receptor abrogates CD28 and IL-2 induced IL-10 secretion by redirected CD4+ T cells. *OncoImmunology* **2012**, *1*, 458–466. [CrossRef] [PubMed]

21. Mata, M.; Gerken, C.; Nguyen, P.; Krenciute, G.; Spencer, D.M.; Gottschalk, S. Inducible Activation of MyD88 and CD40 in CAR T Cells Results in Controllable and Potent Antitumor Activity in Preclinical Solid Tumor Models. *Cancer Discov.* **2017**, *7*, 1306–1319. [CrossRef] [PubMed]

22. Guedan, S.; Chen, X.; Madar, A.; Carpenito, C.; Mcgettigan, S.E.; Frigault, M.J.; Lee, J.; Posey, A.D.; Scholler, J.; Scholler, N.; et al. ICOS-based chimeric antigen receptors program bipolar TH17/TH1 cells. *Blood* **2014**, *124*, 1070–1080. [CrossRef] [PubMed]

23. Roybal, K.T.; Williams, J.Z.; Morsut, L.; Rupp, L.J.; Kolinko, I.; Choe, J.H.; Walker, W.J.; McNally, K.A.; Lim, W.A. Engineering T Cells with Customized Therapeutic Response Programs Using Synthetic Notch Receptors. *Cell* **2016**, *167*, 419–432.e16. [CrossRef] [PubMed]

24. Van Der Stegen, S.J.C.; Hamieh, M.; Sadelain, M. The pharmacology of second-generation chimeric antigen receptors. *Nat. Rev. Drug Discov.* **2015**, *14*, 499–509. [CrossRef] [PubMed]

25. Hyrenius-Wittsten, A.; Roybal, K.T. Paving New Roads for CARs. *Trends Cancer* **2019**, *5*, 583–592. [CrossRef]

26. Chandran, S.S.; Klebanoff, C.A. T cell receptor-based cancer immunotherapy: Emerging efficacy and pathways of resistance. *Immunol. Rev.* **2019**, *290*, 127–147. [CrossRef]

27. Majzner, R.G.; Mackall, C.L. Clinical lessons learned from the first leg of the CAR T cell journey. *Nat. Med.* **2019**, *25*, 1341–1355. [CrossRef]

28. Sadelain, M.; Papapetrou, E.P.; Bushman, F.D. Safe harbours for the integration of new DNA in the human genome. *Nat. Rev. Cancer* **2011**, *12*, 51–58. [CrossRef]

29. Eyquem, J.; Mansilla-Soto, J.; Giavridis, T.; Van Der Stegen, S.J.C.; Hamieh, M.; Cunanan, K.M.; Odak, A.; Gönen, M.; Sadelain, M. Targeting a CAR to the TRAC locus with CRISPR/Cas9 enhances tumour rejection. *Nature* **2017**, *543*, 113–117. [CrossRef]

30. Depil, S.; Duchateau, P.; Grupp, S.A.; Mufti, G.; Poirot, L. 'Off-the-shelf' allogeneic CAR T cells: Development and challenges. *Nat. Rev. Drug Discov.* **2020**, *19*, 185–199. [CrossRef] [PubMed]

31. Anderson, B.D.; Nakamura, T.; Russell, S.J.; Peng, K.W. High CD46 receptor density deter-mines preferential killing of tumor cells by oncolytic measles virus. *Cancer Res.* **2004**, *64*, 4919–4926. [CrossRef]

32. Au, G.G.; Lincz, L.F.; Enno, A.; Shafren, D.R. Oncolytic Coxsackievirus A21 as a novel therapy for multiple myeloma. *Br. J. Haematol.* **2007**, *137*, 133–141. [CrossRef]

33. Martuza, R.L.; Malick, A.; Markert, J.M.; Ruffner, K.L.; Coen, D.M. Experimental therapy of human glioma by means of a genetically engineered virus mutant. *Science* **1991**, *252*, 854–856. [CrossRef] [PubMed]

34. Gammon, D.B.; Gowrishankar, B.; Duraffour, S.; Andrei, G.; Upton, C.; Evans, D.H. Vaccinia Virus–Encoded Ribonucleotide Reductase Subunits Are Differentially Required for Replication and Pathogenesis. *PLoS Pathog.* **2010**, *6*, e1000984. [CrossRef]

35. Peters, C.; Rabkin, S.D. Designing herpes viruses as oncolytics. *Mol. Ther. Oncolytics* **2015**, *2*, 15010. [CrossRef] [PubMed]

36. Parato, K.A.; Breitbach, C.J.; Le Boeuf, F.; Wang, J.; Storbeck, C.; Ilkow, C.; Diallo, J.-S.; Falls, T.; Burns, J.; Garcia, V.; et al. The Oncolytic Poxvirus JX-594 Selectively Replicates in and Destroys Cancer Cells Driven by Genetic Pathways Commonly Activated in Cancers. *Mol. Ther.* **2012**, *20*, 749–758. [CrossRef] [PubMed]

37. Baker, A.T.; Aguirre-Hernández, C.; Halldén, G.; Parker, A.L. Designer Oncolytic Adenovirus: Coming of Age. *Cancers* **2018**, *10*, 201. [CrossRef]

38. Ilkow, C.S.; Swift, S.L.; Bell, J.C.; Diallo, J.-S. From Scourge to Cure: Tumour-Selective Viral Pathogenesis as a New Strategy against Cancer. *PLoS Pathog.* **2014**, *10*, e1003836. [CrossRef]

39. Pikor, L.A.; Bell, J.C.; Diallo, J.-S. Oncolytic Viruses: Exploiting Cancer's Deal with the Devil. *Trends Cancer* **2015**, *1*, 266–277. [CrossRef] [PubMed]

40. Li, S.; Zhu, M.; Pan, R.; Fang, T.; Cao, Y.-Y.; Chen, S.; Zhao, X.; Lei, C.-Q.; Guo, L.; Chen, Y.; et al. The tumor suppressor PTEN has a critical role in antiviral innate immunity. *Nat. Immunol.* **2015**, *17*, 241–249. [CrossRef]

41. Stojdl, D.F.; Lichty, B.D.; Tenoever, B.R.; Paterson, J.M.; Power, A.T.; Knowles, S.; Marius, R.; Reynard, J.; Poliquin, L.; Atkins, H.; et al. VSV strains with defects in their ability to shutdown innate immunity are potent systemic anti-cancer agents. *Cancer Cell* **2003**, *4*, 263–275. [CrossRef]

42. Fiola, C.; Peeters, B.; Fournier, P.; Arnold, A.; Bucur, M.; Schirrmacher, V. Tumor selective replication of Newcastle disease virus: Association with defects of tumor cells in antiviral defence. *Int. J. Cancer* **2006**, *119*, 328–338. [CrossRef]

43. Ahmed, M.; Cramer, S.D.; Lyles, D.S. Sensitivity of prostate tumors to wild type and M protein mutant vesicular stomatitis viruses. *Virology* **2004**, *330*, 34–49. [CrossRef]

44. Obuchi, M.; Fernandez, M.; Barber, G.N. Development of recombinant vesicular stomatitis vi-ruses that exploit defects in host defense to augment specific oncolytic activity. *J. Virol.* **2003**, *77*, 8843–8856. [CrossRef]

45. Willmon, C.L.; Saloura, V.; Fridlender, Z.G.; Wongthida, P.; Diaz, R.M.; Thompson, J.; Kottke, T.; Federspiel, M.; Barber, G.; Vile, R.G.; et al. Expression of IFN-beta enhances both efficacy and safety of oncolytic vesicu-lar stomatitis virus for therapy of mesothelioma. *Cancer Res.* **2009**, *69*, 7713–7720. [CrossRef] [PubMed]

46. Crouse, J.; Kalinke, U.; Oxenius, A. Regulation of antiviral T cell responses by type I interferons. *Nat. Rev. Immunol.* **2015**, *15*, 231–242. [CrossRef]

47. Shaw, A.R.; Porter, C.E.; Watanabe, N.; Tanoue, K.; Sikora, A.; Gottschalk, S.; Brenner, M.K.; Suzuki, M. Adenovirotherapy Delivering Cytokine and Checkpoint Inhibitor Augments CAR T Cells against Metastatic Head and Neck Cancer. *Mol. Ther.* **2017**, *25*, 2440–2451. [CrossRef]

48. Fridman, W.H.; Pagès, F.; Sautès-Fridman, C.; Galon, J. The immune contexture in human tumours: Impact on clinical outcome. *Nat. Rev. Cancer* **2012**, *12*, 298–306. [CrossRef]
49. Kerkar, S.P.; Restifo, N.P. Cellular Constituents of Immune Escape within the Tumor Microenvironment: Figure 1. *Cancer Res.* **2012**, *72*, 3125–3130. [CrossRef] [PubMed]
50. Mullins, I.M.; Slingluff, C.L.; Lee, J.K.; Garbee, C.F.; Shu, J.; Anderson, S.G.; Mayer, M.E.; Knaus, W.A.; Mullins, D.W. CXC Chemokine Receptor 3 Expression by Activated CD8+ T cells Is Associated with Survival in Melanoma Patients with Stage III Disease. *Cancer Res.* **2004**, *64*, 7697–7701. [CrossRef] [PubMed]
51. Harlin, H.; Meng, Y.; Peterson, A.C.; Zha, Y.; Tretiakova, M.; Slingluff, C.; McKee, M.; Gajewski, T.F. Chemokine Expression in Melanoma Metastases Associated with CD8+ T-Cell Recruitment. *Cancer Res.* **2009**, *69*, 3077–3085. [CrossRef]
52. Mikucki, M.E.; Fisher, D.T.; Matsuzaki, J.; Skitzki, J.J.; Gaulin, N.B.; Muhitch, J.B.; Ku, A.W.; Frelinger, J.G.; Odunsi, K.; Gajewski, T.F.; et al. Non-redundant requirement for CXCR3 signalling during tumoricidal T-cell trafficking across tumour vascular checkpoints. *Nat. Commun.* **2015**, *6*, 7458. [CrossRef]
53. Peng, D.; Kryczek, I.; Nagarsheth, N.; Zhao, L.; Wei, S.; Wang, W.; Sun, Y.; Zhao, E.; Vatan, L.; Szeliga, W.; et al. Epigenetic silencing of TH1-type chemokines shapes tumour immunity and immunotherapy. *Nature* **2015**, *527*, 249–253. [CrossRef]
54. Da Silva, R.B.; Laird, M.E.; Yatim, N.; Fiette, L.; Ingersoll, M.A.; Albert, M.L. Dipeptidylpeptidase 4 inhibition enhances lymphocyte trafficking, improving both naturally occurring tumor immunity and immunotherapy. *Nat. Immunol.* **2015**, *16*, 850–858. [CrossRef] [PubMed]
55. Fridlender, Z.G.; Buchlis, G.; Kapoor, V.; Cheng, G.; Sun, J.; Singhal, S.; Crisanti, M.C.; Wang, L.-C.S.; Heitjan, D.; Snyder, L.A.; et al. CCL2 Blockade Augments Cancer Immunotherapy. *Cancer Res.* **2009**, *70*, 109–118. [CrossRef] [PubMed]
56. Lesokhin, A.M.; Hohl, T.M.; Kitano, S.; Cortez, C.; Hirschhorn-Cymerman, D.; Avogadri, F.; Rizzuto, G.A.; Lazarus, J.J.; Pamer, E.G.; Houghton, A.N.; et al. Monocytic CCR2+ Myeloid-Derived Suppressor Cells Promote Immune Escape by Limiting Activated CD8 T-cell Infiltration into the Tumor Microenvironment. *Cancer Res.* **2011**, *72*, 876–886. [CrossRef]
57. Argyle, D.; Kitamura, T. Targeting Macrophage-Recruiting Chemokines as a Novel Therapeutic Strategy to Prevent the Progression of Solid Tumors. *Front. Immunol.* **2018**, *9*, 2629. [CrossRef] [PubMed]
58. Gabrilovich, D.I.; Nagaraj, S. Myeloid-derived suppressor cells as regulators of the immune system. *Nat. Rev. Immunol.* **2009**, *9*, 162–174. [CrossRef] [PubMed]
59. Yu, J.; Du, W.; Yan, F.; Wang, Y.; Li, H.; Cao, S.; Yu, W.; Shen, C.; Liu, J.; Ren, X. Myeloid-Derived Suppressor Cells Suppress Antitumor Immune Responses through IDO Expression and Correlate with Lymph Node Metastasis in Patients with Breast Cancer. *J. Immunol.* **2013**, *190*, 3783–3797. [CrossRef]
60. Han, S.; Toker, A.; Liu, Z.Q.; Ohashi, P.S. Turning the Tide against Regulatory T Cells. *Front. Oncol.* **2019**, *9*, 279. [CrossRef]
61. Rodriguez-Garcia, A.; Palazon, A.; Noguera-Ortega, E.; Powell, D.J.J.; Guedan, S. CAR-T Cells Hit the Tumor Microenvironment: Strategies to Overcome Tumor Escape. *Front. Immunol.* **2020**, *11*, 1109. [CrossRef]
62. Mogensen, T.H. Pathogen Recognition and Inflammatory Signaling in Innate Immune Defenses. *Clin. Microbiol. Rev.* **2009**, *22*, 240–273. [CrossRef]
63. Wongthida, P.; Diaz, R.M.; Galivo, F.; Kottke, T.; Thompson, J.; Melcher, A.; Vile, R. VSV Oncolytic Virotherapy in the B16 Model Depends Upon Intact MyD88 Signaling. *Mol. Ther.* **2011**, *19*, 150–158. [CrossRef] [PubMed]
64. Workenhe, S.T.; Mossman, K.L. Oncolytic Virotherapy and Immunogenic Cancer Cell Death: Sharpening the Sword for Improved Cancer Treatment Strategies. *Mol. Ther.* **2014**, *22*, 251–256. [CrossRef] [PubMed]
65. Van Vloten, J.P.; Workenhe, S.T.; Wootton, S.K.; Mossman, K.L.; Bridle, B.W. Critical Interactions between Immunogenic Cancer Cell Death, Oncolytic Viruses, and the Immune System Define the Rational Design of Combination Immunotherapies. *J. Immunol.* **2018**, *200*, 450–458. [CrossRef] [PubMed]
66. Groom, J.R.; Luster, A.D. CXCR3 ligands: Redundant, collaborative and antagonistic functions. *Immunol. Cell Biol.* **2011**, *89*, 207–215. [CrossRef] [PubMed]
67. Chon, H.J.; Lee, W.S.; Yang, H.; Kong, S.J.; Lee, N.K.; Moon, E.S.; Choi, J.; Han, E.C.; Kim, J.H.; Ahn, J.B.; et al. Tumor microenvironment remodeling by intratumoral oncolytic vaccinia virus enhances the efficacy of immune checkpoint blockade. *Clin. Cancer Res.* **2018**, *25*, 1612–1623. [CrossRef] [PubMed]
68. Durham, N.M.; Mulgrew, K.; McGlinchey, K.; Monks, N.R.; Ji, H.; Herbst, R.; Suzich, J.; Hammond, S.A.; Kelly, E.J. Oncolytic VSV Primes Differential Responses to Immuno-oncology Therapy. *Mol. Ther.* **2017**, *25*, 1917–1932. [CrossRef]
69. Katayama, Y.; Tachibana, M.; Kurisu, N.; Oya, Y.; Terasawa, Y.; Goda, H.; Kobiyama, K.; Ishii, K.J.; Akira, S.; Mizuguchi, H.; et al. Oncolytic Reovirus Inhibits Immunosuppressive Activity of Myeloid-Derived Suppressor Cells in a TLR3-Dependent Manner. *J. Immunol.* **2018**, *200*, 2987–2999. [CrossRef]
70. Engeland, C.E.; Grossardt, C.; Veinalde, R.; Bossow, S.; Lutz, D.; Kaufmann, J.K.; Shevchenko, I.; Umansky, V.; Nettelbeck, D.M.; Weichert, W.; et al. CTLA-4 and PD-L1 Checkpoint Blockade Enhances Oncolytic Measles Virus Therapy. *Mol. Ther.* **2014**, *22*, 1949–1959. [CrossRef]
71. Zamarin, D.; Holmgaard, R.B.; Subudhi, S.K.; Park, J.S.; Mansour, M.; Palese, P.; Merghoub, T.; Wolchok, J.D.; Allison, J.P. Localized Oncolytic Virotherapy Overcomes Systemic Tumor Resistance to Immune Checkpoint Blockade Immunotherapy. *Sci. Transl. Med.* **2014**, *6*, 226ra32. [CrossRef]
72. Saha, D.; Martuza, R.L.; Rabkin, S.D. Macrophage Polarization Contributes to Glioblastoma Eradication by Combination Immunovirotherapy and Immune Checkpoint Blockade. *Cancer Cell* **2017**, *32*, 253–267.e5. [CrossRef]

73. Brown, M.C.; Holl, E.K.; Boczkowski, D.; Dobrikova, E.; Mosaheb, M.; Chandramohan, V.; Bigner, D.D.; Gromeier, M.; Nair, S.K. Cancer immunotherapy with recombinant poliovirus induces IFN-dominant activation of dendritic cells and tumor antigen–specific CTLs. *Sci. Transl. Med.* **2017**, *9*, eaan4220. [CrossRef]

74. Cheema, T.A.; Wakimoto, H.; Fecci, P.E.; Ning, J.; Kuroda, T.; Jeyaretna, D.S.; Martuza, R.L.; Rabkin, S.D. Multifaceted oncolytic virus therapy for glioblastoma in an immunocompetent cancer stem cell model. *Proc. Natl. Acad. Sci. USA* **2013**, *110*, 12006–12011. [CrossRef]

75. Wang, G.; Kang, X.; Chen, K.S.; Jehng, T.; Jones, L.; Chen, J.; Huang, X.F.; Chen, S.-Y. An engineered oncolytic virus expressing PD-L1 inhibitors activates tumor neoantigen-specific T cell responses. *Nat. Commun.* **2020**, *11*, 1–14. [CrossRef] [PubMed]

76. Shen, W.; Patnaik, M.M.; Ruiz, A.; Russell, S.J.; Peng, K.-W. Immunovirotherapy with vesicular stomatitis virus and PD-L1 blockade enhances therapeutic outcome in murine acute myeloid leukemia. *Blood* **2016**, *127*, 1449–1458. [CrossRef] [PubMed]

77. Rajani, K.; Parrish, C.; Kottke, T.; Thompson, J.; Zaidi, S.; Ilett, L.; Shim, K.G.; Diaz, R.-M.; Pandha, H.; Harrington, K.; et al. Combination Therapy With Reovirus and Anti-PD-1 Blockade Controls Tumor Growth Through Innate and Adaptive Immune Responses. *Mol. Ther.* **2016**, *24*, 166–174. [CrossRef]

78. Hardcastle, J.; Mills, L.; Malo, C.S.; Jin, F.; Kurokawa, C.; Geekiyanage, H.; Schroeder, M.; Sarkaria, J.; Johnson, A.J.; Galanis, E. Immunovirotherapy with measles virus strains in combination with anti–PD-1 antibody blockade enhances antitumor activity in glioblastoma treatment. *Neuro Oncol.* **2016**, *19*, 493–502. [CrossRef] [PubMed]

79. Ribas, A.; Dummer, R.; Puzanov, I.; VanderWalde, A.; Andtbacka, R.H.; Michielin, O.; Olszanski, A.J.; Malvehy, J.; Cebon, J.; Fernandez, E.; et al. Oncolytic Virotherapy Promotes Intratumoral T Cell Infiltration and Improves Anti-PD-1 Immunotherapy. *Cell* **2017**, *170*, 1109–1119.e10. [CrossRef]

80. Samson, A.; Scott, K.J.; Taggart, D.; West, E.J.; Wilson, E.; Nuovo, G.J.; Thomson, S.; Corns, R.; Mathew, R.K.; Fuller, M.J.; et al. Intravenous delivery of oncolytic reovirus to brain tumor patients immunologically primes for subsequent checkpoint blockade. *Sci. Transl. Med.* **2018**, *10*, eaam7577. [CrossRef] [PubMed]

81. Packiriswamy, N.; Upreti, D.; Zhou, Y.; Khan, R.; Miller, A.; Diaz, R.M.; Rooney, C.M.; Dispenzieri, A.; Peng, K.-W.; Russell, S.J. Oncolytic measles virus therapy enhances tumor antigen-specific T-cell responses in patients with multiple myeloma. *Leukemia* **2020**, *34*, 3310–3322. [CrossRef]

82. Keppler, S.J.; Rosenits, K.; Koegl, T.; Vucikuja, S.; Aichele, P. Signal 3 Cytokines as Modulators of Primary Immune Responses during Infections: The Interplay of Type I IFN and IL-12 in CD8 T Cell Responses. *PLoS ONE* **2012**, *7*, e40865. [CrossRef] [PubMed]

83. Bahl, K.; Kim, S.-K.; Calcagno, C.; Ghersi, D.; Puzone, R.; Celada, F.; Selin, L.K.; Welsh, R.M. IFN-Induced Attrition of CD8 T Cells in the Presence or Absence of Cognate Antigen during the Early Stages of Viral Infections. *J. Immunol.* **2006**, *176*, 4284–4295. [CrossRef] [PubMed]

84. Bahl, K.; Hübner, A.; Davis, R.J.; Welsh, R.M. Analysis of Apoptosis of Memory T Cells and Dendritic Cells during the Early Stages of Viral Infection or Exposure to Toll-Like Receptor Agonists. *J. Virol.* **2010**, *84*, 6262. [CrossRef]

85. Jangalwe, S.; Kapoor, V.N.; Xu, J.; Girnius, N.; Kennedy, N.J.; Edwards, Y.J.K.; Welsh, R.M.; Davis, R.J.; Brehm, M.A. Cutting Edge: Early Attrition of Memory T Cells during Inflammation and Costimulation Blockade Is Regulated Concurrently by Proapoptotic Proteins Fas and Bim. *J. Immunol.* **2019**, *202*, 647–651. [CrossRef] [PubMed]

86. Welsh, R.M.; Bahl, K.; Marshall, H.D.; Urban, S.L. Type 1 Interferons and Antiviral CD8 T-Cell Responses. *PLoS Pathog.* **2012**, *8*, e1002352. [CrossRef] [PubMed]

87. Evgin, L.; Huff, A.L.; Wongthida, P.; Thompson, J.; Kottke, T.; Tonne, J.; Schuelke, M.; Ayasoufi, K.; Driscoll, C.B.; Shim, K.G.; et al. Oncolytic virus-derived type I interferon restricts CAR T cell therapy. *Nat. Commun.* **2020**, *11*, 1–15. [CrossRef] [PubMed]

88. Crouse, J.; Bedenikovic, G.; Wiesel, M.; Ibberson, M.; Xenarios, I.; Von Laer, D.; Kalinke, U.; Vivier, E.; Jonjic, S.; Oxenius, A. Type I Interferons Protect T Cells against NK Cell Attack Mediated by the Activating Receptor NCR1. *Immunology* **2014**, *40*, 961–973. [CrossRef]

89. Xu, H.C.; Grusdat, M.; Pandyra, A.A.; Polz, R.; Huang, J.; Sharma, P.; Deenen, R.; Köhrer, K.; Rahbar, R.; Diefenbach, A.; et al. Type I Interferon Protects Antiviral CD8+ T Cells from NK Cell Cytotoxicity. *Immunology* **2014**, *40*, 949–960. [CrossRef]

90. Arulanandam, R.; Batenchuk, C.; Angarita, F.A.; Ottolino-Perry, K.; Cousineau, S.; Mottashed, A.; Burgess, E.; Falls, T.J.; De Silva, N.; Tsang, J.; et al. VEGF-Mediated Induction of PRD1-BF1/Blimp1 Expression Sensitizes Tumor Vasculature to Oncolytic Virus Infection. *Cancer Cell* **2015**, *28*, 210–224. [CrossRef]

91. Kottke, T.; Hall, G.; Pulido, J.; Diaz, R.M.; Thompson, J.; Chong, H.; Selby, P.; Coffey, M.; Pandha, H.; Chester, J.; et al. Antiangiogenic cancer therapy combined with oncolytic virotherapy leads to regression of established tumors in mice. *J. Clin. Investig.* **2010**, *120*, 1551–1560. [CrossRef] [PubMed]

92. Breitbach, C.J.; De Silva, N.S.; Falls, T.J.; Aladl, U.; Evgin, L.; Paterson, J.M.; Sun, Y.Y.; Roy, D.G.; Rintoul, J.L.; Daneshmand, M.; et al. Targeting Tumor Vasculature With an Oncolytic Virus. *Mol. Ther.* **2011**, *19*, 886–894. [CrossRef] [PubMed]

93. Breitbach, C.J.; Arulanandam, R.; De Silva, N.; Thorne, S.H.; Patt, R.; Daneshmand, M.; Moon, A.; Ilkow, C.; Burke, J.; Hwang, T.-H.; et al. Oncolytic Vaccinia Virus Disrupts Tumor-Associated Vasculature in Humans. *Cancer Res.* **2013**, *73*, 1265–1275. [CrossRef] [PubMed]

94. Hou, W.; Chen, H.; Rojas, J.; Sampath, P.; Thorne, S.H. Oncolytic vaccinia virus demonstrates antiangiogenic effects mediated by targeting of VEGF. *Int. J. Cancer* **2014**, *135*, 1238–1246. [CrossRef] [PubMed]

95. Matuszewska, K.; Santry, L.A.; Van Vloten, J.P.; Auyeung, A.W.; Major, P.P.; Lawler, J.; Wootton, S.K.; Bridle, B.W.; Petrik, J. Combining Vascular Normalization with an Oncolytic Virus Enhances Immunotherapy in a Preclinical Model of Advanced-Stage Ovarian Cancer. *Clin. Cancer Res.* **2018**, *25*, 1624–1638. [CrossRef] [PubMed]

96. Nishio, N.; Diaconu, I.; Liu, H.; Cerullo, V.; Caruana, I.; Hoyos, V.; Bouchier-Hayes, L.; Savoldo, B.; Dotti, G. Armed Oncolytic Virus Enhances Immune Functions of Chimeric Antigen Receptor–Modified T Cells in Solid Tumors. *Cancer Res.* **2014**, *74*, 5195–5205. [CrossRef]

97. Moon, E.K.; Wang, L.-C.S.; Bekdache, K.; Lynn, R.C.; Lo, A.; Thorne, S.H.; Albelda, S.M. Intra-tumoral delivery of CXCL11 via a vaccinia virus, but not by modified T cells, enhances the efficacy of adoptive T cell therapy and vaccines. *OncoImmunology* **2017**, *7*, e1395997. [CrossRef]

98. Tanoue, K.; Shaw, A.R.; Watanabe, N.; Porter, C.; Rana, B.; Gottschalk, S.; Brenner, M.; Suzuki, M. Armed Oncolytic Adenovirus–Expressing PD-L1 Mini-Body Enhances Antitumor Effects of Chimeric Antigen Receptor T Cells in Solid Tumors. *Cancer Res.* **2017**, *77*, 2040–2051. [CrossRef] [PubMed]

99. Lasek, W.; Zagożdżon, R.; Jakobisiak, M. Interleukin 12: Still a promising candidate for tumor immunotherapy? *Cancer Immunol. Immunother.* **2014**, *63*, 419–435. [CrossRef]

100. Watanabe, K.; Luo, Y.; Da, T.; Guedan, S.; Ruella, M.; Scholler, J.; Keith, B.; Young, R.M.; Engels, B.; Sorsa, S.; et al. Pancreatic cancer therapy with combined mesothelin-redirected chimeric antigen receptor T cells and cytokine-armed oncolytic adenoviruses. *JCI Insight* **2018**, *3*. [CrossRef]

101. Aalipour, A.; Le Boeuf, F.; Tang, M.; Murty, S.; Simonetta, F.; Lozano, A.X.; Shaffer, T.M.; Bell, J.C.; Gambhir, S.S. Viral Delivery of CAR Targets to Solid Tumors Enables Effective Cell Therapy. *Mol. Ther. Oncolytics* **2020**, *17*, 232–240. [CrossRef]

102. Park, A.K.; Fong, Y.; Kim, S.-I.; Yang, J.; Murad, J.P.; Lu, J.; Jeang, B.; Chang, W.-C.; Chen, N.G.; Thomas, S.H.; et al. Effective combination immunotherapy using oncolytic viruses to deliver CAR targets to solid tumors. *Sci. Transl. Med.* **2020**, *12*, eaaz1863. [CrossRef] [PubMed]

103. Orlando, E.J.; Han, X.; Tribouley, C.; Wood, P.A.; Leary, R.J.; Riester, M.; Levine, J.E.; Qayed, M.; Grupp, S.A.; Boyer, M.; et al. Genetic mechanisms of target antigen loss in CAR19 therapy of acute lymphoblastic leukemia. *Nat. Med.* **2018**, *24*, 1504–1506. [CrossRef]

104. O'Rourke, D.M.; Nasrallah, M.P.; Desai, A.; Melenhorst, J.J.; Mansfield, K.; Morrissette, J.J.D.; Martinez-Lage, M.; Brem, S.; Maloney, E.; Shen, A.; et al. A single dose of peripherally infused EGFRvIII-directed CAR T cells mediates antigen loss and induces adaptive resistance in patients with recurrent glioblastoma. *Sci. Transl. Med.* **2017**, *9*, eaaa0984. [CrossRef] [PubMed]

105. Bourgeois-Daigneault, M.-C.; Roy, D.G.; Aitken, A.S.; El Sayes, N.; Martin, N.T.; Varette, O.; Falls, T.; St-Germain, L.E.; Pelin, A.; Lichty, B.D.; et al. Neoadjuvant oncolytic virotherapy before surgery sensitizes triple-negative breast cancer to immune checkpoint therapy. *Sci. Transl. Med.* **2018**, *10*, eaao1641. [CrossRef]

106. Sampson, J.H.; Choi, B.D.; Sanchez-Perez, L.; Suryadevara, C.M.; Snyder, D.J.; Flores, C.T.; Schmittling, R.J.; Nair, S.K.; Reap, E.A.; Norberg, P.K.; et al. EGFRvIII mCAR-Modified T-Cell Therapy Cures Mice with Established Intracerebral Glioma and Generates Host Immunity against Tumor-Antigen Loss. *Clin. Cancer Res.* **2013**, *20*, 972–984. [CrossRef]

107. Beatty, G.L.; O'Hara, M.H.; Lacey, S.F.; Torigian, D.A.; Nazimuddin, F.; Chen, F.; Kulikovskaya, I.M.; Soulen, M.C.; McGarvey, M.; Nelson, A.M.; et al. Activity of Mesothelin-Specific Chimeric Antigen Receptor T Cells Against Pancreatic Carcinoma Metastases in a Phase 1 Trial. *Gastroenterology* **2018**, *155*, 29–32. [CrossRef]

108. Diaz, R.M.; Galivo, F.; Kottke, T.; Wongthida, P.; Qiao, J.; Thompson, J.; Valdes, M.; Barber, G.; Vile, R.G. Oncolytic Immunovirotherapy for Melanoma Using Vesicular Stomatitis Virus. *Cancer Res.* **2007**, *67*, 2840–2848. [CrossRef] [PubMed]

109. Kottke, T.; Errington, F.; Pulido, J.; Galivo, F.; Thompson, J.; Wongthida, P.; Diaz, R.M.; Chong, H.; Ilett, E.; Vile, R.; et al. Broad antigenic coverage induced by viral cDNA library-based vaccination cures established tumors. *Nat. Med.* **2011**, *17*, 854–859. [CrossRef] [PubMed]

110. Pulido, J.; Kottke, T.; Thompson, J.; Galivo, F.; Wongthida, P.; Diaz, R.M.; Rommelfanger, D.; Ilett, E.; Pease, L.; Pandha, H.; et al. Using virally expressed melanoma cDNA libraries to identify tumor-associated antigens that cure melanoma. *Nat. Biotechnol.* **2012**, *30*, 337–343. [CrossRef]

111. Bridle, B.W.; Stephenson, K.B.; Boudreau, J.E.; Koshy, S.; Kazdhan, N.; Pullenayegum, E.; Brunellière, J.; Bramson, J.L.; Lichty, B.D.; Wan, Y. Potentiating Cancer Immunotherapy Using an Oncolytic Virus. *Mol. Ther.* **2010**, *18*, 1430–1439. [CrossRef]

112. Pol, J.G.; Zhang, L.; Bridle, B.W.; Stephenson, K.B.; Rességuier, J.; Hanson, S.; Chen, L.; Kazdhan, N.; Bramson, J.L.; Stojdl, D.F.; et al. Maraba Virus as a Potent Oncolytic Vaccine Vector. *Mol. Ther.* **2014**, *22*, 420–429. [CrossRef] [PubMed]

113. Atherton, M.J.; Stephenson, K.B.; Pol, J.; Wang, F.; Lefebvre, C.; Stojdl, D.F.; Nikota, J.K.; Dvorkin-Gheva, A.; Nguyen, A.; Chen, L.; et al. Customized Viral Immunotherapy for HPV-Associated Cancer. *Cancer Immunol. Res.* **2017**, *5*, 847–859. [CrossRef]

114. Overwijk, W.W.; Theoret, M.R.; Finkelstein, S.E.; Surman, D.R.; De Jong, L.A.; Vyth-Dreese, F.A.; Dellemijn, T.A.; Antony, P.A.; Spiess, P.J.; Palmer, D.C.; et al. Tumor Regression and Autoimmunity after Reversal of a Functionally Tolerant State of Self-reactive CD8+ T Cells. *J. Exp. Med.* **2003**, *198*, 569–580. [CrossRef] [PubMed]

115. Wongthida, P.; Diaz, R.M.; Pulido, C.; Rommelfanger, D.; Galivo, F.; Kaluza, K.; Kottke, T.; Thompson, J.; Melcher, A.; Vile, R. Activating Systemic T-Cell Immunity Against Self Tumor Antigens to Support Oncolytic Virotherapy with Vesicular Stomatitis Virus. *Hum. Gene Ther.* **2011**, *22*, 1343–1353. [CrossRef]

116. Walsh, S.R.; Simovic, B.; Chen, L.; Bastin, D.; Nguyen, A.; Stephenson, K.; Mandur, T.S.; Bramson, J.L.; Lichty, B.D.; Wan, Y. Endogenous T cells prevent tumor immune escape following adoptive T cell therapy. *J. Clin. Investig.* **2019**, *129*, 5400–5410. [CrossRef]

117. Slaney, C.Y.; Von Scheidt, B.; Davenport, A.J.; Beavis, P.A.; Westwood, J.A.; Mardiana, S.; Tscharke, D.C.; Ellis, S.; Prince, H.M.; Trapani, J.A.; et al. Dual-specific Chimeric Antigen Receptor T Cells and an Indirect Vaccine Eradicate a Variety of Large Solid Tumors in an Immunocompetent, Self-antigen Setting. *Clin. Cancer Res.* **2016**, *23*, 2478–2490. [CrossRef]

118. Tanaka, M.; Tashiro, H.; Omer, B.; Lapteva, N.; Ando, J.; Ngo, M.; Mehta, B.; Dotti, G.; Kinchington, P.R.; Leen, A.M.; et al. Vaccination Targeting Native Receptors to Enhance the Function and Proliferation of Chimeric Antigen Receptor (CAR)-Modified T Cells. *Clin. Cancer Res.* **2017**, *23*, 3499–3509. [CrossRef]

119. Lapteva, N.; Gilbert, M.; Diaconu, I.; Rollins, L.A.; Al-Sabbagh, M.; Naik, S.; Krance, R.A.; Tripic, T.; Hiregange, M.; Raghavan, D.; et al. T-Cell Receptor Stimulation Enhances the Expansion and Function of CD19 Chimeric Antigen Receptor–Expressing T Cells. Clin. *Cancer Res.* **2019**, *25*, 7340–7350. [CrossRef]

120. Yu, F.; Wang, X.; Guo, Z.S.; Bartlett, D.L.; Gottschalk, S.M.; Song, X.-T. T-cell Engager-armed Oncolytic Vaccinia Virus Significantly Enhances Antitumor Therapy. *Mol. Ther.* **2014**, *22*, 102–111. [CrossRef] [PubMed]

121. Speck, T.; Heidbuechel, J.P.; Veinalde, R.; Jaeger, D.; Von Kalle, C.; Ball, C.R.; Ungerechts, G.; Engeland, C.E. Targeted BiTE Expression by an Oncolytic Vector Augments Therapeutic Efficacy against Solid Tumors. *Clin. Cancer Res.* **2018**, *24*, 2128–2137. [CrossRef]

122. Freedman, J.D.; Duffy, M.R.; Lei-Rossmann, J.; Muntzer, A.; Scott, E.M.; Hagel, J.; Campo, L.; Bryant, R.J.; Verrill, C.; Lambert, A.; et al. An Oncolytic Virus Expressing a T-cell Engager Simultaneously Targets Cancer and Immunosuppressive Stromal Cells. *Cancer Res.* **2018**, *78*, 6852–6865. [CrossRef]

123. Choi, B.D.; Yu, X.; Castano, A.P.; Bouffard, A.A.; Schmidts, A.; Larson, R.C.; Bailey, S.R.; Boroughs, A.C.; Frigault, M.J.; Leick, M.B.; et al. CAR-T cells secreting BiTEs circumvent antigen escape without detectable toxicity. *Nat. Biotechnol.* **2019**, *37*, 1049–1058. [CrossRef] [PubMed]

124. Wing, A.; Fajardo, C.A.; Posey, A.D.; Shaw, C.; Da, T.; Young, R.M.; Alemany, R.; June, C.H.; Guedan, S. Improving CART-Cell Therapy of Solid Tumors with Oncolytic Virus–Driven Production of a Bispecific T-cell Engager. *Cancer Immunol. Res.* **2018**, *6*, 605–616. [CrossRef]

125. Porter, C.E.; Shaw, A.R.; Jung, Y.; Yip, T.; Castro, P.D.; Sandulache, V.C.; Sikora, A.; Gottschalk, S.; Ittman, M.M.; Brenner, M.K.; et al. Oncolytic Adenovirus Armed with BiTE, Cytokine, and Checkpoint Inhibitor Enables CAR T Cells to Control the Growth of Heterogeneous Tumors. *Mol. Ther.* **2020**, *28*, 1251–1262. [CrossRef]

126. Qiao, J.; Kottke, T.; Willmon, C.; Galivo, F.; Wongthida, P.; Diaz, R.M.; Thompson, J.R.; Ryno, P.; Barber, G.N.; Chester, J.D.; et al. Purging metastases in lymphoid organs using a combination of antigen-nonspecific adoptive T cell therapy, oncolytic virotherapy and immunotherapy. *Nat. Med.* **2007**, *14*, 37–44. [CrossRef] [PubMed]

127. Qiao, J.; Wang, H.; Kottke, T.; Diaz, R.M.; Willmon, C.L.; Hudacek, A.W.; Thompson, J.R.; Parato, K.; Bell, J.C.; Naik, J.D.; et al. Loading of oncolytic vesicular stomatitis virus onto antigen-specific T cells enhances the efficacy of adoptive T-cell therapy of tumors. *Gene Ther.* **2008**, *15*, 604–616. [CrossRef]

128. Kottke, T.; Diaz, R.M.; Kaluza, K.; Pulido, J.; Galivo, F.; Wongthida, P.; Thompson, J.; Willmon, C.; Barber, G.N.; Chester, J.; et al. Use of Biological Therapy to Enhance Both Virotherapy and Adoptive T-Cell Therapy for Cancer. *Mol. Ther.* **2008**, *16*, 1910–1918. [CrossRef] [PubMed]

129. Ilett, E.J.; Prestwich, R.J.; Kottke, T.; Errington, F.; Thompson, J.M.; Harrington, K.J.; Pandha, H.S.; Coffey, M.; Selby, P.J.; Vile, R.G.; et al. Dendritic cells and T cells deliver oncolytic reovirus for tumour killing despite pre-existing anti-viral immunity. *Gene Ther.* **2009**, *16*, 689–699. [CrossRef] [PubMed]

130. Thorne, S.H.; Negrin, R.S.; Contag, C.H. Synergistic Antitumor Effects of Immune Cell-Viral Biotherapy. *Science* **2006**, *311*, 1780–1784. [CrossRef]

131. Thorne, S.H.; Liang, W.; Sampath, P.; Schmidt, T.; Sikorski, R.; Beilhack, A.; Contag, C.H. Targeting Localized Immune Suppression Within the Tumor Through Repeat Cycles of Immune Cell-oncolytic Virus Combination Therapy. *Mol. Ther.* **2010**, *18*, 1698–1705. [CrossRef] [PubMed]

132. VanSeggelen, H.; Tantalo, D.G.; Afsahi, A.; Hammill, J.A.; Bramson, J.L. Chimeric antigen receptor–engineered T cells as oncolytic virus carriers. *Mol. Ther. Oncolytics* **2015**, *2*, 15014. [CrossRef] [PubMed]

133. Huber, J.P.; Farrar, J.D. Regulation of effector and memory T-cell functions by type I interferon. *Immunology* **2011**, *132*, 466–474. [CrossRef] [PubMed]

Review

From Conventional Therapies to Immunotherapy: Melanoma Treatment in Review

Lukasz Kuryk [1,2,*], Laura Bertinato [3], Monika Staniszewska [4,5], Katarzyna Pancer [1], Magdalena Wieczorek [1], Stefano Salmaso [3], Paolo Caliceti [3] and Mariangela Garofalo [3,*]

1. Department of Virology, National Institute of Public Health-National Institute of Hygiene, Chocimska 24, 00-791 Warsaw, Poland; kpancer@pzh.gov.pl (K.P.); mrechnio@pzh.gov.pl (M.W.)
2. Clinical Science, Targovax Oy, Saukonpaadenranta 2, 00180 Helsinki, Finland
3. Department of Pharmaceutical and Pharmacological Sciences, University of Padova, Via F. Marzolo 5, 35131 Padova, Italy; laura.bertinato@studenti.unipd.it (L.B.); stefano.salmaso@unipd.it (S.S.); paolo.caliceti@unipd.it (P.C.)
4. Chair of Drug and Cosmetics Biotechnology, Faculty of Chemistry, Warsaw University of Technology, Noakowskiego 3, 00-664 Warsaw, Poland; mstaniszewska@ch.pw.edu.pl
5. Centre for Advanced Materials and Technologies, Warsaw University of Technology, Poleczki 19, 02-822 Warsaw, Poland
* Correspondence: lkuryk@pzh.gov.pl (L.K.); mariangela.garofalo@unipd.it (M.G.)

Received: 29 September 2020; Accepted: 18 October 2020; Published: 20 October 2020

Simple Summary: Here, we review the current state of knowledge in the field of cancer immunotherapy, focusing on the scientific rationale for the use of oncolytic viruses, checkpoint inhibitors and their combination to combat melanomas. Attention is also given to the immunological aspects of cancer therapy and the shift from conventional therapy towards immunotherapy. This review brings together information on how immunotherapy can be applied to support other cancer therapies in order to maximize the efficacy of melanoma treatment and improve clinical outcomes.

Abstract: In this review, we discuss the use of oncolytic viruses and checkpoint inhibitors in cancer immunotherapy in melanoma, with a particular focus on combinatory therapies. Oncolytic viruses are promising and novel anti-cancer agents, currently under investigation in many clinical trials both as monotherapy and in combination with other therapeutics. They have shown the ability to exhibit synergistic anticancer activity with checkpoint inhibitors, chemotherapy, radiotherapy. A coupling between oncolytic viruses and checkpoint inhibitors is a well-accepted strategy for future cancer therapies. However, eradicating advanced cancers and tailoring the immune response for complete tumor clearance is an ongoing problem. Despite current advances in cancer research, monotherapy has shown limited efficacy against solid tumors. Therefore, current improvements in virus targeting, genetic modification, enhanced immunogenicity, improved oncolytic properties and combination strategies have a potential to widen the applications of immuno-oncology (IO) in cancer treatment. Here, we summarize the strategy of combinatory therapy with an oncolytic vector to combat melanoma and highlight the need to optimize current practices and improve clinical outcomes.

Keywords: oncolytic viruses; melanoma; immunotherapy; checkpoint inhibitors; combinatory therapy

---

1. Introduction

Cancer is one of the three leading causes of death in industrialized countries, along with infectious and cardiovascular diseases. It is caused by the abnormal growth of the progeny of transformed cells, which have previously been subjected to mutations and several other alterations in the cell cycle and metabolism that contributed to giving these cells the typical tumor-like phenotype [1]. One of the most critical aspects in the fight against cancer is the tumor's ability to spread in the patient's body, even in locations far from the primary tumor location, developing metastasis [1]. This event could make the clinical picture significantly more complicated, since in order to cure cancer, all malignant cells in the patient's body need to be destroyed and removed, preferably without side effects for the patients [2].

The immune system (IS) is a complex system which is responsible for the protection of the human body. It consists of many cell types, structures and chemical mediators with different functions that can regulate each other to work effectively and neutralize components recognized as non-self. The idea that our immune system could act as a weapon or a prevention tool against cancer cells has always been particularly attractive, especially because of the specificity of the immune response that could be elicited. The first clue about the host immune system's alleged protective role against cancer emerged from a series of experiments on mice [3], in which it was noticed that mice previously immunized with irradiated tumor cells that were then challenged with an injection of tumor viable cells showed protection against the tumor. The same response was not observed in T cell deficient mice or mice which had been challenged with viable cells from a different tumor than the one used for the immunization process [3]. This evidence led to the discovery of the host immune system involvement in tumor-disruption and tumor prevention mechanisms, suggesting what many years of research have now shown, that is, that the host immune system has a role in the prevention and rejection of tumors [4]. However, since neither the immune system nor the tumor could be defined as simple networks, the relationship between them is obviously complex. This is due to the several factors which are involved in determining the evolution of tumorigenesis, among which there is also the immune system, which can surely exert an anti-tumor effect, but with specific subsets of immune cells, it may also perform a "foster" action on the tumor [5,6]. There are many ways that the IS could carry out its anti-tumor action. First, it protects the host from virus-induced tumors by eliminating or suppressing viral infection [7]; second, it promptly resolves inflammations, avoiding tissue exposure to an inflammatory environment, which is conducive to tumorigenesis [7,8]; third, the immune system is capable of specific recognition and disruption of tumor cells on the basis of their expression of molecules which work as antigens [7,9]. This last specific feature of IS is also known as immunosurveillance, and it is extremely important to guarantee a specific immune reaction which is directed only to tumor cells, sparing healthy tissue and avoiding many side effects [10]. This is possible because tumor cells are antigenic, meaning they express specific antigens usually called tumor associated antigens (TAAs), tumor specific antigens (TSA) or tumor rejection antigens (TRAs) [11–13]. The recognition and identification of these antigens is now a fundamental part in the development of effective immunotherapy, since they represent the main component with which T cells can recognize tumor cells to be activated and trigger the specific immune response. Most of the early efforts in antigens identification focused on shared tumor antigens, which could represent a valid alternative for a wide range of cancers, but these antigens are also expressed in a variety of self-tissues, leading to immunologic tolerance [14,15].

Therefore, the focus of research has slowly shifted to more tumor-specific antigens, usually generated from point mutations in normal genes, known as "neoantigens" [16,17]. Despite advances in conventional cancer therapies including chemotherapy, immuno-oncology is becoming more popular and effective in various cancer indications, including melanoma. Therefore, more conventional modalities seem to be gradually being replaced by more effective IO agents and their combinations (Table 1).

Table 1. The combinatory therapy of oncolytic vectors and CPIs for melanoma treatment.

OV	Checkpoint Inhibitor	Indication	Response Data	ClinicalTrials.gov ID
T-VEC	Ipilimumab	Melanoma	ORR 39% (comb.) vs. 18% (ipi alone)	NCT01740297
T-VEC	Pembrolizumab	Stage IIIB–IV melanoma	48% ORR	NCT02263508
T-VEC	Pembrolizumab	Stage III–IV melanoma	N/A	NCT02965716
HF-10	Ipilimumab	Melanoma	N/A	NCT031530085
HF-10	Ipilimumab	Melanoma	BORR 24% (at 24 weeks); median PFS 19 months; median OS 21.8 months	NCT02272855
HF-10	Nivolumab	Stage IIIB, IIIC, IVM1a melanoma	N/A	NCT03259425
CAVATAK	Ipilimumab	Uveal melanoma with liver metastasis	N/A	NCT03408587
CAVATAK	Pembrolizumab	Melanoma	N/A	NCT02565992
ONCOS-102	Pembrolizumab	Advanced or unresectable melanoma	N/A	NCT03003676

2. Melanoma—Epidemiology and Prevalence

Melanoma is the most aggressive type of skin cancer, and it arises from melanocytes, which are pigment-producing cells in the skin [18]. This type of cancer involves skin (mostly, but not exclusively, sun-exposed skin), but it can also occur in the eye, in the meninges and on gastrointestinal and genital mucosae [7]. In this section, we focus on cutaneous melanoma.

Melanomas can be characterized deeply from a histological point of view, thus leading to the identification of four major subtypes of melanoma [19]: Superficial spreading melanoma, nodular melanoma, lentigo malignant melanoma, and acral lentiginous melanoma. These four subtypes have different patterns of growth and come with different changes in epidermis and dermis [20]. According to a statistic evaluation carried out by the Global Cancer Observatory (GCO), which is part of the International Agency for Research on Cancer (IARC), melanoma incidence is annually increasing worldwide at a very fast rate, which in 2012 was the fastest growing of all types of cancer [21]. In GLOBOCAN 2018, the statistic evaluation of cancer incidence and mortality published by IARC, there were estimated to be approximately 290,000 new cases and 61,000 deaths related to melanoma [22], compared with the 232,000 new cases and 55,000 deaths reported in GLOBOCAN 2012. Melanoma mostly affects young and middle-aged individuals, with a median age at diagnosis of 57 years, while the incidence increases linearly from 25 years until 50 years of age, and then it decreases, especially for females [21]. Overall, the highest incidence is observed in regions with high exposure to solar radiation, such as Australia and New Zealand [23].

There are two types of risk factors for melanoma: (i) Environmental risk factors and (ii) host-related risk factors. Among the environmental risk factors commonly involved in cancer onset, for melanoma there is one particular risk factor which is deeply involved-ultraviolet (UV) light radiation from sunlight [21]. The correlation between sunlight exposure-particularly the UV-B spectrum [24]—and increased risk of melanoma has been deeply investigated, with findings that describe how exposure patterns and timing can contribute to the risk stratification for melanoma [21,25].

Intense and intermittent sun exposure is associated with a higher risk of melanoma, compared with continuous sun exposure, which is more often associated with non-melanoma skin cancers. UV-A exposure from artificial sources, such as sunbeds and devices employed in radiation phototherapy of psoriasis, is associated with a higher risk of melanoma [26]. There are a number of host risk factors related to the patient: (i) The number of congenital and acquired melanocytic nevi, which linearly correlates with melanoma incidence [21]; (ii) pigmentation characteristics of the patient, which are determined by polymorphisms in MC1R gene (melanocortin 1 receptor)—individuals with red hair, light complexion and light eyes exhibit a low pigmentation, and thus an increased risk for melanoma because of their higher sensitivity to UV exposure; (iii) family history of melanoma [21,27]; and (iv) immunosuppression, which is usually caused by comorbidities [28]. Melanoma diagnosis usually comes as an early-stage disease, in which it is possible to proceed with surgical excision and is curable in the majority of cases, while approximately 10% of patients are diagnosed at an advanced stage, which consists of an unresectable and/or metastatic melanoma [21,29]. Furthermore, stage IV melanomas are usually associated with a poor prognosis, lower probability to develop a consistent response to treatments, and, in about 30% of cases, there is brain and visceral involvement [30]. For patients with an advanced-stage melanoma, especially those who cannot undergo excisional surgery or who have metastasis, the wide range of systemic therapies represent the only way to defeat this aggressive type of cancer, which explains their importance and why they are being heavily investigated. This section provides a brief overview of the current available approaches to treat melanoma.

3. Conventional Cancer Therapies

3.1. Excisional Surgery

Surgery is taken into consideration, especially for early-stage melanomas. Excisional surgery is an effective strategy for most patients, but it is not always feasible, and in some cases (approximately 20%) the patient can present a relapse anyway, which is usually associated with a poor prognosis [31].

3.2. Chemotherapy

Chemotherapy for melanoma consists of the following two chemotherapeutics:

- Dacarbazine (DTIC): Approved by the FDA in 1975 for treatment of melanoma, it is an alkylating agent. Like every other chemotherapeutic drug, it is not highly selective for cancer cells over healthy cells, and the high number of clinical trials which have been carried out have reported a modest anti-tumor efficacy. Despite this, dacarbazine remains one of the first-line treatments for metastatic melanoma [32].
- Temozolomide: Despite being considered an analogue of dacarbazine, it has been studied because it has the advantage of oral administration, which is usually more versatile for the patient. Furthermore, temozolomide can reach the central nervous system and, since brain is one of the most common sites for melanoma to metastasize, this represents a crucial point for advanced melanoma treatment [32].

3.3. Targeted Therapies

Targeted therapies revolutionized melanoma treatment in 2011, when the first therapies were approved by FDA. They belong to the following classes:

- BRAF inhibitors: Since BRAF is the most frequently mutated oncogene in melanoma [33], its inhibitors have shown promising results in several clinical trials, with rapid regression of metastasis and positive responses in 50–60% melanoma patients [32,34]. The first drug belonging to this class that has been approved for melanoma is vemurafenib, a selective inhibitor of V600-mutant BRAF [33]. In a randomized phase III clinical trial (BRIM3), vemurafenib showed an objective response rate (ORR) of 48% versus 5% for dacarbazine, and a median progression-free survival (PFS) of 5.3 months versus 1.6 months for dacarbazine [33,35]. The second BRAF inhibitor came soon after the first one, with similar promising results [33]. Toxicities associated with this class of therapeutic agents include rash, arthralgia, fatigue, fever (for dabrafenib only) and photosensitivity (for vemurafenib only), but also the development of secondary non-melanoma cutaneous lesions, such as squamous-cell carcinoma [36,37].

- MEK inhibitors: The development of MEK inhibitors became a priority after the success with BRAF-inhibitors, and it was led by the acknowledgement that BRAF signaling is dependent on MEK1/2 downstream activation [33,38]. Trametinib belongs to this class of new targeted therapies [32], and represents the first drug of its class to be approved by the FDA as a single agent, since in the phase III METRIC clinical trial it showed an ORR of 22% and a median PFS of 4.8 months [39]. Aside from the use of MEK inhibitors to target BRAF-mutated melanomas, there is also preclinical evidence that indicates vulnerability to MEK inhibitors in a not insignificant number of melanomas which do not present BRAF V600 mutations, called wild-type BRAF melanomas (especially in NRAS-Q61-mutant tumors), and also in BRAF/NRAS wild-type melanomas, together with melanomas harboring non-V600 BRAF mutations [33,40].

A translational investigation led to evidence of a possible synergistic relationship between MEK and BRAF inhibitors. Since then, many combinatorial approaches of these two types of inhibitors have been investigated in clinical trials. The combination of vemurafenib and cobimetinib in a phase I study not only resulted in ORR and median PFS values that were very promising, but showed that the incidence of cutaneous hyperproliferative manifestations was substantially lower compared to BRAF inhibitor monotherapy [41]. The combination of BRAF and MEK inhibitors now forms the backbone of advanced BRAF-mutated melanoma treatment [33].

4. Cancer Immunotherapy

The goal of cancer immunotherapy is the stimulation or activation of immune responses against tumor cells, with the ultimate aim of eradicating cancer from the patient's body (Figure 1). In the following sections, we discuss therapeutic treatments falling under the umbrella of the cancer immunotherapy field.

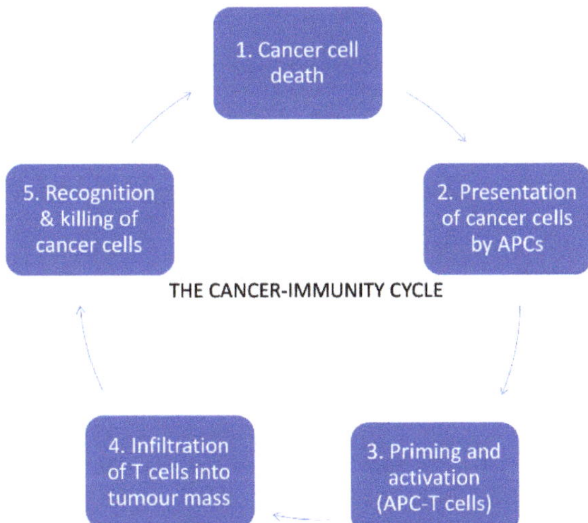

Figure 1. The cancer immunity cycle, modified from [42].

4.1. Immune Checkpoint Inhibitors (ICIs)

Immune checkpoint inhibitors are a new class of cancer therapeutics that have the physiological purpose to negatively regulate the activation of T cells. These checkpoints make it more difficult for T cells to activate, as they need both the interaction with the epitope through the MHC I class, and the presence of co-stimulatory signals to overcome the barrier of negative inhibition. Checkpoint inhibitors (CPIs) are very important to prevent continuous occurrence of immune reactions (Figure 2) [43].

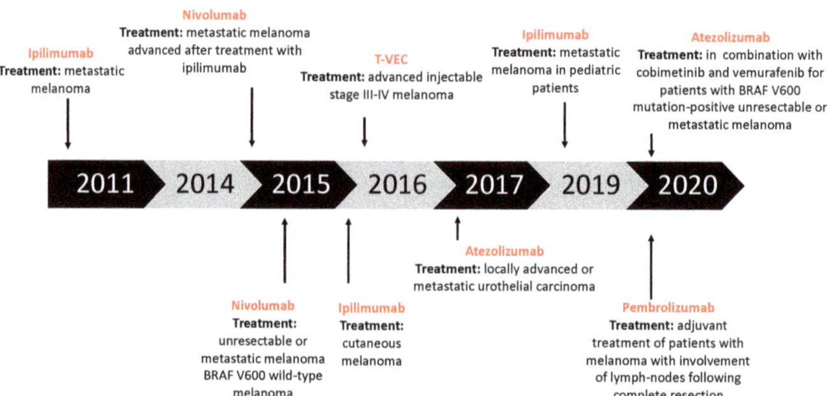

Figure 2. Timeline of immuno-oncology (IO) agents approved for cancer therapies.

The two most important immune checkpoints that have been studied in immunotherapy are the cytotoxic-T lymphocytes antigen 4 (CTLA-4) and the programmed cell death protein 1 (PD-1) [44,45]. CTLA-4 is a receptor and a member of the immunoglobulin superfamily CD28:B7 [46]. It can be found on the surface of both effector T cells and Treg cells, as its function is to regulate the extent of the early stage activation of these two types of immune cells. CTLA-4 binds CD80 and CD86 with higher affinity than CD28 does and blocks the amplification signal that the co-stimulatory binding is supposed to send, in order to trigger T cells expansion. In tumors, CTLA-4 is overexpressed to suppress the activation of immune cells which could have been successful in reaching the tumor site (generally referred to as tumor infiltrating lymphocytes—TILs) [47].

PD-1 is another co-inhibitory molecule expressed in stimulated T cells, Treg cells, B-activated cells and NK cells, and it exerts its function once it is bound to its two ligands, PD-L1 and PD-L2. PD-L1 is expressed more and is found on antigen presenting cells (APCs), dendritic cells (DCs), macrophages and B cells, but it is also expressed in tumor cells which are able to abrogate the lymphocyte response [5]. These two immune checkpoints have been investigated as a target for several monoclonal antibodies, which are already being exploited in cancer therapy for their ability in binding a specific antigen.

The first monoclonal antibody against immune checkpoints to be discovered was ipilimumab, an anti-CTLA-4 antibody that has been firstly approved as a first-line treatment of metastatic melanoma [48]. In the anti-PD-1 group there are other two common ICIs, pembrolizumab and nivolumab, both with indications for metastatic melanoma. Pembrolizumab has been the first anti-PD-1 monoclonal antibody that has been discovered, and with clinical trials KEYNOTE-001, KEYNOTE-002 and KEYNOTE-006 it has gained the first-line therapy indication for metastatic melanoma [49]. In particular, in trials KEYNOTE 006 AND KEYNOTE-002, which both presented comparative arms, patients treated with pembrolizumab significantly improved their progression-free survival (PFS), overall survival (OS) and overall response rates (ORR) relative to ipilimumab in ipilimumab-naive patients (KEYNOTE 006), and significantly improved PFS and ORR, but not OS (although OS data are immature), relative to chemotherapy in ipilimumab-refractory patients, who had also received BRAF/MEK inhibitor therapy if BRAF-mutation positive (KEYNOTE 002) [50].

Pembrolizumab can to be administered as the first line therapy (BRAF wildtype melanoma) or after treatment with ipilimumab, in a combination with anti-CTLA-4 or in patients with BRAF mutations after treatment BRAF inhibitor such as vemurafenib, sorafenib and dabrafenib. Atezolizumab in combination with cobimetinib and vemurafenib is also used for the patients with BRAF V600 mutation-positive unresectable or metastatic melanoma (first line therapy) (Figure 3) (IMspire150, NCT02908672) [51–55]. Therefore, it is reasonable to suppose that atezolizumab could bring some new advantages if compared to the targeting of PD-1 exerted by pembrolizumab, such as the preservation of PD-L2 interactions with PD-1 which carries out the immune checkpoint functions that avoids autoimmune reactions during therapy, allowing for a more tolerable safety profile for this immunotherapeutic new drug [56].

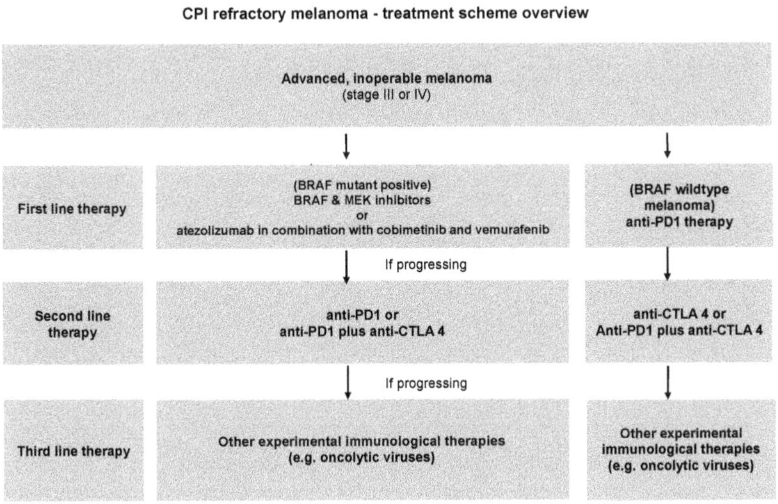

CPI refractory melanoma - treatment scheme overview

Figure 3. CPI refractory melanoma-treatment scheme overview.

Initially, immunotherapy was employed in melanoma treatment with administration of interferon and interleukin cytokines, such as IFN-α and IL-2, which were approved by the FDA with melanoma indications in 1996 and 1998, respectively [32]. Unfortunately, this approach did not show notable benefits for patients, due to the severe side effects associated with systemic administration and to the much poorer therapeutic effects that came with other routes of administration, like the subcutaneous one [32,57]. A modern approach to the immunotherapy of melanoma has grown from elucidations on the role of specific immunomodulatory molecules, and led to a goal shift directed to the enhancement of cell-mediated immunity [33]. To do this, some of the aforementioned ICIs (Figures 2 and 3) have been investigated and were subsequently approved for melanoma therapy:

- Ipilimumab (anti-CTLA-4): Gained regulatory approval by the FDA to treat melanoma after a series of phase III clinical trials (CA184-002 as a single agent, CA184-024 in combination with dacarbazine). The tumor responses according to the Response Criteria in Solid Tumors (RECIST) criteria varied from 5.7% to 11.0% in the anti-CTLA-4 treatment arms. The median overall survival (OS) was improved to 10 months for the anti-CTLA-4 monotherapy arm as compared to 6.4 months for the peptide vaccine-alone arm (HR 0.68; $p < 0.001$ [58], CA184-002, NCT00094653). The five-year survival rate was 18.2% (95% CI, 13.6% to 23.4%) for patients treated with anti-CTLA-4 + dacarbazine vs. 8.8% (95% CI, 5.7% to 12.8%) for patients treated with placebo plus dacarbazine ($p = 0.002$, CA184-024, NCT00324155) [59]. Toxicity associated with ipilimumab includes immune-related symptoms such as dermatitis, colitis, diarrhea and, less commonly, hepatitis, uveitis and hypophysitis [60].

- Pembrolizumab and nivolumab (anti-PD1): After the ipilimumab proof of concept that a checkpoint blockade could actually be an effective strategy to treat melanoma, pembrolizumab and nivolumab were investigated for the same indication, even if (or maybe especially because) they are selective for another receptor which is usually expressed on immune T cell surface—PD-1. Phase III clinical trial reported the median overall survival which has not been reached in the nivolumab-plus-ipilimumab group and was 37.6 months in the nivolumab group, as compared with 19.9 months in the ipilimumab group (hazard ratio for death with nivolumab plus ipilimumab vs. ipilimumab, 0.55 [$p < 0.001$]; hazard ratio for death with nivolumab vs. ipilimumab, 0.65 [$p < 0.001$]). The overall survival rate at 3 years was 58% in the nivolumab-plus-ipilimumab group and 52% in the nivolumab group, as compared with 34% in the ipilimumab group (NCT01844505) [33,61–63].

4.2. Oncolytic Virotherapy

Oncolytic virotherapy [64–70] has elicited increased interest over recent years, even though the first encouraging evidence that led to its development date back to the beginning of the 20th century. It consists of the employment of naturally occurring viruses (e.g., enteroviruses, reoviruses, vaccinia virus) [71–74] or genetically modified viruses (e.g., HSV, adenoviruses). Oncolytic viruses (OVs) [75–77] have the fundamental feature of tumor specificity and many other important advantages, such as the ability to trigger anti-tumor immune responses and the possibility to deliver specific genes in the tumor microenvironment [78].

To avoid damage in healthy tissues, oncolytic viruses are usually genetically modified so that they can replicate only in tumor cells. Their design benefits from a deletion of 24 base pairs in the viral E1A gene which makes the expressed mutated E1A protein unable to bind to retinoblastoma protein (pRb). This interaction is needed in normal cells to activate the E2F transcription factor, which leads to induction of the S-phase of the cell cycle. The deletion-bearing virus is able to infect normal cells but its replication is restricted due to the dysfunctional E1A [74]. The viruses bearing the 24 bp deletion in their E1A gene are commonly tumor-selective and referred to as Δ24-viruses. The only cells in which Δ24-viruses can replicate are tumor cells, which are usually deficient of pRb. Taken together, it is worth highlighting that oncolytic viruses work as anti-tumor agents in a two-step manner: The first is the lysis of tumor cell they have previously infected, but not before they have finished their replication cycle, so that with cell death the release of new progeny occurs. Another feature of OVs is the ability to selectively replicate in cancer cells [79–81]. Even though the virus can enter both healthy and cancerous cells (the selective cell entry must not be confused with exclusive cancer cell entry), there are inherent abnormalities in cancer cell pathways concerning homeostasis, response to stress and their anti-viral machinery, which can give OVs a selective advantage for their replication in these cells [82].

The anti-viral machinery in normal cells is activated by a series of pathways:

- Toll-like receptors (TLRs): This pathway is activated by pathogen-associated molecular patterns (PAMPs), which consist of elements of viral capsid, DNA, RNA and viral proteins. These elements are recognized by TLRs, and they stimulate the innate immune system through a variety of signaling factors (MYD88, TRIF, IRF7, IRF3, NF-kβ), leading to the release of pro-inflammatory cytokines and local type I interferon (IFN-I) [82,83].
- RIG-1 pathway: This pathway is activated by the detection of viral dsDNA and uses some of the same factors exploited by the TLRs pathway, such as IRF3/7. It leads to the release of IFN-I [6].
- IFN-I pathway: This is activated by the local production of type I interferon. After IFN-I binds to its receptor, IFNR, a cascade of signals is triggered and, through the JAK-STAT pathway, it leads to the upregulation of cell-cycle regulators such as PKR and IRF7. These two factors are

important in order to contain viral spread because they induce abortive apoptosis, which blocks the replicative cycle of viruses before the viral progeny is ready to be released [82].

Conventional cytotoxic therapies, as we have already pointed out, are not always effective in melanoma patients, and this statement comes with an even heavier burden when it comes to patients with advanced (unresectable and/or metastatic) melanoma, for whom excisional surgery is not an option [21]. Within this framework, oncolytic viruses pose as a potentially valid therapeutic option for these patients, thanks to their ability to selectively target cancer cells and simultaneously trigger the patient immune system against melanoma cells [84,85]. A decisive role in the efficacy of oncolytic viruses against tumors is covered by their stimulation of immune system, which is triggered to develop a specific anti-tumor immune response by the OV. It is for this reason that the immunogenicity of melanoma as a tumor is an important feature to describe.

Melanoma has been considered an immunogenic malignancy for a long time [86,87]. Virtually all of the major enlightenments concerning tumor immunology have been experimentally observed in melanoma models. When we say that melanoma is a strongly immunogenic malignancy, we refer to the fact that it has a close relationship with host immune cells, which usually infiltrates the tumor microenvironment [87]. The distribution, density, profile and activation state of immune cells which are part of TILs can be variable and modulates the clinical outcome in melanoma patients. TILs are now recognized as an independent prognostic biomarker for melanoma, and the assessment of its composition is even more appealing because it could provide new molecular targets and biomarkers to predict therapeutic responses of immunotherapy drugs [88]. The major components of TIL infiltrate are CD8+ T lymphocytes, Tregs, NK cells, dendritic cells, and macrophages. Furthermore, the high immunogenicity of melanoma also implies the presence of a plethora of tumor antigens, which can be classified as TAAs, which are antigens located on tumor cells' surfaces, and TSAs or neoantigens, which are more specific for a single tumor [87]. Talimogene Laherparepvec (T-VEC), also known as Imlygic or OncoVex^{GM-CSF}, is the first oncolytic virus that has been approved by the FDA and the EMA to treat cancer (Figure 2). The FDA approved T-VEC in 2015, with an indication for local treatment of unresectable, subcutaneous, cutaneous and nodal lesions in patients with melanoma recurrent after initial surgery [89,90]. From this perspective, T-VEC represents a valid second-line treatment for patients with metastatic, unresectable melanoma, especially for those with stage IIIB, IIIC and IV melanoma [89].

5. Combinatorial Approaches with OVs in Melanoma Treatments

5.1. OVs with Immune Checkpoint Inhibitors

The idea behind this combinatorial approach is that these two therapeutic tools can improve each other by addressing one another's shortcomings. Oncolytic viruses present some limitations related mainly to antiviral immunity, which makes it challenging to exploit the bloodstream to reach distant metastatic sites [91]. Therefore, triggering a tumor-specific adaptive immune response is a fundamental feature as OVs cannot travel inside the body to reach other sites, while T cells that have been sensitized to tumor cells surely can, thus assuring an antitumor response even in different sites to that of the primary tumor [91]. From this perspective, ICIs help guarantee the correct activation of the immune system, targeting specific molecules expressed either on the tumor or on immune cells (CTLA-4, PD-1, PD-L1), while viral infections obtained using OVs makes the TME more immunogenic, creating a microenvironment in which ICIs are known to work much better [92].

This combinatorial approach has been explored in numerous clinical trials [91], among which was a phase II clinical trial with 198 stage IIIB–IV melanoma patients, which was organized to evaluate: (1) ipilimumab as a monotherapy and (2) ipilimumab combined with T-VEC [92]. The results showed that the objective response rate of the combination therapy was 39%, while ipilimumab alone had an objective response rate of 18% [91,92]. T-VEC has also been investigated in combination with other ICIs such as pembrolizumab, an anti-PD-1 antibody. In the phase IB portion of the clinical trial

Masterkey-265, T-VEC was administered to 21 patients with stage IIB and IV melanoma in combination with intravenous pembrolizumab. Among the criteria that have been evaluated, the safety profile of the combination was favorable, with no dose-limiting toxicities, and the objective response rate was 62%, while 33% of patients showed a complete response [92].

Multiple adenoviruses are undergoing clinical and preclinical testing in combination with ICIs, both in melanoma and other types of tumor.

The Hemminki group exploited a murine model of melanoma to establish the mechanism under the combination of the anti-PD-1 antibody with the oncolytic viruses encoding for TNFα and IL-2 [93]. What emerged from the combination therapy was a marked increase in intratumoral CD8+ T cells and a statistically significant tumor growth suppression, along with increased survival in animals. Researchers reported complete tumor regression after the course of the combinatory therapy. This preclinical research provides the rationale for a clinical trial where oncolytic adenovirus coding for TNFa and IL-2 (TILT-123) is used in melanoma patients receiving an anti-PD-1 antibody NCT04217473) [92,93].

Thomas et al. reported development of a new fusion-enhanced oncolytic immunotherapy platform based on herpes simplex virus type 1. Researchers developed various oncolytic vectors expressing e.g., GMCSF, an anti-CTLA-4 antibody-like molecule. Anti-cancer assessment was performed in vivo and in nude mouse xenograft models (melanoma, lymphoma, gliosarcoma). The combination therapy with the virus expressing GALV-GP-R- and mGM-CSF and an anti-murine PD1 antibody showed improved anti-tumor effects compared to the control. The treatment of mice with derivatives of this virus coding for anti-mCTLA-4, mCD40L, m4-1BBL, or mOX40L showed enhanced anti-cancer efficacy in un-injected tumors (abscopal effect) [94].

Also, in our previous study we have investigated the anti-cancer potency of ONCOS-102 and pembrolizumab in the humanized melanoma mouse model. Humanized mice engrafted with A2058 melanoma cells showed significant tumor volume reduction after ONCOS-102 treatment. The combination of anti-PD1 with the virus further reduced tumor volume, while pembrolizumab alone did not show therapeutic benefit by itself [45]. Systemic abscopal was also observed when combining oncolytic adenovirus and checkpoint inhibitor in a humanized NOG mouse model of melanoma [44]. These data support the scientific rationale for the ongoing clinical study of combination therapy of ONCOS-102 and pembrolizumab for the treatment of melanoma (NCT03003676).

Currently, there are many oncolytic vectors are under development and investigation in melanoma: coxsackieviruses, HF-10, adenoviruses, reoviruses, echoviruses, and Newcastle disease viruses. Therefore, it is probable that oncolytic vectors will have long-term application in the treatment of advanced melanoma not only as a monotherapy but as a part of combinatory therapies. [95].

T-VEC is the first oncolytic vector approved for the melanoma treatment. Reported data have shown improved therapeutic responses to T-VEC in combination with immune checkpoint blockade in patients with melanoma without additive toxicity [96]. T-VEC combined with anti-PD-1 based immunotherapy for unresectable stage III-IV melanoma showed an overall response rate for on-target lesions of 90%, with 6 patients resulting in a complete response in injected lesions (NCT02263508) [97]. Also, the treatment with T-VEC in patients with advanced melanoma with disease progression following multiple previous systemic therapies (vemurafenib, metformin, ipilimumab, dabrafenib, trametinib, and pembrolizumab) showed signs of anti-cancer effect, and provides potential clinical and immunotherapeutic utility of T-VEC application [98].

CAVATAK, an oncolytic immunotherapy, is an oncolytic strain of Coxsackievirus A21 (CVA21). The virus infects ICAM-1 expressing tumor cells, resulting in cell lysis, and anti-tumor immune response. The Phase II CALM study investigated the efficacy and safety of CVA21 in patients with advanced melanoma (NCT01227551). The treatment with CAVATAK resulted in elevation of the immune CD8+ T cell infiltrates within the tumor (5 of 6 patients), and increased expression of PD-L1+ cells. It was also reported that the virus was able to reconstitute immune cell infiltrates in lesions resistant to immune-checkpoint blockade [99]. The combinatory therapy trials have been conducted where

CAVATAK was administered with ipilimumab (NCT02307149) or pembrolizumab (NCT02565992). The treatment with CAVATAK and anti-CTLA-4 has shown durable response with minimal toxicity. The preliminary ORR rate for the ITT population of 50.0% is higher than published rates for either agent used alone (CAVATAK: ~28% and ipilimumab: ~15–20%) in advanced melanoma patients [100]. Among the evaluable patients (intratumoral CAVATAK and systemic pembrolizumab in advanced melanoma patients), the ORR was 73% (8/11). The DCR (CR + PR + SD) was 91% (10/11). In patients with stage IVM1c disease, the ORR and the DCR is 100% (5/5). Combination therapy of the virus1 and anti-PD1 may present a new strategy for the treatment of patients with injectable advanced melanoma (CAPRA clinical trial) [101].

Another oncolytic adenovirus that has been investigated in combination with pembrolizumab is ONCOS-102 (AdV5/3-Δ24-GM-CSF), which is now under clinical trial (NCT03003676) to investigate its safety and efficacy, supported by preclinical data showing increased CD8+ T cell infiltration in tumor mass upon viral administration [92]. The therapeutics efficacy and safety of the virus was previously tested in C1 study (NCT01598129). The treatment with the virus was safe and well tolerated at the tested doses. Therapy resulted in infiltration of CD8+ T cells to tumors and up-regulation of PD-L1, highlighting the potential of ONCOS-102 as an immunosensitizing agent for combinatory therapies with checkpoint inhibitors [102]. Therefore, providing a scientific rationale for the combinatory therapy with CPIs.

To date, approximately one third of all clinical trials concerning OVs have investigated a combinatorial approach with at least one ICI [91]. Therefore, it is expected that oncolytic viruses have the capability to promote a 'hotter' immune microenvironment which can improve the efficacy of ICI [103,104]. Oncolytic viruses can be thought of as matches—they can light up a fire inside the tumor and this fire will make the TME "hot" enough for ICIs to strike a blow. Many clinical and preclinical models of melanoma and other solid tumors have provided strong evidence that the infection of tumor cells with OVs can result in the creation of a pro-inflammatory tumor microenvironment, which in turn translates into a new influx of T cells that can be protected from inactivation by ICIs [104,105]. Furthermore, some adenoviruses administered in combination with ICIs have been reported to boost release the pool of tumor neoantigens which can be recognized by CD8+ T cells [106]; this is a particularly important finding, because OVs (both as monotherapy and in combination) have most difficulty affecting low mutational burden cancers, which typically have a very small number of TAAs [91].

Nowadays, one of the major challenges for researchers investigating this field is to assess not only which combinations are most effective, but the dosing regimens and schedules to adopt to maximize the synergy and minimize the side effects [91,92]. This is why further clinical trials results are so impatiently awaited. ICIs have contributed to revolutionize cancer treatment. Nevertheless, the best response rates to these agents do not exceed 35% to 40% [107]. Therefore, the goal of combining OVs with ICIs is to enhance clinical efficacy. Oncolytic vectors are used in order attract the immune cells into the lesion, prime anti-tumor immune responses by development of innate and adoptive anticancer immunity. In turn, CPI therapy will prevent inhibition of activates cancer specific T cells. It is expected that those two agents can result in synergistic or additive anti-cancer effect. Interestingly, it has been demonstrated that local OV injection can modulate tumor-specific CD8+ T-cell responses rendering distant tumors susceptible to immune checkpoint inhibitor therapy [108]. Therefore, due to the preclinical success of this combination therapy, there is huge interests in clinical trials: results obtained from patients who have progressed after immune checkpoint inhibition (e.g., NCT 03003676) could shed the light on OV's role in overcoming resistance to immunotherapy. By elucidating the potential of the combination of OVs and checkpoint inhibitors, further development in treatment regimens employing these novel therapeutic agents could be beneficial for patients.

Apart from combinatorial strategies, another aspect concerning the use of ICIs is often investigated to reach some improvement—the response predictions with biomarkers. There are several biomarkers associated with the response of ICIs, some of which have been approved and are currently being exploited to predict the response rate in patients before treatment begins, while others are under further study to establish whether they have a strong correlation with the extent of patients' responses to ICIs. The most important predictive biomarker for anti-PD-1/PD-L1 antibodies is PD-L1 expression [41,82], which is evaluated by immunohistochemistry. PD-L1 expression by cancer cells is recognized as both a prognostic and predictive biomarker in patients with cutaneous melanoma. Approx. 35% of cutaneous melanomas express PD-L1, The PD-L1 immunohistochemistry (IHC) has been approved by FDA as a complement diagnostic to select patients with non-small-cell lung carcinoma (NSCLC) suitable for pembrolizumab therapy. Nevertheless, absence of PD-L1 does not necessarily translates into a poor response to anti-PD-1/PD-L1 inhibitors. Some patients with low PD-L1 expression exhibits clinical efficacy. However, further efforts are still needed to improve the clinical use of PD-L1 expression as biomarkers Apart from combinatorial strategies, another topic concerning the use of ICIs is often investigated to study prediction biomarkers. There are several biomarkers associated with the response of ICIs, some of which have been approved and are currently being exploited to predict the response rate in patients before treatment begins, while others are under further study to establish whether they have a strong correlation with the extent of patients' responses to ICIs. The most important predictive biomarker for anti-PD-1/PD-L1 antibodies is PD-L1 expression [41,109], which is evaluated by immunohistochemistry and is a prerequisite for treatment with drugs such as atezolizumab or pembrolizumab. However, this biomarker may not be enough to identify all of the patients who could benefit from this type of therapy, and this observation led scientists to begin further studies to find more appropriate predictive biomarkers. This biomarker's use is already well established in conventional chemotherapy regimens, but recent studies suggest that it could be exploited to predict the response to immunotherapy and, most importantly, that it could also help discriminate real disease progression from pseudo-progression in patients treated with immunotherapy, avoiding re-biopsy in these patients [109–111].

5.2. OVs with Chemotherapeutic Agents—Future Prospects

In the last decade, many preclinical models have demonstrated that chemotherapeutic agents and OVs could work synergistically [11]. There are two main approaches to setting this combination:

- Use OVs as adjuvant to chemotherapy, which is the most clinically relevant approach, since chemotherapeutic agents represent the cornerstone of almost every cancer standard-of-care therapy [78].
- Use chemotherapeutic drugs to counteract or inhibit factors that limit the effectiveness of oncolytic virotherapy such as large tumor size, poor vasculature, elevated interstitial pressure and other physical barriers [112].

It is important to consider that even if chemotherapy and OVs could seem good partners, not all combinations have showed synergistic effects. In fact, the result depends on different factors including OVs strain, cancer type and the exact drug(s) used, as well as their dosing regimen and schedule [113].

In terms of the two standard-of-care chemotherapeutic agents for melanoma, temozolomide and dacarbazine, there are both in vitro and in vivo studies that explored combinations with various oncolytic vectors. Specifically, an in vitro/in vivo study that tested the combination of dacarbazine with ZD-55-IL18, an oncolytic adenovirus encoding for IL-18, showed that there is a synergy between these two agents, observed in the induction of apoptosis of tumor cells, and inhibition of angiogenesis and metastasis [113]. Another in vitro study conducted on melanoma cell lines treated with temozolomide (TMZ) and another oncolytic adenovirus (Ad5/3.2xTyr) reported that TMZ enhanced the OV's antitumor effect without altering the expression of CAR or other viral receptors on cancer cells, but rather by blocking the tumor cell's cycle in the S/G2 phase, providing a better intracellular environment for the viral replicative cycle to develop. This finding is consistent with the higher number of genome copies detected inside infected tumor cells [114].

By way of conclusion, we could say that the combinatorial approach based on chemotherapy and oncolytic viruses is promising, but like every other approach, it has to face some challenges to push researchers even further. It is clear that there are some incompatibilities between chemotherapy and OVs [114] which must be taken into account when designing new combinations:

- Many chemotherapeutic agents induce apoptosis in cancer cells, while OVs need actively dividing cells to complete their replicative cycle successfully;
- Other chemotherapeutic drugs target angiogenetic mechanisms to impair tumor expansion, but this would also affect viral trafficking inside the tumor mass;
- The immune modulation exerted by some chemotherapeutic drugs could dampen the antitumor immune response triggered by OVs.

We must consider all of these notions to reach a combination that can work effectively with a synergistic interaction.

5.3. OVs with Radiotherapy—Future Prospects

Anticancer synergistic interaction of radiation and OVs therapy has solid backing in the literature, and the enhancement of viral replication due to radiotherapy has been reported in different in vitro and in vivo models such as lung cancer and pleural mesothelioma [115,116]. In melanoma, apart from the clear and demonstrated efficacy of both approaches as a monotherapy to kill cancer cells [32], the synergy is due to three other aspects:

- Radiotherapy may reduce the internal pressure within the tumor mass, making it easier for the OV to penetrate it and work properly.
- Some OVs, such as vesicular stomatitis virus (VSV) or HSV, are able to preferentially target Ras-mutated cancer cells (Ras is one of the driver mutations in melanoma). Since Ras mutations in cancer cells are associated with resistance to radiotherapy, OVs which can target these cells will exert a complementary therapeutic effect to radiotherapy.
- Infection of melanoma cells by OVs will lead to a release of cytokines like TNFα or TRAIL, which can sensitize tumor cells to radiation therapy.

Twigger et al. tried to combine an oncolytic reovirus with radiation therapy in a variety of melanoma cell lines, observing that the combination yielded a statistically significant enhancement of viral cytotoxicity without affecting reoviral replication rates, but with an increase in apoptosis of cancer cells [117]. In another preclinical study, Kyula et al. investigated the combination of an oncolytic *Vaccinia* virus and radiotherapy in BRAF-mutated, Ras-mutated and wild type melanoma cell lines. Results showed that in melanoma cells that carried V600D or V600E BRAF mutations there had been an increased apoptosis [42]. Also, the combination of reovirus and radiation has shown to increase the tumor growth delay of the melanoma xenografts in the treated animals, and significantly improve the overall survival rate compared to the treatment with either of the individual therapies [118]. Importantly, Ras mutation is one of the driver mutations for melanoma and is associated with radio-resistance [58]. However, some viruses like: reovirus, VSV and HSV have been able to selectively target the Ras mutated melanoma cells and mediate cell death [119]. Therefore, oncolytic vectors able to lyse the radiation-resistant melanoma cells can exhibit a complementary therapeutic effect to radiotherapy. There are many ongoing attempts to find the optimal way to combine these two strategies to maximize the antitumor effect preclinically. More investigations are needed to understand how to exploit this combination in the complex context of metastatic unresectable melanomas and their application in clinics.

6. Conclusions

The discovery of T cell checkpoint inhibitors and oncolytic virotherapy has changed the paradigm of oncologic treatment for some cancer types and showed a transition pattern from conventional therapies towards immuno-oncology. Oncolytic viruses can induce anti-tumor immunity and lead to the infiltration of TILs. In turn, a checkpoint blockade can prevent inhibition of T cell activity. Therefore, the combination of those agents seems to be a potent treatment regimen to combat immunogenic cancer types such as melanoma.

Author Contributions: Conceptualization, L.K., M.G.; writing—original draft, M.G., L.K., L.B.; writing—review and editing, L.K., M.G., L.B., M.S., K.P., M.W., S.S., P.C. All authors have read and agreed to the published version of the manuscript.

Funding: L.K. was supported by the National Science Centre, Poland (SONATINA (2019/32/C/NZ7/00156)) and National Institute of Public Health–National Institute of Hygiene in Poland (1BWBW/19). M.G. was supported by PRID-J (Grant Number: GARO_SID19_02) funded by University of Padua.This publication is based upon work from COST Action CA 17140 "Cancer Nanomedicine from the Bench to the Bedside" supported by COST (European Cooperation in Science and Technology) (L.K., M.G.).

Conflicts of Interest: The authors declare no conflict of interest. The funders had no role in the design of the study; in the collection, analyses, or interpretation of data; in the writing of the manuscript, or in the decision to publish the results.

References

1. Loeb, K.R.; Loeb, L.A. Significance of multiple mutations in cancer. *Carcinogenesis* **2000**, *21*, 379–385. [CrossRef] [PubMed]
2. Calì, B.; Molon, B.; Viola, A. Tuning cancer fate: The unremitting role of host immunity. *Open Biol.* **2020**, *7*, 170006. [CrossRef] [PubMed]
3. Pandya, P.H.; Murray, M.E.; Pollok, K.E.; Renbarger, J.L. The Immune System in Cancer Pathogenesis: Potential Therapeutic Approaches. *J. Immunol. Res.* **2016**, *2016*, 4273943. [CrossRef] [PubMed]
4. Dunn, G.P.; Old, L.J.; Schreiber, R.D. The Immunobiology of Cancer Immunosurveillance and Immunoediting. *Immunity* **2004**, *21*, 137–148. [CrossRef] [PubMed]
5. Gun, S.Y.; Lee, S.W.L.; Sieow, J.L.; Wong, S.C. Targeting immune cells for cancer therapy. *Redox Biol.* **2019**, *25*, 101174. [CrossRef]
6. Elion, D.L.; Jacobson, M.E.; Hicks, D.J.; Rahman, B.; Sanchez, V.; Gonzales-Ericsson, P.I.; Fedorova, O.; Pyle, A.M.; Wilson, J.T.; Cook, R.S. Therapeutically Active RIG-I Agonist Induces Immunogenic Tumor Cell Killing in Breast Cancers. *Cancer Res.* **2018**, *78*, 6183–6195. [CrossRef]
7. Lakshmi Narendra, B.; Eshvendar Reddy, K.; Shantikumar, S.; Ramakrishna, S. Immune system: A double-edged sword in cancer. *Inflamm. Res.* **2013**, *62*, 823–834. [CrossRef]
8. Swann, J.B.; Smyth, M.J. Immune surveillance of tumors. *J. Clin. Investig.* **2007**, *117*, 1137–1146. [CrossRef]
9. Chow, M.T.; Möller, A.; Smyth, M.J. Inflammation and immune surveillance in cancer. *Semin. Cancer Biol.* **2012**, *22*, 23–32. [CrossRef]
10. Goldszmid, R.S.; Dzutsev, A.; Trinchieri, G. Host Immune Response to Infection and Cancer: Unexpected Commonalities. *Cell Host Microbe* **2014**, *15*, 295–305. [CrossRef]
11. Corrales, L.; Matson, V.; Flood, B.; Spranger, S.; Gajewski, T.F. Innate immune signaling and regulation in cancer immunotherapy. *Cell Res.* **2017**, *27*, 96–108. [CrossRef] [PubMed]
12. Zamora, A.E.; Crawford, J.C.; Thomas, P.G. Hitting the Target: How T Cells Detect and Eliminate Tumors. *J. Immunol.* **2018**, *200*, 392–399. [CrossRef] [PubMed]
13. Durgeau, A.; Virk, Y.; Corgnac, S.; Mami-Chouaib, F. Recent Advances in Targeting CD8 T-Cell Immunity for More Effective Cancer Immunotherapy. *Front. Immunol.* **2018**, *9*, 14. [CrossRef] [PubMed]
14. Nurieva, R.; Wang, J.; Sahoo, A. T-cell tolerance in cancer. *Immunotherapy* **2013**, *5*, 513–531. [CrossRef]
15. Makkouk, A.; Weiner, G.J. Cancer Immunotherapy and Breaking Immune Tolerance: New Approaches to an Old Challenge. *Cancer Res.* **2015**, *75*, 5–10. [CrossRef]
16. Jiang, T.; Shi, T.; Zhang, H.; Hu, J.; Song, Y.; Wei, J.; Ren, S.; Zhou, C. Tumor neoantigens: From basic research to clinical applications. *J. Hematol. Oncol.* **2019**, *12*, 93. [CrossRef]

17. Peng, M.; Mo, Y.; Wang, Y.; Wu, P.; Zhang, Y.; Xiong, F.; Guo, C.; Wu, X.; Li, Y.; Li, X.; et al. Neoantigen vaccine: An emerging tumor immunotherapy. *Mol. Cancer* **2019**, *18*, 128. [CrossRef]

18. Uong, A.; Zon, L.I. Melanocytes in development and cancer. *J. Cell. Physiol.* **2010**, *222*, 38–41. [CrossRef]

19. Bastian, B.C. The Molecular Pathology of Melanoma: An Integrated Taxonomy of Melanocytic Neoplasia. *Annu. Rev. Pathol. Mech. Dis.* **2014**, *9*, 239–271. [CrossRef]

20. Balois, T.; Chatelain, C.; Ben Amar, M. Patterns in melanocytic lesions: Impact of the geometry on growth and transport inside the epidermis. *J. R. Soc. Interface* **2014**, *11*, 20140339. [CrossRef]

21. Leonardi, G.C.; Falzone, L.; Salemi, R.; Zanghì, A.; Spandidos, D.A.; Mccubrey, J.A.; Candido, S.; Libra, M. Cutaneous melanoma: From pathogenesis to therapy (Review). *Int. J. Oncol.* **2018**, *52*, 1071–1080. [CrossRef] [PubMed]

22. Bray, F.; Ferlay, J.; Soerjomataram, I. Global Cancer Statistics 2018: GLOBOCAN Estimates of Incidence and Mortality Worldwide for 36 Cancers in 185 Countries. *CA Cancer J. Clin.* **2018**, *68*, 394–424. [CrossRef] [PubMed]

23. Ward, W.H.; Farma, J.M. *Cutaneous Melanoma: Etiology and Therapy*; Codon Publication: Wishart, Australia, 2018; ISBN 9780994438140.

24. De Fabo, E.C.; Noonan, F.P.; Fears, T.; Merlino, G. Ultraviolet B but not Ultraviolet a Radiation Initiates Melanoma. *Cancer Res.* **2004**, *64*, 6372–6376. [CrossRef]

25. Gandini, S.; Sera, F.; Cattaruzza, M.S.; Pasquini, P.; Picconi, O.; Boyle, P.; Melchi, C.F. Meta-analysis of risk factors for cutaneous melanoma: II. Sun exposure. *Eur. J. Cancer* **2005**, *41*, 45–60. [CrossRef] [PubMed]

26. Sokolova, A.; Lee, A.; Smith, S.D. The Safety and Efficacy of Narrow Band Ultraviolet B Treatment in Dermatology: A Review. *Am. J. Clin. Dermatol.* **2015**, *16*, 501–531. [CrossRef]

27. Berlin, N.L.; Cartmel, B.; Leffell, D.J.; Bale, A.E.; Mayne, S.T.; Ferrucci, L.M. Family history of skin cancer is associated with early-onset basal cell carcinoma independent of MC1R genotype. *Cancer Epidemiol.* **2015**, *39*, 1078–1083. [CrossRef]

28. Bossen, K.S.; Sørensen, H.T. The impact of comorbidity on cancer survival: A review. *Clin. Epidemiol.* **2013**, *5*, 3–29.

29. Zbytek, B.; Carlson, J.A.; Granese, J.; Ross, J.; Mihm, M.; Slominski, A. Current concepts of metastasis in melanoma. *Expert Rev. Dermatol.* **2008**, *3*, 569–585. [CrossRef]

30. Harries, M.; Malvehy, J.; Lebbe, C.; Heron, L.; Amelio, J.; Szabo, Z.; Schadendorf, D. Treatment patterns of advanced malignant melanoma (stage III–IV)—A review of current standards in Europe. *Eur. J. Cancer* **2016**, *60*, 179–189. [CrossRef]

31. Perera, E.; Gnaneswaran, N.; Jennens, R.; Sinclair, R. Malignant Melanoma. *Healthcare* **2014**, *2*, 1–19. [CrossRef]

32. Mishra, H.; Mishra, P.K.; Ekielski, A.; Jaggi, M.; Iqbal, Z.; Talegaonkar, S. Melanoma treatment: From conventional to nanotechnology. *J. Cancer Res. Clin. Oncol.* **2018**, *144*, 2283–2302. [CrossRef] [PubMed]

33. Luke, J.J.; Flaherty, K.T.; Ribas, A.; Long, G.V. Targeted agents and immunotherapies: Optimizing outcomes in melanoma. *Nat. Rev. Clin. Oncol.* **2017**, *14*, 463–482. [CrossRef]

34. Menzies, A.M.; Long, G.V. Review Systemic treatment for BRAF-mutant melanoma: Where do we go next? *Lancet Oncol.* **2014**, *15*, e371–e381. [CrossRef]

35. Ravnan, M.C.; Matalka, M.S. Vemurafenib in Patients with BRAF V600E Mutation–Positive Advanced Melanoma. *Clin. Ther.* **2012**, *34*, 1474–1486. [CrossRef] [PubMed]

36. Hauschild, A.; Grob, J.-J.; Demidov, L.V.; Jouary, T.; Gutzmer, R.; Millward, M.; Rutkowski, P.; Blank, C.U.; Miller, W.H.; Kaempgen, E.; et al. Dabrafenib in BRAF-mutated metastatic melanoma: A multicentre, open-label, phase 3 randomised controlled trial. *Lancet* **2012**, *380*, 358–365. [CrossRef]

37. Daud, A.; Tsai, K. Management of Treatment-Related Adverse Events with Agents Targeting the MAPK Pathway in Patients with Metastatic Melanoma. *Oncologist* **2017**, *22*, 823–833. [CrossRef] [PubMed]

38. Sanchez, J.N.; Wang, T.; Cohen, M.S. BRAF and MEK Inhibitors: Use and Resistance in BRAF-Mutated Cancers. *Drugs* **2018**, *78*, 549–566. [CrossRef]

39. Chapman, P.B.; Hauschild, A.; Robert, C.; Haanen, J.B.; Ascierto, P.; Larkin, J.; Dummer, R.; Garbe, C.; Testori, A.; Maio, M.; et al. Improved Survival with Vemurafenib in Melanoma with BRAF V600E Mutation. *N. Engl. J. Med.* **2011**, *364*, 2507–2523. [CrossRef]

40. Kudchadkar, R.; Paraiso, K.H.T.; Smalley, K.S.M. Targeting Mutant BRAF in Melanoma: Current Status and Future Development of Combination Therapy Strategies. *Cancer J.* **2012**, *18*, 124. [CrossRef]

41. Ribas, A.; Gonzalez, R.; Pavlick, A.; Hamid, O.; Gajewski, T.F.; Daud, A.; Flaherty, L.; Logan, T.; Chmielowski, B.; Lewis, K.; et al. Combination of vemurafenib and cobimetinib in patients with advanced BRAFV600-mutated melanoma: A phase 1b study. *Lancet Oncol.* **2014**, *15*, 954–965. [CrossRef]

42. Chen, D.S.; Mellman, I. Oncology Meets Immunology: The Cancer-Immunity Cycle. *Immunity* **2013**, *39*, 1–10. [CrossRef]

43. Zappasodi, R.; Merghoub, T.; Wolchok, J.D. Emerging Concepts for Immune Checkpoint Blockade-Based Combination Therapies. *Cancer Cell* **2018**, *33*, 581–598. [CrossRef]

44. Kuryk, L.; Møller, A.-S.W.; Jaderberg, M. Abscopal effect when combining oncolytic adenovirus and checkpoint inhibitor in a humanized NOG mouse model of melanoma. *J. Med. Virol.* **2019**, *91*, 1702–1706. [CrossRef]

45. Kuryk, L.; Møller, A.S.W.; Jaderberg, M. Combination of immunogenic oncolytic adenovirus ONCOS-102 with anti-PD-1 pembrolizumab exhibits synergistic antitumor effect in humanized A2058 melanoma huNOG mouse model. *Oncoimmunology* **2019**, *8*, 1–11. [CrossRef]

46. Sharpe, A.H.; Freeman, G.J. The B7–CD28 superfamily. *Nat. Rev. Immunol.* **2002**, *2*, 116–126. [CrossRef]

47. Seidel, J.A.; Otsuka, A.; Kabashima, K. Anti-PD-1 and Anti-CTLA-4 Therapies in Cancer: Mechanisms of Action, Efficacy, and Limitations. *Front. Oncol.* **2018**, *8*, 86. [CrossRef]

48. Wang, S.; He, Z.; Wang, X.; Li, H.; Liu, X.-S. Antigen presentation and tumor immunogenicity in cancer immunotherapy response prediction. *eLife* **2019**, *8*, e49020. [CrossRef]

49. Cowey, C.L.; Liu, F.X.; Black-Shinn, J.; Stevinson, K.; Boyd, M.; Frytak, J.R.; Ebbinghaus, S.W. Pembrolizumab Utilization and Outcomes for Advanced Melanoma in US Community Oncology Practices. *J. Immunother.* **2018**, *41*, 86. [CrossRef]

50. Deeks, E.D. Pembrolizumab: A Review in Advanced Melanoma. *Drugs* **2016**, *76*, 375–386. [CrossRef]

51. Aris, M.; Barrio, M.M. Combining Immunotherapy with Oncogene-Targeted Therapy: A New Road for Melanoma Treatment. *Front. Immunol.* **2015**, *6*, 46. [CrossRef] [PubMed]

52. Khair, D.O.; Bax, H.J.; Mele, S.; Crescioli, S.; Pellizzari, G.; Khiabany, A.; Nakamura, M.; Harris, R.J.; French, E.; Hoffmann, R.M.; et al. Combining Immune Checkpoint Inhibitors: Established and Emerging Targets and Strategies to Improve Outcomes in Melanoma. *Front. Immunol.* **2019**, *10*, 453. [CrossRef] [PubMed]

53. Larkin, J.; Chiarion-Sileni, V.; Gonzalez, R.; Grob, J.J.; Cowey, C.L.; Lao, C.D.; Schadendorf, D.; Dummer, R.; Smylie, M.; Rutkowski, P.; et al. Combined Nivolumab and Ipilimumab or Monotherapy in Untreated Melanoma. *N. Engl. J. Med.* **2015**, *373*, 23–34. [CrossRef] [PubMed]

54. Zimmer, L.; Apuri, S.; Eroglu, Z.; Kottschade, L.A.; Forschner, A.; Gutzmer, R.; Schlaak, M.; Heinzerling, L.; Krackhardt, A.M.; Loquai, C.; et al. Ipilimumab alone or in combination with nivolumab after progression on anti-PD-1 therapy in advanced melanoma. *Eur. J. Cancer* **2017**, *75*, 47–55. [CrossRef] [PubMed]

55. Constantin Kirchberger, M.; Moreira, A.; Erdmann, M.; Schuler, G.; Heinzerling, L. Real world experience in low-dose ipilimumab in combination with PD-1 blockade in advanced melanoma patients. *Oncotarget* **2018**, *9*, 28903. [CrossRef] [PubMed]

56. Seliger, B. Basis of PD1/PD-L1 Therapies. *J. Clin. Med.* **2019**, *8*, 2168. [CrossRef]

57. Coit, D.G.; Andtbacka, R.; Bichakjian, C.K.; Dilawari, R.A.; DiMaio, D.; Guild, V.; Halpern, A.C.; Hodi, F.S.; Kashani-Sabet, M.; Lange, J.R.; et al. Melanoma: Clinical Practice Guidelines in Oncology. *J. Natl. Compr. Cancer Netw.* **2009**, *7*, 250–275. [CrossRef]

58. Rogiers, A.; Boekhout, A.; Schwarze, J.K.; Awada, G.; Blank, C.U.; Neyns, B. Long-Term Survival, Quality of Life, and Psychosocial Outcomes in Advanced Melanoma Patients Treated with Immune Checkpoint Inhibitors. *J. Oncol.* **2019**, *2019*, 5269062. [CrossRef]

59. Maio, M.; Grob, J.-J.; Aamdal, S.; Bondarenko, I.; Robert, C.; Thomas, L.; Garbe, C.; Chiarion-Sileni, V.; Testori, A.; Chen, T.-T.; et al. Five-Year Survival Rates for Treatment-Naive Patients with Advanced Melanoma Who Received Ipilimumab Plus Dacarbazine in a Phase III Trial. *J. Clin. Oncol.* **2015**, *33*, 1191–1196. [CrossRef]

60. Savoia, P.; Astrua, C.; Fava, P. Ipilimumab (Anti-Ctla-4 Mab) in the treatment of metastatic melanoma: Effectiveness and toxicity management. *Hum. Vaccin. Immunother.* **2016**, *12*, 1092–1101. [CrossRef]

61. Topalian, S.L.; Sznol, M.; McDermott, D.F.; Kluger, H.M.; Carvajal, R.D.; Sharfman, W.H.; Brahmer, J.R.; Lawrence, D.P.; Atkins, M.B.; Powderly, J.D.; et al. Survival, Durable Tumor Remission, and Long-Term Safety in Patients With Advanced Melanoma Receiving Nivolumab. *J. Clin. Oncol.* **2014**, *32*, 1020–1030. [CrossRef]

62. Wolchok, J.D.; Chiarion-Sileni, V.; Gonzalez, R.; Rutkowski, P.; Grob, J.-J.; Cowey, C.L.; Lao, C.D.; Wagstaff, J.; Schadendorf, D.; Ferrucci, P.F.; et al. Overall Survival with Combined Nivolumab and Ipilimumab in Advanced Melanoma. *N. Engl. J. Med.* **2017**, *377*, 1345–1356. [CrossRef] [PubMed]

63. Jessurun, C.A.C.; Vos, J.A.M.; Limpens, J.; Luiten, R.M. Biomarkers for Response of Melanoma Patients to Immune Checkpoint Inhibitors: A Systematic Review. *Front. Oncol.* **2017**, *7*, 233. [CrossRef] [PubMed]

64. Kuryk, L.; Møller, A.S.W. Chimeric oncolytic Ad5/3 virus replicates and lyses ovarian cancer cells through desmoglein-2 cell entry receptor. *J. Med. Virol.* **2020**. [CrossRef] [PubMed]

65. Kuryk, L.; Møller, A.-S.W.; Jaderberg, M. Quantification and functional evaluation of CD40L production from the adenovirus vector ONCOS-401. *Cancer Gene Ther.* **2018**. [CrossRef]

66. Kuryk, L.; Møller, A.-S.; Vuolanto, A.; Pesonen, S.; Garofalo, M.; Cerullo, V.; Jaderberg, M. Optimization of Early Steps in Oncolytic Adenovirus ONCOS-401 Production in T-175 and HYPERFlasks. *Int. J. Mol. Sci.* **2019**, *20*, 621. [CrossRef]

67. Capasso, C.; Magarkar, A.; Cervera-carrascon, V.; Fusciello, M.; Feola, S.; Muller, M. ORIGINAL RESEARCH A novel in silico framework to improve MHC-I epitopes and break the tolerance to melanoma. *Oncoimmunology* **2017**, *6*, e1319028. [CrossRef]

68. Garofalo, M.; Villa, A.; Crescenti, D.; Marzagalli, M.; Kuryk, L.; Limonta, P.; Mazzaferro, V.; Ciana, P. Heterologous and cross-species tropism of cancer- derived extracellular vesicles. *Theranostics* **2019**, *9*, 5681. [CrossRef]

69. Garofalo, M.; Villa, A.; Rizzi, N.; Kuryk, L.; Mazzaferro, V.; Ciana, P. Systemic Administration and Targeted Delivery of Immunogenic Oncolytic Adenovirus Encapsulated in Extracellular Vesicles for Cancer Therapies. *Viruses* **2018**, *10*, 558. [CrossRef]

70. Garofalo, M.; Villa, A.; Rizzi, N.; Kuryk, L.; Rinner, B.; Cerullo, V.; Yliperttula, M. Extracellular vesicles enhance the targeted delivery of immunogenic oncolytic adenovirus and paclitaxel in immunocompetent mice. *J. Control. Release* **2019**, *294*, 165–175. [CrossRef]

71. Jhawar, S.R.; Thandoni, A.; Bommareddy, P.K.; Hassan, S.; Kohlhapp, F.J.; Goyal, S.; Schenkel, J.M.; Silk, A.W.; Zloza, A. Oncolytic Viruses—Natural and Genetically Engineered Cancer Immunotherapies. *Front. Oncol.* **2017**, *7*, 202. [CrossRef]

72. Kuryk, Ł.; Wieczorek, M.; Diedrich, S.; Böttcher, S.; Witek, A.; Litwińska, B. Genetic analysis of poliovirus strains isolated from sewage in Poland. *J. Med. Virol.* **2014**, *86*, 1243–1248. [CrossRef] [PubMed]

73. Wieczorek, M.; Ciąćka, A.; Witek, A.; Kuryk, Ł.; Żuk-Wasek, A. Environmental Surveillance of Non-polio Enteroviruses in Poland, 2011. *Food Environ. Virol.* **2015**, *7*, 224–231. [CrossRef] [PubMed]

74. Pesonen, S.; Kangasniemi, L.; Hemminki, A. Oncolytic Adenoviruses for the Treatment of Human Cancer: Focus on Translational and Clinical Data. *Mol. Pharm.* **2011**, *8*, 12–28. [CrossRef]

75. Garofalo, M.; Saari, H.; Somersalo, P.; Crescenti, D.; Kuryk, L.; Aksela, L.; Capasso, C.; Madetoja, M.; Koskinen, K.; Oksanen, T.; et al. Antitumor effect of oncolytic virus and paclitaxel encapsulated in extracellular vesicles for lung cancer treatment. *J. Control. Release* **2018**, *283*, 223–234. [CrossRef]

76. Kuryk, L.; Møller, A.-S.W.; Garofalo, M.; Cerullo, V.; Pesonen, S.; Alemany, R.; Jaderberg, M. Anti-tumor specific T-cell responses induced by oncolytic adenovirus ONCOS-102 in peritoneal mesothelioma mouse model. *J. Med. Virol.* **2018**, 1669–1673. [CrossRef]

77. Capasso, C.; Hirvinen, M.; Garofalo, M.; Romaniuk, D.; Kuryk, L.; Sarvela, T.; Vitale, A.; Antopolsky, M.; Magarkar, A.; Viitala, T.; et al. Oncolytic adenoviruses coated with MHC-I tumor epitopes increase the antitumor immunity and efficacy against melanoma. *Oncoimmunology* **2016**, *5*, e1105429. [CrossRef]

78. Kuryk, L.; Haavisto, E.; Garofalo, M.; Capasso, C.; Hirvinen, M.; Pesonen, S.; Ranki, T.; Vassilev, L.; Cerullo, V. Synergistic anti-tumor efficacy of immunogenic adenovirus ONCOS-102 (Ad5/3-D24-GM-CSF) and standard of care chemotherapy in preclinical mesothelioma model. *Int. J. Cancer* **2016**, *139*, 1883–1893. [CrossRef] [PubMed]

79. Kuryk, L.; Vassilev, L.; Ranki, T.; Hemminki, A.; Karioja-Kallio, A.; Levälampi, O.; Vuolanto, A.; Cerullo, V.; Pesonen, S. Toxicological and bio-distribution profile of a GM-CSF-expressing, double-targeted, chimeric oncolytic adenovirus ONCOS-102—Support for clinical studies on advanced cancer treatment. *PLoS ONE* **2017**, *12*, e0182715. [CrossRef]

80. Garofalo, M.; Iovine, B.; Kuryk, L.; Capasso, C.; Hirvinen, M.; Vitale, A.; Yliperttula, M.; Bevilacqua, M.A.; Cerullo, V. Oncolytic Adenovirus Loaded with L-carnosine as Novel Strategy to Enhance the Antitumor Activity. *Mol. Cancer Ther.* **2016**, *15*, 651–660. [CrossRef]

81. Hirvinen, M.; Capasso, C.; Guse, K.; Garofalo, M.; Vitale, A.; Ahonen, M.; Kuryk, L.; Vähä-Koskela, M.; Hemminki, A.; Fortino, V.; et al. Expression of DAI by an oncolytic vaccinia virus boosts the immunogenicity of the virus and enhances antitumor immunity. *Mol. Ther. Oncolytics* **2016**, *3*, 16002. [CrossRef]

82. Kaufman, H.L.; Kohlhapp, F.J.; Zloza, A. Oncolytic viruses: A new class of immunotherapy drugs. *Nat. Rev. Drug Discov.* **2015**, *14*, 642. [CrossRef] [PubMed]

83. Shi, M.; Chen, X.; Ye, K.; Yao, Y.; Li, Y. Application potential of toll-like receptors in cancer immunotherapy: Systematic review. *Medicine* **2016**, *95*, e3951. [CrossRef]

84. Marelli, G.; Howells, A.; Lemoine, N.R.; Wang, Y. Oncolytic Viral Therapy and the Immune System: A Double-Edged Sword against Cancer. *Front. Immunol.* **2018**, *9*, 866. [CrossRef]

85. Vähä-Koskela, M.J.V.; Heikkilä, J.E.; Hinkkanen, A.E. Oncolytic viruses in cancer therapy. *Cancer Lett.* **2007**, *254*, 178–216. [CrossRef]

86. Coit, D.G.; Thompson, J.A.; Algazi, A.; Andtbacka, R.; Bichakjian, C.K.; Carson, W.E.; Daniels, G.A.; DiMaio, D.; Ernstoff, M.; Fields, R.C.; et al. Melanoma, Version 2.2016, NCCN Clinical Practice Guidelines in Oncology. *J. Natl. Compr. Cancer Netw.* **2016**, *14*, 450–473. [CrossRef]

87. Antohe, M.; Nedelcu, R.I.; Nichita, L.; Popp, C.G.; Cioplea, M.; Brinzea, A.; Hodorogea, A.; Calinescu, A.; Balaban, M.; Ion, D.A.; et al. Tumor infiltrating lymphocytes: The regulator of melanoma evolution (Review). *Oncol. Lett.* **2019**, *17*, 4155–4161. [CrossRef] [PubMed]

88. Havel, J.J.; Chowell, D.; Chan, T.A. The evolving landscape of biomarkers for checkpoint inhibitor immunotherapy. *Nat. Rev. Cancer* **2019**, *19*, 133–150. [CrossRef]

89. Corrigan, P.A.; Beaulieu, C.; Patel, R.B.; Lowe, D.K. Talimogene Laherparepvec: An Oncolytic Virus Therapy for Melanoma. *Ann. Pharmacother.* **2017**, *51*, 675–681. [CrossRef] [PubMed]

90. O'Donoghue, C.; Doepker, M.P.; Zager, J.S. Talimogene laherparepvec: Overview, combination therapy and current practices. *Melanoma Manag.* **2016**, *3*, 267–272. [CrossRef]

91. Harrington, K.; Freeman, D.J.; Kelly, B.; Harper, J.; Soria, J.-C. Optimizing oncolytic virotherapy in cancer treatment. *Nat. Rev. Drug Discov.* **2019**, *18*, 689–706. [CrossRef]

92. Sivanandam, V.; Larocca, C.J.; Chen, N.G.; Fong, Y.; Warner, S.G. Oncolytic Viruses and Immune Checkpoint Inhibition: The Best of Both Worlds. *Mol. Ther. Oncolytics* **2019**, *13*, 93–106. [CrossRef] [PubMed]

93. Cervera-Carrascon, V.; Siurala, M.; Santos, J.M.; Havunen, R.; Tähtinen, S.; Karell, P.; Sorsa, S.; Kanerva, A.; Hemminki, A. TNFa and IL-2 armed adenoviruses enable complete responses by anti-PD-1 checkpoint blockade. *Oncoimmunology* **2018**, *7*, e1412902. [CrossRef] [PubMed]

94. Thomas, S.; Kuncheria, L.; Roulstone, V.; Kyula, J.N.; Mansfield, D.; Bommareddy, P.K.; Smith, H.; Kaufman, H.L.; Harrington, K.J.; Coffin, R.S. Development of a new fusion-enhanced oncolytic immunotherapy platform based on herpes simplex virus type 1. *J. Immunother. Cancer* **2019**, *7*, 214. [CrossRef] [PubMed]

95. Hromic-jahjefendic, A.; Lundstrom, K. Viral Vector-Based Melanoma Gene Therapy. *Biomedicines* **2020**, *8*, 60. [CrossRef] [PubMed]

96. Larocca, C.A.; LeBoeuf, N.R.; Silk, A.W.; Kaufman, H.L. An Update on the Role of Talimogene Laherparepvec (T-VEC) in the Treatment of Melanoma: Best Practices and Future Directions. *Am. J. Clin. Dermatol.* **2020**. [CrossRef] [PubMed]

97. Sun, L.; Funchain, P.; Song, J.M.; Rayman, P.; Tannenbaum, C.; Ko, J.; Mcnamara, M.; Marcela Diaz-Montero, C.; Gastman, B. Talimogene Laherparepvec combined with anti-PD-1 based immunotherapy for unresectable stage III-IV melanoma: A case series. *J. Immunother. Cancer* **2018**, *6*, 36. [CrossRef] [PubMed]

98. Chesney, J.; Imbert-Fernandez, Y.; Telang, S.; Baum, M.; Ranjan, S.; Fraig, M.; Batty, N. Potential clinical and immunotherapeutic utility of talimogene laherparepvec for patients with melanoma after disease progression on immune checkpoint inhibitors and BRAF inhibitors. *Melanoma Res.* **2018**, *28*, 250. [CrossRef]

99. Andtbacka, R.H.I.; Curti, B.D.; Hallmeyer, S.; Feng, Z.; Paustian, C.; Bifulco, C.; Fox, B.; Grose, M.; Shafren, D. Phase II calm extension study: Coxsackievirus A21 delivered intratumorally to patients with advanced melanoma induces immune-cell infiltration in the tumor microenvironment. *J. Immunother. Cancer* **2015**, *3*, 343. [CrossRef]

100. Curti, B.; Richards, J.; Hallmeyer, S.; Faries, M.; Andtbacka, R.; Daniels, G.; Grose, M.; Shafren, D.R. Abstract CT114: The MITCI (Phase 1b) study: A novel immunotherapy combination of intralesional Coxsackievirus A21 and systemic ipilimumab in advanced melanoma patients with or without previous immune checkpoint therapy treatment. *Cancer Res.* **2017**, *77*, CT114. [CrossRef]

101. Schmid, P.; Cruz, C.; Braiteh, F.S.; Eder, J.P.; Tolaney, S.; Kuter, I.; Nanda, R.; Chung, C.; Cassier, P.; Delord, J.-P.; et al. Abstract 2986: Atezolizumab in metastatic TNBC (mTNBC): Long-term clinical outcomes and biomarker analyses. *Cancer Res.* **2017**, *77*, 2986. [CrossRef]

102. Ranki, T.; Pesonen, S.; Hemminki, A.; Partanen, K.; Kairemo, K.; Alanko, T.; Lundin, J.; Linder, N.; Turkki, R.; Ristimäki, A.; et al. Phase I study with ONCOS-102 for the treatment of solid tumors—An evaluation of clinical response and exploratory analyses of immune markers. *J. Immunother. Cancer* **2016**, *4*, 17. [CrossRef] [PubMed]

103. Chiu, M.; Armstrong, E.J.L.; Jennings, V.; Foo, S.; Crespo-Rodriguez, E.; Bozhanova, G.; Patin, E.C.; McLaughlin, M.; Mansfield, D.; Baker, G.; et al. Combination therapy with oncolytic viruses and immune checkpoint inhibitors. *Expert Opin. Biol. Ther.* **2020**, *20*, 635–652. [CrossRef] [PubMed]

104. Russell, L.; Peng, K.W.; Russell, S.J.; Diaz, R.M. Oncolytic Viruses: Priming Time for Cancer Immunotherapy. *BioDrugs* **2019**, *33*, 485–501. [CrossRef] [PubMed]

105. Marchini, A.; Daeffler, L.; Pozdeev, V.I.; Angelova, A.; Rommelaere, J. Immune Conversion of Tumor Microenvironment by Oncolytic Viruses: The Protoparvovirus H-1PV Case Study. *Front. Immunol.* **2019**, *10*, 1848. [CrossRef]

106. Doloff, J.C.; Waxman, D.J. Dual E1A oncolytic adenovirus: Targeting tumor heterogeneity with two independent cancer-specific promoter elements, DF3/MUC1 and hTERT. *Cancer Gene Ther.* **2011**, *18*, 153–166. [CrossRef]

107. Zamarin, D.; Holmgaard, R.B.; Subudhi, S.K.; Park, J.S.; Mansour, M.; Palese, P.; Merghoub, T.; Wolchok, J.D.; Allison, J.P. Localized Oncolytic Virotherapy Overcomes Systemic Tumor Resistance to Immune Checkpoint Blockade Immunotherapy. *Sci. Transl. Med.* **2014**, *6*, 226ra32. [CrossRef]

108. Woller, N.; Gürlevik, E.; Fleischmann-Mundt, B.; Schumacher, A.; Knocke, S.; Kloos, A.M.; Saborowski, M.; Geffers, R.; Manns, M.P.; Wirth, T.C.; et al. Viral Infection of Tumors Overcomes Resistance to PD-1-immunotherapy by Broadening Neoantigenome-directed T-cell Responses. *Mol. Ther.* **2015**, *23*, 1630–1640. [CrossRef]

109. Kruger, S.; Ilmer, M.; Kobold, S.; Cadilha, B.L.; Endres, S.; Ormanns, S.; Schuebbe, G.; Renz, B.W.; D'Haese, J.G.; Schloesser, H.; et al. Advances in cancer immunotherapy 2019—Latest trends. *J. Exp. Clin. Cancer Res.* **2019**, *38*, 268. [CrossRef]

110. Kaunitz, G.J.; Cottrell, T.R.; Lilo, M.; Muthappan, V.; Esandrio, J.; Berry, S.; Xu, H.; Ogurtsova, A.; Anders, R.A.; Fischer, A.H.; et al. Melanoma subtypes demonstrate distinct PD-L1 expression profiles. *Lab. Investig.* **2017**, *97*, 1063–1071. [CrossRef]

111. Kepp, O.; Senovilla, L.; Vitale, I.; Vacchelli, E.; Adjemian, S.; Agostinis, P.; Apetoh, L.; Aranda, F.; Barnaba, V.; Bloy, N.; et al. Consensus guidelines for the detection of immunogenic cell death. *Oncoimmunology* **2014**, *3*, e955691. [CrossRef]

112. Cerullo, V.; Vähä-Koskela, M.; Hemminki, A. Oncolytic adenoviruses: A potent form of tumor immunovirotherapy. *Oncoimmunology* **2012**, *1*, 979–981. [CrossRef] [PubMed]

113. Nguyen, A.; Ho, L.; Wan, Y. Chemotherapy and Oncolytic Virotherapy: Advanced Tactics in the War against Cancer. *Front. Oncol.* **2014**, *4*, 145. [CrossRef] [PubMed]

114. Quirin, C.; Mainka, A.; Hesse, A.; Nettelbeck, D.M. Combining adenoviral oncolysis with temozolomide improves cell killing of melanoma cells. *Int. J. Cancer* **2007**, *2807*, 2801–2807. [CrossRef] [PubMed]

115. Adusumilli, P.S.; Stiles, B.M.; Chan, M.-K.; Chou, T.-C.; Wong, R.J.; Rusch, V.W.; Fong, Y. Radiation Therapy Potentiates Effective Oncolytic Viral Therapy in the Treatment of Lung Cancer. *Ann. Thorac. Surg.* **2005**, *80*, 409–417. [CrossRef] [PubMed]

116. Adusumilli, P.S.; Chan, M.-K.; Hezel, M.; Yu, Z.; Stiles, B.M.; Chou, T.-C.; Rusch, V.W.; Fong, Y. Radiation-Induced Cellular DNA Damage Repair Response Enhances Viral Gene Therapy Efficacy in the Treatment of Malignant Pleural Mesothelioma. *Ann. Surg. Oncol.* **2007**, *14*, 258–269. [CrossRef]

117. Cheng, L.; Lopez-Beltran, A.; Massari, F.; MacLennan, G.T.; Montironi, R. Molecular testing for BRAF mutations to inform melanoma treatment decisions: A move toward precision medicine. *Mod. Pathol.* **2018**, *31*, 24–38. [CrossRef]

118. Zhang, T.; Suryawanshi, Y.R.; Woyczesczyk, H.M.; Essani, K. Targeting Melanoma with Cancer-Killing Viruses. *Open Virol. J.* **2017**, *11*, 28–47. [CrossRef]

119. Noser, J.A.; Mael, A.A.; Sakuma, R.; Ohmine, S.; Marcato, P.; Lee, P.W.K.; Ikeda, Y. The RAS/Raf1/MEK/ERK Signaling Pathway Facilitates VSV-mediated Oncolysis: Implication for the Defective Interferon Response in Cancer Cells. *Mol. Ther.* **2007**, *15*, 1531–1536. [CrossRef]

Publisher's Note: MDPI stays neutral with regard to jurisdictional claims in published maps and institutional affiliations.

Review

Expanding the Spectrum of Pancreatic Cancers Responsive to Vesicular Stomatitis Virus-Based Oncolytic Virotherapy: Challenges and Solutions

Molly C. Holbrook, Dakota W. Goad and Valery Z. Grdzelishvili *

Department of Biological Sciences, University of North Carolina at Charlotte, Charlotte, NC 28223, USA; mpenton@uncc.edu (M.C.H.); dgoad@uncc.edu (D.W.G.)
* Correspondence: vzgrdzel@uncc.edu

Simple Summary: Pancreatic ductal adenocarcinoma (PDAC) is a devastating malignancy with a poor prognosis and a dismal survival rate. Oncolytic virus (OV) is an anticancer approach that utilizes replication-competent viruses to preferentially infect and kill tumor cells. Vesicular stomatitis virus (VSV), one such OV, is already in several phase I clinical trials against different malignancies. VSV-based recombinant viruses are effective OVs against a majority of tested PDAC cell lines. However, some PDAC cell lines are resistant to VSV. This review discusses multiple mechanisms responsible for the resistance of some PDACs to VSV-based OV therapy, as well multiple rational approaches to enhance permissiveness of PDACs to VSV and expand the spectrum of PDACs responsive to VSV-based oncolytic virotherapy.

Abstract: Pancreatic ductal adenocarcinoma (PDAC) is a devastating malignancy with poor prognosis and a dismal survival rate, expected to become the second leading cause of cancer-related deaths in the United States. Oncolytic virus (OV) is an anticancer approach that utilizes replication-competent viruses to preferentially infect and kill tumor cells. Vesicular stomatitis virus (VSV), one such OV, is already in several phase I clinical trials against different malignancies. VSV-based recombinant viruses are effective OVs against a majority of tested PDAC cell lines. However, some PDAC cell lines are resistant to VSV. Upregulated type I IFN signaling and constitutive expression of a subset of interferon-simulated genes (ISGs) play a major role in such resistance, while other mechanisms, such as inefficient viral attachment and resistance to VSV-mediated apoptosis, also play a role in some PDACs. Several alternative approaches have been shown to break the resistance of PDACs to VSV without compromising VSV oncoselectivity, including (i) combinations of VSV with JAK1/2 inhibitors (such as ruxolitinib); (ii) triple combinations of VSV with ruxolitinib and polycations improving both VSV replication and attachment; (iii) combinations of VSV with chemotherapeutic drugs (such as paclitaxel) arresting cells in the G2/M phase; (iv) arming VSV with p53 transgenes; (v) directed evolution approach producing more effective OVs. The latter study demonstrated impressive long-term genomic stability of complex VSV recombinants encoding large transgenes, supporting further clinical development of VSV as safe therapeutics for PDAC.

Keywords: oncolytic virus; virotherapy; pancreatic cancer; pancreatic ductal adenocarcinoma; vesicular stomatitis virus

Citation: Holbrook, M.C.; Goad, D.W.; Grdzelishvili, V.Z. Expanding the Spectrum of Pancreatic Cancers Responsive to Vesicular Stomatitis Virus-Based Oncolytic Virotherapy: Challenges and Solutions. *Cancers* **2021**, *13*, 1171. https://doi.org/10.3390/cancers13051171

Academic Editors: Antonio Marchini, Carolina S. Ilkow and Alan Melcher

Received: 1 February 2021
Accepted: 5 March 2021
Published: 9 March 2021

Publisher's Note: MDPI stays neutral with regard to jurisdictional claims in published maps and institutional affiliations.

1. Introduction

Pancreatic ductal adenocarcinoma (PDAC) is the most common form of pancreatic neoplasm. It is a highly invasive malignancy, which forms a stromal desmoplastic reaction (desmoplasia), characterized by a dramatic increase in the proliferation of alpha-smooth muscle actin-positive fibroblasts and an increased production of many extracellular matrix components [1]. Family history, diabetes, and smoking are the most well-established risk factors for developing pancreatic cancer. Despite being only the 13th most common type of cancer, PDAC is the fourth-leading cause of cancer-related deaths and is predicted to become the second-leading cause of cancer-related death by 2030, as incidence increases while rates of survivorship remain stagnant due to late diagnosis and limited treatment options [2].

KRAS, CDKN2A, TP53, and *SMAD4* serve as driver genes for PDAC development, and the vast majority of patients with fully established pancreatic cancer carry genetic defects in at least one of these genes [3]. Mutations in KRAS are present in 90% of PDAC tumors, 95% of PDAC tumors have mutations in *CDKN2A* (encodes p16), 50–75% in *TP53*, and *SMAD4* (*DPC4*) is lost in approximately 50% of PDAC tumors [4]. Mutated *KRAS* oncogene leads to an abnormal, constitutively active, Ras protein. This results in aberrant activation of pathways responsible for survival and proliferation [5]. Inactivation of the tumor suppressor gene *CDKN2A* results in the loss of p16, a protein that serves as a regulator of the G1-S checkpoint of the cell cycle. Abnormalities in *TP53* prevent it from acting as a tumor suppressor protein, including its important role as a regulator of DNA-damage checkpoints. Furthermore, many p53 mutants acquire devastating gain-of-function oncogenic activities, actually promoting cell survival, proliferation, invasion, migration, chemoresistance, and chronic inflammation. *SMAD4* (*DPC4*) is related to the TGF-β signaling pathway, but some mutations result in abnormal signaling by TGF-β, a transforming growth factor receptor on the cell surface which can further increase the risk of cancer development by increasing the rate of cell growth and replication. In addition, germline mutations within BRCA2, BRCA1, ATM, and other genes were frequently identified in PDACs as inherited traits increasing susceptibility to PDAC development later in life [6,7]. These genes, especially when identified as being comorbid, are correlated with a significantly higher metastatic burden [8–10].

The primary treatments for PDAC include surgery, chemotherapy, radiotherapy, and palliative care [11]. Surgical resection still retains the greatest chance of success for potentially curing PDAC, however late-stage diagnosis due to ambiguous symptoms often results in tumors that are too-far progressed for surgery alone. Less than 25% of patients that present with PDAC are eligible for surgical resection, and 5-year survivorship of completely resected patients is approximately 37% [4]. In addition, even in patients where surgical resection was performed with either preparatory or subsequent adjuvant chemotherapy, there is a high rate of recurrence, and up to 80% of patients with recurrent PDAC will relapse with local and/or distant disease, which is associated with mortality within 2 years from diagnosis.

Recent advances in the understanding of the molecular biology, diagnosis, and staging of PDAC will hopefully lead to greater progress in the development of novel treatment approaches for PDAC patients. One such approach is oncolytic virus (OV) therapy, which utilizes replication-competent viruses to preferentially infect, replicate in, and kill cancer cells. In this review, we will discuss current advances with OV therapy for PDAC, with a special focus on vesicular stomatitis virus (VSV), the major interest of our laboratory. For comprehensive reviews of gene therapy for pancreatic cancer (unlike oncolytic virotherapy, gene therapy is typically based on replication-defective viral vectors for transgene delivery), we refer to these excellent papers [12,13].

2. Major Challenges with Current PDAC Treatments

Since 1997, gemcitabine-based chemotherapy has been the standard first-line treatment for patients with unresectable locally advanced, or metastatic pancreatic cancer with a median survival rate of 4.4–5.6 months, especially when patients are not healthy enough for combination therapies [14]. Gemcitabine (dFdC) is an analog of deoxycytidine and a pro-drug that, once transported into the cell, must be phosphorylated by cellular deoxycytidine kinase to gemcitabine diphosphate (dFdCTP) and gemcitabine triphosphate (dFdCTP), both of which can inhibit processes required for DNA synthesis. Other commonly used chemotherapies for pancreatic cancer include 5-fluorouracil (5-FU), oxaliplatin, albumin-bound paclitaxel, capecitabine, cisplatin, irinotecan, and docetaxel [15,16].

Although several gemcitabine-based combination treatments exist, most have not considerably improved survival. While some combinatorial chemotherapy treatments, such as gemcitabine with erlotinib, have demonstrated potential for longer patient survival, the majority of patients eventually experience tumor progression due to the development of resistance, and therefore novel therapies are required, especially those that do not rely solely on chemotherapeutic drugs [17,18].

The mechanisms of de novo or inherent resistance of PDACs to chemo- or radiotherapeutics are not well understood. Several factors have been demonstrated to contribute to such resistance, including (i) multiple factors associated with the nature of the PDAC tumor microenvironment (TME) [19,20]; (ii) nucleoside transporters or/and nucleoside enzymes affecting drug uptake and metabolism [21]; (iii) hypoxia-inducible factor-1 alpha (HIF-1α) regulated glucose metabolism [22]; (iv) stromal-derived Insulin-like Growth Factors (IGFs) [23]; (v) abnormal expression of tumor-associated mucin proteins [24]; (vi) IFN-related DNA-damage resistance signature (IRDS) of some tumors [25]. The understanding of chemoresistance of PDACs to chemotherapy is very important, as at least some of these mechanisms could be also contributing to the resistance of PDACs to OV therapy.

The success of any treatment for PDAC is further complicated by the TME of PDAC, which is characterized by dense stroma comprised of abundant fibroblasts, hypoxia, and sparse vasculature. Moreover, the infiltration of tumor-promoting immune cells mediates immune evasion and promotes tumor progression. The stroma surrounding the tumor is primarily composed of pancreatic stellate cells (PSCs) which are activated by secreted factors such as TNFα, TGF-β, and interleukins 1, 2, 10, and themselves secrete mucins, collagen, fibronectin, and laminin in addition to some other factors, forming a thick extracellular matrix (ECM). This composition generates an incredibly dense physical barrier, to both host immune cells and potential therapeutics while also increasing interstitial pressure, which, when combined with sparse vasculature, forms a hypoxic environment, further inhibiting immune cells in terms of recruitment and effectiveness. PI3K/Akt, a key downstream mediator of many receptor tyrosine kinase signaling pathways involved in cell proliferation, migration, and inhibition of apoptosis, is phosphorylated under hypoxic conditions, along with MAPK (Erk), which regulates cell proliferation in response to various growth factors, which have been associated with resistance to gemcitabine [26,27]. The limits on antitumor immune cell recruitment also leads to T-cell exhaustion resulting in loss of cytotoxic effector function and further limits appropriate immune responses. SDF-1α/CXCR4 signaling-induced activation of the intracellular FAK-AKT and ERK1/2 signaling pathways and a subsequent IL-6 autocrine loop in cancer cells can further increase chemoresistance [28].

The low expression of nucleoside transporters (NT) and inactivity of nucleoside enzymes (NE) both affect the activity of gemcitabine. Low expression of a nucleoside transporter hENT1 restricts the uptake of gemcitabine, preventing its incorporation into the DNA of replicating cancer cells, and high expression of hENT1 is related to longer overall survival in pancreatic cancer patients [29,30]. The inactivation of deoxycytidine kinase (dCK), an enzyme responsible for the initial phosphorylation of gemcitabine, also mediates resistance. dCK is often inactivated in gemcitabine-resistant PDAC lines [31], and knockdown of dCK has been shown to lead to the development of resistance [32], while expression of a DCK transgene (along with uridine monophosphate kinase) sensitized pancreatic cancer cells to gemcitabine [33].

Pancreatic cancers metabolize glucose at higher rates and show higher expression of HIF-1α positively correlated with gemcitabine resistance [34,35]. HIF-1α increases glucose uptake and metabolism in the cell and is stabilized by MUC1, a common biomarker for cancers including PDAC [36]. Knockdown of HIF-1α in gemcitabine-resistant cells reduced tumor cell survival following gemcitabine treatment, and treatment with digoxin, and HIF-1α inhibitor, reduced glucose uptake and cell survival in cells treated with gemcitabine [37]. The increased glucose uptake under hypoxic conditions feeds into the glycolysis pathway and increases biomass; however, the exact mechanisms by which HIF-1α reduces sensitivity to chemotherapeutics have yet to be determined.

In addition, stromal-derived IGFs activate the insulin/IGF1R survival signaling pathway, reducing responsiveness to chemotherapeutics [38]. One proposed mechanism describes crosstalk between activated Insulin/IGF signaling pathways in PDAC. IGF-1 and IGF-1R, which are known to be abundantly expressed in the PDAC tissue, can stimulate β-cell proliferation and increase β-cell mass, increasing basal insulin production which may alter the trophic effects of the endocrine cells on the exocrine cells. Endocrine β-cells that express oncogenic K-ras can also be one potential progenitor for PDAC under chronic tissue inflammation [39]. This is further supported by evidence that demonstrates macrophages and myofibroblasts are the two major sources of IGFs within the pancreatic tumor microenvironment, and that chemoresistance is increased when cytotoxic agents increase M2-like macrophage infiltration [23]. For any novel therapies to be effective, they should be able to address most if not all of these challenges.

The structural composition of mucins produced by cells in certain cancers, such as breast and pancreatic cancers, has been suggested to limit immune cell recognition by blocking infiltration [40]. Similarly, the dense mucin mesh prevents cellular uptake of chemotherapeutics like gemcitabine and 5-FU within the tumor. MUC1 and MUV4 are overexpressed and aberrantly glycosylated in the majority of pancreatic tumors [41]. Kalra et al. demonstrated that the inhibition of mucin *O*-glycosylation enhanced the cytotoxic effects of 5-FU against human pancreatic cancer cell lines, but not against the mucin-deficient cell line [40]. They suggest that preventing the formation of the mucin facilitates the diffusion of drugs across the compromised mucus layer, improving intracellular drug uptake and enhancing cytotoxic drug action. Elevated MUC1 and MUC4 expression have also been correlated with greater degrees of resistance to gemcitabine [42]. It was also demonstrated that gemcitabine-resistant cells had accentuated the non-oxidative branch of the pentose phosphate pathway activity and increased pyrimidine biosynthesis, conferring resistance by increased dCTP production. MUC1 and MUC4 overexpression was also shown to upregulate *mdr* genes in pancreatic cancer cells, including *ABCC1*, *ABCC3*, *ABCC5*, and ABCB1 genes [41,43]. MUC4 expression was shown to be conversely correlated with the expression of hCNT1 and hCNT3 transporters, preventing uptake of chemotherapeutic drugs like gemcitabine, and hCNT1 is upregulated when MUC4 is inhibited, resulting in increased drug sensitivity [44]. Finally, MUC4-overexpressing CD18/HPAF-Src were not sensitive to gemcitabine, conferring resistance and survival advantages through erbB2-dependent and anti-apoptotic pathways [45]. Altogether, mucins including MUC1 and MUC4 have been demonstrated to be highly overexpressed and aberrantly glycosylated in pancreatic cancer cells, conferring resistance to various chemotherapies and the downregulation of these oncoproteins may represent a promising therapeutic strategy for reversing chemoresistance and reducing tumor progression and mass.

Type I IFN signaling is upregulated in some tumors responding to chemotherapy and can have antitumor as well as pro-tumor effects. The expression of a type I IFN-related DNA-damage resistance signature (IRDS) was reported to correlate with resistance to chemotherapy and radiotherapy in multiple cancer types. In breast cancer, the IRDS has been implicated in the development of chemoresistance, which may be another potential mechanism of resistance in PDACs as well [25]. The STAT1/IFN pathway transmits a cytotoxic signal either in response to DNA damage or to IFNs, such as in the case of viral infection. Cells with an IRDS (+) profile show constitutive activation of the STAT1/IFN pathway. Interestingly, this chronically activated state of the STAT1/IFN pathway may select against transmission of a cytotoxic signal, instead resulting in pro-survival signals mediated by STAT1 and other IRDS genes [25]. In agreement with this mechanism, STAT1 is highly upregulated in many cancers, including PDAC, and protects SCC-61 cells from ionizing radiation-mediated death [46]. STAT1 may also induce resistance with other DNA damage-based treatments, such as gemcitabine, and may transduce survival/growth signals that enhance tumor survival under some conditions [47]. Sensitivity to DNA damage is coupled with sensitivity to IFNs such that selection for resistance to one may lead to resistance to the other [48], which could prove to be a problem with not only chemo- and radiotherapies, but OV treatments as well.

3. Overview of Common Experimental Models to Study OV Therapy in PDAC

Oncolytic virus (OV) therapy is a relatively novel anticancer approach. Effective OV therapy is dependent on the oncoselectivity of OVs—their ability to preferentially infect, replicate in and kill infected cancer cells without damaging nonmalignant ("normal") cells. The ideal OV therapy not only requires the direct lysis of cancer cells by the virus but also activates innate and adaptive anticancer immune responses [49] (Figure 1).

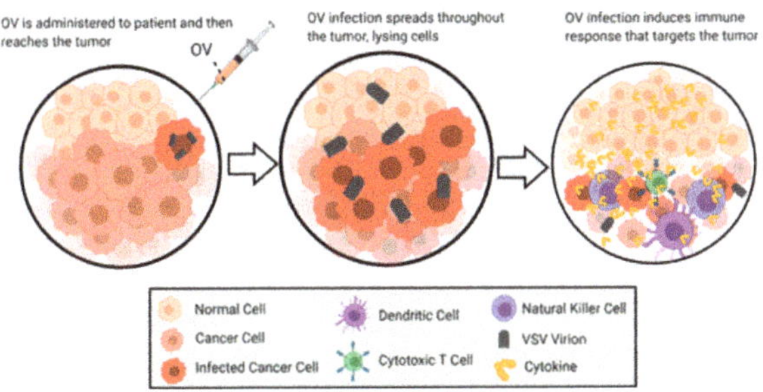

Figure 1. General Overview of Oncolytic Virotherapy. This figure demonstrates the general method of action for the treatment of cancer by oncolytic virotherapy using VSV as an oncolytic virus. The images depict the infection and oncolysis of malignant cells over time, followed by immunos- timulation of cells invading the cleared area. The figure was created by authors with BioRender software (BioRender.com).

Preclinical PDAC models are critical for understanding the biology of PDAC, are platforms for developing novel strategies against PDAC, and are a necessary part of the drug development pipeline. There are several features of an ideal PDAC model system to develop clinically relevant OV therapy against PDAC: (1) the ability to test OV against different PDACs, characterized by various responsiveness to different therapies, including OV therapy; (2) the model should recapitulate a complex TME of PDACs; (3) tractability of the model, including the ability to trace both tumor cells and OV; (4) the ability to deliver OV systemically, as the PDAC are difficult to access; (5) the ability to detect and evaluate innate and adaptive immune responses against both tumor cells and OV. Unfortunately, there is no single PDAC model that successfully recapitulates all these critical features and challenges of the disease. However, there are numerous models for PDAC, each with unique advantages and disadvantages. Here, we will briefly review the advantages and disadvantages of various in vitro and in vivo models of PDAC and how they can contribute to the development of OV therapeutics.

3.1. In Vitro Systems

3.1.1. PDAC Cell Lines

Numerous human PDAC cell lines have been established and can be characterized by their distinctive genotypic and phenotypic variations, including their relative permis- siveness or resistance to OV infection [50–52]. Utilizing cell lines as a model system offers several advantages for studying PDAC, including easy propagation and indefinite growth. These features represent a cost-effective and consistent model that can easily be used to study molecular mechanisms and biomarkers of resistance or permissiveness of PDAC cells to OVs [50,52]. While cell line-based approaches represent quick, straightforward, and consistent models, several features reduce their clinical translatability. First, the homo- geneous nature of cell line models fails to accurately represent the heterogeneous nature of typical in vivo tumors, including PDAC [53]. Indeed, cell lines are under selection for mutations and phenotypes allowing growth advantage in a monolayer, however, the selection mechanisms in vivo are different [54]. In fact, established PDAC cell lines not only lose the heterogeneity present in the primary tumor, but the evolution of these cell lines to grow in culture may obscure genetic aberrations present in the primary tumor [52]. Additionally, many PDAC cell lines are originated from metastasized disease, so the ability to study PDAC progression is severely limited. Secondly, cell lines cultured in a monolayer lack the important three-dimensional structure and function as seen in vivo [54]. Thirdly, the PDAC cell line model fails to represent the TME, which is understood to be a dynamic player in PDAC tumor progression [54]. Lastly, cultured cell lines lack selection pressure from the host adaptive immune system, thus leaving mutations necessary for evading host immunity underrepresented. The outcome of the OV therapy depends on the complex interaction between tumor cells, virus, and innate and adaptive immune systems of the host. One of the desirable outcomes of this interaction is OV-mediated stimulation of immune response against tumor cells. However, normal PDAC stromal cells can induce innate antiviral responses against OV replicating in tumor cells, and adaptive immune response can prematurely clear virus infection instead of targeting tumor cells. Unfortunately, cell culture-based models cannot address these important issues.

Even given the disadvantages of the cell line model, it is a good starting proof-of-principle platform that has allowed our group to investigate mechanisms regarding responsiveness or resistance to OV therapy [50,51,55–62]. For example, our group is interested in understanding why/how certain PDAC cell lines are more resistant to VSV infection than other PDAC cell lines [50]. The cell line model in this aspect allows for reliable comparative measurements of virus replication, spread, and cell lysis. Additionally, the cell line model allows for relatively straightforward screening of both cellular and viral genes and proteins of interest. Cell line models allow for efficient virus tractability through reporter genes such as GFP [63]. Additionally, cell culture-based systems allow innovative imaging approaches for single-cell real-time analysis of OV replication and efficacy in pancreatic cancer cells [64].

Depending on the nature of the investigation, either human or murine PDAC cell lines can be used. Human PDAC cells, derived from primary pancreatic tumors or "cell line-derived xenograft (CDX)" models, have been used since as early as 1963 to characterize and test anti-cancer drugs [65]. The use of human PDAC cells provides the obvious benefit of having the same genetic makeup of the human disease, including key PDAC mutations in KRAS, CDKN2A, p53, and SMAD4 [3]. Although using human PDAC cell lines as a model has numerous informative applications, this model has a limited ability for consequent in vivo studies. If using human PDAC cell lines, researchers are limited to T cell-deficient nude athymic (nude), or B and T cell-deficient severe combined immunodeficient (SCID) mice [66,67]. As will be later discussed in this review, while such in vivo models have many applications, they lack the ability to assess the role of the adaptive immune system against PDAC as well as OV, both important when determining the efficacy of potential OV therapeutics.

To circumvent this caveat, murine PDAC cell lines can be used. Using murine PDAC cells derived from murine PDAC tumors allows researchers to establish PDAC in immunocompetent mice, allowing for the study of OV therapy in the presence of the functional adaptive immune system. One notable drawback to this model is the potential genetic dissimilarity (and thus clinical translatability) between mouse and human PDAC cells.

Generally, murine PDAC cell lines are originated from mice that have PDAC due to either chemical induction or genetic modifications in genetically engineered mouse model (GEMM). Once commonly used PDAC cell line that was cultured from a chemically induced PDAC tumor is Panc02, which has been extensively used for PDAC research [68]. The PDAC tumor from which it was derived was established by implanting 3-methylcholanthrene (3-MCA)-saturated threads of cotton in the pancreas of C57BL/6 mice. Despite its long-term use in evaluating various therapeutic strategies, Panc02 cells lack clinical significance for PDAC due to the absence of some common mutations found in human

PDAC. More relevant murine PDAC cell lines are originated from the KPC mouse model of PDAC (LSL- Kras^{G12D}; LSL-Trp53^{R173H}; *Pdx1-Cre*) [68]. KPC mice develop spontaneous PDAC which closely resemble the genetics, physiology, tumor progression, and metastatic hallmarks of human PDAC [69], and will be described in more detail later in this review.

3.1.2. PDAC Organoid Cultures

To better address the lack of 3D structure and function of 2D models, 3D organoid (organ-like) structures may be used that self-organize into structures that more closely resemble the in vivo tissue structure, composition, and function [70]. To model PDAC, normal or cancerous pancreatic ductal cells are typically embedded in Matrigel™, which contains important components of basement membrane and growth factors. Pancreatic ductal cells form polarized structures due to the cell–cell contacts and cell-matrix interactions, which can greatly influence gene expression when compared to 2D cultures [71]. The ability to better mimic the complex 3D architecture of PDAC in vitro is a valuable platform for testing drug delivery, pharmacokinetics, efficacy, and drug resistance. One of the key advantages of the organoid model is the ability to study PDAC progression by comparing normal pancreatic, preneoplastic, and PDAC cell-based organoids. Better understanding at what phases PDAC is more susceptible or resistant to OV therapy is of particular interest to our group, and such a model represents an unparalleled system for controlled disease progression and visibility. Although organoids have many promising potential applications, there are still limitations. First, this organoid model system is synthetic and the mutational selection is not well understood [72]. Second, many of these models are solely epithelial tissue layers, lacking important elements of the tumor microenvironment such as immune cells, nervous cells, mesenchyme, muscular layers, etc. [73]. This limitation is addressed in more realistic organoid models of PDAC, or, "PDACoids" are additionally co-cultured with stromal components like cancer-associated fibroblasts (CAFs), PSCs, endothelial cells, and immune cells to better mimic the dense stroma which typically represents up to 90% of the tumor volume and is a major player in PDAC tumor progression and therapeutic resistance [74]. Thirdly, the organoid model fails to address the complex immune system dynamics as in the human disease [75]. Unfortunately, even as organoid models continue to progress towards the true complexity of the in vivo tumor, the immune microenvironment around a tumor is exceedingly difficult to truly recapitulate in vitro. Despite the limitations of organoids, this technology has great potential and use to more closely model human tumors. We would like to refer to a more exhaustive review of 3D cell culture approaches, including the spheroid model systems here [76].

PDAC organoid cultures offer a more realistic model to study OV delivery, replication, cell lysis, and oncoselectivity, as they better mimic the 3D organization and complexity of the human disease. Recent studies showed that an adenovirus-based OV exhibited good oncoselectivity, with replication only occurring in organoids from PDAC tumors. The group also concluded that the cytotoxicity observed in PDAC organoids was predictive of antitumor efficacy in both subcutaneous (SC) and orthotopic xenograft models [77]. Although VSV has not yet been tested in a PDAC organoid setting, other OVs used so far [64,78,79] have been promising and provide a more predictive model for in vivo disease.

3.2. In Vivo Murine Model Systems

Murine models for human PDAC research are useful tools as mice and humans have comparable anatomic, cellular, and genomic features, including tumor biology [80]. For the scope of this review, we will focus on murine-based in vivo model systems, however in vivo PDAC models from alternative species are also used, and we refer to these excellent articles [81,82].

Murine models of PDAC can help both researchers and clinicians to better understand the onset, development, and metastatic processes of this disease, as well as to explore new therapeutic modalities such as OV therapy. Ideally, a murine model of PDAC should have the following features: (1) consistent PDAC disease progression similar to that of human disease from precursor lesions to PanIN and then PDAC [83]; (2) Cancerous phenotype similar to that in human disease demonstrating the common hallmarks such as anti-apoptosis, immune evasion and suppression, dense fibrosis/desmoplasia, and metastasis; (3) Should address the phenotypic and genetic heterogeneity as seen in human disease; (4) A reliable, consistent, and relatively quick time to tumor establishment; (5) Ability to study innate immune responses to PDAC cells as wells as OV; (6) Ability to track in vivo both tumor cells and OV. Here, we will briefly review the advantages and disadvantages of common in vivo PDAC models, and how they pertain to the development of OV therapeutics. We will break down these models by the genetic background of the mouse, and how PDAC is established in the mouse. For more comprehensive reviews of in vivo PDAC model systems, we refer to these papers [84,85].

3.2.1. Human Cell Line Derived Xenograft (CDX) and Patient-Derived Xenograft (PDX) Models

Human CDX and PDX xenograft PDAC models are used by introducing PDAC cell lines (CDX) or primary tumor tissues (PDX) into immunocompromised mice (nude or SCID), commonly via SC injection [86]. These are useful models for studies not focused on antitumor immune responses, such as drug screenings as it is procedurally relatively simple and economical [87]. The SC CDX and PDX models have additional advantages: (1) the tumor has good tractability and is relatively easy to measure, even in the absence of reporter genes (e.g., luciferase); (2) depending on the growth rate of the cell line, tumors can be palpable within 2–6 weeks [85], and (3) this model allows for direct intratumoral injection of chemotherapeutics or OVs, and subsequent evaluation. However, these models have serious limitations for studying PDACs, which most often develop metastatic tumors, and SC tumors typically fail to metastasize. Furthermore, the CDX model is characterized by the loss of genetic heterogeneity in culture, whereas the PDX model at least retains some of the patients' original genetic heterogeneity [88].

The orthotopic CDX and PDX xenograft PDAC models are more clinically relevant. In those models, PDAC cells or primary tumor tissues are injected/implanted into the pancreas of nude or SCID mice, which better recapitulates primary human tumors and are more likely to provide metastases and show more relevant tumor microenvironment compared to the SC model [89,90]. However, this approach is more procedurally challenging and requires special imaging techniques such as ultrasonography or an in vivo imaging system (IVIS) in concert with PDAC cells that express a reporter gene such as luciferase [67,89]. Moreover, ideally, the second reporter gene (e.g., for red fluorescent protein) should be encoded by OV to track virus spread in the tumors (primary and metastatic) and potential spread to normal tissues. The major limitation of all human CDX and PDX xenograft PDAC models (SC and orthotopic) is the lack of host immunity that both limits the study of OV-mediated adaptive antitumor and antiviral immunity and the robustness of the host, as immunocompromised mice typically exhibit susceptibility to infections and other health problems [91].

3.2.2. Humanized Murine Model

The use of human CDX and PDX models are severely limited when investigating the dynamic interplay between the tumor, tumor microenvironment, and immune system while studying human PDACs. To compensate for this, researchers have developed humanized murine models, where mice are engineered to express components of the human immune system [92,93]. Humanized mice were created by establishing mutations in the IL2 receptor common chain (IL2rg^{null}) in the non-obese diabetic (NOD)/SCID background [94,95]. With the combined lack of NK cell activity from the NOD background and the impaired B and T cell response from the SCID background, this model can support the implantation of human tissue, peripheral blood mononuclear cells (PBMCs) and hematopoietic stem cells (HSCs), allowing for the modeling of human adaptive immunity in immunocompetent mice [93,96]. It is important to note that, although these models allow for studies involving the adaptive immune system, these mice do not have the complete human immune system. This model has shortcomings such as limited lymph node development, HLA incompatibility between grafted human immune components and PDAC cells/tissue, and limited ability to mimic human immune cell trafficking [97]. Traditionally, human PDAC is characterized as non-immunogenic or "cold" tumors due to its lack of T-cell infiltration and immunosuppressive microenvironment [98]. However, tumor implantations in humanized murine models can cause T-cell infiltrations due to lack of histocompatibility, therefore changing classical cold tumors into artificially hot tumors, which can subsequently lead to false-positive results in immunotherapeutic investigations. Some studies have utilized the humanized murine model to try and better understand the role of the adaptive immune system and its role in anti-tumor immunity in the context of OV therapy [99]. While these studies may offer some insights into the role of anti-tumor immunity during OV therapy, there are many caveats to this model that still need to be addressed.

3.2.3. Genetically Engineered Mouse Models (GEMMs)

The KPC Murine KPC cell lines provide a bridge between the need to recapitulate human PDAC disease phenotype and the use of immunocompetent in vivo murine models. This model was developed when they used a Cre/LoxP approach to express a mutant KRAS allele exclusively in pancreatic progenitor cells, causing the development of PanIN lesions which subsequently progressed to PDAC, but only after a prolonged latency period, as KRAS mutations alone are not sufficient for PDAC development [100]. This GEMM model mimics the classical characteristic of human PDAC, including the pronounced desmoplastic stromal reaction [100,101].

Since, numerous GEMM models have been developed to also include classical human PDAC mutations such as in TP53, SMAD4, CDKN2A, TGF-β, and INK4A. The details of these models are beyond the scope of this review, are more comprehensively described in other reviews [102–104]. As mentioned above, currently, the most commonly used model is LSL-KRAS^{G12D}; LSL-Trp53^{R172H}; PDX-1-Cre (KPC, stands for: Kras, p53, and Cre) mouse (C57BL6 genetic background). Unlike predecessor GEMM models, such as the LSL-KRAS^{G12D}; PDX-1-Cre (KC) mouse, the KPC model creates advanced PDAC with classical human PDAC effects such as cachexia, abdominal distension, bowel and biliary obstruction. PDAC progression in KPC models closely resembles that of human disease as they develop PanIN within 8–10 weeks, followed by invasive and metastatic tumors by 16 weeks, along with the characteristic PDAC desmoplastic stromal reaction [69]. As the disease progresses, the tumor will predictably metastasize to the liver, lung, diaphragm and adrenals, as seen in the human disease. The KPC model, and GEMMs generally, provide the advantage of being able to investigate potential biomarkers, novel diagnostics, and/or therapeutics in the early stages of PDAC, and in the presence of the functional adaptive immune system [105,106]. The KPC model is an attractive platform for investigating the efficacy of OV therapies, due to the high semblance of PDAC disease and immune status of this model compared to that of the human disease.

This model does however have disadvantages. First, although resulting PDAC in the KPC model is very similar to that of the human disease, tumors are of murine origin and will therefore have inherent differences compared to human disease. Additionally, the development of KPC mice is labor-intensive, costly to upkeep, and tumor initiation and formation take up to or greater than one year [107]. As well, tractability is limited as monitoring tumor progression requires specialized equipment that might not be available to all labs [108]. These technical drawbacks of this model make it less than ideal for OV therapy testing.

As the vital and therapeutic role of the immune system continues to be acknowledged during OV therapy, it should be standard to use a murine model with a competent immune system. GEMMs, such as KPC mice, represent such a model but can be limited due to high costs, labor intensity, and long tumor formation periods. Syngeneic murine models are developed by introducing murine tumor cells or tissues into immunocompetent mice of the same or similar genetic background either SC or orthotopically, i.e., implanting PDAC cells or tissue from a C57BL6 background mouse into a "wild-type" (WT) C57BL6 mouse. One of the earliest murine PDAC cell lines cultured, Panc02, was established from chemically induced PDAC in 1984 [68]. However, due to the artificial induction from which these cells arose, they do not harbor classical mutations as in human disease, such as KRAS and p53 [109]. A promising alternative is the KPC cell lines, which originated from the KPC mice and containing clinically relevant genetic mutations [63]. Syngeneic murine models can be established in immunocompetent mice either orthotopically or SC, both having the unique advantages and disadvantages. Both approaches provide the important feature of exhibiting a full immune system. The process of injecting syngeneic PDAC cells SC is procedurally less laborious, and tumor tractability is good, but the SC approach lacks the overall clinical relevance compared to orthotopic due to the tumor not being in the pancreas, and its lack of reliable metastasis. The main limitation in the orthotopic approach is the lack of PDAC cell tractability. However, luciferase can be genetically engineered into the PDAC cell lines to be implanted, allowing for much easier tumor imaging by measuring intensity of bioluminescence [102]. Other methods of syngeneic cell line delivery include intravenous, intraperitoneal, and intrasplenic, which have been used to provide models for lung, peritoneal, lymph node, and liver metastasis, respectively [110–112]. Our group is currently developing the syngeneic KPC cell line model system for studying VSV-based OV therapy against PDAC (i.e., location, tumor microenvironment, immune system), without the high costs and long development times of GEMM models. A key highlight of this system is tractability of both tumor cells as well as VSV via encoded far-red fluorescent protein, and this system utilizes several alternative KPC cell lines that have been engineered to express luciferase, so tumor growth and subsequent metastasis can be tracked and measured easily [113,114].

In conclusion, different preclinical PDAC models provide platforms to study important aspects of PDAC tumor biology, and potential treatments. The in vitro PDAC cell line model allows for large-scale and/or high throughput screenings, as well as determining basic infectivity to OVs and innate immune status but lacks obvious physiological components such as a full immune system, tumor microenvironment, metastasis, and early progression. The in vitro organoid model shares many features of the PDAC cell line model, but better addresses tumor heterogeneity, tumor microenvironment, and disease progression. In vivo PDX models allow similar genetic representation of the human disease by using human-derived tissue, but lack major clinical features of PDAC such as early disease progression, complete immune system, and tumor microenvironment (if implanted SC), as PDXs must typically be implanted into an immunocompromised mouse. The humanized mouse model is a PDX alternative, better allows for studies of immune system interactions with both the tumor and tumor microenvironment, as well as potential immune-modulating treatments such as OV therapy. GEMMs, and in particular the KPC model, best recapitulate human PDAC in full, and are the most ideal systems for research purposes. Unfortunately, GEMM models are time-consuming and costly. The syngeneic model is far less time-consuming than the GEMM KPC model, and shares most of its beneficial features, with the exception of early disease progression studies. Each of these model systems have strengths and weaknesses, and the most suitable model depends on what questions are being asked. It is therefore important to understand the unique advantages and disadvantages of each PDAC model system.

In this review, we will discuss current advances with OV therapy for PDAC, with a special focus on VSV, the major interest of our laboratory. While other OV will be discussed in the current review, we would like to refer to other reviews which give a more general overview of OV therapy for pancreatic cancer [115–118].

4. Overview of the Current Progress in OV Therapy for PDAC

In 2015, the FDA approved Talimogene laherparepvec (T-VEC; Imlygic™), a genetically modified herpes simplex virus, to treat melanoma [119]. T-VEC is the first and still the only FDA-approved OV. However, numerous OVs are currently in preclinical studies and clinical trials for various malignancies, including PDAC. Tables 1 and 2 summarize the results of preclinical studies (Table 1) and clinical trials (Table 2) of various oncolytic viruses against pancreatic cancer.

Preclinical studies demonstrated that a wide range of different viruses could be efficient OVs against PDAC (Table 1).

Additionally, gene therapy targets, such as oncogene knockdown, insertion of functional tumor-suppressor genes, and expression of functional RNAs also demonstrate improved cancer-killing efficacy when combined with OV. One method uses adenoviruses and adeno-associated viruses to deliver apoptotic genes to tumor cells. Such gene therapy using Adenovirus subtype 5 mediates rat insulin promoter directed thymidine kinase (A-5-RIP-TK)/ganciclovir (GCV) gene therapy resulting in significantly enhanced cytotoxicity to both Panc1 and MiaPaCa2 pancreatic cancer cells in vitro [176]. Another review explored the potential use of OV expressing functional p53 [117]. Another method would use OV to deliver siRNA transgenes for oncogenetic knockdown, such as ONYX-411-siRNA^{ras} expressing a mutant K-ras siRNA which significantly reduced K-ras mRNA expression at 48 h posttreatment and improved oncolytic activity [134]. The inclusion of an endostatin-angiostatin fusion gene in VVhEA also showed significant antitumor potency in vivo [152].

There have been many experiments screening for more effective virotherapies within available libraries, and modulated viruses such as the adenovirus AdΔCAR-SYE has been shown to significantly suppress tumor growth, and complete regression of tumors was observed in vivo [129]. In addition, more efficient and tumor-specific targeting peptides and OV could be identified by using additional libraries, and modifications to existing OV based on these findings are also promising. Such modifications have been shown to be effective, with the adenoviruses VCN-01 variants ICOVIR-15K and ICOVIR-17 [130]. As discussed previously, the ability of the OV to modulate the ECM was observed, as tumors treated with VCN-01 showed a dramatic decrease in the intratumoral HA content [130]. Other adenoviral variants such as Ad5PTDf35(pp65) have also demonstrated T-cell stimulation and dendritic cell (DC) modulation to increase efficient transduction within a human context [133].

Combinatorial treatments of chemotherapies, or chemovirotherapy, such as OV paired gemcitabine, have demonstrated improved oncolytic capabilities in vitro and in vivo than either treatment on their own. In vitro and in vivo studies showed that myxoma virus (MYXV) and gemcitabine therapies can be combined sequentially to improve the overall survival in intraperitoneal dissemination (IPD) models of pancreatic cancer [177]. The addition of chemotherapies to OV therapy using a combination of an oncolytic herpes simplex virus-1 mutant NV1066 with 5-FU increased viral replication up to 19-fold compared with cells treated with virus alone, and similar results were achieved by the addition of gemcitabine [125]. Similarly, oVV-Smac combined with gemcitabine greater cytotoxicity and potentiated apoptosis [148]. H-1PV combined with cisplatin, vincristine or sunitinib induced effective immunostimulation via a pronounced DC maturation, better cytokine release and cytotoxic T-cell activation [154]. The addition of gene targets alongside chemovirotherapy has also shown greater cytotoxic efficiency, as with VV-ING4 in combination with gemcitabine [150]. Even in cell lines that demonstrate resistance to viral infection, resistance can be broken with simultaneous treatments. Viruses like VSV rely on nonspecific interactions with the cell surface during the earliest stages of infections, and polycations have been shown to improve viral production and increase oncolysis by increasing the amount of virus interacting with cells during attachment [60]. Additionally, the use of JAK inhibitors like ruxotinilib and IKK inhibitors like TPCA-1 have also been shown to increase viral reproduction and oncolysis [56,60]. Other potential combinatorial treatment regimens could include radiovirotherapy or chemoradiotherapy.

Table 1. Preclinical studies of oncolytic viruses against pancreatic cancer.

Oncolytic Virus Backbone	Oncolytic Virus Name	Brief Description of the Results of the Study
Herpes virus	HSV-GFP (expressing NeoR and EGFP)	The pancreatic cancer lines MIAPACA and PANC-1 exhibited definite cytopathic effects upon infection in vitro (hPDAC cell lines) [120].
	HSV2 L1BR1 (US3-deficient)	US3-deficient HSV virus L1BR1 demonstrates a favorable characteristic regarding the induction of apoptosis in vitro (hPDAC cell lines) [121].
	HSV-1 (DF3gamma134.5)	The DF3/MUC1 promoter is shown to enhance the oncolytic activity of HSV-1 mutants in vitro (hPDAC cell lines) [122].
	R3616: γ134.5 hrR3: UL39	In vivo evaluation of two herpes virus mutants in combination with gemcitabine show complex interactions that can benefit or inhibit oncolytic activity (hPDAC IP into BALB/C nude, immunodeficient) [123].
	FusOn-H2	FusOn-H2 has potent activity against human pancreatic cancer xenografts in vivo (hPDAC SC or OT into Hsd nude, immunodeficient) [124].
	NV1066	5-FU and gemcitabine determined in vivo to potentiate oncolytic herpes viral replication and cytotoxicity across a range of clinically achievable doses in the treatment of human pancreatic cancer (hPDAC cell lines) [125].
	HSV(GM-CSF)	Injection of the recombinant mouse HSV encoding GM-CSF resulted in a significant reduction in tumor growth (mPDAC cells SC into C57BL6, immunocompetent) [126].
	Ad-DHscIL12	Expression of IL-12 in the context of a hypoxia-inducible oncolytic adenovirus is effective against pancreatic cancer in vivo (hamPDAC cells SC into nude, immunodeficient) [127].
	NV1020 G2O7	NV1020 and G2O7 effectively infect and kill human pancreatic cancer cells in vitro and in vitro (hPDAC cell lines) [128].
Adenovirus	AdΔCAR-WT, AdΔCAR-SYE, AdΔCAR-IVR	An in vitro tumor-targeting strategy using an adenovirus library for optimization of oncolytic adenovirus therapy (hPDAC cells SC into BALB/C nude, immunodeficient) [129].
	VCN-01	Oncolytic adenovirus VCN-01 shows an efficacy-toxicity profile in vivo (hPDAC cells SC into BALB/C nude, immunodeficient) [130].
	AdDeltaE1B19K dl337	Novel adenoviral mutants demonstrate the ability to improve efficacy of DNA-damaging drugs such as gemcitabine in vitro and in vivo (hPDAC cells SC into ICRF nude, immunodeficient) [131].
	WtΔE3ADP-Luc, WtΔE3ADP-IFN	Adenoviral death protein and fiber modifications significantly improved oncolysis in vitro and in vivo (hPDAC cells SC into NCr nude, immunodeficient) [132].
	Ad5PTDf35	This vector shows dramatically increased transduction capacity of primary human cell cultures including T cells, monocytes, macrophages, dendritic cells, pancreatic islets and exocrine cells, mesenchymal stem cells and tumor initiating cells (hPancreatic islet cells) [133].
	ONYX-411	These findings indicate that Internavec can generate a two-pronged attack on tumor cells through oncogene knockdown and viral oncolysis, resulting in a significantly enhanced antitumor outcome in vitro and in vivo (hPDAC cells SC into nude, immunodeficient) [134].

Table 1. *Cont.*

Oncolytic Virus Backbone	Oncolytic Virus Name	Brief Description of the Results of the Study
Adenovirus	ZD55-lipocalin-2	ZD55-lipocalin-2 may serve as a potent anticancer drug for pancreatic cancer therapy, especially for patients who have pancreatic adenocarcinoma with KRAS mutations as demonstrated in vitro (hPDAC cell lines) [135].
	LOAd703	LOAd703 is a potent immune activator that modulates the stroma to support antitumor responses in vitro and in vivo (hPDAC cells SC into C57BL/6 nude, immunodeficient) [136].
	OAd-hamIFN	Combination treatment of chemoradiation with IFN-expressing OAd demonstrates enhanced cancer-killing efficacy in vitro and in vivo (hamPDAC cells SC into Golden Syrian hamsters, immunocompetent) [137].
	OAd-TNFa-IL2	Ad-mTNFa-mIL2 increased immune cell infiltration to the tumor and altered host tumor immune status in vivo (hPDAC cells SC into NSG, immunodef and mPDAC cells SC into C57BL/6, immunocompetent) [138].
	YDC002	YDC002 combined with gemcitabine significantly attenuated the expression of major ECM components including collagens, fibronectin, and elastin in tumor spheroids and xenograft tumors compared with gemcitabine alone, resulting in potent induction of apoptosis, gemcitabine-mediated cytotoxicity, and an oncolytic effect through degradation of tumor ECM in vivo (hPDAC cells SC into BALB/C nude, immunodeficient) [139].
	AdΔΔ	AdΔΔ has low toxicity to normal cells while potently sensitizing pancreatic cancer cells to DNA-damaging drugs in vivo (hPDAC cells SC into C57BL/6 nude, immunodeficient) [140].
	dl922-947	dl922-947 is effectively able to elicit an anti-tumoral response in vivo when combined with 5-FU or gemcitabine (hPDAC cells SC into C57BL/6 nude, immunodeficient) [141].
	Delta-24-RGD	Delta-24-RGD significantly inhibited tumor growth in combination with phosphatidylserine targeting antibody in vivo (hPDAC cells SC into nude, immunodeficient) [142].
	Ad5-3Δ-A20T	Ad5-3Δ-A20T is highly selective for $\alpha v \beta 6$ integrin-expressing pancreatic cancer cells for improved targeting of pancreatic cancer in vitro (hPDAC cell lines, 3D culture) [143].
	ICOVIR15	Arming the oncolytic adenovirus ICOVIR15 with miR-99b or miR-485 enhances its fitness and its antitumoral activity in vitro (human lung, breast, colorectal, prostate cancer cell lines) [144]
	Ad5-yCD/mutTK(SR39)rep-ADP	Ad5-yCD/mutTK(SR39)rep-ADP improves oncolysis in vitro and in vivo in combination with radiotherapy (hPDAC cells IM into CD-1 athymic, immunodeficient) [145].
Myxoma Virus	vMyxgfp	vMyxgfp had the ability to infect all pancreatic cancer cell lines tested in vitro (hPDAC cell lines) [146].
Reovirus	Reolysin	Reolysin treatment stimulated selective reovirus replication and decreased cell viability in KRas-transformed immortalized human pancreatic duct epithelial cells and pancreatic cancer cell lines in vitro *and* in vivo (hPDAC cells SC into nude, immunodeficient) [147].

Table 1. *Cont.*

Oncolytic Virus Backbone	Oncolytic Virus Name	Brief Description of the Results of the Study
Vaccinia Virus	oVV-Smac	oVV-Smac is indicated to have a synergistic effect in combinatorial treatment with gemcitabine in vitro and in vivo (hPDAC cells SC into BALB/C nude, immunodeficient) [148].
	VV-HBD2-lacZ	These results indicate that HBD2-expressing VV recruited plasmacytoid DCs (pDCs) to the tumor location, leading to cytotoxic T cell response against the tumor, and thus inhibited tumor growth in vitro and in vivo (murine melanoma cells SC into C57BL/6, immunocompetent) [149].
	GLV-1h68	GLV-1h68 was able to infect, replicate in, and lyse tumor cells in vitro and in vivo (hPDAC cells SC into BALB/C nude, immunodeficient) [150].
	VVLΔTK-IL-10	VV expressing IL-10 demonstrates enhanced anti-tumor efficacy in vivo (GEMM (KPC), immunocompetent) [151].
	VVhEA	The novel Lister strain of vaccinia virus armed with the endostatin-angiostatin fusion gene displayed inherently high selectivity for cancer cells, sparing normal cells both in vitro and in vivo, with effective infection of tumors (hPDAC cells SC into BALB/C nude, immunodeficient) [152].
Parvovirus	H-1PV	H-1PV in combination with gemcitabine enhanced anti-tumor activity of NK cells and effects included reduction in tumor growth, prolonged survival of the animals, and absence of metastases on CT-scans in vitro (hPDAC cell lines) [153].
	H-1PV	In ex vivo human models, H-1PV reinforced drug-induced tumor cell killing and effective immunostimulation (human melanoma cell lines) [154].
	H-1PV	The combination treatment of H-1PV and histone deacetylase inhibitors (HDACIs) such as valproic acid (VPA)acts synergistically to kill a range of human cervical carcinoma and pancreatic carcinoma cell lines by inducing oxidative stress, DNA damage and apoptosis in vitro and in vivo (hPDAC cells SC into NOD/SCID nude, immunodeficient) [155].
Measles Virus	MV-PNP-anti-PSCA	PNP, which activates the prodrug fludarabine effectively, enhanced the oncolytic efficacy of the virus on infected and bystander cells in vitro and in vivo (hPDAC cells SC into NOD/SCID nude, immunodeficient) [156].
	MeV	The chemovirotherapeutic combination of gemcitabine plus oncolytic MeV resulted in improved tumor reduction in vitro (hPDAC cell lines) [157].
Newcastle Disease Virus	NDV	NDV infection was successful in all evaluated PA cell lines in vitro, however the resultant replication kinetics and cytotoxic effects differed (hPDAC cell lines) [158].
	NDV	Infection with NDV activated immune cells which successfully elicited an anti-tumor response in vitro. However, activated NK cells that are abundant in Panc02 tumors lead to outgrowth of nonimmunogenic tumor cells with inhibitory properties (hPDAC cells OT into C57BL/6, immunocompetent) [159].
	MTH-68/H	MTH-68/H selectively kills tumor cell cultures in vitro by inducing endoplasmic reticulum stress leading to p53-independent apoptotic cell death (hPDAC cell lines) [160].

Table 1. *Cont.*

Oncolytic Virus Backbone	Oncolytic Virus Name	Brief Description of the Results of the Study
Poxvirus	CF33	CF33 caused rapid killing of six pancreatic cancer cells lines in vitro, releasing damage-associated molecular patterns, and regression of tumors in vivo (human colorectal cells SC into Hsd nude, immunodeficient) [161].
Influenza Virus	PR8, H5N1, H7N3, H4N8, H7N7, H5N1 HP, H7N1 HP	IAV significantly inhibited tumor growth following intratumoral injection without inducing apoptosis in nonmalignant cells in vivo (hPDAC cells SC into SCID, immunodeficient) [162].
Rhabdovirus	M51R-VSV	M51R-VSV treatment appears to induce antitumor cellular immunity in vivo (hPDAC cells SC into C57BL/6 nude, immunodeficient) [163].
	VSV-FH	VSV-FH can induce potent oncolysis in hepatocellular and pancreatic cancer cell lines in vivo (hPDAC cells SC into athymic nude, immunodeficient) [164].
	VSV-ΔM51	VSV showed oncolytic abilities superior to those of other viruses, and some cell lines that exhibited resistance to other viruses were successfully killed by VSV in vitro (hPDAC cell lines) [51].
	VSV-mp53, VSV-ΔM-mp53	VSV expressing p53 exhibited enhanced oncolytic action, while VSV-ΔM-mp53 was extremely attenuated in vivo due to p53 activating innate immune genes (hPDAC cells IV into BALB/C nude, immunodeficient) [165].
	VSV-ΔM51	VSV recombinants induced robust apoptosis in cells with defective IFN signaling, however cell lines constitutively expressing high levels of IFN-stimulated genes (ISGs) were resistant to apoptosis even when VSV replication levels were dramatically increased by Jak inhibitor I treatment in vitro (hPDAC cell lines) [61].
	VSV-WT, VSV-rM51R-M	Recombinant M51R-M (rM51R-M) virus induces apoptosis much more rapidly in L929 cells than viruses expressing WT M protein by a distinct method in vitro (murine fibroblast cell lines) [166].
	VSV-ΔM51	TPCA-1 (IKK-β inhibitor) and ruxolitinib (JAK1/2 inhibitor), as strong enhancers of VSV-ΔM51 replication and virus-mediated oncolysis in all VSV-resistant cell lines in vitro (hPDAC cell lines) [56].
	VSV	Combining VSV with ruxolitinib and Polybrene or DEAE-dextran successfully broke the resistance of HPAF-II cells to VSV by simultaneously improving VSV attachment and replication in vitro (hPDAC cell lines) [60].
	VSV, VSV-GFP, VSV-ΔM51-GFP	In vivo administration of VSV-ΔM51-GFP resulted in significant reduction in tumor growth for tested mouse PDA xenografts and antitumor efficacy was further improved when the virus was combined with gemcitabine (mPDAC cells SC into C57BL/6, immunocompetent) [55].
	VSV-p53wt, VSV-p53-CC	Two independently evolved VSVs obtained identical glycoprotein mutations, K174E and E238K; these acquired G mutations improved VSV replication, at least in part due to improved virus attachment to SUIT-2 cells, as determined in vitro (hPDAC cell lines) [62].

hPDAC = human PDAC, mPDAC = mouse, SC = subcutaneous injection; OT = orthotopic injection, IP = intraperitoneal injection; IV = intravenous injection.

Table 2. Clinical trials featuring oncolytic viruses against pancreatic cancer.

Oncolytic Virus Backbone	Oncolytic Virus Name	Brief Description of the Clinical Trial
Adenovirus	ONYX-015 (dl1520)	ONYX-015 injection via EUS into pancreatic carcinomas by the transgastric route with prophylactic antibiotics is feasible and generally well tolerated either alone or in combination with gemcitabine [167].
	ONYX-015 (dl1520)	Intratumoral injection of an E1B-55 kDa region-deleted adenovirus into primary pancreatic tumors was feasible and well-tolerated at doses up to 10(11) PFU (2 x 10(12) particles), but viral replication was not detectable [168].
	Ad5-yCD/mutTKSR39rep-ADP	A combination of intratumoral Ad5-DS and gemcitabine is safe and well tolerated in patients with LAPC [169].
	Ad5-yCD/mutTKSR39rep-hIL12	Ongoing clinical trial, no results posted to date. NCT03281382.
	LOAd703	Ongoing clinical trial, no results posted to date. NCT02705196.
	VCN-01	Ongoing clinical trial, no results posted to date. NCT02045589.
	VCN-01	Ongoing clinical trial, no results posted to date. NCT02045602.
Herpesvirus	T-VEC	EUS-guided FNI of T-VEC in advanced pancreatic ca, at initial doses of 104 to 106 PFU/mL followed by up to 107 PFU/mL, was feasible and tolerable. Evidence of biologic activity was observed [170].
	T-VEC	Ongoing clinical trial, no results posted to date. NCT03086642.
	HF10	HF10 direct injection under EUS-guidance in combination with erlotinib and gemcitabine was a safe treatment for locally advanced pancreatic cancer [171].
	HF10	Ongoing clinical trial, no results posted to date. NCT03252808.
	OrienX010	Ongoing clinical trial, no results posted to date. NCT01935453.
Reovirus	Reolysin	Pelareorep was safe but ineffective when administered with carboplatin/paclitaxel, regardless of KRAS mutational status. Immunologic studies suggest that chemotherapy backbone improves immune reconstitution and that targeting remaining immunosuppressive mediators may improve oncolytic virotherapy [172].
	Reolysin	PD analysis revealed reovirus replication within pancreatic tumor and associated apoptosis. Upregulation of immune checkpoint marker PD-L1 suggests future consideration of combining oncolytic virus therapy with anti-PD-L1 inhibitors [173].
	Reolysin	Pelareorep and pembrolizumab added to chemotherapy did not add significant toxicity and showed encouraging efficacy [174].
	Reolysin	Ongoing clinical trial, no results posted to date. NCT01280058.
Parvovirus	ParvOryx	The drug was safe and well-tolerated and showed a promising profile of anti-tumor effects and signs of clinical efficacy, i.e., prolonged survival. However, the optimum dose as well as the most appropriate route and schedule of administration have to be further investigated [175].

PFU = plaque-forming unit.

Some of these treatment methodologies are already being tested in clinical trials. Table 2 describes the OV currently being tested in clinical trials, and while some are still underway, OV including ONYX-15, AD5-yCD, and T-VEC are well-tolerated, and in some cases, biologically active, either alone or in combination chemovirotherapies [167–171,174].

5. Understanding Molecular Mechanisms of Responsiveness and Resistance of PDACs to VSV-Based OV Therapy

VSV is a prototypic nonsegmented negative-strand (NNS) RNA virus (order *Mononegavirales*, family *Rhabdoviridae*). VSV is a promising oncolytic virus against various malignancies, and it has several advantages as an OV [178–180]: (i) its basic biology and interaction with the host have been extensively studied. The oncoselectivity of VSV is mainly based on VSV's high sensitivity to Type I interferon (IFN) mediated antiviral responses (and therefore inability to replicate in healthy cells), while it can specifically infect and kill tumor response cells, most of which lack effective Type I IFN responses; (ii) although WT VSV can cause neurotoxicity in mice, nonhuman primates, several VSV recombinants, including VSV-ΔM51, have been generated which are not neurotropic but retain their OV activity; (iii) VSV has a broad tropism for different types of cancer cells (including PDACs), as its primary mode of entry into a host cell utilizes binding of the VSV-G protein to LDLR, which is ubiquitous, and VSV-G is also capable of using other common surface molecules for cell entry [179]; (iv) there is no preexisting immunity against VSV in most humans; (v) replication occurs in the cytoplasm without risk of host cell transformation; (vi) cellular uptake occurs rapidly; (vii) VSV has a small, easily manipulated genome, and novel VSV-based recombinant viruses can be easily engineered via reverse genetics to improve oncoselectivity, safety, oncotoxicity, and to work synergistically with host immunity and/or other therapies in a specific tumor environment (e.g., PDAC); (viii) as other members of the order *Mononegavirales*, and compared to positive-strand RNA viruses, VSV is less likely to mutate, and our recent study demonstrated long-term genetic stability of VSV recombinants carrying large transgenes [62]. All these and other advantages make VSV a promising candidate OV for PDAC treatment, and we have shown that VSV is effective against the majority of PDAC cell lines in vitro and in vivo [50,51]. Importantly, several phase I clinical trials using VSV against different malignancies are in progress (ClinicalTrials.gov for trials NCT03647163, NCT02923466, NCT03120624, NCT03865212, and NCT03017820).

VSV exhibits inherent oncotropism based largely on defective or reduced type I IFN responses, as specific genes associated with type I IFN responses are downregulated or functionally inactive [165,181]. In addition, IFN signaling can be inhibited by MEK/ERK signaling or by epigenetic silencing of IFN-responsive transcription factors IRF7 or IRF5 [182,183]. However, some PDACs do not have these defects and resist VSV infection like normal cells, which are sensitive to IFN-α treatment and capable of secreting type I IFNs following VSV infection [184].

There has been a demonstration of neurotoxicity in mice infected intranasally or intracranially, demonstrating a need for methods of improvement of VSV oncoselectivity and neurotropic safety without compromising oncolytic ability. There are at least eight approaches demonstrated to address these needs [178–180]: (i) mutating the VSV M protein; (ii) VSV-directed IFN-β expression; (iii) attenuation of VSV through disruption of normal gene order; (iv) mutating the VSV G protein; (v) introducing targets for microRNA from normal cells into the VSV genome; (vi) pseudotyping VSV; (vii) experimental adaptation of VSV to cancer cells; and (viii) using semi-replicative VSV. Most of the studies in our laboratory focus on VSV-ΔM51 recombinants containing a deletion of the methionine residue at position 51 of the M protein, VSV-ΔM51. This mutation results in an inability of VSV-M to inhibit nucleus-to-cytoplasm transport of cellular mRNA, including antiviral transcripts, in normal cells with functional antiviral signaling [185,186].

Our laboratory has characterized numerous human PDAC cells lines and discovered a wide range of susceptibility and permissiveness of different PDAC cell lines to VSV and other tested OVs [50,51,56,58–61]. The range includes "super-permissive" cell lines (such as MIA PaCa-2 and Capan1), "super-resistant" cell lines (such as HPAF-II, Hs766T), and well as many cell lines in between (such as SUIT2 and AsPC-1). Below we describe different mechanisms associated with the resistance of some PDACs to VSV.

5.1. Upregulated Type I IFN Signaling and Constitutive Expression of a Subset of IFN-Stimulated Genes (ISGs)

Our extensive analysis of a large number of human PDAC cell lines demonstrates that PDAC cell lines show surprising diversity with regard to their ability to produce and respond to type I IFNs, and the evaluation of IFN sensitivity and IFN-α and IFN-β production within a cell line may be used to predict its responsiveness to oncolytic treatment [50,51] (Figure 2).

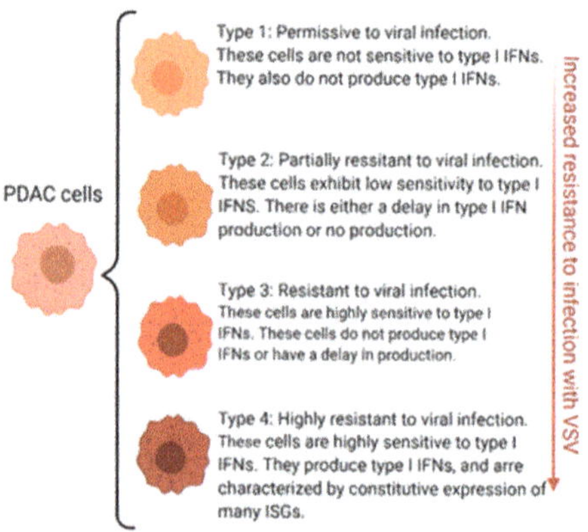

Figure 2. Permissiveness of PDAC to VSV: Four Different Phenotypes. This figure demonstrates the variability across PDAC in regard to permissiveness to infection by VSV. Permissiveness refers to the cells allowance for viral attachment, infection, and replication. The figure was created by authors with BioRender software (BioRender.com).

Upregulated or residual expression of antiviral genes display four unique phenotypes (Figure 2): (i) no type I IFN production and not responsive to type I IFN, (ii) no type I IFN production but responsive to type I IFN, (iii) type I IFN production and responsive to type I IFN, (iv) super resistant PDACs: type I IFN production, responsive to type I IFN and constitutive expression of many antiviral IFN-stimulated genes (ISGs) [51,56,58].

We also conducted a transcriptome analysis to identify biomarkers for resistance of PDAC cell lines to VSV-ΔM51. Of the genes identified, six demonstrate constitutive co-expression in the VSV-resistant cell lines: MX1, EPSTI1, XAF1, GBP1, SAMD9, and SAMD9L [58]. Most of these genes are known to have an antiviral effect. Moreover, shRNA-mediated knockdown of MX1 showed a positive effect on VSV-ΔM51 replication in resistant PDAC cells, suggesting that at least some of the identified ISGs contribute to resistance of PDACs to VSV-ΔM51 [58]. Finally, we demonstrated that JAK inhibitors effectively break resistance to VSV-ΔM51 while affecting very few non-ISGs, suggesting that the constitutive expression of these genes is likely a causative factor for the phenotype of resistance [50,56]. Further evidence that host antiviral response to VSV-ΔM51 infection is the source of resistance has been shown in infection with WT VSV, as even cell lines resistant to VSV-ΔM51 are permissive to at least some degree to the WT-VSV, which is better able to evade antiviral responses in the host [50,51].

5.2. Role of Cell Cycle in Resistance of PDAC Cells to VSV

We have demonstrated that compounds inducing cell cycle arrest in G1/S-phase or S-phase strongly inhibited VSV-ΔM51 replication, while G_2/M phase arrest dramatically enhanced the replication of VSV-ΔM51 in cells with functional antiviral signaling [61]. It was found that G2/M arrest strongly inhibited IFN production and expression of ISGs in response to exogenously added IFN. The replication of IFN-sensitive cytoplasmic viruses can be strongly stimulated during G2/M phase as a result of inhibition of antiviral gene expression, likely due to mitotic inhibition of transcription, a global repression of cellular transcription during G2/M phase. However, G_2/M arrest did not stimulate the replication of VSV-ΔM51 in cells defective in IFN signaling, and it did not stimulate replication of WT VSV, which is more effective at evading antiviral responses [61]. Together, our study suggests that continuous cell cycle transition, a hallmark of cancer cells, could be another factor of oncoselectivity for many viruses, at it would facilitate viral replication via inhibition of antiviral responses in dividing cancer cells during G_2/M phase. It also suggests that slowly dividing PDAC cells could be more resistant to some OVs than faster dividing PDACs.

5.3. Resistance to Virus-Mediated Apoptosis

Inhibition of apoptosis is a hallmark of many malignancies, including PDAC [187], and PDAC with decreased expression or activation of certain apoptotic proteins have the potential to limit/delay cell death following VSV infection [166,188–190]. Our study demonstrated that all three tested VSV recombinants (VSV-GFP (WT M gene), VSV-p1-GFP (WT M gene, GFP in the first position in the genome), and VSV-ΔM51-GFP) induced caspase 3 cleavage following infection, but VSV-ΔM51-GFP induced more caspase 3 cleavage in all cell lines with VSV-inducible Type I IFN responses, despite similar replication levels for the viruses [59]. This indicates a positive role for the ΔM51 mutation, and therefore host antiviral responses, in apoptosis induction, and is unlikely to be simply a result of virus attenuation as VSV-p1-GFP induced caspase cleavage similarly to VSV-GFP. Further, VSV-ΔM51-GFP induces both the extrinsic and intrinsic apoptosis pathways in most PDACs, however, we observed inhibition of VSV-induced and drug-induced apoptosis in some PDAC cells lines, and that was observed even when VSV replication was stimulated using JAK1 inhibitors [59]. In general, our study has demonstrated that resistance of some PDAC cell lines to VSV-mediated oncolysis could be not only due to type I IFN responses that limit virus replication, but also to cellular defects in apoptosis.

5.4. Inefficient Attachment of VSV to PDAC Cells

It is generally accepted that VSV tumor tropism is mainly dependent on the permissiveness of malignant cells to viral replication rather than on receptor specificity. However, when compared VSV attachment to different human PDAC cell lines, we observed a dramatically weaker attachment of VSV to HPAF-II cells, the most resistant human PDAC cell line [60]. Interestingly, although sequence analysis of low-density lipoprotein (LDL) receptor (LDLR) mRNA did not reveal any amino acid substitutions in this cell line, HPAF-II cells displayed the lowest level of LDLR expression and dramatically lower LDL uptake. We also showed that LDLR-independent attachment of VSV to HPAF-II cells and some other PDAC cell lines can be dramatically improved by treating cells with polybrene or DEAE-dextran [60].

6. Enhancing Responsiveness of PDAC Cells to Oncolytic Virotherapy with VSV

6.1. Combination of VSV with Small Molecule Inhibitors

As mentioned above, PDAC cell lines resistant to infection by VSV-ΔM51 demonstrate constitutive expression of numerous ISGs, most notably Mx1 [50,56]. Importantly, a similar expression profile (including upregulation of ISGs such as Mx1) and virus resistance phenotype was demonstrated for primary PDACs isolated from patients [176]. Treatment of resistant cell lines with JAK inhibitor I (a reversible inhibitor of JAK1, JAK2, JAK3 and TYK2) reduced ISG expression and partially overcame resistance to VSV suggesting potential for further improvement by utilizing other inhibitors and/or targeting additional pathways [50]. A similar enhancement of VSV replication was shown for ruxolitinib (JAK1/2 inhibitor), which was previously shown to break resistance of human head and neck cancer cells to VSV [56]. Interestingly, a similar strong inhibition of STAT1 and STAT2 phosphorylation, decreased expression of Mx1 and OAS, and stimulation of VSV-ΔM51 replication was also observed with TPCA-1, a known IKK-β inhibitor. Moreover, using an in situ kinase assay, we demonstrated that TPCA-1 can directly inhibit JAK1 kinase activity [56]. Thus, our study demonstrated that TPCA-1 is a unique dual inhibitor of IKK-β and JAK1 kinase.

6.2. Combination of VSV with Polycations

As mentioned above, some PDAC cell lines, including HPAF-II, are highly resistant to VSV due to a combination of a constitutive antiviral state and type I IFN-independent impaired VSV attachment. It was determined that the source of the impairment to attachment was not a result of mutations to the LDLR or its lowered expression levels, but rather by some other mechanism [60]. We have shown that treatment of cells with polycations improved LDLR-independent virus attachment, as both the cellular membrane and the viral envelope have negative net charges, and the polycations serve to counteract the repulsive electrostatic effects of the lipid barriers. Earlier studies showed a similar effect on cells treated with either DEAE-Dextran or polybrene before infection [191,192]. A novel combinatorial treatment with ruxolitinib and polycations demonstrated improved overall VSV replication and oncolysis and accelerated VSV replication kinetics compared to treatment with ruxolitinib only [60].

6.3. Combination of VSV with FDA-Approved Chemotherapeutic Drugs

VSV is inherently oncoselective due to its high sensitivity to type 1 IFN response, as research indicates that most cancer cells are defective in this type of antiviral signaling, and VSV-ΔM51 is more sensitive than the WT virus, which is better able to inhibit these responses. G2/M arrest stimulates viral replication by inhibiting antiviral responses. Paclitaxel treatment stimulated the replication of VSV-ΔM51 and Sendai virus (another cytoplasmic NNS RNA virus that is also sensitive to type 1 IFN response) in multiple PDAC cell lines via inhibition of antiviral gene expression in treated cells [61]. In cells with functional type I IFN signaling, G2/M arrest inhibited the expression levels of type I and III IFNs, as well as inhibiting the upregulation of ISGs in response to the same amounts of exogenously added type I IFN [61]. A similar effect was also shown for colchicine [193].

6.4. VSV Encoding p53 Transgenes

Studies suggest that p53 enhances type 1 IFN signaling in normal cells and in some cancer cells types. However, our recent study demonstrated that VSV-encoded p53 trasngene actually inhibited antiviral signaling in PDAC cells, while stimulating VSV-ΔM51 replication in those cells [57]. Several potential reasons exist for the contrast to previous reports, including (1) differences between normal cells and different cancer cells in antiviral signaling, (2) constitutive activation of NF-κB pathway in a majority of PDACs, and (3) different timing and level of expression for VSV-encoded p53 in the cell lines used in this study. Future studies with VSVs expressing human p53 in animal models will determine if the oncoselective phenotype seen in PDAC cell lines is also observed in vivo, as it is necessary to see if the benefits are retained in an immunocompetent model without compromising safety.

6.5. Experimental Evolution of VSV

Replication-competent OVs can evolve under natural selection, and reversion to virulence or loss of oncolytic potential can threaten the safety and efficacy of virotherapeutic treatments. Evolutionary risk assessment studies are required to confirm safety, and can also serve the benefit of producing more potent OV, however, the most effective strategy is to combine rational design with evolution, allowing each engineered virus to mutate and fully adapt to its intended target cells. In our recent study, two VSV-ΔM51 variants containing human p53 were serially passaged on Suit-2 cells, resulting in mutations that adapted the viruses to better replicate in multiple PDAC cell lines without developing mutations in the p53 gene or losing oncoselectivity. The mutations of note that were acquired by the viruses include two separate mutations within the G protein sequence; both of these mutations achieved fixation within the span of the experiment in either virus and were identical. It was determined that the acquired G mutations stimulate VSV replication at least in part due to improved virus attachment to PDAC cells [62,194].

7. Future Directions and Conclusions

As discussed previously, several factors pose a series of challenges that will determine whether OV is suitable for cancer treatment, especially for PDAC, where any treatment is further complicated by the TME, which is characterized by dense stroma comprised of abundant fibroblasts, hypoxia, sparse vasculature, as well as infiltration of tumor-promoting immune cells mediating immune evasion and tumor progression. The ideal OV treatment should allow for sufficient delivery and penetration of PDACs with the virus, induction of adaptive antitumor responses, and prevention of premature OV clearance by host antiviral response. It is unlikely that any monotherapy could address all these challenges, and future effective OV-based treatments will likely be combinatorial (chemo-virotherapy, radio-virotherapy, chemo-radio-virotherapy, chemo-radio-immuno-virotherapy, etc.). Many of VSV-based combinatorial approaches have been described in our previously published reviews [178,180], and some additional approaches will be discussed below [195].

First of all, to study and address all these challenges, the ideal model systems should employ immunocompetent animals (to examine antiviral as well as antitumor immune responses) and be able to monitor not only tumor growth and spread, but also OV spread.

Figure 3 illustrates the system that we currently use in our laboratory to investigate various VSV-based OV treatments against PDAC.

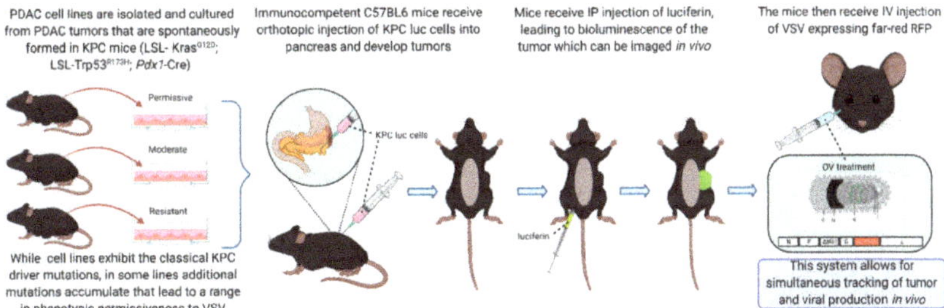

Figure 3. Syngeneic Model of PDAC in Mice. This figure shows the method of development for the murine model of PDAC. In this model, C57BL6 mice are used both for the development of the tumors and to evaluate the treatment in vivo. The figure was created by authors with BioRender software (BioRender.com).

PDAC is a highly heterogenic disease, and our studies have demonstrated dramatic differences between different human PDAC cell lines in their permissiveness to VSV and other OVs [50,51]. Future studies should define distinct subtypes of PDACs to develop personalized treatment strategies for different types of PDACs [3,196,197]. Although there is still no consensus classification for clinical application, some treatments work better against a particular PDAC subtype. For example, patients who had a germline BRCA mutation had significantly longer progression-free survival with maintenance a PARP inhibitor olaparib than with placebo [198]. It could be interesting to test combinations of OVs with olaparib against PDACs that have the BRCAness phenotype [10]. Interestingly, at least two studies showed the increased efficacy of OV therapy for thyroid carcinoma [199] and glioblastoma [200] when OV was combined with olaparib.

One of the major challenges for any PDAC treatment is insufficient drug delivery into the tumors because PDACs are hypovascular, densely packed with ECM components, have a high intratumoral tissue pressure, and very low tumor perfusion. Several previously developed approaches could be used to improve OV delivery into PDAC. For example, administration of a combination of cilengitide (angiogenesis inhibitor) and verapamil (Ca^{2+} channel blocker) promoted tumor angiogenesis, while improving gemcitabine delivery and therapeutic efficacy in mice [201]. Additionally, the angiotensin inhibitor losartan was shown to increase perfusion, drug and oxygen delivery [202]. A more recent study highlighted the potential importance of ROCK inhibition using the oral inhibitor fasudil for dual targeting of tumor tension and vasculature [203]. The administration of fasudil, a Rho-kinase inhibitor, and vasodilator, reduced intratumoral fibrillar collagen, improved sensitivity towards gemcitabine/nab-paclitaxel, and reduced metastasis formation on gemcitabine/abraxane treatment [203]. At least some of these drugs could potentially improve OV therapy when used in combination with VSV or other OVs.

The role of the stromal cells during OV therapy is still unclear, and it is likely dependent on the subtype of the particular PDAC. At least under certain conditions, the stromal cells could play a positive role during OV therapy by dampening antiviral responses within tumor and thus stimulating OV replication and OV-mediated oncolysis [204].

Other areas for development include approaches with a focus on antitumor immune stimulation. The TME of many cancers, including PDAC, is known to be immunosuppressive, due to various factors including a dense, fibrotic composition and a hypoxic environment that prevent access and activation of immune cells within the tumor [205,206]. Adoptive T-cell therapy augments the potency of T-cells by chaperoning virus into the tumor [120], overcoming the stromal barrier. Antigen-specific T-cells that were loaded with VSV-ΔM51 can also be used to produce viral infection, replication, and subsequent oncolysis, as well as producing a proinflammatory environment that helped suppress the immunosuppressive nature of the TME. Immune tolerance mechanisms have been implicated as the main barrier to effective antitumor immunotherapy [207], and the natural flora of the gut has been indicated to possess the ability to exert influence over the immune response of the TME, resulting in immune tolerance that promotes tumor growth and development. Future experiments should examine the role of the natural flora in the efficacy of OV therapy for PDAC.

Author Contributions: M.C.H., D.W.G., and V.Z.G. contributed to the design, writing and editing of this manuscript. All authors have read and agreed to the published version of the manuscript.

Funding: This study was funded by grant 1R15CA238864-01 to V.Z.G. from the National Cancer Institute, National Institutes of Health (Bethesda, MD).

Acknowledgments: All figures were created by authors with BioRender software (BioRender.com).

Conflicts of Interest: The authors declare no conflict of interest.

References

1. Kamisawa, T.; Wood, L.D.; Itoi, T.; Takaori, K. Pancreatic cancer. *Lancet* **2016**, *388*, 73–85. [CrossRef]
2. Aier, I.; Semwal, R.; Sharma, A.; Varadwaj, P.K. A systematic assessment of statistics, risk factors, and underlying features involved in pancreatic cancer. *Cancer Epidemiol.* **2019**, *58*, 104–110. [CrossRef]
3. Orth, M.; Metzger, P.; Gerum, S.; Mayerle, J.; Schneider, G.; Belka, C.; Schnurr, M.; Lauber, K. Pancreatic ductal adenocarcinoma: Biological hallmarks, current status, and future perspectives of combined modality treatment approaches. *Radiat. Oncol.* **2019**, *14*, 1–20. [CrossRef]
4. Siegel, R.L.; Miller, D.K.; Ahmedi, J. Cancer Facts & Figures 2020. *CA Cancer J. Clin.* **2020**, *70*. [CrossRef]
5. Buscail, L.; Bournet, B.; Cordelier, P. Role of oncogenic KRAS in the diagnosis, prognosis and treatment of pancreatic cancer. *Nat. Rev. Gastroenterol. Hepatol.* **2020**, *17*, 153–168. [CrossRef] [PubMed]
6. Salo-Mullen, E.E.; O'Reilly, E.M.; Kelsen, D.P.; Ba, A.M.A.; Lowery, M.A.; Yu, K.H.; Reidy, D.L.; Epstein, A.S.; Lincoln, A.; Bs, A.S.; et al. Identification of germline genetic mutations in patients with pancreatic cancer. *Cancer* **2015**, *121*, 4382–4388. [CrossRef] [PubMed]
7. Shindo, K.; Yu, J.; Suenaga, M.; Fesharakizadeh, S.; Cho, C.; Macgregor-Das, A.; Siddiqui, A.; Witmer, P.D.; Tamura, K.; Song, T.J.; et al. Deleterious Germline Mutations in Patients with Apparently Sporadic Pancreatic Adenocarcinoma. *J. Clin. Oncol.* **2017**, *35*, 3382–3390. [CrossRef]

8. Yachida, S.; White, C.M.; Naito, Y.; Zhong, Y.; Brosnan, J.A.; Macgregor-Das, A.M.; Morgan, R.A.; Saunders, T.; Laheru, D.A.; Herman, J.M.; et al. Clinical Significance of the Genetic Landscape of Pancreatic Cancer and Implications for Identification of Potential Long-term Survivors. *Clin. Cancer Res.* **2012**, *18*, 6339–6347. [CrossRef]

9. Cicenas, J.; Kvederaviciute, K.; Meskinyte, I.; Meskinyte-Kausiliene, E.; Skeberdyte, A.; Cicenas, J. KRAS, TP53, CDKN2A, SMAD4, BRCA1, and BRCA2 Mutations in Pancreatic Cancer. *Cancers* **2017**, *9*, 42. [CrossRef]

10. Wong, W.; Raufi, A.G.; A Safyan, R.; E Bates, S.; A Manji, G. BRCA Mutations in Pancreas Cancer: Spectrum, Current Management, Challenges and Future Prospects. *Cancer Manag. Res.* **2020**, *12*, 2731–2742. [CrossRef]

11. McGuigan, A.; Kelly, P.; Turkington, R.C.; Jones, C.; Coleman, H.G.; McCain, R.S. Pancreatic cancer: A review of clinical diagnosis, epidemiology, treatment and outcomes. *World J. Gastroenterol.* **2018**, *24*, 4846–4861. [CrossRef]

12. Rouanet, M.; Lebrin, M.; Gross, F.; Bournet, B.; Cordelier, P.; Buscail, L. Gene Therapy for Pancreatic Cancer: Specificity, Issues and Hopes. *Int. J. Mol. Sci.* **2017**, *18*, 1231. [CrossRef]

13. Sato-Dahlman, M.; Wirth, K.; Yamamoto, M. Role of Gene Therapy in Pancreatic Cancer—A Review. *Cancers* **2018**, *10*, 103. [CrossRef]

14. Springfeld, C.; Jäger, D.; Büchler, M.W.; Strobel, O.; Hackert, T.; Palmer, D.H.; Neoptolemos, J.P. Chemotherapy for pancreatic cancer. *Presse Médicale* **2019**, *48*, e159–e174. [CrossRef]

15. Turpin, A.; El Amrani, M.; Bachet, J.-B.; Pietrasz, D.; Schwarz, L.; Hammel, P. Adjuvant Pancreatic Cancer Management: Towards New Perspectives in 2021. *Cancers* **2020**, *12*, 3866. [CrossRef] [PubMed]

16. Lambert, A.; Schwarz, L.; Borbath, I.; Henry, A.; Van Laethem, J.-L.; Malka, D.; Ducreux, M.; Conroy, T. An update on treatment options for pancreatic adenocarcinoma. *Ther. Adv. Med Oncol.* **2019**, *11*, 1758835919875568. [CrossRef] [PubMed]

17. Perri, G.; Prakash, L.; Katz, M.H.G. Defining and Treating Borderline Resectable Pancreatic Cancer. *Curr. Treat. Options Oncol.* **2020**, *21*, 1–11. [CrossRef]

18. Parekh, H.D.; Starr, J.; George, T.J. The Multidisciplinary Approach to Localized Pancreatic Adenocarcinoma. *Curr. Treat. Options Oncol.* **2017**, *18*, 73. [CrossRef] [PubMed]

19. Dauer, P.; Nomura, A.; Saluja, A.; Banerjee, S. Microenvironment in determining chemo-resistance in pancreatic cancer: Neighborhood matters. *Pancreatology* **2017**, *17*, 7–12. [CrossRef] [PubMed]

20. Kleeff, J.; Beckhove, P.; Esposito, I.; Herzig, S.; Huber, P.E.; Löhr, J.M.; Friess, H. Pancreatic cancer microenvironment. *Int. J. Cancer* **2007**, *121*, 699–705. [CrossRef] [PubMed]

21. Adamska, A.; Elaskalani, O.; Emmanouilidi, A.; Kim, M.; Razak, N.B.A.; Metharom, P.; Falasca, M. Molecular and cellular mechanisms of chemoresistance in pancreatic cancer. *Adv. Biol. Regul.* **2018**, *68*, 77–87. [CrossRef]

22. Akakura, N.; Kobayashi, M.; Horiuchi, I.; Suzuki, A.; Wang, J.; Chen, J.; Niizeki, H.; Ki, K.; Hosokawa, M.; Asaka, M. Constitutive expression of hypoxia-inducible factor-1alpha renders pancreatic cancer cells resistant to apoptosis induced by hypoxia and nutrient deprivation. *Cancer Res.* **2001**, *61*, 6548–6554.

23. Ireland, L.; Santos, A.; Ahmed, M.S.; Rainer, C.; Nielsen, S.R.; Quaranta, V.; Weyer-Czernilofsky, U.; Engle, D.D.; Perez-Mancera, P.A.; Coupland, S.E.; et al. Chemoresistance in Pancreatic Cancer Is Driven by Stroma-Derived Insulin-Like Growth Factors. *Cancer Res.* **2016**, *76*, 6851–6863. [CrossRef]

24. Tréhoux, S.; Duchêne, B.; Jonckheere, N.; Van Seuningen, I. The MUC1 oncomucin regulates pancreatic cancer cell biological properties and chemoresistance. Implication of p42–44 MAPK, Akt, Bcl-2 and MMP13 pathways. *Biochem. Biophys. Res. Commun.* **2015**, *456*, 757–762. [CrossRef] [PubMed]

25. Weichselbaum, R.R.; Ishwaran, H.; Yoon, T.; Nuyten, D.S.A.; Baker, S.W.; Khodarev, N.; Su, A.W.; Shaikh, A.Y.; Roach, P.; Kreike, B.; et al. An interferon-related gene signature for DNA damage resistance is a predictive marker for chemotherapy and radiation for breast cancer. *Proc. Natl. Acad. Sci. USA* **2008**, *105*, 18490–18495. [CrossRef]

26. Uchida, D.; Shiraha, H.; Kato, H.; Nagahara, T.; Iwamuro, M.; Kataoka, J.; Horiguchi, S.; Watanabe, M.; Takaki, A.; Nouso, K.; et al. Potential of adenovirus-mediated REIC/Dkk-3 gene therapy for use in the treatment of pancreatic cancer. *J. Gastroenterol. Hepatol.* **2013**, *29*, 973–983. [CrossRef] [PubMed]

27. Horioka, K.; Ohuchida, K.; Sada, M.; Zheng, B.; Moriyama, T.; Fujita, H.; Manabe, T.; Ohtsuka, T.; Shimamoto, M.; Miyazaki, T.; et al. Suppression of CD51 in pancreatic stellate cells inhibits tumor growth by reducing stroma and altering tumor-stromal interaction in pancreatic cancer. *Int. J. Oncol.* **2016**, *48*, 1499–1508. [CrossRef]

28. Barnes, C.A.; Chavez, M.I.; Tsai, S.; Aldakkak, M.; George, B.; Ritch, P.S.; Dua, K.; Clarke, C.N.; Tolat, P.; Hagen, C.; et al. Survival of patients with borderline resectable pancreatic cancer who received neoadjuvant therapy and surgery. *Surgery* **2019**, *166*, 277–285. [CrossRef] [PubMed]

29. Giovannetti, E.; Del Tacca, M.; Mey, V.; Funel, N.; Nannizzi, S.; Ricci, S.; Orlandini, C.; Boggi, U.; Campani, D.; Del Chiaro, M.; et al. Transcription Analysis of Human Equilibrative Nucleoside Transporter-1 Predicts Survival in Pancreas Cancer Patients Treated with Gemcitabine. *Cancer Res.* **2006**, *66*, 3928–3935. [CrossRef] [PubMed]

30. Spratlin, J.; Sangha, R.; Glubrecht, D.; Dabbagh, L.; Young, J.D.; Dumontet, C.; Cass, C.; Lai, R.; Mackey, J.R. The Absence of Human Equilibrative Nucleoside Transporter 1 Is Associated with Reduced Survival in Patients with Gemcitabine-Treated Pancreas Adenocarcinoma. *Clin. Cancer Res.* **2004**, *10*, 6956–6961. [CrossRef] [PubMed]

31. Ohhashi, S.; Ohuchida, K.; Mizumoto, K.; Fujita, H.; Egami, T.; Yu, J.; Toma, H.; Sadatomi, S.; Nagai, E.; Tanaka, M. Down-regulation of deoxycytidine kinase enhances acquired resistance to gemcitabine in pancreatic cancer. *Anticancer Res.* **2008**, *28*, 2205–2212.

32. Saiki, Y.; Yoshino, Y.; Fujimura, H.; Manabe, T.; Kudo, Y.; Shimada, M.; Mano, N.; Nakano, T.; Lee, Y.; Shimizu, S.; et al. DCK is frequently inactivated in acquired gemcitabine-resistant human cancer cells. *Biochem. Biophys. Res. Commun.* **2012**, *421*, 98–104. [CrossRef] [PubMed]

33. Vernejoul, F.; Ghénassia, L.; Souque, A.; Lulka, H.; Drocourt, D.; Cordelier, P.; Pradayrol, L.; Pyronnet, S.; Buscail, L.; Tiraby, G.; et al. Gene Therapy Based on Gemcitabine Chemosensitization Suppresses Pancreatic Tumor Growth. *Mol. Ther.* **2006**, *14*, 758–767. [CrossRef] [PubMed]

34. Kasuya, K.; Tsuchida, A.; Nagakawa, Y.; Suzuki, M.; Abe, Y.; Itoi, T.; Serizawa, H.; Nagao, T.; Shimazu, M.; Aoki, T. Hypoxia-inducible factor-1α expression and gemcitabine chemotherapy for pancreatic cancer. *Oncol. Rep.* **2011**, *26*, 1399–1406. [CrossRef] [PubMed]

35. Zhang, Z.; Han, H.; Rong, Y.; Zhu, K.; Zhu, Z.; Tang, Z.; Xiong, C.; Tao, J. Hypoxia potentiates gemcitabine-induced stemness in pancreatic cancer cells through AKT/Notch1 signaling. *J. Exp. Clin. Cancer Res.* **2018**, *37*, 291. [CrossRef] [PubMed]

36. Wang, S.; You, L.; Dai, M.; Zhao, Y. Mucins in pancreatic cancer: A well-established but promising family for diagnosis, prognosis and therapy. *J. Cell. Mol. Med.* **2020**, *24*, 10279–10289. [CrossRef] [PubMed]

37. Shukla, S.K.; Purohit, V.; Mehla, K.; Gunda, V.; Chaika, N.V.; Vernucci, E.; King, R.J.; Abrego, J.; Goode, G.D.; Dasgupta, A.; et al. MUC1 and HIF-1alpha Signaling Crosstalk Induces Anabolic Glucose Metabolism to Impart Gemcitabine Resistance to Pancreatic Cancer. *Cancer Cell* **2017**, *32*, 71–87.e77. [CrossRef]

38. Mutgan, A.C.; Besikcioglu, H.E.; Wang, S.; Friess, H.; Ceyhan, G.O.; Demir, I.E. Insulin/IGF-driven cancer cell-stroma crosstalk as a novel therapeutic target in pancreatic cancer. *Mol. Cancer* **2018**, *17*, 1–11. [CrossRef]

39. Friedlander, S.Y.G.; Chu, G.C.; Snyder, E.L.; Girnius, N.; Dibelius, G.; Crowley, D.; Vasile, E.; DePinho, R.A.; Jacks, T. Context-Dependent Transformation of Adult Pancreatic Cells by Oncogenic K-Ras. *Cancer Cell* **2009**, *16*, 379–389. [CrossRef]

40. Kalra, A.V.; Campbell, R.B. Mucin impedes cytotoxic effect of 5-FU against growth of human pancreatic cancer cells: Overcoming cellular barriers for therapeutic gain. *Br. J. Cancer* **2007**, *97*, 910–918. [CrossRef]

41. Nath, S.K.; Daneshvar, K.; Das Roy, L.; Grover, P.L.; Kidiyoor, A.; Mosley, L.J.; Sahraei, M.; Mukherjee, P.K. MUC1 induces drug resistance in pancreatic cancer cells via upregulation of multidrug resistance genes. *Oncogenesis* **2013**, *2*, e51. [CrossRef]

42. Dang, C.V. MUC-king with HIF May Rewire Pyrimidine Biosynthesis and Curb Gemcitabine Resistance in Pancreatic Cancer. *Cancer Cell* **2017**, *32*, 3–5. [CrossRef]

43. Bafna, S.; Kaur, S.; Momi, N.; Batra, S.K. Pancreatic cancer cells resistance to gemcitabine: The role of MUC4 mucin. *Br. J. Cancer* **2009**, *101*, 1155–1161. [CrossRef]

44. Skrypek, N.; Duchêne, B.; Hebbar, M.; Leteurtre, E.; Van Seuningen, I.; Jonckheere, N. The MUC4 mucin mediates gemcitabine resistance of human pancreatic cancer cells via the Concentrative Nucleoside Transporter family. *Oncogene* **2012**, *32*, 1714–1723. [CrossRef]

45. Mimeault, M.; Johansson, S.L.; Senapati, S.; Momi, N.; Chakraborty, S.; Batra, S.K. MUC4 down-regulation reverses chemoresistance of pancreatic cancer stem/progenitor cells and their progenies. *Cancer Lett.* **2010**, *295*, 69–84. [CrossRef] [PubMed]

46. Amorino, G.P.; Hamilton, V.M.; Valerie, K.; Dent, P.; Lammering, G.; Schmidt-Ullrich, R.K. Epidermal Growth Factor Receptor Dependence of Radiation-induced Transcription Factor Activation in Human Breast Carcinoma Cells. *Mol. Biol. Cell* **2002**, *13*, 2233–2244. [CrossRef] [PubMed]

47. Zeng, S.; Pöttler, M.; Lan, B.; Grützmann, R.; Pilarsky, C.; Yang, H. Chemoresistance in Pancreatic Cancer. *Int. J. Mol. Sci.* **2019**, *20*, 4504. [CrossRef]

48. Khodarev, N.N.; Beckett, M.; Labay, E.; Darga, T.; Roizman, B.; Weichselbaum, R.R. STAT1 is overexpressed in tumors selected for radioresistance and confers protection from radiation in transduced sensitive cells. *Proc. Natl. Acad. Sci. USA* **2004**, *101*, 1714–1719. [CrossRef]

49. Russell, L.; Peng, K.W.; Russell, S.J.; Diaz, R.M. Oncolytic Viruses: Priming Time for Cancer Immunotherapy. *BioDrugs* **2019**, *33*, 485–501. [CrossRef]

50. Moerdyk-Schauwecker, M.; Shah, N.R.; Murphy, A.M.; Hastie, E.; Mukherjee, P.; Grdzelishvili, V.Z. Resistance of pancreatic cancer cells to oncolytic vesicular stomatitis virus: Role of type I interferon signaling. *Virology* **2013**, *436*, 221–234. [CrossRef]

51. Murphy, A.M.; Besmer, D.M.; Moerdyk-Schauwecker, M.; Moestl, N.; Ornelles, D.A.; Mukherjee, P.; Grdzelishvili, V.Z. Vesicular Stomatitis Virus as an Oncolytic Agent against Pancreatic Ductal Adenocarcinoma. *J. Virol.* **2012**, *86*, 3073–3087. [CrossRef]

52. Deer, E.L.; González-Hernández, J.; Coursen, J.D.; Shea, J.E.; Ngatia, J.; Scaife, C.L.; Firpo, M.A.; Mulvihill, S.J. Phenotype and Genotype of Pancreatic Cancer Cell Lines. *Pancreas* **2010**, *39*, 425–435. [CrossRef]

53. Gillet, J.-P.; Varma, S.; Gottesman, M.M. The Clinical Relevance of Cancer Cell Lines. *J. Natl. Cancer Inst.* **2013**, *105*, 452–458. [CrossRef]

54. Froeling, F.E.; Marshall, J.F.; Kocher, H.M. Pancreatic cancer organotypic cultures. *J. Biotechnol.* **2010**, *148*, 16–23. [CrossRef] [PubMed]

55. Hastie, E.; Besmer, D.M.; Shah, N.R.; Murphy, A.M.; Moerdyk-Schauwecker, M.; Molestina, C.; Das Roy, L.; Curry, J.M.; Mukherjee, P.; Grdzelishvili, V.Z. Oncolytic Vesicular Stomatitis Virus in an Immunocompetent Model of MUC1-Positive or MUC1-Null Pancreatic Ductal Adenocarcinoma. *J. Virol.* **2013**, *87*, 10283–10294. [CrossRef] [PubMed]

56. Cataldi, M.; Shah, N.R.; Felt, S.A.; Grdzelishvili, V.Z. Breaking resistance of pancreatic cancer cells to an attenuated vesicular stomatitis virus through a novel activity of IKK inhibitor TPCA-1. *Virology* **2015**, *485*, 340–354. [CrossRef] [PubMed]

57. Hastie, E.; Cataldi, M.; Steuerwald, N.; Grdzelishvili, V.Z. An unexpected inhibition of antiviral signaling by virus-encoded tumor suppressor p53 in pancreatic cancer cells. *Virology* **2015**, *483*, 126–140. [CrossRef]

58. Hastie, E.; Cataldi, M.; Moerdyk-Schauwecker, M.J.; Felt, S.A.; Steuerwald, N.; Grdzelishvili, V.Z. Novel biomarkers of resistance of pancreatic cancer cells to oncolytic vesicular stomatitis virus. *Oncotarget* **2016**, *7*, 61601–61618. [CrossRef]

59. Felt, S.A.; Moerdyk-Schauwecker, M.J.; Grdzelishvili, V.Z. Induction of apoptosis in pancreatic cancer cells by vesicular stomatitis virus. *Virology* **2015**, *474*, 163–173. [CrossRef] [PubMed]

60. Felt, S.A.; Droby, G.N.; Grdzelishvili, V.Z. Ruxolitinib and Polycation Combination Treatment Overcomes Multiple Mechanisms of Resistance of Pancreatic Cancer Cells to Oncolytic Vesicular Stomatitis Virus. *J. Virol.* **2017**, *91*, 91. [CrossRef]

61. Bressy, C.; Droby, G.N.; Maldonado, B.D.; Steuerwald, N.; Grdzelishvili, V.Z. Cell Cycle Arrest in G2/M Phase Enhances Replication of Interferon-Sensitive Cytoplasmic RNA Viruses via Inhibition of Antiviral Gene Expression. *J. Virol.* **2018**, *93*, e01885-18. [CrossRef]

62. Seegers, S.L.; Frasier, C.; Greene, S.; Nesmelova, I.V.; Grdzelishvili, V.Z. Experimental Evolution Generates Novel Oncolytic Vesicular Stomatitis Viruses with Improved Replication in Virus-Resistant Pancreatic Cancer Cells. *J. Virol.* **2019**, *94*, 94. [CrossRef]

63. Torres, M.P.; Rachagani, S.; Souchek, J.J.; Mallya, K.; Johansson, S.L.; Batra, S.K. Novel Pancreatic Cancer Cell Lines Derived from Genetically Engineered Mouse Models of Spontaneous Pancreatic Adenocarcinoma: Applications in Diagnosis and Therapy. *PLoS ONE* **2013**, *8*, e80580. [CrossRef]

64. Quillien, L.; Top, S.; Kappler-Gratias, S.; Redouté, M.A.; Dusetti, N.; Quentin-Froignant, M.C.; Lulka, H.; Camus-Bouclainville, C.; Buscail, L.; Gallardo, F.; et al. A Novel Imaging Approach for Single-Cell Real-Time Analysis of Oncolytic Virus Replication and Efficacy in Cancer Cells. *Hum. Gene Ther.* **2021**, *32*, 166–177. [CrossRef]

65. Dobrynin, Y.V. Establishment and Characteristics of Cell Strains from Some Epithelial Tumors of Human Origin. *J. Natl. Cancer Inst.* **1963**, *31*, 1173–1195. [CrossRef]

66. Walters, D.M.; Stokes, J.B.; Adair, S.J.; Stelow, E.B.; Borgman, C.A.; Lowrey, B.T.; Xin, W.; Blais, E.M.; Lee, J.K.; Papin, J.A.; et al. Clinical, Molecular and Genetic Validation of a Murine Orthotopic Xenograft Model of Pancreatic Adenocarcinoma Using Fresh Human Specimens. *PLoS ONE* **2013**, *8*, e77065. [CrossRef]

67. Fu, X.; Guadagni, F.; Hoffman, R.M. A metastatic nude-mouse model of human pancreatic cancer constructed orthotopically with histologically intact patient specimens. *Proc. Natl. Acad. Sci. USA* **1992**, *89*, 5645–5649. [CrossRef]

68. Corbett, T.H.; Roberts, B.J.; Leopold, W.R.; Peckham, J.C.; Wilkoff, L.J.; Griswold, D.P.; Schabel, F.M. Induction and chemotherapeutic response of two transplantable ductal adenocarcinomas of the pancreas in C57BL/6 mice. *Cancer Res.* **1984**, *44*, 717–726.

69. Lee, J.W.; Komar, C.A.; Bengsch, F.; Graham, K.; Beatty, G.L. Genetically Engineered Mouse Models of Pancreatic Cancer: The KPC Model (LSL-Kras(G12D/+); LSL-Trp53(R172H/+); Pdx-1-Cre), Its Variants, and Their Application in Immuno-oncology Drug Discovery. *Curr. Protoc. Pharmacol.* **2016**, *73*, 14.39.11–14.39.20. [CrossRef] [PubMed]

70. Huch, M.; Koo, B.-K. Modeling mouse and human development using organoid cultures. *Development* **2015**, *142*, 3113–3125. [CrossRef]

71. Ridky, T.W.; Chow, J.M.; Wong, D.J.; Khavari, P.A. Invasive three-dimensional organotypic neoplasia from multiple normal human epithelia. *Nat. Med.* **2010**, *16*, 1450–1455. [CrossRef] [PubMed]

72. Behrens, D.; Walther, W.; Fichtner, I. Pancreatic cancer models for translational research. *Pharmacol. Ther.* **2017**, *173*, 146–158. [CrossRef] [PubMed]

73. Jabs, J.; Zickgraf, F.M.; Park, J.; Wagner, S.; Jiang, X.; Jechow, K.; Kleinheinz, K.; Toprak, U.H.; Schneider, M.A.; Meister, M.; et al. Screening drug effects in patient-derived cancer cells links organoid responses to genome alterations. *Mol. Syst. Biol.* **2017**, *13*, 955. [CrossRef] [PubMed]

74. Hou, S.; Tiriac, H.; Sridharan, B.P.; Scampavia, L.; Madoux, F.; Seldin, J.; Souza, G.R.; Watson, D.; Tuveson, D.; Spicer, T.P. Advanced Development of Primary Pancreatic Organoid Tumor Models for High-Throughput Phenotypic Drug Screening. *SLAS Discov.* **2018**, *23*, 574–584. [CrossRef] [PubMed]

75. Xu, H.; Lyu, X.; Yi, M.; Zhao, W.; Song, Y.; Wu, K. Organoid technology and applications in cancer research. *J. Hematol. Oncol.* **2018**, *11*, 1–15. [CrossRef]

76. Tomás-Bort, E.; Kieler, M.; Sharma, S.; Candido, J.B.; Loessner, D. 3D approaches to model the tumor microenvironment of pancreatic cancer. *Theranostics* **2020**, *10*, 5074–5089. [CrossRef]

77. Raimondi, G.; Mato-Berciano, A.; Pascual-Sabater, S.; Rovira-Rigau, M.; Cuatrecasas, M.; Fondevila, C.; Sánchez-Cabús, S.; Begthel, H.; Boj, S.F.; Clevers, H.; et al. Patient-derived pancreatic tumour organoids identify therapeutic responses to oncolytic adenoviruses. *EBioMedicine* **2020**, *56*, 102786. [CrossRef] [PubMed]

78. Cockle, J.V.; Brüning-Richardson, A.; Scott, K.J.; Thompson, J.; Kottke, T.; Morrison, E.; Ismail, A.; Carcaboso, A.M.; Rose, A.; Selby, P.; et al. Oncolytic Herpes Simplex Virus Inhibits Pediatric Brain Tumor Migration and Invasion. *Mol. Ther. Oncolytics* **2017**, *5*, 75–86. [CrossRef]

79. Wurdinger, T.; Verheije, M.H.; Raaben, M.; Bosch, B.J.; Haan, C.A.M.D.; Van Beusechem, V.W.; Rottier, P.J.M.; Gerritsen, W.R. Targeting non-human coronaviruses to human cancer cells using a bispecific single-chain antibody. *Gene Ther.* **2005**, *12*, 1394–1404. [CrossRef]

80. Li, J.; Qian, W.; Qin, T.; Xiao, Y.; Cheng, L.; Cao, J.; Chen, X.; Ma, Q.; Wu, Z. Mouse-Derived Allografts: A Complementary Model to the KPC Mice on Researching Pancreatic Cancer In Vivo. *Comput. Struct. Biotechnol. J.* **2019**, *17*, 498–506. [CrossRef]

81. Bailey, K.L.; Carlson, M.A. Porcine Models of Pancreatic Cancer. *Front. Oncol.* **2019**, *9*, 144. [CrossRef]

82. Kong, K.; Guo, M.; Liu, Y.; Zheng, J. Progress in Animal Models of Pancreatic Ductal Adenocarcinoma. *J. Cancer* **2020**, *11*, 1555–1567. [CrossRef]

83. Hruban, R.H.; Adsay, N.V.; Albores-Saavedra, J.; Compton, C.; Garrett, E.S.; Goodman, S.N.; Kern, S.E.; Klimstra, D.S.; Kloppel, G.; Longnecker, D.S.; et al. Pancreatic intraepithelial neoplasia: A new nomenclature and classification system for pancreatic duct lesions. *Am. J. Surg. Pathol.* **2001**, *25*, 579–586. [CrossRef] [PubMed]

84. Garcia, P.L.; Miller, A.L.; Yoon, K.J. Patient-Derived Xenograft Models of Pancreatic Cancer: Overview and Comparison with Other Types of Models. *Cancers* **2020**, *12*, 1327. [CrossRef]

85. Krempley, B.D.; Yu, K.H. Preclinical models of pancreatic ductal adenocarcinoma. *Chin. Clin. Oncol.* **2017**, *6*, 25. [CrossRef]

86. Garcia, P.L.; Council, L.N.; Christein, J.D.; Arnoletti, J.P.; Heslin, M.J.; Gamblin, T.L.; Richardson, J.H.; Bjornsti, M.-A.; Yoon, K.J. Development and Histopathological Characterization of Tumorgraft Models of Pancreatic Ductal Adenocarcinoma. *PLoS ONE* **2013**, *8*, e78183. [CrossRef]

87. Rubio-Viqueira, B.; Jimeno, A.; Cusatis, G.; Zhang, X.; Iacobuzio-Donahue, C.; Karikari, C.; Shi, C.; Danenberg, K.; Danenberg, P.V.; Kuramochi, H.; et al. An In Vivo Platform for Translational Drug Development in Pancreatic Cancer. *Clin. Cancer Res.* **2006**, *12*, 4652–4661. [CrossRef]

88. Sereti, E.; Karagianellou, T.; Kotsoni, I.; Magouliotis, D.; Kamposioras, K.; Ulukaya, E.; Sakellaridis, N.; Zacharoulis, D.; Dimas, K. Patient Derived Xenografts (PDX) for personalized treatment of pancreatic cancer: Emerging allies in the war on a devastating cancer? *J. Proteom.* **2018**, *188*, 107–118. [CrossRef]

89. Loukopoulos, P.; Kanetaka, K.; Takamura, M.; Shibata, T.; Sakamoto, M.; Hirohashi, S. Orthotopic Transplantation Models of Pancreatic Adenocarcinoma Derived from Cell Lines and Primary Tumors and Displaying Varying Metastatic Activity. *Pancreas* **2004**, *29*, 193–203. [CrossRef]

90. Morton, C.L.; Houghton, P.J. Establishment of human tumor xenografts in immunodeficient mice. *Nat. Protoc.* **2007**, *2*, 247–250. [CrossRef]

91. Brehm, M.A.; Shultz, L.D.; Luban, J.; Greiner, D.L. Overcoming Current Limitations in Humanized Mouse Research. *J. Infect. Dis.* **2013**, *208*, S125–S130. [CrossRef]

92. Villaudy, J.; Schotte, R.; Legrand, N.; Spits, H. Critical assessment of human antibody generation in humanized mouse models. *J. Immunol. Methods* **2014**, *410*, 18–27. [CrossRef] [PubMed]

93. Walsh, N.C.; Kenney, L.L.; Jangalwe, S.; Aryee, K.-E.; Greiner, D.L.; Brehm, M.A.; Shultz, L.D. Humanized Mouse Models of Clinical Disease. *Annu. Rev. Pathol. Mech. Dis.* **2017**, *12*, 187–215. [CrossRef] [PubMed]

94. Ito, M.; Hiramatsu, H.; Kobayashi, K.; Suzue, K.; Kawahata, M.; Hioki, K.; Ueyama, Y.; Koyanagi, Y.; Sugamura, K.; Tsuji, K.; et al. NOD/SCID/gamma(c)(null) mouse: An excellent recipient mouse model for engraftment of human cells. *Blood* **2002**, *100*, 3175–3182. [CrossRef] [PubMed]

95. Ishikawa, F.; Yasukawa, M.; Lyons, B.; Yoshida, S.; Miyamoto, T.; Yoshimoto, G.; Watanabe, T.; Akashi, K.; Shultz, L.D.; Harada, M. Development of functional human blood and immune systems in NOD/SCID/IL2 receptor {gamma} chain(null) mice. *Blood* **2005**, *106*, 1565–1573. [CrossRef] [PubMed]

96. Shultz, L.D.; Ishikawa, F.; Greiner, D.L. Humanized mice in translational biomedical research. *Nat. Rev. Immunol.* **2007**, *7*, 118–130. [CrossRef]

97. Shultz, L.D.; Brehm, M.A.; Garcia-Martinez, J.V.; Greiner, D.L. Humanized mice for immune system investigation: Progress, promise and challenges. *Nat. Rev. Immunol.* **2012**, *12*, 786–798. [CrossRef]

98. Vareki, S.M. High and low mutational burden tumors versus immunologically hot and cold tumors and response to immune checkpoint inhibitors. *J. Immunother. Cancer* **2018**, *6*, 157. [CrossRef]

99. Tsoneva, D.; Minev, B.; Frentzen, A.; Zhang, Q.; Wege, A.K.; Szalay, A.A. Humanized Mice with Subcutaneous Human Solid Tumors for Immune Response Analysis of Vaccinia Virus-Mediated Oncolysis. *Mol. Ther. Oncolytics* **2017**, *5*, 41–61. [CrossRef]

100. Hingorani, S.R.; Petricoin, E.F.; Maitra, A.; Rajapakse, V.; King, C.; Jacobetz, M.A.; Ross, S.; Conrads, T.P.; Veenstra, T.D.; Hitt, B.A.; et al. Preinvasive and invasive ductal pancreatic cancer and its early detection in the mouse. *Cancer Cell* **2003**, *4*, 437–450. [CrossRef]

101. Neesse, A.; Michl, P.; Frese, K.K.; Feig, C.; Cook, N.; Jacobetz, M.A.; Lolkema, M.P.; Buchholz, M.; Olive, K.P.; Gress, T.M.; et al. Stromal biology and therapy in pancreatic cancer. *Gut* **2011**, *60*, 861–868. [CrossRef]

102. He, M.; Henderson, M.; Muth, S.; Murphy, A.; Zheng, L. Preclinical mouse models for immunotherapeutic and non-immunotherapeutic drug development for pancreatic ductal adenocarcinoma. *Ann. Pancreat. Cancer* **2020**, *3*, 7. [CrossRef]

103. Herreros-Villanueva, M.; Hijona, E.; Cosme, A.; Bujanda, L. Mouse models of pancreatic cancer. *World J. Gastroenterol.* **2012**, *18*, 1286–1294. [CrossRef] [PubMed]

104. Westphalen, C.B.; Olive, K.P. Genetically Engineered Mouse Models of Pancreatic Cancer. *Cancer J.* **2012**, *18*, 502–510. [CrossRef] [PubMed]

105. Fendrich, V.; Schneider, R.; Maitra, A.; Jacobsen, I.D.; Opfermann, T.; Bartsch, D.K. Detection of Precursor Lesions of Pancreatic Adenocarcinoma in PET-CT in a Genetically Engineered Mouse Model of Pancreatic Cancer1. *Neoplasia* **2011**, *13*, 180–186. [CrossRef] [PubMed]

106. Faca, V.M.; Song, K.S.; Wang, H.; Zhang, Q.; Krasnoselsky, A.L.; Newcomb, L.F.; Plentz, R.R.; Gurumurthy, S.; Redston, M.S.; Pitteri, S.J.; et al. A Mouse to Human Search for Plasma Proteome Changes Associated with Pancreatic Tumor Development. *PLoS Med.* **2008**, *5*, e123. [CrossRef]

107. Becher, O.J.; Holland, E.C. Genetically Engineered Models Have Advantages over Xenografts for Preclinical Studies. *Cancer Res.* **2006**, *66*, 3355–3359. [CrossRef]
108. Wang, Y.; Tseng, J.-C.; Sun, Y.; Beck, A.H.; Kung, A.L. Noninvasive Imaging of Tumor Burden and Molecular Pathways in Mouse Models of Cancer. *Cold Spring Harb. Protoc.* **2015**, *2015*, 135–144. [CrossRef]
109. Wang, Y.; Zhang, Y.; Yang, J.; Ni, X.; Liu, S.; Li, Z.; Hodges, S.E.; Fisher, W.E.; Brunicardi, F.C.; Gibbs, R.A.; et al. Genomic Sequencing of Key Genes in Mouse Pancreatic Cancer Cells. *Curr. Mol. Med.* **2012**, *12*, 331–341. [CrossRef]
110. Ijichi, H.; Chytil, A.; Gorska, A.E.; Aakre, M.E.; Fujitani, Y.; Fujitani, S.; Wright, C.V.; Moses, H.L. Aggressive pancreatic ductal adenocarcinoma in mice caused by pancreas-specific blockade of transforming growth factor-beta signaling in cooperation with active Kras expression. *Genes Dev.* **2006**, *20*, 3147–3160. [CrossRef]
111. Soares, K.C.; Foley, K.; Olino, K.; Leubner, A.; Mayo, S.C.; Jain, A.; Jaffee, E.; Schulick, R.D.; Yoshimura, K.; Edil, B.; et al. A Preclinical Murine Model of Hepatic Metastases. *J. Vis. Exp.* **2014**, *10*, e51677. [CrossRef]
112. Yanagihara, K.; Kubo, T.; Mihara, K.; Kuwata, T.; Ochiai, A.; Seyama, T.; Yokozaki, H. Development and Biological Analysis of a Novel Orthotopic Peritoneal Dissemination Mouse Model Generated Using a Pancreatic Ductal Adenocarcinoma Cell Line. *Pancreas* **2019**, *48*, 315–322. [CrossRef]
113. Narayanan, J.S.S.; Ray, P.; Naqvi, I.; White, R. A Syngeneic Pancreatic Cancer Mouse Model to Study the Effects of Irreversible Electroporation. *J. Vis. Exp.* **2018**, e57265. [CrossRef]
114. Manuel, E.R.; Chen, J.; D'Apuzzo, M.; Lampa, M.G.; Kaltcheva, T.I.; Thompson, C.B.; Ludwig, T.; Chung, V.; Diamond, D.J. Salmonella-based therapy targeting indoleamine 2,3-dioxygenase coupled with enzymatic depletion of tumor hyaluronan induces complete regression of aggressive pancreatic tumors. *Cancer Immunol. Res.* **2015**, *3*, 1096–1107. [CrossRef]
115. Wennier, S.; Li, S.; McFadden, G. Oncolytic virotherapy for pancreatic cancer. *Expert Rev. Mol. Med.* **2011**, *13*, e18. [CrossRef]
116. Ahn, D.H.; Bekaii-Saab, T. The Continued Promise and Many Disappointments of Oncolytic Virotherapy in Gastrointestinal Malignancies. *Biomedicines* **2017**, *5*, 10. [CrossRef]
117. Bressy, C.; Hastie, E.; Grdzelishvili, V.Z. Combining Oncolytic Virotherapy with p53 Tumor Suppressor Gene Therapy. *Mol. Ther. Oncolytics* **2017**, *5*, 20–40. [CrossRef]
118. Tassone, E.; Muscolini, M.; Van Montfoort, N.; Hiscott, J. Oncolytic virotherapy for pancreatic ductal adenocarcinoma: A glimmer of hope after years of disappointment? *Cytokine Growth Factor Rev.* **2020**, *56*, 141–148. [CrossRef]
119. Eissa, I.R.; Bustos-Villalobos, I.; Ichinose, T.; Matsumura, S.; Naoe, Y.; Miyajima, N.; Morimoto, D.; Mukoyama, N.; Zhiwen, W.; Tanaka, M.; et al. The Current Status and Future Prospects of Oncolytic Viruses in Clinical Trials against Melanoma, Glioma, Pancreatic, and Breast Cancers. *Cancers* **2018**, *10*, 356. [CrossRef]
120. Stevenson, A.J.; Giles, M.S.; Hall, K.T.; Goodwin, D.J.; A Calderwood, M.; Markham, A.F.; Whitehouse, A. Specific oncolytic activity of herpesvirus saimiri in pancreatic cancer cells. *Br. J. Cancer* **2000**, *83*, 329–332. [CrossRef]
121. Kasuya, H.; Nishiyama, Y.; Nomoto, S.; Goshima, F.; Takeda, S.; Watanabe, I.; Nomura, N.; Shikano, T.; Fujii, T.; Kanazumi, N.; et al. Suitability of a US3-inactivated HSV mutant (L1BR1) as an oncolytic virus for pancreatic cancer therapy. *Cancer Gene Ther.* **2007**, *14*, 533–542. [CrossRef] [PubMed]
122. Gayral, M.; Lulka, H.; Hanoun, N.; Biollay, C.; Sèlves, J.; Vignolle-Vidoni, A.; Berthommé, H.; Trempat, P.; Epstein, A.L.; Buscail, L.; et al. Targeted Oncolytic Herpes Simplex Virus Type 1 Eradicates Experimental Pancreatic Tumors. *Hum. Gene Ther.* **2015**, *26*, 104–113. [CrossRef]
123. Watanabe, I.; Kasuya, H.; Nomura, N.; Shikano, T.; Shirota, T.; Kanazumi, N.; Takeda, S.; Nomoto, S.; Sugimoto, H.; Nakao, A. Effects of tumor selective replication-competent herpes viruses in combination with gemcitabine on pancreatic cancer. *Cancer Chemother. Pharmacol.* **2007**, *61*, 875–882. [CrossRef]
124. Fu, X.; Tao, L.; Li, M.; Fisher, W.E.; Zhang, X. Effective Treatment of Pancreatic Cancer Xenografts with a Conditionally Replicating Virus Derived from Type 2 Herpes Simplex Virus. *Clin. Cancer Res.* **2006**, *12*, 3152–3157. [CrossRef]
125. Eisenberg, D.P.; Adusumilli, P.S.; Hendershott, K.J.; Yu, Z.; Mullerad, M.; Chan, M.-K.; Chou, T.-C.; Fong, Y. 5-Fluorouracil and Gemcitabine Potentiate the Efficacy of Oncolytic Herpes Viral Gene Therapy in the Treatment of Pancreatic Cancer. *J. Gastrointest. Surg.* **2005**, *9*, 1068–1079. [CrossRef]
126. Liu, H.; Yuan, S.-J.; Chen, Y.-T.; Xie, Y.-B.; Cui, L.; Yang, W.-Z.; Yang, D.-X.; Tian, Y.-T. Preclinical evaluation of herpes simplex virus armed with granulocyte-macrophage colony-stimulating factor in pancreatic carcinoma. *World J. Gastroenterol.* **2013**, *19*, 5138–5143. [CrossRef] [PubMed]
127. Bortolanza, S.; Bunuales, M.; Otano, I.; Gonzalez-Aseguinolaza, G.; Ortiz-De-Solorzano, C.; Perez, D.R.; Prieto, J.; Hernandez-Alcoceba, R. Treatment of Pancreatic Cancer with an Oncolytic Adenovirus Expressing Interleukin-12 in Syrian Hamsters. *Mol. Ther.* **2009**, *17*, 614–622. [CrossRef]
128. McAuliffe, P.F.; Jarnagin, W.R.; Johnson, P.; Delman, K.A.; Federoff, H.; Fong, Y. Effective treatment of pancreatic tumors with two multimutated herpes simplex oncolytic viruses. *J. Gastrointest. Surg.* **2000**, *4*, 580–588. [CrossRef]
129. Nishimoto, T.; Yoshida, K.; Miura, Y.; Kobayashi, A.; Hara, H.; Ohnami, S.; Kurisu, K.; Yoshida, T.; Aoki, K. Oncolytic virus therapy for pancreatic cancer using the adenovirus library displaying random peptides on the fiber knob. *Gene Ther.* **2009**, *16*, 669–680. [CrossRef]
130. Rodríguez-García, A.; Giménez-Alejandre, M.; Rojas, J.J.; Moreno, R.; Bazan-Peregrino, M.; Cascalló, M.; Alemany, R. Safety and Efficacy of VCN-01, an Oncolytic Adenovirus Combining Fiber HSG-Binding Domain Replacement with RGD and Hyaluronidase Expression. *Clin. Cancer Res.* **2015**, *21*, 1406–1418. [CrossRef]

131. Leitner, S.; Sweeney, K.; Öberg, D.; Davies, D.; Miranda, E.; Lemoine, N.R.; Halldén, G. Oncolytic Adenoviral Mutants with E1B19K Gene Deletions Enhance Gemcitabine-induced Apoptosis in Pancreatic Carcinoma Cells and Anti-Tumor Efficacy In Vivo. *Clin. Cancer Res.* **2009**, *15*, 1730–1740. [CrossRef]

132. Armstrong, L.; Arrington, A.; Han, J.; Gavrikova, T.; Brown, E.; Yamamoto, M.; Vickers, S.M.; Davydova, J. Generation of a novel, cyclooxygenase-2–targeted, interferon-expressing, conditionally replicative adenovirus for pancreatic cancer therapy. *Am. J. Surg.* **2012**, *204*, 741–750. [CrossRef]

133. Yu, D.; Jin, C.; Ramachandran, M.; Xu, J.; Nilsson, B.; Korsgren, O.; Le Blanc, K.; Uhrbom, L.; Forsberg-Nilsson, K.; Westermark, B.; et al. Adenovirus Serotype 5 Vectors with Tat-PTD Modified Hexon and Serotype 35 Fiber Show Greatly Enhanced Transduction Capacity of Primary Cell Cultures. *PLoS ONE* **2013**, *8*, e54952. [CrossRef] [PubMed]

134. Zhang, Y.-A.; Nemunaitis, J.; Samuel, S.K.; Chen, P.; Shen, Y.; Tong, A.W. Antitumor Activity of an Oncolytic Adenovirus-Delivered Oncogene Small Interfering RNA. *Cancer Res.* **2006**, *66*, 9736–9743. [CrossRef]

135. Xu, B.; Zheng, W.-Y.; Jin, D.-Y.; Wang, D.-S.; Liu, X.-Y.; Qin, X.-Y. Treatment of pancreatic cancer using an oncolytic virus harboring the lipocalin-2 gene. *Cancer* **2012**, *118*, 5217–5226. [CrossRef]

136. Eriksson, E.; Milenova, I.; Wenthe, J.; Ståhle, M.; Leja-Jarblad, J.; Ullenhag, G.; Dimberg, A.; Moreno, R.; Alemany, R.; Loskog, A. Shaping the Tumor Stroma and Sparking Immune Activation by CD40 and 4-1BB Signaling Induced by an Armed Oncolytic Virus. *Clin. Cancer Res.* **2017**, *23*, 5846–5857. [CrossRef]

137. Salzwedel, A.O.; Han, J.; LaRocca, C.J.; Shanley, R.; Yamamoto, M.; Davydova, J. Combination of interferon-expressing oncolytic adenovirus with chemotherapy and radiation is highly synergistic in hamster model of pancreatic cancer. *Oncotarget* **2018**, *9*, 18041–18052. [CrossRef]

138. Watanabe, K.; Luo, Y.; Da, T.; Guedan, S.; Ruella, M.; Scholler, J.; Keith, B.; Young, R.M.; Engels, B.; Sorsa, S.; et al. Pancreatic cancer therapy with combined mesothelin-redirected chimeric antigen receptor T cells and cytokine-armed oncolytic adenoviruses. *JCI Insight* **2018**, *3*. [CrossRef]

139. Jung, K.H.; Choi, I.-K.; Lee, H.-S.; Yan, H.H.; Son, M.K.; Ahn, H.M.; Hong, J.; Yun, C.-O.; Hong, S.-S. Oncolytic adenovirus expressing relaxin (YDC002) enhances therapeutic efficacy of gemcitabine against pancreatic cancer. *Cancer Lett.* **2017**, *396*, 155–166. [CrossRef]

140. Cherubini, G.; Kallin, C.; Mozetic, A.; Hammaren-Busch, K.; Muller, H.; Lemoine, N.R.; Hallden, G. The oncolytic adenovirus AdDeltaDelta enhances selective cancer cell killing in combination with DNA-damaging drugs in pancreatic cancer models. *Gene Ther.* **2011**, *18*, 1157–1165. [CrossRef]

141. Bhattacharyya, M.; Francis, J.; Eddouadi, A.; Lemoine, N.R.; Hallden, G. An oncolytic adenovirus defective in pRb-binding (dl922-947) can efficiently eliminate pancreatic cancer cells and tumors in vivo in combination with 5-FU or gemcitabine. *Cancer Gene Ther.* **2011**, *18*, 734–743. [CrossRef]

142. Dai, B.; Roife, D.; Kang, Y.; Gumin, J.; Perez, M.V.R.; Li, X.; Pratt, M.; Brekken, R.A.; Fueyo-Margareto, J.; Lang, F.F.; et al. Preclinical Evaluation of Sequential Combination of Oncolytic Adenovirus Delta-24-RGD and Phosphatidylserine-Targeting Antibody in Pancreatic Ductal Adenocarcinoma. *Mol. Cancer Ther.* **2017**, *16*, 662–670. [CrossRef] [PubMed]

143. Man, Y.K.S.; Davies, J.A.; Coughlan, L.; Pantelidou, C.; Blazquez-Moreno, A.; Marshall, J.F.; Parker, A.L.; Hallden, G. The Novel Oncolytic Adenoviral Mutant Ad5-3Delta-A20T Retargeted to alphavbeta6 Integrins Efficiently Eliminates Pancreatic Cancer Cells. *Mol. Cancer Ther.* **2018**, *17*, 575–587. [CrossRef]

144. Rovira-Rigau, M.; Raimondi, G.; Marín, M.Á.; Gironella, M.; Alemany, R.; Fillat, C. Bioselection Reveals miR-99b and miR-485 as Enhancers of Adenoviral Oncolysis in Pancreatic Cancer. *Mol. Ther.* **2019**, *27*, 230–243. [CrossRef] [PubMed]

145. Freytag, S.O.; Barton, K.N.; Brown, S.L.; Narra, V.; Zhang, Y.; Tyson, D.; Nall, C.; Lü, M.; Ajlouni, M.; Movsas, B.; et al. Replication-competent Adenovirus-mediated Suicide Gene Therapy with Radiation in a Preclinical Model of Pancreatic Cancer. *Mol. Ther.* **2007**, *15*, 1600–1606. [CrossRef] [PubMed]

146. Woo, Y.; Kelly, K.J.; Stanford, M.M.; Galanis, C.; Chun, Y.S.; Fong, Y.; McFadden, G. Myxoma Virus Is Oncolytic for Human Pancreatic Adenocarcinoma Cells. *Ann. Surg. Oncol.* **2008**, *15*, 2329–2335. [CrossRef]

147. Carew, J.S.; Espitia, C.M.; Zhao, W.; Kelly, K.R.; Coffey, M.; Freeman, J.W.; Nawrocki, S.T. Reolysin is a novel reovirus-based agent that induces endoplasmic reticular stress-mediated apoptosis in pancreatic cancer. *Cell Death Dis.* **2013**, *4*, e728. [CrossRef]

148. Chen, W.; Fan, W.; Ru, G.; Huang, F.; Lu, X.; Zhang, X.; Mou, X.; Wang, S. Gemcitabine combined with an engineered oncolytic vaccinia virus exhibits a synergistic suppressive effect on the tumor growth of pancreatic cancer. *Oncol. Rep.* **2018**, *41*, 67–76. [CrossRef]

149. Sun, T.; Luo, Y.; Wang, M.; Xie, T.; Yan, H. Recombinant Oncolytic Vaccinia Viruses Expressing Human beta-Defensin 2 Enhance Anti-tumor Immunity. *Mol. Ther. Oncolytics* **2019**, *13*, 49–57. [CrossRef]

150. Wu, Y.; Mou, X.; Wang, S.; Liu, X.-E.; Sun, X. ING4 expressing oncolytic vaccinia virus promotes anti-tumor efficiency and synergizes with gemcitabine in pancreatic cancer. *Oncotarget* **2017**, *8*, 82728–82739. [CrossRef]

151. Chard, L.S.; Lemoine, N.R.; Wang, Y. New role of Interleukin-10 in enhancing the antitumor efficacy of oncolytic vaccinia virus for treatment of pancreatic cancer. *OncoImmunology* **2015**, *4*, e1038689. [CrossRef] [PubMed]

152. Tysome, J.R.; Briat, A.; Alusi, G.; Cao, F.; Gao, D.; Yu, J.; Wang, P.; Yang, S.; Dong, Z.; Wang, S.; et al. Lister strain of vaccinia virus armed with endostatin–angiostatin fusion gene as a novel therapeutic agent for human pancreatic cancer. *Gene Ther.* **2009**, *16*, 1223–1233. [CrossRef] [PubMed]

153. Bhat, R.; Dempe, S.; Dinsart, C.; Rommelaere, J. Enhancement of NK cell antitumor responses using an oncolytic parvovirus. *Int. J. Cancer* **2010**, *128*, 908–919. [CrossRef]

154. Moehler, M.; Sieben, M.; Roth, S.; Springsguth, F.; Leuchs, B.; Zeidler, M.; Dinsart, C.; Rommelaere, J.; Galle, P.R. Activation of the human immune system by chemotherapeutic or targeted agents combined with the oncolytic parvovirus H-1. *BMC Cancer* **2011**, *11*, 464. [CrossRef]

155. Li, J.; Bonifati, S.; Hristov, G.; Marttila, T.; Valmary-Degano, S.; Stanzel, S.; Schnölzer, M.; Mougin, C.; Aprahamian, M.; Grekova, S.P.; et al. Synergistic combination of valproic acid and oncolytic parvovirus H-1 PV as a potential therapy against cervical and pancreatic carcinomas. *EMBO Mol. Med.* **2013**, *5*, 1537–1555. [CrossRef] [PubMed]

156. Bossow, S.; Grossardt, C.; Temme, A.; Leber, M.F.; Sawall, S.; Rieber, E.P.; Cattaneo, R.; Von Kalle, C.; Ungerechts, G. Armed and targeted measles virus for chemovirotherapy of pancreatic cancer. *Cancer Gene Ther.* **2011**, *18*, 598–608. [CrossRef]

157. May, V.; Berchtold, S.; Berger, A.; Venturelli, S.; Burkard, M.; Leischner, C.; Malek, N.P.; Lauer, U.M. Chemovirotherapy for pancreatic cancer: Gemcitabine plus oncolytic measles vaccine virus. *Oncol. Lett.* **2019**, *18*, 5534–5542. [CrossRef]

158. Walter, R.J.; Attar, B.M.; Rafiq, A.; Tejaswi, S.; Delimata, M. Newcastle disease virus LaSota strain kills human pancreatic cancer cells in vitro with high selectivity. *JOP* **2012**, *13*, 45–53. [CrossRef]

159. Schwaiger, T.; Knittler, M.R.; Grund, C.; Roemer-Oberdoerfer, A.; Kapp, J.-F.; Lerch, M.M.; Mettenleiter, T.C.; Mayerle, J.; Blohm, U. Newcastle disease virus mediates pancreatic tumor rejection via NK cell activation and prevents cancer relapse by prompting adaptive immunity. *Int. J. Cancer* **2017**, *141*, 2505–2516. [CrossRef]

160. Fábián, Z.; Csatary, C.M.; Szeberényi, J.; Csatary, L.K. p53-Independent Endoplasmic Reticulum Stress-Mediated Cytotoxicity of a Newcastle Disease Virus Strain in Tumor Cell Lines. *J. Virol.* **2007**, *81*, 2817–2830. [CrossRef]

161. O'Leary, M.P.; Choi, A.H.; Kim, S.-I.; Chaurasiya, S.; Lu, J.; Park, A.K.; Woo, Y.; Warner, S.G.; Fong, Y.; Chen, N.G. Novel oncolytic chimeric orthopoxvirus causes regression of pancreatic cancer xenografts and exhibits abscopal effect at a single low dose. *J. Transl. Med.* **2018**, *16*, 1–11. [CrossRef]

162. Kasloff, S.B.; Pizzuto, M.S.; Silic-Benussi, M.; Pavone, S.; Ciminale, V.; Capua, I. Oncolytic Activity of Avian Influenza Virus in Human Pancreatic Ductal Adenocarcinoma Cell Lines. *J. Virol.* **2014**, *88*, 9321–9334. [CrossRef]

163. Blackham, A.U.; Northrup, S.A.; Willingham, M.; Sirintrapun, J.; Russell, G.B.; Lyles, D.S.; Stewart, J.H. Molecular determinants of susceptibility to oncolytic vesicular stomatitis virus in pancreatic adenocarcinoma. *J. Surg. Res.* **2014**, *187*, 412–426. [CrossRef]

164. Nagalo, B.M.; Breton, C.A.; Zhou, Y.; Arora, M.; Bogenberger, J.M.; Barro, O.; Steele, M.B.; Jenks, N.J.; Baker, A.T.; Duda, D.G.; et al. Oncolytic Virus with Attributes of Vesicular Stomatitis Virus and Measles Virus in Hepatobiliary and Pancreatic Cancers. *Mol. Ther. Oncolytics* **2020**, *18*, 546–555. [CrossRef]

165. Heiber, J.F.; Barber, G.N. Vesicular Stomatitis Virus Expressing Tumor Suppressor p53 Is a Highly Attenuated, Potent Oncolytic Agent. *J. Virol.* **2011**, *85*, 10440–10450. [CrossRef] [PubMed]

166. Gaddy, D.F.; Lyles, D.S. Vesicular Stomatitis Viruses Expressing Wild-Type or Mutant M Proteins Activate Apoptosis through Distinct Pathways. *J. Virol.* **2005**, *79*, 4170–4179. [CrossRef] [PubMed]

167. Hecht, J.R.; Bedford, R.; Abbruzzese, J.L.; Lahoti, S.; Reid, T.R.; Soetikno, R.M.; Kirn, D.H.; Freeman, S.M. A phase I/II trial of intratumoral endoscopic ultrasound injection of ONYX-015 with intravenous gemcitabine in unresectable pancreatic carcinoma. *Clin. Cancer Res.* **2003**, *9*, 555–561. [PubMed]

168. Mulvihill, S.; Warren, R.; Venook, A.; Adler, A.; Randlev, B.; Heise, C.; Kirn, D. Safety and feasibility of injection with an E1B-55 kDa gene-deleted, replication-selective adenovirus (ONYX-015) into primary carcinomas of the pancreas: A phase I trial. *Gene Ther.* **2001**, *8*, 308–315. [CrossRef]

169. Lee, J.-C.; Shin, D.W.; Park, H.; Kim, J.; Youn, Y.; Kim, J.H.; Kim, J.; Hwang, J.-H. Tolerability and safety of EUS-injected adenovirus-mediated double-suicide gene therapy with chemotherapy in locally advanced pancreatic cancer: A phase 1 trial. *Gastrointest. Endosc.* **2020**, *92*, 1044–1052.e1. [CrossRef]

170. Chang, K.J.; Senzer, N.N.; Binmoeller, K.; Goldsweig, H.; Coffin, R. Phase I dose-escalation study of talimogene laherparepvec (T-VEC) for advanced pancreatic cancer (ca). *J. Clin. Oncol.* **2012**, *30*, e14546. [CrossRef]

171. Hirooka, Y.; Kasuya, H.; Ishikawa, T.; Kawashima, H.; Ohno, E.; Villalobos, I.B.; Naoe, Y.; Ichinose, T.; Koyama, N.; Tanaka, M.; et al. A Phase I clinical trial of EUS-guided intratumoral injection of the oncolytic virus, HF10 for unresectable locally advanced pancreatic cancer. *BMC Cancer* **2018**, *18*, 1–9. [CrossRef]

172. Noonan, A.M.; Farren, M.R.; Geyer, S.M.; Huang, Y.; Tahiri, S.; Ahn, D.; Mikhail, S.; Ciombor, K.K.; Pant, S.; Aparo, S.; et al. Randomized Phase 2 Trial of the Oncolytic Virus Pelareorep (Reolysin) in Upfront Treatment of Metastatic Pancreatic Adenocarcinoma. *Mol. Ther.* **2016**, *24*, 1150–1158. [CrossRef]

173. Mahalingam, D.; Goel, S.; Aparo, S.; Patel Arora, S.; Noronha, N.; Tran, H.; Chakrabarty, R.; Selvaggi, G.; Gutierrez, A.; Coffey, M.; et al. A Phase II Study of Pelareorep (REOLYSIN((R))) in Combination with Gemcitabine for Patients with Advanced Pancreatic Adenocarcinoma. *Cancers* **2018**, *10*, 160. [CrossRef] [PubMed]

174. Mahalingam, D.; Wilkinson, G.A.; Eng, K.H.; Fields, P.; Raber, P.; Moseley, J.L.; Cheetham, K.; Coffey, M.; Nuovo, G.; Kalinski, P.; et al. Pembrolizumab in Combination with the Oncolytic Virus Pelareorep and Chemotherapy in Patients with Advanced Pancreatic Adenocarcinoma: A Phase Ib Study. *Clin. Cancer Res.* **2019**, *26*, 71–81. [CrossRef] [PubMed]

175. Hajda, J.; Lehmann, M.; Krebs, O.; Kieser, M.; Geletneky, K.; Jäger, D.; Dahm, M.; Huber, B.; Schöning, T.; Sedlaczek, O.; et al. A non-controlled, single arm, open label, phase II study of intravenous and intratumoral administration of ParvOryx in patients with metastatic, inoperable pancreatic cancer: ParvOryx02 protocol. *BMC Cancer* **2017**, *17*, 1–11. [CrossRef] [PubMed]

176. Monsurrò, V.; Beghelli, S.; Wang, R.; Barbi, S.; Coin, S.; Di Pasquale, G.; Bersani, S.; Castellucci, M.; Sorio, C.; Eleuteri, S.; et al. Anti-viral state segregates two molecular phenotypes of pancreatic adenocarcinoma: Potential relevance for adenoviral gene therapy. *J. Transl. Med.* **2010**, *8*, 10. [CrossRef] [PubMed]

177. Wennier, S.T.; Liu, J.; Li, S.; Rahman, M.M.; Mona, M.; McFadden, G. Myxoma Virus Sensitizes Cancer Cells to Gemcitabine and Is an Effective Oncolytic Virotherapeutic in Models of Disseminated Pancreatic Cancer. *Mol. Ther.* **2012**, *20*, 759–768. [CrossRef]

178. Hastie, E.; Grdzelishvili, V.Z. Vesicular stomatitis virus as a flexible platform for oncolytic virotherapy against cancer. *J. Gen. Virol.* **2012**, *93*, 2529–2545. [CrossRef]

179. Hastie, E.; Cataldi, M.; Marriott, I.; Grdzelishvili, V.Z. Understanding and altering cell tropism of vesicular stomatitis virus. *Virus Res.* **2013**, *176*, 16–32. [CrossRef]

180. Felt, S.A.; Grdzelishvili, V.Z. Recent advances in vesicular stomatitis virus-based oncolytic virotherapy: A 5-year update. *J. Gen. Virol.* **2017**, *98*, 2895–2911. [CrossRef]

181. Moussavi, M.; Fazli, L.; Tearle, H.; Guo, Y.; Cox, M.; Bell, J.; Ong, C.; Jia, W.; Rennie, P.S. Oncolysis of Prostate Cancers Induced by Vesicular Stomatitis Virus in PTEN Knockout Mice. *Cancer Res.* **2010**, *70*, 1367–1376. [CrossRef] [PubMed]

182. Zhang, K.-X.; Matsui, Y.; Hadaschik, B.A.; Lee, C.; Jia, W.; Bell, J.C.; Fazli, L.; So, A.I.; Rennie, P.S. Down-regulation of type I interferon receptor sensitizes bladder cancer cells to vesicular stomatitis virus-induced cell death. *Int. J. Cancer* **2010**, *127*, 830–838. [CrossRef]

183. Noser, J.A.; Mael, A.A.; Sakuma, R.; Ohmine, S.; Marcato, P.; Lee, P.W.; Ikeda, Y. The RAS/Raf1/MEK/ERK Signaling Pathway Facilitates VSV-mediated Oncolysis: Implication for the Defective Interferon Response in Cancer Cells. *Mol. Ther.* **2007**, *15*, 1531–1536. [CrossRef] [PubMed]

184. Li, Q.; Tainsky, M.A. Epigenetic Silencing of IRF7 and/or IRF5 in Lung Cancer Cells Leads to Increased Sensitivity to Oncolytic Viruses. *PLoS ONE* **2011**, *6*, e28683. [CrossRef] [PubMed]

185. Stojdl, D.F.; Lichty, B.D.; Tenoever, B.R.; Paterson, J.M.; Power, A.T.; Knowles, S.; Marius, R.; Reynard, J.; Poliquin, L.; Atkins, H.; et al. VSV strains with defects in their ability to shutdown innate immunity are potent systemic anti-cancer agents. *Cancer Cell* **2003**, *4*, 263–275. [CrossRef]

186. Petersen, J.M.; Her, L.-S.; Varvel, V.; Lund, E.; Dahlberg, J.E. The Matrix Protein of Vesicular Stomatitis Virus Inhibits Nucleocytoplasmic Transport When It Is in the Nucleus and Associated with Nuclear Pore Complexes. *Mol. Cell. Biol.* **2000**, *20*, 8590–8601. [CrossRef]

187. Hamacher, R.; Schmid, R.M.; Saur, D.; Schneider, G. Apoptotic pathways in pancreatic ductal adenocarcinoma. *Mol. Cancer* **2008**, *7*, 64. [CrossRef] [PubMed]

188. Cary, Z.D.; Willingham, M.C.; Lyles, D.S. Oncolytic Vesicular Stomatitis Virus Induces Apoptosis in U87 Glioblastoma Cells by a Type II Death Receptor Mechanism and Induces Cell Death and Tumor Clearance In Vivo. *J. Virol.* **2011**, *85*, 5708–5717. [CrossRef] [PubMed]

189. Gaddy, D.F.; Lyles, D.S. Oncolytic Vesicular Stomatitis Virus Induces Apoptosis via Signaling through PKR, Fas, and Daxx. *J. Virol.* **2006**, *81*, 2792–2804. [CrossRef] [PubMed]

190. Sharif-Askari, E.; Nakhaei, P.; Oliere, S.; Tumilasci, V.; Hernández, E.; Wilkinson, P.; Lin, R.; Bell, J.; Hiscott, J. Bax-dependent mitochondrial membrane permeabilization enhances IRF3-mediated innate immune response during VSV infection. *Virology* **2007**, *365*, 20–33. [CrossRef] [PubMed]

191. Conti, C.; Mastromarino, P.; Riccioli, A.; Orsi, N. Electrostatic interactions in the early events of VSV infection. *Res. Virol.* **1991**, *142*, 17–24. [CrossRef]

192. Bailey, C.A.; Miller, D.K.; Lenard, J. Effects of DEAE-dextran on infection and hemolysis by VSV. Evidence that nonspecific electrostatic interactions mediate effective binding of VSV to cells. *Virology* **1984**, *133*, 111–118. [CrossRef]

193. Arulanandam, R.; Batenchuk, C.; Varette, O.; Zakaria, C.; Garcia, V.; Forbes, N.E.; Davis, C.; Krishnan, R.; Karmacharya, R.; Cox, J.; et al. Microtubule disruption synergizes with oncolytic virotherapy by inhibiting interferon translation and potentiating bystander killing. *Nat. Commun.* **2015**, *6*, 6410. [CrossRef] [PubMed]

194. Sanjuán, R.; Grdzelishvili, V.Z. Evolution of oncolytic viruses. *Curr. Opin. Virol.* **2015**, *13*, 1–5. [CrossRef]

195. Olagnier, D.; Lababidi, R.R.; Hadj, S.B.; Sze, A.; Liu, Y.; Naidu, S.D.; Ferrari, M.; Jiang, Y.; Chiang, C.; Beljanski, V.; et al. Activation of Nrf2 Signaling Augments Vesicular Stomatitis Virus Oncolysis via Autophagy-Driven Suppression of Antiviral Immunity. *Mol. Ther.* **2017**, *25*, 1900–1916. [CrossRef]

196. Collisson, E.A.; Sadanandam, A.; Olson, P.; Gibb, W.J.; Truitt, M.; Gu, S.; Cooc, J.; Weinkle, J.; Kim, G.E.; Jakkula, L.; et al. Subtypes of pancreatic ductal adenocarcinoma and their differing responses to therapy. *Nat. Med.* **2011**, *17*, 500–503. [CrossRef]

197. Bailey, P.; Initiative, A.P.C.G.; Chang, D.K.; Nones, K.; Johns, A.L.; Patch, A.-M.; Gingras, M.-C.; Miller, D.K.; Christ, A.N.; Bruxner, T.J.C.; et al. Genomic analyses identify molecular subtypes of pancreatic cancer. *Nature* **2016**, *531*, 47–52. [CrossRef]

198. Golan, T.; Hammel, P.; Reni, M.; Van Cutsem, E.; Macarulla, T.; Hall, M.J.; Park, J.-O.; Hochhauser, D.; Arnold, D.; Oh, D.-Y.; et al. Maintenance Olaparib for Germline BRCA-Mutated Metastatic Pancreatic Cancer. *N. Engl. J. Med.* **2019**, *381*, 317–327. [CrossRef]

199. Passaro, C.; Volpe, M.; Botta, G.; Scamardella, E.; Perruolo, G.; Gillespie, D.; Libertini, S.; Portella, G. PARP inhibitor olaparib increases the oncolytic activity of dl922-947 in In Vitro and In Vivo model of anaplastic thyroid carcinoma. *Mol. Oncol.* **2014**, *9*, 78–92. [CrossRef]

200. Ning, J.; Wakimoto, H.; Peters, C.; Martuza, R.L.; Rabkin, S.D. Rad51 Degradation: Role in Oncolytic Virus—Poly(ADP-Ribose) Polymerase Inhibitor Combination Therapy in Glioblastoma. *J. Natl. Cancer Inst.* **2017**, *109*, 1–13. [CrossRef]

201. Wong, P.-P.; Demircioglu, F.; Ghazaly, E.; Alrawashdeh, W.; Stratford, M.R.; Scudamore, C.L.; Cereser, B.; Crnogorac-Jurcevic, T.; McDonald, S.; Elia, G.; et al. Dual-Action Combination Therapy Enhances Angiogenesis while Reducing Tumor Growth and Spread. *Cancer Cell* **2015**, *27*, 123–137. [CrossRef] [PubMed]

202. Chauhan, V.P.; Martin, J.D.; Liu, H.; Lacorre, D.A.; Jain, S.R.; Kozin, S.V.; Stylianopoulos, T.; Mousa, A.S.; Han, X.; Adstamongkonkul, P.; et al. Angiotensin inhibition enhances drug delivery and potentiates chemotherapy by decompressing tumour blood vessels. *Nat. Commun.* **2013**, *4*, 2516. [CrossRef]

203. Vennin, C.; Chin, V.T.; Warren, S.C.; Lucas, M.C.; Herrmann, D.; Magenau, A.; Melenec, P.; Walters, S.N.; Del Monte-Nieto, G.; Conway, J.R.W.; et al. Transient tissue priming via ROCK inhibition uncouples pancreatic cancer progression, sensitivity to chemotherapy, and metastasis. *Sci. Transl. Med.* **2017**, *9*, eaai8504. [CrossRef]

204. Ilkow, C.S.; Marguerie, M.; Batenchuk, C.; Mayer, J.; Ben Neriah, D.; Cousineau, S.; Falls, T.; Jennings, V.A.; Boileau, M.; Bellamy, D.; et al. Reciprocal cellular cross-talk within the tumor microenvironment promotes oncolytic virus activity. *Nat. Med.* **2015**, *21*, 530–536. [CrossRef] [PubMed]

205. Morrison, A.H.; Byrne, K.T.; Vonderheide, R.H. Immunotherapy and Prevention of Pancreatic Cancer. *Trends Cancer* **2018**, *4*, 418–428. [CrossRef]

206. Bear, A.S.; Vonderheide, R.H.; O'Hara, M.H. Challenges and Opportunities for Pancreatic Cancer Immunotherapy. *Cancer Cell* **2020**, *38*, 788–802. [CrossRef]

207. Wei, M.-Y.; Shi, S.; Liang, C.; Meng, Q.C.; Hua, J.; Zhang, Y.-Y.; Liu, J.; Bo, Z.; Xu, J.; Yu, X.J. The microbiota and microbiome in pancreatic cancer: More influential than expected. *Mol. Cancer* **2019**, *18*, 97. [CrossRef] [PubMed]

Review

Combining Oncolytic Viruses and Small Molecule Therapeutics: Mutual Benefits

Bart Spiesschaert [1,2,3,4], Katharina Angerer [1,2], John Park [4] and Guido Wollmann [1,2,*]

1   Christian Doppler Laboratory for Viral Immunotherapy of Cancer, Medical University Innsbruck, 6020 Innsbruck, Austria; bart.spiesschaert@boehringer-ingelheim.com (B.S.); Katharina.Angerer@i-med.ac.at (K.A.)
2   Institute of Virology, Medical University Innsbruck, 6020 Innsbruck, Austria
3   ViraTherapeutics GmbH, 6063 Rum, Austria
4   Boehringer Ingelheim Pharma GmbH & Co. KG, 88397 Biberach a.d. Riss, Germany; john.park@boehringer-ingelheim.com
*   Correspondence: guido.wollmann@i-med.ac.at

Simple Summary: Oncolytic viruses can be a potent tool in the fight against cancer. However, in clinical settings their ability to replicate in and kill tumors is often limited. Combinations with specific small molecule compounds can address some of these limitations and help oncolytic viruses reach their full potential. The aim of this review is to provide an overview of the different types of small molecules with which oncolytic viruses can achieve therapeutic synergy. We focus on the underlying mechanisms in three functional areas: combinations that increase viral replication, enhance tumor cell killing and improve antitumor immune responses.

Abstract: The focus of treating cancer with oncolytic viruses (OVs) has increasingly shifted towards achieving efficacy through the induction and augmentation of an antitumor immune response. However, innate antiviral responses can limit the activity of many OVs within the tumor and several immunosuppressive factors can hamper any subsequent antitumor immune responses. In recent decades, numerous small molecule compounds that either inhibit the immunosuppressive features of tumor cells or antagonize antiviral immunity have been developed and tested for. Here we comprehensively review small molecule compounds that can achieve therapeutic synergy with OVs. We also elaborate on the mechanisms by which these treatments elicit anti-tumor effects as monotherapies and how these complement OV treatment.

Keywords: oncolytic virus; small molecule; cancer immune therapy; combination therapy; cancer therapy; immunotherapy

Citation: Spiesschaert, B.; Angerer, K.; Park, J.; Wollmann, G. Combining Oncolytic Viruses and Small Molecule Therapeutics: Mutual Benefits. *Cancers* **2021**, *13*, 3386. https://doi.org/10.3390/cancers13143386

Academic Editors: Antonio Marchini, Carolina S. Ilkow and Alan Melcher

Received: 1 May 2021
Accepted: 1 July 2021
Published: 6 July 2021

Publisher's Note: MDPI stays neutral with regard to jurisdictional claims in published maps and institutional affiliations.

1. Introduction

In the course of oncogenic transformation and progression, tumor cells acquire distinct features that have been termed hallmarks of cancer [1,2]. Some of these aberrations form the base for the tumor-preferential infection and propagation of natural or recombinant oncolytic viruses (OVs) [3]. Evasion of growth suppressive mechanisms, continuous proliferative signaling, unrestricted replication machinery and the evasion of innate and adaptive immune control constitute characteristics that can be exploited by OVs. In general, naturally occurring or genetically engineered virotherapy candidate viruses share the core features of tumor-preferential infection, replication, and lysis. Beyond that, they display the diversity of viruses on multiple levels: human pathogen-derived versus animal viruses, DNA versus RNA genome, enveloped versus non-enveloped, nuclear versus cytosolic replication cycle, etc. [4]. Herpes simplex virus (HSV) and adenovirus (AdV) are human pathogenic DNA viruses that have been developed for three decades as oncolytic agents with a plethora of modified variants being tested in preclinical and clinical settings. This resulted in the first regulatory approvals of H101, a genetically engineered adenovirus, in 2005 in China and talimogene laherparepvec (T-VEC), a recombinant attenuated HSV-1 with a transgene encoding for granulocyte-macrophage colony-stimulating factor (GM-CSF), in 2015 in the USA and Europe [5]. Development of oncolytic HSV and AdV variants has continued though with a strong focus on next generation "armed" OVs expressing a multitude of immune modulatory transgenes. Another clinically advanced oncolytic platform is based on the vaccinia virus (VV), a large DNA virus encoding about 200 genes with an exclusive cytosolic replication cycle. Its ability to accommodate up to 40 kb of transgene DNA make VV a prime platform for arming with immune modulatory cargo genes [6]. A related member of the poxvirus family, myxoma virus, has also extensively been explored as an oncolytic agent in pre-clinical settings [7]. H1, a small rat parvovirus, completes the list of the major DNA-based oncolytic agents. This natural onco-preference is in large part based on a dependency on proliferating cells and signaling pathway aberrations [8]. Reovirus, a natural occurring human virus with double stranded RNA genome, is usually not associated with disease in adults and its onco-tropism was originally thought to be linked to RAS transformation in cancer cells, although recent data suggest a more multifactorial relationship [9]. The Edmonston vaccine strain of measles virus, a negative strand RNA paramyxovirus, displays a certain natural onco-selectivity in part due to frequent overexpression of its receptor, CD46, in a range of different cancer types [10]. Newcastle disease virus, an avian paramyxovirus without causing known human disease, harbors a natural onco-selectivity due to interaction with anti-apoptotic proteins and its dependence on a defective antiviral make-up frequently observed in cancer cells [11]. Vesicular stomatitis virus (VSV), a negative strand RNA virus of the rhabdoviridae family, causes mild disease in livestock with clinical symptoms rarely reported in humans. Its ubiquitous receptor entry translates to a pan-tropism for a very broad range of tumor types, but also holds the potential for some neuro-toxicity once it can access the brain. Consequently, VSV development has long been driven by attenuation strategies [12]. As with several other RNA viruses, the primary mode of onco-selectivity is based on reduced antiviral defense mechanisms in certain tumors [13]. In recent years, a large number of VSV variants armed with immunomodulatory transgenes has been tested in preclinical settings and in early phase clinical testing [14]. With few exceptions, most OVs are rather sensitive to innate antiviral control. This increases their safety aspect towards normal cells while letting them take advantage of impaired innate immune signaling in tumors [13]. These OVs are therefore also considerably better suited to be combined with small molecules that counter innate antiviral immunity. During early OV developments, the paradigm was that the efficacy of OV treatment correlated to virus replication. Viral spread throughout the tumor, and subsequent OV-mediated cancer cell lysis, were thought to be the main drivers of OV therapy [15]. According to this thinking, OVs were initially combined with immunosuppressive small molecule compounds in order to limit the antiviral immune response and allow OVs to replicate to higher titers within the treated tumors [16,17]. The different mechanisms and compounds that modulate the innate antiviral immunity are discussed in detail below. Such approaches have yielded promising results mostly in preclinical settings [18]. However, the modes of action by which OVs can be therapeutic are more complex in immunocompetent patients and the immune activating potential of OVs has increasingly dominated the discussion [19–21]. OV treatments are now considered potent partners for immunotherapies [22]. Few treatment modalities inherently hold the potential to simultaneously induce immunogenic cell death (ICD), stimulate innate and adaptive immune responses, enhance T cell infiltration and repolarize an immune-suppressive tumor microenvironment (TME) [23–25]. Immunogenic cell death is associated with the induction and release of pro-inflammatory cytokines and danger-associated molecular patterns (DAMPs) [26]. DAMPs are especially expressed when infected cells die in an immunogenic manner, such as necroptosis. Enhancing these modes of cell death through the combination with tumor cell death enhancing (TCDE) small molecule compounds has therefore become a central focus [27,28] and is also discussed in detail below. The presence of virus related pathogen associated molecular patterns (PAMPs) and DAMPs subsequently facilitates the attraction of immune cells which contribute to the immune-stimulatory state by producing additional inflammatory cytokines [29]. This can eventually shift the immunosuppressive TME allowing a successful antitumor immune response to occur [30,31]. Still, even after induction of an antitumor immune response, the continuous reshaping of the TME at later stages constitutes further challenges [26]. For example, OV treatment commonly induces the expression of programmed cell death ligand 1 (PD-L1). However, this can be successfully countered by immune checkpoint inhibiting antibodies [32]. Small molecule checkpoint inhibitors could contribute to OV treatment in a similar fashion [33]. Other components of the TME, such as tumor growth factor (TGF)-β, epigenetic major histocompatibility complex (MHC) repression, cytotoxic T-lymphocyte-associated Protein 4 (CTLA-4), T-cell immunoglobulin and mucin-domain containing-3 (TIM-3), etc.), regulatory T-cells (Treg), myeloid-derived suppressor cells (MDSC), and M2 tumor associated macrophages (TAMs) can also contribute to an immunosuppressive therapy-resistant state. Some of these factors can be targeted by small molecule therapeutics [34], which will also be discussed in a separate section below. As we show in the following, the different aspects of multimodal OV treatment can be improved by a vast array of small molecule compounds, and a future impact on improving the clinical outcome of such combinations is conceivable.

2. Combinations Affecting Viral Propagation in Tumor Cells

The selectivity of various oncolytic viruses largely depends on defects in the tumor cell's innate ability to fend off viral infections [35]. However, the initial assumption that an impaired interferon (IFN) response is a common feature shared by many tumors [36] may not reflect the clinical reality of solid cancers' heterogeneity [37]. Some tumors, such as pancreas cancer, may even display an upregulated antiviral state leading to primary resistance [38]. A constitutive interferon pathway activation was also described as a main determinant for oncolytic measles virus activity in a human glioblastoma specimen [39]. On the other hand, tumors induced by oncoviruses, such as HPV-associated cervical or head and neck cancers, tend to frequently display strongly impaired antiviral innate responses [40]. However, in light of missing systematic assessments of a large range of tumor types, general conclusions as to what cancer types are more antivirally active and which are not remain to be drawn. Although most viruses have evolved to express proteins that counter antiviral measures [41], engineering of many oncolytic viruses were aimed at abolishing exactly those viral counter measures, generating OVs with a heightened IFN sensitivity [37]. Cornerstones of the antiviral innate immune response are type I (and to a lesser extend type III) interferons [42]. Both IFN types converge in their signaling and induce transcriptional responses through the Janus kinase signal transducers and activators of transcription (JAK/STAT) pathway [43]. Their signaling is associated with downstream expression of interferon stimulated genes (ISGs) which act as antiviral effector proteins countering viral replication. OV replication is impaired when these pathways are still intact in the treated tumor cells [44]. In the following, we will discuss various compound classes involved in inhibiting antiviral signaling pathways and which hold the potential to either enhance replication of OVs or even address OV resistance in cancer cells.

2.1. JAK-STAT Signaling Inhibition

Inhibitors of Janus kinases (JAK), such as JAK inhibitor I (a pan-JAK inhibitor) or ruxolitinib (a specific JAK1/2 inhibitor) (Figure 1), were able to rescue the replication of VSV in several human pancreatic ductal adenocarcinoma (PDA) cells that were otherwise resistant due to constitutive high-level expression of certain interferon stimulated genes (ISGs) [38,45,46]. This effect was improved even further when Polybrene or DEAE-dextran were additionally added, improving VSV attachment and entry and allowing more cells to be infected [47]. A similar effect was seen for refractory human head and neck squamous cell carcinoma (HNSCC) cell lines which owed their VSV resistance to the constitutive expression of a different set of ISGs. Here, JAK inhibitor I and ruxolitinib were also successful in rescuing virus replication with a 100- to 1000-fold increase in yield. Interestingly, other innate immune small molecule compounds, such as histone deacetylase inhibitors (HDI; LBH589), phosphoinositide 3-kinase (PI3K) inhibitors (GDC-0941, LY294002), mammalian target of rapamycin complex 1 (mTORC1) inhibitors (rapamycin) or STAT3 inhibitor VII were not effective [48]. Combination therapy with ruxolitinib and VSV-IFNβ also enhanced viral replication and oncolysis in several non-small cell lung cancer (NSCLC) cell lines [49]. However, several of these compounds were effective in rescuing OV replication in other tumor cell types, as discussed in the sections below underlining the heterogeneity in mechanisms among different tumor cells by which synergy with OVs can occur. In melanoma, mutations in the JAK1/2 signaling pathway as well as JAK1/2 inhibition increase sensitivity to VSV-dM51 [50]. The dual inhibitor of JAK1 and IκB kinase (IKK), TPCA-1 was also shown to improve HSV replication of malignant peripheral nerve sheath tumor (MPNST) cells [51]. OVs that replicate in the cytoplasm, such as RNA viruses and poxviruses, can also trigger direct antiviral effector responses that can hamper their replication and subsequent oncolytic effects. Viral RNA activates the cytosolic PKR by inducing dimerization and subsequent auto-phosphorylation reactions. The protein kinase R (PKR) pathway leads to a stress response by activating other pathways such as the interconnected nuclear factor κ-light chain enhancer of activated B cells (NF-κB) & c-Jun N-terminal kinase (JNK) pathways (Figure 2) [52–54]. JNK are kinases involved in a diverse set of cellular functions, ranging from cell death, survival and proliferation to innate immunity [54]. Specifically, JNK are essential for the expression regulation of many immune mediator genes, such as cytokines (e.g., interleukins (ILs) IL-2, IL-4, IL-8, IL-18, IFN-γ, granulocyte-macrophage colony-stimulating factor (GM-CSF), C-C motif chemokine ligand 5 (CCL5), tumor necrosis factor α (TNF-α)) [55–59] and adhesion molecules (ICAM-1) [53]. While JNK inhibition has been reported to act antivirally on encephalomyocarditis virus, rotavirus and HSV [60–62], a virus promoting effect was seen for vaccinia virus. Here, murine embryonic fibroblasts devoid of JNK showed a significant increase in titer. In line with these results, an increase of apoptosis was seen when wildtype murine embryonic fibroblast cells were co-treated with the JNK-specific inhibitor SP600125 [55,63]. This suggests that JNK inhibition, at least under very specific conditions, can be beneficial for OV therapy [63].

2.2. Inhibition of NF-kB Signaling

Nuclear factor (NF)-κB and inhibitor of NF-κB kinase (IKK) proteins regulate many cellular responses to stimuli, such as innate and adaptive immunity, cell death, and inflammation [64]. NF-κB and IKK therefore play key roles in regulating the innate immune response against OVs. Indeed, two types of compounds enhance OV replication through very distinct mechanisms at different stages of NF-κB-mediated transcription [65]. For instance, fumaric and maleic acid esters, such as dimethyl fumarate (DMF), block the nuclear translocation of NF-κB and have been shown to improve replication of several OVs and subsequent therapeutic outcomes by inhibiting type I IFN [66]. Another point of intervention is in the nucleus after NF-κB has already bound DNA [67]. At this point triptolide blocks transcription, leading to an increase of VSV replication in several VSV-resistant tumor cell types (Figure 2) [68]. Before NF-κB can facilitate transcription of innate immune genes it has to be released from the IκB kinase β (IKKβ) complex. The activation of IKKβ, by the phosphorylation of IκBα and its subsequent proteasomal degradation, allows NF-κB to relocate to the nucleus [64]. Blocking IKKβ can be therapeutically exploited since NF-κB is overexpressed in many cancer types [69]. Consequently, inhibiting IKKβ shows much promise for synergizing with OVs (Figure 2). This would be especially advantageous for OVs, such as VSV and NDV, that rely on defective innate immunity for their onco-selectivity [70]. This was confirmed in studies on malignant peripheral nerve sheath tumor cells and some pancreatic ductal adenocarcinoma cell lines that showed resistance to oncolytic HSV and VSV infection, respectively. In combination with the IKKβ inhibitor TPCA-1, this resistance was overcome and productive infection was achieved [46,51].

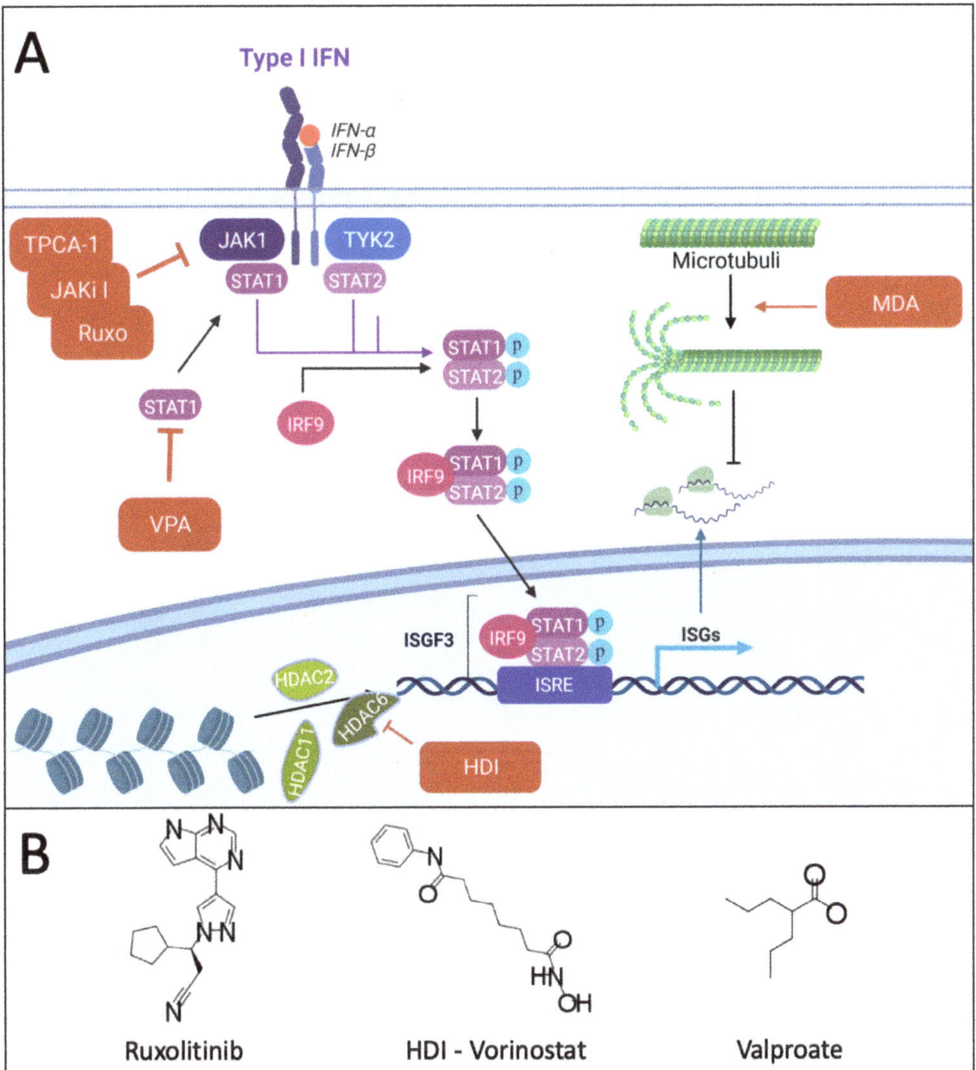

Figure 1. JAK/STAT signaling inhibition for the improvement of OV efficacy. (**A**) IFN binding with its receptor can activate JAK1 and TYK2. This in turn facilitates the phosphorylation of the docking sites of STAT1 and STAT2. Following phosphorylation, both STATs associate with IRF6 to form the transcriptional regulation ISG3. ISG3 trans-locates to the nucleus where it mediates the transcription of ISG mRNAs. The appropriate DNA strains are made accessible for ISGF3 by different histone deacetylases. These mRNAs are in turn transported over microtubules in order to be translated. Targeting these pathways by means of different small molecule inhibitors (red annotated squares) allows OV replication to proceed for longer, resulting in increased viral spread and potentially efficacy. See the main text for more details. Created with biorender.com. (**B**) Selected chemical structures of compounds depicted in panel A. All structures throughout were drawn using MarvinSketch (ChemAxon) from publicly available information. Abbreviations: JAK, Janus kinase; STAT, signal transducers and activators of transcription; IRF9, Interferon regulatory factor 9; ISGF3, Interferon-stimulated gene factor 3; HDAC, histone deacetylase; ISRE, Interferon-sensitive response element; MDA, microtubule destabilizing agent; VPA, Valproate.

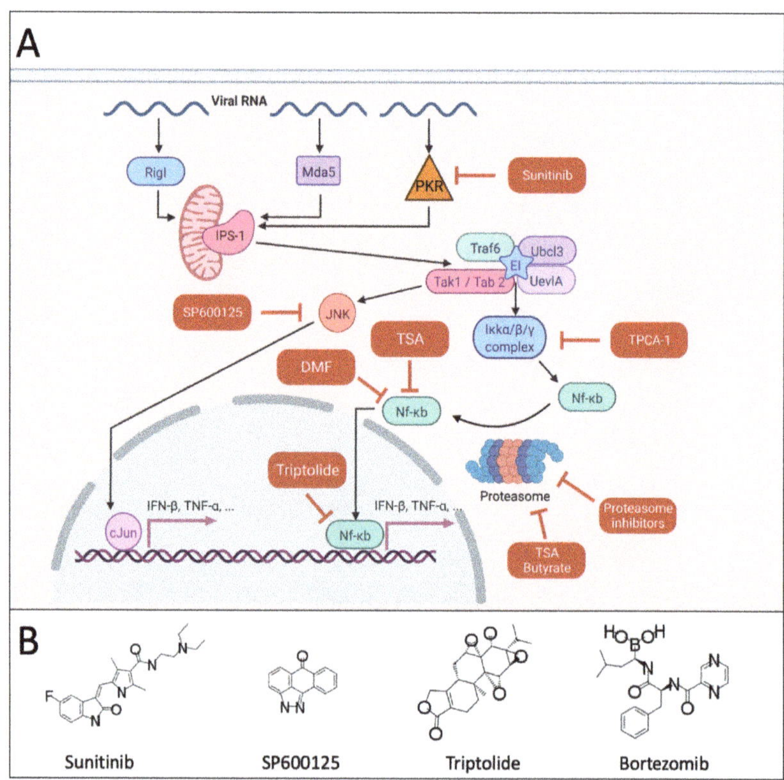

Figure 2. Compound classes that inhibit of NF-kB signaling and synergize with OV treatment. (**A**) Virus replication results in the production of cytosolic DNA and single- and double-stranded RNA. This triggers multiple signaling cascades, including the recruitment of RIG-I and Mda5 to the adaptor IPS-1 on the membrane of the mitochondria. This in turn leads to kinase activation through TRAF family members. More specifically, this activates the IKK complex, which phosphorylates IκB proteins. Phosphorylation of IκB leads to its ubiquitination and proteasomal degradation, freeing NF-κB complexes for transcription induction. TRAF6 signaling also leads to JNK activation. Activated JNK trans-locates to the nucleus and activates c-Jun and other target transcription factors. These transcription factors, such as cJun and NF-κB lead to the transcription of numerous proteins involved in innate immunity and cells death, including IFN-β. Interfering with the different steps of signaling pathways using different classes of compounds (red annotated red squares) have resulted in increased viral replication and subsequent efficacy. See the main text for more details. Created with biorender.com. (**B**) Selected chemical structures of compounds depicted in panel A. All structures throughout were drawn using MarvinSketch (ChemAxon) from publicly available information. Abbreviations: TRAF, TNF Receptor Associated Factor; JNK, c-Jun N-terminal kinase; Atf2, Activating transcription factor 2; IPS-1, interferon-β promoter stimulator 1; TSA, Trichostatin A; DMF, dimethyl fumarate; RigI, retinoic acid-inducible gene I; mda5, melanoma differentiation-associated protein 5; PRK, protein kinase R; Ubcl3, ubiquitin-conjugating enzyme 13; ubiquitin-conjugating enzyme E2 variant 1; Tak1, transforming growth factor-6-activated kinase 1; IKK, IκB kinase β.

2.3. PI3K/AKT/mTOR Pathway Antagonists

Important for cell survival and growth, the phosphoinositide 3-kinase (PI3K)/Ak strain transforming (AKT)/mTOR pathway is also crucially involved in the induction of type 1 interferons (Figure 3) [71]. It is commonly activated in numerous types of cancer [72] via mutations or amplification of genes encoding receptor tyrosine kinases, subunits of

PI3K, AKT or activating isoforms of rat sarcoma (Ras) [73]. The first agents, targeting the PI3K pathway with the specific purpose of treating cancer, were analogues of rapamycin, namely everolimus (RAD 001) and temsirolimus [73]. Hence, inhibition of mTOR is expected to augment the oncolytic activity particularly of those viruses depending on impaired antiviral responses within a tumor cell. The macrolide compound rapamycin is a prototypical inhibitor of the serine/threonine protein kinase mTOR. Combining rapamycin with the highly IFN-sensitive VSV-mutant strain (VSVΔM51) led to significant increase of the oncolytic effect [74]. In addition other oncolytic RNA viruses, such as NDV and reovirus, showed improved oncolytic effect in mice when co-treated with rapamycin [75,76]. Oncolytic DNA viruses also benefit from co-treatment with rapamycin. The yield and dissemination of an HSV-derived oncolytic virus was markedly increased in semi-permissive tumor cell lines [77]. An oncolytic vaccinia virus (VACV) only achieved complete remission in in vivo models when it was combined with rapamycin [78]. A key restriction factor for myxoma virus in human cells is its dependence on AKT activation [79]. By inhibiting mTORC1, AKT becomes hyperactivated through the release from the negative feedback loop between ribosomal protein S6 kinase beta-1 (S6K1) and insulin receptor substrate 1 (IRS-1) [80]. This subsequently enhances myxoma virus replication which also translates to increased survival in vivo [81–83]. mTOR inhibition can also lead to a decrease in phosphorylation of the effector proteins, eukaryotic translation initiation factor 4E-binding protein 1 (4E-BPs) and S6Ks, which are essential for type I interferon (IFN) production (Figure 3) [84,85]. This inhibition of the type I interferon response also contributes to a more pronounced replication of myxoma virus in vitro and increased efficacy in vivo [86]. Everolimus was tested in combination with an oncolytic adenovirus. Even though, in vitro, RAD001 seemed to interfere with the viral replication, potent anti-glioma effects were seen in vivo. This was presumably due to the induction of autophagic cell death [87,88]. Increased efficacy through modulation of autophagy in similar settings is also described for other OVs [75,76]. The hyperactivation of AKT during mTORC1 inhibition might have benefits when combined with myxoma virus [81–83], but in other settings can have a negative effect on survival. In phosphate and tensin homolog (PTEN)-deficient glioblastoma patients, for instance, hyperactivation of AKT, following rapamycin treatment, was associated with more rapid onset of tumor progression [89]. The mTORC2 complex, which is insensitive to rapamycin and its analogues, activates AKT and has a distinct role in tumor maintenance and progression [90]. For OVs with a dependency on a weakened antiviral state within the tumor, mTORC2 antagonists that also inhibit mTORC1 would be a superior option. ATP-competitive mTOR kinase inhibitors (TKIs) achieve this by targeting the kinase domain of mTOR, thereby also blocking the activation feedback of PI3K/Akt signaling (Figure 3) [91]. Indeed, mTORC1/2 inhibitors, such as PP242, INK1341, INK128 or Torin1, were also able to increase HSV replication and oncolysis by altering eIF4E/4E-BPs expression [77]. Specific inhibitors, such as rapamycin and TKIs, are prone to trigger the development of secondary resistance after prolonged treatment [92]. Consequently, inhibitors were developed that target the same signaling pathway but at multiple sites. Dual PI3K/mTOR inhibitors, such as voxtalisib [93], target the p110α, β, and γ isoforms of PI3K as well as the ATP-binding sites of both mTORC1 and mTORC2, completely suppressing PI3K/Akt signaling [91]. Combinations with OVs have yet to be reported. BKM120, another pan-class PI3K inhibitor, targeting all four catalytic isoforms, in combination with oncolytic HSV-1, was effective in the treatment of Du145 prostate cancer sphere forming cells (PCSCs) [94]. Finally, the benefits of combining PI3K/Akt signaling blockade and OVs can also work in the opposite direction, demonstrated by the combination of an oncolytic HSV and PI3K/Akt inhibitors (LY294002, triciribine, GDC-0941, BEZ235). Here, treatment with the OV sensitized the tumor cells to the inhibitors through enhanced Akt activation [95,96]. Indirectly, PI3K inhibitors, more specifically PI3Kδ-selective inhibitors, could improve systemic OV delivery to tumors through attachment inhibition of systemic macrophages [97].

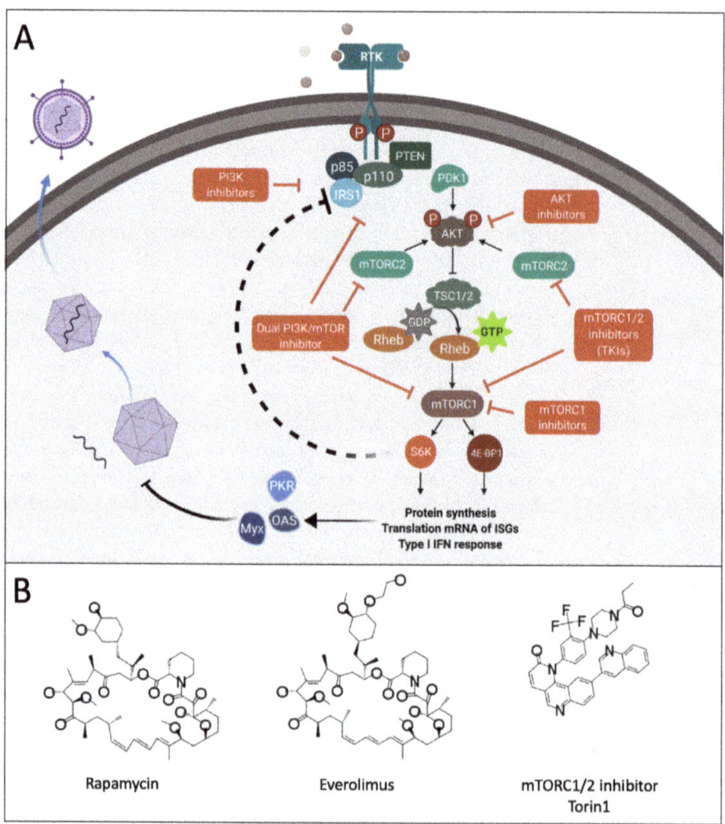

Figure 3. Overview of the PI3K(p85/p110)/AKT/mTOR pathway and small molecule compounds that target this pathway in synergy with OV therapy. (**A**) Activating (PI3K, AKT, PDK1, mTORC1 and mTORC2) and inhibiting proteins (PTEN, TSC1/2) of the signaling pathway are shown. PI3K consists of catalytic subunit p110 and the regulatory subunit p85. PI3K phosphorylates phosphatidylinositol bisphosphate, which in turn activates PDK1 and AKT. PTEN negatively regulates the activation of AKT, which can inhibit TSC1/2, a negative regulator of mTOR. Active mTOR phosphorylates S6K1 and 4EBP1 leading to increased translation and synthesis of, among others, ISGs [73]. Targeting this process by means of different small molecule inhibitors (red annotated squares) allows OV replication to proceed for longer, resulting in increased viral spread and efficacy. See the main text for more details. Created with biorender.com. (**B**) Selected chemical structures of compounds depicted in panel A. All structures throughout were drawn using MarvinSketch (ChemAxon) from publicly available information. Abbreviations: RTK, receptor tyrosine kinase; PDK1, phosphoinositide-dependent kinase 1; IRS1, insulin receptor substrate 1; PTEN, phosphatase and tensin homologue; mTOR, mammalian target of rapamycin. PKR, protein kinase R; Myx, GTP-binding protein MX; AOS, oligoadenylate synthetase; S6K, S6 kinase; 4E-BP1, Eukaryotic translation initiation factor 4E-binding protein 1; Rheb, Ras homolog enriched in brain; IRS1, insulin receptor substrate 1.

2.4. Proteasome Inhibitors

Another approach to indirectly inhibit NF-κB is by blocking proteasomal degradation. The rationale is that proteasome inhibition blocks NF-kBs release from the IKKβ complex (Figure 2). Indeed, the proteasome inhibitor bortezomib improved the viral replication of oncolytic HSV and also enhanced necroptotic tumor cell death through increased endoplasmatic reticulum (ER) stress and unfolded protein response (UPR) (Figure 4C) [98–100].

However, when bortezomib was combined with VSV, a reduction in replication and spread was seen in myeloma cells despite NF-κB activation being blocked. Interestingly, despite these antagonistic effects in vitro, co-treatment in vivo did improve the antitumor efficacy [101]. Similarly, another proteasome inhibitor PS-341 blocked the replication of VSV in human adenocarcinoma A549 cells [102] and infection with HSV strains. These seemingly contradictory studies make the combination of proteasome inhibitors and OVs a treatment option that needs to be further elucidated.

2.5. Tankyrase Inhibition

Resistance to PI3/AktT inhibitors is linked to Wnt/b-catenin signaling hyperactivation [103] and can be countered by the Wnt/tankyrase inhibitor NVP-TNKS656 [104]. Hence a direct synergy between tankyrase inhibitors (TNKSi) and OVs might be possible. Tankyrases play a role in the replication of different herpes viruses. The inhibition of tankyrase has been shown to promote replication of beta- (cytomegalovirus) and gamma-herpesvirus (Epstein-Barr virus), with the underlying mechanism via which this benefits the virus still to be elucidated [105,106]. In contrast, TNKS inhibition acts suppressive on the alpha-herpesvirus, HSV-1 [104]. However, direct combination regimens of TNKSi and OVs have not yet been published, but such studies might be merited.

2.6. Receptor Tyrosine Kinase Inhibitor

In the antiviral context, direct inhibition of PKR and Rnase was also achieved by another class of small molecule compounds. The ATP-competitive inhibitor of vascular endothelial growth factor (VEGF) and platelet derived growth factor (PDGF) receptors, sunitinib, was reported also to be a strong inhibitor for both PKR and RnaseL [107] (Figure 2). These compounds also have more direct impact on tumor growth through their negative regulation of tumor vascularization. Due to their broader mode of action this group of inhibitors can be referred to in more general terms as receptor tyrosine kinase inhibitors (RTKIs). These compounds proved to be very beneficial when combined with oncolytic VSV, leading to the elimination of prostate, breast, and kidney malignant tumors in mice [108]. Synergistic effects with RTKIs were also shown for vaccinia and reovirus in pancreatic neuroendocrine tumors and renal cell carcinoma, respectively [109,110], as well as for the combination with HSV in glioblastoma [111]. However, vaccine virus is also connected to the activation of the epidermal growth factor receptor (EGFR) pathway for their replication and spread. Here, simultaneous administration of RTKIs, such as imatinib and sorafenib, resulted in the inhibition of vaccinia virus replication [112,113]. Nonetheless, oncolytic vaccinia virotherapy, followed by sorafenib treatment, showed enhanced efficacy compared to either monotherapy. This is most likely due to OV-mediated sensitization of the tumor cells and tumor vasculature to VEGF/VEGFR inhibitors [112]. Part of these reported benefits are also achieved through modulation of the tumor microenvironment. When MC38 tumor bearing mice were pretreated with sunitinib, the anti-tumor response, induced by a tumor associated antigen (TAA)-armed virus, was markedly improved through a decrease in inhibitory regulatory T cells (Tregs) and myeloid-derived suppressor cells (MDSCs) after sunitinib treatment [114]. This adaptive immune modulation is achieved by interacting with RTKs expressed on regulatory immune cell populations, such as c-KIT and VEGFR-1 [115,116]. In a similar setting, the more broad-range RTK inhibitor cabozantinib also showed a more diverse and potent effect and immunomodulatory effects with additional expression of MHC-I molecules, ICAM-1, Fas, and calreticulin on tumor cells. Modulation of antigen expression is most likely to be facilitated by its hepatocyte growth factor receptor (MET) inhibition [117]. Another more specific EGFR inhibitor, erlotinib, also seems to enhance the oncolytic effect in some human pancreatic cancer cells through a similar mechanism for oncolytic HSV. Here, prolonged viral presence was reported [118]. On the other hand, in tumors, characterized by upregulated EGFR signaling, the synergism seemed predominantly driven by a concerted antiangiogenic effect [119].

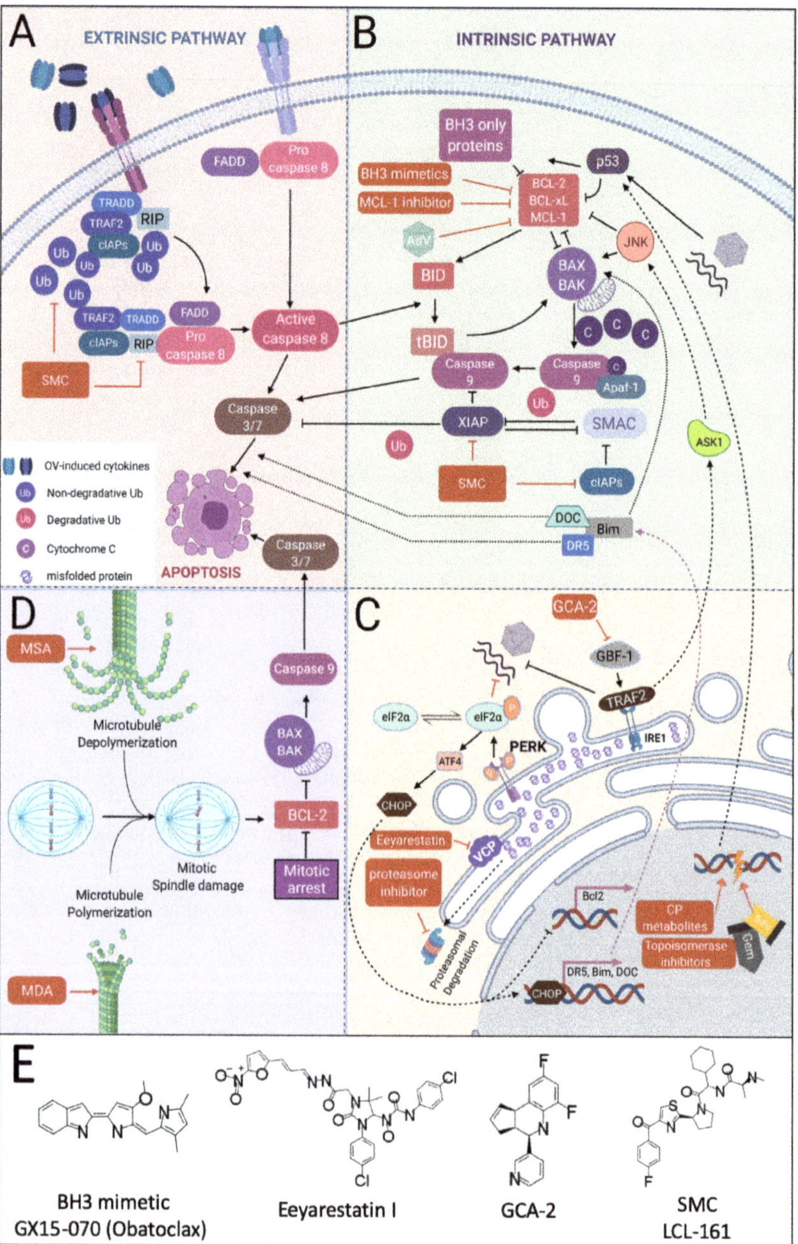

Figure 4. Increasing bystander killing of tumor cells by small molecules after OV treatment. (**A**) Cytokines produced in response to OV treatment of the tumor can activate the extrinsic pathway for apoptosis through binding with death receptors such as Fas and TNF-α receptor. Oligomerization of these receptors in turn facilitates the recruitment of adaptor proteins, for example, binding of Fas ligand with Fas recruits caspase-8 through the adaptor protein FADD. Cleaved caspase-8 can directly

activate caspase-3 and result in cell death. (**B**) Additionally, cleaved caspase-8 connects to the pathways of intrinsic apoptosis. This occurs when it cleaves Bid. Truncated Bid subsequently translocates to the mitochondria where it induces cytochrome release leading to activation of caspase-9 and caspase-3. This cytochrome c release is facilitated by the oligomerization of the pro-apoptotic Bax and Bak proteins at the outer mitochondria membrane. This process stands under the control of several proteins including Bcl-2, Bcl-xL and MCL-1. These pro-survival proteins in turn are inhibited by "BH3 only" proteins. (**C**) Intrinsic apoptosis can also be additionally stimulated through compounds that induce DNA damage, since this leads to p53 upregulation, resulting in indirectly Bax/Bak activation. ER stress signaling, caused by the accumulation of misfolded protein in the ER, can also facilitate this effect through ASK1 with the activation and subsequent translocation of JNK to the mitochondrial membrane. In addition, ER stress can also promote cell death through the activation of MAPK-mediated activation of eIF2α and ATF4 leading to the nuclear translocation of CHOP where it promotes transcription of pro-apoptotic genes. Apart from promoting cell death, eIF2α and TRAF2 also attenuates protein translation when misfolded protein accumulate in the ER. Since this is often the case during OV replication, the inhibition of these mechanisms can improve the efficacy of OV treatment. (**D**) Also the stabilizing or destabilizing of microtubules can trigger apoptosis. More specially, when cells are arrested G2/M phase, this can lead to the activation of intrinsic apoptosis. Targeting these pathways can improve oncolysis, tumor immunogenicity and viral replication depending on what aspect of cell death is targeted. Small molecule compounds targeting different stages of this process are presented by red annotated squares. See the main text for more information. Created with biorender.com. (**E**) Selected chemical structures of compounds depicted in panels A-D. All structures throughout were drawn using MarvinSketch (ChemAxon) from publicly available information. Abbreviations: TRADD, TNFR1-associated death domain protein; TRAF2, TNF receptor-associated factor 2; cIAP, cellular inhibitor of apoptosis; RIP, receptor interacting protein; FADD, fas-associated death domain; BH3, BCL-2 homology domain 3; SMC, Second mitochondria-derived activator of caspase mimetic compounds; Ub, ubiquitin; MCL-1; myeloid cell leukemia 1; XIAP, X-linked inhibitor of apoptosis protein; BID, BH3 interacting-domain death agonist; tBID, truncated Bid; AdV, adenovirus; JNK, c-Jun NH2-terminal kinase; BCL-xL, B-cell lymphoma, extra-large; BCL-2, B-cell lymphoma 2; BAX, BCL2 associated X; BAK, Bcl-2 homologous antagonist killer; Apaf-1, apoptotic protease activating factor-1; ASK1, Apoptosis signal-regulating kinase 1; CHOP, CCAAT-enhancer-binding protein homologous protein; DOX, downstream of CHOP; DR5, death receptor 5 (DR5); MDA, microtubule-destabilizing agents; MSA, microtubule-stabilizing agent; ATF4, Activating transcription factor 4; PERK, PRKR-like endoplasmic reticulum kinase; IRE1, inositol-requiring enzyme; CP, cyclophosphamide; Gem, gemcitabine; 5-Fu, fluorouracil; GBF-1, Golgi-specific brefeldin A-resistant guanine nucleotide exchange factor 1; GCA-2, GBF-1 inhibitor golgicide A; P, phosphorylated; VCP, valosin-containing protein; eIF2α, eukaryotic translation initiation factor 2α.

2.7. Histone Deacetylase Inhibitors (HDIs)

Transcription regulation requires deacetylase activity [120]. Histone deacetylase inhibitor compounds (HDIs) were found to rescue viral replication in resistant cells [120–122], which led to several investigations into the potential to augment OV replication. Interestingly, the blunting of the antiviral response (Figure 1) seemed to be limited to tumor cells, leaving the inhibition of viral replication in normal tissue intact [17]. However, an enhanced effect was also seen in proliferating endothelial cells [123]. The mechanism by which this specificity occurs remains unclear. It is suggested that this might be due to either an inherent preference of OVs for tumor cells or an enhanced susceptibility of tumor cells for these small molecules [124]. This enhanced susceptibility could be caused by the aberrant activity of histone deacetylases (HDACs), documented for several types of cancers [125–127]. Numerous HDI/OV combinations were tested in different tumor models showing the therapeutic benefit of blunting the innate antiviral response during OV treatment (Table 1). Some HDIs, such as butyrate and trichostatin A (TSA), can also indirectly inhibit the innate immune signaling through the inhibition of NF-κB activation by reducing proteasome subunit expression [128]. Apart from inhibiting the innate immune response, the adaptive immune response was also beneficially influenced with entinostat resulting in prolonged lymphopenia and depletion of Tregs [129–131]. Another HDI, valproate, was shown to suppress production of IFN-γ, and immune cell infiltration including NK cells, macrophages and lymphocytes, which helped promote virus growth but also has the potential to dampen anti-tumor immune responses [130,132–134]. This discrepancy in modulating the adaptive immune response can be related to the differences in HDAC targets of the different HDIs. Trichostatin A inhibits class I and II HDACS [135], Entinostat inhibits class I HDACs [136], whereas vorinostat and to a lesser extent valproate are pan-HDAC inhibitors [137,138]. Among the HDIs vorinostat is considered the more potent candidate for combination with OVs. However, more recent screenings have uncovered an even more potent compound to promote viral replication in less permissive tumors, namely viral sensitizer 1 and analog 28 (VSe1-28). This increased viral yield of VSV up to 2000 fold in vitro [124]. Further, reovirus has recently been described to synergize with HDAC inhibitor belinostat in both sensitive and belinostat-resistant T cell lymphoma cells [139].

Table 1. Synergy of HDIs and OVs.

HDI	OV	Tumor	References
entinostat	VSV	B16-F10, CT26, L363(MM), HT29, M14, PC3, SW620, 4T1	[17,129,130,140]
vorinostat	VSV	B16-F10	[130,141]
trichostatin	HSV, vaccinia	SAS, Ca9-22, HSC, HCT116, B26-F10, U87, SW480, HeLa	[123,142,143]
valproate	HSV, H1	U87, AGS1, U251, Gli36, HeLa	[132–134,144]
Scriptaid & LBH589	Adenovirus	Glioblastoma	[145]

In addition, the HDI trichostatin has been reported to increase expression of MHC-I molecules on the cell surface [146]. This is of particular interest for OVs used in a cancer vaccine setting, where downregulation of MHC-I expression can result in a relapse [147]. This increased MHC-I expression was further improved when trichostatin was combined with the hypomethylation agent, 5-azacytidine [146,148]. Beyond the interference with the innate antiviral activity and stimulating effects on the adaptive immune responses, HDIs have also been shown to enhance the direct tumor cell killing and replication of H1 parvovirus by increasing the acetylation of the viral NS-1 protein [144].

3. Combinations Enhancing Tumor Cell Death

Evasion of cell death is one of the main hallmarks of cancer. Apoptosis resistance develops frequently by either upregulation of anti-apoptotic elements or countering pro-apoptotic stimuli [149]. Though less prominent, other forms of programmed cell death can be similarly overridden, such as necroptosis [150]. Of note, some viruses employ analogous strategies to counter cell death as an archetypal cellular defense mechanism against viral infection, exemplified by the oncolytic HSV [151] and vaccinia virus [152]. Consequently, viral oncolysis alone rarely leads to widespread and complete cell death, opening the door for a combination approach with cell death sensitizers. Another aspect of such combinations links the aforementioned often limited intra-tumoral spread of OVs with the potential of bystander killing of uninfected cells [153]. Sensitizing a tumor mass with agents promoting cell death has been shown to significantly increase the kill zone of oncolytic viruses beyond the infected areas, yet still confined to the tumor [154]. The following section gives an overview of small molecule compounds that augment tumor cell killing and thus hold promise to synergize with oncolytic virotherapy.

3.1. ER Stress Inducers

One approach to promote tumor cell death is by amplifying ER stress. When cells synthesize secretory proteins in amounts that exceed the processing machinery, proteins are accumulated in the ER. Because this setting is linked to cells with high protein synthesis levels such as cancer cells and virally infected cells [155,156], OV-infected tumor cells would be particularly sensitive to disruption of ER homeostasis. The protein accumulation triggers the unfolded protein response (UPR) which tries to alleviate the ER by increasing ER chaperone gene transcription, lowering protein synthesis, and, if all else fails, inducing cell death (Figure 4C) [157]. Inhibiting these adaptive UPR measures has been studied in combination with the oncolytic M1- and adenovirus using the valosin-containing protein (VCP) inhibitor Eeyarestatin I and the Golgi-specific brefeldin A-resistant guanine nucleotide exchange factor 1 (GBF-1) inhibitor golgicide A (GCA-2), respectively. These combinations resulted in the significantly enhanced anticancer efficacy of the OV treatment [158,159]. The fine balance between homeostasis and apoptotic induction by the UPRER, now requires more mechanistic knowledge of virus interactions with the UPRER and drug synergy experiments, before this field is ripe for clinical applications [160]. Indirect effects of ER stress inducers, such as thapsigargin (Tg) and ionomycin (Im), can also enhance the activity of oncolytic adenoviruses through an alteration in Ca2+ flux and protein kinase C signaling [161].

3.2. Analogues of DNA Building Blocks

Pyrimidine analogues, such as Gemcitabine and 5-fluorouracil, are common chemotherapeutic compounds used for treating various types of malignancies. By interfering with DNA replication these antimetabolites induce inhibition of DNA synthesis with subsequent p53 upregulation, which ultimately can lead to cell death (Figure 4C) [162]. Naturally, these cytotoxic compounds combine well with several OVs [163–169]. However, these antimetabolites can also induce senescence of tumor cells which can regain proliferative activity after treatment cessation [170]. Here certain OVs, like oncolytic measles virus, have been shown to contribute to eliminating these senescent cells, thereby avoiding relapse [167]. Specific pyrimidine analogues can also have immune modulating effects. These have been suggested to positively affect the antitumor immune response over the antiviral one [166].

3.3. Antagonizing Inhibitors of Apoptosis (IAPs)

One major barrier to effective OV therapy is virus-induced expression of type I IFN and nuclear factor kappa B (NF-κB)-responsive cytokines, which can orchestrate an antiviral state in tumors. On the other hand, the subsequently produced cytokines (TNF-α, Fas ligand (FasL), TNF-related apoptosis-inducing ligand (TRAIL), etc.) can also be exploited to induce tumor cell killing beyond the zone of initial infection, facilitated via co-treatment with a number of different pharmaceutical agents, such as SMAC-mimetic compounds (SMCs) [154,171,172] and B cell lymphoma-2 (BCL-2) homology domain 3 (BH3) mimetics [173,174]. Of note, tumor cells are often more sensitive to these chemical compounds than normal cells since NF-κB signaling is frequently constitutively activated [175], leading to elevated expression of proteins participating in cell death pathways [176].

The second mitochondria-derived activator of caspase (SMAC) is a pro-apoptotic factor released from the mitochondria during the process of cell death. Cytosolic SMAC can potentiate the activity of different caspases by inhibiting X-linked inhibitor of apoptosis protein (XIAP) and cellular inhibitors of apoptosis (cIAPs) (Figure 4B), which otherwise antagonize caspase cleavage [177]. SMAC mimetic compounds (SMCs) are small molecule mimetics of this cellular factor that can potentiate TRAIL- and TNF-α-mediated cell death (Figure 4A,B), especially in tumor cells where theses signaling pathways are aberrant [178]. Despite their potent effects on certain cell lines as a single agent due to the presence of endogenous TNF-α, SMAC mimetics are ineffective as a monotherapy in most tumor cell lines. In addition, drug resistance mechanisms include a SMC-induced upregulation of cIAP2 [179] and LRIG1 [180]. As enhancers of pro-apoptotic stimuli, however, they act as strong enhancers of the cytotoxicity of many apoptosis-inducing therapies, such as OVs [181]. This synergy has been described for several SMCs and viruses (see Table 2) and is mainly facilitated by the cytokines produced in response to OV infection. The most important cytokines involved are TRAIL [178,182,183], IL-8 [183], IL-1A [183], IL-1β [184] and TNF-α [176,185]. To improve the synergy between SMC and OVs even further, OVs have been armed with exogenous tumor cell death enhancing (TCDE) cytokines, like TNF-α [186], which also addresses toxicity issues commonly associated with their systemic delivery. In an armed OV setting, production of these cytokines is largely limited to the tumor [187].

Table 2. Selected SMC/OV combinations.

SMC	OV	Tumor Model	References
LCL-161	VSV, M1	EMT-6, CT26, MOC-11, SNB75, SG539, BTIC, HCT-116, Kym-1, M-3	[154,183,186,188–191]
Birinapant	M1	HCT-116, Huh-7	[183,191]

Apart from enhanced cytotoxic effects, SMC/OV combinations can also improve the antitumor response by modulating the adaptive immune response. Exhaustion of CD8$^+$ T-cells was reduced by an SMC-induced tumor macrophage M2 to M1 repolarization, an effect that could be further enhanced by PD-1 checkpoint blockade [190].

B cell lymphoma-2 (BCL-2) homology domain 3 (BH3) mimetics are antagonists that can bind with the hydrophobic Bcl-2 homology (BH) groove of Bcl-2 family proteins, thereby inhibiting these pro-survival proteins and restoring the apoptotic processes in tumor cells (Figure 4B) [192]. Several BH3 mimetics, namely GX15-070 (Obatoclax), EM20-25, BI-97D6 were shown to synergistically increase tumor cell death when combined with oncolytic vaccinia virus, VSV and AdV, respectively [173,174,193,194]. BH3 mimetics also could have a place in the cancer vaccine setting where treatment with GX15-070 (Obatoclax) increased intra-tumoral activated CD8$^+$ T-cells while reducing Treg activity [193].

3.4. Microtubule Targeting Compounds

Taxane compounds achieve their therapeutic effect through stabilizing the spindle microtubule dynamics resulting in inhibited cell division (Figure 4D) [195]. In combination with OVs, the microtubule stabilizing agents (MSAs), docetaxel and paclitaxel, were able to sensitize a variety of tumor types to cell death following stimulation by a subset of OV infection-induced cytokines [196–202]. In combination with reovirus, even tumor cells not sensitive to paclitaxel alone showed a strongly enhanced cell death, which was less due to increased oncolyis but, rather, resulted from activation of cell death programs prior to viral assembly [203]. OVs, armed with pro-apoptotic cargos, could sensitized the cancer cells even further to combination treatment [204]. More out-of-the-box ideas, such as encapsulating paclitaxel and oncolytic adenovirus, together in extracellular vesicles with improved transduction and efficacy, show that there new modes of synergy still to be elucidated [205].

Another way of interfering with the tubuline network is through destabilization. Indeed, microtubule-destabilizing agents (MDAs), such as vinca alkaloids, colchicine and platinum compounds, have long been used as cancer chemotherapeutics. These compounds can also increase cell death through bystander killing after exposure to OV-induced cytokines [206–208]. The synergy of these types of compounds have been described in numerous animal and human settings [200,201,203,208–213]. In addition, MDAs were able to increase OV replication through a previously unappreciated role of microtubule structures in regulating type I IFN translation (Figure 1). A colchicine-induced drop in IFN and ISG expression allowed for a more robust replication of an oncolytic VSV variant with a heightened IFN sensitivity [206,214]. On the other hand, HSV-induced cisplatin retention was reported, resulting in increased DNA damage and anti-tumor immunity [215]. An additional route through which OV treatment can facilitate cell death in combination with chemotherapeutics, more specifically platinum compounds, is by downregulating myeloid cell leukemia 1 (MCL-1) (Figure 4B). MCL-1 is an anti-apoptotic member of the BCL-2 protein family that is more strongly degraded during oncolytic adenovirus infection. Its elimination in turn allows compounds like cisplatin to push tumor cells more efficiently towards cell death [216].

3.5. Topoisomerase Inhibitors

DNA topoisomerases are enzymes that solve topological problems associated with DNA replication, transcription, recombination, and chromatin remodeling by introducing temporary single- or double-strand breaks in the DNA [217]. Topoisomerase inhibitors are small molecules that interfere with the function of these enzymes through either intercalation or alkylation, leading to single and double stranded DNA breaks (Figure 4C). When the integrity of the genome is sufficiently compromised, apoptosis and cell death will follow, particularly in fast dividing cells, such as tumor cells, which are especially sensitive to this [218,219]. Improving the potency of these inhibitors, specifically in tumor cells, could allow lower dosing of these compounds, thereby limiting their adverse effects. This is of special importance for these therapeutics, since their use has been linked to the development of leukemia later in life [220,221]. An important mode of action of the reported synergy between OV treatment and doxorubicin is believed to be both treatments pushing the tumor cells in conflicting states of mitotic progression, resulting in higher tumor cell death than either monotherapy could achieve [222]. In addition, the effect of doxorubicin can be augmented by OV-mediated MCL-1 downregulation with co-treatment significantly increasing tumor cell death (Figure 4B,C) [223].For several cancer types, doxorubicin-treated senescent tumor cells, which are resistant to more classical methods of treatment, were efficiently killed by an oncolytic measles virus [167]. The combination of doxorubicin with an oncolytic adenovirus improved cell death in a more immunogenic fashion. This was further enhanced with additional co-treatment of the cyclophosphamide analogue ifosfamide [224]. Alternatively, the co-application of doxorubicin can also promote an increased infectivity of tumor cells by oncolytic viruses such as certain reovirus strains [225,226]. A more complex interplay has also been reported, where OV treatment induces the nuclear translocation of the cytoplasmic transcription factor cAMP response element-binding protein 3-like 1 (CREB3L1) [227], which in turn is associated with augmented doxorubicin-mediated cell death [228].

4. Combinations Improving the Antitumor Immune Response

Although initially envisioned to act primarily via their tumoricidal actions, over the last decade oncolytic viruses have emerged as potent immune activators and promising partners for cancer immunotherapies. The potential and promising preclinical and clinical findings of combinations of OVs with major immunotherapeutic approaches such as immune checkpoint inhibitors, T cell therapies, and cancer vaccines are beyond the scope of this small molecule themed review but are extensively discussed in recent publications [22,229–233]. Small molecule compounds that augment the antitumor immune response can modulate the tumor microenvironment or affect the adaptive immunity arm. The natural immune-activating characteristics renders OVs as the ideal platform to work in conjunction with small molecule immunotherapies. The TME consists of extracellular matrix (ECM), stromal and immune cells. Some of these cells such as Tregs, MDSCs and M2 macrophages drive an immunosuppressive environment by the secretion of cytokines such as IL-10 or TGF-β [234,235]. Within the TME many human tumors are infiltrated by Tregs [236], with preclinical data indicating that their depletion can enhance or restore anti-tumor immunity [237]. This makes Treg-depleting small molecules attractive candidates to counter cancer relapses caused by these immunosuppressive cells after OV treatment.

4.1. Cyclophosphamide (CP)

CP was extensively tested in combination with OVs, where synergy was described mostly through CPs immunosuppressive effects which allowed the OVs to replicate longer, thereby prolonging and enhancing their therapeutic efficacy [238–241]. However, CP can also play a role in improving the anti-tumor immune response elicited by initial OV treatment. Low-dose CP does not have the same immunosuppressive and toxic effects that allow increased OV replication, but does decrease the number of Tregs without compromising induction of antitumor or antiviral T-cell responses [242,243]. This selective sensitivity of Tregs to CP, comprehensively reviewed by Madondo et al. [244], works through several mechanisms. Combined, these mechanisms allow for depletion or reduced activity of Tregs, while leaving other cell populations intact [244]. This approach shows great promise, especially in combination with oncolytic virus-based cancer vaccination [245].

4.2. Inhibitors of VEGF and PDGF Signaling

VEGF-targeting agents such as sunitinib and cabozantinib can modulate the composition of immune cell subpopulations in the tumor and have been shown to enhance the efficacy of OV treatment. These agents, in combination with OVs, also act on several other aspects of the tumor adaptive immunity and TME, but mainly act through reducing the function of immunosuppressive cells, such as MDSCs, which in turn change cytokine levels (IL-1b, IL-6 and C-X-C motif chemokine ligand 1 (CXCL1)) and amplify the CD4$^+$ and CD8$^+$-mediated tumor regression [109,110,117,246]. The molecular mechanism underlying this MDSC depletion is believed to relate to inhibition of STAT3, which blocks the development of immature myeloid cells into MDSCs, and VEGFR blockade, which results in a lower capacity of MDSCs to migrate to the TME [247].

4.3. Transforming Growth Factor-β TGF-β Inhibition

During cancer progression, cross-talk of EGFR signaling occurs with another important signaling cascade, which is centered around the cytokine family of TGF-β [248,249]. The effects of TGF-β are very diverse and affect many signaling pathways of numerous cell types in vivo, including cancer cells [249]. Due to the interaction complexity, the effect of TGF-β evolves throughout the progression of cancer. Initially, it has a suppressing effect by triggering cell cycle arrest [250]. However, as cancer progresses, tumor cells become resistant to this response and TGF-β signaling results in epithelial–to-mesenchymal transition and increased cell migration with subsequent metastases [250,251]. TGF-β also contributes to an immunosuppressive TME [252], which impedes any anti-tumor immune response that is elicited during OV treatment [253]. Indeed, when a small-molecule inhibitor of TGF-β receptor 1 (TGF-βR1), known as A8301 [254], was combined with oncolytic HSV as treatment for murine rhabdomyosarcoma, an increased efficacy was seen due to an improved anti-tumor T cell response [255]. During non-canonical TGF-β signaling, crosstalk occurs with numerous other signaling pathways, such as PI3K, JNK and NF-κB [249]. As described above, these signaling pathways can have inhibiting effects on the replication and potency of OVs. In certain tumor settings an indirect inhibition of the pathways through TGF-β blockage could also promote OV replication. Indeed, in glioblastoma (GBM) the TGF-βRI kinase inhibitors, galunisertib [256], SB431542 and LY2109761 facilitated an increase in HSV replication through indirect inhibition of JNK-MAPK signaling [257]. Interestingly, SB431542 also inhibited oncolytic reovirus-mediated cell lysis, contrary to A8301 and galunisertib (LY2157299), indicating TGF-β signaling independent mechanisms further to be elucidated [258].

4.4. Topoisomerase Inhibitors

The cytotoxicity of some topoisomerase inhibitor compounds has been shown to be associated with enhanced immunogenicity of dying cells, in part due to the widespread genomic damages [259]. In addition, topoisomerase inhibitors can also improve tumor immunogenicity by upregulating antigen presentation as shown for a variety of melanoma cell lines and gliomas in response to nanomolar levels of DNA intercalating daunorubicin [260]. These immune activating characteristics could be synergistically enhanced by a combination of an oncolytic herpesvirus and adenovirus with mitoxantrone [261] and temozolomide [27,262,263].

4.5. Novel Compounds Targeting Adaptive Treatment Resistance of the Tumor

There are also numerous other small molecule inhibitors that counteract different aspects of immunosuppressive adaptive-mediated treatment resistance. However, these compounds have yet to be tested in combination with OVs and will therefore only be mentioned briefly, for example, inhibition of ubiquitin-specific peptidase 7 (USP7) [264,265], PI3Kdelta [266], the CBP/EP300. In addition, topoisomerase inhibitors can also improve tumor immunogenicity by upregulating antigen presentation as shown for a variety of melanoma cell lines and gliomas in response to nanomolar levels of DNA intercalating daunorubicin [260] or bromodomain [267]; all have been shown to inhibit Treg function, subsequently allowing for a more potent antitumor immune response to arise.

4.6. Checkpoint Inhibitors (CPIs)

The benefits of combining antibody-based CPIs with OVs are well-known and have been comprehensively reviewed elsewhere [22,268–270]. Naturally, upregulation of immune checkpoints is a common result after OV treatment, leading to an increase in immune suppression and subsequent tumor relapse [32]. This can be countered by macromolecule CPIs. However, small molecule CPIs have also been developed and hold several benefits over their antibody counterparts. This upcoming class of small molecules has been extensively reviewed [271–274]. However, combinations with OVs have not yet been described for small molecule CPIs.

4.7. Stimulator of Interferon Genes (STING)

The cyclic guanosine monophosphate–adenosine monophosphate (GMP-AMP) synthase (cGAS)-stimulator of the interferon genes (STING) signaling pathway has recently been described as playing an important role, not only in the innate response to infection [275–278], but also in cancer immune surveillance. STING activation initiates a type I interferon (IFN)-driven pro-inflammatory program that stimulates basic leucine zipper transcriptional factor ATF-like 3 (BATF3)-dependent dendritic cell (DC) cross-presentation and promotes CD8$^+$ T cell-mediated anti-tumor immune responses [279–282]. STING agonists have thus emerged as a class of promising new therapeutics that may enhance tumor immunogenicity and several candidates are being evaluated in pre-clinical and clinical contexts [283–285]. However, STING deficiency is common in several cancer entities due to the anti-tumorigenic and immune-activating role of STING signaling [286–288] and data suggest that, consequently, oncolytic viruses benefit from STING loss due to a decreased antiviral IFN response [287,288]. Several OVs also encode gene products that interfere with the cGAS–STING signaling pathway [289,290]. These considerations make a potential combination of OV with STING agonists at first look counterintuitive. However, STING deficiency or dysfunction has been associated with an exclusion of lymphoid cells from the TME [279] and, while viral replication may be enhanced in STING loss tumors, an optimal induction of an adaptive anti-tumor immune response could be hindered. Indeed, OVs that induce an IFN response via cGAS-STING signaling may have an advantage due to the involvement of this pathway in the bridging of innate and adaptive immunity [291]. Hence, the combination of small molecule STING agonists with certain oncolytic viruses may represent an interesting novel approach to enhance anti-tumor immune responses in OV therapy, although careful assessment of the co-treatment regimen to balance the antiviral and antitumoral effects of STING will be paramount.

5. Safety Considerations

To date, clinical experience with virotherapy-enhancing combinations is limited and our current understanding on the synergism of select combinations has been based on extensive preclinical studies. Twenty years of clinical testing of OV's in monotherapy settings have underlined their excellent safety profile with grade 1 and 2 being the most commonly reported adverse events [5]. To what extent some small molecule combinations may compromise such a safety profile or adversely affect the overall therapeutic efficacy of oncolytic viruses is currently, in large part, subject to conjecture and should therefore be carefully addressed in pre-clinical settings. For example, dimethyl fumarate potentiates replication and oncolysis induced by VSVΔM51 [66], but lowers leukocyte counts and can result in reactivation of JC virus, leading to multifocal leukoencephalopathy (PML). Some HDIs have also been shown to reactivate latent HIV [292], EBV and HSV-1 [293]. The risk that such compounds may reactivate a second virus, with that virus' interactions with the initial oncolytic virus being unknown, should not be underestimated. The specific inhibition profiles of the particular small molecule, as well as the OV in question, will also determine the outcome of an OV/drug combination. While enhancing OV replication, inhibition of certain HDACs (HDAC 2, 6, 11) may enhance Treg function [294], so choosing a drug with a favorable profile, selection of patients with low tumor Treg counts or careful scheduling of the drug and OV may enhance the final anti-tumor synergy. In addition, some virotherapy-enhancing combinations may also potentially enhance the safety profile. For example, ruxolitinib has long been proposed to enhance activity of numerous OVs due to countering the antiviral JAK/STAT signaling and no toxicities have been reported in different preclinical studies [44,49]. However, its combination with an interferon-armed VSV-hIFN-NIS in two current clinical trials (see Table 3) may also act to offset potential toxicities caused by excessive production of the interferon transgene in particularly permissive tumors.

Table 3. Currently active * clinical trials with oncolytic virus and small molecule compound combinations.

Virus Family	Oncolytic Virus Design	Small Molecule Compound	Indication	Phase/Status	CinicalTrials.gov Reference
HSV	rQNestin34.5v.2 HSV-1 with viral gene ICP34.5 under glioma specific nestin promoter control	Cyclophosphamide	Glioma	I recruiting	NCT03152318
	TBI-1401(HF10) naturally attenuated HSV-1	Gemcitabine + nab-pactitaxel	Pancreatic cancer	I not recruiting	NCT03252808
AdV	ONCOS-102 Ad5/3-24 expressing a GM-CSF transgene	Cyclophosphamide	Melanoma	I not recruiting	NCT03003676
	ONCOS-102 Ad5/3-24 expressing a GM-CSF transgene	Cyclophosphamide	Mesothelioma	II not recruiting	NCT02879669
	LOAd703 AdV5/35 expressing TMZ-CD40L and 4-1BBL transgenes	Gemcitabine + nab-pactitaxel	Pancreatic cancer	I/IIa recruiting	NCT02705196
RV	Pelareorep Unmodified human reovirus typ 3 (Dearing strain)	Paclitaxel	Breast cancer	II recruiting	NCT04215146
	Pelareorep Unmodified human reovirus typ 3 (Dearing strain)	Carfilzomib	Multiple myeloma	I recruiting	NCT03605719
VV	JX-594 (Pexa-Vec) Wyeth strain VV expressing a GM-CSF transgene	Cyclophosphamide	Sarcoma, breast cancer	II recruiting	NCT02630368

Table 3. *Cont.*

Virus Family	Oncolytic Virus Design	Small Molecule Compound	Indication	Phase/Status	CinicalTrials.gov Reference
VSV	VSV-hIFN-NIS VSV expressing an interferon and a sodium iodide symporter transgene	Ruxolitinib	Multiple myeloma, AML, T-cell lymphoma	I recruiting	NCT03017820
	VSV-hIFN-NIS VSV expressing an interferon and a sodium iodide symporter transgene	Ruxolitinib	Endometrial cancer	I recruiting	NCT03120624

AdV, adenovirus; GM-CSF, granulocyte-macrophage colony-stimulating factor; hIFN, human interferon; HSV-1, herpes simplex virus type 1; ICP, infected cell protein; TMZ-CD40L, trimerized membrane-bound CD40 ligand; VSV, vesicular stomatitis virus; VV, vaccinia virus. * clinicaltrials.org accessed on 23 June 2021; search term "oncolytic"; filters "recruiting" and "active, not recruiting".

6. Conclusions

While our understanding of how to capture the full potential of oncolytic virotherapy continues to evolve, it appears clear that release of tumor associated antigens and activation of the immune system is crucial for these anti-oncolytic agents. Consequently, combinations of oncolytic viruses with immune checkpoint inhibitors are dominating the current clinical trial landscape [295,296]. However, combinations with select small molecule compounds can address some of the limitations of the oncolytic core features and improve oncolysis, intra-tumoral spread, immunogenicity of tumor cell killing, as well as improving antigen processing and the regulation of immune cell populations. Such combinations have now also entered clinical testing [18] (for currently active trials, see Table 3).

In conclusion, there are many potent compounds available to counter most immunosuppressive mechanisms a tumor can display. The big challenge will be to develop methods to efficiently and affordably determine which combination to use when, and for which patients.

Author Contributions: Conceptualization, B.S. and G.W.; writing–original draft preparation, B.S., K.A., J.P. and G.W.; writing–review and editing, B.S. and G.W.; visualization, B.S. All authors have read and agreed to the published version of the manuscript.

Funding: B.S., K.A. and G.W. were supported by funding from the Christian Doppler Research Association.

Conflicts of Interest: B.S. and J.P. are employees of ViraTherapeutics GmbH and Boehringer Ingelheim Pharma, respectively. G.W. serves as scientific advisor for Boehringer Ingelheim. ViraTherapeutics or Boehringer Ingelheim had no role in the conceptualization or writing of the manuscript. The other authors declare no competing interest.

References

1. Hanahan, D.; Weinberg, R.A. The hallmarks of cancer. *Cell* **2000**, *100*, 57–70. [CrossRef]
2. Hanahan, D.; Weinberg, R.A. Hallmarks of cancer: The next generation. *Cell* **2011**, *144*, 646–674. [CrossRef]
3. Kaufman, H.L.; Kohlhapp, F.J.; Zloza, A. Oncolytic viruses: A new class of immunotherapy drugs. *Nat. Rev. Drug Discov.* **2016**, *15*, 660. [CrossRef]
4. Maroun, J.; Muñoz-Alía, M.; Ammayappan, A.; Schulze, A.; Peng, K.W.; Russell, S. Designing and building oncolytic viruses. *Future Virol.* **2017**, *12*, 193–213. [CrossRef] [PubMed]
5. Macedo, N.; Miller, D.M.; Haq, R.; Kaufman, H.L. Clinical landscape of oncolytic virus research in 2020. *J. Immunother. Cancer* **2020**, *8*. [CrossRef] [PubMed]
6. Guo, Z.S.; Lu, B.; Guo, Z.; Giehl, E.; Feist, M.; Dai, E.; Liu, W.; Storkus, W.J.; He, Y.; Liu, Z.; et al. Vaccinia virus-mediated cancer immunotherapy: Cancer vaccines and oncolytics. *J. Immunother. Cancer* **2019**, *7*, 1–21. [CrossRef]
7. Rahman, M.M.; McFadden, G. Oncolytic Virotherapy with Myxoma Virus. *J. Clin. Med.* **2020**, *9*, 171. [CrossRef]
8. Bretscher, C.; Marchini, A. H-1 parvovirus as a cancer-killing agent: Past, present, and future. *Viruses* **2019**, *11*, 562. [CrossRef]
9. Müller, L.; Berkeley, R.; Barr, T.; Ilett, E.; Errington-Mais, F. Past, present and future of oncolytic reovirus. *Cancers* **2020**, *12*, 3219. [CrossRef] [PubMed]

10. Pidelaserra-Martí, G.; Engeland, C.E. Mechanisms of measles virus oncolytic immunotherapy. *Cytokine Growth Factor Rev.* **2020**, *56*, 28–38. [CrossRef]
11. Tayeb, S.; Zakay-Rones, Z.; Panet, A. Therapeutic potential of oncolytic Newcastle disease virus: A critical review. *Oncolytic Virother.* **2015**, *4*, 49–62. [CrossRef]
12. Felt, S.A.; Grdzelishvili, V.Z. Ecent advances in vesicular stomatitis virus-based oncolytic virotherapy: A 5-year update. *J. Gen. Virol.* **2017**, *98*, 2895–2911. [CrossRef] [PubMed]
13. Stojdl, D.F.; Lichty, B.; Knowles, S.; Marius, R.; Atkins, H.; Sonenberg, N.; Bell, J.C. Exploiting tumor-specific defects in the interferon pathway with a previously unknown oncolytic virus. *Nat. Med.* **2000**, *6*, 821–825. [CrossRef] [PubMed]
14. Melzer, M.; Lopez-Martinez, A.; Altomonte, J. Oncolytic Vesicular Stomatitis Virus as a Viro-Immunotherapy: Defeating Cancer with a "Hammer" and "Anvil". *Biomedicines* **2017**, *5*, 8. [CrossRef]
15. Kirn, D.; Martuza, R.L.; Zwiebel, J. Replication-selective virotherapy for cancer: Biological principles, risk management and future directions. *Nat. Med.* **2001**, *7*, 781–787. [CrossRef]
16. Wakimoto, H.; Fulci, G.; Tyminski, E.; Antonio Chiocca, E. Altered expression of antiviral cytokine mRNAs associated with cyclophosphamide's enhancement of viral oncolysis. *Gene Ther.* **2004**, *11*, 214–223. [CrossRef]
17. Nguyên, T.; Abdelbary, H.; Arguello, M.; Breitbach, C.; Leveille, S.; Diallo, J.-S.; Yasmeen, A.; Bismar, T.A.; Kirn, D.; Falls, T.; et al. Chemical targeting of the innate antiviral response by histone deacetylase inhibitors renders refractory cancers sensitive to viral oncolysis. *Proc. Natl. Acad. Sci. USA* **2008**, *105*, 14981–14986. [CrossRef]
18. Phan, M.; Watson, M.F.; Alain, T.; Diallo, J.S. Oncolytic Viruses on Drugs: Achieving Higher Therapeutic Efficacy. *ACS Infect. Dis.* **2018**, *4*, 1448–1467. [CrossRef]
19. Prestwich, R.J.; Errington, F.; Diaz, R.M.; Pandha, H.S.; Harrington, K.J.; Melcher, A.A.; Vile, R.G. The case of oncolytic viruses versus the immune system: Waiting on the judgment of Solomon. *Hum. Gene Ther.* **2009**, *20*, 1119–1132. [CrossRef]
20. Workenhe, S.T.; Mossman, K.L. Oncolytic virotherapy and immunogenic cancer cell death: Sharpening the sword for improved cancer treatment strategies. *Mol. Ther.* **2014**, *22*, 251–256. [CrossRef]
21. Workenhe, S.T.; Verschoor, M.L.; Mossman, K.L. The role of oncolytic virus immunotherapies to subvert cancer immune evasion. *Future Oncol.* **2015**, *11*, 675–689. [CrossRef]
22. Harrington, K.; Freeman, D.J.; Kelly, B.; Harper, J.; Soria, J.-C. Optimizing oncolytic virotherapy in cancer treatment. *Nat. Rev. Drug Discov.* **2019**, *18*, 689–706. [CrossRef]
23. Van Vloten, J.P.; Workenhe, S.T.; Wootton, S.K.; Mossman, K.L.; Bridle, B.W. Critical Interactions between Immunogenic Cancer Cell Death, Oncolytic Viruses, and the Immune System Define the Rational Design of Combination Immunotherapies. *J. Immunol.* **2018**, *200*, 450–458. [CrossRef]
24. Choi, A.; O'Leary, M.; Fong, Y.; Chen, N. From Benchtop to Bedside: A Review of Oncolytic Virotherapy. *Biomedicines* **2016**, *4*, 18. [CrossRef]
25. Gujar, S.; Pol, J.G.; Kroemer, G. Heating it up: Oncolytic viruses make tumors 'hot' and suitable for checkpoint blockade immunotherapies. *Oncoimmunology* **2018**, *7*, e1442169. [CrossRef]
26. Krysko, D.V.; Garg, A.D.; Kaczmarek, A.; Krysko, O.; Agostinis, P.; Vandenabeele, P. Immunogenic cell death and DAMPs in cancer therapy. *Nat. Rev. Cancer* **2012**, *12*, 860–875. [CrossRef]
27. Liikanen, I.; Ahtiainen, L.; Hirvinen, M.L.; Bramante, S.; Cerullo, V.; Nokisalmi, P.; Hemminki, O.; Diaconu, I.; Pesonen, S.; Koski, A.; et al. Oncolytic Adenovirus with Temozolomide Induces Autophagy and Antitumor Immune Responses in Cancer Patients. *Mol. Ther.* **2013**, *21*, 1212–1223. [CrossRef]
28. Komorowski, M.P.; McGray, A.R.; Kolakowska, A.; Eng, K.; Gil, M.; Opyrchal, M.; Litwinska, B.; Nemeth, M.J.; Odunsi, K.O.; Kozbor, D. Reprogramming antitumor immunity against chemoresistant ovarian cancer by a CXCR4 antagonist-armed viral oncotherapy. *Mol. Ther. Oncolytics* **2016**, *3*, 16034. [CrossRef]
29. Errington, F.; Steele, L.; Prestwich, R.; Harrington, K.J.; Pandha, H.S.; Vidal, L.; de Bono, J.; Selby, P.; Coffey, M.; Vile, R.; et al. Reovirus Activates Human Dendritic Cells to Promote Innate Antitumor Immunity. *J. Immunol.* **2014**, *180*, 6018–6026. [CrossRef]
30. Gujar, S.; Dielschneider, R.; Clements, D.; Helson, E.; Shmulevitz, M.; Marcato, P.; Pan, D.; Pan, L.Z.; Ahn, D.G.; Alawadhi, A.; et al. Multifaceted therapeutic targeting of ovarian peritoneal carcinomatosis through virus-induced immunomodulation. *Mol. Ther.* **2013**, *21*, 338–347. [CrossRef]
31. Gujar, S.A.; Lee, P.W.K. Oncolytic Virus-Mediated Reversal of Impaired Tumor Antigen Presentation. *Front. Oncol.* **2014**, *4*, 77. [CrossRef]
32. Liu, Z.; Ravindranathan, R.; Kalinski, P.; Guo, Z.S.; Bartlett, D.L. Rational combination of oncolytic vaccinia virus and PD-L1 blockade works synergistically to enhance therapeutic efficacy. *Nat. Commun.* **2017**, *8*, 14754. [CrossRef] [PubMed]
33. Huck, B.R.; Kötzner, L.; Urbahns, K. Small Molecules Drive Big Improvements in Immuno-Oncology Therapies. *Angew. Chem. Int. Ed.* **2018**, *57*, 4412–4428. [CrossRef]
34. Sharma, P.; Hu-Lieskovan, S.; Wargo, J.A.; Ribas, A. Primary, Adaptive, and Acquired Resistance to Cancer Immunotherapy. *Cell* **2017**, *168*, 707–723. [CrossRef]
35. Matveeva, O.V.; Chumakov, P.M. Defects in interferon pathways as potential biomarkers of sensitivity to oncolytic viruses. *Rev. Med. Virol.* **2018**, *28*, e2008. [CrossRef] [PubMed]

36. Stojdl, D.F.; Lichty, B.D.; TenOever, B.R.; Paterson, J.M.; Power, A.T.; Knowles, S.; Marius, R.; Reynard, J.; Poliquin, L.; Atkins, H.; et al. VSV strains with defects in their ability to shutdown innate immunity are potent systemic anti-cancer agents. *Cancer Cell* **2003**, *4*, 263–275. [CrossRef]

37. Kurokawa, C.; Galanis, E. Interferon signaling predicts response to oncolytic virotherapy. *Oncotarget* **2019**, *10*, 1544–1545. [CrossRef] [PubMed]

38. Hastie, E.; Cataldi, M.; Moerdyk-Schauwecker, M.J.; Felt, S.A.; Steuerwald, N.; Grdzelishvili, V.Z.; Hastie, E.; Cataldi, M.; Moerdyk-Schauwecker, M.J.; Felt, S.A.; et al. Novel biomarkers of resistance of pancreatic cancer cells to oncolytic vesicular stomatitis virus. *Oncotarget* **2016**, *7*, 61601–61618. [CrossRef]

39. Kurokawa, C.; Iankov, I.D.; Anderson, S.K.; Aderca, I.; Leontovich, A.A.; Maurer, M.J.; Oberg, A.L.; Schroeder, M.A.; Giannini, C.; Greiner, S.M.; et al. Constitutive Interferon Pathway Activation in Tumors as an Efficacy Determinant Following Oncolytic Virotherapy. *JNCI J. Natl. Cancer Inst.* **2018**, *110*, 1–10. [CrossRef]

40. Vähä-Koskela, M.; Hinkkanen, A. Tumor Restrictions to Oncolytic Virus. *Biomedicines* **2014**, *2*, 163–194. [CrossRef]

41. Rojas, J.M.; Alejo, A.; Martín, V.; Sevilla, N. Viral pathogen-induced mechanisms to antagonize mammalian interferon (IFN) signaling pathway. *Cell. Mol. Life Sci.* **2021**, *78*, 1423–1444. [CrossRef]

42. Lin, F.; Young, H.A. Interferons: Success in anti-viral immunotherapy. *Cytokine Growth Factor Rev.* **2014**, *25*, 369–376. [CrossRef] [PubMed]

43. Leonard, W.J.; O'Shea, J.J. JAKS and STATS: Biological Implications. *Annu. Rev. Immunol.* **2002**, *16*, 293–322. [CrossRef]

44. Dold, C.; Rodriguez Urbiola, C.; Wollmann, G.; Egerer, L.; Muik, A.; Bellmann, L.; Fiegl, H.; Marth, C.; Kimpel, J.; von Laer, D. Application of interferon modulators to overcome partial resistance of human ovarian cancers to VSV-GP oncolytic viral therapy. *Mol. Ther. Oncolytics* **2016**, *3*, 16021. [CrossRef]

45. Moerdyk-Schauwecker, M.; Shah, N.R.; Murphy, A.M.; Hastie, E.; Mukherjee, P.; Grdzelishvili, V.Z. Resistance of pancreatic cancer cells to oncolytic vesicular stomatitis virus: Role of type I interferon signaling. *Virology* **2013**, *436*, 221–234. [CrossRef]

46. Cataldi, M.; Shah, N.R.; Felt, S.A.; Grdzelishvili, V.Z. Breaking resistance of pancreatic cancer cells to an attenuated vesicular stomatitis virus through a novel activity of IKK inhibitor TPCA-1. *Virology* **2015**, *485*, 340–354. [CrossRef]

47. Felt, S.A.; Droby, G.N.; Grdzelishvili, V.Z. Ruxolitinib and Polycation Combination Treatment Overcomes Multiple Mechanisms of Resistance of Pancreatic Cancer Cells to Oncolytic Vesicular Stomatitis Virus. *J. Virol.* **2017**, *91*, e00461-17. [CrossRef] [PubMed]

48. Escobar-Zarate, D.; Liu, Y.-P.; Suksanpaisan, L.; Russell, S.J.; Peng, K.-W. Overcoming cancer cell resistance to VSV oncolysis with JAK1/2 inhibitors. *Cancer Gene Ther.* **2013**, *20*, 582–589. [CrossRef]

49. Patel, M.R.; Dash, A.; Jacobson, B.A.; Ji, Y.; Baumann, D.; Ismail, K.; Kratzke, R.A. JAK/STAT inhibition with ruxolitinib enhances oncolytic virotherapy in non-small cell lung cancer models. *Cancer Gene Ther.* **2019**, *26*, 411–418. [CrossRef] [PubMed]

50. Nguyen, T.-T.; Ramsay, L.; Ahanfeshar-Adams, M.; Lajoie, M.; Schadendorf, D.; Alain, T.; Watson, I.R. Mutations in the IFNγ-JAK-STAT Pathway Causing Resistance to Immune Checkpoint Inhibitors in Melanoma Increase Sensitivity to Oncolytic Virus Treatment. *Clin. Cancer Res.* **2021**. [CrossRef] [PubMed]

51. Jackson, J.D.; Markert, J.M.; Li, L.; Carroll, S.L.; Cassady, K.A. STAT1 and NF- B Inhibitors Diminish Basal Interferon-Stimulated Gene Expression and Improve the Productive Infection of Oncolytic HSV in MPNST Cells. *Mol. Cancer Res.* **2016**, *14*, 482–492. [CrossRef] [PubMed]

52. Domingo, E.; Guerra, S.; Esteban, M.; Gil, J.; Rivas, C.; Garcia, M.A.; Ventoso, I. Impact of Protein Kinase PKR in Cell Biology: From Antiviral to Antiproliferative Action. *Microbiol. Mol. Biol. Rev.* **2006**, *70*, 1032–1060. [CrossRef]

53. Tamanini, A.; Rolfini, R.; Nicolis, E.; Melotti, P.; Cabrini, G. MAP kinases and NF-κB collaborate to induce ICAM-1 gene expression in the early phase of adenovirus infection. *Virology* **2003**, *307*, 228–242. [CrossRef]

54. Vlahopoulos, S.; Zoumpourlis, V.C. JNK: A key modulator of intracellular signaling. *Biochemistry* **2004**, *69*, 844–854. [CrossRef] [PubMed]

55. Motiwala, A.; Pierce, S.; Satoh, Y.; Bhagwat, S.S.; Bennett, B.L.; Sakata, S.T.; O'Leary, E.C.; Manning, A.M.; Leisten, J.C.; Xu, W.; et al. SP600125, an anthrapyrazolone inhibitor of Jun N-terminal kinase. *Proc. Natl. Acad. Sci. USA* **2002**, *98*, 13681–13686. [CrossRef]

56. Li, B.; Tournier, C.; Davis, R.J.; Flavell, R.A. Regulation of IL-4 expression by the transcription factor JunB during T helper cell differentiation. *EMBO J.* **1999**, *18*, 420–432. [CrossRef]

57. Li, J.; Manaligod, J.M.; Bhat, R.K.; Brasier, A.R.; Hershenson, M.B.; Page, K.; Tan, A.; Iasvovskaia, S.; Kartha, S. Regulation of human airway epithelial cell IL-8 expression by MAP kinases. *Am. J. Physiol. Cell. Mol. Physiol.* **2015**, *283*, L690–L699. [CrossRef]

58. Wang, Y.; Li, C.; Wang, X.; Zhang, J.; Chang, Z. Heat shock response inhibits IL-18 expression through the JNK pathway in murine peritoneal macrophages. *Biochem. Biophys. Res. Commun.* **2002**, *296*, 742–748. [CrossRef]

59. Oltmanns, U.; Issa, R.; Sukkar, M.B.; John, M.; Chung, K.F. Role of c-jun N-terminal kinase in the induced release of GM-CSF, RANTES and IL-8 from human airway smooth muscle cells. *Br. J. Pharmacol.* **2003**, *139*, 1228–1234. [CrossRef]

60. Holloway, G.; Coulson, B.S. Rotavirus Activates JNK and p38 Signaling Pathways in Intestinal Cells, Leading to AP-1-Driven Transcriptional Responses and Enhanced Virus Replication. *J. Virol.* **2006**, *80*, 10624–10633. [CrossRef]

61. Hirasawa, K.; Kim, A.; Han, H.-S.; Han, J.; Jun, H.-S.; Yoon, J.-W. Effect of p38 mitogen-activated protein kinase on the replication of encephalomyocarditis virus. *J. Virol.* **2003**, *77*, 5649–5656. [CrossRef] [PubMed]

62. McLean, T.; Bachenmeier, S.L. Activation of cJUN N-Terminal Kinase by Herpes Simplex Virus Type 1 Enhances Viral Replication. *J. Virol.* **1999**, *73*, 8415–8426. [CrossRef]

63. Hu, W.; Hofstetter, W.; Guo, W.; Li, H.; Pataer, A.; Peng, H.H.; Guo, Z.S.; Bartlett, D.L.; Lin, A.; Swisher, S.G.; et al. JNK-deficiency enhanced oncolytic vaccinia virus replication and blocked activation of double-stranded RNA-dependent protein kinase. *Cancer Gene Ther.* **2008**, *15*, 616–624. [CrossRef] [PubMed]

64. Perkins, N.D. Integrating cell-signalling pathways with NF-κB and IKK function. *Nat. Rev. Mol. Cell Biol.* **2007**, *8*, 49–62. [CrossRef]

65. Struzik, J.; Szulc-Dąbrowska, L. NF-κB signaling in targeting tumor cells by oncolytic viruses—Therapeutic perspectives. *Cancers* **2018**, *10*, 426. [CrossRef] [PubMed]

66. Selman, M.; Ou, P.; Rousso, C.; Bergeron, A.; Krishnan, R.; Pikor, L.; Chen, A.; Keller, B.A.; Ilkow, C.; Bell, J.C.; et al. Dimethyl fumarate potentiates oncolytic virotherapy through NF-B inhibition. *Sci. Transl. Med.* **2018**, *10*, eaao1613. [CrossRef]

67. Qiu, D.; Kao, P.N. Immunosuppressive and Anti-Inflammatory Mechanisms of Triptolide, the Principal Active Diterpenoid from the Chinese Medicinal Herb *Tripterygium wilfordii* Hook. f. *Drugs R & D* **2006**, *4*, 1–18. [CrossRef]

68. Ben Yebdri, F.; Van Grevenynghe, J.; Tang, V.A.; Goulet, M.L.; Wu, J.H.; Stojdl, D.F.; Hiscott, J.; Lin, R. Triptolide-mediated inhibition of interferon signaling enhances vesicular stomatitis virus-based oncolysis. *Mol. Ther.* **2013**, *21*, 2043–2053. [CrossRef]

69. Baumann, J.; Garner, J.M.; Davidoff, A.M.; Morton, C.L.; Du, Z.; Whitt, M.A.; Pfeffer, L.M. Inhibition of Type I Interferon-Mediated Antiviral Action in Human Glioma Cells by the IKK Inhibitors BMS-345541 and TPCA-1. *J. Interferon Cytokine Res.* **2012**, *32*, 368–377. [CrossRef]

70. Wollmann, G.; Ozduman, K.; van den Pol, A.N. Oncolytic virus therapy for glioblastoma multiforme: Concepts and candidates. *Cancer J.* **2012**, *18*, 69–81. [CrossRef]

71. Cao, W.; Manicassamy, S.; Tang, H.; Kasturi, S.P.; Pirani, A.; Murthy, N.; Pulendran, B. Toll-like receptor-mediated induction of type I interferon in plasmacytoid dendritic cells requires the rapamycin-sensitive PI(3)K-mTOR-p70S6K pathway. *Nat. Immunol.* **2008**, *9*, 1157–1164. [CrossRef] [PubMed]

72. Weigelt, B.; Downward, J. Genomic Determinants of PI3K Pathway Inhibitor Response in Cancer. *Front. Oncol.* **2012**, *2*, 109. [CrossRef]

73. Dienstmann, R.; Rodon, J.; Serra, V.; Tabernero, J. Picking the Point of Inhibition: A Comparative Review of PI3K/AKT/mTOR Pathway Inhibitors. *Mol. Cancer Ther.* **2014**, *13*, 1021–1031. [CrossRef]

74. Alain, T.; Lun, X.; Martineau, Y.; Sean, P.; Pulendran, B.; Petroulakis, E.; Zemp, F.J.; Lemay, C.G.; Roy, D.; Bell, J.C.; et al. Vesicular stomatitis virus oncolysis is potentiated by impairing mTORC1-dependent type I IFN production. *Proc. Natl. Acad. Sci. USA* **2010**, *107*, 1576–1581. [CrossRef]

75. Li, Y.; Meng, S.; Xu, J.; Zhu, Q.; Liu, Q.; Jiang, K.; Zhang, G.; Deng, W.; Wang, Y. Pharmacological modulation of autophagy enhances Newcastle disease virus-mediated oncolysis in drug-resistant lung cancer cells. *BMC Cancer* **2014**, *14*, 551. [CrossRef]

76. Comins, C.; Simpson, G.R.; Rogers, W.; Relph, K.; Harrington, K.; Melcher, A.; Roulstone, V.; Kyula, J.; Pandha, H. Synergistic antitumour effects of rapamycin and oncolytic reovirus. *Cancer Gene Ther.* **2018**, *25*, 148–160. [CrossRef] [PubMed]

77. Zakaria, C.; Sean, P.; Hoang, H.D.; Leroux, L.P.; Watson, M.; Workenhe, S.T.; Hearnden, J.; Pearl, D.; Truong, V.T.; Robichaud, N.; et al. Active-site mTOR inhibitors augment HSV1-dICP0 infection in cancer cells via dysregulated eIF4E/4E-BP axis. *PLoS Pathog.* **2018**, *14*. [CrossRef]

78. Lun, X.; Chan, J.; Zhou, H.; Sun, B.; Kelly, J.J.; Stechishin, O.O.; Bell, J.C.; Parato, K.; Hu, K.; Vaillant, D.; et al. Efficacy and safety/toxicity study of recombinant vaccinia virus JX-594 in two immunocompetent animal models of glioma. *Mol. Ther.* **2010**, *18*, 1927–1936. [CrossRef]

79. Wang, G.; Barrett, J.W.; Stanford, M.; Werden, S.J.; Johnston, J.B.; Gao, X.; Sun, M.; Cheng, J.Q.; McFadden, G. Infection of human cancer cells with myxoma virus requires Akt activation via interaction with a viral ankyrin-repeat host range factor. *Proc. Natl. Acad. Sci. USA* **2006**, *103*, 4640–4645. [CrossRef] [PubMed]

80. O'Reilly, K.E.; Rojo, F.; She, Q.-B.; Solit, D.; Mills, G.B.; Smith, D.; Lane, H.; Hofmann, F.; Hicklin, D.J.; Ludwig, D.L.; et al. mTOR Inhibition Induces Upstream Receptor Tyrosine Kinase Signaling and Activates Akt. *Cancer Res.* **2006**, *66*, 1500–1508. [CrossRef] [PubMed]

81. Xue, Q.L.; Zhou, H.; Alain, T.; Sun, B.; Wang, L.; Barrett, J.W.; Stanford, M.M.; McFadden, G.; Bell, J.; Senger, D.L.; et al. Targeting human medulloblastoma: Oncolytic virotherapy with myxoma virus is enhanced by rapamycin. *Cancer Res.* **2007**, *67*, 8818–8827. [CrossRef]

82. Stanford, M.M.; Shaban, M.; Barrett, J.W.; Werden, S.J.; Gilbert, P.A.; Bondy-Denomy, J.; MacKenzie, L.; Graham, K.C.; Chambers, A.F.; McFadden, G. Myxoma virus oncolysis of primary and metastatic B16F10 mouse tumors in vivo. *Mol. Ther.* **2008**, *16*, 52–59. [CrossRef]

83. Zemp, F.J.; Lun, X.; McKenzie, B.A.; Zhou, H.; Maxwell, L.; Sun, B.; Kelly, J.J.P.; Stechishin, O.; Luchman, A.; Weiss, S.; et al. Treating brain tumor-initiating cells using a combination of myxoma virus and rapamycin. *Neuro Oncol.* **2013**, *15*, 904–920. [CrossRef]

84. Costa-Mattioli, M.; Sonenberg, N. RAPping production of type I interferon in pDCs through mTOR. *Nat. Immunol.* **2008**, *9*, 1097–1099. [CrossRef]

85. Livingstone, M.; Sikström, K.; Robert, P.A.; Uzé, G.; Larsson, O.; Pellegrini, S.P. Assessment of mTOR-Dependent translational regulation of interferon stimulated genes. *PLoS ONE* **2015**, *10*, e0133482. [CrossRef]

86. Lun, X.; Alain, T.; Zemp, F.J.; Zhou, H.; Rahman, M.M.; Hamilton, M.G.; McFadden, G.; Bell, J.; Senger, D.L.; Forsyth, P.A. Myxoma virus virotherapy for glioma in immunocompetent animal models: Optimizing administration routes and synergy with rapamycin. *Cancer Res.* **2010**, *70*, 598–608. [CrossRef]

87. Alonso, M.M.; Jiang, H.; Yokoyama, T.; Xu, J.; Bekele, N.B.; Lang, F.F.; Kondo, S.; Gomez-Manzano, C.; Fueyo, J. Delta-24-RGD in Combination with RAD001 Induces Enhanced Anti-glioma Effect via Autophagic Cell Death. *Mol. Ther.* **2008**, *16*, 487–493. [CrossRef] [PubMed]

88. Alonso, M.M.; Gomez-Manzano, C.; Jiang, H.; Bekele, N.B.; Piao, Y.; Yung, W.K.A.; Alemany, R.; Fueyo, J. Combination of the oncolytic adenovirus ICOVIR-5 with chemotherapy provides enhanced anti-glioma effect in vivo. *Cancer Gene Ther.* **2007**, *14*, 756–761. [CrossRef]

89. Cloughesy, T.F.; Yoshimoto, K.; Nghiemphu, P.; Brown, K.; Dang, J.; Zhu, S.; Hsueh, T.; Chen, Y.; Wang, W.; Youngkin, D.; et al. Antitumor Activity of Rapamycin in a Phase I Trial for Patients with Recurrent PTEN-Deficient Glioblastoma. *PLoS Med.* **2008**, *5*, e8. [CrossRef]

90. Bhagwat, S.V.; Gokhale, P.C.; Crew, A.P.; Cooke, A.; Yao, Y.; Mantis, C.; Kahler, J.; Workman, J.; Bittner, M.; Dudkin, L.; et al. Preclinical Characterization of OSI-027, a Potent and Selective Inhibitor of mTORC1 and mTORC2: Distinct from Rapamycin. *Mol. Cancer Ther.* **2011**, *10*, 1394–1406. [CrossRef]

91. Zaytseva, Y.Y.; Valentino, J.D.; Gulhati, P.; Mark Evers, B. mTOR inhibitors in cancer therapy. *Cancer Lett.* **2012**, *319*, 1–7. [CrossRef]

92. Rodrik-Outmezguine, V.S.; Okaniwa, M.; Yao, Z.; Novotny, C.J.; McWhirter, C.; Banaji, A.; Won, H.; Wong, W.; Berger, M.; de Stanchina, E.; et al. Overcoming mTOR resistance mutations with a new-generation mTOR inhibitor. *Nature* **2016**, *534*, 272–276. [CrossRef] [PubMed]

93. Wen, P.Y.; Omuro, A.; Ahluwalia, M.S.; Fathallah-Shaykh, H.M.; Mohile, N.; Lager, J.J.; Laird, A.D.; Tang, J.; Jiang, J.; Egile, C.; et al. Phase I dose-escalation study of the PI3K/mTOR inhibitor voxtalisib (SAR245409, XL765) plus temozolomide with or without radiotherapy in patients with high-grade glioma. *Neuro Oncol.* **2015**, *17*, 1275–1283. [CrossRef] [PubMed]

94. Wang, L.; Ning, J.; Wakimoto, H.; Wu, S.; Wu, C.L.; Humphrey, M.R.; Rabkin, S.D.; Martuza, R.L. Oncolytic Herpes Simplex Virus and PI3K Inhibitor BKM120 Synergize to Promote Killing of Prostate Cancer Stem-like Cells. *Mol. Ther. Oncolytics* **2019**, *13*, 58–66. [CrossRef] [PubMed]

95. Kanai, R.; Wakimoto, H.; Martuza, R.L.; Rabkin, S.D. A novel oncolytic herpes simplex virus that synergizes with phosphoinositide 3-kinase/Akt pathway inhibitors to target glioblastoma stem cells. *Clin. Cancer Res.* **2011**, *17*, 3686–3696. [CrossRef] [PubMed]

96. Liu, T.C.; Wakimoto, H.; Martuza, R.L.; Rabkin, S.D. Herpes simplex virus Us3(-) mutant as oncolytic strategy and synergizes with phosphatidylinositol 3-kinase-Akt-targeting molecular therapeutics. *Clin. Cancer Res.* **2007**, *13*, 5897–5902. [CrossRef] [PubMed]

97. Ferguson, M.S.; Chard Dunmall, L.S.; Gangeswaran, R.; Marelli, G.; Tysome, J.R.; Burns, E.; Whitehead, M.A.; Aksoy, E.; Alusi, G.; Hiley, C.; et al. Transient Inhibition of PI3Kδ Enhances the Therapeutic Effect of Intravenous Delivery of Oncolytic Vaccinia Virus. *Mol. Ther.* **2020**, *28*, 1263–1275. [CrossRef]

98. Yoo, J.Y.; Hurwitz, B.S.; Bolyard, C.; Yu, J.G.; Zhang, J.; Selvendiran, K.; Rath, K.S.; He, S.; Bailey, Z.; Eaves, D.; et al. Bortezomib-induced unfolded protein response increases oncolytic HSV-1 replication resulting in synergistic antitumor effects. *Clin. Cancer Res.* **2014**, *20*, 3787–3798. [CrossRef] [PubMed]

99. Yoo, J.Y.; Jaime-Ramirez, A.C.; Bolyard, C.; Dai, H.; Nallanagulagari, T.; Wojton, J.; Hurwitz, B.S.; Relation, T.; Lee, T.J.; Lotze, M.T.; et al. Bortezomib treatment sensitizes oncolytic HSV-1-treated tumors to NK cell immunotherapy. *Clin. Cancer Res.* **2016**, *22*, 5265–5276. [CrossRef]

100. Kim, Y.; Yoo, J.Y.; Lee, T.J.; Liu, J.; Yu, J.; Caligiuri, M.A.; Kaur, B.; Friedman, A. Complex role of NK cells in regulation of oncolytic virus–bortezomib therapy. *Proc. Natl. Acad. Sci. USA* **2018**, *115*, 4927–4932. [CrossRef]

101. Yarde, D.N.; Nace, R.A.; Russell, S.J. Oncolytic vesicular stomatitis virus and bortezomib are antagonistic against myeloma cells invitro but have additive anti-myeloma activity in vivo. *Exp. Hematol.* **2013**, *41*, 1038–1049. [CrossRef]

102. Dudek, S.E.; Luig, C.; Pauli, E.-K.; Schubert, U.; Ludwig, S. The Clinically Approved Proteasome Inhibitor PS-341 Efficiently Blocks Influenza A Virus and Vesicular Stomatitis Virus Propagation by Establishing an Antiviral State. *J. Virol.* **2010**, *84*, 9439–9451. [CrossRef] [PubMed]

103. Tenbaum, S.P.; Ordóñez-Morán, P.; Puig, I.; Chicote, I.; Arqués, O.; Landolfi, S.; Fernández, Y.; Herance, J.R.; Gispert, J.D.; Mendizabal, L.; et al. β-catenin confers resistance to PI3K and AKT inhibitors and subverts FOXO3a to promote metastasis in colon cancer. *Nat. Med.* **2012**, *18*, 892–901. [CrossRef] [PubMed]

104. Arqués, O.; Chicote, I.; Puig, I.; Tenbaum, S.P.; Argilés, G.; Dienstmann, R.; Fernández, N.; Caratù, G.; Matito, J.; Silberschmidt, D.; et al. Tankyrase Inhibition Blocks Wnt/b-Catenin Pathway and Reverts Resistance to PI3K and AKT Inhibitors in the Treatment of Colorectal Cancer. *Clin. Cancer Res.* **2016**, *22*, 644–656. [CrossRef]

105. Deng, Z.; Atanasiu, C.; Zhao, K.; Marmorstein, R.; Sbodio, J.I.; Chi, N.-W.; Lieberman, P.M. Inhibition of Epstein-Barr virus OriP function by tankyrase, a telomere-associated poly-ADP ribose polymerase that binds and modifies EBNA1. *J. Virol.* **2005**, *79*, 4640–4650. [CrossRef] [PubMed]

106. Roy, S.; Liu, F.; Arav-Boger, R. Human cytomegalovirus inhibits the PARsylation activity of tankyrase—A potential strategy for suppression of the Wnt pathway. *Viruses* **2015**, *8*, 8. [CrossRef] [PubMed]

107. Jha, B.K.; Polyakova, I.; Kessler, P.; Dong, B.; Dickerman, B.; Sen, G.C.; Silverman, R.H. Inhibition of RNase L and RNA-dependent protein kinase (PKR) by sunitinib impairs antiviral innate immunity. *J. Biol. Chem.* **2011**, *286*, 26319–26326. [CrossRef] [PubMed]

108. Jha, B.K.; Dong, B.; Nguyen, C.T.; Polyakova, I.; Silverman, R.H. Suppression of antiviral innate immunity by sunitinib enhances oncolytic virotherapy. *Mol. Ther.* **2013**, *21*, 1749–1757. [CrossRef]

109. Lawson, K.A.; Mostafa, A.A.; Shi, Z.Q.; Spurrell, J.; Chen, W.; Kawakami, J.; Gratton, K.; Thakur, S.; Morris, D.G. Repurposing sunitinib with oncolytic reovirus as a novel immunotherapeutic strategy for renal cell carcinoma. *Clin. Cancer Res.* **2016**, *22*, 5839–5850. [CrossRef] [PubMed]

110. Kim, M.; Nitschke, M.; Sennino, B.; Murer, P.; Schriver, B.J.; Bell, A.; Subramanian, A.; McDonald, C.E.; Wang, J.; Cha, H.; et al. Amplification of oncolytic vaccinia virus widespread tumor cell killing by sunitinib through multiple mechanisms. *Cancer Res.* **2018**, *78*, 922–937. [CrossRef]

111. Saha, D.; Wakimoto, H.; Peters, C.W.; Antoszczyk, S.J.; Rabkin, S.D.; Martuza, R.L. Combinatorial effects of vegfr kinase inhibitor axitinib and oncolytic virotherapy in mouse and human glioblastoma stem-like cell models. *Clin. Cancer Res.* **2018**, *24*, 3409–3422. [CrossRef]

112. Heo, J.; Breitbach, C.J.; Moon, A.; Kim, C.W.; Patt, R.; Kim, M.K.; Lee, Y.K.; Oh, S.Y.; Woo, H.Y.; Parato, K.; et al. Sequential therapy with JX-594, a targeted oncolytic poxvirus, followed by sorafenib in hepatocellular carcinoma: Preclinical and clinical demonstration of combination efficacy. *Mol. Ther.* **2011**, *19*, 1170–1179. [CrossRef] [PubMed]

113. Reeves, P.M.; Bommarius, B.; Lebeis, S.; McNulty, S.; Christensen, J.; Swimm, A.; Chahroudi, A.; Chavan, R.; Feinberg, M.B.; Veach, D.; et al. Disabling poxvirus pathogenesis by inhibition of Abl-family tyrosine kinases. *Nat. Med.* **2005**, *11*, 731–739. [CrossRef] [PubMed]

114. Farsaci, B.; Higgins, J.P.; Hodge, J.W. Consequence of dose scheduling of sunitinib on host immune response elements and vaccine combination therapy. *Int. J. Cancer* **2012**, *130*, 1948–1959. [CrossRef]

115. Ozao-Choy, J.; Ge, M.; Kao, J.; Wang, G.X.; Meseck, M.; Sung, M.; Schwartz, M.; Divino, C.M.; Pan, P.Y.; Chen, S.H. The novel role of tyrosine kinase inhibitor in the reversal of immune suppression and modulation of tumor microenvironment for immune-based cancer therapies. *Cancer Res.* **2009**, *69*, 2514–2522. [CrossRef] [PubMed]

116. Kao, J.; Ko, E.C.; Eisenstein, S.; Sikora, A.G.; Fu, S.; Chen, S. Targeting immune suppressing myeloid-derived suppressor cells in oncology. *Crit. Rev. Oncol. Hematol.* **2011**, *77*, 12–19. [CrossRef]

117. Ardiani, A.; Donahue, R.N.; Aftab, D.T.; Hodge, J.W.; Kwilas, A.R. Dual effects of a targeted small-molecule inhibitor (cabozantinib) on immune-mediated killing of tumor cells and immune tumor microenvironment permissiveness when combined with a cancer vaccine. *J. Transl. Med.* **2014**, *12*, 294. [CrossRef]

118. Yamamura, K.; Kasuya, H.; Sahin, T.T.; Tan, G.; Hotta, Y.; Tsurumaru, N.; Fukuda, S.; Kanda, M.; Kobayashi, D.; Tanaka, C.; et al. Combination treatment of human pancreatic cancer xenograft models with the epidermal growth factor receptor tyrosine kinase inhibitor erlotinib and oncolytic herpes simplex virus hf10. *Ann. Surg. Oncol.* **2014**, *21*, 691–698. [CrossRef] [PubMed]

119. Mahller, Y.Y.; Vaikunth, S.S.; Currier, M.A.; Miller, S.J.; Ripberger, M.C.; Hsu, Y.H.; Mehrian-Shai, R.; Collins, M.H.; Crombleholme, T.M.; Ratner, N.; et al. Oncolytic HSV and erlotinib inhibit tumor growth and angiogenesis in a novel malignant peripheral nerve sheath tumor xenograft model. *Mol. Ther.* **2007**, *15*, 279–286. [CrossRef]

120. Stark, G.R.; Kerr, I.M.; Leung, S.; Muzaffar, R.; Vlieststra, R.J.; Levy, D.E.; Trapman, J.; Bluyssen, H.A.; van der Made, A.C. Combinatorial association and abundance of components of interferon-stimulated gene factor 3 dictate the selectivity of interferon responses. *Proc. Natl. Acad. Sci. USA* **2006**, *92*, 5645–5649. [CrossRef]

121. Nusinzon, I.; Horvath, C.M. Positive and Negative Regulation of the Innate Antiviral Response and Beta Interferon Gene Expression by Deacetylation. *Mol. Cell. Biol.* **2006**, *26*, 3106–3113. [CrossRef]

122. Genin, P.; Morin, P.; Civas, A. Impairment of Interferon-Induced IRF-7 Gene Expression due to Inhibition of ISGF3 Formation by Trichostatin A. *J. Virol.* **2003**, *77*, 7113–7119. [CrossRef]

123. Liu, T.C.; Castelo-Branco, P.; Rabkin, S.D.; Martuza, R.L. Trichostatin A and oncolytic HSV combination therapy shows enhanced antitumoral and antiangiogenic effects. *Mol. Ther.* **2008**, *16*, 1041–1047. [CrossRef]

124. Dornan, M.H.; Krishnan, R.; MacKlin, A.M.; Selman, M.; El Sayes, N.; Son, H.H.; Davis, C.; Chen, A.; Keillor, K.; Le, P.J.; et al. First-in-class small molecule potentiators of cancer virotherapy. *Sci. Rep.* **2016**, *6*, 26786. [CrossRef] [PubMed]

125. Gilbert, J.; Gore, S.D.; Herman, J.G.; Carducci, M.A. The clinical application of targeting cancer through histone acetylation and hypomethylation. *Clin. Cancer Res.* **2004**, *10*, 4589–4596. [CrossRef]

126. Archer, S.Y.; Hodin, R.A. Histone acetylation and cancer. *Curr. Opin. Genet. Dev.* **1999**, *9*, 171–174. [CrossRef]

127. Jacobson, S.; Pillus, L. Modifying chromatin and concepts of cancer. *Curr. Opin. Genet. Dev.* **1999**, *9*, 175–184. [CrossRef]

128. Place, R.F.; Noonan, E.J.; Giardina, C. HDAC inhibition prevents NF-κB activation by suppressing proteasome activity: Downregulation of proteasome subunit expression stabilizes IκBα. *Biochem. Pharmacol.* **2005**, *70*, 394–406. [CrossRef] [PubMed]

129. Bridle, B.W.; Stephenson, K.B.; Boudreau, J.E.; Koshy, S.; Kazdhan, N.; Pullenayegum, E.; Brunellière, J.; Bramson, J.L.; Lichty, B.D.; Wan, Y. Potentiating cancer immunotherapy using an oncolytic virus. *Mol. Ther.* **2010**, *18*, 1430–1439. [CrossRef]

130. Bridle, B.W.; Chen, L.; Lemay, C.G.; Diallo, J.-S.; Pol, J.; Nguyen, A.; Capretta, A.; He, R.; Bramson, J.L.; Bell, J.C.; et al. HDAC inhibition suppresses primary immune responses, enhances secondary immune responses, and abrogates autoimmunity during tumor immunotherapy. *Mol. Ther.* **2013**, *21*, 887–894. [CrossRef] [PubMed]

131. Shen, L.; Shrikant, P.; Ciesielski, M.; Ellis, L.; Fenstermaker, R.; Miles, K.M.; Ramakrishnan, S.; Pili, R.; Sotomayor, P. Class I Histone Deacetylase Inhibitor Entinostat Suppresses Regulatory T Cells and Enhances Immunotherapies in Renal and Prostate Cancer Models. *PLoS ONE* **2012**, *7*, e30815. [CrossRef] [PubMed]

132. Otsuki, A.; Patel, A.; Kasai, K.; Suzuki, M.; Kurozumi, K.; Chiocca, E.A.; Saeki, Y. Histone deacetylase inhibitors augment antitumor efficacy of herpes-based oncolytic viruses. *Mol. Ther.* **2008**, *16*, 1546–1555. [CrossRef]

133. Alvarez-Breckenridge, C.A.; Yu, J.; Price, R.; Wei, M.; Wang, Y.; Nowicki, M.O.; Ha, Y.P.; Bergin, S.; Hwang, C.; Fernandez, S.A.; et al. The histone deacetylase inhibitor valproic acid lessens NK cell action against oncolytic virus-infected glioblastoma cells by inhibition of STAT5/T-BET signaling and generation of gamma interferon. *J. Virol.* **2012**, *86*, 4566–4577. [CrossRef] [PubMed]

134. Jennings, V.A.; Scott, G.B.; Rose, A.M.S.; Scott, K.J.; Migneco, G.; Keller, B.; Reilly, K.; Donnelly, O.; Peach, H.; Dewar, D.; et al. Potentiating Oncolytic Virus-Induced Immune-Mediated Tumor Cell Killing Using Histone Deacetylase Inhibition. *Mol. Ther.* **2019**, *27*, 1139–1152. [CrossRef]

135. Vanhaecke, T.; Papeleu, P.; Elaut, G.; Rogiers, V. Trichostatin A-like Hydroxamate Histone Deacetylase Inhibitors as Therapeutic Agents: Toxicological Point of View. *Curr. Med. Chem.* **2004**, *11*, 1629–1643. [CrossRef]

136. Bracker, T.U.; Sommer, A.; Fichtner, I.; Faus, H.; Haendler, B.; Hess-Stumpp, H. Efficacy of MS-275, a selective inhibitor of class I histone deacetylases, in human colon cancer models. *Int. J. Oncol.* **2009**, *35*, 909–920. [CrossRef] [PubMed]

137. McKinsey, T.A. Isoform-selective HDAC inhibitors: Closing in on translational medicine for the heart. *J. Mol. Cell. Cardiol.* **2011**, *51*, 491–496. [CrossRef] [PubMed]

138. Richon, V.M. Cancer biology: Mechanism of antitumour action of vorinostat (suberoylanilide hydroxamic acid), a novel histone deacetylase inhibitor. *Br. J. Cancer* **2006**, *95*, S2–S6. [CrossRef]

139. Islam, S.; Espitia, C.M.; Persky, D.O.; Carew, J.S.; Nawrocki, S.T. Resistance to histone deacetylase inhibitors confers hypersensitivity to oncolytic reovirus therapy. *Blood Adv.* **2020**, *4*, 5297–5310. [CrossRef]

140. Stiff, A.; Caserta, E.; Sborov, D.W.; Nuovo, G.J.; Mo, X.; Schlotter, S.Y.; Canella, A.; Smith, E.; Badway, J.; Old, M.; et al. Histone Deacetylase Inhibitors Enhance the Therapeutic Potential of Reovirus in Multiple Myeloma. *Mol. Cancer Ther.* **2016**, *15*, 830–841. [CrossRef]

141. Bartlett, D.L.; Liu, Z.; Sathaiah, M.; Ravindranathan, R.; Guo, Z.; He, Y.; Guo, Z.S. Oncolytic viruses as therapeutic cancer vaccines. *Mol. Cancer* **2013**, *12*, 1–16. [CrossRef] [PubMed]

142. Katsura, T.; Iwai, S.; Ota, Y.; Shimizu, H.; Ikuta, K.; Yura, Y. The effects of trichostatin A on the oncolytic ability of herpes simplex virus for oral squamous cell carcinoma cells. *Cancer Gene Ther.* **2009**, *16*, 237–245. [CrossRef] [PubMed]

143. MacTavish, H.; Diallo, J.; Huang, B.; Stanford, M.; Le Boeuf, F.; De Silva, N.; Cox, J.; Simmons, J.G.; Guimond, T.; Falls, T.; et al. Enhancement of vaccinia virus based oncolysis with histone deacetylase inhibitors. *PLoS ONE* **2010**, *5*, e14462. [CrossRef] [PubMed]

144. Li, J.; Bonifati, S.; Hristov, G.; Marttila, T.; Valmary-Degano, S.; Stanzel, S.; Schnölzer, M.; Mougin, C.; Aprahamian, M.; Grekova, S.P.; et al. Synergistic combination of valproic acid and oncolytic parvovirus H-1PV as a potential therapy against cervical and pancreatic carcinomas. *EMBO Mol. Med.* **2013**, *5*, 1537–1555. [CrossRef]

145. Leenstra, S.; Kloezeman, J.J.; de Vrij, J.; Dirven, C.M.F.; Kleijn, A.; van den Bossche, W.; Berghauser Pont, L.M.E.; Lamfers, M.L.M.; Kaufmann, J.K. The HDAC Inhibitors Scriptaid and LBH589 Combined with the Oncolytic Virus Delta24-RGD Exert Enhanced Anti-Tumor Efficacy in Patient-Derived Glioblastoma Cells. *PLoS ONE* **2015**, *10*, e0127058. [CrossRef]

146. Manning, J.; Indrova, M.; Lubyova, B.; Pribylova, H.; Bieblova, J.; Hejnar, J.; Simova, J.; Jandlova, T.; Bubenik, J.; Reinis, M. Induction of MHC class I molecule cell surface expression and epigenetic activation of antigen-processing machinery components in a murine model for human papilloma virus 16-associated tumours. *Immunology* **2008**, *123*, 218–227. [CrossRef]

147. Leone, P.; Shin, E.C.; Perosa, F.; Vacca, A.; Dammacco, F.; Racanelli, V. MHC class I antigen processing and presenting machinery: Organization, function, and defects in tumor cells. *J. Natl. Cancer Inst.* **2013**, *105*, 1172–1187. [CrossRef]

148. Marincola, F.M.; Nicolay, H.J.M.; Altomonte, M.; Guidoboni, M.; Colizzi, F.; Fonsatti, E.; Calabro, L.; Pezzani, L.; Maio, M.; Sigalotti, L. Functional Up-regulation of Human Leukocyte Antigen Class I Antigens Expression by 5-aza-2′-deoxycytidine in Cutaneous Melanoma: Immunotherapeutic Implications. *Clin. Cancer Res.* **2007**, *13*, 3333–3338. [CrossRef]

149. Fernald, K.; Kurokawa, M. Evading apoptosis in cancer. *Trends Cell Biol.* **2013**, *23*, 620–633. [CrossRef]

150. Gong, Y.; Fan, Z.; Luo, G.; Yang, C.; Huang, Q.; Fan, K.; Cheng, H.; Jin, K.; Ni, Q.; Yu, X.; et al. The role of necroptosis in cancer biology and therapy. *Mol. Cancer* **2019**, *18*, 100. [CrossRef]

151. You, Y.; Cheng, A.-C.; Wang, M.-S.; Jia, R.-Y.; Sun, K.-F.; Yang, Q.; Wu, Y.; Zhu, D.; Chen, S.; Liu, M.-F.; et al. The suppression of apoptosis by α-herpesvirus. *Cell Death Dis.* **2017**, *8*, e2749. [CrossRef] [PubMed]

152. Veyer, D.L.; Carrara, G.; Maluquer de Motes, C.; Smith, G.L. Vaccinia virus evasion of regulated cell death. *Immunol. Lett.* **2017**, *186*, 68–80. [CrossRef]

153. Sprague, L.; Braidwood, L.; Conner, J.; Cassady, K.A.; Benencia, F.; Cripe, T.P. Please stand by: How oncolytic viruses impact bystander cells. *Future Virol.* **2018**, *13*, 671–680. [CrossRef]

154. Beug, S.T.; Tang, V.A.; LaCasse, E.C.; Cheung, H.H.; Beauregard, C.E.; Brun, J.; Nuyens, J.P.; Earl, N.; St-Jean, M.; Holbrook, J.; et al. Smac mimetics and innate immune stimuli synergize to promote tumor death. *Nat. Biotechnol.* **2014**, *32*, 182–190. [CrossRef]

155. Yoshida, H. ER stress and diseases. *FEBS J.* **2007**, *274*, 630–658. [CrossRef]

156. Lee, A.S.; Hendershot, L.M. ER stress and cancer. *Cancer Biol. Ther.* **2006**, *5*, 721–722. [CrossRef]

157. Gugliotta, G.; Sudo, M.; Cao, Q.; Lin, D.C.; Sun, H.; Takao, S.; Le Moigne, R.; Rolfe, M.; Gery, S.; Müschen, M.; et al. Valosin-Containing Protein/p97 as a Novel Therapeutic Target in Acute Lymphoblastic Leukemia. *Neoplasia* **2017**, *19*, 750–761. [CrossRef]

158. Prasad, V.; Suomalainen, M.; Pennauer, M.; Yakimovich, A.; Andriasyan, V.; Hemmi, S.; Greber, U.F. Chemical Induction of Unfolded Protein Response Enhances Cancer Cell Killing through Lytic Virus Infection. *J. Virol.* **2014**, *88*, 13086–13098. [CrossRef] [PubMed]

159. Zhang, H.; Li, K.; Lin, Y.; Xing, F.; Xiao, X.; Cai, J.; Zhu, W.; Liang, J.; Tan, Y.; Fu, L.; et al. Targeting VCP enhances anticancer activity of oncolytic virus M1 in hepatocellular carcinoma. *Sci. Transl. Med.* **2017**, *9*, eaam7996. [CrossRef]

160. Prasad, V.; Greber, U.F. The endoplasmic reticulum unfolded protein response—Homeostasis, cell death and evolution in virus infections. *FEMS Microbiol. Rev.* **2021**, *16*, 1–19. [CrossRef]

161. Taverner, W.K.; Jacobus, E.J.; Christianson, J.; Champion, B.; Paton, A.W.; Paton, J.C.; Su, W.; Cawood, R.; Seymour, L.W.; Lei-Rossmann, J. Calcium Influx Caused by ER Stress Inducers Enhances Oncolytic Adenovirus Enadenotucirev Replication and Killing through PKCα Activation. *Mol. Ther. Oncolytics* **2019**, *15*, 117–130. [CrossRef] [PubMed]

162. Galmarini, C.M.; Mackey, J.R.; Dumontet, C. Nucleoside analogues and nucleobases in cancer treatment. *Lancet Oncol.* **2002**, *3*, 415–424. [CrossRef]

163. Nakano, K.; Todo, T.; Zhao, G.; Yamaguchi, K.; Kuroki, S.; Cohen, J.B.; Glorioso, J.C.; Tanaka, M. Enhanced efficacy of conditionally replicating herpes simplex virus (G207) combined with 5-fluorouracil and surgical resection in peritoneal cancer dissemination models. *J. Gene Med.* **2005**, *7*, 638–648. [CrossRef]

164. Leitner, S.; Sweeney, K.; Öberg, D.; Davies, D.; Miranda, E.; Lemoine, N.R.; Halldén, G. Oncolytic adenoviral mutants with E1B19KGene deletions enhance gemcitabine-induced apoptosis in pancreatic carcinoma cells and anti-tumor efficacy in vivo. *Clin. Cancer Res.* **2009**, *15*, 1730–1740. [CrossRef]

165. Gutermann, A.; Mayer, E.; von Dehn-Rothfelser, K.; Breidenstein, C.; Weber, M.; Muench, M.; Gungor, D.; Suehnel, J.; Moebius, U.; Lechmann, M. Efficacy of oncolytic herpesvirus NV1020 can be enhanced by combination with chemotherapeutics in colon carcinoma cells. *Hum. Gene Ther.* **2006**, *17*, 1241–1253. [CrossRef]

166. Angelova, A.L.; Aprahamian, M.; Grekova, S.P.; Hajri, A.; Leuchs, B.; Giese, N.A.; Dinsart, C.; Herrmann, A.; Balboni, G.; Rommelaere, J.; et al. Improvement of gemcitabine-based therapy of pancreatic carcinoma by means of oncolytic parvovirus H-1PV. *Clin. Cancer Res.* **2009**, *15*, 511–519. [CrossRef]

167. Weiland, T.; Lampe, J.; Essmann, F.; Venturelli, S.; Berger, A.; Bossow, S.; Berchtold, S.; Schulze-Osthoff, K.; Lauer, U.M.; Bitzer, M. Enhanced killing of therapy-induced senescent tumor cells by oncolytic measles vaccine viruses. *Int. J. Cancer* **2014**, *134*, 235–243. [CrossRef]

168. Chen, W.; Fan, W.; Ru, G.; Huang, F.; Lu, X.; Zhang, X.; Mou, X.; Wang, S. Gemcitabine combined with an engineered oncolytic vaccinia virus exhibits a synergistic suppressive effect on the tumor growth of pancreatic cancer. *Oncol. Rep.* **2019**, *41*, 67–76. [CrossRef]

169. Liu, X.; Yang, Z.; Li, Y.; Zhu, Y.; Li, W.; Li, S.; Wang, J.; Cui, Y.; Shang, C.; Liu, Z.; et al. Chemovirotherapy of Lung Squamous Cell Carcinoma by Combining Oncolytic Adenovirus with Gemcitabine. *Front. Oncol.* **2020**, *10*, 229. [CrossRef]

170. May, V.; Berchtold, S.; Berger, A.; Venturelli, S.; Burkard, M.; Leischner, C.; Malek, N.P.; Lauer, U.M. Chemovirotherapy for pancreatic cancer: Gemcitabine plus oncolytic measles vaccine virus. *Oncol. Lett.* **2019**, *18*, 5534–5542. [CrossRef] [PubMed]

171. Mannhold, R.; Fulda, S.; Carosati, E. IAP antagonists: Promising candidates for cancer therapy. *Drug Discov. Today* **2010**, *15*, 210–219. [CrossRef]

172. Michie, J.; Kearney, C.J.; Hawkins, E.D.; Silke, J.; Oliaro, J. The Immuno-Modulatory Effects of Inhibitor of Apoptosis Protein Antagonists in Cancer Immunotherapy. *Cells* **2020**, *9*, 207. [CrossRef]

173. Tumilasci, V.F.; Oliere, S.; Nguyen, T.L.-A.; Shamy, A.; Bell, J.; Hiscott, J. Targeting the Apoptotic Pathway with BCL-2 Inhibitors Sensitizes Primary Chronic Lymphocytic Leukemia Cells to Vesicular Stomatitis Virus-Induced Oncolysis. *J. Virol.* **2008**, *82*, 8487–8499. [CrossRef]

174. Sarkar, S.; Quinn, B.A.; Shen, X.-N.; Dash, R.; Das, S.K.; Emdad, L.; Klibanov, A.L.; Wang, X.-Y.; Pellecchia, M.; Sarkar, D.; et al. Therapy of prostate cancer using a novel cancer terminator virus and a small molecule BH-3 mimetic. *Oncotarget* **2015**, *6*, 10712–10727. [CrossRef] [PubMed]

175. Karin, M. Nuclear factor-κB in cancer development and progression. *Nature* **2006**, *441*, 431–436. [CrossRef]

176. Petersen, S.L.; Wang, L.; Yalcin-Chin, A.; Li, L.; Peyton, M.; Minna, J.; Harran, P.; Wang, X. Autocrine TNFα Signaling Renders Human Cancer Cells Susceptible to Smac-Mimetic-Induced Apoptosis. *Cancer Cell* **2007**, *12*, 445–456. [CrossRef]

177. Du, C.; Fang, M.; Li, Y.; Li, L.; Wang, X. Smac, a mitochondrial protein that promotes cytochrome c-dependent caspase activation by eliminating IAP inhibition. *Cell* **2000**, *102*, 33–42. [CrossRef]

178. Li, L.; Thomas, R.M.; Suzuki, H.; De Brabander, J.K.; Wang, X.; Harran, P.G. A small molecule Smac mimic potentiates TRAIL- and TNFα-mediated cell death. *Science* **2004**, *305*, 1471–1474. [CrossRef]

179. Petersen, S.L.; Peyton, M.; Minna, J.D.; Wang, X. Overcoming cancer cell resistance to Smac mimetic induced apoptosis by modulating cIAP-2 expression. *Proc. Natl. Acad. Sci. USA* **2010**, *107*, 11936–11941. [CrossRef]

180. Bai, L.; McEachern, D.; Yang, C.Y.; Lu, J.; Sun, H.; Wang, S. LRIG1 modulates cancer cell sensitivity to Smac mimetics by regulating TNFα expression and receptor tyrosine kinase signaling. *Cancer Res.* **2012**, *72*, 1229–1238. [CrossRef]

181. Fulda, S. Smac mimetics as IAP antagonists. *Semin. Cell Dev. Biol.* **2015**, *39*, 132–138. [CrossRef]

182. Fulda, S.; Wick, W.; Weller, M.; Debatin, K.M. Smac agonists sensitize for Apo2L/TRAIL-or anticancer drug-induced apoptosis and induce regression of malignant glioma in vivo. *Nat. Med.* **2002**, *8*, 808–815. [CrossRef]

183. Cai, J.; Lin, Y.; Zhang, H.; Liang, J.; Tan, Y.; Cavenee, W.K.; Yan, G. Selective replication of oncolytic virus M1 results in a bystander killing effect that is potentiated by Smac mimetics. *Proc. Natl. Acad. Sci. USA* **2017**, *114*, 201701002. [CrossRef] [PubMed]

184. Cheung, H.H.; Beug, S.T.; St Jean, M.; Brewster, A.; Kelly, N.L.; Wang, S.; Korneluk, R.G. Smac mimetic compounds potentiate interleukin-1β-mediated cell death. *J. Biol. Chem.* **2010**, *285*, 40612–40623. [CrossRef] [PubMed]

185. Lalaoui, N.; Hänggi, K.; Brumatti, G.; Chau, D.; Nguyen, N.Y.N.; Vasilikos, L.; Spilgies, L.M.; Heckmann, D.A.; Ma, C.; Ghisi, M.; et al. Targeting p38 or MK2 Enhances the Anti-Leukemic Activity of Smac-Mimetics. *Cancer Cell* **2016**, *29*, 145–158. [CrossRef]

186. Beug, S.T.; Pichette, S.J.; St-Jean, M.; Holbrook, J.; Walker, D.E.; LaCasse, E.C.; Korneluk, R.G. Combination of IAP Antagonists and TNF-α-Armed Oncolytic Viruses Induce Tumor Vascular Shutdown and Tumor Regression. *Mol. Ther. Oncolytics* **2018**, *10*, 28–39. [CrossRef]

187. Beug, S.T.; Conrad, D.P.; Alain, T.; Korneluk, R.G.; Lacasse, E.C. Combinatorial cancer immunotherapy strategies with proapoptotic small-molecule IAP antagonists. *Int. J. Dev. Biol.* **2015**, *59*, 141–147. [CrossRef]

188. Beug, S.T.; Beauregard, C.E.; Healy, C.; Sanda, T.; St-Jean, M.; Chabot, J.; Walker, D.E.; Mohan, A.; Earl, N.; Lun, X.; et al. Smac mimetics synergize with immune checkpoint inhibitors to promote tumour immunity against glioblastoma. *Nat. Commun.* **2017**, *8*, 14278. [CrossRef]

189. Dobson, C.C.; Naing, T.; Beug, S.T.; Faye, M.D.; Chabot, J.; St-Jean, M.; Walker, D.E.; LaCasse, E.C.; Stojdl, D.F.; Korneluk, R.G.; et al. Oncolytic virus synergizes with Smac mimetic compounds to induce rhabdomyosarcoma cell death in a syngeneic murine model. *Oncotarget* **2016**. [CrossRef]

190. Kim, D.-S.; Dastidar, H.; Zhang, C.; Zemp, F.J.; Lau, K.; Ernst, M.; Rakic, A.; Sikdar, S.; Rajwani, J.; Naumenko, V.; et al. Smac mimetics and oncolytic viruses synergize in driving anticancer T-cell responses through complementary mechanisms. *Nat. Commun.* **2017**, *8*, 344. [CrossRef] [PubMed]

191. Cai, J.; Yan, G. The identification and development of a novel oncolytic virus: Alphavirus M1. *Hum. Gene Ther.* **2021**, *32*, 138–149. [CrossRef]

192. Delbridge, A.R.D.; Strasser, A. The BCL-2 protein family, BH3-mimetics and cancer therapy. *Cell Death Differ.* **2015**, *22*, 1071–1080. [CrossRef] [PubMed]

193. Takai, S.; Hodge, J.W.; Farsaci, B.; Schlom, J.; Sabzevari, H.; Higgins, J.P.; Di Bari, M.G. Effect of a small molecule BCL-2 inhibitor on immune function and use with a recombinant vaccine. *Int. J. Cancer* **2010**, *127*, 1603–1613. [CrossRef]

194. Samuel, S.; Tumilasci, V.F.; Oliere, S.; Nguyên, T.L.; Shamy, A.; Bell, J.; Hiscott, J. VSV oncolysis in combination with the BCL-2 inhibitor obatoclax overcomes apoptosis resistance in chronic lymphocytic leukemia. *Mol. Ther.* **2010**, *18*, 2094–2103. [CrossRef]

195. Zhou, J.; Giannakakou, P. Targeting microtubules for cancer chemotherapy. *Curr. Med. Chem. Anticancer Agents* **2005**, *5*, 65–71. [CrossRef]

196. Zeng, W.G.; Li, J.J.; Hu, P.; Lei, L.; Wang, J.N.; Liu, R. Bin An oncolytic herpes simplex virus vector, G47Δ, synergizes with paclitaxel in the treatment of breast cancer. *Oncol. Rep.* **2013**, *29*, 2355–2361. [CrossRef]

197. Lin, S.F.; Gao, S.P.; Price, D.L.; Li, S.; Chou, T.C.; Singh, P.; Huang, Y.Y.; Fong, Y.; Wong, R.J. Synergy of a herpes oncolytic virus and paclitaxel for anaplastic thyroid cancer. *Clin. Cancer Res.* **2008**, *14*, 1519–1528. [CrossRef]

198. Passer, B.J.; Castelo-Branco, P.; Buhrman, J.S.; Varghese, S.; Rabkin, S.D.; Martuza, R.L. Oncolytic herpes simplex virus vectors and taxanes synergize to promote killing of prostate cancer cells. *Cancer Gene Ther.* **2009**, *16*, 551–560. [CrossRef]

199. Heinemann, L.; Simpson, G.R.; Boxall, A.; Kottke, T.; Relph, K.L.; Vile, R.; Melcher, A.; Prestwich, R.; Harrington, K.J.; Morgan, R.; et al. Synergistic effects of oncolytic reovirus and docetaxel chemotherapy in prostate cancer. *BMC Cancer* **2011**, *11*, 221. [CrossRef]

200. Karapanagiotou, E.M.; Roulstone, V.; Twigger, K.; Ball, M.; Tanay, M.A.; Nutting, C.; Newbold, K.; Gore, M.E.; Larkin, J.; Syrigos, K.N.; et al. Phase I/II trial of carboplatin and paclitaxel chemotherapy in combination with intravenous oncolytic reovirus in patients with advanced malignancies. *Clin. Cancer Res.* **2012**, *18*, 2080–2089. [CrossRef] [PubMed]

201. Fujiwara, T.; Kagawa, S.; Kishimoto, H.; Endo, Y.; Hioki, M.; Ikeda, Y.; Sakai, R.; Urata, Y.; Tanaka, N.; Fujiwara, T. Enhanced antitumor efficacy of telomerase-selective oncolytic adenoviral agent OBP-401 with docetaxel: Preclinical evaluation of chemovirotherapy. *Int. J. Cancer* **2006**, *119*, 432–440. [CrossRef] [PubMed]

202. Bourgeois-Daigneault, M.-C.; St-Germain, L.E.; Roy, D.G.; Pelin, A.; Aitken, A.S.; Arulanandam, R.; Falls, T.; Garcia, V.; Diallo, J.-S.; Bell, J.C. Combination of Paclitaxel and MG1 oncolytic virus as a successful strategy for breast cancer treatment. *Breast Cancer Res.* **2016**, *18*, 83. [CrossRef] [PubMed]

203. Sei, S.; Mussio, J.K.; Yang, Q.E.; Nagashima, K.; Parchment, R.E.; Coffey, M.C.; Shoemaker, R.H.; Tomaszewski, J.E. Synergistic antitumor activity of oncolytic reovirus and chemotherapeutic agents in non-small cell lung cancer cells. *Mol. Cancer* **2009**, *8*, 47. [CrossRef]

204. Lal, G.; Rajala, M.S. Combination of oncolytic measles virus armed with BNiP3, a pro-apoptotic gene and paclitaxel induces breast cancer cell death. *Front. Oncol.* **2019**, *9*, 676. [CrossRef]

205. Garofalo, M.; Saari, H.; Somersalo, P.; Crescenti, D.; Kuryk, L.; Aksela, L.; Capasso, C.; Madetoja, M.; Koskinen, K.; Oksanen, T.; et al. Antitumor effect of oncolytic virus and paclitaxel encapsulated in extracellular vesicles for lung cancer treatment. *J. Control. Release* **2018**, *283*, 223–234. [CrossRef]

206. Arulanandam, R.; Batenchuk, C.; Varette, O.; Zakaria, C.; Garcia, V.; Forbes, N.E.; Davis, C.; Krishnan, R.; Karmacharya, R.; Cox, J.; et al. Microtubule disruption synergizes with oncolytic virotherapy by inhibiting interferon translation and potentiating bystander killing. *Nat. Commun.* **2015**, *6*, 6410. [CrossRef]

207. Ziauddin, M.F.; Guo, Z.S.; O'Malley, M.E.; Austin, F.; Popovic, P.J.; Kavanagh, M.A.; Li, J.; Sathaiah, M.; Thirunavukarasu, P.; Fang, B.; et al. TRAIL gene-armed oncolytic poxvirus and oxaliplatin can work synergistically against colorectal cancer. *Gene Ther.* **2010**, *17*, 550–559. [CrossRef]

208. Pan, Q.; Huang, Y.; Chen, L.; Gu, J.F.; Zhou, X. SMAC-armed vaccinia virus induces both apoptosis and necroptosis and synergizes the efficiency of vinblastine in HCC. *Hum. Cell* **2014**, *27*, 162–171. [CrossRef] [PubMed]

209. Cinatl, J.; Cinatl, J.; Michaelis, M.; Kabickova, H.; Kotchetkov, R.; Vogel, J.U.; Doerr, H.W.; Klingebiel, T.; Hernáiz Driever, P. Potent oncolytic activity of multimutated herpes simplex virus G207 in combination with vincristine against human rhabdomyosarcoma. *Cancer Res.* **2003**, *63*, 1508–1514. [CrossRef] [PubMed]

210. Moehler, M.; Sieben, M.; Roth, S.; Springsguth, F.; Leuchs, B.; Zeidler, M.; Dinsart, C.; Rommelaere, J.; Galle, P.R. Activation of the human immune system by chemotherapeutic or targeted agents combined with the oncolytic parvovirus H-1. *BMC Cancer* **2011**, *11*, 464. [CrossRef]

211. Choi, C.-H.; Cha, Y.-J.; An, C.-S.; Kim, K.-J.; Kim, K.-C.; Moon, S.-P.; Lee, Z.; Min, Y.-D. Molecular mechanisms of heptaplatin effective against cisplatin-resistant cancer cell lines: Less involvement of metallothionein. *Cancer Cell Int.* **2004**, *4*, 6. [CrossRef]

212. Khuri, F.R.; Nemunaitis, J.; Ganly, I.; Arseneau, J.; Tannock, I.F.; Romel, L.; Gore, M.; Ironside, J.; MacDougall, R.H.; Heise, C.; et al. A controlled trial of intratumoral ONYX-015, a selectively-replicating adenovirus, in combination with cisplatin and 5-fluorouracil in patients with recurrent head and neck cancer. *Nat. Med.* **2000**, *6*, 879–885. [CrossRef]

213. Mell, L.K.; Brumund, K.T.; Daniels, G.A.; Advani, S.J.; Zakeri, K.; Wright, M.E.; Onyeama, S.J.; Weisman, R.A.; Sanghvi, P.R.; Martin, P.J.; et al. Phase I trial of intravenous oncolytic vaccinia virus (GL-ONC1) with cisplatin and radiotherapy in patients with locoregionally advanced head and neck carcinoma. *Clin. Cancer Res.* **2017**, *23*, 5696–5702. [CrossRef]

214. Arulanandam, R.; Taha, Z.; Garcia, V.; Selman, M.; Chen, A.; Varette, O.; Jirovec, A.; Sutherland, K.; Macdonald, E.; Tzelepis, F.; et al. The strategic combination of trastuzumab emtansine with oncolytic rhabdoviruses leads to therapeutic synergy. *Commun. Biol.* **2020**. [CrossRef]

215. Hong, B.; Chapa, V.; Saini, U.; Modgil, P.; Cohn, D.E.; He, G.; Siddik, Z.H.; Sood, A.K.; Yan, Y.; Selvendiran, K.; et al. Oncolytic HSV therapy modulates vesicular trafficking inducing cisplatin sensitivity and antitumor immunity. *Clin. Cancer Res.* **2021**, *27*, 542–553. [CrossRef]

216. You, L.; Wang, Y.; Jin, Y.; Qian, W. Downregulation of Mcl-1 synergizes the apoptotic response to combined treatment with cisplatin and a novel fiber chimeric oncolytic adenovirus. *Oncol. Rep.* **2012**, *27*, 971–978. [CrossRef]

217. Champoux, J.J. DNA Topoisomerases: Structure, Function, and Mechanism. *Annu. Rev. Biochem.* **2002**, *70*, 369–413. [CrossRef]

218. Hande, K.R. Etoposide: Four decades of development of a topoisomerase II inhibitor. *Eur. J. Cancer* **1998**, *34*, 1514–1521. [CrossRef]

219. Ewesuedo, R.B.; Ratain, M.J. Topoisomerase I Inhibitors. *Oncologist* **1997**, *2*, 359–364. [CrossRef]

220. Ezoe, S. Secondary leukemia associated with the anti-cancer agent, etoposide, a topoisomerase II inhibitor. *Int. J. Environ. Res. Public Health* **2012**, *9*, 2444–2453. [CrossRef]

221. Felix, C.A. Leukemias related to treatment with DNA topoisomerase II inhibitors. *Med. Pediatr. Oncol. Off. J. SIOP Int. Soc. Pediatr. Oncol.* **2001**, *36*, 525–535. [CrossRef] [PubMed]

222. Skelding, K.A.; Barry, R.D.; Shafren, D.R. Enhanced oncolysis mediated by Coxsackievirus A21 in combination with doxorubicin hydrochloride. *Investig. New Drugs* **2012**, *30*, 568–581. [CrossRef] [PubMed]

223. Schache, P.; Gürlevik, E.; Strüver, N.; Woller, N.; Malek, N.; Zender, L.; Manns, M.; Wirth, T.; Kühnel, F.; Kubicka, S. VSV virotherapy improves chemotherapy by triggering apoptosis due to proteasomal degradation of Mcl-1. *Gene Ther.* **2009**, *16*, 849–861. [CrossRef]

224. Siurala, M.; Bramante, S.; Vassilev, L.; Hirvinen, M.; Parviainen, S.; Tähtinen, S.; Guse, K.; Cerullo, V.; Kanerva, A.; Kipar, A.; et al. Oncolytic adenovirus and doxorubicin-based chemotherapy results in synergistic antitumor activity against soft-tissue sarcoma. *Int. J. Cancer* **2015**, *136*, 945–954. [CrossRef]

225. Rodríguez Stewart, R.M.; Berry, J.T.L.; Berger, A.K.; Yoon, S.B.; Hirsch, A.L.; Guberman, J.A.; Patel, N.B.; Tharp, G.K.; Bosinger, S.E.; Mainou, B.A. Enhanced Killing of Triple-Negative Breast Cancer Cells by Reassortant Reovirus and Topoisomerase Inhibitors. *J. Virol.* **2019**, *93*. [CrossRef] [PubMed]

226. Berry, J.T.L.; Muñoz, L.E.; Rodríguez Stewart, R.M.; Selvaraj, P.; Mainou, B.A. Doxorubicin Conjugation to Reovirus Improves Oncolytic Efficacy in Triple-Negative Breast Cancer. *Mol. Ther. Oncolytics* **2020**, *18*, 556–572. [CrossRef]

227. Denard, B.; Seemann, J.; Chen, Q.; Gay, A.; Huang, H.; Chen, Y.; Ye, J. The membrane-bound transcription factor CREB3L1 is activated in response to virus infection to inhibit proliferation of virus-infected cells. *Cell Host Microbe* **2011**, *10*, 65–74. [CrossRef] [PubMed]

228. Mistarz, A.; Graczyk, M.; Winkler, M.; Singh, P.K.; Cortes, E.; Miliotto, A.; Liu, S.; Long, M.; Yan, L.; Stablewski, A.; et al. Induction of Cell Death in Ovarian Cancer Cells by Doxorubicin and Oncolytic Vaccinia Virus is Associated with CREB3L1 Activation. *Mol. Ther. Oncolytics* **2021**. [CrossRef]

229. Aitken, A.S.; Roy, D.G.; Bourgeois-Daigneault, M.-C. Taking a Stab at Cancer; Oncolytic Virus-Mediated Anti-Cancer Vaccination Strategies. *Biomedicines* **2017**, *5*, 3. [CrossRef]

230. Watanabe, N.; McKenna, M.K.; Rosewell Shaw, A.; Suzuki, M. Clinical CAR-T Cell and Oncolytic Virotherapy for Cancer Treatment. *Mol. Ther.* **2021**, *29*, 505–520. [CrossRef] [PubMed]

231. Mondal, M.; Guo, J.; He, P.; Zhou, D. Recent advances of oncolytic virus in cancer therapy. *Hum. Vaccines Immunother.* **2020**, *16*, 2389–2402. [CrossRef] [PubMed]

232. Sivanandam, V.; LaRocca, C.J.; Chen, N.G.; Fong, Y.; Warner, S.G. Oncolytic Viruses and Immune Checkpoint Inhibition: The Best of Both Worlds. *Mol. Ther. Oncolytics* **2019**, *13*, 93–106. [CrossRef] [PubMed]

233. Zou, P.; Tang, R.; Luo, M. Oncolytic virotherapy, alone or in combination with immune checkpoint inhibitors, for advanced melanoma: A systematic review and meta-analysis. *Int. Immunopharmacol.* **2020**, *78*, 106050. [CrossRef] [PubMed]

234. Roma-Rodrigues, C.; Mendes, R.; Baptista, P.; Fernandes, A. Targeting Tumor Microenvironment for Cancer Therapy. *Int. J. Mol. Sci.* **2019**, *20*, 840. [CrossRef]

235. Zhang, B.; Wang, X.; Cheng, P. Remodeling of Tumor Immune Microenvironment by Oncolytic Viruses. *Front. Oncol.* **2021**, *10*, 3478. [CrossRef] [PubMed]

236. Chaudhary, B.; Elkord, E. Regulatory T Cells in the Tumor Microenvironment and Cancer Progression: Role and Therapeutic Targeting. *Vaccines* **2016**, *4*, 28. [CrossRef] [PubMed]

237. Viehl, C.T.; Moore, T.T.; Liyanage, U.K.; Frey, D.M.; Ehlers, J.P.; Eberlein, T.J.; Goedegebuure, P.S.; Linehan, D.C. Depletion of CD4+CD25+ regulatory T cells promotes a tumor-specific immune response in pancreas cancer-bearing mice. *Ann. Surg. Oncol.* **2006**, *13*, 1252–1258. [CrossRef]

238. Lun, X.Q.; Jang, J.-H.; Tang, N.; Deng, H.; Head, R.; Bell, J.C.; Stojdl, D.F.; Nutt, C.L.; Senger, D.L.; Forsyth, P.A.; et al. Efficacy of systemically administered oncolytic vaccinia virotherapy for malignant gliomas is enhanced by combination therapy with rapamycin or cyclophosphamide. *Clin. Cancer Res.* **2009**, *15*, 2777–2788. [CrossRef]

239. Fulci, G.; Breymann, L.; Gianni, D.; Kurozomi, K.; Rhee, S.S.; Yu, J.; Kaur, B.; Louis, D.N.; Weissleder, R.; Caligiuri, M.A.; et al. Cyclophosphamide enhances glioma virotherapy by inhibiting innate immune responses. *Proc. Natl. Acad. Sci. USA* **2006**, *103*, 12873–12878. [CrossRef]

240. Ikeda, K.; Ichikawa, T.; Wakimoto, H.; Silver, J.S.; Deisboeck, T.S.; Finkelstein, D.; Harsh, G.R.; Louis, D.N.; Bartus, R.T.; Hochberg, F.H.; et al. Oncolytic virus therapy of multiple tumors in the brain requires suppression of innate and elicited antiviral responses. *Nat. Med.* **1999**, *5*, 881–887. [CrossRef]

241. Peng, K.-W.; Myers, R.; Greenslade, A.; Mader, E.; Greiner, S.; Federspiel, M.J.; Dispenzieri, A.; Russell, S.J. Using clinically approved cyclophosphamide regimens to control the humoral immune response to oncolytic viruses. *Gene Ther.* **2013**, *20*, 255–261. [CrossRef] [PubMed]

242. Cerullo, V.; Diaconu, I.; Kangasniemi, L.; Rajecki, M.; Escutenaire, S.; Koski, A.; Romano, V.; Rouvinen, N.; Tuuminen, T.; Laasonen, L.; et al. Immunological Effects of Low-dose Cyclophosphamide in Cancer Patients Treated with Oncolytic Adenovirus. *Mol. Ther.* **2011**, *19*, 1737–1746. [CrossRef]

243. Ghiringhelli, F.; Larmonier, N.; Schmitt, E.; Parcellier, A.; Cathelin, D.; Garrido, C.; Chauffert, B.; Solary, E.; Bonnotte, B.; Martin, F. CD4+CD25+ regulatory T cells suppress tumor immunity but are sensitive to cyclophosphamide which allows immunotherapy of established tumors to be curative. *Eur. J. Immunol.* **2004**, *34*, 336–344. [CrossRef] [PubMed]

244. Madondo, M.T.; Quinn, M.; Plebanski, M. Low dose cyclophosphamide: Mechanisms of T cell modulation. *Cancer Treat. Rev.* **2016**, *42*, 3–9. [CrossRef] [PubMed]

245. Pol, J.G.; Atherton, M.J.; Stephenson, K.B.; Bridle, B.W.; Workenhe, S.T.; Kazdhan, N.; McGray, A.J.R.; Wan, Y.; Kroemer, G.; Lichty, B.D. Enhanced immunotherapeutic profile of oncolytic virus-based cancer vaccination using cyclophosphamide preconditioning. *J. Immunother. Cancer* **2020**, *8*. [CrossRef]

246. Roland, C.L.; Lynn, K.D.; Toombs, J.E.; Dineen, S.P.; Udugamasooriya, D.G.; Brekken, R.A. Cytokine levels correlate with immune cell infiltration after anti-VEGF therapy in preclinical mouse models of breast cancer. *PLoS ONE* **2009**, *4*, e7669. [CrossRef]

247. Zhao, Q.; Guo, J.; Wang, G.; Chu, Y.; Hu, X. Suppression of immune regulatory cells with combined therapy of celecoxib and sunitinib in renal cell carcinoma. *Oncotarget* **2017**, *8*, 1668–1677. [CrossRef]

248. Zhao, Y.; Ma, J.; Fan, Y.; Wang, Z.; Tian, R.; Ji, W.; Zhang, F.; Niu, R. TGF-β transactivates EGFR and facilitates breast cancer migration and invasion through canonical Smad3 and ERK/Sp1 signaling pathways. *Mol. Oncol.* **2018**, *12*, 305–321. [CrossRef]

249. Akhurst, R.J.; Hata, A. Targeting the TGFβ signalling pathway in disease. *Nat. Rev. Drug Discov.* **2012**, *11*, 790–811. [CrossRef]

250. Pickup, M.; Novitskiy, S.; Moses, H.L. The roles of TGFβ in the tumour microenvironment. *Nat. Rev. Cancer* **2013**, *13*, 788–799. [CrossRef]

251. Padua, D.; Massagué, J. Roles of TGFβ in metastasis. *Cell Res.* **2009**, *19*, 89–102. [CrossRef] [PubMed]

252. Yang, L.; Pang, Y.; Moses, H.L. TGF-β and immune cells: An important regulatory axis in the tumor microenvironment and progression. *Trends Immunol.* **2010**, *31*, 220–227. [CrossRef] [PubMed]

253. Groeneveldt, C.; van Hall, T.; van der Burg, S.H.; ten Dijke, P.; van Montfoort, N. Immunotherapeutic Potential of TGF-β Inhibition and Oncolytic Viruses. *Trends Immunol.* **2020**, *41*, 406–420. [CrossRef] [PubMed]

254. Hamashima, Y.; Saitoh, M.; Node, M.; Hanyu, A.; Tojo, M.; Imamura, T.; Miyazono, K.; Kajimoto, T. The ALK-5 inhibitor A-83-01 inhibits Smad signaling and epithelial-to-mesenchymal transition by transforming growth factor-beta. *Cancer Sci.* **2005**, *96*, 791–800. [CrossRef]

255. Hutzen, B.; Chen, C.Y.; Wang, P.Y.; Sprague, L.; Swain, H.M.; Love, J.; Conner, J.; Boon, L.; Cripe, T.P. TGF-β Inhibition Improves Oncolytic Herpes Viroimmunotherapy in Murine Models of Rhabdomyosarcoma. *Mol. Ther. Oncolytics* **2017**, *7*, 17–26. [CrossRef] [PubMed]

256. Yingling, J.M.; McMillen, W.T.; Yan, L.; Huang, H.; Sawyer, J.S.; Graff, J.; Clawson, D.K.; Britt, K.S.; Anderson, B.D.; Beight, D.W.; et al. Preclinical assessment of galunisertib (LY2157299 monohydrate), a first-in-class transforming growth factor-β receptor type I inhibitor. *Oncotarget* **2018**, *9*, 6659–6677. [CrossRef]

257. Esaki, S.; Nigim, F.; Moon, E.; Luk, S.; Kiyokawa, J.; Curry, W.; Cahill, D.P.; Chi, A.S.; Iafrate, A.J.; Martuza, R.L.; et al. Blockade of transforming growth factor-β signaling enhances oncolytic herpes simplex virus efficacy in patient-derived recurrent glioblastoma models. *Int. J. Cancer* **2017**, *141*, 2348–2358. [CrossRef]

258. Ishigami, I.; Shuwari, N.; Kaminade, T.; Mizuguchi, H.; Sakurai, F. A TGFβ Signaling Inhibitor, SB431542, inhibits reovirus-mediated lysis of human hepatocellular carcinoma cells in a TGFβ-independent Manner. *Anticancer Res.* **2021**, *41*, 2431–2440. [CrossRef]

259. Huang, K.C.-Y.; Chiang, S.-F.; Yang, P.-C.; Ke, T.-W.; Chen, T.-W.; Hu, C.-H.; Huang, Y.-W.; Chang, H.-Y.; Chen, W.T.-L.; Chao, K.S.C. Immunogenic Cell Death by the Novel Topoisomerase I Inhibitor TLC388 Enhances the Therapeutic Efficacy of Radiotherapy. *Cancers* **2021**, *13*, 1218. [CrossRef]

260. Haggerty, T.J.; Dunn, I.S.; Rose, L.B.; Newton, E.E.; Martin, S.; Riley, J.L.; Kurnick, J.T. Topoisomerase inhibitors modulate expression of melanocytic antigens and enhance T cell recognition of tumor cells. *Cancer Immunol. Immunother.* **2011**, *60*, 133–144. [CrossRef]

261. Workenhe, S.T.; Pol, J.G.; Lichty, B.D.; Cummings, D.T.; Mossman, K.L. Combining Oncolytic HSV-1 with Immunogenic Cell Death-Inducing Drug Mitoxantrone Breaks Cancer Immune Tolerance and Improves Therapeutic Efficacy. *Cancer Immunol. Res.* **2013**, *1*, 309–319. [CrossRef]

262. Fan, J.; Jiang, H.; Cheng, L.; Ma, B.; Liu, R. Oncolytic herpes simplex virus and temozolomide synergistically inhibit breast cancer cell tumorigenesis invitro and in vivo. *Oncol. Lett.* **2021**, *21*, 1-1. [CrossRef]

263. Saha, D.; Rabkin, S.D.; Martuza, R.L. Temozolomide antagonizes oncolytic immunovirotherapy in glioblastoma. *J. Immunother. Cancer* **2020**, *8*, 345. [CrossRef]

264. Kumar, S.; Wu, J.; Wang, F.; Wang, L.; Chen, L.; Sokirniy, I.; Wang, H.; Sterner, D.; Grove, C.; Cunnion, B.; et al. Abstract B202: Covalent irreversible usp7 inhibitors for cancer immunotherapy. In: Proceedings of the AACR-NCI-EORTC International Conference: Molecular Targets and Cancer Therapeutics. *Mol. Cancer Ther.* **2018**, *17*, B202.

265. Fu, C.; Zhu, X.; Xu, P.; Li, Y. Pharmacological inhibition of USP7 promotes antitumor immunity and contributes to colon cancer therapy. *OncoTargets Ther.* **2019**, *12*, 609–617. [CrossRef] [PubMed]

266. Hanna, B.S.; Roessner, P.M.; Scheffold, A.; Jebaraj, B.M.C.; Demerdash, Y.; Öztürk, S.; Lichter, P.; Stilgenbauer, S.; Seiffert, M. PI3Kδ inhibition modulates regulatory and effector T-cell differentiation and function in chronic lymphocytic leukemia. *Leukemia* **2018**, *33*, 1427–1438. [CrossRef]

267. De Almeida Nagata, D.E.; Chiang, E.Y.; Jhunjhunwala, S.; Caplazi, P.; Arumugam, V.; Modrusan, Z.; Chan, E.; Merchant, M.; Jin, L.; Arnott, D.; et al. Regulation of Tumor-Associated Myeloid Cell Activity by CBP/EP300 Bromodomain Modulation of H3K27 Acetylation. *Cell Rep.* **2019**, *27*, 269–281. [CrossRef] [PubMed]

268. Shi, T.; Song, X.; Wang, Y.; Liu, F.; Wei, J. Combining Oncolytic Viruses with Cancer Immunotherapy: Establishing a New Generation of Cancer Treatment. *Front. Immunol.* **2020**, *11*, 683. [CrossRef]

269. Marchini, A.; Scott, E.M.; Rommelaere, J. Overcoming barriers in oncolytic virotherapy with HDAC inhibitors and immune checkpoint blockade. *Viruses* **2016**, *8*, 9. [CrossRef] [PubMed]

270. Rojas, J.J.; Sampath, P.; Hou, W.; Thorne, S.H. Defining effective combinations of immune checkpoint blockade and oncolytic virotherapy. *Clin. Cancer Res.* **2015**, *21*, 5543–5551. [CrossRef] [PubMed]

271. Sasikumar, P.G.; Ramachandra, M. Small-molecule antagonists of the immune checkpoint pathways: Concept to clinic. *Future Med. Chem.* **2017**, *9*, 1305–1308. [CrossRef] [PubMed]

272. Sasikumar, P.G.; Ramachandra, M. Small-Molecule Immune Checkpoint Inhibitors Targeting PD-1/PD-L1 and Other Emerging Checkpoint Pathways. *BioDrugs* **2018**, *32*, 481–497. [CrossRef] [PubMed]

273. Chen, S.; Song, Z.; Zhang, A. Small-Molecule Immuno-Oncology Therapy: Advances, Challenges and New Directions. *Curr. Top. Med. Chem.* **2019**, *19*, 180–185. [CrossRef]

274. Park, J.-J.; Thi, E.P.; Carpio, V.H.; Bi, Y.; Cole, A.G.; Dorsey, B.D.; Fan, K.; Harasym, T.; Iott, C.L.; Kadhim, S.; et al. Checkpoint inhibition through small molecule-induced internalization of programmed death-ligand 1. *Nat. Commun.* **2021**, *12*, 1222. [CrossRef] [PubMed]

275. Ishikawa, H.; Ma, Z.; Barber, G.N. STING regulates intracellular DNA-mediated, type I interferon-dependent innate immunity. *Nature* **2009**, *461*, 788–792. [CrossRef] [PubMed]

276. Ishikawa, H.; Barber, G.N. STING is an endoplasmic reticulum adaptor that facilitates innate immune signalling. *Nature* **2008**, *455*, 674–678. [CrossRef] [PubMed]

277. Lau, A.; Gray, E.E.; Brunette, R.L.; Stetson, D.B. DNA tumor virus oncogenes antagonize the cGAS-STING DNA-sensing pathway. *Science* **2015**, *350*, 568–571. [CrossRef] [PubMed]

278. Lam, E.; Stein, S.; Falck-Pedersen, E. Adenovirus Detection by the cGAS/STING/TBK1 DNA Sensing Cascade. *J. Virol.* **2014**, *88*, 974–981. [CrossRef] [PubMed]

279. Woo, S.R.; Fuertes, M.B.; Corrales, L.; Spranger, S.; Furdyna, M.J.; Leung, M.Y.K.; Duggan, R.; Wang, Y.; Barber, G.N.; Fitzgerald, K.A.; et al. STING-dependent cytosolic DNA sensing mediates innate immune recognition of immunogenic tumors. *Immunity* **2014**, *41*, 830–842. [CrossRef]

280. Corrales, L.; McWhirter, S.M.; Dubensky, T.W.; Gajewski, T.F. The host STING pathway at the interface of cancer and immunity. *J. Clin. Investig.* **2016**, *126*, 2404–2411. [CrossRef] [PubMed]

281. Deng, L.; Liang, H.; Xu, M.; Yang, X.; Burnette, B.; Arina, A.; Li, X.D.; Mauceri, H.; Beckett, M.; Darga, T.; et al. STING-dependent cytosolic DNA sensing promotes radiation-induced type I interferon-dependent antitumor immunity in immunogenic tumors. *Immunity* **2014**, *41*, 843–852. [CrossRef] [PubMed]

282. Demaria, O.; De Gassart, A.; Coso, S.; Gestermann, N.; Di Domizio, J.; Flatz, L.; Gaide, O.; Michielin, O.; Hwu, P.; Petrova, T.V.; et al. STING activation of tumor endothelial cells initiates spontaneous and therapeutic antitumor immunity. *Proc. Natl. Acad. Sci. USA* **2015**, *112*, 15408–15413. [CrossRef]

283. Jiang, M.; Jiang, M.; Chen, P.; Chen, P.; Wang, L.; Li, W.; Chen, B.; Liu, Y.; Liu, Y.; Wang, H.; et al. CGAS-STING, an important pathway in cancer immunotherapy. *J. Hematol. Oncol.* **2020**, *13*, 1–11. [CrossRef] [PubMed]

284. Pan, B.S.; Perera, S.A.; Piesvaux, J.A.; Presland, J.P.; Schroeder, G.K.; Cumming, J.N.; Wesley Trotter, B.; Altman, M.D.; Buevich, A.V.; Cash, B.; et al. An orally available non-nucleotide STING agonist with antitumor activity. *Science* **2020**, *369*. [CrossRef]

285. Chin, E.N.; Yu, C.; Vartabedian, V.F.; Jia, Y.; Kumar, M.; Gamo, A.M.; Vernier, W.; Ali, S.H.; Kissai, M.; Lazar, D.C.; et al. Antitumor activity of a systemic STING-activating non-nucleotide cGAMP mimetic. *Science* **2020**, *369*, 993–999. [CrossRef] [PubMed]

286. Xia, T.; Konno, H.; Ahn, J.; Barber, G.N. Deregulation of STING Signaling in Colorectal Carcinoma Constrains DNA Damage Responses and Correlates with Tumorigenesis. *Cell Rep.* **2016**, *14*, 282–297. [CrossRef]

287. Xia, T.; Konno, H.; Barber, G.N. Recurrent loss of STING signaling in melanoma correlates with susceptibility to viral oncolysis. *Cancer Res.* **2016**, *76*, 6747–6759. [CrossRef]

288. De Queiroz, N.M.G.P.; Xia, T.; Konno, H.; Barber, G.N. Ovarian cancer cells commonly exhibit defective STING signaling which affects sensitivity to viral oncolysis. *Mol. Cancer Res.* **2019**, *17*, 974–986. [CrossRef] [PubMed]

289. Eaglesham, J.B.; Kranzusch, P.J. Conserved strategies for pathogen evasion of cGAS–STING immunity. *Curr. Opin. Immunol.* **2020**, *66*, 27–34. [CrossRef]

290. Ahn, J.; Barber, G.N. STING signaling and host defense against microbial infection. *Exp. Mol. Med.* **2019**, *51*, 1–10. [CrossRef]

291. Russell, L.; Peng, K.W.; Russell, S.J.; Diaz, R.M. Oncolytic Viruses: Priming Time for Cancer Immunotherapy. *BioDrugs* **2019**, *33*, 485–501. [CrossRef] [PubMed]

292. Zaikos, T.D.; Painter, M.M.; Sebastian Kettinger, N.T.; Terry, V.H.; Collins, K.L. Class 1-Selective Histone Deacetylase (HDAC) Inhibitors Enhance HIV Latency Reversal while Preserving the Activity of HDAC Isoforms Necessary for Maximal HIV Gene Expression. *J. Virol.* **2018**, *92*. [CrossRef] [PubMed]

293. Nehme, Z.; Pasquereau, S.; Herbein, G. Control of viral infections by epigenetic-targeted therapy. *Clin. Epigenet.* **2019**, *11*, 1–17. [CrossRef]

294. Wang, L.; Beier, U.H.; Akimova, T.; Dahiya, S.; Han, R.; Samanta, A.; Levine, M.H.; Hancock, W.W. Histone/protein deacetylase inhibitor therapy for enhancement of Foxp3+ T-regulatory cell function posttransplantation. *Am. J. Transplant.* **2018**, *18*, 1596–1603. [CrossRef] [PubMed]

295. Chiu, M.; Armstrong, E.J.L.; Jennings, V.; Foo, S.; Crespo-Rodriguez, E.; Bozhanova, G.; Patin, E.C.; McLaughlin, M.; Mansfield, D.; Baker, G.; et al. Combination therapy with oncolytic viruses and immune checkpoint inhibitors. *Expert Opin. Biol. Ther.* **2020**, *20*, 635–652. [CrossRef] [PubMed]

296. Hwang, J.K.; Hong, J.; Yun, C.O. Oncolytic viruses and immune checkpoint inhibitors: Preclinical developments to clinical trials. *Int. J. Mol. Sci.* **2020**, *21*, 8627. [CrossRef] [PubMed]

Review

Tackling HLA Deficiencies Head on with Oncolytic Viruses

Kerry Fisher *, Ahmet Hazini and Leonard W. Seymour *

Department of Oncology, University of Oxford, Oxford OX3 7DQ, UK; Ahmet.Hazini@oncology.ox.ac.uk
* Correspondence: Kerry.Fisher@oncology.ox.ac.uk (K.F.); len.seymour@oncology.ox.ac.uk (L.W.S.)

Simple Summary: Oncolytic viruses show great promise as anticancer agents by simultaneously lysing cancer cells while stimulating innate and adaptive immune responses. However, the extent to which the adaptive immune system contributes to overall efficacy alongside oncolytic viruses is, in part, dependent on the compliance of cancer cells to present antigens correctly. Dysregulation of any part of the antigen presentation machinery provides a strong selection pressure for immune escape. In this review, we consider the key immunological factors that might be measured to allow for the optimum deployment of oncolytic viruses for effective cancer therapy.

Abstract: Dysregulation of HLA (human leukocyte antigen) function is increasingly recognized as a common escape mechanism for cancers subject to the pressures exerted by immunosurveillance or immunotherapeutic interventions. Oncolytic viruses have the potential to counter this resistance by upregulating HLA expression or encouraging an HLA-independent immunological responses. However, to achieve the best therapeutic outcomes, a prospective understanding of the HLA phenotype of cancer patients is required to match them to the characteristics of different oncolytic strategies. Here, we consider the spectrum of immune competence observed in clinical disease and discuss how it can be best addressed using this novel and powerful treatment approach.

Keywords: oncolytic virus; class I HLA; immunosurveillance; immunotherapy

Citation: Fisher, K.; Hazini, A.; Seymour, L.W. Tackling HLA Deficiencies Head on with Oncolytic Viruses. *Cancers* **2021**, *13*, 719. https://doi.org/10.3390/cancers13040719

Academic Editors: Antonio Marchini, Carolina S. Ilkow and Alan Melcher
Received: 12 December 2020
Accepted: 3 February 2021
Published: 10 February 2021

Publisher's Note: MDPI stays neutral with regard to jurisdictional claims in published maps and institutional affiliations.

1. Introduction

Cancer is a disease that constantly adapts to remain hidden beneath the immune surveillance radar. Cells acquiring mutations or changes that are noticed by the immune system can be pruned away by T-cell immunosurveillance, effectively providing a threshold governing which changes are sufficiently stealthy to persist. However, under constant selective pressure, tumors often develop a capacity for subversion, for example by the upregulation of TGF-β or immune checkpoints, effectively raising the immunosurveillance threshold which allows a greater range of genetic and epigenetic changes (tumor associated antigens (TAAs)) to persist without detection [1].

We can counter this immune subversion strategy in a significant minority of patients using immune checkpoint inhibitors that can reveal cancer cells, once again, as legitimate targets for destruction by the immune system. However, the majority of patients do not respond to checkpoint inhibition, and those that do can eventually develop resistance [2]. This suggests there are multiple mechanisms of immune evasion and a constantly evolving battleground between tumor and immune cells. In many cases, cancer cells may eventually acquire the capacity to turn off the immune surveillance radar completely by usurping the class I HLA system to protect themselves from detection (Figure 1).

Without class I HLA presentation, T-cells cannot recognize and kill target cells, even if all other aspects of the immune system are fully functional [3,4]. The majority of immunotherapy approaches currently under development—including cancer vaccines, checkpoint blockade, adoptive T-cells and STING (stimulator of interferon genes) agonists—are rendered completely obsolete without functional class I HLA presentation by cancer cells. Any aspirations for encouraging epitope spreading [5] or inciting a systemic ab-scopal effect are likely to become redundant if key elements of the HLA pathway are sufficiently compromised.

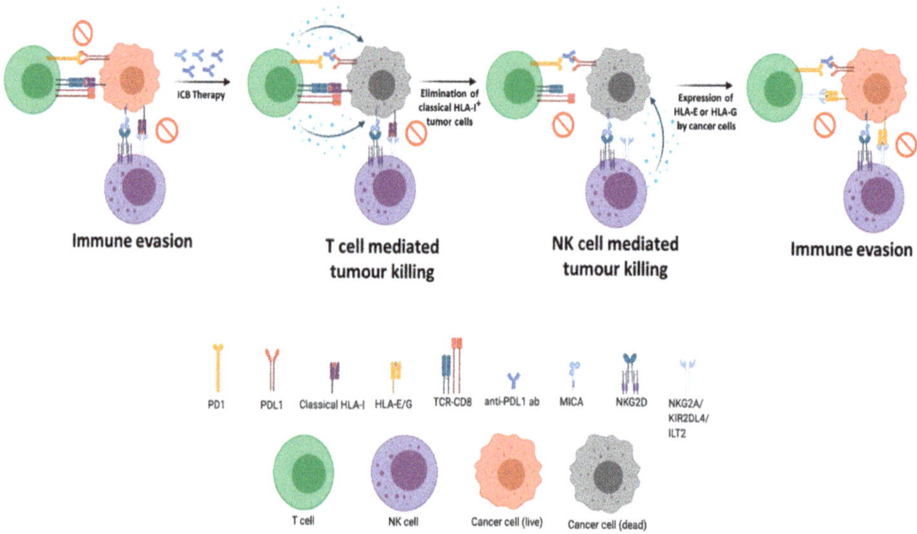

Figure 1. Overview of immune evasion strategies employed by tumor cells. T-cell mediated killing requires functional class I HLA and can be inhibited by tumor expression of immune checkpoints such as PDL1 (programmed death ligand 1). Immune activity in this situation can be restored using checkpoint inhibitor antibodies such as anti-PDL1. Similarly, NK (natural killer) cell-mediated cytotoxicity requires engagement via an NK cell ligand such as MIC-A (major histocompatibility complex class I polypeptide—related sequence A), coupled with the absence of class I HLA. Expression of the non-classical HLA haplotypes HLA-E or HLA-G (which are not generally recognized by the TCR (T cell receptor)) provide a simple mechanism for tumor cells to evade killing by both T-cells and NK cells. Additional abbreviations: ICB (immune checkpoint blockade), NKG2D (natural killer cell receptor G2 type D), KIR2DL4 (Killer cell immunoglobulin-like receptor 2DL4) and ILT2 (immunoglobulin-like transcript 2).

2. The Murky and Complex World of HLA Deregulation

There are a myriad of ways that class I HLA expression can be downregulated, corrupted or made dysfunctional [6–10]. A frequently observed event in cancer is loss of heterozygosity (LOH) in the HLA locus, meaning that one allele is lost, effectively providing cancer cells with a wider operational envelope to accumulate TAAs without detection. For example, LOH in the HLA locus of lung cancer can cause the loss of HLA-C*08:02, meaning that driver mutations like *K-RAS* (Kirsten rat sarcoma) G12D are no longer presented and G12D-specific T-cells are no longer effective [11]. The frequency of this change in cancer in early-stage lung cancer is reported to be 40% [12], emphasizing the importance of this escape mechanism to mask immunogenic TAA.

Non-silent genetic mutations in class I HLA genes (including *HLA-A, -B, -C* and *beta-microglobulin (β2M)*) are reported at low frequencies in many cancers, typically <5% [13], but in microsatellite unstable colorectal cancer (MSI-H CRC), where there is a strong immune pressure, this can be as high as 30% [14]. However, focusing on mutations directly in HLA genes alone will only reveal a small fraction of the problem because any gene involved in the antigen presentation pathway (e.g., proteasomal processing, peptide loading in the ER or regulatory genes such as *NLRC5*) can result in deficient class I HLA function [14]. The cumulative impact of all these molecular changes, together with heterogeneity between clones of tumor cells, adds up to a complex picture. Given the lack of any established biomarkers for HLA functionality in patients, the importance of genetic mutations is likely not fully appreciated.

Defects arising from genetic mutations or allelic loss are often referred to as irreversible or hard defects [9]. In contrast, soft defects are epigenetic in nature and are potentially reversible through pharmacological intervention. Causes of soft HLA defects are varied and include deficiencies in interferon pathways [15] or the hypermethylation of key HLA regulatory elements [16,17]. They are much harder to study and quantify than genetic changes and may reflect responses to applied therapeutic immune pressure or be related to other features of cancer progression and the tumor microenvironment including TGF-β signaling [18], ER stress [19] and hypoxia [20]. Upregulation of class I HLA can be achieved, in principle, through interferon signaling due to an interferon response element (ISRE) in the promoter region of all the classical HLA molecules [15,21–23], although this approach may not be effective in all cancers due to frequent deregulation of interferon pathways [24–26].

The loss of type I HLA expression in cancer cells is often seen as an invitation for elimination by NK cells. However, in the complex evolutionary environment of the tumor, NK cells provide just one more challenge to work around. In consequence, the simple absence of class I HLA is not likely to be a common occurrence; tumor cells that lose their classical class I function alongside the upregulation of the non-classical HLA molecules HLA-G and/or E, which do not provide classical antigen presentation but can inhibit the activation of NK cells [27,28], are likely to be more successful. Tumor heterogeneity may apply to HLA function in the same way that it applies to mutational load, meaning that different tumor cells and their progeny may acquire different HLA deregulation strategies, particularly in the face of immunological therapies. In other words, we should not think in terms of binary HLA loss, but of a constant bio-selection to maintain a balance of HLA expression appropriate for the continued existence of a population of cancer cells. For further details on HLA loss and the underlying mechanisms, please refer to an earlier review in this journal [6].

3. Treating HLA-Competent Tumors with Oncolytic Viruses or Immunotherapy

Cancers with functional HLA should be relatively amenable to a wide range of immunotherapies, although this can probably occur only if they are infiltrated with functional CD8+ T-cells (Figure 2, category A). In these patients, a single adjustment to the immune system through either checkpoint inhibition or oncolysis could arguably be sufficient to reach a tipping point that enables the immune system to mount an effective response. The first approved oncolytic viruses for human use, Imlygic, lends support to this concept with significant numbers of responses that appear to be immunological in nature [29]. For tumors showing an abscopal response (>50% shrinkage of non-injected lesions), comprising of about 34% of superficial lesions and 15% of visceral lesions, it would be reasonable to hypothesize that the HLA system remains at least partially functional. The addition of checkpoint inhibitors with Imlygic has further increased the response rate [30,31], perhaps reflecting their independent mechanisms of action that may synergize to allow HLA-mediated cytotoxicity. Whether this is sufficient to extend patient survival compared to either treatment alone is currently being evaluated in a phase 3 clinical trial (MASTERKEY-256/KEYNOTE-034). Meanwhile several other oncolytic agents are being explored in combination with checkpoint inhibitors, including Cavatak, Reolysin, MG1-MAGEA3, ONCOS-102, DNX-2401, HF-10, Pexa-Vec and Enadenotucirev [32]. For all these trials, it would be very helpful to prospectively correlate patient HLA expression with clinical observations, although we are not aware that any are planning to do so.

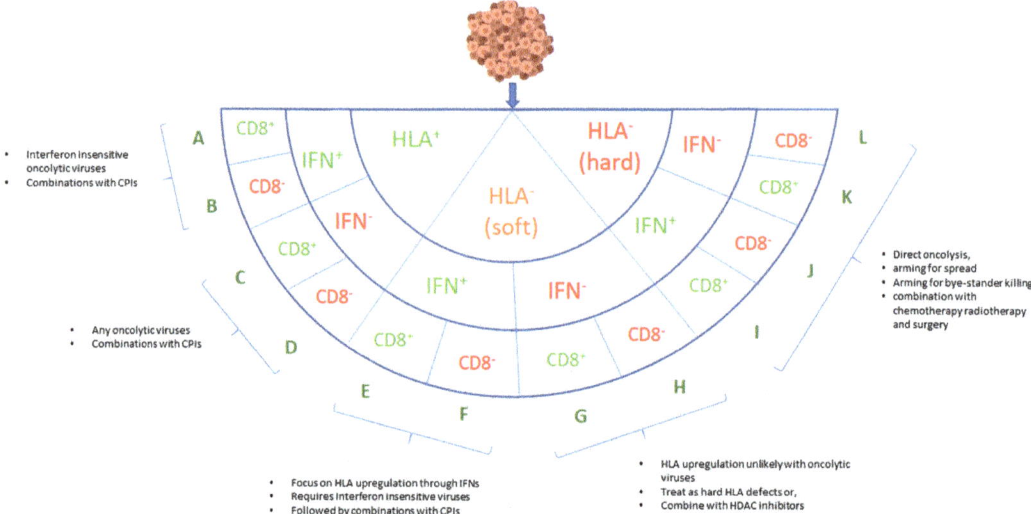

Figure 2. Simplified model of immunological phenotypes from the perspective of oncolytic therapy. HLA$^+$ cancers (A–D) have a high potential for immune responses including oncolytic strategies to foster in situ vaccination and combinations with checkpoint inhibitors (CPIs). Cancers with a HLA$^-$ (soft) phenotype (E–H) require conversion to HLA$^+$ before an adaptive immune response is likely to succeed. One of the most effective ways to restore HLA expression is through interferons (IFN), which are usually expected to be an inherent byproduct of oncolysis and immunogenic cell death. Understanding the IFN status of cancer cells is of particular importance in the context of reversible HLA defects and oncolytic viruses. Cancer cells need to be IFN-competent to allow HLA upregulation, but these cells are a poor target for oncolytic viruses that are designed to selectively replicate in IFN-defective cells. With functional HLA status restored, CD8$^+$ infiltration becomes an important variable. Some viruses are vulnerable to lymphocytes by design, exploiting HLA defects or immune exclusion in order to spread. Drawing lymphocytes into the tumor is an essential part of the adaptive immune response and a forte of oncolytic viruses, however lymphocytes may also eliminate some oncolytic vectors prematurely. Cancers with molecular hard defects (HLA$^-$; categories I–L) or indeed soft HLA deficiencies that cannot be restored, may be difficult to treat with any therapy that relies on the adaptive immune response for efficacy. For these cancers, direct oncolysis combined with conventional cytotoxic chemoradiotherapy or the use of HLA independent killing strategies, for example encoding bi-specific T-cell engagers, may be more appropriate. Finally, it is worth noting that tumor heterogeneity means that different phenotypes may well occur within different regions of the same tumor. HDAC: histone deacetylase.

Where functional HLA presentation on cancer cells is confirmed, it is compelling to pursue oncolytic viruses as in situ personalized vaccines, releasing and exposing TAAs upon lysis [33–35]. This effect is amplified by triggering immunogenic cell death (ICD) pathways that induce wide-ranging immunological consequences, including the maturation of dendritic cells and activation of T-cells [36–40]. On top of the inherent ability of viruses to induce ICD, arming them with transgenes can further stimulate different arms of the immune system through the careful choice of cytokines or checkpoint inhibitors, among myriad other possibilities [41].

4. Boosting Lymphocyte Uptake into HLA Competent Tumors

An important and unusual benefit of oncolytic viruses is the potential to attract CD8$^+$ cells that may otherwise be excluded from tumors, a phenomenon particularly apparent in carcinomas [42–44]. Promoting lymphocyte engraftment into tumors could be essential to allow for the initiation of an effective immune response, although it can only be useful if type I HLA expression is functional (Figure 2, categories A–D). Lymphocytes are attracted towards a chemokine concentration gradient and replicating lytic viruses, established within the tumor, are well placed to become a homing beacon for immune cells [31,45–47]. It is also possible that new CD8$^+$ cells entering the tumor may have greater activation potential than the endogenous T-cells that may have become anergic following exposure to the tumor microenvironment (TME) for an extended period. This phenomenon should place oncolytic viruses as ideal partners for other immunotherapy approaches, since mobilizing immune cells in cancer patients by other approaches is far more challenging. For example, administering immune stimulants like STING agonists, chemokines, or interferons directly into the blood stream is likely to provide only a relatively short period of activation at the expense of systemic side effects, without any locoregional information to guide immune cells towards the tumor [48,49].

To further augment the capacity of oncolytic viruses to promote immune engraftment, it is possible to arm them to express chemokines within the tumor [50–52]. In this context, particularly desirable oncolytic viruses might be those that that persist locally and express chemokines for extended periods of time. However, attracting cytotoxic T-cells into the locality of a replicating oncolytic virus brings with it the capacity for rapid recognition and elimination of the virus itself, and a consequent premature end to the therapy [53]. The subtle interplay between viruses and the HLA system has been evolving for millennia, and the implications for oncolytic virus design are considered in the next section.

5. The Consequences of Removing HLA-Manipulating Viral Proteins from Oncolytic Viruses

Many viruses contain proteins, such as E3 19k in adenovirus or ICP47 in HSV, that are capable of corrupting antigen presentation inside infected cells using strategies as diverse as blocking the TAP transporter or pulling HLA molecules away from the cell surface [54]. These elements have been removed from many oncolytic virus candidates with the intention of making oncolysis more immunogenic, and to improve safety by accelerating viral clearance in normal tissues [55,56]. However, in hot tumors, broadcasting a cancer cell as virus-infected in this way may lead to premature virus elimination by the immune system, necessitating repeated virus delivery in the face of increasing levels of neutralizing antibodies or restricting treatment options to direct intratumoral injection. Conversely, in cold tumors that lack CD8$^+$ cells, this may be less of an issue, at least initially, with virus infected cells only becoming targets after immune cell infiltration is restored.

Removing HLA-inactivating genes from oncolytic viruses might also be counterproductive to creating a durable anticancer immune response. One of the benefits of oncolytic viruses is the capacity to cause immunogenic cancer cell death, shedding adenosine triphosphate, heat shock proteins and other immune-provoking signals alongside cancer antigens into the interstitial space [37,40,57]. This could create a proinflammatory environment, potentially leading to an anticancer immune response. In contrast, allowing the presentation of virus antigens on HLA will likely lead to efficient T-cell-mediated killing of infected cancer cells by caspase-mediated apoptosis, regarded as a less inflammatory death mechanism [58]. Arguably, this could be less effective at priming new anticancer immune responses than virus mediated lysis with the simultaneous presentation of pathogen or damage associated molecular patterns (PAMPs and DAMPs).

Cancer cells infected with an oncolytic virus are likely to be destined for eventual elimination, either through lysis or following the exhaustion of metabolites. Rather than killing them rapidly via T-cell cytotoxicity, it may be preferable to allow virus infected cells to persist as a factory, modifying the tumor microenvironment and releasing TAAs, while T-cells kill any residual cells not infected by the virus [59]. Accordingly, cancer selective viruses that can avoid T-cell killing may have the advantage, especially when relying on the adaptive immune response as the major driver for efficacy.

6. Treating Cancers with Reversible HLA Defects

Where a cancer cell lacks functional class I HLA, little benefit can accrue from checkpoint inhibition or attracting lymphocytes into the tumor microenvironment. Efforts to express or expose TAA are likely to be fruitless until HLA function is restored. Turning HLA defective (soft) cancers (Figure 2, categories E–H) into HLA competent tumors (Figure 2, categories A–D) is therefore an attractive proposition and can exploit several strengths of the oncolytic approach, particularly where those soft mutations are mediated through deficient interferon pathways. In particular, viruses have an intrinsic capability to induce interferons during infection and lysis by triggering pathogen recognition receptors such as cGAS (cyclic GMP-AMP synthase) or RIG-I (retinoic acid-inducible gene I) in cancer cells or adjacent stromal cells [60,61]. Viruses can also be armed to express additional levels of interferons selectively in the tumor microenvironment [62–64].

Although soft HLA defects are not understood in detail in patients, they feature strongly in laboratory models. For example, B16 cancer cells are a classic example of soft class I HLA defects, with low levels of HLA that can be fully restored by treatment with gamma interferon [65]. It is intriguing to ponder whether it is interferon-induced class I HLA upregulation rather than anything else that leads to efficacy with oncolytic viruses in this model, given that persistent lysis would not be required in the presence of an anticancer immune response [66].

This potential for the locoregional upregulation of class I HLA expression in tumors, using interferon either encoded within the virus or produced naturally following virus infection, should synergize with the ability of oncolytic viruses to attract and activate T-cells. In turn, this could condition the tumor microenvironment to support additional immune therapies like checkpoint inhibitors that have little chance otherwise of having an impact. Accordingly, oncolytic viruses could be the key to enabling the effective immunotherapy of cancers with reversible HLA defects (E to H).

While upregulating classical HLA-A, -B and -C in tumors with soft mutations may be highly desirable to engender an adaptive immune effect, care must be taken not to upregulate non-classical HLA-G/E, which have a variety of mechanisms to inhibit classical antigen presentation. Variations in the promoter regions of HLA-A/B/C vs. HLA-G/E may allow pharmacological approaches to achieve this [67]. Careful analysis of the regulation of classical and non-classical HLA molecules, together with the versatility of oncolytic viruses, could permit the subtle manipulation of HLA expression to achieve the greatest therapeutic impact.

7. Implications of Interferon Competency in Cancer Cells with HLA Reversible Cancers

Cancer cells with defects in interferon pathways or NF-κB signaling pathways may not be amenable to the correction of HLA function via the expression of interferons (Figure 2, categories G,H). This is crucially important for some oncolytic approaches that are designed to exploit cancer cells that have deregulated interferon pathways [68]. At face value, these oncolytic agents would perhaps not be the first choice for patients with interferon-reversible soft HLA defects because, by definition, these tumors are likely to be interferon responsive and therefore the activity of these viruses would be limited.

However, the reality is more nuanced, because defects in interferon signaling can be at different stages of the pathway, with some cancer cells not able to express interferons and others not able to respond to them. In a stromal-rich tumor, fibroblasts and macrophages are likely to have the full capacity to detect oncolytic viruses and express interferons even if the cancer cells are defective [69]. Cells that can be stimulated by external interferon (with intact JAK/STAT signaling) would often have functional interferon regulatory factors (IRFs), and therefore be able to trigger ISRE to upregulate classical class I HLA expression. Conversely, cells with deficient JAK/STAT pathways could likely still upregulate HLA, but only if their own virus-sensing pathways remain intact. Consequently, a major subset of cancer cells with functional JAK/STAT pathways but defects in virus sensing and interferon expression, could likely upregulate HLA in response to interferons generated within the TME by an oncolytic virus interacting with stromal cells. However, this may be to the detriment of further oncolytic virus spread and lysis. Accordingly, matching the patient population to the oncolytic strategy is very important.

Some DNA viruses have sophisticated mechanisms to overcome interferon responses, for example the E1A protein and the VA RNAs in adenovirus, B8R and B18R/B19R in vaccinia, and ICP34.5, US11 and others in HSV-1 [70]. Adenoviruses, for example, can continue replicating despite triggering interferon and STING pathways, allowing both immune stimulation and lysis to happen concurrently [71]. When DNA viruses are genetically attenuated to render them interferon sensitive, the innate immune response works to limit viral replication [72]. Consequently, patients that are considered to have interferon-inducible HLA may be preferentially matched with viruses that can operate in a wider variety of interferon competent environments. This may be particularly true in heterogeneous tumors where there is likely to be a variety of different interferon defects in different populations of cancer cells, and also in those cancers with a high stromal cell content. In contrast, for viruses that are unavoidably dependent on dysfunctional interferon, it may be possible to achieve upregulation of HLA via interferon-independent pharmacological means such as HDAC inhibition [73].

8. Oncolytic Viruses for the Treatment of Irreversible Hard Defects

In cancers where there are genetic defects in the HLA pathway, attempts to mount a comprehensive adaptive immune response are likely to be futile. In these cancers (Figure 2, categories I–L), decisions over how best to invoke ICD or stimulate antigen presenting cells becomes straightforward because none are likely to be successful. For these cancers, the full spectrum of oncolytic mechanisms and arming strategies will need to focus on achieving cytotoxicity independent of HLA expression, and this embodies the essence of the whole concept of direct virolysis. Prior to discussing the range of different HLA-independent cytotoxic strategies available to oncolytic viruses, it is worth challenging the very notion that hard HLA defects are definitively irreversible. Oncolytic viruses have the capacity, in principle, to normalize HLA expression in any cell they infect.

8.1. Gene Supplementation Therapies

Wild type copies of each of the components of the antigen presentation pathway can in theory be encoded within oncolytic viruses and expressed locally within tumor cells to restore pathway function. Although this approach could be adopted for any pathway component, perhaps the most widely studied is β2M [74], where gene replacement with adenovirus has been shown to restore cell surface expression of classical class I HLA molecules, implying that β2M function has been restored [75]. Importantly, the restoration of β2M was also shown to lead to peptide-specific immune recognition by cognate antigen-specific T-cells [76].

Unfortunately, the therapeutic benefits of direct class I HLA gene replacement therapy will likely be very limited, because current understanding suggests that the expressed transgene product would be restricted only to cells that are actively infected with the oncolytic virus. This raises two immediate concerns—first, that functional antigen presentation would not be restored more broadly within the tumor, but only in cells that would be expected to be killed by direct oncolysis, and secondly that the restored HLA function in those infected cells might begin to present viral epitopes rather than TAAs, giving rise to an augmented antiviral T-cell response rather than stimulating an anticancer response. This might accelerate the immune-mediated clearance of the virus rather than empowering a cancer vaccination effect. Hence, the concept of replacing mutated components of the antigen presentation pathway appears flawed unless it can be somehow delivered more broadly within the tumor and not restricted just to virus-infected cells. One possible approach that might be worth exploring is the use of exosomes. Although still in its infancy, the potential for programming viruses to manipulate exosomes to distribute functional HLA molecules amongst cancer cells may be an effective way of restoring the presentation of TAAs to allow renewed immunosurveillance.

8.2. Turning Lymphocytes into HLA-Independent Killers with Virus-Deployed Bispecific T-Cell Engagers

Bispecific T-cell engagers (BiTEs) crosslink endogenous CD3 on T-cells to surface targets on cancer cells, creating an activating pseudosynapse through clustering and leading to rapid and efficient target-specific cytotoxicity. The T-cells then detach from the target cell and can bind to a new target cell, earning them the epithet serial killer T-cells. This powerful approach is reminiscent of converting endogenous T-cells into antigen-specific chimaeric antigen receptor (CAR) T cell-like cells in situ, recognising any chosen surface antigen.

BiTEs are difficult to deploy through conventional intravenous delivery because of their short plasma half-life, and fine tuning of affinity of the CD3 binding arm is crucial to prevent sequestration to T-cells outside the tumor. When it comes to the choice of the antigen-binding arm for a systemic therapy, there is a trade-off between using broadly expressed antigens that could mediate on-target off-disease side effects, or highly cancer-selective antigens that may not be universally expressed throughout the tumor. Expressing BiTES locally and exclusively in the tumor microenvironment from an oncolytic virus may avoid these complications, allowing for a far broader range of target and effector arms to be considered. A constantly renewed supply of BiTEs might be expected to synergize with newly arriving T-cells encouraged to the tumor through chemokines or oncolysis.

Several groups have now demonstrated the successful delivery of BiTEs using a range of oncolytic viruses. The earliest approaches used oncolytic vaccinia and adenovirus to express BiTEs targeting receptors on the cancer cell surface, allowing cancer cells to be targeted simultaneously by two distinct cytotoxic entities—the oncolytic virus and the BiTE-targeted T-cell [77–79]. It would be difficult to use any other strategy to target these surface markers with the anatomical selectivity provided by oncolytic viruses. However, despite the potency and selectivity of the approach, it was soon considered suboptimal because the rapid cytotoxicity can decrease the pool of therapeutic viruses by eradicating the tumor cells that produce them. Accordingly, more recent innovations have explored combining the direct cytotoxicity of oncolytic viruses against cancer cells with BiTEs that retarget endogenous T-cells to attack other cellular components of the tumor microenvironment. To date, these have included tumor-associated macrophages via targeting to folate receptor b or CD206 [80], or tumor-associated fibroblasts via fibroblast activation protein (FAP) [81,82]. This latter approach forms the basis of an ongoing clinical trial which uses an oncolytic adenovirus to express a FAP BiTE alongside alpha interferon (which should increase immune stimulation and may restore HLA function), together with two chemokines intended to recruit T-cells into the tumor (Clinical Trials identifier NCT04053283). Finally, harnessing endogenous T-cells in this way should provide greater therapeutic potential for heterogeneous solid tumors than exogenously applied CAR T-cells, since every T-cell within the tumor may be redeployed against the chosen target antigen, whereas CAR T-cells will always be a subset of those engrafting into solid tumors.

9. Orchestrating Anti-Cancer Innate Immune Cells in the Absence of Functional HLA

A great strength of oncolytic viruses is that they are fully customizable drugs that can be fine-tuned to accomplish the tasks at hand. Where adaptive immunity is unlikely to be harnessed, arming elements targeting cancer vaccine approaches can be swapped out for those targeting the innate immune system. NK cells have been directly implicated in enhancing the efficacy of oncolytic viruses in experimental models [83]. Arming oncolytic viruses with IL-15 [84] is likely to augment NK activation, while strategies based on CCL5 chemotaxis [85] have been used to attract them in to tumors. Expression of SIRPa-Fc antagonists from oncolytic viruses can block the CD47 "don't eat me" signals on cancer cells and facilitate macrophages to attack them directly [86]. Building on the BiTE principle, it is now possible to develop bispecific killer cell engagers (BiKEs) that crosslink cancer cell surface antigens with CD16 on NK cells, or better still, trispecific TriKEs that included further functionality through integrating IL-15 [87,88]. Equivalent approaches using BiMEs (bis-pecific macrophage engagers) to activate macrophages have also been reported [89]. Like all arming approaches to activate the adaptive immune system, strategies for exploiting innate cells will benefit from selective expression in the tumor microenvironment provided by oncolytic viruses. Careful design of the innate immune cell stimulation is required in order to avoid the premature and unwanted elimination of virus-infected cells.

10. Strategies to Maximize Cell Killing by Oncolytic Viruses

Oncolysis is an HLA independent killing mechanism that should be exploited in cancers with severe immunological defects. Several hallmarks of cancer, including immune deregulation, appear to provide an advantageous niche for virus activity [90]. These also include the intrinsic resistance of cancer cells to apoptosis and their reprogrammed energy metabolism that provides biosynthetic intermediates to support macromolecular synthesis. When this is combined with hard HLA defects and little to no prospect of a functional cytotoxic T lymphocyte response, a virus might be expected to remain on station for as long as there are substrate cancer cells to infect. In xenografts, efficacy through oncolysis can be demonstrated without help or hindrance from T-cells [91–93].

The combination of virotherapy with chemotherapy can also lead to enhanced virus spread and cell lysis. Synergy with a wide range of chemotherapeutic agents, including platinum drugs, taxanes and topoisomerase inhibitors, among others, have been reported and reviewed in detail elsewhere [94]. Arguably, the most convincing anticancer efficacy in solid carcinomas with oncolytic viruses was in combination with chemotherapy [95]. In this study, patients refractory to several lines of previous treatment were injected intratumorally with ONYX-015 concurrently with intravenous cisplatin and 5FU. Only one of the tumor nodules (the largest) was injected, with uninfected tumors acting as internal controls. In 19 evaluable patients, eight had a complete response in the injected tumor and the remainder had a partial response with signs of extensive lysis. That this effect was mediated by direct oncolysis without adaptive immunity was evidenced by the lack of any effect on non-injected tumors.

Arming viruses to treat HLA compromised tumors can be fully focused on promoting intratumoral virus spread and direct oncolysis. This is made easier because virus gene products that evade premature CTL killing like E3 19k in adenovirus or ICP47 in HSV-1 become redundant in this setting and removing them makes space for additional transgenes. Matrix degrading enzymes or cell fusion peptides have both demonstrated enhanced spread of oncolytic viruses in vitro and in vivo. Enzyme prodrug therapy using thymidine kinase (TK) or cytosine deaminase (CD) [96] is an alternative approach that combines many of the benefits of targeted chemotherapy with the anatomical selectivity of a virus. An innocuous prodrug given systemically that is activated into a cytotoxic entity only by the enzyme within the tumor restricts cytotoxicity to the environs of the infected tumor cells and spares normal tissues, including the bone marrow. In the case of TK, the prodrug is usually ganciclovir, which is metabolized to yield a potent DNA polymerase inhibitor, whereas CD metabolism 5-fluorocytosine to become 5-fluorouracil (5FU), which is a widely used inhibitor of thymidylate synthase. In particular, 5FU is an attractive therapeutic for this approach because it diffuses widely and will mediate a considerable bystander toxicity against cells that are not directly virus infected [97]. Future strategies to enhance virus spread or cytotoxicity could involve transgene products, including proteins or nucleic acids packaged into exosomes and other membrane vesicles that are shed from the infected cell, and thereby allow their transfer into the cytosol of uninfected cancer cells nearby [98].

11. Conclusions

Progress in cancer immunotherapy over the past two decades has been extraordinary, transforming the lives of many patients that were previously incurable. Intense ongoing research is focused on the limits of our ability to re-educate the immune system and turn it against tumor antigens. Success in the clinic, together with more sophisticated animal models that allow for the demonstration of novel immunotherapies, has attracted much of the global capacity for oncology research and development.

Yet despite this, the majority of patients do not respond to immunotherapy. This failure may continue unless we tackle the underlying features of immune evasion head on. HLA deficiency is arguably the primary culprit because deregulation in any of the genes or pathways involved in antigen processing or presentation will thwart any attempts at perpetuating the clearance of cancer cells via the adaptive immune system. Current laboratory tests for class I HLA expression do not take into account the presence of non-classical HLA molecules (notably HLA-E and -G), and hence cannot reliably assess HLA function. It follows that a deeper understanding of a patient's capacity for adaptive immune responses, including HLA functionality, will be essential to allow as much clinical impact as possible.

Oncolytic viruses are supremely versatile anticancer agents that have the capacity to address some of the greatest therapeutic challenges and unmet needs for patients. This includes exploiting class I HLA where it is functional, or inducing it where it is not, provided we can accurately identify patients who would benefit in those scenarios. Despite the inherent ability of viruses to induce interferons and potentially restore or upregulate HLA, surprisingly no primary oncolytic paper has focused on this area. In many cases, this might be because cancers with functional interferon pathways have been avoided for pre-clinical research using those oncolytic agents that are interferon sensitive. On a broader point, a greater awareness of the HLA status of animal tumour models could be very useful to help interpret preclinical therapeutic activity and might contribute towards stratification of patients suitable for different types of treatment.

Perhaps the most powerful approach of all is to use the characteristics of oncolytic viruses to invoke HLA independent immunotherapy or mediate direct cytotoxicity alone, or perhaps in combination with chemo/radiotherapy. Although this is particularly relevant to cancers with severe immune deregulation, such a strategy could be applied more broadly without having to be overly concerned with HLA status.

It follows that oncolytic viruses have great potential to contribute to meaningful therapies for patients with any status of immune function. However, which oncolytic approach will be most successful depends on both the class I HLA status and the interferon competence of the tumor, including whether there is clonal heterogeneity between metastases or even within individual tumors. Knowledge of HLA functionality and interferon status in individual patients is essential to guide the choice of optimal treatment strategy, but that information is currently very hard to obtain. A laboratory test for HLA function that could be performed on biopsies would revolutionize our ability to deploy oncolytic and other immune stimulatory strategies effectively.

Funding: We are grateful to Cancer Research UK for their support in Program C552/A29106.

Conflicts of Interest: L.W.S. and K.F. own equity in Psioxus Therapeutics and Theolytics Ltd.; K.F. is an employee of Theolytics and L.W.S. consults for both companies. The funders had no role in the design of the study; in the collection, analyses, or interpretation of data; in the writing of the manuscript; or in the decision to publish the results.

References

1. Dunn, G.P.; Bruce, A.T.; Ikeda, H.; Old, L.J.; Schreiber, R.D. Cancer immunoediting: From immunosurveillance to tumor escape. *Nat. Immunol.* **2002**, *3*, 991–998. [CrossRef]
2. Gettinger, S.; Choi, J.; Hastings, K.; Truini, A.; Datar, I.; Sowell, R.; Wurtz, A.; Dong, W.; Cai, G.; Melnick, M.A.; et al. Impaired HLA Class I Antigen Processing and Presentation as a Mechanism of Acquired Resistance to Immune Checkpoint Inhibitors in Lung Cancer. *Cancer Discov.* **2017**, *7*, 1420–1435. [CrossRef] [PubMed]
3. Rodig, S.J.; Gusenleitner, D.; Jackson, D.G.; Gjini, E.; Giobbie-Hurder, A.; Jin, C.; Chang, H.; Lovitch, S.B.; Horak, C.; Weber, J.S.; et al. MHC proteins confer differential sensitivity to CTLA-4 and PD-1 blockade in untreated metastatic melanoma. *Sci. Transl. Med.* **2018**, *10*, eaar3342. [CrossRef] [PubMed]
4. Sade-Feldman, M.; Jiao, Y.J.; Chen, J.H.; Rooney, M.S.; Barzily-Rokni, M.; Eliane, J.P.; Bjorgaard, S.L.; Hammond, M.R.; Vitzthum, H.; Blackmon, S.M.; et al. Resistance to checkpoint blockade therapy through inactivation of antigen presentation. *Nat. Commun.* **2017**, *8*, 1–11. [CrossRef] [PubMed]
5. Brossart, P. The Role of Antigen Spreading in the Efficacy of Immunotherapies. *Clin. Cancer Res.* **2020**, *26*, 4442–4447. [CrossRef] [PubMed]
6. Cornel, A.M.; Mimpen, I.L.; Nierkens, S. MHC Class I Downregulation in Cancer: Underlying Mechanisms and Potential Targets for Cancer Immunotherapy. *Cancers* **2020**, *12*, 1760. [CrossRef]

7. Rodriguez, J.A. HLA-mediated tumor escape mechanisms that may impair immunotherapy clinical outcomes via T-cell activation. *Oncol. Lett.* **2017**, *14*, 4415–4427. [CrossRef] [PubMed]

8. Garrido, F.; Perea, F.; Bernal, M.; Sanchez-Palencia, A.; Aptsiauri, N.; Ruiz-Cabello, F. The Escape of Cancer from T Cell-Mediated Immune Surveillance: HLA Class I Loss and Tumor Tissue Architecture. *Vaccines* **2017**, *5*, 7. [CrossRef]

9. Garrido, F. MHC/HLA Class I Loss in Cancer Cells. *Adv. Exp. Med. Biol.* **2019**, *1151*, 15–78. [CrossRef]

10. Garrido, F. HLA Class-I Expression and Cancer Immunotherapy. *Adv. Exp. Med. Biol.* **2019**, *1151*, 79–90. [CrossRef]

11. Tran, E.; Robbins, P.F.; Lu, Y.C.; Prickett, T.D.; Gartner, J.J.; Jia, L.; Pasetto, A.; Zheng, Z.; Ray, S.; Groh, E.M.; et al. T-Cell Transfer Therapy Targeting Mutant KRAS in Cancer. *N. Engl. J. Med.* **2016**, *375*, 2255–2262. [CrossRef]

12. McGranahan, N.; Rosenthal, R.; Hiley, C.T.; Rowan, A.J.; Watkins, T.B.K.; Wilson, G.A.; Birkbak, N.J.; Veeriah, S.; Van Loo, P.; Herrero, J.; et al. Allele-Specific HLA Loss and Immune Escape in Lung Cancer Evolution. *Cell* **2017**, *171*, 1259–1271.e1211. [CrossRef] [PubMed]

13. Shukla, S.A.; Rooney, M.S.; Rajasagi, M.; Tiao, G.; Dixon, P.M.; Lawrence, M.S.; Stevens, J.; Lane, W.J.; Dellagatta, J.L.; Steelman, S.; et al. Comprehensive analysis of cancer-associated somatic mutations in class I HLA genes. *Nat. Biotechnol.* **2015**, *33*, 1152–1158. [CrossRef] [PubMed]

14. Ozcan, M.; Janikovits, J.; von Knebel Doeberitz, M.; Kloor, M. Complex pattern of immune evasion in MSI colorectal cancer. *Oncoimmunology* **2018**, *7*, e1445453. [CrossRef]

15. Seliger, B.; Ruiz-Cabello, F.; Garrido, F. IFN inducibility of major histocompatibility antigens in tumors. *Adv. Cancer Res.* **2008**, *101*, 249–276. [CrossRef]

16. Fonsatti, E.; Sigalotti, L.; Coral, S.; Colizzi, F.; Altomonte, M.; Maio, M. Methylation-regulated expression of HLA class I antigens in melanoma. *Int. J. Cancer* **2003**. [CrossRef] [PubMed]

17. Jung, H.; Kim, H.S.; Kim, J.Y.; Sun, J.M.; Ahn, J.S.; Ahn, M.J.; Park, K.; Esteller, M.; Lee, S.H.; Choi, J.K. DNA methylation loss promotes immune evasion of tumours with high mutation and copy number load. *Nat. Commun.* **2019**, *10*, 1–12. [CrossRef] [PubMed]

18. Chen, X.H.; Liu, Z.C.; Zhang, G.; Wei, W.; Wang, X.X.; Wang, H.; Ke, H.P.; Zhang, F.; Wang, H.S.; Cai, S.H.; et al. TGF-beta and EGF induced HLA-I downregulation is associated with epithelial-mesenchymal transition (EMT) through upregulation of snail in prostate cancer cells. *Mol. Immunol.* **2015**, *65*, 34–42. [CrossRef] [PubMed]

19. Granados, D.P.; Tanguay, P.L.; Hardy, M.P.; Caron, E.; de Verteuil, D.; Meloche, S.; Perreault, C. ER stress affects processing of MHC class I-associated peptides. *BMC Immunol.* **2009**, *10*, 1–15. [CrossRef] [PubMed]

20. Sethumadhavan, S.; Silva, M.; Philbrook, P.; Nguyen, T.; Hatfield, S.M.; Ohta, A.; Sitkovsky, M.V. Hypoxia and hypoxia-inducible factor (HIF) downregulate antigen-presenting MHC class I molecules limiting tumor cell recognition by T cells. *PLoS ONE* **2017**, *12*, e0187314. [CrossRef]

21. Jongsma, M.L.M.; Guarda, G.; Spaapen, R.M. The regulatory network behind MHC class I expression. *Mol. Immunol.* **2019**, *113*, 16–21. [CrossRef]

22. Cangemi, G.; Morandi, B.; D'Agostino, A.; Peri, C.; Conte, R.; Damonte, G.; Ferlazzo, G.; Biassoni, R.; Melioli, G. IFN-alpha mediates the up-regulation of HLA class I on melanoma cells without switching proteasome to immunoproteasome. *Int. Immunol.* **2003**, *15*, 1415–1421. [CrossRef] [PubMed]

23. Lorenzi, S.; Forloni, M.; Cifaldi, L.; Antonucci, C.; Citti, A.; Boldrini, R.; Pezzullo, M.; Castellano, A.; Russo, V.; van der Bruggen, P.; et al. IRF1 and NF-kB restore MHC class I-restricted tumor antigen processing and presentation to cytotoxic T cells in aggressive neuroblastoma. *PLoS ONE* **2012**, *7*, e46928. [CrossRef]

24. Katsoulidis, E.; Kaur, S.; Platanias, L.C. Deregulation of Interferon Signaling in Malignant Cells. *Pharmaceuticals* **2010**, *3*, 406–418. [CrossRef] [PubMed]

25. Stojdl, D.F.; Lichty, B.; Knowles, S.; Marius, R.; Atkins, H.; Sonenberg, N.; Bell, J.C. Exploiting tumor-specific defects in the interferon pathway with a previously unknown oncolytic virus. *Nat. Med.* **2000**, *6*, 821–825. [CrossRef] [PubMed]

26. Danziger, O.; Shai, B.; Sabo, Y.; Bacharach, E.; Ehrlich, M. Combined genetic and epigenetic interferences with interferon signaling expose prostate cancer cells to viral infection. *Oncotarget* **2016**, *7*, 52115–52134. [CrossRef] [PubMed]

27. Lin, A.; Yan, W.H.; Xu, H.H.; Gan, M.F.; Cai, J.F.; Zhu, M.; Zhou, M.Y. HLA-G expression in human ovarian carcinoma counteracts NK cell function. *Ann. Oncol.* **2007**, *18*, 1804–1809. [CrossRef] [PubMed]

28. Lo Monaco, E.; Tremante, E.; Cerboni, C.; Melucci, E.; Sibilio, L.; Zingoni, A.; Nicotra, M.R.; Natali, P.G.; Giacomini, P. Human leukocyte antigen E contributes to protect tumor cells from lysis by natural killer cells. *Neoplasia* **2011**, *13*, 822–830. [CrossRef]

29. Andtbacka, R.H.; Ross, M.; Puzanov, I.; Milhem, M.; Collichio, F.; Delman, K.A.; Amatruda, T.; Zager, J.S.; Cranmer, L.; Hsueh, E.; et al. Patterns of Clinical Response with Talimogene Laherparepvec (T-VEC) in Patients with Melanoma Treated in the OPTiM Phase III Clinical Trial. *Ann. Surg. Oncol.* **2016**, *23*, 4169–4177. [CrossRef]

30. Chesney, J.; Puzanov, I.; Collichio, F.; Singh, P.; Milhem, M.M.; Glaspy, J.; Hamid, O.; Ross, M.; Friedlander, P.; Garbe, C.; et al. Randomized, Open-Label Phase II Study Evaluating the Efficacy and Safety of Talimogene Laherparepvec in Combination With Ipilimumab Versus Ipilimumab Alone in Patients With Advanced, Unresectable Melanoma. *J. Clin. Oncol.* **2018**, *36*, 1658–1667. [CrossRef] [PubMed]

31. Ribas, A.; Dummer, R.; Puzanov, I.; VanderWalde, A.; Andtbacka, R.H.I.; Michielin, O.; Olszanski, A.J.; Malvehy, J.; Cebon, J.; Fernandez, E.; et al. Oncolytic Virotherapy Promotes Intratumoral T Cell Infiltration and Improves Anti-PD-1 Immunotherapy. *Cell* **2017**, *170*, 1109–1119.e1110. [CrossRef] [PubMed]

32. Eissa, I.R.; Bustos-Villalobos, I.; Ichinose, T.; Matsumura, S.; Naoe, Y.; Miyajima, N.; Morimoto, D.; Mukoyama, N.; Zhiwen, W.; Tanaka, M.; et al. The Current Status and Future Prospects of Oncolytic Viruses in Clinical Trials against Melanoma, Glioma, Pancreatic, and Breast Cancers. *Cancers* **2018**, *10*, 356. [CrossRef]
33. Russell, S.J.; Barber, G.N. Oncolytic Viruses as Antigen-Agnostic Cancer Vaccines. *Cancer Cell* **2018**, *33*, 599–605. [CrossRef]
34. Nguyen, T.; Avci, N.G.; Shin, D.H.; Martinez-Velez, N.; Jiang, H. Tune Up In Situ Autovaccination against Solid Tumors with Oncolytic Viruses. *Cancers* **2018**, *10*, 171. [CrossRef]
35. Guo, Z.S.; Liu, Z.; Bartlett, D.L. Oncolytic Immunotherapy: Dying the Right Way is a Key to Eliciting Potent Antitumor Immunity. *Front. Oncol.* **2014**, *4*, 74. [CrossRef]
36. Workenhe, S.T.; Mossman, K.L. Oncolytic virotherapy and immunogenic cancer cell death: Sharpening the sword for improved cancer treatment strategies. *Mol. Ther.* **2014**, *22*, 251–256. [CrossRef]
37. Dyer, A.; Di, Y.; Calderon, H.; Illingworth, S.; Kueberuwa, G.; Tedcastle, A.; Jakeman, P.; Chia, S.L.; Brown, A.; Silva, M.A.; et al. Oncolytic Group B Adenovirus Enadenotucirev Mediates Non-apoptotic Cell Death with Membrane Disruption and Release of Inflammatory Mediators. *Mol. Ther. Oncolytics* **2017**, *4*, 18–30. [CrossRef] [PubMed]
38. Di Somma, S.; Iannuzzi, C.A.; Passaro, C.; Forte, I.M.; Iannone, R.; Gigantino, V.; Indovina, P.; Botti, G.; Giordano, A.; Formisano, P.; et al. The Oncolytic Virus dl922-947 Triggers Immunogenic Cell Death in Mesothelioma and Reduces Xenograft Growth. *Front. Oncol.* **2019**, *9*, 564. [CrossRef]
39. De Munck, J.; Binks, A.; McNeish, I.A.; Aerts, J.L. Oncolytic virus-induced cell death and immunity: A match made in heaven? *J. Leukoc. Biol.* **2017**, *102*, 631–643. [CrossRef] [PubMed]
40. Bommareddy, P.K.; Zloza, A.; Rabkin, S.D.; Kaufman, H.L. Oncolytic virus immunotherapy induces immunogenic cell death and overcomes STING deficiency in melanoma. *Oncoimmunology* **2019**, *8*, e1591875. [CrossRef]
41. De Graaf, J.F.; de Vor, L.; Fouchier, R.A.M.; van den Hoogen, B.G. Armed oncolytic viruses: A kick-start for anti-tumor immunity. *Cytokine Growth Factor Rev.* **2018**, *41*, 28–39. [CrossRef] [PubMed]
42. Joyce, J.A.; Fearon, D.T. T cell exclusion, immune privilege, and the tumor microenvironment. *Science* **2015**, *348*, 74–80. [CrossRef] [PubMed]
43. De Olza, M.O.; Navarro Rodrigo, B.; Zimmermann, S.; Coukos, G. Turning up the heat on non-immunoreactive tumours: Opportunities for clinical development. *Lancet Oncol.* **2020**, *21*, e419–e430. [CrossRef]
44. Naito, Y.; Saito, K.; Shiiba, K.; Ohuchi, A.; Saigenji, K.; Nagura, H.; Ohtani, H. CD8+ T cells infiltrated within cancer cell nests as a prognostic factor in human colorectal cancer. *Cancer Res.* **1998**, *58*, 3491–3494. [PubMed]
45. Alayo, Q.A.; Ito, H.; Passaro, C.; Zdioruk, M.; Mahmoud, A.B.; Grauwet, K.; Zhang, X.; Lawler, S.E.; Reardon, D.A.; Goins, W.F.; et al. Glioblastoma infiltration of both tumor- and virus-antigen specific cytotoxic T cells correlates with experimental virotherapy responses. *Sci. Rep.* **2020**, *10*, 1–11. [CrossRef] [PubMed]
46. Garcia-Carbonero, R.; Salazar, R.; Duran, I.; Osman-Garcia, I.; Paz-Ares, L.; Bozada, J.M.; Boni, V.; Blanc, C.; Seymour, L.; Beadle, J.; et al. Phase 1 study of intravenous administration of the chimeric adenovirus enadenotucirev in patients undergoing primary tumor resection. *J. Immunother. Cancer* **2017**, *5*, 1–13. [CrossRef] [PubMed]
47. Samson, A.; Scott, K.J.; Taggart, D.; West, E.J.; Wilson, E.; Nuovo, G.J.; Thomson, S.; Corns, R.; Mathew, R.K.; Fuller, M.J.; et al. Intravenous delivery of oncolytic reovirus to brain tumor patients immunologically primes for subsequent checkpoint blockade. *Sci. Transl. Med.* **2018**, *10*, eaar3342. [CrossRef] [PubMed]
48. Waldmann, T.A. Cytokines in Cancer Immunotherapy. *Cold Spring Harb. Perspect. Biol.* **2018**, *10*, a028472. [CrossRef]
49. Le Naour, J.; Zitvogel, L.; Galluzzi, L.; Vacchelli, E.; Kroemer, G. Trial watch: STING agonists in cancer therapy. *Oncoimmunology* **2020**, *9*, 1777624. [CrossRef]
50. Eckert, E.C.; Nace, R.A.; Tonne, J.M.; Evgin, L.; Vile, R.G.; Russell, S.J. Generation of a Tumor-Specific Chemokine Gradient Using Oncolytic Vesicular Stomatitis Virus Encoding CXCL9. *Mol. Ther. Oncolytics* **2020**, *16*, 63–74. [CrossRef]
51. Li, J.; O'Malley, M.; Urban, J.; Sampath, P.; Guo, Z.S.; Kalinski, P.; Thorne, S.H.; Bartlett, D.L. Chemokine expression from oncolytic vaccinia virus enhances vaccine therapies of cancer. *Mol. Ther.* **2011**, *19*, 650–657. [CrossRef]
52. Lapteva, N.; Aldrich, M.; Weksberg, D.; Rollins, L.; Goltsova, T.; Chen, S.Y.; Huang, X.F. Targeting the intratumoral dendritic cells by the oncolytic adenoviral vaccine expressing RANTES elicits potent antitumor immunity. *J. Immunother.* **2009**, *32*, 145–156. [CrossRef]
53. Wojton, J.; Kaur, B. Impact of tumor microenvironment on oncolytic viral therapy. *Cytokine Growth Factor Rev.* **2010**, *21*, 127–134. [CrossRef]
54. Ambagala, A.P.; Solheim, J.C.; Srikumaran, S. Viral interference with MHC class I antigen presentation pathway: The battle continues. *Vet. Immunol. Immunopathol.* **2005**, *107*, 1–15. [CrossRef]
55. Liu, B.L.; Robinson, M.; Han, Z.Q.; Branston, R.H.; English, C.; Reay, P.; McGrath, Y.; Thomas, S.K.; Thornton, M.; Bullock, P.; et al. ICP34.5 deleted herpes simplex virus with enhanced oncolytic, immune stimulating, and anti-tumour properties. *Gene Ther.* **2003**, *10*, 292–303. [CrossRef] [PubMed]
56. Bortolanza, S.; Bunuales, M.; Alzuguren, P.; Lamas, O.; Aldabe, R.; Prieto, J.; Hernandez-Alcoceba, R. Deletion of the E3-6.7K/gp19K region reduces the persistence of wild-type adenovirus in a permissive tumor model in Syrian hamsters. *Cancer Gene Ther.* **2009**, *16*, 703–712. [CrossRef] [PubMed]

57. Ma, J.; Ramachandran, M.; Jin, C.; Quijano-Rubio, C.; Martikainen, M.; Yu, D.; Essand, M. Characterization of virus-mediated immunogenic cancer cell death and the consequences for oncolytic virus-based immunotherapy of cancer. *Cell Death Dis.* **2020**, *11*, 1–15. [CrossRef] [PubMed]

58. Green, D.R.; Ferguson, T.; Zitvogel, L.; Kroemer, G. Immunogenic and tolerogenic cell death. *Nat. Rev. Immunol.* **2009**, *9*, 353–363. [CrossRef]

59. Davola, M.E.; Mossman, K.L. Oncolytic viruses: How "lytic" must they be for therapeutic efficacy? *OncoImmunology* **2019**, *8*, e1581528. [CrossRef] [PubMed]

60. Kell, A.M.; Gale, M., Jr. RIG-I in RNA virus recognition. *Virology* **2015**, *479*, 110–121. [CrossRef] [PubMed]

61. Ishikawa, H.; Barber, G.N. STING is an endoplasmic reticulum adaptor that facilitates innate immune signalling. *Nature* **2008**, *455*, 674–678. [CrossRef]

62. Bourgeois-Daigneault, M.C.; Roy, D.G.; Falls, T.; Twumasi-Boateng, K.; St-Germain, L.E.; Marguerie, M.; Garcia, V.; Selman, M.; Jennings, V.A.; Pettigrew, J.; et al. Oncolytic vesicular stomatitis virus expressing interferon-gamma has enhanced therapeutic activity. *Mol. Ther. Oncolytics* **2016**, *3*, 16001. [CrossRef] [PubMed]

63. LaRocca, C.J.; Han, J.; Gavrikova, T.; Armstrong, L.; Oliveira, A.R.; Shanley, R.; Vickers, S.M.; Yamamoto, M.; Davydova, J. Oncolytic adenovirus expressing interferon alpha in a syngeneic Syrian hamster model for the treatment of pancreatic cancer. *Surgery* **2015**, *157*, 888–898. [CrossRef] [PubMed]

64. Patel, M.R.; Jacobson, B.A.; Ji, Y.; Drees, J.; Tang, S.; Xiong, K.; Wang, H.; Prigge, J.E.; Dash, A.S.; Kratzke, A.K.; et al. Vesicular stomatitis virus expressing interferon-beta is oncolytic and promotes antitumor immune responses in a syngeneic murine model of non-small cell lung cancer. *Oncotarget* **2015**, *6*, 33165–33177. [CrossRef]

65. Seliger, B.; Wollscheid, U.; Momburg, F.; Blankenstein, T.; Huber, C. Characterization of the major histocompatibility complex class I deficiencies in B16 melanoma cells. *Cancer Res.* **2001**, *61*, 1095–1099.

66. Galivo, F.; Diaz, R.M.; Wongthida, P.; Thompson, J.; Kottke, T.; Barber, G.; Melcher, A.; Vile, R. Single-cycle viral gene expression, rather than progressive replication and oncolysis, is required for VSV therapy of B16 melanoma. *Gene Ther.* **2010**, *17*, 158–170. [CrossRef]

67. Castelli, E.C.; Veiga-Castelli, L.C.; Yaghi, L.; Moreau, P.; Donadi, E.A. Transcriptional and posttranscriptional regulations of the HLA-G gene. *J. Immunol. Res.* **2014**, *2014*, 734068. [CrossRef]

68. Aref, S.; Castleton, A.Z.; Bailey, K.; Burt, R.; Dey, A.; Leongamornlert, D.; Mitchell, R.J.; Okasha, D.; Fielding, A.K. Type 1 Interferon Responses Underlie Tumor-Selective Replication of Oncolytic Measles Virus. *Mol. Ther.* **2020**, *28*, 1043–1055. [CrossRef] [PubMed]

69. Arwert, E.N.; Milford, E.L.; Rullan, A.; Derzsi, S.; Hooper, S.; Kato, T.; Mansfield, D.; Melcher, A.; Harrington, K.J.; Sahai, E. STING and IRF3 in stromal fibroblasts enable sensing of genomic stress in cancer cells to undermine oncolytic viral therapy. *Nat. Cell Biol.* **2020**, *22*, 758–766. [CrossRef] [PubMed]

70. Katze, M.G.; He, Y.; Gale, M., Jr. Viruses and interferon: A fight for supremacy. *Nat. Rev. Immunol.* **2002**, *2*, 675–687. [CrossRef] [PubMed]

71. Lam, E.; Falck-Pedersen, E. Unabated adenovirus replication following activation of the cGAS/STING-dependent antiviral response in human cells. *J. Virol.* **2014**, *88*, 14426–14439. [CrossRef]

72. Xia, T.; Konno, H.; Ahn, J.; Barber, G.N. Deregulation of STING Signaling in Colorectal Carcinoma Constrains DNA Damage Responses and Correlates With Tumorigenesis. *Cell Rep.* **2016**, *14*, 282–297. [CrossRef] [PubMed]

73. Mora-Garcia Mde, L.; Duenas-Gonzalez, A.; Hernandez-Montes, J.; De la Cruz-Hernandez, E.; Perez-Cardenas, E.; Weiss-Steider, B.; Santiago-Osorio, E.; Ortiz-Navarrete, V.F.; Rosales, V.H.; Cantu, D.; et al. Up-regulation of HLA class-I antigen expression and antigen-specific CTL response in cervical cancer cells by the demethylating agent hydralazine and the histone deacetylase inhibitor valproic acid. *J. Transl. Med.* **2006**, *4*, 1–14. [CrossRef]

74. Salamone, F.N.; Gleich, L.L.; Li, Y.Q.; Stambrook, P.J. Major histocompatibility gene therapy: The importance of haplotype and beta 2-microglobulin. *Laryngoscope* **2004**, *114*, 612–615. [CrossRef]

75. Del Campo, A.B.; Aptsiauri, N.; Mendez, R.; Zinchenko, S.; Vales, A.; Paschen, A.; Ward, S.; Ruiz-Cabello, F.; Gonzalez-Aseguinolaza, G.; Garrido, F. Efficient recovery of HLA class I expression in human tumor cells after beta2-microglobulin gene transfer using adenoviral vector: Implications for cancer immunotherapy. *Scand. J. Immunol.* **2009**, *70*, 125–135. [CrossRef] [PubMed]

76. Del Campo, A.B.; Carretero, J.; Munoz, J.A.; Zinchenko, S.; Ruiz-Cabello, F.; Gonzalez-Aseguinolaza, G.; Garrido, F.; Aptsiauri, N. Adenovirus expressing beta2-microglobulin recovers HLA class I expression and antitumor immunity by increasing T-cell recognition. *Cancer Gene Ther.* **2014**, *21*, 317–332. [CrossRef]

77. Freedman, J.D.; Hagel, J.; Scott, E.M.; Psallidas, I.; Gupta, A.; Spiers, L.; Miller, P.; Kanellakis, N.; Ashfield, R.; Fisher, K.D.; et al. Oncolytic adenovirus expressing bispecific antibody targets T-cell cytotoxicity in cancer biopsies. *EMBO Mol. Med.* **2017**, *9*, 1067–1087. [CrossRef]

78. Fajardo, C.A.; Guedan, S.; Rojas, L.A.; Moreno, R.; Arias-Badia, M.; de Sostoa, J.; June, C.H.; Alemany, R. Oncolytic Adenoviral Delivery of an EGFR-Targeting T-cell Engager Improves Antitumor Efficacy. *Cancer Res.* **2017**, *77*, 2052–2063. [CrossRef] [PubMed]

79. Yu, F.; Wang, X.; Guo, Z.S.; Bartlett, D.L.; Gottschalk, S.M.; Song, X.T. T-cell engager-armed oncolytic vaccinia virus significantly enhances antitumor therapy. *Mol. Ther.* **2014**, *22*, 102–111. [CrossRef]

80. Scott, E.M.; Jacobus, E.J.; Lyons, B.; Frost, S.; Freedman, J.D.; Dyer, A.; Khalique, H.; Taverner, W.K.; Carr, A.; Champion, B.R.; et al. Bi- and tri-valent T cell engagers deplete tumour-associated macrophages in cancer patient samples. *J. Immunother. Cancer* **2019**, *7*, 1–18. [CrossRef]

81. Freedman, J.D.; Duffy, M.R.; Lei-Rossmann, J.; Muntzer, A.; Scott, E.M.; Hagel, J.; Campo, L.; Bryant, R.J.; Verrill, C.; Lambert, A.; et al. An Oncolytic Virus Expressing a T-cell Engager Simultaneously Targets Cancer and Immunosuppressive Stromal Cells. *Cancer Res.* **2018**, *78*, 6852–6865. [CrossRef]

82. De Sostoa, J.; Fajardo, C.A.; Moreno, R.; Ramos, M.D.; Farrera-Sal, M.; Alemany, R. Targeting the tumor stroma with an oncolytic adenovirus secreting a fibroblast activation protein-targeted bispecific T-cell engager. *J. Immunother. Cancer* **2019**, *7*, 1–15. [CrossRef]

83. Leung, E.Y.L.; Ennis, D.P.; Kennedy, P.R.; Hansell, C.; Dowson, S.; Farquharson, M.; Spiliopoulou, P.; Nautiyal, J.; McNamara, S.; Carlin, L.M.; et al. NK Cells Augment Oncolytic Adenovirus Cytotoxicity in Ovarian Cancer. *Mol. Ther. Oncolytics* **2020**, *16*, 289–301. [CrossRef] [PubMed]

84. Stephenson, K.B.; Barra, N.G.; Davies, E.; Ashkar, A.A.; Lichty, B.D. Expressing human interleukin-15 from oncolytic vesicular stomatitis virus improves survival in a murine metastatic colon adenocarcinoma model through the enhancement of anti-tumor immunity. *Cancer Gene Ther.* **2012**, *19*, 238–246. [CrossRef]

85. Li, F.; Sheng, Y.; Hou, W.; Sampath, P.; Byrd, D.; Thorne, S.; Zhang, Y. CCL5-armed oncolytic virus augments CCR5-engineered NK cell infiltration and antitumor efficiency. *J. Immunother. Cancer* **2020**, *8*, e000131. [CrossRef] [PubMed]

86. Huang, Y.; Lv, S.Q.; Liu, P.Y.; Ye, Z.L.; Yang, H.; Li, L.F.; Zhu, H.L.; Wang, Y.; Cui, L.Z.; Jiang, D.Q.; et al. A SIRPalpha-Fc fusion protein enhances the antitumor effect of oncolytic adenovirus against ovarian cancer. *Mol. Oncol.* **2020**, *14*, 657–668. [CrossRef] [PubMed]

87. Vallera, D.A.; Ferrone, S.; Kodal, B.; Hinderlie, P.; Bendzick, L.; Ettestad, B.; Hallstrom, C.; Zorko, N.A.; Rao, A.; Fujioka, N.; et al. NK-Cell-Mediated Targeting of Various Solid Tumors Using a B7-H3 Tri-Specific Killer Engager In Vitro and In Vivo. *Cancers* **2020**, *12*, 2659. [CrossRef]

88. Felices, M.; Lenvik, T.R.; Davis, Z.B.; Miller, J.S.; Vallera, D.A. Generation of BiKEs and TriKEs to Improve NK Cell-Mediated Targeting of Tumor Cells. *Methods Mol. Biol.* **2016**, *1441*, 333–346. [CrossRef] [PubMed]

89. Ring, N.G.; Herndler-Brandstetter, D.; Weiskopf, K.; Shan, L.; Volkmer, J.P.; George, B.M.; Lietzenmayer, M.; McKenna, K.M.; Naik, T.J.; McCarty, A.; et al. Anti-SIRPalpha antibody immunotherapy enhances neutrophil and macrophage antitumor activity. *Proc. Natl. Acad. Sci. USA* **2017**, *114*, E10578–E10585. [CrossRef]

90. Seymour, L.W.; Fisher, K.D. Oncolytic viruses: Finally delivering. *Br. J. Cancer* **2016**, *114*, 357–361. [CrossRef] [PubMed]

91. Heise, C.C.; Williams, A.M.; Xue, S.; Propst, M.; Kirn, D.H. Intravenous administration of ONYX-015, a selectively replicating adenovirus, induces antitumoral efficacy. *Cancer Res.* **1999**, *59*, 2623–2628.

92. Kuhn, I.; Harden, P.; Bauzon, M.; Chartier, C.; Nye, J.; Thorne, S.; Reid, T.; Ni, S.; Lieber, A.; Fisher, K.; et al. Directed evolution generates a novel oncolytic virus for the treatment of colon cancer. *PLoS ONE* **2008**, *3*, e2409. [CrossRef] [PubMed]

93. Brun, J.; McManus, D.; Lefebvre, C.; Hu, K.; Falls, T.; Atkins, H.; Bell, J.C.; McCart, J.A.; Mahoney, D.; Stojdl, D.F. Identification of genetically modified Maraba virus as an oncolytic rhabdovirus. *Mol. Ther.* **2010**, *18*, 1440–1449. [CrossRef] [PubMed]

94. Wennier, S.T.; Liu, J.; McFadden, G. Bugs and drugs: Oncolytic virotherapy in combination with chemotherapy. *Curr. Pharm. Biotechnol.* **2012**, *13*, 1817–1833. [CrossRef] [PubMed]

95. Khuri, F.R.; Nemunaitis, J.; Ganly, I.; Arseneau, J.; Tannock, I.F.; Romel, L.; Gore, M.; Ironside, J.; MacDougall, R.H.; Heise, C.; et al. A controlled trial of intratumoral ONYX-015, a selectively-replicating adenovirus, in combination with cisplatin and 5-fluorouracil in patients with recurrent head and neck cancer. *Nat. Med.* **2000**, *6*, 879–885. [CrossRef]

96. Connors, T.A.; Knox, R.J. Prodrugs in cancer chemotherapy. *Stem Cells* **1995**, *13*, 501–511. [CrossRef]

97. Dachs, G.U.; Hunt, M.A.; Syddall, S.; Singleton, D.C.; Patterson, A.V. Bystander or no bystander for gene directed enzyme prodrug therapy. *Molecules* **2009**, *14*, 4517–4545. [CrossRef]

98. Sterzenbach, U.; Putz, U.; Low, L.H.; Silke, J.; Tan, S.S.; Howitt, J. Engineered Exosomes as Vehicles for Biologically Active Proteins. *Mol. Ther.* **2017**, *25*, 1269–1278. [CrossRef]

Review

Oncolytic Virotherapy: The Cancer Cell Side

Marcelo Ehrlich * and Eran Bacharach *

Shmunis School of Biomedicine and Cancer Research, George S. Wise Faculty of Life Sciences, Tel Aviv University, Tel Aviv 6997801, Israel
* Correspondence: marceloe@tauex.tau.ac.il (M.E.); eranba@tauex.tau.ac.il (E.B.)

Simple Summary: Oncolytic viruses (OVs) are a promising immunotherapy that specifically target and kill cancer cells and stimulate anti-tumor immunity. While different OVs are endowed with distinct features, which enhance their specificity towards tumor cells; attributes of the cancer cell also critically contribute to this specificity. Such features comprise defects in innate immunity, including antiviral responses, and the metabolic reprogramming of the malignant cell. The tumorigenic features which support OV replication can be intrinsic to the transformation process (e.g., a direct consequence of the activity of a given oncogene), or acquired in the course of tumor immunoediting—the selection process applied by antitumor immunity. Oncogene-induced epigenetic silencing plays an important role in negative regulation of immunostimulatory antiviral responses in the cancer cells. Reversal of such silencing may also provide a strong immunostimulant in the form of viral mimicry by activation of endogenous retroelements. Here we review features of the cancer cell that support viral replication, tumor immunoediting and the connection between oncogenic signaling, DNA methylation and viral oncolysis. As such, this review concentrates on the malignant cell, while detailed description of different OVs can be found in the accompanied reviews of this issue.

Citation: Ehrlich, M.; Bacharach, E. Oncolytic Virotherapy: The Cancer Cell Side. *Cancers* **2021**, *13*, 939. https://doi.org/10.3390/cancers13050939

Academic Editor: G. Eric Blair

Received: 19 January 2021
Accepted: 12 February 2021
Published: 24 February 2021

Abstract: Cell autonomous immunity genes mediate the multiple stages of anti-viral defenses, including recognition of invading pathogens, inhibition of viral replication, reprogramming of cellular metabolism, programmed-cell-death, paracrine induction of antiviral state, and activation of immunostimulatory inflammation. In tumor development and/or immunotherapy settings, selective pressure applied by the immune system results in tumor immunoediting, a reduction in the immunostimulatory potential of the cancer cell. This editing process comprises the reduced expression and/or function of cell autonomous immunity genes, allowing for immune-evasion of the tumor while concomitantly attenuating anti-viral defenses. Combined with the oncogene-enhanced anabolic nature of cancer-cell metabolism, this attenuation of antiviral defenses contributes to viral replication and to the selectivity of oncolytic viruses (OVs) towards malignant cells. Here, we review the manners by which oncogene-mediated transformation and tumor immunoediting combine to alter the intracellular milieu of tumor cells, for the benefit of OV replication. We also explore the functional connection between oncogenic signaling and epigenetic silencing, and the way by which restriction of such silencing results in immune activation. Together, the picture that emerges is one in which OVs and epigenetic modifiers are part of a growing therapeutic toolbox that employs activation of anti-tumor immunity for cancer therapy.

Keywords: oncolytic viruses; immunoediting; oncogenic signaling; RAS; DNA methyltransferase inhibitor (DNMTi); viral mimicry; epigenetic silencing

1. Introduction

The present review focuses on the differential and enhanced susceptibility of cancer cells to oncolytic viruses (OVs). We propose that such hyper-susceptibility of the malignant cells stems from unique features of the cancer-cell milieu, including defective antiviral responses and metabolic reprograming. The sources of such tumor-cell specific alterations comprise a combination of factors, which are intrinsic to the tumor cell—e.g., oncogene-stimulated signaling, and/or extrinsic ones; e.g., selective pressure applied by the tumor immune microenvironment. We begin by focusing on the cancer-cell *per se*, analyzing how oncogene-induced modifications serve to optimize the intracellular environment towards OV replication. To this end, we employ RAS-activated pathways as a pivot, exemplifying how this intrinsic oncogenic pathway modulates antiviral responses. We then proceed to focus on the immunoediting of tumors, as this provides a critical extrinsic (selective) source of alterations to cancer-cell autonomous immune functions. Given the overlap in the immune-activation-potential of a cancer cell and its ability to raise antiviral responses, the selective pressure applied by anti-tumor immunity results in both decreased immunogenicity and in defective antiviral responses. We finalize our review by focusing on oncogene-stimulated DNA methylation in the context of immune evasion, as an example of how the two processes (oncogenic signaling and immunoediting) converge to influence OVs-cancer-cell interactions. Our focus on DNA methylation stems from its prominence as a molecular mechanism for silencing of cell-autonomous immune responses. In this context, we also discuss the reversal of this form of epigenetic silencing, which may elicit tumor immunogenicity through the expression of endogenous retroelements, thus generating a "viral mimicry" state, emulating the immune-stimulatory potential of OVs.

2. Defects to Cell Autonomous Immunity and Metabolic Reprogramming Optimize the Cancer Cell Milieu towards Viral Infection

2.1. Cell Autonomous Immunity: The Antiviral Response

The cell autonomous immune response provides the first line of defense against cellular pathogens, including viruses [1]. To deal with a wide variety of pathogens, activation of cell autonomous immunity occurs in an antigen-independent fashion. Instead, it relies on the ability of the cell to recognize molecular patterns which are abundant in pathogens (pathogen-associated-molecular patterns, PAMPs), yet relatively absent in healthy cells. These molecular patterns are recognized by pattern recognition receptors (PRRs), which survey distinct cellular compartments for the presence of PAMPs. In addition, aberrant intracellular localization of nucleic acids (e.g., intra-endosomal localization of RNA or DNA, or cytoplasmic localization of DNA) also serves to discern between nucleic acids of cellular vs. pathogen origin, and when detected, stimulates cell autonomous immune responses (reviewed in [2–4]). A prototypic PAMP is double stranded RNA (dsRNA), an obligatory molecular pattern of viral infection, which may be recognized by toll-like receptor 3 (TLR3) upon exposure to the endosomal lumen, or by RNA helicases—the retinoic acid-inducible gene I (RIG-I) and the melanoma differentiation-associated gene 5 (MDA5) upon exposure in the cytoplasm [1,5]. DNA too can serve as a PAMP, depending on its composition or intracellular localization. In these contexts, TLR9 recognizes DNA molecules rich in unmethylated CpG sequences, as commonly occurs in genomes of viruses and bacteria [6]; while cytoplasm-localized DNA is recognized by cyclic GMP-AMP Synthase (cGAS) [7]. In a typical case, exemplified here by the cellular response to RNA virus infections, PAMP-induced PRR signals are transduced through mitochondrial antiviral-signaling protein (MAVS), Tank-binding kinase 1 (TBK1) and IKKs; resulting in the activation of nuclear factor kappa-light-chain-enhancer of activated B cells (NF-kB) and interferon (IFN)-regulatory factors (IRFs) 3 and 7. These in turn translocate to the nucleus and mediate the transcriptional activation of type I or type III IFNs (e.g., IFN-β). Following synthesis and secretion, IFNs activate Janus kinase (JAK)- signal transducer and activator of transcription (STAT) signaling, resulting in STAT-mediated massive amplification of the cell autonomous immune response via the induction of IFN-stimulated-genes (ISGs) [1,5,8].

2.2. Oncogene-Induced Perturbations to Antiviral Responses: A Reduction in Impediments to Viral Replication

Oncogene-induced perturbations to antiviral responses are prominent molecular mechanisms by which the cancer-cell milieu becomes optimized towards OV replication. To exemplify this concept, we focus on such effects related to oncogenic RAS. Oncogenic mutations in RAS, a GTP-activated molecular switch, ensue exposure to genotoxic agents, and are estimated to occur in 16–30% of all human cancers, with highest incidence in pancreatic (90%) and colon (50%) cancers; and considerable portions of melanoma and lung adenocarcinoma [9–11]. Activated RAS (either because of oncogenic mutations or following stimulation of upstream growth receptors) stimulates downstream signaling pathways mediated by phosphatidylinositol 3 (OH)-kinase (PI3K), RAL guanine nucleotide dissociation stimulator (RALGDS) family members, and members of the RAF family, which activate the RAF/MEK/ERK pathway [12]. Thus, RAS functions as a multi-pronged signaling node, which upon activation, endows tumor cells with multiple malignancy-associated features. Multiple lines of evidence place RAS, and its associated signaling pathways, as negative regulators cell autonomous immunity.

2.2.1. RAS-Mediated Regulation of Immune Transcription Factors

In accord with oncogene-mediated regulation of gene expression programs, a critical mechanism by which they modify immune/antiviral functions of tumor cells is through regulation of the expression of immunity-related transcription factors. In HRAS transformed murine fibroblasts, and RAS-transformed human cancer cells, MEK-ERK signaling was shown to negatively regulate IRF-1-dependent transcription of IRF1 and STAT2 [13,14], thus hampering IFN responses, and supporting the replication of oncolytic vesicular stomatitis virus (VSV). In addition to immune-related functions (e.g., as antiviral gene, master regulator of acute inflammation, and main effector of IFNγ signaling), IRF1 was also characterized as a tumor-suppressor [15–19]. Thus, IRF1 inhibition by RAS-MEK is predicted to concomitantly promote tumorigenicity, alter the interactions between tumor- and immune cells and enhance the susceptibility of cancer cells to OVs. Of note, the antagonism of IRF1 function by mitogenic pathways is not restricted to cancer settings. For example, in airway epithelial cells, influenza A virus (IAV) and rhinovirus activate the epidermal growth factor receptor (EGFR, [20])—an upstream activator of the RAS/RAF/MEK/ERK pathway [21]. Activated EGFR diminishes both IRF1 expression and induction of IFN-λ production, thus increasing viral infection. Oncogenic KRAS was shown to inhibit the expression of STAT1, STAT2 and IRF9 (members of the ISGF3 transcription-promoting complex); thus, hampering the basal and IFN-induced expression of ISGs in colorectal cancer cell lines [22]. This effect was proposed to be mediated (at least in part) through the PI3K pathway. Moreover, a recent study employing a murine model of colorectal cancer combining oncogenic KRAS expression with conditional null alleles of adenomatous polyposis coli (*APC*) and *TRP53*, identified repression of IRF2 as a key mechanism for KRAS-induced immune-suppression in colorectal cancer [23]. It should be noted that the roles of IRF2 in cancer are controversial. Thus, while IRF2 expression is downregulated in many different tumor types [24] suggesting potential tumor suppressor roles, other studies proposed pro-tumorigenic functions for IRF2, including via antagonism of IRF1 functions [15,25]. Similarly, while IRF2 was proposed to antagonize IRF1 antiviral responses [26], more recent studies suggest complementary roles for IRF1 and IRF2 in IFN-induced gene expression.

2.2.2. Inhibition of PKR Licenses Cells for Viral Infection

A major antiviral signaling node, which is targeted by RAS-induced signaling, is the dsRNA-activated protein kinase, PKR, which following the binding of dsRNA inhibits protein synthesis via phosphorylation of the eukaryotic initiation factor 2 α (eIF2α) [27,28]. In accord with the enhanced protein synthesis requirements of cancer cells, PKR has been identified as a tumor suppressor in different malignancy settings [28–30]; inducing apoptosis upon its activation [31,32]. The notion of PKR as a main antiviral gene is underscored by the numerous inhibitory mechanisms against PKR which are encoded/induced by different viruses [33–37], and by the enhancement of viral replication and viral-induced lethality in PKR-null cells and mice, respectively [38]. Based on this dual role of tumor suppressor and antiviral effector, oncogene-mediated targeting of PKR in general, and its inhibition by the RAS/RAF/MEK/ERK pathway in particular, can be exploited by

OVs. For example, wild-type IAV counters PKR via its NS1 protein [39], and via activation of mitogen-activated protein kinase-activated protein kinases (MAPKAPKs) MK2 and MK3 [40]. In accord with PKR being an ISG [38], mutant IAV lacking NS1 replicate only in interferon-deficient systems [41] and perturbation of expression of MK2 or MK3 reduces IAV titers, and enhances PKR activation and eIF2α phosphorylation by the dsRNA mimic polyI:C [40]. In accord with RAF/MEK/ERK-mediated licensing of cells towards IAV infection, IAV shows a strong tropism towards cells expressing active RAF both in vitro and in vivo [42]. Similarly, expression of oncogenic NRAS in melanoma cells, suffices to make them selectively susceptible to oncolysis by IAV lacking NS1 [43]. The centrality of PKR inhibition by the RAS/RAF/MEK/ERK signaling axis in determining susceptibility of cancer cells to OVs is further exemplified by: (i) the requirements of herpes simplex virus 1 (HSV1) $\Delta\gamma$(1)34.5 mutants for MEK-mediated PKR inhibition [44], (ii) the oncotropism of VAI mutant adenovirus towards cells in which RAS inactivates PKR [45], and (iii) the selectivity of the mammalian reovirus towards RAS-transformed cells, which was initially identified as dependent on PKR inactivation [46,47]. This latter tropism has been further dissected and was shown to involve additional mechanisms, including: activation of RAL-GTP exchange factor (RAL-GEF) and the p38 kinase, downstream of RAS [48]; the RAS-mediated enhancement of multiple reovirus infection features including uncoating, particle infectivity, and apoptosis-dependent virion release [49]; and the RAS-mediated inhibition of RIG-I expression/function [50]. In line with the latter inhibitory mechanism, RAF/MEK/ERK activation also hampers RIG-I- and IFN-mediated restriction of VSV replication [51].

2.2.3. Inhibition of Antiviral Responses by RAS-Regulated Factors

Oncogenic RAS may also regulate OV replication through effects on additional oncogenes. For example, the enhanced replication of oncolytic Newcastle disease virus (NDV) depends on RAC1 in highly-malignant RAS-transformed keratinocytes [52]; and RAC1 is a downstream effector of oncogenic RAS [53,54]. In addition, the CDC25 phosphatase, a RAF-regulated oncogene [55], negatively regulates TBK1 through dephosphorylation, inhibiting RIG-I-mediated induction of IFN [56]. Moreover, while oncogenic KRAS increases PKC-βII expression in a murine colon-cancer model [57], this enzyme phosphorylates and inhibits RIG-I, and enhances VSV replication in different cellular settings [58]. The notion of a functional interaction between MDA-5 and oncogenic-RAS is exemplified by the suppression of pro-apoptotic effects of MDA-5 overexpression by either oncogenic RAS or RAF [59]. An additional mode of action is observed for the MYC oncogene, which functions as a crucial effector of oncogenic KRAS, [60,61] and represses, together with the transcriptional repressor MIZ, the type I IFN pathway [61]. Interestingly, inactivation of the tumor suppressor phosphatase and tensin homologue (PTEN), which among its well-documented malignancy-promoting activities [62] accelerates tumorigenesis induced by KRAS [63], results in increased phosphorylation of Ser97 in IRF3, in the negative regulation of IRF-mediated IFN induction upon viral challenge, and in increased viral (VSV) replication [64].

Together, the above-mentioned examples (Section 2.2) demonstrate the ability of oncogenic signaling to interfere with all steps of the antiviral response continuum, including PRR-mediated PAMP recognition, IFN induction, JAK/STAT signaling and ISG expression.

2.3. Oncogene-Mediated Stimulation of Anabolism: Supplying the Metabolic Needs of Replicating Viruses

Both viral replication and tumor-cell growth are anabolic processes, i.e., dependent on the biosynthesis of macromolecules (nucleic acids, proteins, lipids and oligosaccharides). As such, both oncogenic transformation and viral infection optimize the cell's metabolic regulation towards their anabolic needs. The efficiency and extent by which oncogene-induced processes carry out such reprograming is predicted to support enhanced replication of OVs. For example, oncogenic KRAS stimulates anabolic metabolism to maintain pancreatic tumors through activation of MAPK and MYC pathways and the ensuing increased expression of genes which regulate sterol biosynthesis, pyrimidine metabolism and glycosylation [65]. Such metabolically reprogramed cells are characterized by increased glycolytic flux (Warburg effect, [66]) and by glutamine serving as a major carbon source for the tri-carboxylic acid (TCA) cycle [67]. Multiple lines of evidence support the notion that viruses benefit from analogous metabolic reprograming, as different viruses manipulate cell metabolism towards aerobic glycolysis (reviewed in [68,69]) and reprogram glutamine catabolism to optimize virus replication [70]. Similarly, fatty acid synthase (FASN), which regulates the production of long-chain fatty acids [71], is overexpressed in different tumors [71,72], and induced upon oncogenic-RAS-mediated cell transformation [73,74]. Analogous to its role in tumorigenesis, FASN-mediated lipogenesis is required for infection with diverse viruses [75–79]. The similitude of the metabolic requirements of KRAS-transformed tumors and viruses is further exemplified by the effects of inhibitors of dihydroorotate dehydrogenase (DHODH), which perturb *de novo* pyrimidine biosynthesis, selectively inhibit the growth of KRAS mutant cell lines [80] and exhibit broad-range antiviral activity against RNA viruses [81].

The multiple effects of oncogenic RAS, which promote viral replication and reduce tumor-cell immunogenicity are schematically depicted in Figure 1.

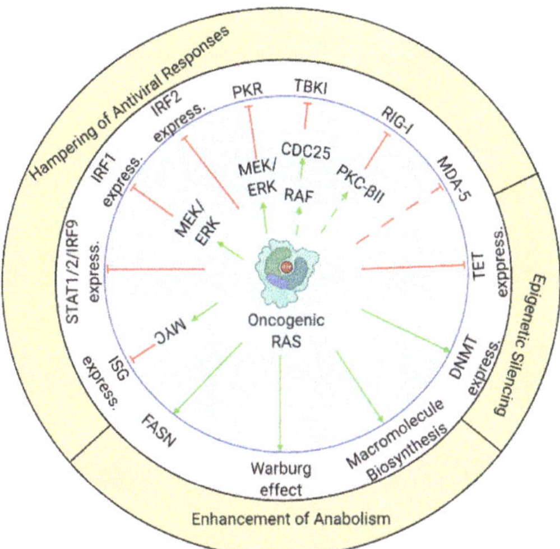

Figure 1. Oncogenic RAS supports viral infection through multiple molecular mechanisms. Scheme depicts mechanisms described throughout review. Green arrows or blunt red arrows denote stimulation or inhibition, respectively. Dashed arrows indicate cases where one source of information supports the connection between oncogenic RAS and its effector, and another source supports the link between the effector and the oncolysis-regulating mechanism. The figure was created with BioRender.com (accessed on 12 February 2021).

3. Immunoediting Selects for Cancer Cells with Defects in Immune-Stimulatory Abilities

Immunosurveillance and tumor immunoediting are complementary and consecutive processes involving the interaction of a competent immune system with developing tumors. The former refers to the continuous recognition and targeting of malignant cells as a result immune activity. Contrastingly, immunoediting results in the selection of tumor cells with reduced immunogenicity as consequence of selective pressures applied by innate and adaptive immunity. Tumor immunoediting is commonly divided into three phases (the "three E's"): (i) elimination, where cancer cells are destroyed by immunosurveillance mechanisms; (ii) equilibrium, where cells surviving the initial immune onslaught undergo consecutive rounds of functional, epigenetic and genetic changes. These result in adaptation, i.e., improved fitness of the malignant cells within the tumor microenvironment (TME) co-populated by immune cells; (iii) escape, where outgrowth of resistant clones induces and supports an immunosuppressive microenvironment (reviewed in [82,83], schematically depicted in Figure 2).

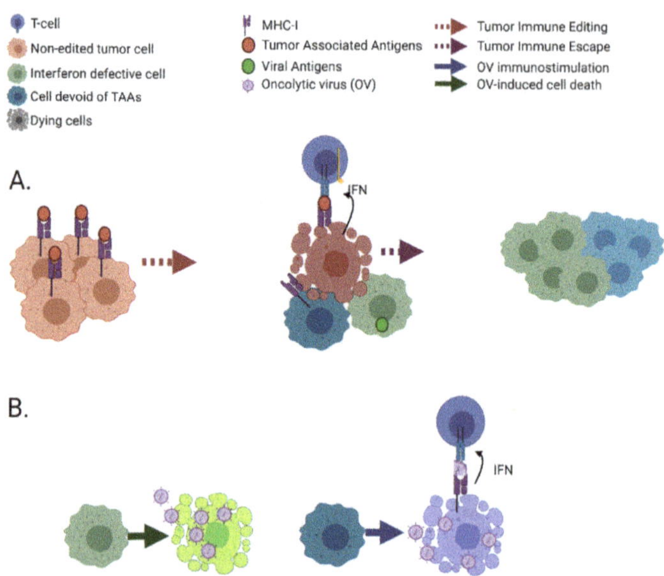

Figure 2. Tumor immunoediting and treatment of escape mutants with oncolytic viruses. (**A**) Tumor cells prior to editing are depicted (in pink) at the left side. Anti-tumor immunity kills a portion of susceptible tumor cells while selecting for escape mutants (middle), allowing their subsequent clonal expansion (right). Two types of escape mutants are depicted: green—IFN-defective cells, blue—cells devoid of tumor-associated antigens. (**B**) OV treatments (e.g., by naturally oncolytic viruses, see Section 5 for definition) of the immunoedited tumors (described in A). Direct cell killing by OVs (left), immune-mediated killing of infected cells (right). A number of such naturally oncolytic viruses are now under clinical trials for treatment of diverse cancer types. The figure was created with BioRender.com (accessed on 12 February 2021).

3.1. Molecular Mechanisms of Immunoediting: Optimization of the Cancer Cell towards Viral Oncolysis

The molecular mechanisms underpinning immunoediting are multifold and include: (i) Increased ability of cancer cells to survive immune-cell-induced death. This occurs through multiple mechanisms including: inactivating mutations, epigenetic silencing or sequestration of components of cell death pathways induced by immune cells [84–90], overexpression of decoy receptors (reviewed in [91]), or interference with the cancer-cell apoptotic machineries [92]. While, in theory, such interference may make it more difficult for OVs to kill cancer cells by apoptosis, it may also allow for an extension of the period during which the virus replicates, increasing thus the viral titer within the tumor. Of note, OVs have been shown to kill cancer cells via multiple pathways (in addition to apoptosis), including necrosis, necroptosis, pyroptosis, and autophagic cell death (reviewed in [93]), suggesting their ability to circumvent the enhanced resistance to apoptosis of cancer cells. (ii) Reduced immunogenicity of cancer cells. A main mode of loss of immunogenicity are acquired defects to the expression and/or function of the cell's antigen processing and presentation machineries [94]. This occurs via a broad range of processes including inactivating mutations or epigenetic silencing of MHC-I *per se* or of co-factors required for its expression [95–97]; inhibition of signaling pathways that promote MHC-I expression [98–100]; or activation of pathways that inhibit MHC-I expression [101–103]). Additionally, cancer cells also decrease expression of pro-inflammatory cytokines, such as in the epigenetic silencing of IFN-γ or IFN-κ in cervical cancer and Human Papillomavirus Type 16 (HPV-16)-positive cells, respectively [104,105]; or the reduced expression of pro-inflammatory cytokines in non-small cell lung cancers (NSCLC) [106]. The overlap in the genetic/signaling programs which mediate MHC-I expression, inflammation and antiviral responses, suggests that the downregulation of the former programs in the context of immunoediting should diminish cancer-cell resistance to OV infection. For immune evasion, the reduction in immune stimuli is complemented through increased expression of negative regulators of immune cell function (e.g., programmed cell death-ligand 1 (PD-L1) [107–110]. In accord with its function as an effector of negative feedback of inflammatory responses, PD-L1 expression is stimulated by IFN-γ, JAK/STAT signaling, and IRF1 [111]; and by TNFα and NF-κB [112]. Given that these pathways mediate cell autonomous immunity, this would suggest that PD-L1 upregulation can be associated with increased resistance to OV infection. However, PD-L1 expression is also upregulated by variety of tumorigenesis-related factors, including: EGFR in NSCLC [113]; the oncogenic BRAF V600E mutant in colorectal cancer [114]; or the loss of PTEN and activation of the PI3K pathway in glioma [115]. As mentioned above, activation of mitogenic pathways (e.g., EGFR, BRAF, or PI3K) entail modifications of the cancer cell milieu, making it more prone to OV infection.

Tumor-induced defects to IFN signaling form a class of mechanisms for altering the interactions of immune cells and malignant cells, with unique implications for oncolytic virotherapy. The uniqueness of such defects stems from the breadth of the IFN response that concomitantly regulates hundreds of immune-mediators [1], many of which directly inhibit different stages of viral infection. In light of the multiple steps involved in the induction, signal transduction and cellular response to IFNs, cancer-induced defects to IFN signaling occur through a plethora of molecular mechanisms including: (i) perturbations to the expression of the IFN receptor; e.g., the ubiquitination and downregulation of the type I IFN receptor (IFNAR1) following inflammatory signaling, nutrient deprivation or hypoxia (all conditions prevalent in the TME) [116,117]. Such down regulation, which was observed in melanoma and colorectal cancer [118,119], is associated with increased metastatic propensity and with the generation of an immune-privileged TME; (ii) perturbations to JAK/STAT1 signaling including epigenetic silencing and inactivating mutations in JAK1 [120–122]. In this context, whole-exome and RNA sequencing, and reverse-phase protein array data from different the Cancer Genome Atlas (TCGA) datasets (skin cutaneous melanoma, breast invasive carcinoma, lung adenocarcinoma, and colorectal adenocarcinoma) revealed alterations in *JAK1* or *JAK2* in 5–12 % of the samples, with dependence on cancer type [123]; (iii) crosstalk of JAK/STAT1 signaling with pro-tumorigenic signaling pathways; such as the inhibition of IFN-induced expression of inflammatory genes following STAT3 activation [124].

An interesting aspect of the interactions between immune and malignant cells pertains to the identity (source) of cancer-cell derived immune stimuli. In this context, viruses cause ~15 percent of cancer cases [125], and may thus supply PAMPs for immune-stimulation in virus-transformed cancer cells. However, the majority of tumors do not necessarily encounter pathogens in the course of their developments. A major additional source of stimuli are mutations, which are recognized as tumor-associated antigens and play a prominent immunostimulatory role [126]. Additionally, damage (or danger) associated molecular patterns (DAMPs), which activate PRRs, may also contribute immune-activating stimuli. Thus, DNA fragments generated as a result of genomic instability [127] or upon therapeutic induction of double-stranded DNA breaks [128], activate cGAS/IFN-mediated responses [129], serving thus as a source of immunostimulatory cytokines. Similarly, cytoplasmic exposure of mtDNA [130], resulting from inhibition of the tumor suppressor ataxia telangiectasia mutated (ATM) protein, entails PRR-mediated activation of type I IFN responses [131]. These scenarios support the notion that PRR-mediated activation of type I-IFN responses occurs throughout tumorigenesis, and may force the cancer cell to hamper such responses in order to escape the anti-proliferative and the immune-stimulatory effects of IFN signaling. As mentioned above, such hampered responses optimize the cancer cell milieu towards OV replication.

3.2. Acquired Resistance to Immunotherapy, An Additional Source of Modifications to Tumors Which Can Be Exploited by OVs

Acquired resistance to immunotherapy can be viewed as an acute case of tumor immunoediting. In the context of immunotherapy, the release from the constraints imposed by the immune checkpoints, enforces high selective pressure applied on cancer cells by TME-localized immune cells. Thus, clustered regularly interspaced short palindromic repeat (CRISPR)/CRISPR associated protein 9 (Cas9)—mediated knockout screens identified genes related to IFN-γ, in addition to TNF-α and antigen presentation pathways as required for the T-cell mediated killing and its enhancement by anti-PD1 antibodies [132–134]. Similarly, truncation in the β2-microglobulin gene resulting in defects in MHC-I-mediated antigen presentation and loss-of-function mutations to JAK1 or JAK2, implying defects to the transduction of antiviral IFN signals; mediate resistance to PD-1 blockade in melanoma [135]. Given that immunostimulatory roles for PRRs have been identified in immunotherapy settings [136–140], they may also be targeted in acquired resistance to this form of therapy, with profound implications to the susceptibility of such edited tumors to OVs. Together, these studies show how escape from immune pressure, in the context of immunoediting in the course of tumor progression, or in the context of immunotherapy; can directly contribute to reduced resistance to infection of cancer cells with OVs.

4. Oncogene-Induced Silencing of Immune Genes by DNA Methylation

Methylation of cytosines within CpG dinucleotides is a highly abundant epigenetic modification of mammalian genomes [141]. Methylation patterns, which regulate gene expression, are dynamically regulated via the opposing activities of enzymes that introduce or remove this modification, known as 'writers' and 'erasers', respectively. This regulatory apparatus is complemented by chromatin 'readers', i.e., protein modules that recognize histone and DNA modifications [142]. In accord with the deregulation of methylation in cancer development, DNA methyl transferases (DNMTs, 1, 3A and 3B) are overexpressed in many tumors [143–146]. A connection between tumorigenic features of cancer cells, epigenetic silencing and defects in antiviral responses is already observed upon spontaneous immortalization of fibroblasts which results in epigenetic silencing of ISGs [147]. Numerous studies reported on promoter methylation and down regulation of different IRFs (e.g., different combinations of IRF4, IRF5, IRF6, IRF7) in cancers, including fibrosarcoma [148], melanoma [149], lung cancer [150], and gastric cancer [151]. Similarly, the promoter of IFN-γ was shown to be methylated in cervical cancer [104]. Moreover, our analysis of the TCGA skin cutaneous melanoma (SKCM) database revealed significantly higher methylation of promoters of genes presenting highly-correlated expression with STAT1 (a gene group that is enriched for cell autonomous immunity genes), as compared to randomly selected genes [152]. In accord with its tumor-promoting functions, RAS was termed as "silent assassin", due to its gene silencing abilities in cancer cells [153]. In this context, DNMT1 expression is transcriptionally regulated by RAS-induced signaling pathways [143,154,155]. RAS-mediated transformation also modulates the function of DNA-methylation readers such as MBD2 [156]. Furthermore, the expression of enzymes that revert DNA methylation (ten-eleven translocation (TET) methylcytosine dioxygenases) is also regulated by oncogenic signaling in general, and RAS signaling in particular; and ERK-mediated suppression of TET1 is required for K-RAS-induced cellular transformation and hypermethylation of DNA [157]. In accord with a functional interaction between RAS and DNMTs in mediating pro-tumorigenic features, a genome-wide RNA interference (RNAi) screen in K-RAS-transformed NIH-3T3 cells identified DNMT1 and members of the RAS/MEK pathway (ERK2 and MAP3K9) as required for the silencing of the pro-apoptotic FAS gene [158]. The role of such functional interaction in mediating the suppression of immune responses is observed in the downregulation of the RAS-effector MYC and the up-regulation of ISGs in lung cancer, following DNMT inhibition [159]. Interestingly, promoter methylation of IRF7 and enhancement of viral infection was observed in nasal epithelial cells exposed to cigarette smoke [160], suggesting that exposure to carcinogens may already set the stage for the silencing of immune genes observed in malignant cells.

5. Naturally Oncolytic Viruses Exploit the Altered Cancer-Cell Milieu

The combined metabolic and defense-defective features of the cancer cell milieu (see schematic depiction in Figure 1) can be exploited in the context of oncolytic virotherapy. This is particularly relevant for viruses that are naturally devoid of human disease-causing potential but retain the potential to replicate in, and kill, malignant cells. Such viruses are referred to here as "naturally oncolytic" viruses, to differentiate them from "armed/engineered oncolytic viruses". Examples of "naturally oncolytic" viruses include attenuated clones of human pathogens (e.g., vaccine clones of measles and mumps viruses, [161]), viruses of veterinary origin (e.g., Newcastle disease virus (NDV), VSV, rat parvovirus (H-1PV), [162–165]) or the mammalian reovirus, a virus naturally devoid of disease-causing potential [47]. Indeed, we explored the complete absence of IFN signaling in LNCaP prostate cancer cells [120,121], which also present oncogenic KRAS mutation [166], to select an oncolytic mutant of the epizootic hemorrhagic disease virus (EHDV), an orbivirus (arbovirus of the Reoviridae family) that naturally targets ruminants, and that we named EHDV-Tel Aviv University (EHDV-TAU) [120]. Our studies demonstrate productive infection of EHDV-TAU in cells with defective IFN/antiviral responses, e.g. the absence of JAK1 expression/function in LNCaP prostate cancer cells [120,167], or the low basal expression levels of PRRs and defective induction of IFN (following viral infection) by B16F10 murine melanoma cells [152]. Moreover, in the latter case, treatment with inhibitors of epigenetic silencing restored PRR expression and viral induction of IFN responses in the B16F10 cells; exemplifying the role of epigenetic silencing of IFN/ISGs in the cancer cell, as a mechanism for OV selectivity. Additionally, our studies revealed that while productive infection was inhibited upon treatment with IFN, EHDV-TAU retained its cell killing potential of LNCaP cells engineered to express JAK1 (LNCaP-JAK1), when infection was carried out in presence of interleukin-6 (IL-6), an inflammatory cytokine and strong activator of cell autonomous immunity [167]. Thus, with dependence on the cellular setting, OVs may also exploit antiviral responses for induction of cancer cell death.

6. Endogenous Retroviruses, Viral Mimicry That Elicits Anti-Tumor Immunity

Tumor cells often show enhanced DNA methylation at CpG-rich sites, located in endogenous retroelements (reviewed in [168,169]). These elements, which make up more than 40% of the human genome, consist of repetitive sequences that belong to three major classes: endogenous retroviruses (ERVs), short interspersed nuclear elements (SINEs) and long interspersed nuclear elements (LINEs). Endogenous retroelements have originated from ancient infections by exogenous retroviruses, which integrated their genomes into the genome of germ cells of the host. This allowed for the vertical transmission of these elements to the offspring of the infected host. During evolution, the majority of such elements have accumulated excessive DNA mutations that inactivated their genes. However, a minority (thousands) retained some of their protein coding potential. Importantly, peptides that are derived from human endogenous retroviruses (hERVs) can be recognized by immune cells. This is exemplified by the infiltration of T cells with receptors specific for hERVs-derived epitopes, into hERVs-expressing clear cell renal cell carcinoma tumors [170]. Furthermore, endogenous retroelements may express additional immunostimulators since transcription of these elements may generate dsRNA molecules (by bidirectional transcription, as well as by sense–antisense pairing); and if reverse transcription follows, complementary DNA (cDNA) and double-stranded DNA (dsDNA) may be created too. These products, which mimic viral infection, may then be sensed by endosomal TLR3, 7, 8 or 9, and/or by cytoplasmic PRRs, including RIG- I, MDA5, cGAS [168,169]. Sensing this 'viral mimicry', activates antiviral signaling cascades, including an IFN response (see [171] and additional examples below). ERVs are repressed by variety of mechanisms, including epigenetic silencing through DNA methylation and histone modifications (reviewed in [172–175]).

Given the potential immunogenicity of endogenous retroelements and their epigenetic suppression, reactivation of these elements by epigenetic modifiers in cancer cells may results in the abovementioned viral mimicry, leading to an anti-cancerous state. For example, treatment of colorectal or ovarian cancer cells with DNMT inhibitors (DNMTis) results in induction of transcription from otherwise suppressed ERVs, the subsequent formation of dsRNA from specific ERV elements, recognition of these dsRNA molecules by MDA5/TLR3 sensors, activation of the mitochondrial antiviral-signaling protein (MAVS)-IRF7 axis and induction of IFN. Together, these result in enhanced anti-proliferative/apoptotic responses [176,177].

The complex interactions among oncogenic signaling, epigenetics and viral mimicry can be further demonstrated by the effects of the cyclin-dependent kinases 4 and 6 (CDK4/6) on cancer immunity [178]. CDK4/6, which interact with D-type cyclins, are central drivers of the cell cycle at the G1-S transition, transduce variety of mitogenic signals and their activity is associated with oncogenesis of several types of cancer (recently reviewed in [179]). Upon the induction of mitogenic signal, cyclin D-CDK4/6 complex promotes retinoblastoma (Rb) phosphorylation, leading to the release of transcription factor E2F from the Rb-E2F complex, and entry into S phase and DNA replication. One of the many targets of E2F is the *Dnmt1* gene [178,180]. Accordingly, CDK4/6 inhibition reduces DNMT1 activity, which leads to activation of ERVs expression, formation of ERVs dsRNA and IFN responses to this viral mimicry. Overall, this increases tumor antigen presentation and, together with additional effects of the CDK4/6 inhibitors, leading to cytotoxic T-lymphocytes (CTL)-mediated clearance of the tumor cells in mouse models [178]. Thus, mitogenic signals suppress ERVs expression via DNA methylation, mediated by the CDK4/6-Cyclin D-Rb-DNMT1 axis, and inhibition of this axis results in ERVs activation followed by enhanced anti-tumor immunity.

7. Concluding Remarks

IFNs and ISGs mediate antiviral and tumor-suppressor functions, via cell-autonomous and non-cell autonomous mechanisms. Tumor cells silence IFNs and ISGs along tumorigenesis, and in pronounced fashion in the context of immunoediting. OVs exploit the IFN/ISG-silenced cellular context for replication, and exert part of their therapeutic benefit through stimulation of anti-tumor immunity. Similar to what is observed in OV-infected cells, reversal of DNA methylation-mediated epigenetic silencing of hERVs stimulates anti-tumor immunity through viral mimicry. While the possibility OV/DNMTi combinations may be attractive due to their immunostimulatory potential, the activation of cell autonomous immunity by DNMTi is predicted to be inhibitory towards viral replication. Indeed, our studies showed inhibition of productive infection of EHDV-TAU and oncolytic VSV following DNMTi treatment of murine melanoma cells. However, while the cell-killing potential of oncolytic VSV was diminished in presence of DNMTi, EHDV-TAU retained its cell-killing potential under these conditions ([152], see schematic depiction in Figure 3). This difference in outcome of combined OV/DNMTi treatment, supports the notion of tailoring therapy combinations to the distinct proprieties of different OVs.

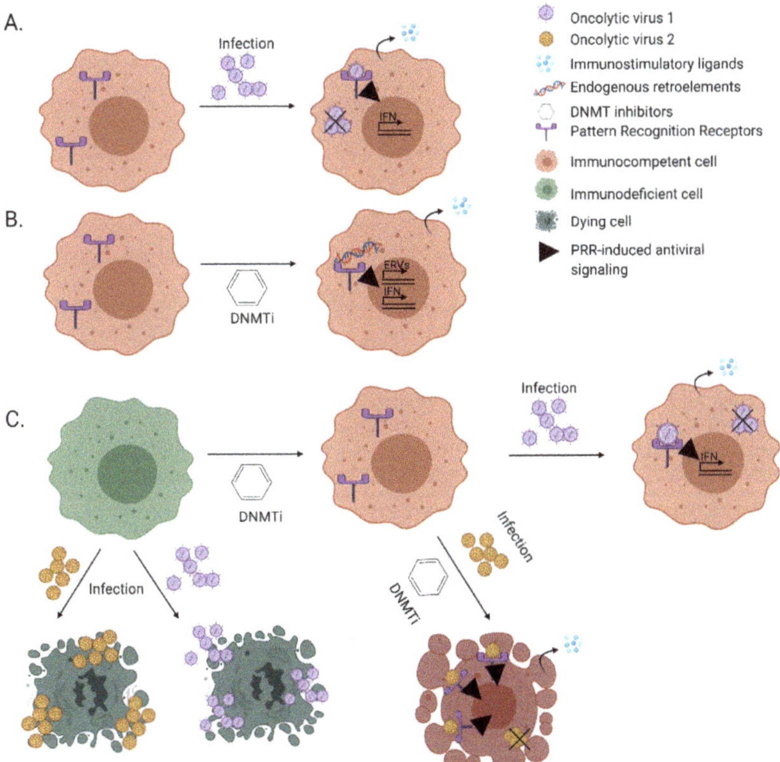

Figure 3. Immuno-stimulation: the interplay between OVs and DNMTis. (**A**)—Infection of oncolytic virus is aborted in an immunocompetent tumor cell in a process involving PRR-mediated PAMP sensing and secretion of immunostimulatory ligands, e.g., interferons. (**B**)—Immunocompetent tumor cells, when treated with DNMTis, may express endogenous retroelements and sense their products. This process is termed viral mimicry. The secreted immunostimulatory cytokines (in A or B) may act in autocrine or paracrine fashions, to induce anti-proliferative/cell death effects at the level of the cancer cell, and/or immune-mediated anti-tumorigenic effects (not depicted for sake of simplicity). (**C**)—The subset of immunodeficient cancer cells (green cells) can be killed directly by replicating oncolytic virus. DNMTi treatment renders them immunocompetent (green to pink shift). When combined with oncolytic viruses, the outcome of such treatment depends on the identity of the oncolytic virus. In a subset of cases (e.g. with EHDV-TAU, yellow viruses) the combined treatment induces cell death, in spite of the DNMTi-mediated reduction in viral replication (as in [152]). The figure was created with BioRender.com (accessed on 12 February 2021).

Author Contributions: M.E. and E.B. conceived and wrote the manuscript. All authors have read and agreed to the published version of the manuscript.

Funding: The authors acknowledge by Israel Science Foundation (1966/18; M.E.; 470/17, E.B.), Israel Cancer Association (20200132; M.E.), German Cancer Research Fund-Israel Ministry of Science and Technology (DKFZ-MOST, 0123955, M.E. and E.B.), Rosetrees Foundation (PGS19-2/10160, M.E.), Emerson Collective Cancer Research Fund (M.E., E.B.).

Conflicts of Interest: The authors declare no conflict of interests.

References

1. Schneider, W.M.; Chevillotte, M.D.; Rice, C.M. Interferon-Stimulated Genes: A Complex Web of Host Defenses. *Annu. Rev. Immunol.* **2014**, *32*, 513–545. [CrossRef] [PubMed]
2. Takeuchi, O.; Akira, S. Pattern Recognition Receptors and Inflammation. *Cell* **2010**, *140*, 805–820. [CrossRef] [PubMed]

3. Amarante-Mendes, G.P.; Adjemian, S.; Branco, L.M.; Zanetti, L.C.; Weinlich, R.; Bortoluci, K.R. Pattern Recognition Receptors and the Host Cell Death Molecular Machinery. *Front. Immunol.* **2018**, *9*, 2379. [CrossRef] [PubMed]

4. Palm, N.W.; Medzhitov, R. Pattern recognition receptors and control of adaptive immunity. *Immunol. Rev.* **2008**, *227*, 221–233. [CrossRef] [PubMed]

5. Schoggins, J.W.; Rice, C.M. Interferon-stimulated genes and their antiviral effector functions. *Curr. Opin. Virol.* **2011**, *1*, 519–525. [CrossRef]

6. Pisetsky, D.S. The origin and properties of extracellular DNA: From PAMP to DAMP. *Clin. Immunol.* **2012**, *144*, 32–40. [CrossRef] [PubMed]

7. Motwani, M.; Pesiridis, S.; Fitzgerald, K.A. DNA sensing by the cGAS–STING pathway in health and disease. *Nat. Rev. Genet.* **2019**, *20*, 657–674. [CrossRef] [PubMed]

8. Odendall, C.; Kagan, J.C. The unique regulation and functions of type III interferons in antiviral immunity. *Curr. Opin. Virol.* **2015**, *12*, 47–52. [CrossRef] [PubMed]

9. Prior, I.A.; Lewis, P.D.; Mattos, C. A Comprehensive Survey of Ras Mutations in Cancer. *Cancer Res.* **2012**, *72*, 2457–2467. [CrossRef] [PubMed]

10. Chin, L.; Tam, A.; Pomerantz, J.; Wong, M.; Holash, J.; Bardeesy, N.; Shen, Q.; O'Hagan, R.; Pantginis, J.; Zhou, H.; et al. Essential role for oncogenic Ras in tumour maintenance. *Nat. Cell Biol.* **1999**, *400*, 468–472. [CrossRef] [PubMed]

11. Malumbres, M.; Barbacid, M. RAS oncogenes: The first 30 years. *Nat. Rev. Cancer* **2003**, *3*, 459–465. [CrossRef] [PubMed]

12. Yang, J.-J.; Kang, J.-S.; Krauss, R.S. Ras Signals to the Cell Cycle Machinery via Multiple Pathways To Induce Anchorage-Independent Growth. *Mol. Cell. Biol.* **1998**, *18*, 2586–2595. [CrossRef] [PubMed]

13. Christian, S.L.; Collier, T.W.; Zu, D.; Licursi, M.; Hough, C.M.; Hirasawa, K. Activated Ras/MEK Inhibits the Antiviral Response of Alpha Interferon by Reducing STAT2 Levels. *J. Virol.* **2009**, *83*, 6717–6726. [CrossRef] [PubMed]

14. Komatsu, Y.; Christian, S.L.; Ho, N.; Pongnopparat, T.; Licursi, M.; Hirasawa, K. Oncogenic Ras inhibits IRF1 to promote viral oncolysis. *Oncogene* **2014**, *34*, 3985–3993. [CrossRef] [PubMed]

15. Harada, H.; Kitagawa, M.; Tanaka, N.; Yamamoto, H.; Harada, K.; Ishihara, M.; Taniguchi, T. Anti-oncogenic and oncogenic potentials of interferon regulatory factors-1 and -2. *Science* **1993**, *259*, 971–974. [CrossRef] [PubMed]

16. Tanaka, N.; Ishihara, M.; Lamphier, M.S.; Nozawa, H.; Matsuyama, T.; Mak, T.W.; Aizawa, S.; Tokino, T.; Oren, M.; Taniguchi, T. Cooperation of the tumour suppressors IRF-1 and p53 in response to DNA damage. *Nat. Cell Biol.* **1996**, *382*, 816–818. [CrossRef] [PubMed]

17. Taniguchi, T.; Ogasawara, K.; Takaoka, A.; Tanaka, N. IRF Family of Transcription Factors as Regulators of Host Defense. *Ann. Rev. Immunol.* **2001**, *19*, 623–655. [CrossRef] [PubMed]

18. Kroger, A.; Koster, M.; Schroeder, K.; Hauser, H.; Mueller, P.P. Activities of IRF-1. *J. Interferon. Cytokine Res.* **2002**, *22*, 5–14. [CrossRef] [PubMed]

19. Kimura, T.; Nakayama, K.; Penninger, J.; Kitagawa, M.; Harada, H.; Matsuyama, T.; Tanaka, N.; Kamijo, R.; Vilcek, J.; Mak, T.W.; et al. Involvement of the IRF-1 transcription factor in antiviral responses to interferons. *Science* **1994**, *264*, 1921–1924. [CrossRef] [PubMed]

20. Ueki, I.F.; Min-Oo, G.; Kalinowski, A.; Ballon-Landa, E.; Lanier, L.L.; Nadel, J.A.; Koff, J.L. Respiratory virus–induced EGFR activation suppresses IRF1-dependent interferon λ and antiviral defense in airway epithelium. *J. Exp. Med.* **2013**, *210*, 1929–1936. [CrossRef]

21. Wee, P.; Wang, Z. Epidermal Growth Factor Receptor Cell Proliferation Signaling Pathways. *Cancers* **2017**, *9*, 52. [CrossRef]

22. Klampfer, L.; Huang, J.; Corner, G.; Mariadason, J.; Arango, D.; Sasazuki, T.; Shirasawa, S.; Augenlicht, L. Oncogenic Ki-Ras Inhibits the Expression of Interferon-responsive Genes through Inhibition of STAT1 and STAT2 Expression. *J. Biol. Chem.* **2003**, *278*, 46278–46287. [CrossRef] [PubMed]

23. Liao, W.; Overman, M.J.; Boutin, A.T.; Shang, X.; Zhao, D.; Dey, P.; Li, J.; Wang, G.; Lan, Z.; Li, J.; et al. KRAS-IRF2 Axis Drives Immune Suppression and Immune Therapy Resistance in Colorectal Cancer. *Cancer Cell* **2019**, *35*, 559–572.e7. [CrossRef] [PubMed]

24. Kriegsman, B.A.; Vangala, P.; Chen, B.J.; Meraner, P.; Brass, A.L.; Garber, M.; Rock, K.L. Frequent Loss of IRF2 in Cancers Leads to Immune Evasion through Decreased MHC Class I Antigen Presentation and Increased PD-L1 Expression. *J. Immunol.* **2019**, *203*, 1999–2010. [CrossRef]

25. Wang, Y.; Liu, D.-P.; Chen, P.-P.; Koeffler, H.P.; Tong, X.-J.; Xie, D. Involvement of IFN Regulatory Factor (IRF)-1 and IRF-2 in the Formation and Progression of Human Esophageal Cancers. *Cancer Res.* **2007**, *67*, 2535–2543. [CrossRef]

26. Harada, H.; Willison, K.; Sakakibara, J.; Miyamoto, M.; Fujita, T.; Taniguchi, T. Absence of the type I IFN system in EC cells: Transcriptional activator (IRF-1) and repressor (IRF-2) genes are developmentally regulated. *Cell* **1990**, *63*, 303–312. [CrossRef]

27. Garciía, M.A.; Gil, J.; Ventoso, I.; Guerra, S.; Domingo, E.; Rivas, C.; Esteban, M. Impact of Protein Kinase PKR in Cell Biology: From Antiviral to Antiproliferative Action. *Microbiol. Mol. Biol. Rev.* **2006**, *70*, 1032–1060. [CrossRef] [PubMed]

28. Gal-Ben-Ari, S.; Barrera, I.; Ehrlich, M.; Rosenblum, K. PKR: A Kinase to Remember. *Front. Mol. Neurosci.* **2019**, *11*, 480. [CrossRef] [PubMed]

29. Yoon, C.-H.; Lee, E.-S.; Lim, D.-S.; Bae, Y.-S. PKR, a p53 target gene, plays a crucial role in the tumor-suppressor function of p53. *Proc. Natl. Acad. Sci. USA* **2009**, *106*, 7852–7857. [CrossRef] [PubMed]

30. Cuddihy, A.R.; Li, S.; Tam, N.W.N.; Wong, A.H.-T.; Taya, Y.; Abraham, N.; Bell, J.C.; Koromilas, A.E. Double-Stranded-RNA-Activated Protein Kinase PKR Enhances Transcriptional Activation by Tumor Suppressor p53. *Mol. Cell. Biol.* **1999**, *19*, 2475–2484. [CrossRef] [PubMed]

31. Balachandran, S.; Kim, C.N.; Yeh, W.-C.; Mak, T.W.; Bhalla, K.; Barber, G.N. Activation of the dsRNA-dependent protein kinase, PKR, induces apoptosis through FADD-mediated death signaling. *EMBO J.* **1998**, *17*, 6888–6902. [CrossRef]

32. Gil, J.; Esteban, M. The interferon-induced protein kinase (PKR), triggers apoptosis through FADD-mediated activation of caspase 8 in a manner independent of Fas and TNF-α receptors. *Oncogene* **2000**, *19*, 3665–3674. [CrossRef] [PubMed]

33. Li, S.; Min, J.-Y.; Krug, R.M.; Sen, G.C. Binding of the influenza A virus NS1 protein to PKR mediates the inhibition of its activation by either PACT or double-stranded RNA. *J. Virol.* **2006**, *349*, 13–21. [CrossRef] [PubMed]

34. Carroll, K.; Elroy-Stein, O.; Moss, B.; Jagus, R. Recombinant vaccinia virus K3L gene product prevents activation of double-stranded RNA-dependent, initiation factor 2 alpha-specific protein kinase. *J. Biol. Chem.* **1993**, *268*, 12837–12842. [CrossRef]

35. Budt, M.; Niederstadt, L.; Valchanova, R.S.; Jonjić, S.; Brune, W. Specific Inhibition of the PKR-Mediated Antiviral Response by the Murine Cytomegalovirus Proteins m142 and m143. *J. Virol.* **2008**, *83*, 1260–1270. [CrossRef] [PubMed]

36. Gale, M.; Blakely, C.M.; Kwieciszewski, B.; Tan, S.-L.; Dossett, M.; Tang, N.M.; Korth, M.J.; Polyak, S.J.; Gretch, D.R.; Katze, M.G. Control of PKR Protein Kinase by Hepatitis C Virus Nonstructural 5A Protein: Molecular Mechanisms of Kinase Regulation. *Mol. Cell. Biol.* **1998**, *18*, 5208–5218. [CrossRef] [PubMed]

37. Peters, G.A.; Khoo, D.; Mohr, I.; Sen, G.C. Inhibition of PACT-Mediated Activation of PKR by the Herpes Simplex Virus Type 1 Us11 Protein. *J. Virol.* **2002**, *76*, 11054–11064. [CrossRef] [PubMed]

38. Balachandran, S.; Roberts, P.C.; Brown, L.E.; Truong, H.; Pattnaik, A.K.; Archer, D.R.; Barber, G.N. Essential role for the dsRNA-dependent protein kinase PKR in innate immunity to viral infection. *Immunity* **2000**, *13*, 129–141. [CrossRef]

39. Bergmann, M.; Garcia-Sastre, A.; Carnero, E.; Pehamberger, H.; Wolff, K.; Palese, P.; Muster, T. Influenza Virus NS1 Protein Counteracts PKR-Mediated Inhibition of Replication. *J. Virol.* **2000**, *74*, 6203–6206. [CrossRef] [PubMed]

40. Luig, C.; Köther, K.; Dudek, S.E.; Gaestel, M.; Hiscott, J.; Wixler, V.; Ludwig, S. MAP kinase-activated protein kinases 2 and 3 are required for influenza A virus propagation and act via inhibition of PKR. *FASEB J.* **2010**, *24*, 4068–4077. [CrossRef]

41. García-Sastre, A.; Egorov, A.; Matassova, D.; Brandtbc, S.; Levy, D.E.; Durbin, J.E.; Palese, P.; Musterbc, T. Influenza A Virus Lacking the NS1 Gene Replicates in Interferon-Deficient Systems. *J. Virol.* **1998**, *252*, 324–330. [CrossRef] [PubMed]

42. Ölschläger, V.; Pleschka, S.; Fischer, T.; Rziha, H.-J.; Wurzer, W.; Stitz, L.; Rapp, U.R.; Ludwig, S.; Planz, O. Veronika Lung-specific expression of active Raf kinase results in increased mortality of influenza A virus-infected mice. *Oncogene* **2004**, *23*, 6639–6646. [CrossRef] [PubMed]

43. Bergmann, M.; Romirer, I.; Sachet, M.; Fleischhacker, R.; García-Sastre, A.; Palese, P.; Wolff, K.; Pehamberger, H.; Jakesz, R.; Muster, T. A genetically engineered influenza A virus with ras-dependent oncolytic properties. *Cancer Res.* **2001**, *61*, 8188–8193. [PubMed]

44. Smith, K.D.; Mezhir, J.J.; Bickenbach, K.; Veerapong, J.; Charron, J.; Posner, M.C.; Roizman, B.; Weichselbaum, R.R. Activated MEK suppresses activation of PKR and enables efficient replication and in vivo oncolysis by Deltagamma(1)34.5 mutants of herpes simplex virus 1. *J. Virol.* **2006**, *80*, 1110–1120. [CrossRef] [PubMed]

45. Cascalló, M.; Capellà, G.; Mazo, A.; Alemany, R. Ras-dependent oncolysis with an adenovirus VAI mutant. *Cancer Res.* **2003**, *63*, 5544–5550. [PubMed]

46. Coffey, M.C.; Strong, J.E.; Forsyth, P.A.; Lee, P.W. Reovirus therapy of tumors with activated Ras pathway. *Science* **1998**, *282*, 1332–1334. [CrossRef] [PubMed]

47. Strong, J.E.; Coffey, M.C.; Tang, D.; Sabinin, P.; Lee, P.W. The molecular basis of viral oncolysis: Usurpation of the Ras signaling pathway by reovirus. *EMBO J.* **1998**, *17*, 3351–3362. [CrossRef]

48. Norman, K.L.; Hirasawa, K.; Yang, A.-D.; Shields, M.A.; Lee, P.W.K. Reovirus oncolysis: The Ras/RalGEF/p38 pathway dictates host cell permissiveness to reovirus infection. *Proc. Natl. Acad. Sci. USA* **2004**, *101*, 11099–11104. [CrossRef]

49. Marcato, P.; Shmulevitz, M.; Pan, D.; Stoltz, D.; Lee, P.W. Ras Transformation Mediates Reovirus Oncolysis by Enhancing Virus Uncoating, Particle Infectivity, and Apoptosis-dependent Release. *Mol. Ther.* **2007**, *15*, 1522–1530. [CrossRef] [PubMed]

50. Shmulevitz, M.; Pan, L.-Z.; Garant, K.; Pan, D.; Lee, P.W.K. Oncogenic Ras Promotes Reovirus Spread by Suppressing IFN-β Production through Negative Regulation of *RIG-I* Signaling. *Cancer Res.* **2010**, *70*, 4912–4921. [CrossRef] [PubMed]

51. Noser, J.A.; Mael, A.A.; Sakuma, R.; Ohmine, S.; Marcato, P.; Lee, P.W.; Ikeda, Y. The RAS/Raf1/MEK/ERK Signaling Pathway Facilitates VSV-mediated Oncolysis: Implication for the Defective Interferon Response in Cancer Cells. *Mol. Ther.* **2007**, *15*, 1531–1536. [CrossRef]

52. Puhlmann, J.; Puehler, F.; Mumberg, D.; Boukamp, P.; Beier, R. Rac1 is required for oncolytic NDV replication in human cancer cells and establishes a link between tumorigenesis and sensitivity to oncolytic virus. *Oncogene* **2010**, *29*, 2205–2216. [CrossRef] [PubMed]

53. Walsh, A.B.; Bar-Sagi, D. Differential Activation of the Rac Pathway by Ha-Ras and K-Ras. *J. Biol. Chem.* **2001**, *276*, 15609–15615. [CrossRef] [PubMed]

54. Lambert, J.M.; Lambert, Q.T.; Reuther, G.W.; Malliri, A.; Siderovski, D.P.; Sondek, J.; Collard, J.G.; Der, C.J. Tiam1 mediates Ras activation of Rac by a PI(3)K-independent mechanism. *Nat. Cell Biol.* **2002**, *4*, 621–625. [CrossRef] [PubMed]

55. Galaktionov, K.; Jessus, C.; Beach, D. Raf1 interaction with Cdc25 phosphatase ties mitogenic signal transduction to cell cycle activation. *Genes Dev.* **1995**, *9*, 1046–1058. [CrossRef] [PubMed]

56. Qi, D.; Hu, L.; Jiao, T.; Zhang, T.; Tong, X.; Ye, X. Phosphatase Cdc25A Negatively Regulates the Antiviral Immune Response by Inhibiting TBK1 Activity. *J. Virol.* **2018**, *92*, e01118-18. [CrossRef] [PubMed]

57. Calcagno, S.R.; Li, S.; Colon, M.; Kreinest, P.A.; Thompson, E.A.; Fields, A.P.; Murray, N.R. Oncogenic K-ras promotes early carcinogenesis in the mouse proximal colon. *Int. J. Cancer* **2008**, *122*, 2462–2470. [CrossRef] [PubMed]

58. Maharaj, N.P.; Wies, E.; Stoll, A.; Gack, M.U. Conventional PKC-α/β Negatively Regulate RIG-I Antiviral Signal Transduction. *J. Virol.* **2012**, *86*, 1358–1371. [CrossRef] [PubMed]

59. Lin, L.; Su, Z.; Lebedeva, I.V.; Gupta, P.; Boukerche, H.; Rai, T.; Barber, G.N.; Dent, P.; Sarkar, D.; Fisher, P.B. Activation of Ras/Raf protects cells from melanoma differentiation-associated gene-5-induced apoptosis. *Cell Death Differ.* **2006**, *13*, 1982–1993. [CrossRef]

60. Wirth, M.; Schneider, G. MYC: A Stratification Marker for Pancreatic Cancer Therapy. *Trends Cancer* **2016**, *2*, 1–3. [CrossRef] [PubMed]

61. Muthalagu, N.; Monteverde, T.; Raffo-Iraolagoitia, X.; Wiesheu, R.; Whyte, D.; Hedley, A.; Laing, S.; Kruspig, B.; Upstill-Goddard, R.; Shaw, R.; et al. Repression of the Type I Interferon Pathway Underlies MYC- and KRAS-Dependent Evasion of NK and B Cells in Pancreatic Ductal Adenocarcinoma. *Cancer Discov.* **2020**, *10*, 872–887. [CrossRef] [PubMed]

62. Steck, P.A.; Pershouse, M.A.; Jasser, S.A.; Yung, W.K.; Lin, H.; Ligon, A.H.; Langford, L.A.; Baumgard, M.L.; Hattier, T.; Davis, T.; et al. Identification of a candidate tumour suppressor gene, MMAC1, at chromosome 10q23.3 that is mutated in multiple advanced cancers. *Nat. Genet.* **1997**, *15*, 356–362. [CrossRef] [PubMed]

63. Iwanaga, K.; Yang, Y.; Raso, M.G.; Ma, L.; Hanna, A.E.; Thilaganathan, N.; Moghaddam, S.; Evans, C.M.; Li, H.; Cai, W.-W.; et al. Pten Inactivation Accelerates Oncogenic K-ras–Initiated Tumorigenesis in a Mouse Model of Lung Cancer. *Cancer Res.* **2008**, *68*, 1119–1127. [CrossRef] [PubMed]

64. Li, S.; Zhu, M.; Pan, R.; Fang, T.; Cao, Y.-Y.; Chen, S.; Zhao, X.; Lei, C.-Q.; Guo, L.; Chen, Y.; et al. The tumor suppressor PTEN has a critical role in antiviral innate immunity. *Nat. Immunol.* **2016**, *17*, 241–249. [CrossRef] [PubMed]

65. Ying, H.; Kimmelman, A.C.; Lyssiotis, C.A.; Hua, S.; Chu, G.C.; Fletcher-Sananikone, E.; Locasale, J.W.; Son, J.; Zhang, H.; Coloff, J.L.; et al. Oncogenic Kras Maintains Pancreatic Tumors through Regulation of Anabolic Glucose Metabolism. *Cell* **2012**, *149*, 656–670. [CrossRef]

66. Warburg, O. On the Origin of Cancer Cells. *Science* **1956**, *123*, 309–314. [CrossRef] [PubMed]

67. Weinberg, F.; Hamanaka, R.; Wheaton, W.W.; Weinberg, S.; Joseph, J.; Lopez, M.; Kalyanaraman, B.; Mutlu, G.M.; Budinger, G.R.S.; Chandel, N.S. Mitochondrial metabolism and ROS generation are essential for Kras-mediated tumorigenicity. *Proc. Natl. Acad. Sci. USA* **2010**, *107*, 8788–8793. [CrossRef] [PubMed]

68. Polcicova, K.; Badurova, L.; Tomaskova, J. Metabolic reprogramming as a feast for virus replication. *Acta Virol.* **2020**, *64*, 201–215. [CrossRef]

69. Sanchez, E.L.; Lagunoff, M. Viral activation of cellular metabolism. *Virology* **2015**, *479–480*, 609–618. [CrossRef]

70. Thai, M.; Thaker, S.K.; Feng, J.; Du, Y.; Hu, H.; Wu, T.T.; Graeber, T.G.; Braas, D.; Christofk, H.R. MYC-induced reprogramming of glutamine catabolism supports optimal virus replication. *Nat. Commun.* **2015**, *6*, 8873. [CrossRef] [PubMed]

71. Flavin, R.; Peluso, S.; Nguyen, P.L.; Loda, M. Fatty acid synthase as a potential therapeutic target in cancer. *Futur. Oncol.* **2010**, *6*, 551–562. [CrossRef] [PubMed]

72. Menendez, J.A.; Lupu, R. Fatty acid synthase and the lipogenic phenotype in cancer pathogenesis. *Nat. Rev. Cancer* **2007**, *7*, 763–777. [CrossRef] [PubMed]

73. Gouw, A.M.; Eberlin, L.S.; Margulis, K.; Sullivan, D.K.; Toal, G.G.; Tong, L.; Zare, R.N.; Felsher, D.W. Oncogene KRAS activates fatty acid synthase, resulting in specific ERK and lipid signatures associated with lung adenocarcinoma. *Proc. Natl. Acad. Sci. USA* **2017**, *114*, 4300–4305. [CrossRef] [PubMed]

74. Smith, B.; Schafer, X.L.; Ambeskovic, A.; Spencer, C.M.; Land, H.; Munger, J. Addiction to Coupling of the Warburg Effect with Glutamine Catabolism in Cancer Cells. *Cell Rep.* **2016**, *17*, 821–836. [CrossRef] [PubMed]

75. Munger, J.; Bennett, B.D.; Parikh, A.; Feng, X.-J.; McArdle, J.; Rabitz, H.A.; Shenk, T.; Rabinowitz, J.D. Systems-level metabolic flux profiling identifies fatty acid synthesis as a target for antiviral therapy. *Nat. Biotechnol.* **2008**, *26*, 1179–1186. [CrossRef] [PubMed]

76. Greseth, M.D.; Traktman, P. De novo Fatty Acid Biosynthesis Contributes Significantly to Establishment of a Bioenergetically Favorable Environment for Vaccinia Virus Infection. *PLoS Pathog.* **2014**, *10*, e1004021. [CrossRef] [PubMed]

77. Martín-Acebes, M.A.; Blázquez, A.-B.; De Oya, N.J.; Escribano-Romero, E.; Saiz, J.-C. West Nile Virus Replication Requires Fatty Acid Synthesis but Is Independent on Phosphatidylinositol-4-Phosphate Lipids. *PLoS ONE* **2011**, *6*, e24970. [CrossRef]

78. Lee, M.; Sugiyama, M.; Mekhail, K.; Latreille, E.; Khosraviani, N.; Wei, K.; Lee, W.L.; Antonescu, C.; Fairn, G.D. Fatty Acid Synthase inhibition prevents palmitoylation of SARS-CoV2 Spike Protein and improves survival of mice infected with murine hepatitis virus. *bioRxiv* **2020**. [CrossRef]

79. Ohol, Y.M.; Wang, Z.; Kemble, G.; Duke, G. Direct Inhibition of Cellular Fatty Acid Synthase Impairs Replication of Respiratory Syncytial Virus and Other Respiratory Viruses. *PLoS ONE* **2015**, *10*, e0144648. [CrossRef] [PubMed]

80. Koundinya, M.; Sudhalter, J.; Courjaud, A.; Lionne, B.; Touyer, G.; Bonnet, L.; Menguy, I.; Schreiber, I.; Perrault, C.; Vougier, S.; et al. Dependence on the Pyrimidine Biosynthetic Enzyme DHODH Is a Synthetic Lethal Vulnerability in Mutant KRAS-Driven Cancers. *Cell Chem. Biol.* **2018**, *25*, 705–717. [CrossRef]

81. Xiong, R.; Zhang, L.; Li, S.; Sun, Y.; Ding, M.; Wang, Y.; Zhao, Y.; Wu, Y.; Shang, W.; Jiang, X.; et al. Novel and potent inhibitors targeting DHODH are broad-spectrum antivirals against RNA viruses including newly-emerged coronavirus SARS-CoV-2. *Protein Cell* **2020**, *11*, 723–739. [CrossRef]

82. Vesely, M.D.; Schreiber, R.D. Cancer immunoediting: Antigens, mechanisms, and implications to cancer immunotherapy. *Ann. N. Y. Acad. Sci.* **2013**, *1284*, 1–5. [CrossRef] [PubMed]

83. O'Donnell, J.S.; Teng, M.W.L.; Smyth, M.J. Cancer immunoediting and resistance to T cell-based immunotherapy. *Nat. Rev. Clin. Oncol.* **2019**, *16*, 1511–1567. [CrossRef]

84. Twomey, J.D.; Zhang, B. Circulating Tumor Cells Develop Resistance to TRAIL-Induced Apoptosis Through Autophagic Removal of Death Receptor 5: Evidence from an In Vitro Model. *Cancers* **2019**, *11*, 94. [CrossRef] [PubMed]

85. Hopkins-Donaldson, S.; Ziegler, A.; Kurtz, S.; Bigosch, C.; Kandioler, D.; Ludwig, C.; Zangemeister-Wittke, U.; Stahel, R. Silencing of death receptor and caspase-8 expression in small cell lung carcinoma cell lines and tumors by DNA methylation. *Cell Death Differ.* **2003**, *10*, 356–364. [CrossRef]

86. Müschen, M.; Warskulat, U.; Beckmann, M. Defining CD95 as a tumor suppressor gene. *J. Mol. Med.* **2000**, *78*, 312–325. [CrossRef] [PubMed]

87. Soung, Y.H.; Lee, J.W.; Kim, S.Y.; Jang, J.; Park, Y.G.; Park, W.S.; Nam, S.W.; Lee, J.Y.; Yoo, N.J.; Lee, S.H. CASPASE-8 gene is inactivated by somatic mutations in gastric carcinomas. *Cancer Res.* **2005**, *65*, 815–821. [PubMed]

88. Kim, H.S.; Lee, J.W.; Soung, Y.H.; Park, W.S.; Kim, S.Y.; Lee, J.H.; Park, J.Y.; Cho, Y.G.; Kim, C.J.; Jeong, S.W.; et al. Inactivating mutations of caspase-8 gene in colorectal carcinomas. *Gastroenterology* **2003**, *125*, 708–715. [CrossRef]

89. Debatin, K.-M.; Krammer, P.H. Death receptors in chemotherapy and cancer. *Oncogene* **2004**, *23*, 2950–2966. [CrossRef] [PubMed]

90. Rooney, M.S.; Shukla, S.A.; Wu, C.J.; Getz, G.; Hacohen, N. Molecular and Genetic Properties of Tumors Associated with Local Immune Cytolytic Activity. *Cell* **2015**, *160*, 48–61. [CrossRef] [PubMed]

91. O' Reilly, E.; Tirincsi, A.; Logue, S.E.; Szegezdi, E. The Janus Face of Death Receptor Signaling during Tumor Immunoediting. *Front. Immunol.* **2016**, *7*, 446. [CrossRef] [PubMed]

92. Lee, J.H.; Soung, Y.H.; Lee, J.W.; Park, W.S.; Kim, S.Y.; Cho, Y.G.; Kim, C.J.; Seo, S.H.; Kim, H.S.; Nam, S.W.; et al. Inactivating mutation of the pro-apoptotic geneBID in gastric cancer. *J. Pathol.* **2004**, *202*, 439–445. [CrossRef] [PubMed]

93. Guo, Z.S.; Liu, Z.; Bartlett, D.L. Oncolytic Immunotherapy: Dying the Right Way is a Key to Eliciting Potent Antitumor Immunity. *Front. Oncol.* **2014**, *4*, 74. [CrossRef]

94. Leone, P.; Shin, E.-C.; Perosa, F.; Vacca, A.; Dammacco, F.; Racanelli, V. MHC Class I Antigen Processing and Presenting Machinery: Organization, Function, and Defects in Tumor Cells. *J. Natl. Cancer Inst.* **2013**, *105*, 1172–1187. [CrossRef] [PubMed]

95. Sucker, A.; Zhao, F.; Real, B.; Heeke, C.; Bielefeld, N.; Maßen, S.; Horn, S.; Moll, I.; Maltaner, R.; Horn, P.A.; et al. Genetic Evolution of T-cell Resistance in the Course of Melanoma Progression. *Clin. Cancer Res.* **2014**, *20*, 6593–6604. [CrossRef]

96. Restifo, N.P.; Marincola, F.M.; Kawakami, Y.; Taubenberger, J.; Yannelli, J.R.; Rosenberg, S.A. Loss of Functional Beta2-Microglobulin in Metastatic Melanomas From Five Patients Receiving Immunotherapy. *J. Natl. Cancer Inst.* **1996**, *88*, 100–108. [CrossRef] [PubMed]

97. Burr, M.L.; Sparbier, C.E.; Chan, K.L.; Chan, Y.-C.; Kersbergen, A.; Lam, E.Y.; Azidis-Yates, E.; Vassiliadis, D.; Bell, C.C.; Gilan, O.; et al. An Evolutionarily Conserved Function of Polycomb Silences the MHC Class I Antigen Presentation Pathway and Enables Immune Evasion in Cancer. *Cancer Cell* **2019**, *36*, 385–401. [CrossRef] [PubMed]

98. Hobart, M.; Ramassar, V.; Goes, N.; Urmson, J.; Halloran, P.F. IFN regulatory factor-1 plays a central role in the regulation of the expression of class I and II MHC genes in vivo. *J. Immunol.* **1997**, *158*, 4260–4269.e8. [PubMed]

99. Lorenzi, S.; Forloni, M.; Cifaldi, L.; Antonucci, C.; Citti, A.; Boldrini, R.; Pezzullo, M.; Castellano, A.; Russo, V.; van der Bruggen, P.; et al. IRF1 and NF-kB restore MHC class I-restricted tumor antigen processing and presentation to cytotoxic T cells in aggressive neuroblastoma. *PLoS ONE* **2012**, *7*, e46928. [CrossRef] [PubMed]

100. Kobayashi, K.S.; Elsen, P.J.V.D. NLRC5: A key regulator of MHC class I-dependent immune responses. *Nat. Rev. Immunol.* **2012**, *12*, 813–820. [CrossRef] [PubMed]

101. Jones, L.M.; Broz, M.L.; Ranger, J.J.; Ozcelik, J.; Ahn, R.; Zuo, D.; Ursini-Siegel, J.; Hallett, M.T.; Krummel, M.; Muller, W.J. STAT3 Establishes an Immunosuppressive Microenvironment during the Early Stages of Breast Carcinogenesis to Promote Tumor Growth and Metastasis. *Cancer Res.* **2016**, *76*, 1416–1428. [CrossRef] [PubMed]

102. Kosack, L.; Wingelhofer, B.; Popa, A.; Orlova, A.; Agerer, B.; Vilagos, B.; Majek, P.; Parapatics, K.; Lercher, A.; Ringler, A.; et al. The ERBB-STAT3 Axis Drives Tasmanian Devil Facial Tumor Disease. *Cancer Cell* **2019**, *35*, 125–139.e9. [CrossRef] [PubMed]

103. Liu, C.; Peng, W.; Xu, C.; Lou, Y.; Zhang, M.; Wargo, J.A.; Chen, J.Q.; Li, H.S.; Watowich, S.S.; Yang, Y.; et al. BRAF Inhibition Increases Tumor Infiltration by T cells and Enhances the Antitumor Activity of Adoptive Immunotherapy in Mice. *Clin. Cancer Res.* **2013**, *19*, 393–403. [CrossRef] [PubMed]

104. Ma, D.; Jiang, C.; Hu, X.; Liu, H.; Li, Q.; Li, T.; Yang, Y.; Li, O. Methylation Patterns of the IFN-γ Gene in Cervical Cancer Tissues. *Sci. Rep.* **2014**, *4*, 6331. [CrossRef] [PubMed]

105. Rincon-Orozco, B.; Halec, G.; Rosenberger, S.; Muschik, D.; Nindl, I.; Bachmann, A.; Ritter, T.M.; Dondog, B.; Ly, R.; Bosch, F.X.; et al. Epigenetic Silencing of Interferon-κ in Human Papillomavirus Type 16–Positive Cells. *Cancer Res.* **2009**, *69*, 8718–8725. [CrossRef] [PubMed]

106. Tekpli, X.; Landvik, N.E.; Anmarkud, K.H.; Skaug, V.; Haugen, A.; Zienolddiny, S. DNA methylation at promoter regions of interleukin 1B, interleukin 6, and interleukin 8 in non-small cell lung cancer. *Cancer Immunol. Immunother.* **2012**, *62*, 337–345. [CrossRef] [PubMed]

107. Yu, J.; Wang, X.; Teng, F.; Kong, L. PD-L1 expression in human cancers and its association with clinical outcomes. *OncoTargets Ther.* **2016**, *9*, 5023–5039. [CrossRef] [PubMed]

108. Nakanishi, J.; Wada, Y.; Matsumoto, K.; Azuma, M.; Kikuchi, K.; Ueda, S. Overexpression of B7-H1 (PD-L1) significantly associates with tumor grade and postoperative prognosis in human urothelial cancers. *Cancer Immunol. Immunother.* **2007**, *56*, 1173–1182. [CrossRef] [PubMed]

109. Nomi, T.; Sho, M.; Akahori, T.; Hamada, K.; Kubo, A.; Kanehiro, H.; Nakamura, S.; Enomoto, K.; Yagita, H.; Azuma, M.; et al. Clinical Significance and Therapeutic Potential of the Programmed Death-1 Ligand/Programmed Death-1 Pathway in Human Pancreatic Cancer. *Clin. Cancer Res.* **2007**, *13*, 2151–2157. [CrossRef]

110. Wilmotte, R.; Burkhardt, K.; Kindler, V.; Belkouch, M.-C.; Dussex, G.; De Tribolet, N.; Walker, P.R.; Dietrich, P.-Y. B7-homolog 1 expression by human glioma: A new mechanism of immune evasion. *NeuroReport* **2005**, *16*, 1081–1085. [CrossRef]

111. Lee, S.J.; Jang, B.C.; Lee, S.W.; Yang, Y.I.; Suh, S.I.; Park, Y.M.; Oh, S.; Shin, J.G.; Yao, S.; Chen, L.; et al. Interferon regulatory factor-1 is prerequisite to the constitutive expression and IFN-gamma-induced upregulation of B7-H1 (CD274). *FEBS Lett.* **2006**, *580*, 755–762. [CrossRef]

112. Antonangeli, F.; Natalini, A.; Garassino, M.C.; Sica, A.; Santoni, A.; Di Rosa, F. Regulation of PD-L1 Expression by NF-κB in Cancer. *Front Immunol.* **2020**, *11*, 584626. [CrossRef] [PubMed]

113. Azuma, K.; Ota, K.; Kawahara, A.; Hattori, S.; Iwama, E.; Harada, T.; Matsumoto, K.; Takayama, K.; Takamori, S.; Kage, M.; et al. Association of PD-L1 overexpression with activating EGFR mutations in surgically resected nonsmall-cell lung cancer. *Ann. Oncol.* **2014**, *25*, 1935–1940. [CrossRef] [PubMed]

114. Feng, D.; Qin, B.; Pal, K.; Sun, L.; Dutta, S.; Dong, H.; Liu, X.; Mukhopadhyay, D.; Huang, S.; Sinicrope, F.A. BRAFV600E-induced, tumor intrinsic PD-L1 can regulate chemotherapy-induced apoptosis in human colon cancer cells and in tumor xenografts. *Oncogene* **2019**, *38*, 6752–6766. [CrossRef] [PubMed]

115. Parsa, A.T.; Waldron, J.S.; Panner, A.; Crane, C.A.; Parney, I.F.; Barry, J.J.; Cachola, K.E.; Murray, J.C.; Tihan, T.; Jensen, M.C.; et al. Loss of tumor suppressor PTEN function increases B7-H1 expression and immunoresistance in glioma. *Nat. Med.* **2007**, *13*, 84–88. [CrossRef] [PubMed]

116. Bhattacharya, S.; Katlinski, K.V.; Reichert, M.; Takano, S.; Brice, A.; Zhao, B.; Yu, Q.; Zheng, H.; Carbone, C.J.; Katlinskaya, Y.V.; et al. Triggering ubiquitination of IFNAR 1 protects tissues from inflammatory injury. *EMBO Mol. Med.* **2014**, *6*, 384–397. [CrossRef] [PubMed]

117. Huangfu, W.-C.; Qian, J.; Liu, C.; Liu, J.; Lokshin, A.E.; Baker, D.P.; Rui, H.; Fuchs, S.Y. Inflammatory signaling compromises cell responses to interferon alpha. *Oncogene* **2011**, *31*, 161–172. [CrossRef] [PubMed]

118. Katlinski, K.V.; Gui, J.; Katlinskaya, Y.V.; Ortiz, A.; Chakraborty, R.; Bhattacharya, S.; Carbone, C.J.; Beiting, D.P.; Girondo, M.A.; Peck, A.R.; et al. Inactivation of Interferon Receptor Promotes the Establishment of Immune Privileged Tumor Microenvironment. *Cancer Cell* **2017**, *31*, 194–207. [CrossRef] [PubMed]

119. Katlinskaya, Y.V.; Katlinski, K.V.; Yu, Q.; Ortiz, A.; Beiting, D.P.; Brice, A.; Davar, D.; Sanders, C.; Kirkwood, J.M.; Rui, H.; et al. Suppression of Type I Interferon Signaling Overcomes Oncogene-Induced Senescence and Mediates Melanoma Development and Progression. *Cell Rep.* **2016**, *15*, 171–180. [CrossRef] [PubMed]

120. Danziger, O.; Shai, B.; Sabo, Y.; Bacharach, E.; Ehrlich, M. Combined genetic and epigenetic interferences with interferon signaling expose prostate cancer cells to viral infection. *Oncotarget* **2016**, *7*, 52115–52134. [CrossRef] [PubMed]

121. Dunn, G.P.; Sheehan, K.C.; Old, L.J.; Schreiber, R.D. IFN Unresponsiveness in LNCaP Cells Due to the Lack of JAK1 Gene Expression. *Cancer Res.* **2005**, *65*, 3447–3453. [CrossRef] [PubMed]

122. Rossi, M.R.; Hawthorn, L.; Platt, J.; Burkhardt, T.; Cowell, J.K.; Ionov, Y. Identification of inactivating mutations in the JAK1, SYNJ2, and CLPTM1 genes in prostate cancer cells using inhibition of nonsense-mediated decay and microarray analysis. *Cancer Genet. Cytogenet.* **2005**, *161*, 97–103. [CrossRef] [PubMed]

123. Shin, D.S.; Zaretsky, J.M.; Escuin-Ordinas, H.; Garcia-Diaz, A.; Hu-Lieskovan, S.; Kalbasi, A.; Grasso, C.S.; Hugo, W.; Sandoval, S.; Torrejon, D.Y.; et al. Primary Resistance to PD-1 Blockade Mediated by JAK1/2 Mutations. *Cancer Discov.* **2017**, *7*, 188–201. [CrossRef] [PubMed]

124. Ho, H.H.; Ivashkiv, L.B. Role of STAT3 in type I interferon responses. Negative regulation of STAT1-dependent inflammatory gene activation. *J. Biol. Chem.* **2006**, *281*, 14111–14118. [CrossRef] [PubMed]

125. zur Hausen, H. Viruses in human cancers. *Science* **1991**, *254*, 1167–1173. [CrossRef] [PubMed]

126. Chalmers, Z.R.; Connelly, C.F.; Fabrizio, D.; Gay, L.; Ali, S.M.; Ennis, R.; Schrock, A.; Campbell, B.; Shlien, A.; Chmielecki, J.; et al. Analysis of 100,000 human cancer genomes reveals the landscape of tumor mutational burden. *Genome Med.* **2017**, *9*, 1–14. [CrossRef] [PubMed]

127. Negrini, S.; Gorgoulis, V.G.; Halazonetis, T.D. Genomic instability—An evolving hallmark of cancer. *Nat. Rev. Mol. Cell Biol.* **2010**, *11*, 220–228. [CrossRef] [PubMed]

128. Harding, S.M.; Benci, J.L.; Irianto, J.; Discher, D.E.; Minn, A.J.; Greenberg, R.A. Mitotic progression following DNA damage enables pattern recognition within micronuclei. *Nat. Cell Biol.* **2017**, *548*, 466–470. [CrossRef] [PubMed]

129. MacKenzie, K.J.; Carroll, P.; Martin, C.-A.; Murina, O.; Fluteau, A.; Simpson, D.J.; Olova, N.; Sutcliffe, H.; Rainger, J.K.; Leitch, A.; et al. cGAS surveillance of micronuclei links genome instability to innate immunity. *Nat. Cell Biol.* **2017**, *548*, 461–465. [CrossRef] [PubMed]
130. McArthur, K.; Whitehead, L.W.; Heddleston, J.M.; Li, L.; Padman, B.S.; Oorschot, V.; Geoghegan, N.D.; Chappaz, S.; Davidson, S.; Chin, H.S.; et al. BAK/BAX macropores facilitate mitochondrial herniation and mtDNA efflux during apoptosis. *Science* **2018**, *359*, eaao6047. [CrossRef]
131. Hu, M.; Zhou, M.; Bao, X.; Pan, D.; Jiao, M.; Liu, X.; Li, F.; Li, C.-Y. ATM inhibition enhances cancer immunotherapy by promoting mtDNA leakage and cGAS/STING activation. *J. Clin. Investig.* **2021**, *131*. [CrossRef] [PubMed]
132. Kearney, C.J.; Vervoort, S.J.; Hogg, S.J.; Ramsbottom, K.M.; Freeman, A.J.; Lalaoui, N.; Pijpers, L.; Michie, J.; Brown, K.K.; Knight, D.A.; et al. Tumor immune evasion arises through loss of TNF sensitivity. *Sci. Immunol.* **2018**, *3*, eaar3451. [CrossRef] [PubMed]
133. Manguso, R.T.; Pope, H.W.; Zimmer, M.D.; Brown, F.D.; Yates, K.B.; Miller, B.C.; Collins, N.B.; Bi, K.; LaFleur, M.W.; Juneja, V.R.; et al. In vivo CRISPR screening identifies Ptpn2 as a cancer immunotherapy target. *Nat. Cell Biol.* **2017**, *547*, 413–418. [CrossRef] [PubMed]
134. Lawson, K.A.; Sousa, C.M.; Zhang, X.; Kim, E.; Akthar, R.; Caumanns, J.J.; Yao, Y.; Mikolajewicz, N.; Ross, C.; Brown, K.R.; et al. Functional genomic landscape of cancer-intrinsic evasion of killing by T cells. *Nat. Cell Biol.* **2020**, *586*, 120–126. [CrossRef] [PubMed]
135. Zaretsky, J.M.; Garcia-Diaz, A.; Shin, D.S.; Escuin-Ordinas, H.; Hugo, W.; Hu-Lieskovan, S.; Torrejon, D.Y.; Abril-Rodriguez, G.; Sandoval, S.; Barthly, L.; et al. Mutations Associated with Acquired Resistance to PD-1 Blockade in Melanoma. *N. Engl. J. Med.* **2016**, *375*, 819–829. [CrossRef] [PubMed]
136. Heidegger, S.; Wintges, A.; Stritzke, F.; Bek, S.; Steiger, K.; Koenig, P.-A.; Göttert, S.; Engleitner, T.; Öllinger, R.; Nedelko, T.; et al. RIG-I activation is critical for responsiveness to checkpoint blockade. *Sci. Immunol.* **2019**, *4*, eaau8943. [CrossRef]
137. Ruzicka, M.; Koenig, L.M.; Formisano, S.; Boehmer, D.F.R.; Vick, B.; Heuer, E.-M.; Meinl, H.; Kocheise, L.; Zeitlhöfler, M.; Ahlfeld, J.; et al. RIG-I-based immunotherapy enhances survival in preclinical AML models and sensitizes AML cells to checkpoint blockade. *Leukemia* **2020**, *34*, 1017–1026. [CrossRef]
138. Such, L.; Zhao, F.; Liu, D.; Thier, B.; Le-Trilling, V.T.K.; Sucker, A.; Coch, C.; Pieper, N.; Howe, S.; Bhat, H.; et al. Targeting the innate immunoreceptor RIG-I overcomes melanoma-intrinsic resistance to T cell immunotherapy. *J. Clin. Investig.* **2020**, *130*, 4266–4281. [CrossRef] [PubMed]
139. Duewell, P.; Beller, E.; Kirchleitner, S.V.; Adunka, T.; Bourhis, H.; Siveke, J.; Mayr, D.; Kobold, S.; Endres, S.; Schnurr, M. Targeted activation of melanoma differentiation-associated protein 5 (MDA5) for immunotherapy of pancreatic carcinoma. *OncoImmunology* **2015**, *4*, e1029698. [CrossRef]
140. Ishizuka, J.J.; Manguso, R.T.; Cheruiyot, C.K.; Bi, K.; Panda, A.; Iracheta-Vellve, A.; Miller, B.C.; Du, P.P.; Yates, K.B.; Dubrot, J.; et al. Loss of ADAR1 in tumours overcomes resistance to immune checkpoint blockade. *Nat. Cell Biol.* **2019**, *565*, 43–48. [CrossRef] [PubMed]
141. Greenberg, M.V.C.; Bourc'His, D. The diverse roles of DNA methylation in mammalian development and disease. *Nat. Rev. Mol. Cell Biol.* **2019**, *20*, 590–607. [CrossRef] [PubMed]
142. Torres, I.O.; Fujimori, D.G. Functional coupling between writers, erasers and readers of histone and DNA methylation. *Curr. Opin. Struct. Biol.* **2015**, *35*, 68–75. [CrossRef] [PubMed]
143. Lin, R.-K.; Wang, Y.-C. Dysregulated transcriptional and post-translational control of DNA methyltransferases in cancer. *Cell Biosci.* **2014**, *4*, 46. [CrossRef] [PubMed]
144. Lin, R.-K.; Hsu, H.-S.; Chang, J.-W.; Chen, C.-Y.; Chen, J.-T.; Wang, Y.-C. Alteration of DNA methyltransferases contributes to 5′CpG methylation and poor prognosis in lung cancer. *Lung Cancer* **2007**, *55*, 205–213. [CrossRef] [PubMed]
145. Rahman, M.; Qian, Z.R.; Wang, E.L.; Yoshimoto, K.; Nakasono, M.; Sultana, R.; Yoshida, T.; Hayashi, T.; Haba, R.; Ishida, M.; et al. DNA methyltransferases 1, 3a, and 3b overexpression and clinical significance in gastroenteropancreatic neuroendocrine tumors. *Hum. Pathol.* **2010**, *41*, 1069–1078. [CrossRef] [PubMed]
146. Qu, Y.; Mu, G.; Wu, Y.; Dai, X.; Zhou, F.; Xu, X.; Wang, Y.; Wei, F. Overexpression of DNA Methyltransferases 1, 3a, and 3b Significantly Correlates With Retinoblastoma Tumorigenesis. *Am. J. Clin. Pathol.* **2010**, *134*, 826–834. [CrossRef]
147. Kulaeva, O.I.; Draghici, S.; Tang, L.; Kraniak, J.M.; Land, S.J.; Tainsky, M.A. Epigenetic silencing of multiple interferon pathway genes after cellular immortalization. *Oncogene* **2003**, *22*, 4118–4127. [CrossRef] [PubMed]
148. Lu, R.; Au, W.-C.; Yeow, W.-S.; Hageman, N.; Pitha, P.M. Regulation of the Promoter Activity of Interferon Regulatory Factor-7 Gene: ACTIVATION BY INTERFERON AND SILENCING BY HYPERMETHYLATION*. *J. Biol. Chem.* **2000**, *275*, 31805–31812. [CrossRef]
149. Nobeyama, Y.; Nakagawa, H. Silencing of interferon regulatory factor gene 6 in melanoma. *PLoS ONE* **2017**, *12*, e0184444. [CrossRef] [PubMed]
150. Li, Q.; Tainsky, M.A. Epigenetic Silencing of IRF7 and/or IRF5 in Lung Cancer Cells Leads to Increased Sensitivity to Oncolytic Viruses. *PLoS ONE* **2011**, *6*, e28683. [CrossRef] [PubMed]
151. Yamashita, M.; Toyota, M.; Suzuki, H.; Nojima, M.; Yamamoto, E.; Kamimae, S.; Watanabe, Y.; Kai, M.; Akashi, H.; Maruyama, R.; et al. DNA methylation of interferon regulatory factors in gastric cancer and noncancerous gastric mucosae. *Cancer Sci.* **2010**, *101*, 1708–1716. [CrossRef] [PubMed]

152. Dellac, S.; Ben-Dov, H.; Raanan, A.; Saleem, H.; Zamostiano, R.; Semyatich, R.; Lavi, S.; Witz, I.P.; Bacharach, E.; Ehrlich, M. Constitutive low expression of antiviral effectors sensitizes melanoma cells to a novel oncolytic virus. *Int. J. Cancer* **2020**. [CrossRef] [PubMed]

153. Cheng, X. Silent Assassin: Oncogenic Ras Directs Epigenetic Inactivation of Target Genes. *Sci. Signal.* **2008**, *1*, pe14. [CrossRef] [PubMed]

154. Bakin, A.V.; Curran, T. Role of DNA 5-methylcytosine transferase in cell transformation by fos. *Science* **1999**, *283*, 387–390. [CrossRef] [PubMed]

155. Bigey, P.; Ramchandani, S.; Theberge, J.; Araujo, F.D.; Szyf, M. Transcriptional regulation of the human DNA Methyltransferase (dnmt1) gene. *Gene* **2000**, *242*, 407–418. [CrossRef]

156. Devailly, G.; Grandin, M.; Perriaud, L.; Mathot, P.; Delcros, J.-G.; Bidet, Y.; Morel, A.-P.; Bignon, J.-Y.; Puisieux, A.; Mehlen, P.; et al. Dynamics of MBD2 deposition across methylated DNA regions during malignant transformation of human mammary epithelial cells. *Nucleic Acids Res.* **2015**, *43*, 5838–5854. [CrossRef]

157. Wu, B.-K.; Brenner, C. Suppression of TET1-Dependent DNA Demethylation Is Essential for KRAS-Mediated Transformation. *Cell Rep.* **2014**, *9*, 1827–1840. [CrossRef]

158. Gazin, C.; Wajapeyee, N.; Gobeil, S.; Virbasius, C.-M.A.; Green, M.R. An elaborate pathway required for Ras-mediated epigenetic silencing. *Nat. Cell Biol.* **2007**, *449*, 1073–1077. [CrossRef] [PubMed]

159. Topper, M.J.; Vaz, M.; Chiappinelli, K.B.; Shields, C.E.D.; Niknafs, N.; Yen, R.-W.C.; Wenzel, A.; Hicks, J.; Ballew, M.; Stone, M.; et al. Epigenetic Therapy Ties MYC Depletion to Reversing Immune Evasion and Treating Lung Cancer. *Cell* **2017**, *171*, 1284–1300.e21. [CrossRef] [PubMed]

160. Jaspers, I.; Horvath, K.M.; Zhang, W.; Brighton, L.E.; Carson, J.L.; Noah, T.L. Reduced Expression of IRF7 in Nasal Epithelial Cells from Smokers after Infection with Influenza. *Am. J. Respir. Cell Mol. Biol.* **2010**, *43*, 368–375. [CrossRef]

161. Myers, R.; Greiner, S.; Harvey, M.; Soeffker, D.; Frenzke, M.; Abraham, K.; Shaw, A.; Rozenblatt, S.; Federspiel, M.J.; Russell, S.J.; et al. Oncolytic activities of approved mumps and measles vaccines for therapy of ovarian cancer. *Cancer Gene Ther.* **2005**, *12*, 593–599. [CrossRef] [PubMed]

162. Stojdl, D.F.; Lichty, B.D.; Knowles, S.; Marius, R.; Atkins, H.; Sonenberg, N.; Bell, J.C. Exploiting tumor-specific defects in the interferon pathway with a previously unknown oncolytic virus. *Nat. Med.* **2000**, *6*, 821–825. [CrossRef] [PubMed]

163. Stojdl, D.F.; Lichty, B.D.; Tenoever, B.R.; Paterson, J.M.; Power, A.T.; Knowles, S.; Marius, R.; Reynard, J.; Poliquin, L.; Atkins, H.; et al. VSV strains with defects in their ability to shutdown innate immunity are potent systemic anti-cancer agents. *Cancer Cell* **2003**, *4*, 263–275. [CrossRef]

164. Li, J.; Bonifati, S.; Hristov, G.; Marttila, T.; Valmary-Degano, S.; Stanzel, S.; Schnölzer, M.; Mougin, C.; Aprahamian, M.; Grekova, S.P.; et al. Synergistic combination of valproic acid and oncolytic parvovirus H-1 PV as a potential therapy against cervical and pancreatic carcinomas. *EMBO Mol. Med.* **2013**, *5*, 1537–1555. [CrossRef] [PubMed]

165. Zakay-Rones, Z.; Tayeb, S.; Panet, A. Therapeutic potential of oncolytic Newcastle disease virus a critical review. *Oncol. Virother.* **2015**, *4*, 49–62. [CrossRef]

166. Pergolizzi, R.G.; Kreis, W.; Rottach, C.; Susin, M.; Broome, J.D. Mutational Status of Codons 12 and 13 of the N- and K-ras Genes in Tissue and Cell Lines Derived from Primary and Metastatic Prostate Carcinomas. *Cancer Investig.* **1993**, *11*, 25–32. [CrossRef] [PubMed]

167. Danziger, O.; Pupko, T.; Bacharach, E.; Ehrlich, M. Interleukin-6 and Interferon-α Signaling via JAK1–STAT Differentially Regulate Oncolytic versus Cytoprotective Antiviral States. *Front. Immunol.* **2018**, *9*, 94. [CrossRef]

168. Jones, P.A.; Ohtani, H.; Chakravarthy, A.; De Carvalho, D.D. Epigenetic therapy in immune-oncology. *Nat. Rev. Cancer* **2019**, *19*, 151–161. [CrossRef]

169. Kassiotis, G.; Stoye, J.P. Immune responses to endogenous retroelements: Taking the bad with the good. *Nat. Rev. Immunol.* **2016**, *16*, 207–219. [CrossRef] [PubMed]

170. Smith, C.C.; Beckermann, K.E.; Bortone, D.S.; De Cubas, A.A.; Bixby, L.M.; Lee, S.J.; Panda, A.; Ganesan, S.; Bhanot, G.; Wallen, E.M.; et al. Endogenous retroviral signatures predict immunotherapy response in clear cell renal cell carcinoma. *J. Clin. Investig.* **2018**, *128*, 4804–4820. [CrossRef] [PubMed]

171. Tunbak, H.; Enriquez-Gasca, R.; Tie, C.H.C.; Gould, P.A.; Mlcochova, P.; Gupta, R.K.; Fernandes, L.; Holt, J.; Van Der Veen, A.G.; Giampazolias, E.; et al. The HUSH complex is a gatekeeper of type I interferon through epigenetic regulation of LINE-1s. *Nat. Commun.* **2020**, *11*, 5387. [CrossRef]

172. Goodier, J.L. Restricting retrotransposons: A review. *Mobile DNA* **2016**, *7*, 16. [CrossRef] [PubMed]

173. Groh, S.; Schotta, G. Silencing of endogenous retroviruses by heterochromatin. *Cell Mol. Life Sci.* **2017**, *74*, 2055–2065. [CrossRef] [PubMed]

174. Fukuda, K.; Shinkai, Y. SETDB1-Mediated Silencing of Retroelements. *Viruses* **2020**, *12*, 596. [CrossRef] [PubMed]

175. Geis, F.K.; Goff, S.P. Silencing and Transcriptional Regulation of Endogenous Retroviruses: An Overview. *Viruses* **2020**, *12*, 884. [CrossRef] [PubMed]

176. Chiappinelli, K.B.; Strissel, P.L.; Desrichard, A.; Li, H.; Henke, C.; Akman, B.; Hein, A.; Rote, N.S.; Cope, L.M.; Snyder, A.; et al. Inhibiting DNA Methylation Causes an Interferon Response in Cancer via dsRNA Including Endogenous Retroviruses. *Cell* **2015**, *162*, 974–986. [CrossRef] [PubMed]

177. Roulois, D.; Yau, H.L.; Singhania, R.; Wang, Y.; Danesh, A.; Shen, S.Y.; Han, H.; Liang, G.; Jones, P.A.; Pugh, T.J.; et al. DNA-Demethylating Agents Target Colorectal Cancer Cells by Inducing Viral Mimicry by Endogenous Transcripts. *Cell* **2015**, *162*, 961–973. [CrossRef] [PubMed]
178. Goel, S.; DeCristo, M.J.; Watt, A.C.; BrinJones, H.; Sceneay, J.; Li, B.B.; Khan, N.; Ubellacker, J.M.; Xie, S.; Metzger-Filho, O.; et al. CDK4/6 inhibition triggers anti-tumour immunity. *Nature* **2017**, *548*, 471–475. [CrossRef] [PubMed]
179. Du, Q.; Guo, X.; Wang, M.; Li, Y.; Sun, X.; Li, Q. The application and prospect of CDK4/6 inhibitors in malignant solid tumors. *J. Hematol. Oncol.* **2020**, *13*, 41. [CrossRef] [PubMed]
180. Kimura, H.; Nakamura, T.; Ogawa, T.; Tanaka, S.; Shiota, K. Transcription of mouse DNA methyltransferase 1 (Dnmt1) is regulated by both E2F-Rb-HDAC-dependent and -independent pathways. *Nucleic Acids Res.* **2003**, *31*, 3101–3113. [CrossRef]

Review

Personalizing Oncolytic Virotherapy for Glioblastoma: In Search of Biomarkers for Response

Eftychia Stavrakaki, Clemens M. F. Dirven and Martine L. M. Lamfers *

Department of Neurosurgery, Brain Tumor Center, Erasmus University Medical Center,
3015 CN Rotterdam, The Netherlands; e.stavrakaki@erasmusmc.nl (E.S.); c.dirven@erasmusmc.nl (C.M.F.D.)
* Correspondence: m.lamfers@erasmusmc.nl

Simple Summary: Glioblastoma (GBM) is the most frequent and aggressive primary brain tumor. Despite multimodal treatment, the prognosis of GBM patients remains very poor. Oncolytic virotherapy is being evaluated as novel treatment for this patient group and clinical trials testing oncolytic viruses have shown impressive responses, albeit in a small subset of GBM patients. Obtaining insight into specific tumor- or patient-related characteristics of the responding patients, may in the future improve response rates. In this review we discuss factors related to oncolytic activity of the most widely applied oncolytic virus strains as well as potential biomarkers and future assays that may allow us to predict response to these agents. Such biomarkers and tools may in the future enable personalizing oncolytic virotherapy for GBM patients.

Abstract: Oncolytic virus (OV) treatment may offer a new treatment option for the aggressive brain tumor glioblastoma. Clinical trials testing oncolytic viruses in this patient group have shown promising results, with patients achieving impressive long-term clinical responses. However, the number of responders to each OV remains low. This is thought to arise from the large heterogeneity of these tumors, both in terms of molecular make-up and their immune-suppressive microenvironment, leading to variability in responses. An approach that may improve response rates is the personalized utilization of oncolytic viruses against Glioblastoma (GBM), based on specific tumor- or patient-related characteristics. In this review, we discuss potential biomarkers for response to different OVs as well as emerging ex vivo assays that in the future may enable selection of optimal OV for a specific patient and design of stratified clinical OV trials for GBM.

Keywords: oncolytic viruses; glioblastoma; clinical trials; biomarkers; personalized oncolyticvirotherapy

Citation: Stavrakaki, E.; Dirven, C.M.F.; Lamfers, M.L.M. Personalizing Oncolytic Virotherapy for Glioblastoma: In Search of Biomarkers for Response. *Cancers* **2021**, *13*, 614. https://doi.org/10.3390/cancers13040614

Academic Editors: Antonio Marchini, Carolina S. Ilkow and Alan Melcher

Received: 28 December 2020
Accepted: 29 January 2021
Published: 4 February 2021

Publisher's Note: MDPI stays neutral with regard to jurisdictional claims in published maps and institutional affiliations.

1. Introduction

Oncolytic viral therapy or virotherapy is a form of immunotherapy showing promising results for cancers with poor prognosis [1]. In this approach, oncolytic viruses (OVs) are employed to kill tumor cells, while in parallel stimulating an anti-tumor immune response [2]. OVs exhibit either natural tropism to malignant cells or their genome is altered to confer them higher specificity for malignant cells [3]. Viruses from ten different families (Adenoviridae, Herpesviridae, Paramyxoviridae, Reoviridae, Retroviridae, Picornaviridae, Parvoviridae, Poxviridae, Rhabdoviridae, Alphaviruses) have thus far been utilized as oncolytic virus platforms in clinical trials for various cancer types [2].

One deadly type of cancer is glioblastoma multiforme (GBM), the most common and aggressive primary brain tumor [4]. The standard treatment consists of maximal safe surgical resection followed by radiotherapy plus concomitant and adjuvant temozolomide chemotherapy. However, the median overall survival among all GBM patients is less than one year, and only 15 months in patients receiving complete standard treatment with 3-year survival being less than 10% [5,6]. In the past decades, numerous therapeutic approaches have been tested in clinical trials, with disappointing outcomes. The main obstacles in treating GBM include its infiltrative growth, its intrinsic resistance to chemo- and radiotherapy, its notorious intratumoral heterogeneity with dynamic changes in subclones facilitating treatment escape, its protected location behind the blood-brain-barrier and the immunological 'cold' microenvironment of these tumors. These hurdles to more conventional therapies, as well as the dismal prognosis of GBM patients, have encouraged scientists and clinicians to develop and evaluate the local application of various types of oncolytic viruses in this patient group. Table 1 summarizes the most commonly applies OVs in GBM trials. The OVs differ in their primary attachment molecules to host receptors as well as in the source of their tumor selectivity, which may be derived from a natural tropism to cancer cells or by genetic engineering.

Table 1. Characteristics of the most commonly used Oncolytic viruses (OVs) in glioblastoma multiforme (GBM) clinical trials.

Family	Genome	OV Examples	Genetic Engineering	Entry Receptor	Tumor Specificity
		HSV1716	ICP34.5-deleted	HVEM, 3-O-sulfated heparin sulfate and nectin-2	Defects in the p16/Rb, PKR or interferon pathways [7]
		G207	ICP34.5 and ICP6 -deleted mutant oHSV	HVEM, 3-O-sulfated heparin sulfate and nectin-2	Defects in the p16/Rb, PKR or interferon pathways [8]
Herpesvirus	dsDNA	G47Δ	ICP34.5, ICP6 and α47 -deleted mutant oHSV	HVEM, 3-O-sulfated heparin sulfate and nectin-2	Defects in the p16/Rb, PKR or interferon pathways [9]
		rQnestinHSV-1	ICP34.5-deleted mutant oHSV, in which γ134.5 gene was reinserted under control of nestin promoter	HVEM, 3-O-sulfated heparin sulfate and nectin-2	Expression of nestin [10]
Adenovirus	dsDNA	Onyx-015	E1B-55k and E3B -deleted mutant group C adenovirus	CAR	Defects in p53 pathway, defects in cell cycle, late viral RNA export [11]
		delta24-RGD	24-base pair deletion in the E1A gene and insertion of an RGD sequence in the viral knob	CAR, αvβ3 and αvβ5 integrins	Defects in Rb pathway [12]
Paramyxoviridae	(−) ssRNA	MV-CEA	Edmonston (MV-Edm) vaccine strain with insertion of the human carcinoembryonic antigen gene	CD46, nectin-4, SLAM	Overexpression of CD46, defects in the interferon pathway [13]
		NDV	Natural tropism	Sialic acids	Defects in the interferon pathway [14]
Reovirus	dsRNA	R124	Natural tropism	JAM-A, Nogo Receptor NgR1	Defects in the Ras signaling pathway [15]
Picornaviridae	(+) ssRNA	PVSRIPO	Poliovirus type 1 (Sabin) vaccine with replacement of the internal ribosomal entry site (IRES) with the human rhinovirus type 2 IRES	CD155	Overexpression of CD155 [16,17]
Parvovirus H1	ssDNA	Parvovirus H-1PV	Natural tropism	Sialic acids	Defects in interferon pathway, defects in cell proliferation pathways [18]

In a recent review, Chiocca et al. summarized the findings [19] from all the recent GBM oncolytic virotherapy trials and illustrated that a subgroup of GBM patients responds exceptionally well to OV treatments, with survivors at 36-months, and with some patients exhibiting long term remission [20,21]. This phenomenon has also been observed in OV trials for other cancer types. For instance, a phase II clinical trial employing an oncolytic herpes simplex virus 1 for stage IIIC or IV melanoma showed 26% overall response [22].

These observations raise the question: would the responding patients have been the same individuals if they had been treated with any other OV, or are we looking at responders to a specific OV? In other words, is the elicited immune response a generalized one for all types of OVs, or does each OV elicit a specific anti-tumor immune response? The latter would suggest that response rates may be significantly increased if we are able to define which OV is best suited for a particular patient. Identification of robust predictive biomarkers for OV response would allow future design of stratified clinical trials employing multiple OV strains. The replication efficiency of the virus is thought to be of importance for generation of the subsequent inflammatory and anti-tumor responses. Moreover, host immune status is also expected to contribute to the efficacy of OV treatment. This review, therefore, focuses on tumor and host resistance mechanisms to viral infection, replication and oncolysis and discusses potential biomarkers that have previously been reported in relation to sensitivity or resistance to the most frequently employed OVs in preclinical and clinical GBM research.

2. Glioblastoma

2.1. Heterogeneity, Stem Cells and Therapy Resistance

Common molecular abnormalities involved in the evolution of glioblastomas include aberrations in the oncogenes (EGFR, PDGF and its receptors) and tumor suppressor genes (p16INK4a, p14ARF, PTEN, RB1, and TP53), which are often observed in other human cancers as well [23]. GBM is also characterized by inter-tumoral heterogeneity, which is highlighted by the classification of GBMs into three subgroups: proneural, classical and mesenchymal [24,25]. Each subtype is characterized by specific gene expression patterns and molecular abnormalities, resulting in different clinical treatment outcomes [25,26]. Proneural subtype has the most favorable prognosis among the three subtypes; aberrations in the isocitrate dehydrogenase 1 (IDH1) gene and the platelet-derived growth factor receptor A (PDGFRA) define this subgroup. The classical subgroup is characterized by the amplification of EGFR, lack of TP53 mutations and often with homozygous CDKN2A deletions [26]. Lastly, the mesenchymal subtype is the most aggressive and it is characterized by aberrations in the neurofibromin 1 (NF1) and PTEN genes [23]. It is also characterized by a pro-inflammatory environment compared with the other subtypes [27]. It was hypothesized that one underlying cause for this was the higher incidence of tumor-associated antigens (TAAs), however this could not be proven, as specific tumor antigens are expressed in each subtype [27]. Nevertheless, this classification has not led to altered or adapted treatment approaches [28,29].

Apart from intertumoral heterogeneity, intra-tumoral heterogeneity poses another therapeutic obstacle in treatment of GBM, allowing escape of subclones from (targeted) therapies and driving treatment resistance. This heterogeneity was captured by genome-wide and single cell RNA studies, which showed tumor cells with different transcriptional profiles within the same tumor [30,31]. In addition, it was shown that within the same tumor, different subtypes can coexist, highlighting the heterogeneity that characterizes GBM [31]. In another study, paired primary and recurrent tumor tissue samples were analyzed to determine the persistence of possible drug targets. The results showed that the molecular targets between primary and recurrent tumors changed by 90% [32]. This may explain the failure of drugs that target specific molecular mutations in GBM, such as the EGFR [33].

Eventually, most of the patients experience tumor relapse due to therapeutic resistance [29]. This therapeutic resistance is mainly attributed to glioblastoma stem cells (GSCs), which activate DNA repair mechanisms to promote survival after chemo- and radiotherapy [34]. Additionally, outgrowth of resistant subclones and downregulation of targeted molecules contribute to drug resistance. Furthermore, the highly infiltrative nature of GSCs makes total surgical resection of the tumor impossible [35]. The remaining and/or treatment-resistant clones will eventually generate functional vessels for the nutrient transport and develop tumor recurrence [34].

2.2. GBM Microenvironment: Local Immunosuppressive Mechanisms

Glioblastoma arises in the central nervous system (CNS) [36], which is an immunologically distinct site. In the past, the CNS was considered an immune privileged site, due to its unique properties. For instance, the blood brain barrier, which tightly regulates the transportation of the immune cells from the periphery to the CNS; the lack of antigen presenting cells in a non-inflamed state; and more importantly the lack of a classic lymphatic system [37–39]. The concept of CNS being immune privileged has now been revised. Recent studies have shown that antigens derived from the CNS can efficiently elicit an immune response [40]. More importantly, Louveau et al. [41] discovered a functional lymphatic system, parallel to the dural sinuses, a possible route of transportation of antigen-presenting cells to the deep cervical lymph nodes, where they can present CNS-derived antigens and prime T cells. These recent studies have provided evidence that CNS-derived antigens can mount a vigorous immune response, offering ground to investigate immunotherapy approaches for GBM.

The GBM environment is characterized by the high influx of tumor-associated macrophages (TAMs). In a non-inflamed state, the myeloid composition of the CNS consists of the tissue-resident macrophages that arise from the yolk sac, the microglia [42]. However, in GBM, the microenvironment is comprised mainly of a mixture of microglia and infiltrating monocytes from the periphery. Glioma cells produce a milieu of monocyte chemoattractant proteins along with other factors, leading to disruption of the blood-brain barrier and facilitating recruitment of monocytes from the periphery [43]. When monocytes arrive at the tumor site, glioma cells drive their polarization to an immunosuppressive M2 phenotype [44,45]. These M2-like TAMs promote tumor growth and migration as well as the immune invasion by hampering the adaptive immunity [44,46,47]. TAMs are the most abundant immune cell population in GBM and can consist up to 50% of the GBM tumor mass. Their importance in tumor growth is highlighted by the correlation between increased TAM numbers and worse prognosis in GBM patients; furthermore, TAM infiltration has been associated with the mesenchymal subtype of GBM, being the most aggressive one [48,49].

Another feature that facilitates the local immune suppression in GBM is T cell dysfunction. Severe T cell exhaustion is observed in GBM, which is characterized by upregulation of expression of co-inhibitory molecules like PD-1, LAG-3 and TIM-3 [50]. Furthermore, an increase in numbers of the regulatory T cells (Tregs), which can suppress the antigen-specific T cells, was found in high grade gliomas compared to low grade gliomas [51]. The recruitment of Tregs at the tumor site is mainly facilitated by the production of the attractant indoleamine 2,3 dioxygenase (IDO) by gliomas [52]. Another facet that contributes to the "cold" tumor microenvironment is the relatively low mutational burden of GBM cells, associated with limited expression of neoantigens [53,54]. Taken together, GBM has all the characteristics of a tumor with low immunogenicity. The M2-like macrophages that are abundant at the tumor site, the dysfunctional T cells and the low neoantigen expression are some of the barriers that we need to overcome to design successful immunotherapies.

Considering all of the above, a therapeutic strategy that is not hindered by specificity for a single molecular target or differentiation state of tumor cells, that is delivered locally in a single surgical intervention, hence bypassing the BBB, that is self-perpetuating in its anti-tumor activity, and which can overcome the immune-suppressive tumor microenvironment, may offer opportunities for achieving therapeutic responses in glioblastoma patients. Oncolytic viruses offer such a treatment strategy.

3. Factors Affecting OV Therapy in GBM

Several oncolytic virotherapy clinical trials have shown impressive and durable responses in a subset of patients, indicating that OVs might be a very promising therapeutic tool for treating GBM. The establishment of an efficient viral infection, lysis of tumor cells, viral spreading and anti-tumor immune activation, all depends on multiple factors (Figure 1). It is therefore conceivable that we may improve OV efficacy if we take these factors into account when selecting patients for treatment.

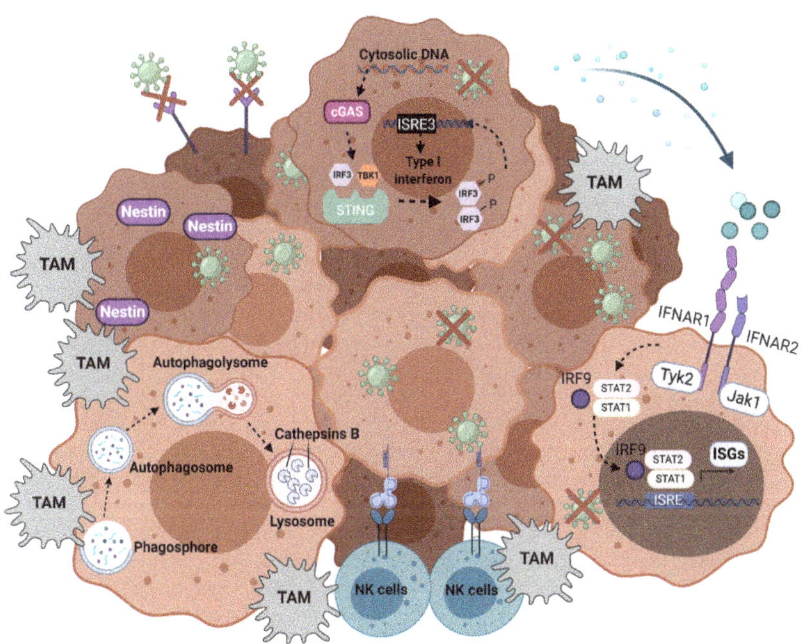

Figure 1. OV restriction mechanisms of GBM tumors. Infiltration of NK cells and tumor-associated macrophages (TAMs) at the tumor site, activation status of autophagy, expression of viral entry molecules and viral sensors (e.g., cGAS-STING) that lead to constitutive active type I interferon pathways, all could hamper the OV replication and oncolysis. Furthermore, cathepsin B expression and expression of specific proteins that drive specific tumor replication (e.g., nestin) could determine the OV efficacy. Created with BioRender.com.

3.1. Viral Entry Molecule Expression

Tumor cell infection and oncolysis are a prerequisite for mounting an inflammatory response in the tumor microenvironment and ultimately generating an anti-tumor immune response. This depends on the cell entry possibilities for the virus. As GBM cells are not the natural host cells for entry of most viruses, low levels or even lack of receptor molecules on these cells can form the first obstacle to virotherapy. It has been shown that tremendous inter-tumoral variability exists in expression levels of specific adenovirus and reovirus entry molecules on patient-derived GBM cells [55,56]. As a result, various retargeting strategies have been applied to overcome such limitations, including EGFR and integrin retargeting of OVs [57]. Therefore, OV efficacy could potentially be enhanced by stratification of patients based on expression of specific viral receptor molecules in their tumors.

3.2. Status of Oncogenic Signaling Pathways Affected in Glioblastoma

Many OVs applied in glioma studies are genetically engineered or have naturally evolved to exploit oncogenic signaling pathways in cancer cells, such as the Ras, Rb, p53 or nucleotide synthesis pathways [58]. Therefore, OV efficacy could potentially be enhanced by stratification of patients based on activation status or presence of mutations in targeted pathways.

Another targeting approach is by the insertion of tumor-specific promoters to drive specific viral replication in tumor cells and avoid toxicity to normal tissue [59]. Various promoter candidates have been applied to design tumor-specific promoter-driven OVs, including nestin, survivin, cyclooxygenase-2 (COX-2), C-X-C chemokine receptor type 4 (CXCR4), hypoxia inducible factor-1 (HIF-1) and telomerase [10,60–63]. Considering the intertumoral heterogeneity in transcription profiles of GBM, it would be expected that the response to such OVs might vary between GBM subtypes. One could hypothesize that GBM with proneural features might be more sensitive to viruses targeting cells expressing neuronal progenitor genes (e.g., nestin), whereas tumors of mesenchymal subtype may be more sensitive to viruses in which replication is driven by the inflammation-activated COX-2 or CXCR4 promoter [62–64].

3.3. Innate Anti-Viral Responses

Upon cell entry, antiviral host defenses may be activated that counteract a productive lytic cycle and progeny production. There is a plethora of innate sensors that could lead to clearance of the infected cell and halt the viral spreading, leading to resistance to OV therapy. As soon as the cell is infected, the viral pathogen-associated molecular patterns (PAMPs) are sensed by pattern recognition receptors (PRRs). These PRRs include Toll-like receptors (TLRs) [65], RIG-I-like receptors (retinoic acid-inducible gene-I-like receptors, RLRs) [66], C-type lectin receptors (CLRs) [66], oligomerization domain containing receptors {(NOD-like receptors (NLRs)} [66], cyclic GMP-AMP synthase (cGAS) [67] and absent in melanoma 2 (AIM2)-like receptors (ALRs) [68]. The recognition of the viral PAMPs from the host PRRs results in interferon type I (IFNα, IFNβ, IFN-ε, IFN-κ, IFN-ω, IFN-δ, IFN-ζ and IFN-τ) and interferon type III (IFN-$\lambda\varsigma$) production, as well as the expression of interferon stimulated genes (ISGs) and other proinflammatory cytokines and chemokines [69,70].

Although the aforementioned PRRs have been extensively studied, not many studies have attempted to correlate their overexpression in tumor cells with OV resistance; but rather with the aftermath of the recognition, the antiviral interferon pathways. However, a few studies have implicated the cytosolic DNA sensing pathway to oncolytic herpes virus-1 resistance (see below). The main sensor of dsDNA in the cytoplasm is the cGAS, which recognizes dsDNA viruses and reverse transcribing RNA viruses like HIV-1. As soon as cGAS is activated, it synthesizes cGAMP which activates the adaptor protein stimulator of interferon genes (STING) [71]. Stimulation of STING leads to the activation of IRF3 and NFκB [71]. Interferon gamma inducible protein 16 (IFI16) is another sensor of dsDNA that signals via STING to activate IRF3 and NFκB resulting in IFNβ production [72].

Ultimately, viral detection by the aforementioned sensors will lead to the activation of host defenses such as the production of type I and type III interferons. These have distinct receptors, however, both activate a signaling cascade via receptor-associated protein tyrosine kinases Janus kinase 1 (JAK1) and tyrosine kinase 2 (TYK2), which activate the activator of transcription 1 (STAT1) and STAT2, which subsequently form a complex with the IFN regulatory factor 9 (IRF9), the ISGF3 complex [73]. This complex translocates to the nucleus, resulting in the expression of more than 300 ISGs and pro-inflammatory molecules and establishing an anti-viral state in the infected cell [74,75]. The cytokine and chemokine milieu produced by the infected cell also acts in a paracrine manner to induce an ISG-mediated anti-viral state in the (uninfected) adjacent cells. Some of these ISGs, such as GTPase myxovirus resistance 1 (MxA), ribonuclease L (RNaseL) and protein kinase R (PKR) have direct antiviral activity. For instance, MxA monomers reside in the cytoplasm and upon binding to viral components can degrade them [76]. PKR regulates a plethora of signaling pathways and its role in antiviral response and inhibition of host translation is considered crucial upon virus infection [76].

The anti-viral IFNs are major determinants of OV efficacy. Many OVs exploit the IFN pathway defects to successfully replicate in tumor cells. For instance, it has been shown that STING pathway is correlated with oncolytic herpes virus-1 resistance (see below). However, new evidence shows that this advantage in viral replication may not correlate with tumor eradication in vivo [77]. This may be explained by the inability of oncolytic viruses to induce immunogenic cell death in STING-deficient tumor cells, thus hampering the induction of innate and adaptive immunity [78]. Identifying specific defects in the IFN pathway that may 'assist' the viral replication without harming the induction of antitumor immunity, could lead to identification of predictive biomarkers for OV sensitivity.

3.4. The Autophagic Response to Viral Infections

There is growing evidence that the role of autophagy in the infectious cycle of many viruses is critical. Autophagy is an evolutionary conserved adaptive process in which the cells attempt to maintain their homeostasis [79,80]. It can be triggered by different types of stress, such as hypoxia, nutrient deprivation and infection [81]. The role of autophagy is to remove detrimental cytosolic material such as protein aggregates and damaged organelles. During this process, a phagophore engulfs cytosolic material to form an autophagosome, which subsequently fuses with a lysosome to degrade its cytoplasmic content [81].

Autophagy is activated by the infected cell to degrade and remove the virions from the cell, a process called xenophagy [82]. TLRs, RLRs and cGAS signaling pathways, among others, lead to autophagy activation to enhance the interferon production and create an anti-viral milieu [83]. Notably, plasmacytoid dendritic cells (pDCs) which lack the autophagy protein 5 (Atg5) showed decreased TLR7-dependent IFNα and IL-12 production after VSV and Sendai virus infection, indicating the importance of autophagy for mounting an anti-viral response [84]. Additionally, autophagy was shown to have anti-viral effects against Sindbis virus and Rift Valley Fever Virus infection [85,86]. On the other hand, many viruses have developed mechanisms to exploit autophagy in favor of their viral replication. For example, herpesvirus and dengue virus were shown to enhance autophagy to promote cell survival in order to establish a successful infection and enhanced viral replication [87,88]. Moreover, rapamycin, an autophagy inducer, was shown to increase the viral replication of various oncolytic viruses in tumor cells, including adenovirus, reovirus, poliovirus, herpes virus, NDV and myxoma virus [89–94]. In line with these findings, it was shown that knocking out two key autophagy genes (ATG5 or ATG10) impaired virus-induced lysis of cancer cells by a modified oncolytic adenovirus (delta24-RGD) [95]. Furthermore, co-treatment with Everolimus, a rapamycin derivative, and delta24-RGD enhanced autophagic dependent cell death in an in vivo glioma model [96].

The double-sided role of autophagy in oncolytic virus efficacy has captured the attention of the research community and it has been extensively reviewed [97,98]. The results thus far suggest that, for certain OVs, the tumor cells' ability to activate autophagy can contribute to the degree of viral replication, and ultimately the therapeutic efficacy of the viral treatment.

4. Potential Biomarkers for Sensitivity to Oncolytic Viruses

The different OV strains employed in GBM immunotherapy utilize different cell entry receptors of entry and their cell killing mechanisms are distinct from each other. Furthermore, each OV strain triggers the host responses in diverse ways. Available in vitro and in vivo data provides numerous leads to pathways and molecules involved in OV sensitivity or resistance. OV trials are increasingly incorporating trial-associated (immune) monitoring studies to gain insight into the in situ mechanisms involved in clinical OV therapy. Such valuable data may yield relevant information for identifying potential biomarkers related to response to OV therapy in GBM.

4.1. Oncolytic Herpes Simplex Virus

Oncolytic HSV-1 (oHSV) is an enveloped double-stranded DNA virus that belongs to the alpha-herpesvirus subfamily [99]. It is a neurotropic virus and therefore requires engineering for tumor-restricted replication [100]. Modified oHSV-1 variants that have been tested in glioma patients are G207, G47Δ, HSV1716 and rQNestin-34.5 [9]. Safety and feasibility of local oHSV injection in GBM was shown in two phase 1 trials testing G207 and HSV1716 [7,8]. However, in both studies viral replication was detected in only a few patients; 3 out of 6 and 2 out of 12 patients, respectively. These results suggest that the replication of the virus was restricted in some patients. In addition, seroconversion was observed in some patients indicating that the antiviral immune response may have contributed to the rapid clearance of the virus [7,8]. In an effort to understand these restricting mechanisms, Peters et al. studied G207 infection in vitro and found glioma stem cells (GSCs) to be non-permissive to infection [101]. G207 virus contains mutations in both copies of the γ34.5 gene to prevent neurovirulence, however, in GSCs this deletion results in a translational shut down preventing the production of progeny virions [101]. Other oHSV-1 variants designed to express the γ34.5 protein under a tumor-specific promoter such as the rQNestin-34.5, might enhance the oncolytic activity of the virus in GSCs [10].

Another modified HSV-1 is Talimogene laherparepvec (T-VEC) which is the first Food and Drug Administration (FDA) approved oncolytic virus and is indicated for treatment of patients with advanced melanoma [102]. Numerous clinical trials have employed T-VEC, however, no biomarkers for response have been described thus far. In a clinical study in melanoma patients, a favorable outcome was observed in a subgroup of patients with unresectable Stage III or IV M1a disease [103]. Recently, an in vitro study in melanoma cell lines revealed that STING expression can restrict T-VEC-mediated oncolysis and loss of its expression may confer sensitivity to oncolysis [104].

4.2. Oncolytic Adenovirus

Adenoviruses are non-enveloped double-stranded DNA viruses [59]. Oncolytic adenoviruses have been extensively explored and utilized in many clinical trials against several cancers [105]. Conditionally replicating adenoviruses (CRAds) have been modified in diverse ways to target oncogenic pathways frequently mutated cancers such as the retinoblastoma (Rb) or the p53 pathway [58,106]. Dl1520 (ONYX-015) was the first CRAd tested in a phase I clinical trial for recurrent gliomas, in which the safety of local peritumoral injection of the virus was shown [107]. Several possible selectivity mechanisms have been proposed for dl1520, including p53/p14ARF defects, aberrant late mRNA transport and cell cycle disruption, all of which may relate to the functions of the early viral E1B-55k gene which is deleted in this virus [11]. Whether any of these factors can serve as biomarkers for dl1520 response has not been evaluated.

The results of another phase I clinical study against recurrent malignant gliomas using the CRAd delta24-RGD (DNX-2401) were recently published [20]. This OV was engineered to selectively replicate in tumor cells with dysfunctional Rb pathway, which is the case in approximately 80% of GBM tumors [12,58]. Impressive anti-tumor effects were found, with 17% of the treated patients surviving beyond 3 years [20]. Another early clinical trial with this CRAd was conducted in our institute in patients with recurrent GBM (NCT01582516). A subgroup of patients revealed high concentrations of different cytokines in post treatment CSF samples, indicating that delta24-RGD can induce an inflammatory microenvironment, which is potentially key for its therapeutic efficacy [108]. Whether specific cytokines can serve as biomarkers for response requires further investigation.

Very few studies have focused on elucidating the resistance mechanisms in the non-responder patients in oncolytic adenoviral trials. In one study, it was demonstrated that the IFN signaling pathway was upregulated in Ad5/3-Δ24-resistant ovarian tumors compared to untreated tumors [109]. Moreover, the authors showed that the MxA, an ISG which is induced by IFN type I or type III signaling could provide a predictive marker for resistance to oncolytic adenoviral therapies [109,110].

Of great interest are the studies that have focused on the pre-treatment immune status of patients receiving oncolytic adenovirus. Specifically, it was shown that chronic inflammation was a negative predictive marker for response to oncolytic adenovirus therapy in different types of cancer [111]. Furthermore, high-mobility group box 1 (HMGB1), a nuclear protein secreted by immune cells and which is associated with a pro-inflammatory state and immunological cell death, could serve as a predictive and prognostic marker for oncolytic virotherapy with adenoviruses. The study suggested that patients with low serum HMGB1 have more robust anti-tumor responses after oncolytic adenovirus therapy [112]. It was hypothesized that a higher pro-inflammatory state as measured by HMGB1, leads to inhibition of viral replication. These results may suggest that the use of immunosuppressants for a limited amount of time during post virus administration, may improve response rates in this subgroup of patients.

Lastly, in a study using pancreatic cell lines it was shown that high expression of cyclin D1 enhances delta24-RGD-induced cytotoxicity [113]. Cyclin D1 activates the cyclin-dependent kinase 4 and 6 (CDK4 and CDK6), which then phosphorylates the Rb protein resulting in cell cycle progression. Over-expression of cyclin D1 has been observed in many cancer types like head and neck squamous cell carcinomas (HNSCC), pancreatic and breast cancer [114]. In GBM, the highest expression of cyclin D1 was observed in the proneural subtype, which may suggest that this subtype would benefit more from delta24-RGD treatment.

4.3. Oncolytic Retrovirus

Replication-competent retroviruses are a relative newcomer to the OV field. These are single-stranded RNA viruses. Originally, retroviruses were applied in gene therapy approaches, however, it was later shown that replication competence of retroviruses can provide a powerful tool for gene delivery of anticancer agents in tumors [115]. Vocimagene amiretrorepvec (Toca 511) is a replicating γ-retrovirus derived from murine leukemia virus and is engineered to encode a yeast cytosine deaminase (CD) gene [116]. In the presence of the prodrug 5-fluorocytosin (5-FC), CD converts 5-FC to the potent anti-cancer drug 5-fluorouracil (5-FU). The results from a phase I clinical trial for recurrent high-grade gliomas showed that the median survival of the patients ($n = 53$) was 13.6 months with six patients showing complete response [21]. However, in a subsequent phase III study in 403 patients, clinical endpoints were not met [117].

New results from the earlier phase I clinical trial (NCT01470794) were recently published demonstrating that 86% of the patients that lived >2 years had neoantigens deriving from IDH1, PI3K3CA, EGFR, SYNE1 genes. Interestingly, only 26% of the patients with <2 years survival had neoantigens arising from these genes, suggesting that neoantigens arising from driver genes may support Toca+5-FC therapy response. Moreover, the numbers of M0 macrophages and NK cells at the tumor site at the time of treatment were associated with poor response [118]. It is conceivable that these cells contributed to clearance of the virus before its therapeutic effect could take place. Further investigation is needed to establish if the immune composition could serve as a predictive marker and whether this is also the case for other oncolytic viruses.

4.4. Oncolytic Measles Virus

Oncolytic measles virus (oMV) has been applied in many Phase I/II clinical trials against numerous types of cancers including ovarian cancer, pancreatic cancer and glioblastoma (GBM) [119]. MV is a single-stranded, negative-sense RNA virus that belongs to the Paramyxoviridae family [120]. The entry of MV is mediated by the attachment of the viral Hemagglutinin (H) protein to three known cell surface receptors; the complement regulatory protein CD46, the signaling lymphocyte activating molecule (SLAM) or nectin-4 [121]. The wild type strains of MV mainly bind to the SLAM receptor, the attenuated MV Edmonston's (MV-Edm) vaccine strains enter through CD46 receptor, while nectin-4 can be used by both wild type and Edm strains [121].

The attenuated MV strains have revealed tropism for infecting and killing glioma cells, due to the overexpression of the entry receptor CD46 on the cell surface of these cells [13,122]. Although CD46 is abundantly expressed on glioma cells, facilitating efficient infection, some glioma cell lines show resistance to oncolysis after the viral entry, indicating that other processes can affect its oncolytic efficacy [123].

Indeed, a recent study pinpointed the expression of the interferon-induced transmembrane protein 1 (IFITM1) gene as the responsible ISG for restricting oMV replication in transformed human mesenchymal stromal cells [124]. Additionally, in another study researchers screened eight sarcoma cell lines and found that five of them were susceptible to the oMV. The resistance in the three remaining sarcoma cell lines was attributed to the upregulation of the RIG-I and IFITM1 mRNA expression. Interestingly, it was found that resistance could be broken by increasing the multiplicity of infection (MOI) in combination with the pro-drug 5-FC [125].

The fact that oMVs are sensitive to antiviral responses was also highlighted by a translational study from Kurokawa et al. [123]. Specifically, it was shown that mice bearing GBM tumors with defective interferon pathway were more responsive to oMV treatment, producing 387-fold higher infectious progeny virions compared to mice bearing GBM with intact interferon pathway. Moreover, gene expression analysis of tumor samples from GBM patients treated with oMV (NCT00390299) showed an inverse correlation of ISGs expression and viral replication [123].

Taken together, evidence suggests that IFITM1 expression may be a biomarker for resistance to oMV virotherapy for GBM patients and could help stratify the patients in oMV trials. Further research needs to be performed in other tumor types to investigate if this is a pan-cancer biomarker for oMV response.

4.5. Newcastle Disease Virus (NDV)

NDV is an avian paramyxovirus with a negative-sense, single-stranded RNA enclosed in its viral envelope. It was thought that some NDV strains have oncolytic properties by taking advantage of the inability of the tumor cells to elicit an anti-viral response due to deficiencies in IFN pathway [14]. Therefore, the search for markers of resistance to NDV-mediated oncolysis have focused on the antiviral pathways.

Krishnamurthy et al. showed that NDV susceptibility was linked to impairment of the type I interferon pathway [126]. Fibrosarcoma cells that were susceptible to NDV infection were unable to induce IFN-β production [126]. Specifically, the STAT1 and STAT2 phosphorylation was significantly reduced in the permissive tumor cells, resulting in reduced expression of ISG mRNAs and IFN-β [126]. In an approach to overcome resistance, Zamarin et al. engineered an NDV variant expressing an IFNα-antagonist, which demonstrated enhanced oncolytic activity in melanoma cell lines compared to the NDV strain without the IFN-antagonist [127].

In contrast with these findings, Mansour et al. revealed that the human non-small-cell lung cancer cell line A549 was susceptible to NDV oncolysis, despite the production of high levels of type I IFN response. It was proposed that the restriction mechanism in NDV oncolysis was based on the expression level of the anti-apoptotic protein Bcl-xl, where over-expression of Bcl-xl correlates with increased sensitivity to NDV [128]. Moreover, it was recently demonstrated that STAT3 inhibition suppresses immunogenic cell death (ICD) by NDV in melanoma tumor cells [129]. Interestingly, Bcl-xl is one of the target genes of active STAT3 [129].

Another potential marker for NDV cytotoxicity is the status of the autophagy pathway. Meng et al. showed in U251 glioma cells that NDV exploits the autophagic machinery to increase its replication. In this study, inhibition of BECLIN-1 or ATG5 gene, which are critical for the autophagosome formation, led to reduced production of NDV [130]. A more recent study showed autophagy modulators act as sensitizers for NDV in drug-resistant lung cancers [131]. On a different note, Puhlmann et al. showed that the oncogenic protein, Rac 1, is essential for NDV replication and it could confer sensitivity to viral replication [132]. Rac1 is a Rho GTPase protein and is involved in processes like cell proliferation and cytoskeleton organization [133]. In GBM, it has been shown that Rac1 is important for maintaining the stemness of GSCs, which may suggest that NDV could potentially be utilized to target the treatment-resistant cancer stem cell clones within GBM [134].

4.6. Mammalian Orthoreovirus (Reovirus)

Wild type reovirus has been tested clinically for various types of cancer including GBM, both by local and systemic administration [135,136]. Reovirus is a non-enveloped virus with 10 segments of dsRNA enclosed in its capsid. The main receptor that wild type reoviruses use to enter the cells is the junctional adhesion molecule A (JAM-A) [15]. Van den Hengel et al. tested a panel of primary GBM cell cultures, showing that they exhibit a large intertumoral variability in JAM-A expression, suggesting that reovirus efficacy may be hampered in low JAM-A expressing tumors [56]. Recently a second receptor was shown to mediate reovirus infection in the central nervous system (CNS), the Nogo receptor NgR1 [137]. The NgR1 expression in GBM cell lines has been established in various studies, however in a recent study it was shown that the NgR maturation, and thus expression to the cell membrane, is inhibited by transforming growth factor-β (TGF-β) 1, which is highly expressed by GBM cells [138–140]. The cell surface expression of NgR1 in primary GBM cells is yet to be elucidated.

Reovirus' natural tropism to replicate and kill tumor cells makes it an ideal candidate for oncolytic virotherapy. Early on, researchers attempted to elucidate the mechanism of tumor selectivity. It was found that the constitutively active Ras signaling pathway potentiates the reovirus replication via inhibition of PKR [141]. Additional studies have supported that Ras pathway activating mutations enhance reovirus replication [142,143]. Contrary to these findings, Twigger et al. showed that reovirus oncolysis does not depend on the status of EGFR/Ras/MAPK pathway in squamous cell carcinoma primary cell lines [144]. Another factor involved in host cell sensitivity to reovirus is determined by the levels of inhibitors of proteases, such as cathepsin B, that are required to disassemble the virus in the cytoplasm. Inhibition of these proteases restricts disassembly and inhibits viral replication [145].

In a clinical phase I trial for high grade gliomas and brain metastasis, the safety of reovirus after intravenous injection prior to brain surgery was demonstrated. The reovirus capsid σ3 protein and the virus RNA detection in tumors varied between the nine trial patients and correlated with the high proliferation index of the tumor cells (Ki67 expression) [136]. The efficacy of reovirus was also evaluated in a clinical trial for non-small cell lung cancer (NCT01708993), in which reovirus (Reolysin) was injected intravenously. A post-hoc analysis of the data obtained from this clinical trial showed a favorable trend for patients with p53 and EGFR mutations [146]. Another clinical trial (NCT01199263) testing oncolytic reovirus in ovarian cancer patients did not show any clinical benefit and authors stated that one explanation could be that only 20% of the ovarian cancer patients harbor Ras mutations [147].

4.7. Oncolytic Poliovirus

Poliovirus is an enterovirus in the Picornaviridae family of single-stranded RNA vi-ruses. A non-pathogenic oncolytic variant was engineered by replacing the internal ribo-somal entry site (IRES) of the poliovirus type 1 (Sabin) vaccine strain with the human rhi-novirus type 2 IRES (PVS-RIPO). The affinity of poliovirus for its cellular receptor CD155, which is upregulated on GBM cells, provides a unique opportunity target these tumors [16]. This approach demonstrated potent anti-glioma activity in mouse models and led to translation to clinical investigation for GBM [17,148]. Impressive results were obtained in a clinical phase II trial testing the recombinant nonpathogenic polio–rhinovirus chimera (PVSRIPO), oncolytic virus in 61 patients with recurrent GBM. Twenty-one percent of patients were still alive 36 months after initiation of treatment [149]. The source of this striking response in a subgroup of patients has yet to be elucidated. However, intriguingly, transcriptomic analysis revealed a correlation between low tumor mutational burden, tumor-intrinsic inflammation, and improved survival after PVSRIPO or anti-PD1 immunotherapy in recurrent GBM patients [150]. Further studies are required to confirm whether these characteristics hold predictive biomarker potential.

4.8. Oncolytic Parvovirus H-1

Oncolytic parvovirus H-1 is a single-stranded DNA rodent protoparvovirus 1 virus [18]. The lack of pre-existing immunity in humans makes parvovirus an interesting oncolytic virus to explore in the clinic [151]. Oncolytic H-1 parvovirus was administered intratumorally and intravenously in recurrent GBM patients in a phase I/IIa trial [152]. Robust immune response in terms of high infiltration of cytotoxic T cells was observed in the oncolytic parvovirus patients in this first study. Furthermore, in peripheral blood of the patients, specific T cell responses against glioma and viral antigens were detected [152]. Unlike for many other oncolytic viruses, antiviral type I interferon responses are not evoked by parvovirus infection [152]. Interestingly, parvovirus infection of GBM cells is followed by cathepsin B upregulation [152]. Inhibition of cathepsin B protects the glioma cells from parvovirus oncolysis, highlighting cathepsin B's importance in the parvovirus infection [153].

4.9. Oncolytic Vaccinia Virus

Vaccinia virus is a double-stranded DNA virus that belongs to poxvirus family [154]. Its natural tropism to enter the CNS has made it an attractive candidate for systemic delivery in GBM patients [155]. A phase I/II clinical study for GBM patients is currently ongoing (NCT03294486), in which patients are treated systemically with the vaccinia virus TG6002 [156]. Different strains of oncolytic vaccinia virus have been applied for other types of cancer as well and researchers tried to identify predictive response markers [157]. Zloza et al. identified the inhibitory molecule immunoglobulin-like transcript 2 (ILT2) on the cell surface of T cells, as a potential biomarker for vaccinia virus immunotherapy in melanoma patients. Increased expression of ILT2 on T cells was associated with poor response to oncolytic virotherapy using vaccinia virus [157]. Another study attempted to identify biomarkers associated with resistance to vaccinia virus therapy in hematological malignancies [158]. Genes involved in the ubiquitination pathway, DNA damage response and antigen presentation, among others, were identified and associated with resistance to vaccinia virus-induced oncolysis [158].

5. Development of Personalized in Vitro Models for OV Selection

It can be concluded that the interplay between the OVs and tumor cells is very complex and that defining a single biomarker or set of biomarkers predicting efficacy of a specific OV may not be achievable. An alternative approach is the development of patient-specific assays to screen a set of OVs and identify the optimal OV for a particular patient. Such a predictive assay would need to provide information on the efficacy of OV infection, replication, and oncolysis as well as on the immune response that is mounted by the OVs.

At present, in vitro culture models from patient-derived tumors have become the gold standard in drug development research for GBM. Our group has developed a preclinical screening system based on patient-derived low-passage cell cultures under serum-free conditions for preserving the molecular genetic make-up of the parental tumors [159,160]. Such screening activities have for example led to the identification of viral sensitizers which enhance the oncolytic activity of delta24-RGD in GBM cells [161]. The screening system was also applied to assess the efficiency of infection, replication and cell killing by four different OVs on a panel of primary glioma cell lines, which revealed tremendous intertumoral heterogeneity in viral sensitivities [56] (and unpublished data). Such panels of molecularly characterized cell cultures may also help identify new markers of sensitivity or resistance to tested OVs.

To gain insight in the relationship between oncolytic efficacy and immune stimulation, a co-culture model of glioma cells and (autologous) immune cell populations, could potentially provide useful insight on the immune response that is triggered after treatment with different OV candidates, as well as on the relationship between infectivity, oncolysis and immune activation. For instance, a co-culture of macrophages and delta24-RGD-infected (and permissive) GSCs revealed a shift of the tumor-supportive macrophages M2 to the pro-inflammatory M1 [108]. Such approaches are also being taken for other forms of cancer. A platform has been established for cancers like colorectal and non-small cell lung in which tumor organoids were co-cultured with autologous T cells derived from the peripheral blood of the patient [162]. With such model systems, the T cell-mediated cell killing could be evaluated for individual patients after infection with different OVs [162].

However, the establishment of such primary GSC or organoid-immune cell co-cultures is time-consuming and may not yield a robust OV therapy recommendation within the required timeframe. Furthermore, with serial passaging of primary glioma cells, the diverse clones that characterize the GBM tumor cannot be maintained [59]. The establishment of ex vivo 3D tumor model systems directly from fresh tumor tissue may therefore offer a more attractive approach for performing OV screens on the heterogeneous landscape of GBM as such models still retain architecture and cellular composition of the original tumor, including the presence of immune cell infiltrates. We previously reported that fresh tissue derived organotypic multicellular spheroids (OMS) offer a versatile system for studying OV infection, replication and tissue penetration [163]. Similarly, fresh GBM tumor slices have been employed to assess oncolytic myxoma virus efficacy [164]. Other approaches being developed include the culture of fresh tumor cells in slices, on matrices and in microfluidic systems [165–167]. Culturing fresh tissue also has limitations, since the culturing methods generally favor the tumor cells and not the immune cells [168]. Identification of culture conditions supporting all cell populations would also offer an improvement to these models.

Efforts in the field to generate 3D models from fresh tissue under culture conditions that support and recapitulate the unique immune tumor microenvironment are expected to facilitate investigations into both the dynamics of viral infection and replication in tumor cells as well as the effects thereof on local immune responses [169]. Such systems may in the future offer a tool to screen multiple OVs for a specific patient and select the optimal viral treatment within a clinically-relevant timeframe.

6. Conclusions

With the translation of oncolytic virotherapy to clinical trials for GBM patients, impressive responses have been documented in small subgroups of patients. To increase these response rates, better understanding of factors affecting viral replication, oncolysis and subsequent immune activation is required for each of the OVs under development. Studies in clinically-relevant in vitro and in vivo models as well as trial-associated immune and tumor monitoring studies are crucial for defining these factors and are expected to offer leads for stratification of patients in future OV trials.

Based on current literature reviews, we have identified factors related to the sensitivity of GBM tumors for specific OVs including the expression of the viral entry molecules, activation state of (cell cycle) signaling or autophagy pathways and the induction of specific antiviral signaling pathways that clear the virus infection (Table 2). These individual markers of sensitivity or resistance may together yield predictive profiles. However, further investigations are needed to shed light on the interplay between oncolytic activity on the tumor cells and the immune system. Ultimately, the complexity of these interactions in a background of a heterogeneous tumor and interpatient immune status variations, may require development of personalized ex vivo models to aid in identifying the most promising OV for a specific patient. The convergence of these developments toward applicable tools will enable classification of each GBM patient as sensitive or resistant to specific OVs. Ideally, future clinical trial design will incorporate more than one OV in parallel arms, such that patients can be stratified to the OV that best matches their tumor properties and/or immune status (Figure 2). Such a selection and stratification approach is expected to significantly improve response rates in OV trials for GBM patients.

Table 2. Potential predictive markers for sensitivity or resistance to OVs.

OV	Potential Predictive Marker	Effect	Tumor Type/ Cell Line
T-VEC	STING	Resistance	Melanoma cell lines [104]
Adenovirus (Ad5/3-Δ24)	MxA	Resistance	Ovarian carcinoma cell line [110]
	High HMGB1	Resistance	Patients with advanced metastatic solid tumors [112]
(Δ24-RGD)	Cyclin D1	Sensitivity	Pancreatic cell line [113]
Vocimagene amiretrorepvec	IDH1, PI3K3CA, EGFR, SYNE1 Neo-epitopes	Sensitivity	Patients with recurrent high-grade gliomas [118]
	NK cells and M0 macrophage tumor infiltration	Resistance	Patients with recurrent high-grade gliomas [118]
MV	IFITM1	Resistance	Transformed mesenchymal stromal cells [124]
	RIG-I	Resistance	Sarcoma cells [125]
	ISG15	Resistance	Primary GBM cells [123]
NDV	Bcl-xl	Sensitivity	A549 cell line [128]
	Rac-1	Sensitivity	HaCaT A5-RT3 [132]
	STAT3	Sensitivity	Melanoma cell lines [129]
Reovirus (R124)	JAM-A	Sensitivity	Primary GBM cells [56]
	Cathepsin B	Sensitivity	Glioma cell line [144]
	P53 and EGFR mutations	Sensitivity	Patients with non-small cell lung cancer [145]
	Ki-67	Sensitivity	Patients with high grade gliomas and metastatic brain tumors [136]
Oncolytic parvovirus H-1	Cathepsin B	Sensitivity	Glioma cell line [18,151]
Vaccinia virus	Expression of ILT2 on T cells	Resistance	Melanoma patients [155]
	LEF1, STAMBPL1, and SLFN11	Sensitivity	myeloid and lymphoid leukemia cell lines [156]
	PVRIG, LPP, CECR1, Arhgef6, IRX3, IGFBP2, and CD1d	Resistance	myeloid and lymphoid leukemia cell lines [156]

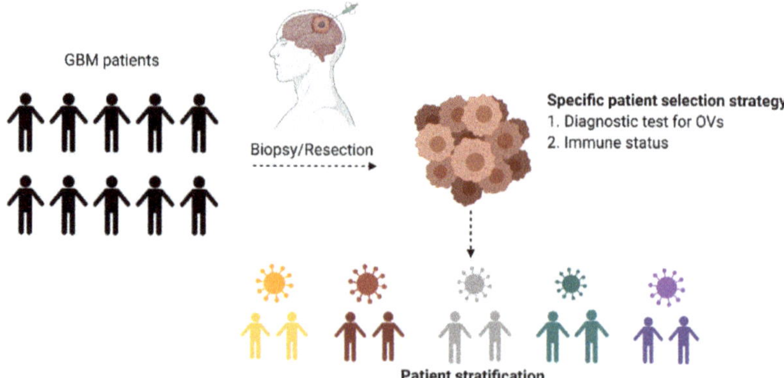

Figure 2. Personalized oncolytic virotherapy. In an envisioned personalized OV trial, resected tumor material from each patient would be used for diagnostic tests to identify the most potent OV, in terms of oncolysis for the particular tumor. Additional tests to determine local and/or systemic immune status after OV infection could further aid in stratifying the patients based on their unique tumor microenvironment and immune status. Created with BioRender.com.

Funding: The authors gratefully acknowledge the support from Foundation Overleven met Alvleesklierkanker and Foundation Support Casper.

Conflicts of Interest: No conflict of interest to disclose.

References

1. Zhang, S.; Rabkin, S.D. The discovery and development of oncolytic viruses: Are they the future of cancer immunotherapy? *Expert Opin. Drug Discov.* **2020**. [CrossRef]
2. Harrington, K.; Freeman, D.J.; Kelly, B.; Harper, J.; Soria, J.C. Optimizing oncolytic virotherapy in cancer treatment. *Nat. Rev. Drug Discov.* **2019**, *18*, 689–706. [CrossRef] [PubMed]
3. Kaufman, H.L.; Kohlhapp, F.J.; Zloza, A. Oncolytic viruses: A new class of immunotherapy drugs. *Nat. Rev. Drug Discov.* **2015**, *14*, 642–662. [CrossRef] [PubMed]
4. Taylor, O.G.; Brzozowski, J.S.; Skelding, K.A. Glioblastoma multiforme: An overview of emerging therapeutic targets. *Front. Oncol.* **2019**, *9*, 963. [CrossRef]
5. Stupp, R.; Mason, W.P.; van den Bent, M.J.; Weller, M.; Fisher, B.; Taphoorn, M.J.; Belanger, K.; Brandes, A.A.; Marosi, C.; Bogdahn, U.; et al. Radiotherapy plus concomitant and adjuvant temozolomide for glioblastoma. *N. Engl. J. Med.* **2005**. [CrossRef]
6. De Witt Hamer, P.C.; Ho, V.K.Y.; Zwinderman, A.H.; Zwinderman, A.H.; Ackermans, L.; Ardon, H.; Boomstra, S.; Bouwknegt, W.; van den Brink, W.A.; Dirven, C.M.; et al. Between-hospital variation in mortality and survival after glioblastoma surgery in the Dutch Quality Registry for Neuro Surgery. *J. Neurooncol.* **2019**, *144*, 313–323. [CrossRef] [PubMed]
7. Papanastassiou, V.; Rampling, R.; Fraser, M.; Petty, R.; Hadley, D.; Nicoll, J.; Harland, J.; Mabbs, R.; Brown, M. The potential for efficacy of the modified (ICP 34.5) herpes simplex virus HSV1716 following intratumoural injection into human malignant glioma: A proof of principle study. *Gene Ther.* **2002**, *9*, 398–406. [CrossRef] [PubMed]
8. Markert, J.M.; Liechty, P.G.; Wang, W.; Gaston, S.; Braz, E.; Karrasch, M.; Nabors, L.B.; Markiewicz, M.; Lakeman, A.D.; Palmer, C.A.; et al. Phase Ib trial of mutant herpes simplex virus G207 inoculated pre-and post-tumor resection for recurrent GBM. *Mol. Ther.* **2009**, *17*, 199–207. [CrossRef] [PubMed]
9. Ning, J.; Wakimoto, H. Oncolytic herpes simplex virus-based strategies: Toward a breakthrough in glioblastoma therapy. *Front. Microbiol.* **2014**, *5*, 303. [CrossRef]
10. Kambara, H.; Okano, H.; Chiocca, E.A.; Saeki, Y. An Oncolytic HSV-1 Mutant Expressing ICP34. 5 under Control of a Nestin Promoter Increases Survival of Animals even when Symptomatic from a Brain Tumor. *Cancer Res.* **2005**, *67*, 8980–8984.
11. Cheng, P.H.; Wechman, S.L.; McMasters, K.M.; Zhou, H.S. Oncolytic replication of E1b-deleted adenoviruses. *Viruses* **2015**, *7*, 5767–5779. [CrossRef] [PubMed]
12. Suzuki, K.; Fueyo, J.; Krasnykh, V.; Reynolds, P.N.; Curiel, D.T.; Alemany, R. A conditionally replicative adenovirus with enhanced infectivity shows improved oncolytic potency. *Clin. Cancer Res.* **2001**, *7*, 120–126. [PubMed]
13. Allen, C.; Opyrchal, M.; Aderca, I.; Schroeder, M.A.; Sarkaria, J.N.; Domingo, E.; Federspiel, M.J.; Galanis, E. Oncolytic measles virus strains have significant antitumor activity against glioma stem cells. *Gene Ther.* **2013**, *20*, 444–449. [CrossRef] [PubMed]
14. Schirrmacher, V.; van Gool, S.; Stuecker, W. Breaking therapy resistance: An update on oncolytic newcastle disease virus for improvements of cancer therapy. *Biomedicines* **2019**, *7*, 66. [CrossRef] [PubMed]
15. Gong, J.; Mita, M.M. Activated Ras signaling pathways and reovirus oncolysis: An update on the mechanism of preferential reovirus replication in cancer cells. *Front. Oncol.* **2014**, *4*, 167. [CrossRef]
16. Gromeier, M.; Lachmann, S.; Rosenfeld, M.R.; Gutin, P.H.; Wimmer, E. Intergeneric poliovirus recombinants for the treatment of malignant glioma. *Proc. Natl. Acad. Sci. USA.* **2000**. [CrossRef] [PubMed]
17. Dobrikova, E.Y.; Broadt, T.; Poiley-Nelson, J.; Yang, X.; Soman, G.; Giardina, S.; Harris, R.; Gromeier, M. Recombinant oncolytic poliovirus eliminates glioma in vivo without genetic adaptation to a pathogenic phenotype. *Mol. Ther.* **2008**. [CrossRef]
18. Nüesch, J.P.F.; Lacroix, J.; Marchini, A.; Rommelaere, J. Molecular pathways: Rodent parvoviruses—Mechanisms of oncolysis and prospects for clinical cancer treatment. *Clin. Cancer Res.* **2012**, *18*, 3516–3523. [CrossRef]
19. Chiocca, E.A.; Nassiri, F.; Wang, J.; Peruzzi, P.; Zadeh, G. Viral and other therapies for recurrent glioblastoma: Is a 24-month durable response unusual? *Neuro-Oncology* **2019**, *21*, 14–25. [CrossRef]
20. Lang, F.F.; Conrad, C.; Gomez-Manzano, C.; Alfred Yung, W.K.; Sawaya, R.; Weinberg, J.S.; Prabhu, S.S.; Rao, G.; Fuller, G.N.; Aldape, K.D.; et al. Phase I study of DNX-2401 (delta-24-RGD) oncolytic adenovirus: Replication and immunotherapeutic effects in recurrent malignant glioma. *J. Clin. Oncol.* **2018**, *36*, 1419–1427. [CrossRef]
21. Cloughesy, T.F.; Landolfi, J.; Vogelbaum, M.A.; Ostertag, D.; Elder, J.B.; Bloomfield, S.; Carter, B.; Chen, C.C.; Kalkanis, S.N.; Kesari, S.; et al. Durable complete responses in some recurrent high-grade glioma patients treated with Toca 511 + Toca FC. *Neuro-Oncology* **2018**, *20*, 1383–1392. [CrossRef]
22. Senzer, N.N.; Kaufman, H.L.; Amatruda, T.; Nemunaitis, M.; Reid, T.; Daniels, G.; Gonzalez, R.; Glaspy, J.; Whitman, E.; Harrington, K.; et al. Phase II Clinical Trial of a Granulocyte-Macrophage Colony-Stimulating Factor—Encoding, Second-Generation Oncolytic Herpesvirus in Patients With Unresectable Metastatic Melanoma. *J. Clin. Oncol.* **2009**, *27*, 5763–5767. [CrossRef] [PubMed]
23. Crespo, I.; Vital, A.L.; Gonzalez-Tablas, M.; Patino, M.D.C.; Otero, A.; Lopes, M.C.; De Oliveira, C.; Domingues, P.; Orfao, A.; Tabernero, M.D. Molecular and Genomic Alterations in Glioblastoma Multiforme. *Am. J. Pathol.* **2015**, *185*, 1820–1833. [CrossRef] [PubMed]

24. Wang, Q.; Hu, X.; Muller, F.; Kim, H.; Squatrito, M.; Mikkelsen, T.; Scarpace, L.; Barthel, F.; Lin, Y.-H.; Satani, N.; et al. Tumor Evolution of Glioma Intrinsic Gene Expression Subtype Associates With Immunological Changes in the Microenvironment. *Neuro-Oncology* **2016**, *18*, vi202. [CrossRef]

25. Gill, B.J.; Pisapia, D.J.; Malone, H.R.; Goldstein, H.; Lei, L.; Sonabend, A.; Yun, J.; Samanamud, J.; Sims, J.S.; Banu, M.; et al. MRI-localized biopsies reveal subtype-specific differences in molecular and cellular composition at the margins of glioblastoma. *Proc. Natl. Acad. Sci. USA* **2014**, *111*, 12550–12555. [CrossRef]

26. Verhaak, R.G.; Hoadley, K.A.; Purdom, E.; Wang, V.; Qi, Y.; Wilkerson, M.D.; Miller, C.R.; Ding, L.; Golub, T.; Mesirov, J.P.; et al. Integrated Genomic Analysis Identifies Clinically Relevant Subtypes of Glioblastoma Characterized by Abnormalities in PDGFRA, IDH1, EGFR, and NF1. *Cancer Cell.* **2010**, *17*, 98–110. [CrossRef] [PubMed]

27. Doucette, T.; Rao, G.; Rao, A.; Shen, L.; Aldape, K.; Wei, J.; Dziurzynski, K.; Gilbert, M.; Heimberger, A.B. Immune heterogeneity of glioblastoma subtypes: Extrapolation from the cancer genome atlas. *Cancer Immunol. Res.* **2013**. [CrossRef]

28. Weller, M.; Cloughesy, T.; Perry, J.R.; Wick, W. Standards of care for treatment of recurrent glioblastoma-are we there yet? *Neuro-Oncology* **2013**, *15*, 4–27. [CrossRef]

29. Weller, M.; van den Bent, M.; Preusser, M.; Le Rhun, E.; Tonn, J.C.; Minniti, G.; Bendszus, M.; Balana, C.; Chinot, O.; Dirven, L.; et al. EANO guidelines on the diagnosis and treatment of diffuse gliomas of adulthood. *Nat. Rev. Clin. Oncol.* **2020**. [CrossRef] [PubMed]

30. Patel, A.P.; Tirosh, I.; Trombetta, J.J.; Shalek, A.K.; Gillespie, S.M.; Wakimoto, H.; Cahill, D.P.; Nahed, B.V.; Curry, W.T.; Martuza, R.L.; et al. Single-cell RNA-seq highlights intratumoral heterogeneity in primary glioblastoma. *Science* **2014**, *344*, 1396–1402. [CrossRef]

31. Sottoriva, A.; Spiteri, I.; Piccirillo, S.G.M.; Touloumis, A.; Collins, V.P.; Marioni, J.C.; Curtis, C.; Watts, C.; Tavaré, S. Intratumor heterogeneity in human glioblastoma reflects cancer evolutionary dynamics. *Proc. Natl. Acad. Sci. USA* **2013**, *110*, 4009–4014. [CrossRef]

32. Schäfer, N.; Gielen, G.H.; Rauschenbach, L.; Kebir, S.; Till, A.; Reinartz, R.; Simon, M.; Niehusmann, P.; Kleinschnitz, C.; Herrlinger, U.; et al. Longitudinal heterogeneity in glioblastoma: Moving targets in recurrent versus primary tumors. *J. Transl. Med.* **2019**, *17*, 1–9. [CrossRef]

33. Lee, A.; Arasaratnam, M.; Chan, D.L.H.; Khasraw, M.; Howell, V.M.; Wheeler, H. Anti-epidermal growth factor receptor therapy for glioblastoma in adults. *Cochrane Database Syst. Rev.* **2020**. [CrossRef]

34. Cheng, L.; Huang, Z.; Zhou, W.; Wu, Q.; Donnola, S.; James, K.; Fang, X.; Sloan, A.E.; Mao, Y.; Lathia, J.D.; et al. Glioblastoma Stem Cells Generate Vascular Pericytes to Support Vessel Function and Tumor Growth. *Cell* **2013**, *153*, 139–152. [CrossRef]

35. Darmanis, S.; Sloan, S.A.; Croote, D.; Mignardi, M.; Chernikova, S.; Samghababi, P.; Zhang, Y.; Neff, N.; Kowarsky, M.; Caneda, C.; et al. Single-Cell RNA-Seq Analysis of Infiltrating Neoplastic Cells at the Migrating Front. of Human Glioblastoma. *Cell Rep.* **2017**, *21*, 1399–1410. [CrossRef] [PubMed]

36. Alcantara Llaguno, S.R.; Parada, L.F. Cell of origin of glioma: Biological and clinical implications. *Br. J. Cancer* **2016**. [CrossRef] [PubMed]

37. Engelhardt, B.; Carare, R.O.; Bechmann, I.; Flügel, A.; Laman, J.D.; Weller, R.O. Vascular, glial, and lymphatic immune gateways of the central nervous system. *Acta Neuropathol.* **2016**. [CrossRef] [PubMed]

38. Ransohoff, R.M.; Engelhardt, B. The anatomical and cellular basis of immune surveillance in the central nervous system. *Nat. Publ. Gr.* **2012**, *12*. [CrossRef] [PubMed]

39. Ludewig, P.; Gallizioli, M.; Urra, X.; Behr, S.; Brait, V.H.; Gelderblom, M.; Magnus, T.; Planas, A.M. Biochimica et Biophysica Acta Dendritic cells in brain diseases. *BBA Mol. Basis Dis.* **2016**, *1862*, 352–367. [CrossRef]

40. Miller, S.D.; Mahon, E.J.M.C.; Schreiner, B.; Bailey, S.L. Antigen Presentation in the CNS by Myeloid Dendritic Cells Drives Progression of Relapsing Experimental Autoimmune Encephalomyelitis. *Ann. N. Y. Acad. Sci.* **2007**, *191*, 179–191. [CrossRef]

41. Louveau, A.; Smirnov, I.; Keyes, T.J.; Eccles, J.D.; Rouhani, S.J.; Peske, J.D.; Derecki, N.C.; Castle, D.; Mandell, J.W.; Lee, K.S.; et al. Structural and functional features of central nervous system lymphatic vessels. *Nature* **2015**, *523*, 337–341. [CrossRef] [PubMed]

42. Sevenich, L. Brain-Resident Microglia and Blood-Borne Macrophages Orchestrate Central Nervous System Inflammation in Neurodegenerative Disorders and Brain Cancer. *Front. Immunol.* **2018**, *9*, 697. [CrossRef] [PubMed]

43. Hambardzumyan, D.; Gutmann, D.H.; Kettenmann, H. The role of microglia and macrophages in glioma maintenance and progression. *Nat. Neurosci.* **2016**, *19*, 20–27. [CrossRef]

44. Takenaka, M.C.; Gabriely, G.; Rothhammer, V.; Mascanfroni, I.D.; Wheeler, M.A.; Chao, C.; Gutiérrez-vázquez, C.; Kenison, J.; Tjon, E.C.; Barroso, A.; et al. Control of tumor-associated macrophages and T cells in glioblastoma via AHR and CD39. *Nat. Neurosci.* **2019**, *22*, 729–740. [CrossRef]

45. Zhou, W.; Ke, S.Q.; Huang, Z.; Flavahan, W.; Fang, X.; Paul, J.; Wu, L.; Sloan, A.E.; Mclendon, R.E.; Li, X.; et al. Periostin secreted by glioblastoma stem cells recruits M2 tumour-associated macrophages and promotes malignant growth. *Nat. Cell Biol.* **2015**, *17*, 170–182. [CrossRef] [PubMed]

46. Mantovani, A.; Marchesi, F.; Malesci, A.; Laghi, L.; Allavena, P. Tumour-associated macrophages as treatment targets in oncology. *Nat. Rev. Clin. Oncol.* **2017**, *14*, 399–416. [CrossRef] [PubMed]

47. Shi, Y.; Ping, Y.; Zhou, W.; He, Z.; Chen, C.; Bian, B.; Zhang, L.; Chen, L.; Lan, X.; Zhang, X.; et al. Tumour-associated macrophages secrete pleiotrophin to promote PTPRZ1 signalling in glioblastoma stem cells for tumour growth. *Nat. Commun.* **2017**, *8*, 1–17. [CrossRef]

48. Sørensen, M.D. Tumour-associated microglia/macrophages predict poor prognosis in high-grade gliomas and correlate with an aggressive tumour subtype. *Neuropathol. Appl. Neurobiol.* **2018**, *44*, 185–206. [CrossRef] [PubMed]

49. Lu-emerson, C.; Snuderl, M.; Kirkpatrick, N.D.; Goveia, J.; Davidson, C.; Huang, Y.; Riedemann, L.; Taylor, J.; Ivy, P.; Duda, G.; et al. Increase in tumor-associated macrophages after antiangiogenic therapy is associated with poor survival among patients with recurrent glioblastoma. *Neuro-Oncology* **2013**, *15*, 1079–1087. [CrossRef]

50. Woroniecka, K.; Chongsathidkiet, P.; Rhodin, K.; Kemeny, H.; Dechant, C.; Farber, S.H.; Elsamadicy, A.A.; Cui, X.; Koyama, S.; Jackson, C.; et al. T-Cell Exhaustion Signatures Vary with Tumor Type and Are Severe in Glioblastoma. *Clin. Cancer Res.* **2018**, *44*, 4175–4187. [CrossRef] [PubMed]

51. Heimberger, A.B.; Abou-ghazal, M.; Reina-ortiz, C.; Yang, D.S.; Sun, W.; Qiao, W.; Hiraoka, N.; Fuller, G.N. Incidence and Prognostic Impact of FoxP3 + Regulatory T Cells in Human Gliomas. *Clin. Cancer Res.* **2008**, *14*, 5166–5173. [CrossRef] [PubMed]

52. Martinez-lage, M.; Lynch, T.M.; Bi, Y.; Cocito, C.; Way, G.P.; Pal, S.; Haller, J.; Yan, R.E.; Ziober, A.; Nguyen, A.; et al. Immune landscapes associated with different glioblastoma molecular subtypes. *Acta Neuropathol. Commun.* **2019**, *6*, 1–12. [CrossRef]

53. Büttner, R.; Longshore, J.W.; López-Ríos, F.; Vandergrift, W.A., III.; Patel, S.J.; Cachia, D.; Bartee, E. Implementing TMB measurement in clinical practice: Considerations on assay requirements. *ESMO Open* **2019**, *4*, e00044. [CrossRef]

54. Alexandrov, L.B.; Nik-Zainal, S.; Wedge, D.C.; Aparicio, S.A.; Behjati, S.; Biankin, A.V.; Bignell, G.R.; Bolli, N.; Borg, A.; Børresen-Dale, A.L.; et al. Signatures of mutational processes in human cancer. *Nature* **2013**, *500*, 415–421. [CrossRef]

55. Lamfers, M.L.M.; Idema, S.; Bosscher, L.; Heukelom, S.; Moeniralm, S.; Van Der Meulen-Muileman, I.H.; Overmeer, R.M.; Van Der Valk, P.; Van Beusechem, V.W.; Gerritsen, W.R. Differential effects of combined Ad5-Δ24RGD and radiation therapy in in vitro versus in vivo models of malignant glioma. *Clin. Cancer Res.* **2007**, *13*, 7451–7458. [CrossRef] [PubMed]

56. Van Den Hengel, S.K.; Balvers, R.K.; Dautzenberg, I.J.C.; Van Den Wollenberg, D.J.M.; Kloezeman, J.J.; Lamfers, M.L.; Sillivis-Smit, P.A.E.; Hoeben, R.C. Heterogeneous reovirus susceptibility in human glioblastoma stem-like cell cultures. *Cancer Gene Ther.* **2013**, *20*, 507–513. [CrossRef] [PubMed]

57. Grill, J.; Van Beusechem, V.W.; Van Der Valk, P.; Dirven, C.M.F.; Leonhart, A.; Pherai, D.S.; Haisma, H.J.; Pinedo, H.M.; Curiel, D.T.; Gerritsen, W.R. Combined Targeting of Adenoviruses to Integrins and Epidermal Growth Factor Receptors Increases Gene Transfer into Primary Glioma Cells and Spheroids. *Clin. Cancer Res.* **2001**, *7*, 641–650.

58. Sonabend, A.M.; Ulasov, I.V.; Han, Y.; Lesniak, M.S. Oncolytic adenoviral therapy for glioblastoma multiforme. *Neurosurg. Focus* **2006**, *20*, E19. [CrossRef]

59. Balvers, R.K.; Gomez-Manzano, C.; Jiang, H.; Piya, S.; Klein, S.R.; Lamfers, M.L.M.; Dirven, C.M.F.; Fueyo, J. *Advances in Oncolytic Virotherapy for Brain Tumors*; Elsevier Inc.: Amsterdam, The Netherlands, 2013. [CrossRef]

60. Oh, E.; Hong, J.; Kwon, O.J.; Yun, C.O. A hypoxia- and telomerase-responsive oncolytic adenovirus expressing secretable trimeric TRAIL triggers tumour-specific apoptosis and promotes viral dispersion in TRAIL-resistant glioblastoma. *Sci. Rep.* **2018**, *8*, 1–13. [CrossRef] [PubMed]

61. Post, D.E.; Sandberg, E.M.; Kyle, M.M.; Devi, N.S.; Brat, D.J.; Xu, Z.; Tighiouart, M.; Van Meir, E.G. Targeted cancer gene therapy using a hypoxia inducible factor-dependent oncolytic adenovirus armed with interleukin-4. *Cancer Res.* **2007**, *67*, 6872–6881. [CrossRef]

62. Ulasov, I.V.; Rivera, A.A.; Sonabend, A.M.; Rivera, L.B.; Wang, M.; Zhu, Z.B.; Lesniak, M.S. Comparative evaluation of survivin, midkine and CXCR4 promoters for transcriptional targeting of glioma gene therapy. *Cancer Biol. Ther.* **2007**. [CrossRef] [PubMed]

63. Hoffmann, D.; Wildner, O. Efficient generation of double heterologous promoter controlled oncolytic adenovirus vectors by a single homologous recombination step in Escherichia coli. *BMC Biotechnol.* **2006**. [CrossRef] [PubMed]

64. Saito, N.; Hirai, N.; Aoki, K.; Sato, S.; Suzuki, R.; Hiramoto, Y.; Fujita, S.; Nakayama, H.; Hayashi, M.; Sakurai, T.; et al. Genetic and lineage classification of glioma-initiating cells identifies a clinically relevant glioblastoma model. *Cancers* **2019**, *11*, 1564. [CrossRef] [PubMed]

65. Hardison, S.E.; Brown, G.D. C-type Lectin Receptors Orchestrate Anti-Fungal Immunity. *Nat. Immunol.* **2012**, *13*, 817–822. [CrossRef] [PubMed]

66. Goubau, D.; Deddouche, S.; Reis e Sousa, C. Cytosolic Sensing of Viruses. *Immunity* **2013**, *38*, 855–869. [CrossRef] [PubMed]

67. Lee, H.C.; Chathuranga, K.; Lee, J.S. Intracellular sensing of viral genomes and viral evasion. *Exp. Mol. Med.* **2019**, *51*. [CrossRef]

68. Caneparo, V.; Landolfo, S.; Gariglio, M.; De Andrea, M. The absent in melanoma 2-like receptor IFN-inducible protein 16 as an inflammasome regulator in systemic lupus erythematosus: The dark side of sensing microbes. *Front. Immunol.* **2018**, *9*, 1–14. [CrossRef]

69. Lazear, H.M.; Nice, T.J.; Diamond, M.S. Interferon-λ: Immune Functions at Barrier Surfaces and Beyond. *Immunity* **2015**, *43*, 15–28. [CrossRef] [PubMed]

70. Levy, D.E.; Marié, I.J.; Durbin, J.E. Induction and function of type I and III interferon in response to viral infection. *Curr. Opin. Virol.* **2012**, *1*, 476–486. [CrossRef] [PubMed]

71. Hornung, V.; Hartmann, R.; Ablasser, A.; Hopfner, K.P. OAS proteins and cGAS: Unifying concepts in sensing and responding to cytosolic nucleic acids. *Nat. Rev. Immunol.* **2014**, *14*, 521–528. [CrossRef]

72. Unterholzner, L.; Keating, S.E.; Baran, M.; Horan, K.A.; Jensen, S.B.; Sharma, S.; Sirois, C.M.; Jin, T.; Latz, E.; Xiao, T.S.; et al. IFI16 is an innate immune sensor for intracellular DNA. *Nat. Immunol.* **2010**, *11*, 997–1004. [CrossRef] [PubMed]

73. Lazear, H.M.; Schoggins, J.W.; Diamond, M.S.; Hill, C.; Andrew, M.; Programs, I.; Louis, S. Shared and Distinct Functions of Type I and Type III Interferons. *Immunity* **2019**, *50*, 907–923. [CrossRef] [PubMed]

74.   Lionel BIvashkivand Laura, T. Donlin1 Regulation of type I interferon responses Lionel. *Bone* **2008**, *23*, 1–7. [CrossRef]

75.   Ma, F.; Li, B.; Liu, S.; Iyer, S.S.; Yu, Y.; Wu, A.; Cheng, G. Positive Feedback Regulation of Type I IFN Production by the IFN-Inducible DNA Sensor cGAS. *J. Immunol.* **2015**, *194*, 1545–1554. [CrossRef]

76.   Sadler, A.J.; Williams, B.R.G. Interferon-inducible antiviral effectors. *Nat. Rev. Immunol.* **2008**, *8*, 559–568. [CrossRef]

77.   Froechlich, G.; Caiazza, C.; Gentile, C.; D'alise, A.M.; De Lucia, M.; Langone, F.; Leoni, G.; Cotugno, G.; Scisciola, V.; Nicosia, A.; et al. Integrity of the antiviral STING-mediated DNA sensing in tumor cells is required to sustain the immunotherapeutic efficacy of herpes simplex oncolytic virus. *Cancers* **2020**, *12*, 3407. [CrossRef]

78.   Barber, G.N. STING: Infection, inflammation and cancer. *Nature* **2015**, *15*, 11–13. [CrossRef] [PubMed]

79.   Wang, Y.; Jiang, K.; Zhang, Q.; Meng, S.; Ding, C. Autophagy in negative-strand RNA virus infection. *Front. Microbiol.* **2018**, *9*, 206. [CrossRef] [PubMed]

80.   Dikic, I.; Elazar, Z. Mechanism and medical implications of mammalian autophagy. *Nat. Rev. Mol. Cell Biol.* **2018**, *19*, 349–364. [CrossRef] [PubMed]

81.   Mizushima, N.; Yoshimori, T.; Ohsumi, Y. The Role of Atg Proteins in Autophagosome Formation. *Annu. Rev. Cell Dev. Biol.* **2011**, *27*, 107–132. [CrossRef] [PubMed]

82.   Sharma, V.; Verma, S.; Seranova, E.; Sarkar, S.; Kumar, D. Selective autophagy and xenophagy in infection and disease. *Front. Cell Dev. Biol.* **2018**, *6*, 47. [CrossRef]

83.   Choi, Y.; Bowman, J.W.; Jung, J.U. Autophagy during viral infection—A double-edged sword. *Nat. Rev. Microbiol.* **2018**, *16*, 341–354. [CrossRef]

84.   Lee, H.K.; Lund, J.M.; Ramanathan, B.; Mizushima, N.; Iwasaki, A. Autophagy-dependent viral recognition by plasmacytoid dendritic cells. *Science* **2007**, *315*, 1398–1401. [CrossRef] [PubMed]

85.   Orvedahl, A.; MacPherson, S.; Sumpter, R.; Tallóczy, Z.; Zou, Z.; Levine, B. Autophagy Protects against Sindbis Virus Infection of the Central Nervous System. *Cell Host Microbe* **2010**, *7*, 115–127. [CrossRef] [PubMed]

86.   Moy, R.H.; Gold, B.; Molleston, J.M.; Schad, V.; Yanger, K.; Salzano, M.V.; Yagi, Y.; Fitzgerald, K.A.; Stanger, B.Z.; Soldan, S.S.; et al. Antiviral autophagy restricts rift valley fever virus infection and is conserved from flies to mammals. *Immunity* **2014**, *40*, 51–65. [CrossRef] [PubMed]

87.   Lussignol, M.; Esclatine, A. Herpesvirus and autophagy: "All right, everybody be cool, this is a robbery!". *Viruses* **2017**, *9*, 372. [CrossRef] [PubMed]

88.   Heaton, N.S.; Randall, G. Dengue virus-induced autophagy regulates lipid metabolism. *Cell Host Microbe.* **2010**, *8*, 422–432. [CrossRef] [PubMed]

89.   Cheng, P.H.; Lian, S.; Zhao, R.; Rao, X.M.; McMasters, K.M.; Zhou, H.S. Combination of autophagy inducer rapamycin and oncolytic adenovirus improves antitumor effect in cancer cells. *Virol. J.* **2013**, *10*, 293. [CrossRef] [PubMed]

90.   Comins, C.; Simpson, G.R.; Rogers, W.; Relph, K.; Harrington, K.; Melcher, A.; Roulstone, V.; Kyula, J.; Pandha, H. Synergistic antitumour effects of rapamycin and oncolytic reovirus. *Cancer Gene Ther.* **2018**, *25*, 148–160. [CrossRef]

91.   Fu, X.; Tao, L.; Rivera, A.; Zhang, X. Rapamycin enhances the activity of oncolytic herpes simplex virus against tumor cells that are resistant to virus replication. *Int. J. Cancer* **2011**, *129*, 1503–1510. [CrossRef] [PubMed]

92.   Stanford, M.M.; Barrett, J.W.; Nazarian, S.H.; Werden, S.; McFadden, G. Oncolytic Virotherapy Synergism with Signaling Inhibitors: Rapamycin Increases Myxoma Virus Tropism for Human Tumor Cells. *J. Virol.* **2007**, *81*, 1251–1260. [CrossRef] [PubMed]

93.   Goetz, C.; Dobrikova, E.; Shveygert, M.; Dobrikov, M.; Gromeier, M. Oncolytic poliovirus against malignant glioma. *Future Virol.* **2011**, *6*, 1045–1058. [CrossRef] [PubMed]

94.   Hu, L.; Jiang, K.; Ding, C.; Meng, S. Targeting autophagy for oncolytic immunotherapy. *Biomedicines* **2017**, *5*, 1–9. [CrossRef] [PubMed]

95.   Jiang, H.; White, E.J.; Rios-Vicil, C.I.; Xu, J.; Gomez-Manzano, C.; Fueyo, J. Human Adenovirus Type 5 Induces Cell Lysis through Autophagy and Autophagy-Triggered Caspase Activity. *J. Virol.* **2011**, *85*, 4720–4729. [CrossRef] [PubMed]

96.   Alonso, M.M.; Jiang, H.; Yokoyama, T.; Xu, J.; Bekele, N.B.; Lang, F.F.; Kondo, S.; Gomez-Manzano, C.; Fueyo, J. Delta-24-RGD in combination with RAD001 induces enhanced anti-glioma effect via autophagic cell death. *Mol. Ther.* **2008**, *16*, 487–493. [CrossRef] [PubMed]

97.   Tazawa, H.; Kagawa, S.; Fujiwara, T. Oncolytic adenovirus-induced autophagy: Tumor-suppressive efect and molecular basis. *Acta Med. Okayama* **2013**, *67*, 333–342. [CrossRef] [PubMed]

98.   Tazawa, H.; Kuroda, S.; Hasei, J.; Kagawa, S.; Fujiwara, T. Impact of autophagy in oncolytic adenoviral therapy for cancer. *Int. J. Mol. Sci.* **2017**, *18*, 1479. [CrossRef]

99.   Ma, W.; He, H.; Wang, H. Oncolytic herpes simplex virus and immunotherapy. *BMC Immunol.* **2018**, *19*, 1–11. [CrossRef]

100. Totsch, S.K.; Schlappi, C.; Kang, K.D.; Ishizuka, A.S.; Lynn, G.M.; Fox, B.; Beierle, E.A.; Whitley, R.J.; Markert, J.M.; Gillespie, G.Y.; et al. Oncolytic herpes simplex virus immunotherapy for brain tumors: Current pitfalls and emerging strategies to overcome therapeutic resistance. *Oncogene* **2019**, *38*, 6159–6171. [CrossRef]

101. Peters, C.; Paget, M.; Tshilenge, K.T.; Saha, D.; Antoszczyk, S.; Baars, A.; Frost, T.; Martuza, R.L.; Wakimoto, H.; Rabkin, S.D. Restriction of Replication of Oncolytic Herpes Simplex Virus with a Deletion of γ34.5 in Glioblastoma Stem-Like Cells. *J. Virol.* **2018**. [CrossRef]

102. Lalu, M.; Leung, G.J.; Dong, Y.Y.; Montroy, J.; Butler, C.; Auer, R.C.; Fergusson, D.A. Mapping the preclinical to clinical evidence and development trajectory of the oncolytic virus talimogene laherparepvec (T-VEC): A systematic review. *BMJ Open* **2019**, *9*, e029475. [CrossRef]

103. Harrington, K.J.; Andtbacka, R.H.I.; Collichio, F.; Downey, G.; Chen, L.; Szabo, Z.; Kaufman, L.K. Efficacy and safety of talimogene laherparepvec versus granulocyte-macrophage colony-stimulating factor in patients with stage IIIB/C and IVM1a melanoma: Subanalysis of the Phase III OPTiM trial. *Onco Targets Ther.* **2016**, *9*, 7081–7093. [CrossRef]

104. Bommareddy, P.K.; Zloza, A.; Rabkin, S.D.; Kaufman, H.L. Oncolytic virus immunotherapy induces immunogenic cell death and overcomes STING deficiency in melanoma. *Oncoimmunology* **2019**, *8*, e1591875. [CrossRef]

105. Rosewell Shaw, A.; Suzuki, M. Recent advances in oncolytic adenovirus therapies for cancer. *Curr. Opin. Virol.* **2016**, *21*, 9–15. [CrossRef] [PubMed]

106. Larson, C.; Oronsky, B.; Scicinski, J.; Fanger, G.R.; Stirn, M.; Oronsky, A.; Reid, T.R. Going viral: A review of replication-selective oncolytic adenoviruses. *Oncotarget* **2015**, *6*, 19976–19989. [CrossRef] [PubMed]

107. Chiocca, E.A.; Abbed, K.M.; Tatter, S.; Louis, D.N.; Hochberg, F.H.; Barker, F.; Kracher, J.; Grossman, S.A.; Fisher, J.D.; Carson, K.; et al. A phase I open-label, dose-escalation, multi-institutional trial of injection with an E1B-attenuated adenovirus, ONYX-015, into the peritumoral region of recurrent malignant gliomas, in the adjuvant setting. *Mol. Ther.* **2004**, *10*, 958–966. [CrossRef] [PubMed]

108. Bossche WBLVan Den Kleijn, A.; Teunissen, C.E.; Voerman, J.S.A.; Teodosio, C.; David, P.; Dongen, J.J.M.; Van Dirven, C.M.F.; Lamfers, M.L.M. Oncolytic virotherapy in glioblastoma patients induces a tumor macrophage phenotypic shift leading to an altered glioblastoma microenvironment. *Neuro-Oncology* **2018**, *20*, 1494–1504. [CrossRef] [PubMed]

109. Liikanen, I.; Monsurrò, V.; Ahtiainen, L.; Raki, M.; Hakkarainen, T.; Diaconu, I.; Escutenaire, S.; Hemminki, O.; Dias, J.D.; Cerullo, V.; et al. Induction of interferon pathways mediates in vivo resistance to oncolytic adenovirus. *Mol. Ther.* **2011**, *19*, 1858–1866. [CrossRef] [PubMed]

110. Haller, O.; Kochs, G. Human MxA protein: An interferon-induced dynamin-like GTPase with broad antiviral activity. *J. Interf Cytokine Res.* **2011**, *31*, 79–87. [CrossRef]

111. Taipale, K.; Liikanen, I.; Juhila, J.; Turkki, R.; Tähtinen, S.; Kankainen, M.; Vassilev, L.; Ristimäki, A.; Koski, A.; Kanerva, A.; et al. Chronic activation of innate immunity correlates with poor prognosis in cancer patients treated with oncolytic adenovirus. *Mol. Ther.* **2016**, *24*, 175–183. [CrossRef]

112. Liikanen, I.; Koski, A.; Merisalo-Soikkeli, M.; Hemminki, O.; Oksanen, M.; Kairemo, K.; Joensuu, T.; Kanerva, A.; Hemminki, A. Serum HMGB1 is a predictive and prognostic biomarker for Oncolytic immunotherapy. *Oncoimmunology* **2015**, *4*, e989771. [CrossRef] [PubMed]

113. Dai, B.; Roife, D.; Kang, Y.; Gumin, J.; Perez, M.V.R.; Li, X.; Pratt, M.; Brekken, R.A.; Fueyo-margareto, J.; Lang, F.F.; et al. Preclinical Evaluation of Sequential Combination of Oncolytic Adenovirus Delta-24-RGD and Phosphatidylserine-Targeting Antibody in Pancreatic Ductal Adenocarcinoma. *Mol. Cancer Ther.* **2017**, *16*, 662–671. [CrossRef] [PubMed]

114. Musgrove, E.A.; Caldon, C.E.; Barraclough, J.; Stone, A.; Sutherland, R.L. Cyclin D as a therapeutic target in cancer. *Nat. Rev. Cancer* **2011**, *11*. [CrossRef] [PubMed]

115. Logg, C.R.; Robbins, J.M.; Jolly, D.J.; Gruber, H.E.; Kasahara, N. *Retroviral Replicating Vectors in Cancer*, 1st ed.; Elsevier Inc.: Amsterdam, The Netherlands, 2012; Volume 507. [CrossRef]

116. Perez, O.D.; Logg, C.R.; Hiraoka, K.; Diago, O.; Burnett, R.; Inagaki, A.; Jolson, D.; Amundson, K.; Buckley, T.; Lohse, D.; et al. Design and selection of toca 511 for clinical use: Modified retroviral replicating vector with improved stability and gene expression. *Mol. Ther.* **2012**, *20*, 1689–1698. [CrossRef] [PubMed]

117. Cloughesy, T.F.; Petrecca, K.; Walbert, T.; Butowski, N.; Salacz, M.; Perry, J.; Damek, D.; Bota, D.; Bettegowda, C.; Zhu, J.J.; et al. Effect of Vocimagene Amiretrorepvec in Combination with Flucytosine vs Standard of Care on Survival following Tumor Resection in Patients with Recurrent High-Grade Glioma: A Randomized Clinical Trial. *JAMA Oncol.* **2020**, *6*, 1939–1946. [CrossRef] [PubMed]

118. Accomando, W.P.; Rao, A.R.; Hogan, D.J.; Newman, A.M.; Nakao, A.; Alizadeh, A.A.; Diehn, M.; Diago, O.R.; Gammon, D.; Haghighi, A.; et al. Molecular and Immunologic Signatures are Related to Clinical Benefit from Treatment with Vocimagene Amiretrorepvec (Toca 511) and 5-Fluorocytosine (Toca FC) in Patients with Glioma. *Clin. Cancer Res.* **2020**, *26*, 6176–6186. [CrossRef] [PubMed]

119. Msaouel, P.; Opyrchal, M.; Dispenzieri, A.; Peng, K.W.; Federspiel, M.J.; Russell, S.J.; Galanis, E. Clinical Trials with Oncolytic Measles Virus: Current Status and Future Prospects. *Curr. Cancer Drug Targets* **2017**, *18*, 177–187. [CrossRef]

120. Laksono, B.M.; de Vries, R.D.; McQuaid, S.; Duprex, W.P.; de Swart, R.L. Measles virus host invasion and pathogenesis. *Viruses* **2016**, *8*, 210. [CrossRef] [PubMed]

121. Lin, L.T.; Richardson, C.D. The host cell receptors for measles virus and their interaction with the viral Hemagglutinin (H) Protein. *Viruses.* **2016**, *8*, 250. [CrossRef]

122. Ulasov, I.V.; Tyler, M.A.; Zheng, S.; Han, Y.; Lesniak, M.S. CD46 represents a target for adenoviral gene therapy of malignant glioma. *Hum Gene Ther.* **2006**, *17*, 556–564. [CrossRef] [PubMed]

123. Kurokawa, C.; Iankov, I.D.; Anderson, S.K.; Aderca, I.; Leontovich, A.A.; Maurer, M.J.; Oberg, A.L.; Schroeder, M.A.; Giannini, C.; Greiner, S.M.; et al. Constitutive interferon pathway activation in tumors as an efficacy determinant following oncolytic virotherapy. *J. Natl. Cancer Inst.* **2018**, *110*, 1123–1132. [CrossRef] [PubMed]

124. Aref, S.; Castleton, A.Z.; Bailey, K.; Burt, R.; Dey, A.; Leongamornlert, D.; Mitchell, R.J.; Okasha, D.; Fielding, A.K. Type 1 Interferon Responses Underlie Tumor-Selective Replication of Oncolytic Measles Virus. *Mol. Ther.* **2020**, *28*, 1–13. [CrossRef] [PubMed]

125. Berchtold, S.; Lampe, J.; Weiland, T.; Smirnow, I.; Schleicher, S.; Handgretinger, R.; Kopp, H.-G.; Reiser, J.; Stubenrauch, F.; Mayer, N.; et al. Innate Immune Defense Defines Susceptibility of Sarcoma Cells to Measles Vaccine Virus-Based Oncolysis. *J. Virol.* **2013**, *87*, 3484–3501. [CrossRef] [PubMed]

126. Krishnamurthy, S.; Takimoto, T.; Scroggs, R.A.; Portner, A. Differentially Regulated Interferon Response Determines the Outcome of Newcastle Disease Virus Infection in Normal and Tumor Cell Lines. *J. Virol.* **2006**, *80*, 5145–5155. [CrossRef]

127. Zamarin, D.; Martínez-Sobrido, L.; Kelly, K.; Mansour, M.; Sheng, G.; Vigila, A.; García-Sastre, A.; Palese, P.; Fong, Y. Enhancement of oncolytic properties of recombinant newcastle disease virus through antagonism of cellular innate immune responses. *Mol. Ther.* **2009**, *17*, 697–706. [CrossRef] [PubMed]

128. Mansour, M.; Palese, P.; Zamarin, D. Oncolytic Specificity of Newcastle Disease Virus Is Mediated by Selectivity for Apoptosis-Resistant Cells. *J. Virol.* **2011**, *85*, 6015–6023. [CrossRef]

129. Shao, X.; Wang, X.; Guo, X.; Jiang, K.; Ye, T.; Chen, J.; Fang, J.; Gu, L.; Wang, S.; Zhang, G.; et al. STAT3 contributes to oncolytic newcastle disease virus-induced immunogenic cell death in melanoma cells. *Front. Oncol.* **2019**, *9*, 436. [CrossRef]

130. Meng, C.; Zhou, Z.; Jiang, K.; Yu, S.; Jia, L.; Wu, Y.; Liu, Y.; Meng, S.; Ding, C. Newcastle disease virus triggers autophagy in U251 glioma cells to enhance virus replication. *Arch. Virol.* **2012**, *157*, 1011–1018. [CrossRef] [PubMed]

131. Jiang, K.; Li, Y.; Zhu, Q.; Xu, J.; Wang, Y.; Deng, W.; Liu, Q.; Zhang, G.; Meng, S. Pharmacological modulation of autophagy enhances Newcastle disease virus-mediated oncolysis in drug-resistant lung cancer cells. *BMC Cancer* **2014**, *14*, 551. [CrossRef] [PubMed]

132. Puhlmann, J.; Puehler, F.; Mumberg, D.; Boukamp, P.; Beier, R. Rac1 is required for oncolytic NDV replication in human cancer cells and establishes a link between tumorigenesis and sensitivity to oncolytic virus. *Oncogene* **2010**, *29*, 2205–2216. [CrossRef] [PubMed]

133. Van den Broeke, C.; Jacob, T.; Favoreel, H.W. Rho'ing in and out of cells: Viral interactions with Rho GTPase signaling. *Small GTPases* **2014**, *5*, e28318. [CrossRef] [PubMed]

134. Lai, Y.J.; Tsai, J.C.; Tseng, Y.T.; Wu, M.S.; Liu, W.S.; Lam, H.I.; Yu, J.H.; Nozell, S.E.; Benveniste, E.N. Small G protein Rac GTPases regulate the maintenance of glioblastoma stem-like cells in vitro and in vivo. *Oncotarget* **2017**, *8*, 18031–18049. [CrossRef]

135. Forsyth, P.; Roldán, G.; George, D.; Wallace, C.; Palmer, C.A.; Morris, D.; Cairncross, G.; Matthews, M.V.; Markert, J.; Gillespie, Y.; et al. A phase I trial of intratumoral administration of reovirus in patients with histologically confirmed recurrent malignant gliomas. *Mol. Ther.* **2008**, *16*, 627–632. [CrossRef] [PubMed]

136. Samson, A.; Scott, K.J.; Taggart, D.; West, E.J.; Wilson, E.; Nuovo, G.J.; Thomson, S.; Corns, R.; Mathew, R.K.; Fuller, M.J.; et al. Intravenous delivery of oncolytic reovirus to brain tumor patients immunologically primes for subsequent checkpoint blockade. *Sci. Transl. Med.* **2018**, *10*, eaam7577. [CrossRef] [PubMed]

137. Konopka-Anstadt, J.L.; Mainou, B.A.; Sutherland, D.M.; Sekine, Y.; Strittmatter, S.M.; Dermody, T.S. The Nogo receptor NgR1 mediates infection by mammalian reovirus. *Cell Host Microbe* **2014**, *15*, 681–691. [CrossRef] [PubMed]

138. Liao, H.; Duka, T.; Teng, F.Y.H.; Sun, L.; Bu, W.Y.; Ahmed, S.; Tang, B.L.; Xiao, Z.C. Nogo-66 and myelin-associated glycoprotein (MAG) inhibit the adhesion and migration of Nogo-66 receptor expressing human glioma cells. *J. Neurochem.* **2004**, *90*, 1156–1162. [CrossRef] [PubMed]

139. Kang, Y.H.; Han, S.R.; Jeon, H.; Lee, S.; Lee, J.; Yoo, S.M.; Park, J.B.; Park, M.J.; Kim, J.T.; Lee, H.G.; et al. Nogo receptor–vimentin interaction: A novel mechanism for the invasive activity of glioblastoma multiforme. *Exp. Mol. Med.* **2019**, *51*, 1–15. [CrossRef] [PubMed]

140. Papachristodoulou, A.; Silginer, M.; Weller, M.; Schneider, H.; Hasenbach, K.; Janicot, M.; Roth, P. Therapeutic targeting of TGFb ligands in glioblastoma using novel antisense oligonucleotides reduces the growth of experimental gliomas. *Clin. Cancer Res.* **2019**, *25*, 7189–7201. [CrossRef]

141. Strong, J.E.; Coffey, M.C.; Tang, D.; Sabinin, P.; Lee, P.W.K. The molecular basis of viral oncolysis: Usurpation of the Ras signaling pathway by reovirus. *EMBO J.* **1998**, *17*, 3351–3362. [CrossRef]

142. Norman, K.L.; Hirasawa, K.; Yang, A.D.; Shields, M.A.; Lee, P.W.K. Reovirus oncolysis: The Ras/RalGEF/p38 pathway dictates host cell permissiveness to reovirus infection. *Proc. Natl. Acad. Sci. USA* **2004**, *101*, 11099–11104. [CrossRef]

143. Marcato, P.; Shmulevitz, M.; Pan, D.; Stoltz, D.; Lee, P.W.K. Ras transformation mediates reovirus oncolysis by enhancing virus uncoating, particle infectivity, and apoptosis-dependent release. *Mol. Ther.* **2007**, *15*, 1522–1530. [CrossRef] [PubMed]

144. Twigger, K.; Roulstone, V.; Kyula, J.; Karapanagiotou, E.M.; Syrigos, K.N.; Morgan, R.; White, C.; Bhide, S.; Nuovo, G.; Coffey, M.; et al. Reovirus exerts potent oncolytic effects in head and neck cancer cell lines that are independent of signalling in the EGFR pathway. *BMC Cancer* **2012**, *12*, 368. [CrossRef] [PubMed]

145. Alain, T.; Kim, T.S.Y.; Lun, X.Q.; Liacini, A.; Schiff, L.A.; Senger, D.L.; Forsyth, P.A. Proteolytic disassembly is a critical determinant for reovirus oncolysis. *Mol. Ther.* **2007**, *15*, 1512–1521. [CrossRef] [PubMed]

146. Morris, D.; Tu, D.; Tehfe, M.A.; Nicholas, G.A.; Goffin, J.R.; Gregg, R.W.; Shepherd, F.A.; Murray, N.; Wierzbicki, R.; Lee, C.W.; et al. A Randomized Phase II study of Reolysin in Patients with Previously Treated Advanced or Metatstatic Non Small Cell Lung Cancer (NSCLC) receiving Standard Salvage Chemotherapy—Canadian Cancer Trials Group IND 211. *J. Clin. Oncol.* **2016**. [CrossRef]

147. Cohn, D.E.; Sill, M.W.; Walker, J.L.; Malley, D.O.; Nagel, C.I.; Rutledge, T.L.; Bradley, W.; Richardson, D.L.; Moxley, K.M.; Aghajanian, C. Randomized phase IIB evaluation of weekly paclitaxel versus weekly paclitaxel with oncolytic reovirus (Reolysin®) in recurrent ovarian, tubal, or peritoneal cancer: An NRG Oncology/Gynecologic Oncology Group study. *Gynecol. Oncol.* **2017**, *146*, 477–483. [CrossRef] [PubMed]

148. Barton, K.L.; Misuraca, K.; Cordero, F.; Dobrikova, E.; Min, H.D.; Gromeier, M.; Kirsch, D.G.; Becher, O.J. PD-0332991, a CDK4/6 inhibitor, significantly prolongs survival in a genetically engineered mouse model of brainstem glioma. *PLoS ONE* **2013**. [CrossRef]

149. Desjardins, A.; Gromeier, M.; Herndon, J.E.; Beaubier, N.; Bolognesi, D.P.; Friedman, A.H.; Friedman, H.S.; McSherry, F.; Muscat, A.M.; Nair, S.; et al. Recurrent glioblastoma treated with recombinant poliovirus. *N. Engl. J. Med.* **2018**, *379*, 150–161. [CrossRef] [PubMed]

150. Gromeier, M.; Brown, M.C.; Zhang, G.; Lin, X.; Chen, Y.; Wei, Z.; Beaubier, N.; Yan, H.; He, Y.; Desjardins, A.; et al. Very low mutation burden is a feature of inflamed recurrent glioblastomas responsive to cancer immunotherapy. *Nat. Commun.* **2021**. [CrossRef]

151. Geletneky, K.; Huesing, J.; Rommelaere, J.; Leuchs, B.; Capper, D.; Bartsch, A.J.; Neumann, J.O.; Schöning, T.; Hüsing, J.; Beelte, B.; et al. Phase I/IIa study of intratumoral/intracerebral or intravenous/intracerebral administration of Parvovirus H-1 (ParvOryx) in patients with progressive primary or recurrent glioblastoma multiforme: ParvOryx01 protocol. *BMC Cancer* **2012**, *12*, 3516–3523. [CrossRef]

152. Geletneky, K.; Hajda, J.; Angelova, A.L.; Leuchs, B.; Capper, D.; Bartsch, A.J.; Neumann, J.O.; Schöning, T.; Hüsing, J.; Beelte, B.; et al. Oncolytic H-1 Parvovirus Shows Safety and Signs of Immunogenic Activity in a First Phase I/IIa Glioblastoma Trial. *Mol. Ther.* **2017**, *25*, 2620–2634. [CrossRef] [PubMed]

153. Di Piazza, M.; Mader, C.; Geletneky, K.; Herrero y Calle, M.; Weber, E.; Schlehofer, J.; Deleu, L.; Rommelaere, J. Cytosolic Activation of Cathepsins Mediates Parvovirus H-1-Induced Killing of Cisplatin and TRAIL-Resistant Glioma Cells. *J. Virol.* **2007**, *81*, 4186–4198. [CrossRef]

154. Guo, Z.S.; Lu, B.; Guo, Z.; Giehl, E.; Feist, M.; Dai, E.; Liu, W.; Storkus, W.J.; He, Y.; Liu, Z.; et al. Vaccinia virus-mediated cancer immunotherapy: Cancer vaccines and oncolytics. *J. Immuno. Ther. Cancer* **2019**, *7*, 1–21. [CrossRef] [PubMed]

155. Advani, S.J.; Buckel, L.; Chen, N.G.; Scanderbeg, D.J.; Geissinger, U.; Zhang, Q.; Yu, Y.A.; Aguilar, R.J.; Mundt, A.J.; Szalay, A.A. Preferential replication of systemically delivered oncolytic vaccinia virus in focally irradiated glioma xenografts. *Clin. Cancer Res.* **2012**, *18*, 2579–2590. [CrossRef] [PubMed]

156. Foloppe, J.; Kempf, J.; Futin, N.; Wallace, C.; Palmer, C.A.; Morris, D.; Cairncross, G.; Matthews, M.V.; Markert, J.; Gillespie, Y.; et al. The Enhanced Tumor Specificity of TG6002, an ArMed. Oncolytic Vaccinia Virus Deleted in Two Genes Involved in Nucleotide Metabolism. *Mol. Ther. Oncolytics* **2019**, *14*, 1–14. [CrossRef] [PubMed]

157. Zloza, A.; Kim, D.W.; Kim-Schulze, S.; Jagoda, M.C.; Monsurro, V.; Marincola, F.M.; Kaufman, H.L. Immunoglobulin-like transcript 2 (ILT2) is a biomarker of therapeutic response to oncolytic immunotherapy with vaccinia viruses. *J. Immunother. Cancer* **2014**, *2*, 1–8. [CrossRef]

158. Lee, N.H.; Kim, M.; Oh, S.Y.; Kim, S.G.; Kwon, H.C.; Hwang, T.H. Gene expression profiling of hematologic malignant cell lines resistant to oncolytic virus treatment. *Oncotarget* **2017**, *8*, 1213–1225. [CrossRef]

159. Balvers, R.K.; Kleijn, A.; Kloezeman, J.J.; French, P.J.; Kremer, A.; Bent MJVan Den Dirven, C.M.F.; Leenstra, S.; Lamfers, M.L.M. Serum-free culture success of glial tumors is related to specific molecular profiles and expression of extracellular matrix–associated gene modules. *Neuro-Oncology* **2013**, *15*, 1684–1695. [CrossRef] [PubMed]

160. Kleijn, A.; Kloezeman, J.J.; Balvers, R.K.; Kaaij MVan Der Dirven, C.M.F.; Leenstra, S.; Lamfers, M.L.M. A Systematic Comparison Identifies an ATP-Based Viability Assay as Most Suitable Read-Out for Drug Screening in Glioma Stem-Like Cells. *Stem Cells Int.* **2016**, *2016*. [CrossRef]

161. Pont, L.B.; Balvers, R.K.; Kloezeman, J.J.; Nowicki, M.O.; Van Den Bossche, W.; Kremer, A.; Wakimoto, H.; Van Den Hoogen, B.G.; Leenstra, S.; Dirven, C.M.F.; et al. In vitro screening of clinical drugs identifies sensitizers of oncolytic viral therapy in glioblastoma stem-like cells. *Gene Ther.* **2015**, *22*, 947–959. [CrossRef] [PubMed]

162. Dijkstra, K.K.; Cattaneo, C.M.; Weeber, F.; Chalabi, M.; van de Haar, J.; Fanchi, L.F.; Slagter, M.; van der Velden, D.L.; Kaing, S.; Kelderman, S.; et al. Generation of Tumor-Reactive T Cells by Co-culture of Peripheral Blood Lymphocytes and Tumor Organoids. *Cell* **2018**, *174*, 1586–1598. [CrossRef]

163. Grill, J.; Lamfers, M.L.M.; van Beusechem, V.W.; Dirven, C.M.; Pherai, D.S.; Kater, M.; Van der Valk, P.; Vogels, R.; Vandertop, W.P.; Pinedo, H.M.; et al. The Organotypic Multicellular Spheroid Is a Relevant Three-Dimensional Model to Study Adenovirus Replication and Penetration in Human Tumors in Vitro. *Mol. Ther.* **2002**, *6*, 609–614. [CrossRef] [PubMed]

164. Burton, C.; Das, A.; McDonald, D.; Vandergrift, W.A., III.; Patel, S.J.; Cachia, D.; Bartee, E. Oncolytic myxoma virus synergizes with standard of care for treatment of glioblastoma multiforme. *Oncolytic Virotherapy* **2018**, *7*, 107–116. [CrossRef] [PubMed]

165. Sood, D.; Tang-Schomer, M.; Pouli, D.; Mizzoni, C.; Raia, N.; Tai, A.; Arkun, K.; Wu, J.; Black, L.D.; Scheffle, B.; et al. 3D extracellular matrix microenvironment in bioengineered tissue models of primary pediatric and adult brain tumors. *Nat. Commun.* **2019**, *10*, 1–14. [CrossRef] [PubMed]

166. Robertson, F.L.; Marqués-Torrejón, M.A.; Morrison, G.M.; Pollard, S.M. Experimental models and tools to tackle glioblastoma. *DMM Dis. Model Mech.* **2019**, *12*. [CrossRef]

167. Cai, X.; Briggs, R.G.; Homburg, H.B.; Young, I.M.; Davis, E.J.; Lin, Y.H.; Battiste, J.D.; Sughrue, M.E. Application of microfluidic devices for glioblastoma study: Current status and future directions. *Biomed. Microdevices* **2020**, *22*, 1–10. [CrossRef]
168. Jacob, F.; Salinas, R.D.; Zhang, D.Y.; Nguyen, P.T.T.; Schnoll, J.G.; Wong, S.Z.H.; Thokala, R.; Sheikh, S.; Saxena, D.; Prokop, S.; et al. A Patient-Derived Glioblastoma Organoid Model and Biobank Recapitulates Inter- and Intra-tumoral Heterogeneity. *Cell* **2020**, *180*, 188–204. [CrossRef] [PubMed]
169. Kemp, V.; Lamfers, M.L.M.; Pluijm, G.; Van Der Hoogen, B.G.; Van Den Hoeben, R.C. Developing oncolytic viruses for clinical use: A consortium approach. *Cytokine Growth Factor Rev.* **2020**. [CrossRef]

Review

HSV-1 Oncolytic Viruses from Bench to Bedside: An Overview of Current Clinical Trials

Marilin S. Koch, Sean E. Lawler * and E. Antonio Chiocca

Harvey Cushing Neurooncology Research Laboratories, Department of Neurosurgery, Brigham and Women's Hospital, Harvard Medical School, Boston, MA 02115, USA; eachiocca@bwh.harvard.edu (E.A.C.)
* Correspondence: slawler@bwh.harvard.edu

Received: 22 October 2020; Accepted: 23 November 2020; Published: 26 November 2020

Simple Summary: Oncolytic Herpes simplex virus-1 (HSV-1) offers the dual potential of both lytic tumor-specific cell killing and inducing anti-tumor immune responses. The HSV-1 genome can be altered to enhance both components and this may be applicable for the treatment of a broad range of cancers. Several engineered oncolytic viruses based on the HSV-1 backbone are currently under investigation in various clinical trials, both as single agents and in combination with various immunomodulatory drugs.

Abstract: Herpes simplex virus 1 (HSV-1) provides a genetic chassis for several oncolytic viruses (OVs) currently in clinical trials. Oncolytic HSV1 (oHSV) have been engineered to reduce neurovirulence and enhance anti-tumor lytic activity and immunogenicity to make them attractive candidates in a range of oncology indications. Successful clinical data resulted in the FDA-approval of the oHSV talimogene laherparepvec (T-Vec) in 2015, and several other variants are currently undergoing clinical assessment and may expand the landscape of future oncologic therapy options. This review offers a detailed overview of the latest results from clinical trials as well as an outlook on newly developed HSV-1 oncolytic variants with improved tumor selectivity, replication, and immunostimulatory capacity and related clinical studies.

Keywords: HSV-1; oncolytic virus; immunotherapy; clinical trials

---

1. Introduction

In the past decade, immunotherapeutic drugs for oncology have revolutionized the field. The landscape of immunotherapeutic drugs has been spearheaded by immune checkpoint inhibition [1–3], as well as CAR (chimeric antigen receptor)-T-cell therapy [4,5], suicide-gene approaches [6], and a range of other agents, e.g., tumor antigen vaccinations [7]. In addition to these, oncolytic viruses (OVs) have emerged as an important part of the immunotherapeutic armory (Figure 1).

OVs infect tumor cells and cause their lysis leading to a release of tumor-specific antigens as well as neoantigens. Antigen presentation and virus induced activation of the innate immune cells in turn trigger the activation of tumor-specific T-cells.

Among OVs in clinical trials, Herpes simplex virus 1 (HSV-1)-derived agents are some of the most widely tested viral vectors and have also been thoroughly investigated in numerous pre-clinical studies [8]. HSV-1 is a double-stranded neurotropic DNA-virus [9,10]; the wild-type virus in humans can cause mucocutaneous lesions, keratoconjunctivitis, encephalitis, and respiratory infections [10]. Its large genome of 150 kb [11], infectivity, and lytic activity present ideal properties for a potent engineerable OV: HSV-1 can infect a variety of cell types and cause lysis; its comparatively large genome facilitates modifications that can enhance anti-tumorigenic features and reduce neurovirulence [12] and it can easily be inactivated by the anti-herpetic drugs ganciclovir, acyclovir, or valacyclovir. To date,

17 strains of HSV-1 are known [11]. Multiple genetic modifications of HSV-1 have been described that alter infectiousness, neurovirulence, and lytic activity (Table 1). Engineering strategies aim at (a) preventing infection of the nervous system, e.g., by deleting the neurovirulence gene *γ34.5/RL1* [13] (b) enhancing tumor-selectivity, e.g., by deleting the ribonucleotide reductase expressing gene *ICP6* [14] and (c) increasing immunogenicity by adding genes to express immunostimulatory mediators, such as GM-CSF [15] and IL-12 [16,17] or counteract T-cell exhaustion by arming the HSV-genome with anti-CTLA-4 and anti-PD-1 targeting antibody sequences [18]. Current oHSVs tested in published clinical trials include HSV1716, G207, HF10, NV1020, and talimogene laherparepvec (T-Vec), which is until now the most thoroughly investigated HSV-1 related OV and in 2015 became the first OV to gain FDA-approval, after a successful trial in advanced melanoma [13]. There are several additional oHSVs that are currently under clinical and re-clinical investigation. This review aims to give an overview over the state of clinical applications of oncolytic viral therapy with oHSV-1 and future directions.

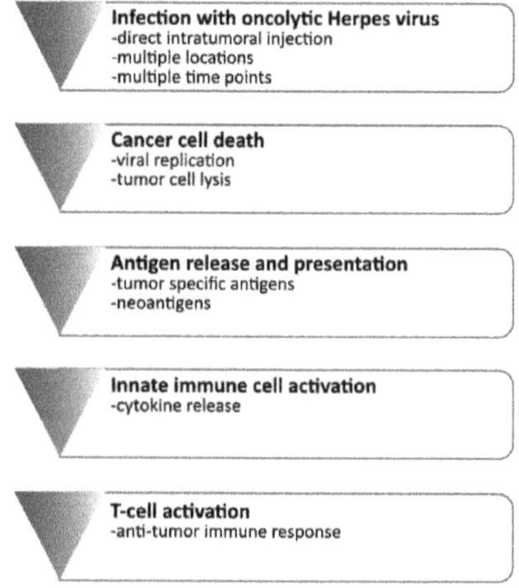

Figure 1. Mechanism of oncolytic virus therapy and interaction with the immune system.

Table 1. Oncolytic Herpes viruses tested in oncology clinical trials to date.

Virus Strain	Modifications	Aim
G207	insertion of the Escherichia coli *lacZ* sequence at *ICP6/UL39*	reducing ribonucleotide reductase activity [14]
	deletion of *γ34.5/RL1*	reducing neurovirulence [15]
1716	deletion of *γ34.5/RL1*	reducing neurovirulence [15]
HF10	deletion in the Bam HI-B fragment	unknown
	two incomplete *UL56* copies without promoter	possibly reducing neurovirulence [16]
	reduced expression of UL43, UL49.5, UL55, LAT	possible influence on immunogenicity (UL43), unknown (UL49.5), reduced virus reactivation (LAT) [17]
	increased expression of UL53 and UL54	reduced viral shedding (UL53) [17]
NV1020	deletion of one allele of *α0*, *α4*, *γ34.5* and *UL56*	reducing infectivity, viral replication and neuroinvasiveness [18]

Table 1. *Cont.*

Virus Strain	Modifications	Aim
Talimogene laherparepvec (T-Vec)	deletion of *ICP34.5*	reducing neurovirulence [15]
	deletion of *ICP47*	augment immune response [19]
	insertion of *GM-CSF* gene	augment immune response [19]

2. HSV-1-Derived Oncolytic Viruses in Clinical Trials

A number of oHSVs have been developed and tested in clinical trials so far. Overall they have shown efficacy, and encouraging responses as exemplified by T-Vec. For clinical trials, GMP-grade virus stocks are injected intratumorally following biosafety procedures. Depending on the trial, the virus may be injected at multiple areas within the same tumor or by repeated intratumoral injections over time; intravenous virus administration has also been evaluated [20].

2.1. HSV-1716

HSV-1716 (Seprehvir by Virttu Biologics/Sorrento Therapeutics Inc. San Diego, CA, USA) has deletions of both copies of $\gamma 34.5/RL1$ that mitigate neurovirulence [15]. This variant has been tested for the treatment of recurrent malignant glioma [21] and stage IV melanoma [22] in phase I studies. Toxicity was the primary endpoint in both studies. Rampling et al. injected HSV-1716 stereotactically into the tumor of patients with recurrent anaplastic astrocytoma and glioblastoma. No encephalitis or virus shedding could be detected, thereby demonstrating safe delivery [21]. Mackie et al. conducted a pilot study with the same construct for malignant melanoma. HSV-1716 was applied subcutaneously into melanoma nodules. No toxicity or virus shedding was observed. Pathological workup showed necrosis within excised tumor tissue from three patients. Further, signs of viral replication within the samples were observed [22]. Intravenous injections in pediatric and young adult patients (11–30 years) with recurrent or progressive non-CNS solid tumors were also well tolerated, as no dose-limiting toxicities or shedding of the virus (monitored with HSV-1 cultures and PCR from patient samples) were observed. Due to the small cohort size of nine patients and varying therapy regimens pre- and post-virus treatment, no conclusion regarding the efficacy of HSV-1716 could be drawn [20].

2.2. G207

G207 is an attenuated HSV-1 variant that contains an insertion of the *Escherichia coli lacZ* sequence in the *ICP6* gene and deletions at both $\gamma 34.5$ loci [23], aiming at diminishing viral growth and neurovirulence [14,15]. Deletion of the ribonucleotide reductase encoding *ICP6* gene allows for selective viral replication in dividing (tumor) cells [23]. Markert and colleagues tested the safety of G207 in several phase I studies in recurrent or residual anaplastic astrocytoma, glioblastoma, and gliosarcoma. The initial phase I study [24] evaluated the safety profile of intratumorally inoculated G207 in a dose-escalation scheme. While it was demonstrated that the virus could be safely administered without the development of encephalitis, other potential adverse events (AEs) were difficult to distinguish from disease-related symptoms. MRI (magnetic resonance imaging) confirmed a decrease in enhancement volume in 40% of the patients; two patients tested positive for the HSV-1 and lacZ sequence in the tissue analysis, suggesting successful inoculation of G207. A follow-up phase Ib study investigated the safety profile of two inoculations each before and after tumor resection in patients with recurrent glioblastoma [25]. Again, no signs of encephalitis were detected and the therapy was well tolerated. Every patient experienced at least one AE with 13% being possibly associated with G207, but an ameliorated Karnofsky Performance Score (KPS) was noticed in 50% of the patients. Another subsequent phase I study focused on the combination of G207 with radiation in patients with recurrent or residual anaplastic astrocytoma, glioblastoma, and gliosarcoma [26]. Patients were treated with G207 via stereotactic inoculation and subsequent radiation with 5 Gy. As in the other two studies, no patient developed encephalitis; in some cases, seizures were classified as possible G207-related adverse events. Overall, the treatment combination was assessed as safe. The secondary endpoint of this study was efficacy: The median progression-free survival was stated with 2.5 months, the median survival from G207 inoculation added up to 7.5 months. Signs of therapy response in MRI were noticed in two patients on three occasions.

2.3. HF10

HF10 (Canerpaturev, C-REV by Takara Bio Inc. Mountain View, CA, USA) is a HSV-1 strain with a deletion in the Bam HI-B fragment [16,27,28] and additional alterations resulting in defective expression of UL43, UL49.5, UL55, UL56, LAT genes, and increased expression of UL53 and UL54 [17]. In contrast to other oHSVs, HF10 was not engineered—the mutations that define this strain occurred spontaneously [17]. Preclinical evaluation of this construct presented promising results in a syngeneic immunocompetent mouse model for peritoneally disseminated fibrosarcoma with the HF-10-treated animals showing prolonged survival. The development of anti-tumor immunity was also shown in the mice since they rejected a tumor rechallenge [16]. HF10 was first tested in humans in a pilot study to assess toxicity and efficacy in patients with recurrent metastatic breast cancer and (sub)cutaneous metastases [29,30]. One nodule per patient was injected with HF10 for up to three days, while another was injected with saline. No macroscopic reduction of tumors was observed, but histological analysis showed 30–100% tumor cell death and signs of viral infection of breast cancer cells. No shedding or reactivation of HSV-1 was detected. There were no therapy-related adverse effects. A follow-up phase I dose-escalation study examined possible toxicity and efficacy of HF10 in patients with non-resectable pancreatic cancer [31]. HF10 was injected intratumorally at several locations during laparotomy and via catheter for three days in a row. The primary endpoints were assessed 30 days after virus inoculation. No adverse events were registered and approximately 66% of the patients presented with stable disease or even partial response. Furthermore, the tumor marker CA19-9 (cancer antigen 19.9) decreased in 50% of the patients. All of the patients were HSV1 antibody positive from the beginning and no virus shedding could be detected, either in the abdomen or in the blood. Histopathological analysis found scar tissue at the HF10 injection site with virus-specific patterns (inclusion bodies, small segmented nuclei), corresponding with the results of the previous study conducted for breast cancer, suggesting viral replication [30,31]. In comparison to normal tumor tissue, HF10-injected tumors showed a significantly higher rate of CD8$^+$-T-cell and macrophage infiltration. A follow-up phase I study combined ultrasound guided HF-10 injections with erlotinib and gemcitabine chemotherapy in unresectable locally advanced pancreatic cancer [32]. After an initial chemotherapy cycle, patients received intratumoral endoscopic ultrasound (EUS)-guided HF10 injections every two weeks with a total of four injections. While a chemotherapy-related grade III myelosuppression was noticed in 50% of the patients, no HF10-specific adverse events occurred. 90% of the patients received all planned treatments and were assessed for therapy response in accordance with RECIST criteria, with >70% of the patients showing either stable disease or partial response overall. Analysis of target lesion response even showed a partial response in 33% and a stable disease in 66% of the cases. A complete surgical response was noted in two patients who underwent surgery after therapy. An infiltration of CD8$^+$ T cells was observed in the resected tissue from both patients. Another small pilot study conducted by Fujimoto et al. [33] investigated the effects HF10 in subcutaneous metastases of head and neck squamous cell carcinoma in two patients; the authors admittedly described no macroscopic changes two weeks after virus inoculation, but report tumor cell death and fibrosis as well as an enrichment of CD4$^+$- and CD8$^+$-T-cells in the histopathological analyses of resected tumor specimens.

2.4. NV1020

NV1020 is a derivative of the HSV-1 strain R7020 that was initially developed as a vaccine against HSV-2 and has been attenuated by several genetic modifications including deletions of one allele of the genes for *ICP0, ICP4,* and *γ34.5,* as well as *UL56,* thereby reducing infectiousness, viral replication, and neuroinvasiveness; additionally, NV1020 has been altered by a deletion in the region of the thymidine kinase (*tk*) gene and insertions of a fragment of HSV-2 DNA and the *tk* gene [18]. NV1020 has been shown to be successful in the treatment of various preclinical cancer models such as pleural, gastric, and hepatic cancer as well as head and neck squamous cell carcinoma [18,34–36]. Combined treatment of NV1020 with 5-FU, SN38 and oxaliplatin proved to act additively or synergistically in the treatment of colon cancer models [37]. It was first applied in a clinical setting in a phase I study for liver metastases of colorectal cancer to evaluate safety and tolerability [38]. Patients received a single dose of NV1020 via hepatic arterial infusion followed by implantation of a hepatic arterial infusion pump for local delivery of chemotherapy. Virus-associated adverse events that appeared directly after administration of NV1020 included pyrexia, headache, and muscle stiffness. NV1020-related individual cases of increased GGT (gamma glutamyl transferase) levels, gastroenteritis, and leukocytosis were registered. Analysis of cytokine and T-cell serum levels did not indicate a measurable immunogenic effect of NV1020 and evaluation of anti-tumor efficacy with CT scans 28 days after treatment showed tumor reduction in 17% and stable disease in 58% of the patients, while 25% were diagnosed with further progression. Radiologic assessment up to 12 months after treatment showed partial responses to chemotherapy after NV1020 in all patients; the authors also observed a 24% median decrease of the tumor marker CEA (carcinoembryonic antigen) [39]. The median survival was 25 months; after 62 months of observation, one patient was still alive. A follow-up study by Geevarghese et al. [40] examined safety and efficacy of NV1020 for the same disease type. NV1020 was administered into the hepatic artery weekly in four fixed doses, followed by adjuvant treatment at the physician's discretion. Similar to the first study by Kemeny et al., pyrexia, chills, headache, nausea, myalgia, and fatigue were registered as adverse events within 24 h after NV1020 infusion. Although no shedding of NV1020 could be detected, infrequent HSV-1 shedding was observed. Higher doses of NV1020 were associated with stable disease in 50% of the patients and additional chemotherapy resulted in a clinical control rate of 68%. Immunologically, a dose-associated increase in levels of IL-6, TNF-α, and IFN-γ was noted by the authors and therefore 1×10^8 pfu (plaque forming units) was defined as the optimal biological dose.

2.5. Talimogene Laherparepvec

Talimogene laherparepvec (IMLYGIC^{TM}, T-Vec, OncoVEX$^{\text{GM-CSF}}$ by Amgen Inc. Cambridge, MA, USA) is a genetically engineered OV based on the HSV-1 strain JS1, which has been modified by deletion of $\gamma 34.5$ and *ICP47* as well as an insertion of the gene for GM-CSF [19] to render the virus more immunogenic. The first phase I clinical trial was performed in patients with (sub)cutaneous metastases of breast, gastrointestinal adenocarcinoma, malignant melanoma, and epithelial cancer of the head and neck to determine safety, biological activity and adequate dosing [41]. For the first part of the study, patients were categorized in three cohorts with the HSV-seropositive patients receiving the highest dose. The second part of the study focused on evaluating three dose regimens with the HSV-seropositive patients receiving the highest doses. The authors recorded pyrexia, low-grade anorexia, nausea and vomiting, fatigue, and reaction at the injection site as the main adverse events. 1×10^7 pfu/mL was declared as the maximum-tolerated dose (MTD) for seronegative patients, while no MTD for seropositive patients could be stated. All HSV-seronegative patients seroconverted, whereas in the seropositive cohort, an increase in HSV antibody titer was noted. No treatment-associated effects on cytokines were recorded. Histological analyses of tumor tissue frequently showed necrotic areas and positive HSV1 staining primarily in necrotic tumor tissues suggesting a correlation. In three patients, stable disease was achieved and in some cases size reductions of the injected tumor was seen.

Further studies on the effects of T-Vec on clinical response and survival were conducted by Senzer et al. in a phase II study for patients with unresectable stage IIIc and stage IV melanoma [42]. The patients each received initial intratumoral injections, followed three weeks later by injections every two weeks for a possible total of 24 treatments. All seronegative patients seroconverted. Eighty-five percent of the patients experienced grade I/II adverse effects with the most common being fever, chills, fatigue, nausea, and vomiting, as well as headache. Treatment was associated with local as well as distant responses in lung, liver, pancreas, lymph nodes, and soft tissue. Clinical response assessment resulted in 20% complete responses; 13% of the patients were classified as having "no evidence of disease" with some cases involving additional surgery. Overall median survival was 16 months, and the one-year survival rate of patients with complete or partial response totaled 93%. Kaufman et al. [43] further analyzed local and distant immune responses of this patient cohort. The authors used peripheral blood mononuclear cells (PBMCs) from study patients, non-study patients, and healthy donors as well as tumor tissue from study patients and non-study melanoma patients to compare the immune cell status. Higher amounts of activated $CD8^+$-T-cells expressing Perforin and Granzyme B as well as PD-1 expressing T-cells and Tregs in the local tumor tissue compared to the periphery in non-study melanoma patients were observed. Functional analysis of tumor infiltrating lymphocytes (TILs) and PBMCs from a study patient showed an enrichment of MART-1-specific T-cells, indicating the development of a T-Vec-mediated systemic anti-tumor immunity. Moreover, a decrease of $CD4^+$-T-cells, Tregs, T-suppressor cells, and myeloid-derived suppressor cells (MDSCs) within TILs of study patients compared to non-study patients was noted. A comparison of immune cell populations between treated tumor sites and peripheral tumor sites showed more distinct local responses but still provided evidence for the induction of a systemic anti-tumor immunity.

A randomized phase III trial of T-Vec compared to GM-CSF in patients with unresected stage IIIB-IV melanoma [13] showed that T-Vec treatment resulted in a prolonged median overall survival (23.3. vs. 18.9 months T-Vec vs. GM-CSF) and an improved durable response rate in T-Vec patients (16.3%) in contrast to GM-CSF-treated patients (2.1%). The T-Vec dosing scheme followed previous strategies [42], while GM-CSF was administered daily for two weeks in 28-day cycles. The most common adverse events in the T-Vec cohort included chills, pyrexia, pain at the injection site, nausea, influenza-like symptoms, and fatigue, therefore matching the profile of adverse events seen in preceding studies. In the T-Vec group, the authors further observed decreased size of more than 50% in injected as well as in uninjected lesions [44], which points to the development of a systemic anti-tumor response as previously reported [43].

Additional clinical data show that oHSV therapy appears to work well with immune checkpoint blockade. Combined treatment of T-Vec (1×10^6–1×10^8 pfu/mL) with the CTLA-4 blocking monoclonal antibody ipilimumab in 19 patients with stage III and IV melanoma did not lead to dose-limiting toxicities [45]. Moreover, Puzanov et al. [45] reported 22% complete responses, 28% partial responses, and 22% stable disease and an objective response rate of 50% referring to immune-related response criteria. As already noted in previous studies with T-Vec monotherapy, both injected and uninjected tumor lesions showed a size reduction after treatment with T-Vec and ipilimumab. Significant enrichment of total $CD8^+$ and activated $CD8^+$-T-cells during T-Vec monotherapy as well as a gain of ICOS-expressing $CD4^+$-T-cells during combination therapy was observed.

3. Future Directions for Next Generation oHSVs

Currently, more than 20 clinical trials on already tested, but newly developed HSV-1 related OVs are also underway (Table 2). Further studies on known compounds such as G207, HF10, and T-Vec are designed to determine safety and tolerability for either different malignancies or combinations with chemotherapy (NCT03252808, NCT02779855, NCT03300544, NCT03554044), radiotherapy (NCT03911388, NCT04482933, NCT03300544, NCT02923778), or checkpoint inhibition (NCT03153085, NCT04185311, NCT02978625, NCT02965716, NCT04163952).

Newly developed candidates include G47Δ, rQNestin, M032, RP1, RP2, Rrp450, ONCR-177, and C134. As many of the initial trials had shown safety but no efficacy as described above, subsequent trials were designed to answer remaining questions.

Table 2. Outlook on ongoing and future clinical trials on oncolytic Herpes viruses.

Virus	Study Title	Study Type	Disease Type	Study Aim	Status	NCT/UMIN #
HF10	A study of TBI-1401(HF10) in patients with solid tumors with superficial lesions	phase I	solid tumors	safety and tolerability of repeated intratumoral injections	completed	NCT02428036
	Phase I Study of TBI-1401(HF10) plus chemotherapy in patients with unresectable pancreatic cancer	phase I	stage III/IV unresectable pancreatic cancer	dose determination of combined treatment of HF10 with Gemcitabine+Nab-paclitaxel or TS-1	active, not recruiting	NCT03252808
	Study of HF10 in patients with refractory head and neck cancer or solid tumors with cutaneous and/or superficial lesions	phase I	refractory head and neck cancer, squamous cell carcinoma, skin carcinoma of the breast, malignant melanoma	dose escalation study for single and repeated intratumoral injections, assessment of local tumor response	completed	NCT01017185
	A study of combination with TBI-1401(HF10) and ipilimumab in Japanese patients with unresectable or metastatic melanoma	phase II	stage IIIB, IIIC, or IV unresectable or metastatic malignant melanoma	safety and efficacy of repeated administration of intratumoral injections of HF10 in combination with ipilimumab, best overall response rate	completed	NCT03153085
	A study of combination treatment with HF10 and ipilimumab in patients with unresectable or metastatic melanoma	phase II	stage IIIB, IIIC, or IV unresectable or metastatic melanoma	efficacy of the combination of HF10 with ipilimumab, best overall response rate	completed	NCT02272855
G207	HSV G207 alone or with a single radiation dose in children with progressive or recurrent supratentorial brain tumors	phase I	recurrent or progressive supratentorial neoplasms, malignant glioma, glioblastoma, anaplastic astrocytoma, PNET, cerebral primitive neuroectodermal tumor, embryonal tumor	safety and tolerability of intratumoral injection, also in combination with a single low dose of radiation	active, not recruiting	NCT02457845
	HSV G207 in children with recurrent or refractory cerebellar brain tumors	phase I	recurrent or refractory medulloblastoma, glioblastoma multiforme, giant cell glioblastoma, anaplastic astrocytoma, primitive neuroectodermal tumor, ependymoma, atypical teratoid/rhabdoid tumor, germ cell tumor, other high-grade malignant tumor	safety and tolerability of intratumoral injection, also in combination with a single low dose of radiation	recruiting	NCT03911388
	HSV G207 with a single radiation dose in children with recurrent high-grade glioma	phase II	recurrent/progressive high grade glioma including glioblastoma multiforme, giant cell glioblastoma, anaplastic astrocytoma, midline diffuse glioma	efficacy and safety of intratumoral inoculation of G207 combined with a single radiation dose	not yet recruiting	NCT04482933

Table 2. *Cont.*

Virus	Study Title	Study Type	Disease Type	Study Aim	Status	NCT/UMIN #
G47Δ	A clinical study of G47delta oncolytic virus therapy for progressive glioblastoma	phase I/II	recurrent/progressive glioblastoma	safety and efficacy of intratumoral inoculation of G47Δ	completed	UMIN000002661
	A clinical study of an oncolytic HSV-1 G47delta for patients with castration resistant prostate cancer	phase I	castration resistant prostate cancer	safety and efficacy of intratumoral inoculation of G47Δ	completed	UMIN000010463
	A clinical study of G47delta oncolytic virus therapy for progressive olfactory neuroblastoma	n/a	recurrent olfactory neuroblastoma	safety and efficacy of intratumoral inoculation of G47Δ	recruiting	UMIN000011636
	A clinical study of G47delta oncolytic virus therapy for progressive malignant pleural mesothelioma	phase I	inoperable/recurrent/progressive malignant pleural mesothelioma	safety and efficacy of inoculation of G47Δ into the pleural cavity	recruiting	UMIN000034063
Talimogene laherparepvec	Talimogene laherparepvec in combination with neoadjuvant chemotherapy in triple negative breast cancer	phase I/II	triple negative breast carcinoma	determination of the maximum tolerated dose of talimogene laherparepvec administered with paclitaxel-doxorubicin/cyclophosphamide, pathological complete response rate	active, not recruiting	NCT02779855
	T-VEC in non-melanoma skin cancer	phase I	locally advanced squamous cell carcinoma, basal cell, carcinoma, Merkel cell carcinoma or cutaneous T cell lymphoma	detection of local immune effects after talimogene laherparepvec injection	recruiting	NCT03458117
	Ipilimumab, nivolumab, and talimogene laherparepvec before surgery in treating participants with localized, triple- negative or estrogen receptor positive, HER2 negative breast cancer	phase I	triple negative or ER positive HER2 negative infiltrating ductal breast cancer	safety of combined treatment of talimogene laherparepvec with nivolumab and ipilimumab	recruiting	NCT04185311
	Talimogene laherparepvec in treating patients with recurrent breast cancer that cannot be removed by surgery	phase II	recurrent stage IV breast cancer	determination of talimogene laherparepvec efficacy with overall response rate (ORR)	active, not yet recruiting	NCT02658812
	Talimogene laherparepvec and nivolumab in treating patients with refractory lymphomas or advanced or refractory non-melanoma skin cancers	phase II	T cell and NK cell lymphomas, Merkel cell carcinoma, Squamous cell carcinoma of the skin, Other non-melanoma skin cancers	response rate to talimogene laherparepvec, also in combination with nivolumab	recruiting	NCT02978625

Table 2. *Cont.*

Virus	Study Title	Study Type	Disease Type	Study Aim	Status	NCT/UMIN #
Talimogene laherparepvec	Talimogene laherparepvec and pembrolizumab in treating patients with stage III-IV melanoma	phase II	stage IV or unresectable stage III melanoma	response rate to talimogene laherparepvec in combination with pembrolizumab	recruiting	NCT02965716
	Talimogene laherparepvec, chemotherapy, and radiation therapy before surgery in treating patients with locally advanced or metastatic rectal cancer	phase I	stage III/IV rectal adenocarcinoma	dose determination and toxicity of talimogene laherparepvec in combination with capecitabibe, 5-fluoruracil, leucovorin, oxaliplatin, radiation	recruiting	NCT03300544
	Talimogene laherparepvec with chemotherapy or endocrine therapy in treating participants with metastatic, unresectable, or recurrent HER2- negative breast cancer	phase Ib	HER2-negative, estrogen receptor positive stage III/IV breast carcinoma	safety and tolerability of talimogene laherparepvec in combination with either chemotherapy (paclitaxel, nab-paclitaxel, or gemcitabine/carboplatin) or endocrine therapy	recruiting	NCT03554044
	Talimogene laherparepvec and panitumumab for the treatment of locally advanced or metastatic squamous cell carcinoma of the skin	phase I	locally advanced or metastatic squamous cell carcinoma of the skin	safety and efficacy of combined talimogene laherparepvec and panitumumab	recruiting	NCT04163952
	Talimogene laherparepvec and radiation therapy in treating patients with newly diagnosed soft tissue sarcoma that can be removed by surgery	phase II	liposarcoma, leiomyosarcoma, undifferentiated pleomorphic sarcoma (UPS)/ malignant fibrous histiosarcoma (MFH)	evaluation of the pathologic complete necrosis rate and safety following neoadjuvant treatment with talimogene laherparepvec and radiation	recruiting	NCT02923778
	A Phase 1, multi-center, open-label, dose de-escalation study to evaluate the safety and efficacy of Talimogene laherparepvec in pediatric subjects with advanced non-CNS tumors that are amenable to direct injection	phase I	recurring non-CNS solid tumor	safety and efficacy	recruiting	NCT02756845
ONCR-177	Study of ONCR-177 alone and in combination with PD-1 blockade in adult subjects with advanced and/or refractory cutaneous, subcutaneous or metastatic nodal solid tumors	phase I	advanced or metastatic solid tumors	determination of the maximum tolerated dose as well as preliminary efficacy of ONCR-177 in combination with pembrolizumab	recruiting	NCT04348916
RP2	Study of RP2 monotherapy and RP2 in combination with nivolumab in patients with solid tumors	phase I	advanced or metastatic non-neurological solid tumors	safety and tolerability of RP2, also in combination with nivolumab	recruiting	NCT04336241

Table 2. *Cont.*

Virus	Study Title	Study Type	Disease Type	Study Aim	Status	NCT/UMIN #
RP1	Study evaluating cemiplimab alone and combined with RP1 in treating advanced squamous skin cancer	phase II	locally advanced or metastatic cutaneous squamous cell carcinoma	determination of the clinical response rate/overall response rate of cemiplimab monotherapy versus combination with RP1	recruiting	NCT04050436
	Study of RP1 monotherapy and RP1 in combination with nivolumab	phase I/II	advanced and/or refractory solid tumors	determination of the maximum tolerated dose as well as preliminary efficacy of RP1 in combination with nivolumab	recruiting	NCT03767348
	A Phase 1b study of RP1 in transplant patients with advanced cutaneous squamous cell carcinoma	phase I	recurrent, locally advanced or metastatic cutaneous squamous cell carcinoma	safety and tolerability	recruiting	NCT04349436
rQNestin	A study of the treatment of recurrent malignant glioma with rQNestin34.5v.2	phase I	astrocytoma, malignant astrocytoma, oligodendroglioma, anaplastic oligodendroglioma, mixed oligo-astrocytoma	safety and dose determination of rQNestin with or without previous immunomodulation with cyclophosphamide	recruiting	NCT03152318
M032	Genetically engineered HSV-1 Phase 1 study for the treatment of recurrent malignant glioma	phase I	recurrent or progressive glioblastoma multiforme, anaplastic astrocytoma, gliosarcoma	safety and tolerability	recruiting	NCT02062827
C134	Trial of C134 in patients with recurrent GBM	phase I	recurrent or progressive glioblastoma multiforme, anaplastic astrocytoma, gliosarcoma	safety and tolerability	recruiting	NCT03657576
Rrp450	rRp450-Phase I trial in liver metastases and primary liver tumors	phase I	liver metastases or primary liver cancer	safety and tolerability	recruiting	NCT01071941

3.1. G47Δ

G47Δ was first described by Todo et al. in 2001: It is based on the G207 virus and contains an additional deletion in the region of the ICP47 gene, which eventually mitigates enhanced expression of MHC I on virus-infected cells [46]. Preclinical evaluation indeed showed positive effects on MHC I expression, T-cell stimulation of melanoma cells as well as increased cytolytic potency in melanoma and glioblastoma cell lines in vitro and survival in a immunocompetent neuroblastoma model in vivo [46]. Promising results with this agent have also been obtained for the treatment of breast cancer cell lines [47]. G47Δ has been tested for safety and efficacy in patients with recurrent or progressive glioblastoma (UMIN000002661) and castration resistant prostate cancer (UMIN000010463) in Japan. An interim analysis of the phase 2 glioblastoma study in 2019 presented with encouraging data, i.e., a one-year-survival rate of 92.3% compared to control (15%) [48]. Currently, this agent is also being tested in recurrent olfactory neuroblastoma (UMIN000011636) and malignant pleural mesothelioma (UMIN000034063).

3.2. rQNestin34.5

rQNestin34.5 is a an engineered oHSV based on F-strain HSV1 that expresses the neurovirulence factor ICP34.5 under a synthetic nestin promoter to drive robust tumor-selective viral replication [49]. In vivo experiments showed that the survival after symptom-onset of glioma-bearing animals was significantly prolonged after treatment with rQNestin34.5 compared to controls including the previous generation of oHSV [49]. rQNestin34.5v2 is a derivative that lacks a fusion ICP6-GFP transcript [50] and is currently under investigation in a phase I clinical trial for recurrent glioblastoma in combination with cyclophosphamide (NCT03152318). Chiocca et al. [50] showed that rQNestin34.5v2 is selectively cytotoxic for glioma cells and conducted toxicologic analyses to determine a starting dose of 1×10^6 pfu for use in humans.

3.3. M032

M032 is derived from the HSV-1 F-strain, containing deletions for both alleles of the neurovirulence factor $\gamma 34.5$ and armed to express the stimulatory cytokine IL-12 [51,52]. The murine variant of this construct–M002–has been well characterized by Parker et al. [53]: In vitro data support its toxicity against human glioblastoma and murine neuroblastoma cell lines, and in vivo survival data from neuroblastoma-bearing mice indicate a significant increase of median survival compared to control; immunohistologic workups of murine brain sections revealed an increase of CD4$^+$- and CD8$^+$-T-cells. A phase-1 trial (NCT02062827) is investigating safety and tolerability of M032 in patients with recurrent or progressive high-grade glioma.

3.4. ONCR-177

ONCR-177 (by Oncorus Inc. Cambridge, MA, USA) is a recombinant HSV-1 virus construct that is a derivative of ONCR-159 [54], which contains a *UL37* and *ICP47* mutation and 4 miR-T cassettes that were inserted into the gene regions of *ICP4*, *ICp27*, *UL8* and $\gamma 34.5$, thereby diminishing viral replication and mitigating reduced neurovirulence and also resistance to shut down by host interferon responses [55]. Based on this, the authors state that ONCR-177 has been further modified by expression for IL-12, CCL4, FLT3LG, and blocking antibody sequences for CTLA-4 and PD-1 to increase NK- and T-cell activation, dendritic cell availability, and antagonize T-cell exhaustion [54]. ONCR-177 monotherapy as well as combined treatment with pembrolizumab is being tested for the maximum tolerated dose and preliminary efficacy in advanced and metastatic solid cancers in a phase I study (NCT04348916).

3.5. C134

C134 is a chimeric oHSV that was altered by deletion of the neurovirulence factor $\gamma34.5$ and expression of the human cytomegalovirus gene IRS1, with the latter preserving late viral protein synthesis, which is disabled by deletion of $\gamma34.5$ [56]. Preclinical studies proved that C134, compared to $\gamma34.5$ deleted HSV-1 variants, had a higher replication potential in glioblastoma in in vivo models and was able to increase survival in glioma and neuroblastoma-bearing mice in contrast to $\gamma34.5$ deleted controls [57]. Safety and tolerability of C134 for treatment of advanced or progressive gliomas is currently investigated in a phase I clinical trial (NCT03657576).

3.6. RP1/2

RP-1 (by Replimmune Group Inc. Woburn, MA, USA) is a derivative of a wild-type HSV1 isolate containing deletions of $\gamma34.5$ and *ICP47* and expresses GM-CSF and GALV-GP-R⁻-a fusogenic membrane glycoprotein from gibbon ape leukemia virus that was shown to increase tumor-cell killing potential and immunogenic effects [58]. The viral construct is in clinical trials for recurrent or advanced squamous cell carcinoma (NCT04349436), combinations with the anti-PD-1 antibody cemiplimab (NCT04050436) in recurrent or advanced squamous cell carcinoma, and nivolumab in advanced or refractory solid tumors (NCT03767348) are also under evaluation. A further development on this backbone is the RP2 oHSV, which additionally expresses anti-CTLA-4 [59] and is being tested in combination with nivolumab for advanced or metastatic solid tumors in a phase I study (NCT04336241).

3.7. Rrp450

Rrp450 is a genetically engineered oHSV with a deletion of ribonucleotide reductase gene *ICP6* as well as an insertion of the *CYP2B1* gene, thereby diminishing replication potency in non-dividing cells and encoding for a cytochrome of the P450 family that activates the prodrug cyclophosphamide [60]. Pawlik et al. demonstrated in in vitro and in vivo models of hepatocellular carcinoma that Rrp450 causes tumor cell death, which is augmented by additional administration of cyclophosphamide [61]. These results were confirmed in preclinical models for sarcoma, high-grade medulloblastoma, and atypical teratoid/rhabdoid tumors [62,63]. The first phase I study for assessment of safety and tolerability of Rrp450 in liver metastases or primary liver cancer is currently in the recruitment phase (NCT01071941).

4. Conclusions

HSV-1 based OVs have shown promising results in various preclinical studies regarding efficacy based on combined tumor cell killing abilities and immunostimulation in a broad range of cancers. Attempts at clinical translation have often not been successful due to lack of efficacy, although safety has been good even at the maximum achievable doses of these agents. The success of T-Vec in melanoma leading to FDA approval has provided great impetus to the field, proving for the first time that this approach can provide durable clinical benefit. However, melanoma is known to be responsive to immunotherapies, and therefore the challenge now is to come up with approaches that may be broadly applicable in more tumor types, by engineering more potent viruses, with enhanced tumor cell killing and immunogenic responses. As described in this review, treatment with oHSV-1 proved to be safe throughout the various different viruses tested so far. oHSVs have the potential to be an efficient weapon in anti-cancer treatment and qualify as a potent combination partner with chemotherapeutic as well as immunotherapeutic regimens—this possibility has been recognized as several studies on combinatorial treatment are underway. Although effects on the immune system and prolonged survival were observed in some cases, these results have to be critically reviewed since the majority of the studies discussed were phase I clinical trials, designed for evaluation of safety and tolerability. It is therefore of the utmost importance to acquire reliable and detailed clinical data on the influence of oHSVs on the immune response and overall survival in follow-up studies to further characterize efficacy and find the most suitable combination partners. Better understanding the factors involved in response and resistance will lead to improved application of these agents in future trials.

Funding: The authors were supported by NCI P01 CA069246 and P01 CA163205 (E.A.C).

Conflicts of Interest: E.A.C. is currently an advisor to Advantagene Inc., Insightec, Inc., Seneca Therapeutics, Immunomic Therapeutics and DNAtrix Inc. and has equity interest in Immunomic Therapeutics, Seneca Therapeutics and DNAtrix; he has also advised Alcyone Biosciences, Voyager Therapeutics, Sangamo Therapeutics, Oncorus, Merck, Tocagen, Ziopharm, Stemgen, NanoTx., Ziopharm Oncology, Cerebral Therapeutics, Genenta. Merck, Janssen, Karcinolysis, Shanaghai Biotech, Sigilon Therapeutics. He has received research support from NIH, US Department of Defense, American Brain Tumor Association, National Brain Tumor Society, Alliance for Cancer Gene Therapy, Neurosurgical Research Education Foundation, Advantagene, NewLink Genetics and Amgen. He also is a named inventor on patents related to oncolytic HSV1.

References

1. Phan, G.Q.; Yang, J.C.; Sherry, R.M.; Hwu, P.; Topalian, S.L.; Schwartzentruber, D.J.; Restifo, N.P.; Haworth, L.R.; Seipp, C.A.; Freezer, L.J.; et al. Cancer regression and autoimmunity induced by cytotoxic T lymphocyte-associated antigen 4 blockade in patients with metastatic melanoma. *Proc. Natl. Acad. Sci. USA* **2003**, *100*, 8372–8377. [CrossRef] [PubMed]

2. Brahmer, J.R.; Tykodi, S.S.; Chow, L.Q.M.; Hwu, W.J.; Topalian, S.L.; Hwu, P.; Drake, C.G.; Camacho, L.H.; Kauh, J.; Odunsi, K.; et al. Safety and activity of anti-PD-L1 antibody in patients with advanced cancer. *N. Engl. J. Med.* **2012**, *366*, 2455–2465. [CrossRef] [PubMed]

3. Topalian, S.L.; Hodi, F.S.; Brahmer, J.R.; Gettinger, S.N.; Smith, D.C.; McDermott, D.F.; Powderly, J.D.; Carvajal, R.D.; Sosman, J.A.; Atkins, M.B.; et al. Safety, activity, and immune correlates of anti-PD-1 antibody in cancer. *N. Engl. J. Med.* **2012**, *366*, 2443–2454. [CrossRef] [PubMed]

4. Brown, C.E.; Alizadeh, D.; Starr, R.; Weng, L.; Wagner, J.R.; Naranjo, A.; Ostberg, J.R.; Blanchard, M.S.; Kilpatrick, J.; Simpson, J.; et al. Regression of glioblastoma after chimeric antigen receptor T-cell therapy. *N. Engl. J. Med.* **2016**, *375*, 2561–2569. [CrossRef] [PubMed]

5. Ali, S.A.; Shi, V.; Maric, I.; Wang, M.; Stroncek, D.F.; Rose, J.J.; Brudno, J.N.; Stetler-Stevenson, M.; Feldman, S.A.; Hansen, B.G.; et al. T cells expressing an anti-B-cell maturation antigen chimeric antigen receptor cause remissions of multiple myeloma. *Blood* **2016**, *128*, 1688–1700. [CrossRef] [PubMed]

6. Wheeler, L.A.; Manzanera, A.G.; Bell, S.D.; Cavaliere, R.; McGregor, J.M.; Grecula, J.C.; Newton, H.B.; Lo, S.S.; Badie, B.; Portnow, J.; et al. Phase II multicenter study of gene-mediated cytotoxic immunotherapy as adjuvant to surgical resection for newly diagnosed malignant glioma. *Neuro. Oncol.* **2016**, *18*, 1137–1145. [CrossRef]

7. Hilf, N.; Kuttruff-Coqui, S.; Frenzel, K.; Bukur, V.; Stevanović, S.; Gouttefangeas, C.; Platten, M.; Tabatabai, G.; Dutoit, V.; van der Burg, S.H.; et al. Actively personalized vaccination trial for newly diagnosed glioblastoma. *Nature* **2019**, *565*, 240–245. [CrossRef]

8. Lawler, S.E.; Speranza, M.C.; Cho, C.F.; Chiocca, E.A. Oncolytic viruses in cancer treatment a review. *JAMA Oncol.* **2017**, *3*, 841–849. [CrossRef]

9. Kristensson, K. Morphological studies of the neural spread of herpes simplex virus to the central nervous system. *Acta Neuropathol.* **1970**, *16*, 54–63. [CrossRef]

10. Studahl, M.; Cinque, P.; Bergström, T. *Herpes Simplex Viruses*; CRC Press: Boca Raton, FL, USA, 2005; ISBN 9780849355806.

11. Watson, G.; Xu, W.; Reed, A.; Babra, B.; Putman, T.; Wick, E.; Wechsler, S.L.; Rohrmann, G.F.; Jin, L. Sequence and comparative analysis of the genome of HSV-1 strain McKrae. *Virology* **2012**, *433*, 528–537. [CrossRef]

12. Ma, W.; He, H.; Wang, H. Oncolytic herpes simplex virus and immunotherapy. *BMC Immunol.* **2018**, *19*. [CrossRef] [PubMed]

13. Andtbacka, R.H.I.; Kaufman, H.L.; Collichio, F.; Amatruda, T.; Senzer, N.; Chesney, J.; Delman, K.A.; Spitler, L.E.; Puzanov, I.; Agarwala, S.S.; et al. Talimogene laherparepvec improves durable response rate in patients with advanced melanoma. *J. Clin. Oncol.* **2015**, *33*, 2780–2788. [CrossRef] [PubMed]

14. Goldstein, D.J.; Weller, S.K. Herpes simplex virus type 1-induced ribonucleotide reductase activity is dispensable for virus growth and DNA synthesis: Isolation and characterization of an ICP6 lacZ insertion mutant. *J. Virol.* **1988**, *62*, 196–205. [CrossRef] [PubMed]

15. MacLean, A.R.; Ul-Fareed, M.; Robertson, L.; Harland, J.; Brown, S.M. Herpes simplex virus type 1 deletion variants 1714 and 1716 pinpont neurovirulence-related sequences in Glasgow strain 17+ between immediate early gene 1 and the "a" sequence. *J. Gen. Virol.* **1991**, *72*, 631–639. [CrossRef]

16. Takakuwa, H.; Goshima, F.; Nozawa, N.; Yoshikawa, T.; Kimata, H.; Nakao, A.; Nawa, A.; Kurata, T.; Sata, T.; Nishiyama, Y. Oncolytic viral therapy using a spontaneously generated herpes simplex virus type 1 variant for disseminated peritoneal tumor in immunocompetent mice. *Arch. Virol.* **2003**, *148*, 813–825. [CrossRef]

17. Eissa, I.R.; Naoe, Y.; Bustos-Villalobos, I.; Ichinose, T.; Tanaka, M.; Zhiwen, W.; Mukoyama, N.; Morimoto, T.; Miyajima, N.; Hitoki, H.; et al. Genomic signature of the natural oncolytic herpes simplex virus HF10 and its therapeutic role in preclinical and clinical trials. *Front. Oncol.* **2017**, *7*, 149. [CrossRef]

18. Kelly, K.J.; Wong, J.; Fong, Y. Herpes simplex virus NV1020 as a novel and promising therapy for hepatic malignancy. *Expert Opin. Investig. Drugs* **2008**, *17*, 1105–1113. [CrossRef]

19. Liu, B.L.; Robinson, M.; Han, Z.Q.; Branston, R.H.; English, C.; Reay, P.; McGrath, Y.; Thomas, S.K.; Thornton, M.; Bullock, P.; et al. ICP34.5 deleted herpes simplex virus with enhanced oncolytic, immune stimulating, and anti-tumour properties. *Gene Ther.* **2003**, *10*, 292–303. [CrossRef]

20. Streby, K.A.; Currier, M.A.; Triplet, M.; Ott, K.; Dishman, D.J.; Vaughan, M.R.; Ranalli, M.A.; Setty, B.; Skeens, M.A.; Whiteside, S.; et al. First-in-Human Intravenous Seprehvir in Young Cancer Patients: A Phase 1 Clinical Trial. *Mol. Ther.* **2019**, *27*, 1930–1938. [CrossRef]

21. Rampling, R.; Cruickshank, G.; Papanastassiou, V.; Nicoll, J.; Hadley, D.; Brennan, D.; Petty, R.; MacLean, A.; Harland, J.; McKie, E.; et al. Toxicity evaluation of replication-competent herpes simplex virus (ICP 34.5 null mutant 1716) in patients with recurrent malignant glioma. *Gene Ther.* **2000**, *7*, 859–866. [CrossRef]

22. Mackie, R.M.; Stewart, B.; Brown, S.M. Intralesional injection of herpes simplex virus 1716 in metastatic melanoma. *Lancet* **2001**, *357*, 525–526. [CrossRef]

23. Mineta, T.; Rabkin, S.D.; Yazaki, T.; Hunter, W.D.; Martuza, R.L. Attenuated multi-mutated herpes simplex virus-1 for the treatment of malignant gliomas. *Nat. Med.* **1995**, *1*, 938–943. [CrossRef] [PubMed]

24. Markert, J.M.; Medlock, M.D.; Rabkin, S.D.; Gillespie, G.Y.; Todo, T.; Hunter, W.D.; Palmer, C.A.; Feigenbaum, F.; Tornatore, C.; Tufaro, F.; et al. Conditionally replicating herpes simplex virus mutant G207 for the treatment of malignant glioma: Results of a phase I trial. *Gene Ther.* **2000**, *7*, 867–874. [CrossRef] [PubMed]

25. Markert, J.M.; Liechty, P.G.; Wang, W.; Gaston, S.; Braz, E.; Karrasch, M.; Nabors, L.B.; Markiewicz, M.; Lakeman, A.D.; Palmer, C.A.; et al. Phase Ib trial of mutant herpes simplex virus G207 inoculated pre-and post-tumor resection for recurrent GBM. *Mol. Ther.* **2009**, *17*, 199–207. [CrossRef] [PubMed]

26. Markert, J.M.; Razdan, S.N.; Kuo, H.C.; Cantor, A.; Knoll, A.; Karrasch, M.; Nabors, L.B.; Markiewicz, M.; Agee, B.S.; Coleman, J.M.; et al. A phase 1 trial of oncolytic HSV-1, g207, given in combination with radiation for recurrent GBM demonstrates safety and radiographic responses. *Mol. Ther.* **2014**, *22*, 1048–1055. [CrossRef]

27. Umene, K.; Eto, T.; Mori, R.; Takagi, Y.; Enquist, L.W. Herpes simplex virus type 1 restriction fragment polymorphism determined using southern hybridization. *Arch. Virol.* **1984**, *80*, 275–290. [CrossRef]

28. Nishiyama, Y.; Kimura, H.; Daikoku, T. Complementary lethal invasion of the central nervous system by nonneuroinvasive herpes simplex virus types 1 and 2. *J. Virol.* **1991**, *65*, 4520–4524. [CrossRef]

29. Nakao, A.; Kimata, H.; Imai, T.; Kikumori, T.; Teshigahara, O. Intratumoral injection of herpes simplex virus HF10 in recurrent breast cancer. *Ann. Oncol.* **2004**, *15*, 987–988. [CrossRef]

30. Kimata, H.; Imai, T.; Kikumori, T.; Teshigahara, O.; Nagasaka, T.; Goshima, F.; Nishiyama, Y.; Nakao, A. Pilot study of oncolytic viral therapy using mutant herpes simplex virus (HF10) against recurrent metastatic breast cancer. *Ann. Surg. Oncol.* **2006**, *13*, 1078–1084. [CrossRef]

31. Nakao, A.; Kasuya, H.; Sahin, T.T.; Nomura, N.; Kanzaki, A.; Misawa, M.; Shirota, T.; Yamada, S.; Fujii, T.; Sugimoto, H.; et al. A phase I dose-escalation clinical trial of intraoperative direct intratumoral injection of HF10 oncolytic virus in non-resectable patients with advanced pancreatic cancer. *Cancer Gene Ther.* **2011**, *18*, 167–175. [CrossRef]

32. Hirooka, Y.; Kasuya, H.; Ishikawa, T.; Kawashima, H.; Ohno, E.; Villalobos, I.B.; Naoe, Y.; Ichinose, T.; Koyama, N.; Tanaka, M.; et al. A Phase I clinical trial of EUS-guided intratumoral injection of the oncolytic virus, HF10 for unresectable locally advanced pancreatic cancer. *BMC Cancer* **2018**, *18*. [CrossRef] [PubMed]

33. Fujimoto, Y.; Mizuno, T.; Sugiura, S.; Goshima, F.; Kohno, S.I.; Nakashima, T.; Nishiyama, Y. Intratumoral injection of herpes simplex virus HF10 in recurrent head and neck squamous cell carcinoma. *Acta Otolaryngol.* **2006**, *126*, 1115–1117. [CrossRef] [PubMed]

34. Ebright, M.I.; Zager, J.S.; Malhotra, S.; Delman, K.A.; Weigel, T.L.; Rusch, V.W.; Fong, Y. Replication-competent herpes virus NV1020 as direct treatment of pleural cancer in a rat model. *J. Thorac. Cardiovasc. Surg.* **2002**, *124*, 123–129. [CrossRef] [PubMed]

35. Wong, R.J.; Kim, S.H.; Joe, J.K.; Shah, J.P.; Johnson, P.A.; Fong, Y. Effective treatment of head and neck squamous cell carcinoma by an oncolytic herpes simplex virus. *J. Am. Coll. Surg.* **2001**, *193*, 12–21. [CrossRef]

36. Bennett, J.J.; Delman, K.A.; Burt, B.M.; Mariotti, A.; Malhotra, S.; Zager, J.; Petrowsky, H.; Mastorides, S.; Federoff, H.; Fong, Y. Comparison of safety, delivery, and efficacy of two oncolytic herpes viruses (G207 and NV1020) for peritoneal cancer. *Cancer Gene Ther.* **2002**, *9*, 935–945. [CrossRef]

37. Gutermann, A.; Mayer, E.; Von Dehn-Rothfelser, K.; Breidenstein, C.; Weber, M.; Muench, M.; Gungor, D.; Suehnel, J.; Moebius, U.; Lechmann, M. Efficacy of oncolytic herpesvirus NV1020 can be enhanced by combination with chemotherapeutics in colon carcinoma cells. *Hum. Gene Ther.* **2006**, *17*, 1241–1253. [CrossRef]

38. Kemeny, N.; Brown, K.; Covey, A.; Kim, T.; Bhargava, A.; Brody, L.; Guilfoyle, B.; Haag, N.P.; Karrasch, M.; Glasschroeder, B.; et al. Phase I, open-label, dose-escalating study of a genetically engineered herpes simplex virus, NV1020, in subjects with metastatic colorectal carcinoma to the liver. *Hum. Gene Ther.* **2006**, *17*, 1214–1224. [CrossRef]

39. Fong, Y.; Kim, T.; Bhargava, A.; Schwartz, L.; Brown, K.; Brody, L.; Covey, A.; Karrasch, M.; Getrajdman, G.; Mescheder, A.; et al. A herpes oncolytic virus can be delivered via the vasculature to produce biologic changes in human colorectal cancer. *Mol. Ther.* **2009**, *17*, 389–394. [CrossRef]

40. Geevarghese, S.K.; Geller, D.A.; De Haan, H.A.; Hörer, M.; Knoll, A.E.; Mescheder, A.; Nemunaitis, J.; Reid, T.R.; Sze, D.Y.; Tanabe, K.K.; et al. Phase I/II study of oncolytic herpes simplex virus NV1020 in patients with extensively pretreated refractory colorectal cancer metastatic to the liver. *Hum. Gene Ther.* **2010**, *21*, 1119–1128. [CrossRef]

41. Hu, J.C.C.; Coffin, R.S.; Davis, C.J.; Graham, N.J.; Groves, N.; Guest, P.J.; Harrington, K.J.; James, N.D.; Love, C.A.; McNeish, I.; et al. A phase I study of OncoVEXGM-CSF, a second-generation oncolytic herpes simplex virus expressing granulocyte macrophage colony-stimulating factor. *Clin. Cancer Res.* **2006**, *12*, 6737–6747. [CrossRef]

42. Senzer, N.N.; Kaufman, H.L.; Amatruda, T.; Nemunaitis, M.; Reid, T.; Daniels, G.; Gonzalez, R.; Glaspy, J.; Whitman, E.; Harrington, K.; et al. Phase II clinical trial of a granulocyte-macrophage colony-stimulating factor-encoding, second-generation oncolytic herpesvirus in patients with unresectable metastatic melanoma. *J. Clin. Oncol.* **2009**, *27*, 5763–5771. [CrossRef]

43. Kaufman, H.L.; Kim, D.W.; Deraffele, G.; Mitcham, J.; Coffin, R.S.; Kim-Schulze, S. Local and distant immunity induced by intralesional vaccination with an oncolytic herpes virus encoding GM-CSF in patients with stage IIIc and IV melanoma. *Ann. Surg. Oncol.* **2010**, *17*, 718–730. [CrossRef] [PubMed]

44. Andtbacka, R.H.I.; Ross, M.; Puzanov, I.; Milhem, M.; Collichio, F.; Delman, K.A.; Amatruda, T.; Zager, J.S.; Cranmer, L.; Hsueh, E.; et al. Patterns of Clinical Response with Talimogene Laherparepvec (T-VEC) in Patients with Melanoma Treated in the OPTiM Phase III Clinical Trial. *Ann. Surg. Oncol.* **2016**, *23*, 4169–4177. [CrossRef] [PubMed]

45. Puzanov, I.; Milhem, M.M.; Minor, D.; Hamid, O.; Li, A.; Chen, L.; Chastain, M.; Gorski, K.S.; Anderson, A.; Chou, J.; et al. Talimogene laherparepvec in combination with ipilimumab in previously untreated, unresectable stage IIIB-IV melanoma. *J. Clin. Oncol. Am. Soc.* **2016**, *34*, 2619–2626. [CrossRef]

46. Todo, T.; Martuza, R.L.; Rabkin, S.D.; Johnson, P.A. Oncolytic herpes simplex virus vector with enhanced MHC class I presentation and tumor cell killing. *Proc. Natl. Acad. Sci. USA* **2001**, *98*, 6396–6401. [CrossRef]

47. Zeng, W.; Hu, P.; Wu, J.; Wang, J.; Li, J.; Lei, L.; Liu, R. The oncolytic herpes simplex virus vector G47Δ effectively targets breast cancer stem cells. *Oncol. Rep.* **2013**, *29*, 1108–1114. [CrossRef] [PubMed]

48. Todo, T. ATIM-14. Results of phase II clinical trial of oncolytic herpes virus G47Δ in patients with glioblastoma. *Neuro. Oncol.* **2019**, *21*, vi4. [CrossRef]

49. Kambara, H.; Okano, H.; Chiocca, E.A.; Saeki, Y. An oncolytic HSV-1 mutant expressing ICP34.5 under control of a nestin promoter increases survival of animals even when symptomatic from a brain tumor. *Cancer Res.* **2005**, *65*, 2832–2839. [CrossRef]

50. Chiocca, E.A.; Nakashima, H.; Kasai, K.; Fernandez, S.A.; Oglesbee, M. Preclinical Toxicology of rQNestin34.5v.2: An Oncolytic Herpes Virus with Transcriptional Regulation of the ICP34.5 Neurovirulence Gene. *Mol. Ther. Methods Clin. Dev.* **2020**, *17*, 871–893. [CrossRef]
51. Cassady, K.A.; Parker, J.N. Herpesvirus Vectors for Therapy of Brain Tumors. *Open Virol. J.* **2010**, *4*, 103–108. [CrossRef]
52. Patel, D.M.; Foreman, P.M.; Nabors, L.B.; Riley, K.O.; Gillespie, G.Y.; Markert, J.M. Design of a Phase i Clinical Trial to Evaluate M032, a Genetically Engineered HSV-1 Expressing IL-12, in Patients with Recurrent/Progressive Glioblastoma Multiforme, Anaplastic Astrocytoma, or Gliosarcoma. *Hum. Gene Ther. Clin. Dev.* **2016**, *27*, 69–78. [CrossRef] [PubMed]
53. Parker, J.N.; Gillespie, G.Y.; Love, C.E.; Randall, S.; Whitley, R.J.; Markert, J.M. Engineered herpes simplex virus expressing IL-12 in the treatment of experimental murine brain tumors. *Proc. Natl. Acad. Sci. USA* **2000**, *97*, 2208–2213. [CrossRef] [PubMed]
54. Kennedy, E.M.; Farkaly, T.; Behera, P.; Colthart, A.; Goshert, C.; Jacques, J.; Grant, K.; Grzesik, P.; Lerr, J.; Salta, L.V.; et al. Abstract 1455: Design of ONCR-177 base vector, a next generation oncolytic herpes simplex virus type-1, optimized for robust oncolysis, transgene expression and tumor-selective replication. *Immunol. Am. Assoc. Cancer Res.* **2019**, *79*, 1455.
55. Kennedy, E.M.; Farkaly, T.; Grzesik, P.; Lee, J.; Denslow, A.; Hewett, J.; Bryant, J.; Behara, P.; Goshert, C.; Wambua, D.; et al. Design of an interferon-resistant oncolytic HSV-1 incorporating redundant safety modalities for improved tolerability. *Mol. Ther. Oncol.* **2020**. [CrossRef]
56. Cassady, K.A. Human Cytomegalovirus TRS1 and IRS1 Gene Products Block the Double-Stranded-RNA-Activated Host Protein Shutoff Response Induced by Herpes Simplex Virus Type 1 Infection. *J. Virol.* **2005**, *79*, 8707–8715. [CrossRef]
57. Shah, A.C.; Parker, J.N.; Gillespie, G.Y.; Lakeman, F.D.; Meleth, S.; Markert, J.M.; Cassady, K.A. Enhanced antiglioma activity of chimeric HCMV/HSV-1 oncolytic viruses. *Gene Ther.* **2007**, *14*, 1045–1054. [CrossRef]
58. Thomas, S.; Kuncheria, L.; Roulstone, V.; Kyula, J.N.; Mansfield, D.; Bommareddy, P.K.; Smith, H.; Kaufman, H.L.; Harrington, K.J.; Coffin, R.S. Development of a new fusion-enhanced oncolytic immunotherapy platform based on herpes simplex virus type 1. *J. Immunother. Cancer* **2019**, *7*. [CrossRef]
59. Thomas, S.; Kuncheria, L.; Roulstone, V.; Kyula, J.N.; Kaufman, H.L.; Harrington, K.J.; Coffin, R.S. Abstract 1470: Development & characterization of a new oncolytic immunotherapy platform based on herpes simplex virus type 1. In Proceedings of the AACR Annual Meeting 2019, Atlanta, GA, USA, 29 March–3 April 2019; Volume 79, p. 1470.
60. Chase, M.; Chung, R.Y.; Antonio Chiocca, E. An oncolytic viral mutant that delivers the CYP2B1 transgene and augments cyclophosphamide chemotherapy. *Nat. Biotechnol.* **1998**, *16*, 444–448. [CrossRef]
61. Pawlik, T.M.; Nakamura, H.; Yoon, S.S.; Mullen, J.T.; Chandrasekhar, S.; Chiocca, E.A.; Tanabe, K.K. Oncolysis of diffuse hepatocellular carcinoma by intravascular administration of a replication-competent, genetically engineered herpesvirus. *Cancer Res.* **2000**, *60*, 2790–2795.
62. Currier, M.A.; Gillespie, R.A.; Sawtell, N.M.; Mahller, Y.Y.; Stroup, G.; Collins, M.H.; Kambara, H.; Chiocca, E.A.; Cripe, T.P. Efficacy and safety of the oncolytic herpes simplex virus rRp450 alone and combined with cyclophosphamide. *Mol. Ther.* **2008**, *16*, 879–885. [CrossRef]
63. Studebaker, A.W.; Hutzen, B.J.; Pierson, C.R.; Haworth, K.B.; Cripe, T.P.; Jackson, E.M.; Leonard, J.R. Oncolytic Herpes Virus rRp450 Shows Efficacy in Orthotopic Xenograft Group 3/4 Medulloblastomas and Atypical Teratoid/Rhabdoid Tumors. *Mol. Ther. Oncol.* **2017**, *6*, 22–30. [CrossRef] [PubMed]

Publisher's Note: MDPI stays neutral with regard to jurisdictional claims in published maps and institutional affiliations.

MDPI

St. Alban-Anlage 66

4052 Basel

Switzerland

Tel. +41 61 683 77 34

Fax +41 61 302 89 18

www.mdpi.com

Cancers Editorial Office

E-mail: cancers@mdpi.com

www.mdpi.com/journal/cancers